International Conference on Computation in Science and Engineering (ICCSE 2017)

AA006037

Journal of Physics: Conference Series Volume 1090

Bandung, Indonesia
10 – 12 July 2017

ISBN: 978-1-5108-7300-1
ISSN: 1742-6588

Table of contents

Volume 1090

International Conference on Computation in Science and Engineering
10-12 July 2017, Bandung, Indonesia

Accepted papers received: 20 August 2018
Published online: 28 September 2018

Preface

Papers

IOP Conf. Series: Journal of Physics: Conf. Series **1090** (2018) 011001 doi:10.1088/1742-6596/1090/1/011001

PREFACE

The role of computation and simulation to support research activities in sciences and technologies has become more important in recent decades. The International Conference on Computation in Science and Engineering (ICCSE) 2017 has been conducted in Bandung Institute of Technology (ITB) campus, Bandung, Indonesia, on 10-12 July 2017. This conference is organized by Bandung Institute of Technology (ITB, Indonesia) and supported by UIN Sunan Gunung Jati University. This conference aims at summarizing recent research activities relevant to the computation application in science and engineering and facilitate communication among relevant experts.

More than 150 people from Indonesia, Japan, Malaysia, India, Australia, Korea, and some other countries have participated in this conference. About 150 presentations including 9 plenary talks are presented. The presentations are grouped into 12 areas of particular interest: (1): Complex system Modelling, (2): Energy System Computation, (3): Fluid Dynamics Computation, (4): Artificial Intelligent and Soft Computing, (5): High Energy Physics, (6): Simulation in general, (7): Quantum Computation, (8) HPC, (9): Material Computation, (10): Sub surface modelling and computation, (11): Nuclear and Radiation Computation, and (12): Atmospheric and meteorological computation.

From about 100 full papers submitted after the conference for publication in IOP conference proceedings, after peer review process by relevant experts, 83 papers were accepted for publication in this proceeding. We are indebted to all of authors for submitting their original papers.

We would like to thank all participants, all members of the International Board, members of the Organizing Committee, and I express my gratitude to all those who helped the success of this conference.

Zaki Su'ud

ICCSE 2017 Chairman

Dwi Irwanto

ICCSE 2017 Co-Chairman

Content from this work may be used under the terms of the Creative Commons Attribution 3.0 licence. Any further distribution of this work must maintain attribution to the author(s) and the title of the work, journal citation and DOI.
Published under licence by IOP Publishing Ltd

IOP Conf. Series: Journal of Physics: Conf. Series **1090** (2018) 011002 doi:10.1088/1742-6596/1090/1/011002

Peer review statement

All papers published in this volume of *Journal of Physics: Conference Series* have been peer reviewed through processes administered by the proceedings Editors. Reviews were conducted by expert referees to the professional and scientific standards expected of a proceedings journal published by IOP Publishing.

Content from this work may be used under the terms of the Creative Commons Attribution 3.0 licence. Any further distribution of this work must maintain attribution to the author(s) and the title of the work, journal citation and DOI.
Published under licence by IOP Publishing Ltd

IOP Conf. Series: Journal of Physics: Conf. Series **1090** (2018) 012001 doi:10.1088/1742-6596/1090/1/012001

Current and Future Computation in Nuclear Engineering

Hiroshi Sekimoto

Tokyo Institute of Technology

Abstract. This paper consists of two parts. In the first part, current computations in nuclear engineering are presented, where areas of nuclear engineering, procedures for solution, limits of analysis, and increases of computer performance are explained. In the second part, future computations in nuclear engineering are presented, where two examples are discussed. One example is computations for innovative nuclear reactors. The other is computations for accident management.

1. Introduction

From the start of nuclear energy development, computation was a key technology for nuclear reactor design and operation. The computer development is one of the largest development among of scientific technologies. And the nuclear engineering computation becomes more and more accurate and sophisticated along with this development.

In the first part of this paper, current computations in nuclear engineering are presented. It covers the areas of nuclear engineering, procedures for solution, limits of analysis, and development of computer power.

In the second part, computations in future nuclear engineering are discussed, where two examples are presented. One example is computations for innovative nuclear reactors. The other is computations for accident managements which should be implemented in actual reactor operation system

2. Current Computation in Nuclear Engineering

2.1. Areas of Numerical Analyses in Nuclear Engineering

The important areas of numerical analyses in nuclear engineering are shown in table 1. They covers so many different kinds of calculation techniques. The present paper discusses mainly on nuclear reactors, since we have only a limited number of pages.

For each area we have many sub-areas to be analyzed by calculation, for example, the fuel cycle, which covers mining, enrichment, fuel fabrication, reprocessing, waste disposal, and fuel transportation. On the other hand some areas may be overlapped, for example, the cost is often investigated with optimization.

Required computation method usually differs in different areas.

Content from this work may be used under the terms of the Creative Commons Attribution 3.0 licence. Any further distribution of this work must maintain attribution to the author(s) and the title of the work, journal citation and DOI.
Published under licence by IOP Publishing Ltd

IOP Conf. Series: Journal of Physics: Conf. Series **1090** (2018) 012001 doi:10.1088/1742-6596/1090/1/012001

Table 1. Areas of numerical analyses in nuclear engineering

cross section evaluations,
reactor physics analysis (mainly neutron transport),
thermal aspects including fluid behaviors,
instrumentations and control,
stability, operations research,
stress analyses of solid components,
diffusion of materials especially in fuel elements,
fuel cycle including wastes disposal,
safety and accidents,
seculity,
environmental effects of radioactive materials from facilities, environment dynamics,
effects of irradiations on materials and human body,
designs of reactor system and elements,
costs,
optimization,
others.

2.2. *Procedures for Solution*

The general procedures for solution is shown in figure 1. Usually problem definition and formulation are combined and called as problem formulation.

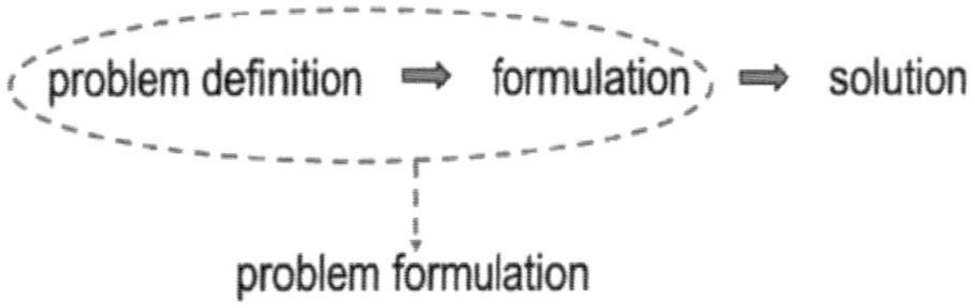

Figure 1. Procedures for solution

Albert Einstein said "The formulation of a problem is often far more essential than its solution. [1]" This is true especially for physics. However, for the present nuclear engineering the problem is already presented and our jobs are usually solutions.

Many calculation methods are available for solving these problems. The problems for neutron transport are given by differential and/or integral equations. Usually these equations are transformed to matrix equations by employing discretization of continuous variables such as spatial coordinates, time, energy and angle. Electric computer is powerful for solving these equations. The neutron transport is also solved by Monte Carlo method, where discretization of continuous variables is not necessary. Since each method has its original merits and demerits, we should be careful for choosing proper method for each problem.

For safety analysis, we usually employ probabilistic risk analysis (PRA) which include event tree analysis and fault tree analysis. These analyses are entirely different from the neutron transport analysis. For this kind of problem, difficult part is not solving equation, but obtaining proper inputs.

We meet often optimization and stability problems. These analyses are also very different from the neutron transport analysis. Especially we have so many unique optimization methods from analytical to heuristic. We need to be familiar to these methods.

Figure 1 shows the only one direction from the problem formulation to solution, but after obtaining a solution we should check the whole procedure and change the problem formulation and/or solution, if necessary.

IOP Conf. Series: Journal of Physics: Conf. Series **1090** (2018) 012001 doi:10.1088/1742-6596/1090/1/012001

2.3. Limits of Analysis

When more contents are included in the analysis of problem, we can get more informative results. More general results can be obtained by treating wider problem region. More accurate results can be obtained by changing the assumption employed for the calculation to less approximated. More accurate results are obtained when number of calculation meshes is increased and convergence criteria are changed to be smaller. These improvements of analysis increase the computation size. The computations are usually performed by using computer. Both of the accuracy of results and problem size are limited by computer performance as shown in figure 2.

(Accuracy of Results) x (Problem Size) < (Computer Performance)

Figure 2. Limits of analysis

2.4. Increase of Computer Performance

The computer performance is increasing according to Moor's law [2] "the number of transistors in a dense integrated circuit doubles approximately every two years" for many years as shown in figure 3.

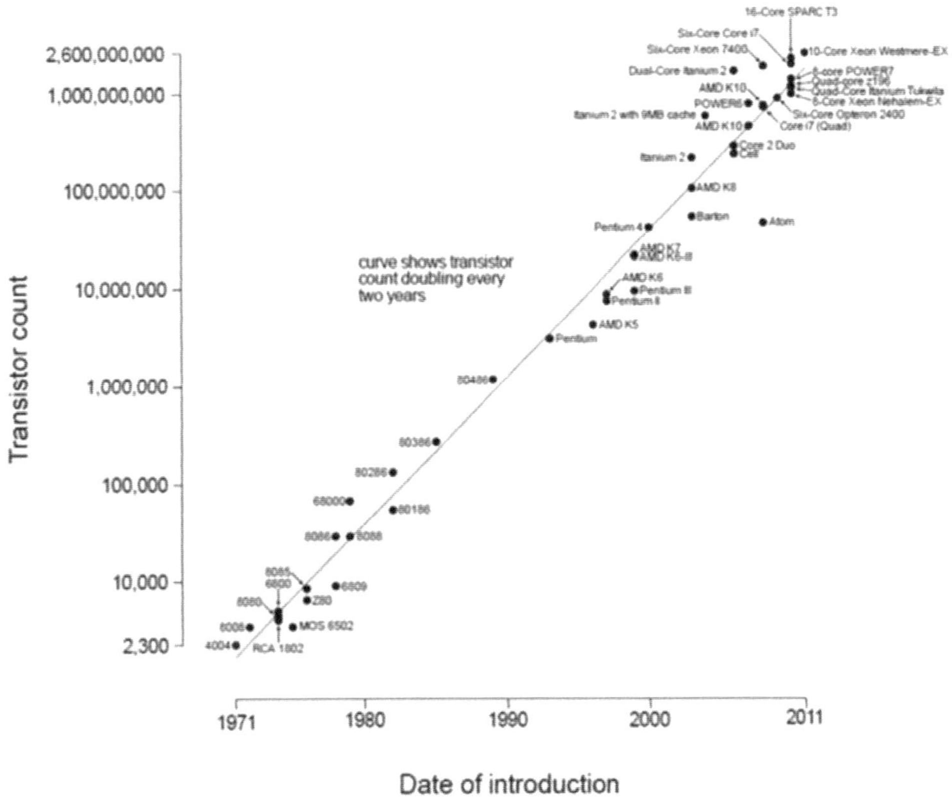

Figure 3. Increase of computer performance: microprocessor transistor counts 1971-2011 [2]

The change of this computer performance enables the larger computation in the later years. One example is shown in figure 4, which is in-core fuel management problem.

IOP Conf. Series: Journal of Physics: Conf. Series **1090** (2018) 012001 doi:10.1088/1742-6596/1090/1/012001

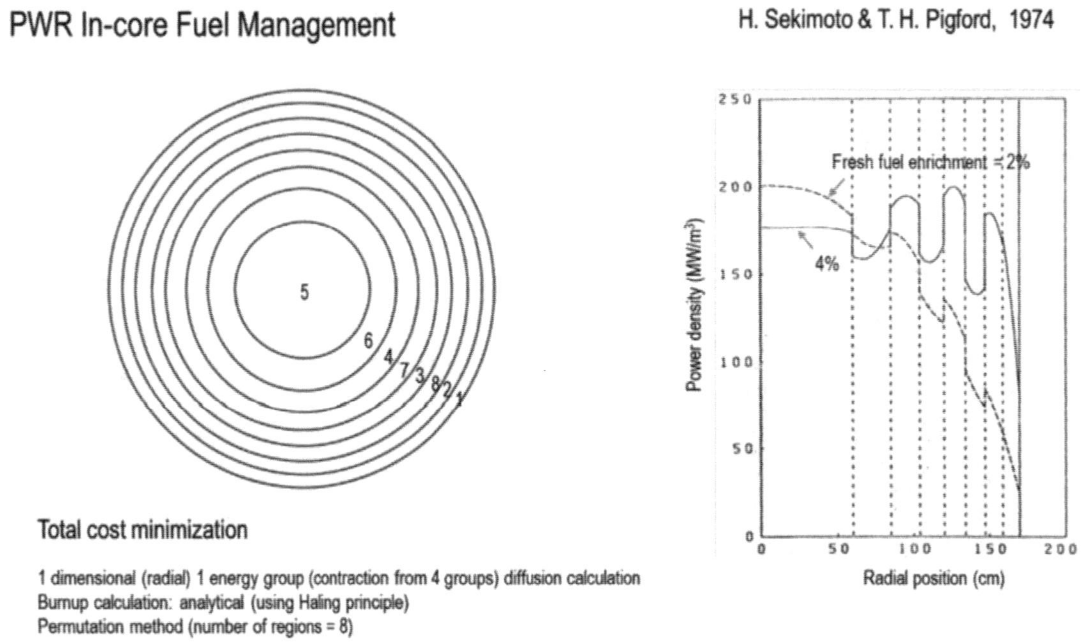

(a) Calculation done in 1973 [3]

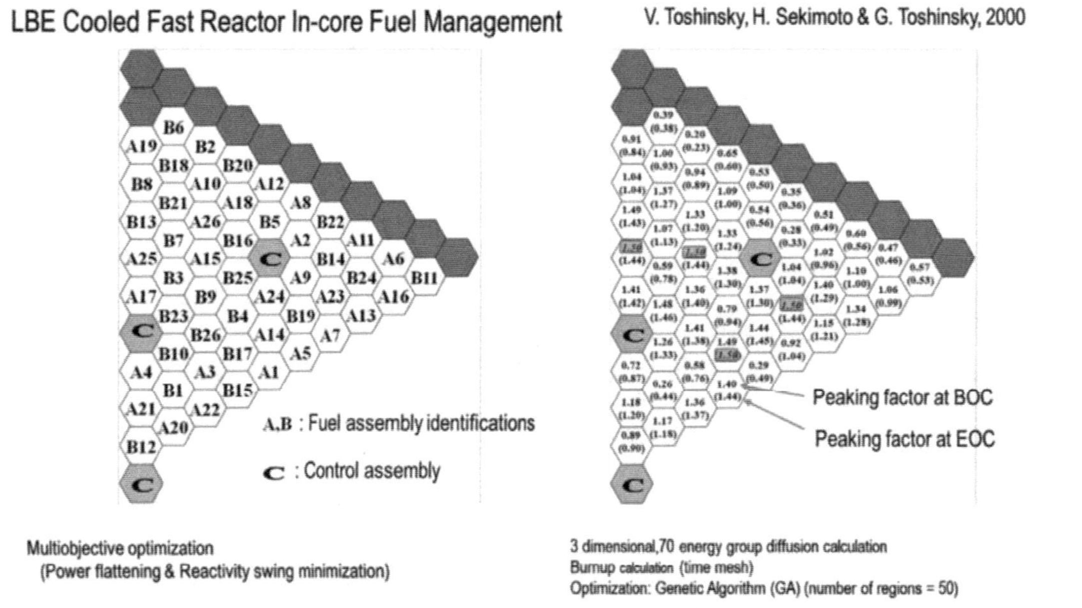

(b) Calculation done in 1999 [4]

Figure 4. Effects of increase of computer performance on actual reactor analyses

In-core fuel management needs to treat a lot of space meshes for treating fine fuel assembly positions and a lot of cases for different fuel shuffling scheme. In 1973 it was impossible to treat actual fuel shuffling scheme of each fuel assembly. The calculation presented in figure 4 (a) treated cylindrically divided core which is not realistic [3]. In 1999 treatment of each fuel assembly became possible as shown in figure 4 (b) [4]. The total mesh numbers including energy groups increased by more than 10^4,

IOP Conf. Series: Journal of Physics: Conf. Series **1090** (2018) 012001 doi:10.1088/1742-6596/1090/1/012001

and the number of regions increased 5.6 times which increased the number of shuffling patterns so drastically. The used computers were the large computers in university computer center: University of California, Berkeley for the calculation (a) and Tokyo Institute of Technology for the calculation (b). Both machines had the world highest level of performance at each time. This situation is shown in figure 5. The treatable problem size increased so much by the increase of computer performance. However, not only computer performance but developments of analytical method also contributed to the great increase of problem size.

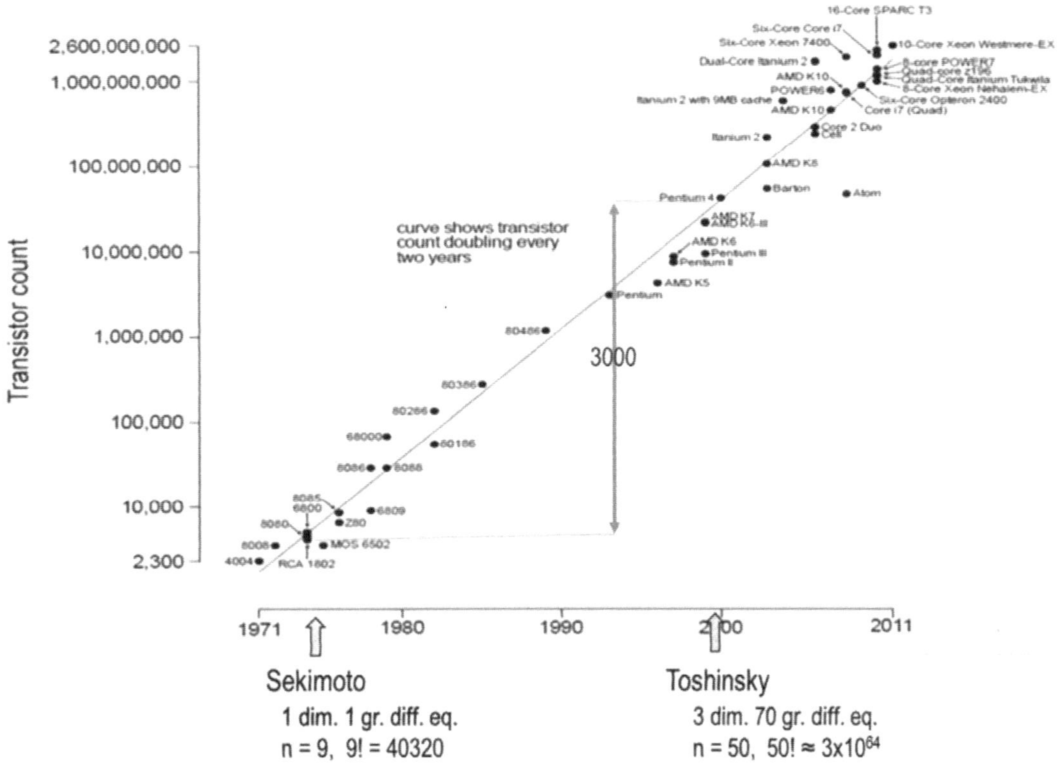

Figure 5. Increase of computer performance and problem size with year

3. Future Computation in Nuclear Engineering

3.1. Computation in Future Nuclear Engineering

The computer performance will continue to increase as mentioned in Section 2.4 in the future. Several innovative computers such as quantum computers[5] are expected to be put to practical use in the near future. It will be able to handle larger and more complicated systems faster and more accurately. In the future of nuclear power, the role of computation should increase more and more.

One case of the computation in future nuclear engineering is for a new field of innovative nuclear engineering under development. In another case, the development of the computer enables calculation that was impossible to date. Here in this paper an interesting example of computation is shown for each case: innovative nuclear reactors for the fast case and accident managements for the second case.

3.2. Computation for Innovative Nuclear Reactor System

The computation technique will be improved even for the basic problems of conventional reactors. However, developments of computation technique are considered much more for innovative reactors[6]. Some of the important innovative reactors are shown in table 2 with their unique characteristics.

Table 2. Several unique characteristics for some innovative nuclear reactors

reactor types	unique characteristics
Fast Reactor	hard neutron spectrum re-criticality accidents
High Temperature Gas Cooled Reactor	neutron self shielding for double heterogeneity depressurizing accident
Molten Salt Reactor	liquid fuel (FP flow, on-line reprocessing)
Accelerator Driven System	sub-criticality unisotropic neutron field hard neutron spectrum
Fusion Reactor Blanket	sub-criticality unisotropic neutron field hard neutron spectrum

Each reactor type in this table contains many different subtypes. The fast reactor has many options for different fuel: oxide, metal, nitride, et al., and for different coolants: sodium, lead (or lead bismuth), gas, et al. The high temperature gas cooled reactor has typical two options: block-type fuel reactor and pebble bed reactor. The molten salt reactor has two types: thermal thorium reactor and fast reactor. The accelerator driven system uses solid core or liquid core. The fusion reactor blanket has two options; simple tritium breeding blanket and complicated fusion-fission hybrid blanket. It has also many options for different plasma core design, and many subtypes of design like the fast reactor.

Many innovative reactors other than those listed in this table have been proposed.

The unique characteristics shown in table 2 are only those related to reactor physics. The FP flow for the molten salt reactor includes both delayed neutron precursor and high neutron absorbers like Xe-135. They affect greatly the reactor physics characteristics.

Proper computation method should be developed to treat these characteristics shown in table 2.

3.3. Computation for Accident management

Experiences of Fukushima-Daiichi accident provided a lot of suggestions for protecting and mitigating such a severe accident. Quick proper response to the accident is one of the most important requirements. The director of reactor has to know all details of the reactor and should make an appropriate judgment on what to do when problems arise. However, this ability has been greatly exceeding that of ordinary human beings. Furthermore, at the accident all of necessary data cannot be obtained in most cases, and human error easily occur. Mental stress of the director and operators become too strong.

Computer can make a more calm judgment as compared to humans. It can make a proper judgment instantaneously from a lot of information. In case of an accident it is conceivable that the computer makes a decision on behalf of human beings.

All of the reactor information must be given to the computer. In an accident, some information may be impossible to be given due to disconnection of the signal cable or some other reasons. We have to build a system so that this kind of thing does not happen, but even if it happens, the computer system must make appropriate judgments with available information.

Rapid judgment is inevitable though the problem is so large and complicated. The artificial intelligence (AI)[7] seems very efficient tool for this problem. The AI programs for games such as chess and go-game has been developed drastically for recent years. Some of the technology of game programs may be employed in our program.

To err is human, and the responsibility for a nuclear accident is too heavy for the humans. The author considers humans should not make such judgements. However, according to usual human rules, it may appear that there is a doubt about the judgment of the computer. In such a case, it may become necessary to make a system directed by humans that can communicate with the computer.

References

[1] https://todayinsci.com/E/Einstein_Albert/EinsteinAlbert-ProblemQuote800px.htm.
[2] Wikipedia 2017 *Moore's law*, https://en.wikipedia.org/wiki/Moore%27s_law.
[3] Sekimoto H, Pigford TH 1974 A New Method for Calculating Space-Dependent Burnup *Trans. American Nuclear Society* **18** 137.
[4] Toshinsky VG, Sekimoto H, Toshinsky G 2000 A Method to Improve Multiobjective Genetic Algorithm Optimization of a Self-Fuel-Providing LMFBR by Niche Induction among Nondominated Solutions *Ann. Nucl. Energy* **27** 397.
[5] Wikipedia 2018 *Quantum computing*, https://en.wikipedia.org/wiki/Quantum_computing.
[6] OECD/NEA 2002 *innovative nuclear reactor development*, https://www.oecd-nea.org/ndd/reports/2002/nea3969-innovative-reactor.pdf.
[7] Wikipedia 2018 *Artificial intelligence*, https://en.wikipedia.org/wiki/Artificial_intelligence.

By-product group benefits of non-kin resource-sharing in vampire bats

R Donepudi[1] and R Ramaswamy[1,2]

[1] Department of Biotechnology and Bioinformatics, School of Life Sciences, University of Hyderabad, Hyderabad 500046, India
[2] School of Computational & Integrative Sciences, Jawaharlal Nehru University, New Delhi 110067, India

E-mail: donepudiraviteja@gmail.com

Abstract. We develop an agent based model (ABM) to simulate the behaviour of a colony of vampire bats (Order: *Chiroptera*) and study the by-product group benefits that result from resource-sharing among related as well as unrelated members of the colony. Such cooperative behaviour can lead to unexpected group benefits; there is an increase the inclusive fitness of related members of the colony (namely *kin*) and can have direct benefit when shared with unrelated members (namely *non-kin*). Sharing can also provides by-product benefits when individuals have a shared (or *group*) interest.

Our study focuses on the contrast in the group estimates between sharing and non-sharing populations. For constant ecological resources, sharing behaviour can increase the sustainable population size, increase the total resource stored in the population, and reduce the average resource required per individual, compared to a non-sharing population. (The extent of the increase or decrease will depend on the parameters of the model). This increased carrying capacity due to resource sharing can increase the fitness of individuals in the group. The increase in cooperativity has a nonlinear effect on group benefits: Substantial group benefits are shown only after a cooperativity threshold, and it increases exponentially to a maximum thereafter.

1. Introduction

Agent based modeling (ABM) techniques are known to provide considerable insight into a number of different problems in different areas of enquiry, ranging from biology, physics, and chemistry, to economics and the social sciences [1, 2, 3]. In order to understand the emergence of properties in a complex system, its parts are modelled as interacting agents with a specified minimal set of properties and behaviours. As an example of such an approach, the properties of an ecology can be seen to derive from the populations of its constituent species [4, 5] and agent-based modelling has been usefully applied in to understand the complex process of the emergence of cooperation [6] among the species. In general, ABMs using simple local interactions can give insight into complex global patterns [7]. Our aim in this work is to use autonomous behaviours of vampire bats (such as foraging, starvation, death, breeding and blood sharing) in an ABM framework in order to understand the effect at a population level, particularly in respect of the inclusive fitness of the population.

Content from this work may be used under the terms of the Creative Commons Attribution 3.0 licence. Any further distribution of this work must maintain attribution to the author(s) and the title of the work, journal citation and DOI.
Published under licence by IOP Publishing Ltd

Cooperating organisms often invest in partners preferentially so as to increase the inclusive fitness benefit [12]. Inclusive fitness is the sum of direct fitness and indirect fitness [8, 9, 10, 11]. The ability to identify close relatives, namely the phenomenon of kin discrimination, and reciprocity (a tit for tat strategy) are mechanisms that ensures that so–called cheaters do not benefit from cooperation. On the other hand, when the act of cooperation automatically provides so-called by-product benefits (which in general are shared group benefits) no specific enforcement may be required [9, 13]. In the present paper we explore such group benefits.

In the common vampire bat (*Desmodus rotundus*) food or blood sharing is a cooperative behaviour [14]. Vampire bats are obligate blood feeders and can store very limited resources for survival; a 72 hour starvation will kill the bat [15]. At the same time, bats regurgitate in order to share blood with kin, namely related members of the colony, as well as with others [16]. Cheaters, namely those who do not reciprocate help can be detected by social grooming [12, 17]. It has been observed that food-sharing with unrelated members of the colony, namely non-kin, occurs preferentially with individuals having high past reciprocation [14, 18].

Simulations by Wilkinson [16] have shown that the direct fitness benefit is low compared to indirect fitness benefit. A bat with 90% success rate of foraging takes 1110 days on average to miss three consecutive meals. Thus a typical bat may need no more than 3 to 5 donations of food through sharing in its entire lifespan. Considerable attention has been given to examine both primary (direct) fitness and indirect fitness benefits and mechanisms to maintain coopera-tion [12, 14, 16, 18]. By-product group benefits of food sharing in vampire bats have also been studied, although to a lesser extent. This is useful in understanding cooperation among non-kin individuals, and as has been seen in simulations, energy sharing as in vampire bats can bring substantial benefits to the group as a whole [19].

We present an agent-based model (ABM) of food sharing in vampire bats. Our simulations explore the group benefits of resource sharing with all individuals in the group. We find that within our model, for a given constant rate of ecological resources, both the carrying capacity as well as the total resources stored with the individuals in the population increase significantly. This increase in the sustainable population size with a small increase in resources gathered can reduce the resources required per individual.The increase in fitness due to the increased group size can result in the increased reproductive capacity of individuals.

We have also examined the effect of the rate of cooperativity on group size and find that considerable group benefits can occur only for large cooperativity. This suggests that such by-product benefits might be of use in maintaining the cooperativity, although this cannot explain the origin of cooperativity itself. Another factor in the model is the capacity to store food by an individual: this is varied from sufficient food for 3 days to sufficient food for 12 days. The increase in capacity to store food reduces the sustainable size for both cooperating and non-cooperating populations, but the population size ratio (sharing to non sharing) is nonmonotonic, increasing first and then decreasing. Our present results suggest that significant additional by-product group benefits accrue from food-sharing behaviour in vampire bats, and this aspect needs to be explicitly included in any estimation of the total fitness benefit of such cooperative behaviour.

2. Materials and Methods

The agent based model that we employ here is as follows. Each individual bat forages every day, and if successful will store three units of food resource. The success is probabilistic and depends on its own efficiency and resources available per individual. The rate of food resource in the habitat (total resource per day) is taken to be constant and thus the resources available per bat per day is inversely related to the population size. If foraging is unsuccessful the bat's stored resource is reduced by one unit, and the particular bat survives only if the stored resources have a non-zero value. This is in keeping with the observation that a bat that is unsuccessful in foraging for three consecutive days is not likely to survive [14, 15]. The foraging efficiencies for initial individuals is randomly chosen in a range (for convenience between 1/2 and 1) and we further assume that each bat reproduces once every year at a random time [20], offspring inheriting the mother's foraging efficiency.

In a population with resources-sharing behaviour, individuals that are successful in foraging are considered donors and are denoted D: these have at least 3 units of stored food. Bats with only one unit of stored resources, namely those that are unsuccessful in foraging, are denoted N (for needy). So long as the total number of N bats is lower than the D, each will receive one unit of resource randomly from a donor. If N exceeds D, then each donor randomly selects one N bat to donate one unit of resources.

We find that in our simulations, the model bat populations stabilize after a transient time (which depends on the efficiency range chosen). Quantities such as the stable size of the population, the total food stored with individuals in each population and the resources available per individual can be measured, and compared between populations that indulge in food-sharing versus those that do not. For proper comparison, we take both populations to feed on the same amount of external resource, to have the same foraging efficiencies, and further, that both groups start with the same number of individuals. The efficiency of a strategy is measured by the size of the stable population that is eventually achieved.

In order to understand the effect of partial cooperativity, each resource-sharing event is taken to be probabilistic. If the index of cooperativity is w, only that fraction of resource-sharing acts will be successful, while others maintain the *status quo* by not sharing. Another parameter of interest is c, the capacity of an individual to store resources, and in our simulations we have varied its value from three to twelve. The model, data and NETLOGO code for reproducing the results can be downloaded from the https://github.com/Donepudiraviteja/Resource-sharing/.

3. Results

Results of our simulations for populations of equal size with each habitat having the same rate of resource availability are shown in Fig. 1. Transients are discarded and the different measures are calculated after the populations have stabilized.

The sustainable surviving population on the given constant food resource in the habitat for population with resources sharing behaviour is 417, where as for non-sharing behaviour it is 277 for this set of parameters, indicating that on average, the sustainable population size increases by approximately 50% due to resource-sharing, since this is the only difference between the two models. These results are typical: simulations for various levels of resource availability per day showed similar trends: both sharing and non-sharing populations have a linear population growth as a function of the resource availability, and with a similar size ratio (data not shown).

The total resources assimilated from the habitat in a single day is calculated by adding the resources available to all individuals. In our simulations, sharing behaviour actually increased the total resources assimilated by about 11% over non-sharing population since the cooperative behaviour allowed for more foragers and consequently, fewer N individuals. The resource required per individual, namely the total food resources assimilated divided by the population

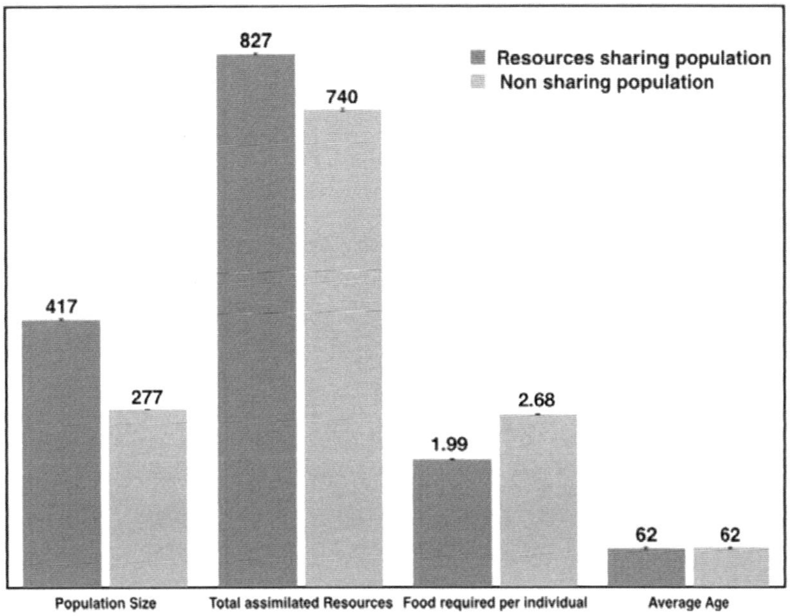

Figure 1. Comparison of measured quantities for the sharing and non-sharing populations, starting with 200 individuals and 600 units of ecological resources. A transient time of 100 days is taken, and simulations have been averaged and rounded over an ensemble of 12 realizations. The sample to sample variation is small and does not show up on this scale.

size is about one fourth *less* for food sharing population.

We also considered the case when individuals were pre-classified as resource sharing or non-sharing, and allowed them to compete for resources. An equal number of both types of individuals was considered, and in steady state, we find (see Fig. 2) that resources sharing individuals eventually take over the population: this is the dominant strategy. We keep the populations distinct, namely there are no resource transfers between sharing and non sharing individuals, and there are no defectors between the types in the population. Simulations show that an increase in cooperativity leads to a nonlinear increases in group benefits: substantial benefits are shown only after the cooperativity parameter is 60%, and it increases exponentially to a maximum thereafter.

The sustainable group size is strongly dependent on the maximum capacity c that an individual possesses to store resources. Group sizes were measured for c varying between 3 and 12. The effect of increasing capacity is, paradoxically, to *reduce* group size in both sharing and non-sharing populations (see Fig. 3). The ratio between sharing and non sharing populations has a nonmonotonic dependence on c whereas the total resources stored by both sharing and non sharing populations has the inverse effect.

Finally, to see whether the increased population size effects the lifespan, we measure the average age and the average age at death for both the populations. (In this simplified model, we only consider death to occur via resource-deprivation). For both average age and average age at death the food-sharing population is statistically similar to the non-sharing population. And

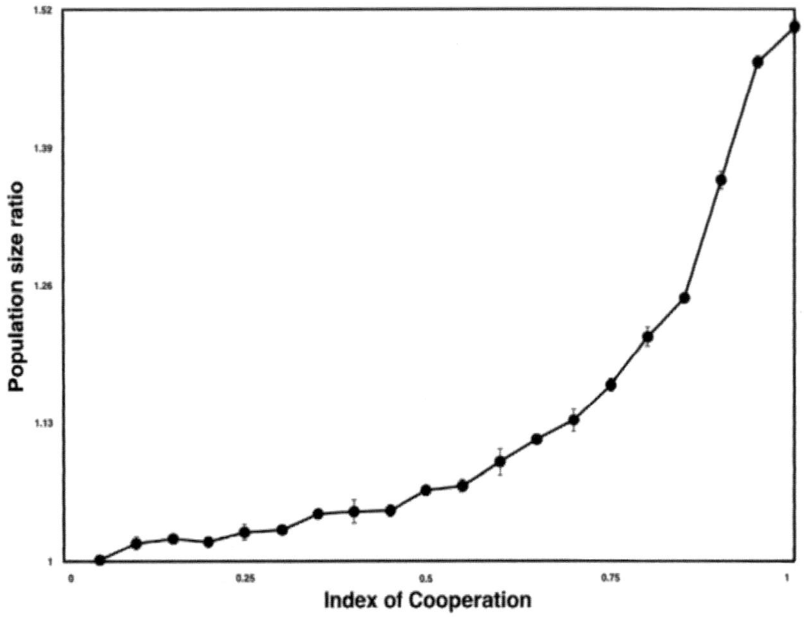

Figure 2. Group benefits with increasing cooperativity. The cooperativity of the population is varied on the abscissa, and the resulting population size ratio (between populations of sharers versus non-sharers) is on the ordinate. All results are averaged over 5 realizations, and the error bars are shown.

the average age for both food-sharing population and non-sharing population are not affected by the amount of ecological resources available or the intitial population size.

If p is the probability of unsuccessful forages the average number of days before there are three consecutive unsuccessful forages (TCUF), is clearly given by

$$\text{TCUF} = \frac{1 + p + p^2}{p^3}.$$

For the reported probability of 10% unsuccessful forages [14, 16], therefore, on average TCUF will occur after more than three years; the average lifespan of a vampire bat being about 10 years.

4. Summary and Discussion

Pseudo-reciprocity [13, 21] results when the cheaters (in the sense used here) indirectly benefit those individuals who share resources. This is a *by-product* group benefit, and in the present work we study a model of resource-sharing between individuals in a bat colony where such behaviour can lead to substantial group benefits. Mechanisms such as kin recognition and reciprocality ensure and reinforce cooperative behaviour. The by-product group benefits that result in our simulations has the effect of increasing the effective total fitness of individuals and thereby might also help in maintaining the fitness through pseudo-reciprocity.

Increasing the maximum capacity of resources that can be stored affects the quantifiers we have focussed upon for both sharing and non-sharing populations. However, the ratio of their

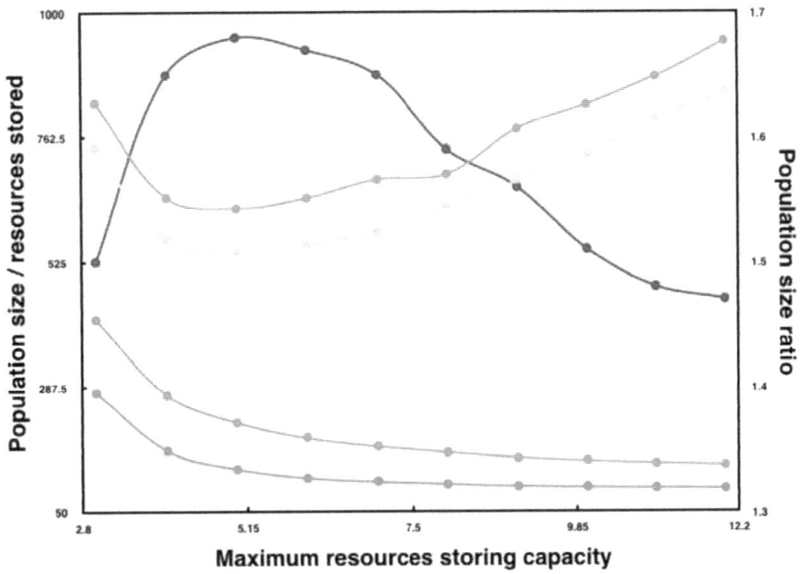

Figure 3. Group benefits with increasing maximum resource storing capacity, c plotted along the abscissa. Red curve represents population size ratio between sharing and non-sharing populations, blue and yellow curves represent the population size and total resources stored for non sharing population. The green and orange curves represents the population size and total resources stored for a sharing population. All results are averaged over 5 realizations.

respective population sizes is a nonmontonic function of storage capacity, suggesting that for a species that can store resources for sufficiently long, there is unlikely to be any added benefits due to cooperation.

Cooperativity has a nonlinear effect on benefits. For low levels of cooperativity the benefits are also low, but after the cooperativity increases to about 60% benefits rapidly increase. In the evolution of food-sharing behaviour, in the initial stages, when the co-operativity is low, the by-product benefits would be minimal, but once significant co-operativity is established, the consequent benefits help in its maintenance.

In summary, our simple application of agent based models to understand cooperative behaviour of vampire bats suggests that resource-sharing between unrelated individuals in vampire bat colonies can yield substantial by-product benefits. Such behaviour has not been explored in previous studies, and this suggests that more specifically defined resource-sharing models need to be developed and studied, particularly to understand better the differential advantages of various strategies of foraging and resource allocation.

Acknowledgment
RR thanks the Department of Science and Technology for support through the J C Bose National Fellowship, and DR is grateful to the CSIR, India, for the award of the SPM Junior Research fellowship.

IOP Conf. Series: Journal of Physics: Conf. Series **1090** (2018) 012002 doi:10.1088/1742-6596/1090/1/012002

References

[1] Tesfatsion L 2002 Agent-based computational economics: growing economies from the bottom Up *Artificial Life* **8** 1 55–82
[2] An G, Mi Q, Dutta-Moscato J and Vodovotz Y 2009 Agent-based models in translational systems biology *Wiley Interdiscip Rev. Syst. Biol. Med.* **1** 159–71
[3] Joshua E M 1999 Agent-based computational models and generative social science *Complexity* **4** 5 41–60
[4] Grimm V and Railsback S F 2005 *Individual-Based Modeling and Ecology* (Princeton and Oxford: Princeton University Press) p 437
[5] Lomnicki A 1992 Population Ecology from the Individual Perspective *Individual-Based Models and Approaches Ecology: Populations, Communities and Ecosystem* ed. DeAngelis D L and Gross L J (New York: Chapman and Hall) pp 3–17
[6] Axelrod R 1997 *The Complexity of Cooperation: Agent-Based Models of Competition and Collaboration* (Princeton: Princeton University Press) p 248
[7] Macy M W and Willer R 2002 From factors to factors: computational sociology and agent-based modeling *Annual Review of Sociology* **28** 143–66
[8] Hamilton W 1964 The genetical evolution of social behaviour I *J. Theor. Biol.* **7** 1–16
[9] West S A, Griffin A S and Gardner A 2007 Evolutionary Explanations for Cooperation *Curr. Biol.* **17** R661–R672
[10] Grafen A 1984 Natural selection, kin selection and group selection *Behavioural Ecology: An Evolutionary Approach* 2nd Edition ed Krebs J R and Davies N B (Oxford: Blackwell Scientific Publications) pp 62–84
[11] Lehmann L and Keller L 2006 The evolution of cooperation and altruism–a general framework and a classification of models *J. Evol. Biol.* **19** 1365–76
[12] Carter G G and Wilkinson G S 2015 Social benefits of non-kin food sharing by female vampire bats *Proc. R. Soc. B* **282**, (2015)2524
[13] Connor R C 2010 Cooperation beyond the dyad: on simple models and a complex society *Phil. Trans. R. Soc. B* **365** 2687–97
[14] Wilkinson G S 1984 Reciprocal food sharing in the vampire bat *Nature* **308** 181–84
[15] McNab B K 1973 Energetics and the distribution of vampires *J. Mammal.* **54** 131–44
[16] Wilkinson G S 1988 Reciprocal altruism in bats and other mammals *Ethol. Sociobiol.* **9** 85–100
[17] Wilkinson G S 1986 Social grooming in the common vampire bat, Desmodus rotundus *Anim. Behav.* **34** 1880–9
[18] Carter G G and Wilkinson G S 2013 Food sharing in vampire bats: reciprocal help predicts donations more than relatedness or harassment *Proc. R. Soc. B* **280** (2012)2573
[19] Witkowski M 2007 Energy sharing for swarms modeled on the Common vampire bat *Adapt. Behav.* **15** 307–328
[20] Greenhall A M, Joermann G and Schmidt U 1983 *Desmodu rotundus* (Mammalian Species) *The American Society of Mammalogists* **202** 1–6
[21] Clutton-Brock T 2009 Cooperation between non-kin in animal societies *Nature* **462** 51–57

IOP Conf. Series: Journal of Physics: Conf. Series **1090** (2018) 012003 doi:10.1088/1742-6596/1090/1/012003

AMIR CPU: World's First and Only 32-bit Softcore Processor in Schematic on Freeware Platform

Dr. Muhammad Nasir Ibrahim[1], Namazi Azhari[2], Adam Baharum[2], Dr. Mariani Idroas[1], Uswah Khairudin[1], Johari Kassim[1], Md Mohar Muhammad[1]

[1]Universiti Teknologi Malaysia,
[2]Pahlawan Mikropemproses Sdn Bhd

Abstract. The developing world has always been a spectator and consumer participant in the ICT technology since the pioneering computing years of post second world war of world economic growth. The developed world led by the US, Europe, Japan, and later South Korea have surged far ahead in ICT being represented by their own creation of microprocessor platforms evolving into an exclusive club which they locked with licensing. Their early advantage of mostly trial and error processor designs have proven to be a market success that discouraged new aspiring contenders. New CAD tools have arrived along with the cloud computing model presenting new opportunities. Major ICT developments such as the Microsoft Catapult project in 2012, and the Intel purchase of Altera for USD16.7 billion in 2016 should be perceived as alert warnings. To us this is the perfect time to capitalize on the mistakes of the earlier processor designs to create a new and novel 32-bit processor for the rest of the free world to hop along the ICT revolution into the cloud and IoT applications on *Freeware* platform as active creative participants. These aspirations have to begin at the education level with an easy to learn 32-bit processor model.

Keywords: Freeware, Softcore, RISC/CISC, FPGA, HDL/VHDL/Verilog

1. Introduction

We believe that for the developing countries to enter the playing field in ICT in order to create technological innovations of their own, it is imperative to have control over one platform that is a major building block of ICT: the microprocessor which is free from licensing beginning at the education level. ICT education tools have advanced so far as to be able to leapfrog the traditional methods of circuit wiring of breadboarding by supplanting with *Freeware* development tools and Intellectual-Property (IP) components. This freeware IP is the AMIR 32-bit softcore processor which is also multi-platform able to target all similarly capable devices from different vendors.

The processor was created in schematic so as to be able to reach a wide user-base of science and engineering professionals including occasional users such as scientists, students, and hobbyists. Unlike HDL (Hardware Description Language) design entry such as VHDL and Verilog which are favored by electronic design engineers which require steep learning curve, the schematic design entry requires virtually zero learning curve.

Content from this work may be used under the terms of the Creative Commons Attribution 3.0 licence. Any further distribution of this work must maintain attribution to the author(s) and the title of the work, journal citation and DOI.
Published under licence by IOP Publishing Ltd

We define softcore processor as:

- Processor distribution that exists only in software, i.e., entity can be emailed as files in a folder
- On a board, without power there is no ready processor hardware; processor comes to life after a software image file of it is uploaded to an FPGA (Field-Programmable-Gate-Array).

We define AMIR freeware platform as:

- All required software tools; development environment, components library used are free from IP license as they can be legitimately downloaded from provider's website as well as freedom to make copies, and distribute
- An additional freedom is added specifically for the AMIR platform is that any AMIR derivative designs have the same *Freeware* rights with the rights as proprietary products without returning source codes. The last requirement is what differentiates between *Open-Source.*
- All this applies only in an FPGA target.

The basic software tools & components under the AMIR softcore freeware platform are listed below:

- Quartus II Web edition or any other freely downloadable FPGA tools
- Built-in libraries consisting primitive logic gates, 74hc series ICs, FPGA SRAM
- AMIR softcore processor component
- AMIR assembler
- GCC (GNU compiler)
- Editors using built-in Microsoft such as: Notepad or WordPad
- Editors under Ubuntu such as: gedit, LibreOffice
- Schematic+PCB CAD using KiCad in Ubuntu and Windows.

2. History of Freeware and Softcore Processor

An essentially "free software system" or "Freeware" built around the Unix platform was started by Richard Stallman who announced the GNU project in 1983 [1]. By 1990 all major components of GNU was completed except the kernel. Linus Torvalds happened to create the Linux kernel around this time, releasing it as freeware in 1992 [1]. Thus the GNU/Linux freeware platform was born and influenced the trajectory of computing within two decades. Microprocessors followed a course of gradual improvements beginning from 4-bit, 8-bit, 16-bit before hitting to the mature 32-bit in which Linux were required. When Linus Torvalds started Linux in 1991 as a hobby he chose a readily available platform: a 32-bit i386 PC thus unknowingly chosen the bit-width of his free OS based on Unix in which Posix was based [2,3]. This already set the data size that a new 21st century CPU design should begin. A 32-bit CPU can now take place jobs done by 16-bit, 8-bit, 4-bit easily and in the same footprint.

Xilinx introduced the first FPGA in 1985 with the XC2064 chip having 1200 programmable logic gates, and 64 CLBs (*Configurable-Logic-Blocks*) in a 2-micron process CMOS silicon [4,5]. This was the first technology that enabled simple 8-bit softcore processors to be implemented in a number of this part. In 1991 Xilinx made a major improvement to the FPGA by including SRAM blocks in the new XC4000 chip [6]. The SRAM blocks opened the possibility to create a single-chip softcore processor. The same year Philip Freidin created the first general purpose softcore CPU, the R16 or RISC4005 a 16-bit RISC processor with 16 registers running at 20 Mhz [7].
From there onwards new softcore processors were being created every year including soft versions of hard processor architectures from Intel, ARM, and others.

Currently as of March 2018 there are more than a hundred softcore processors consisting of 4-bit, 8-bit, 16-bit, 32-bit, 64-bit, and variable-bit data path. Roughly 40 of them are 32-bit [8]. Most are from the US and Europe. Some are freeware while some require licensing. All exist in high-level description language (HDL) with only one done in schematic by Jan Gray which is a 16-bit softcore processor [8]. This AMIR is the first and only one in the world that is 32-bit done in schematic and available as a freeware component for all target platforms.

We list below [9] some of the 32-bit softcore processors available as freeware :

- ZPU is a 32-bit stack-based from Norway created in 2008 by Zylin which has the copyright
- F32C is a 32-bit RISC V implementation subset from the University of Zagreb, Croatia
- F-CPU (Freedom CPU linked to YASEP)started in 1999, restarted in 2004 was once an ambitious effort which is still unfinished
- YASEP 16/32 bits; RTL in VHDL, & asm in Java, microcontroller subset: ready
- OpenRISC from a loose-knit organization called OpenCores in 2002 offered a free 32-bit processor IP whose architecure is similar to Sparc
- RISC V from Berkeley, a creation of the RISC inventor Prof David Patterson is a 32/64/128-bit softcore based on the pioneering quantitative study
- ZIP CPU is a 32-bit integer RISC processor created by Dan Gisselquist from Virginia US in 2015

3. Why Another 32-Bit Processor and Soft-Core

This section highlights some of the motivations for a new 32-bit processor architecture beginning with a softcore implementation. Selection of 32-bit as the data size covers 8-bit, and 16-bit applications for biggest market share presently, and the future especially for IoT devices. Processors have evolved to two main application groups: 32-bit and 64-bit. Currently PCs are slowly migrating to 64-bit which is the domain of servers. 32-bit now covers many application areas in embedded systems and user/work devices. Embedded applications in IoT, Broadband equipment, Robotics, Instrumentation which are dominated by ARM. User/Work applications in PC, Internet access, Smartphones which are dominated by both Intel and ARM. The technical advantages of 32-bit over 8/16-bit rests on the following four main points:

- Addressing; 16-bit addressing of 64kB is now too small
- Instructions; wider bit-fields preserve operands, larger immediate data and offset branch
- Data processing; minimum floating point data definition is 32-bit. Smaller width requires complex multi-byte operation
- Counters; necessary part of hardware and software housekeeping; 8-bit counters overflow easily, 16-bit inadequate

When we asked a number of instructors with regard to teaching the architecture of major 32-bit processors which inherently must include the assembly language, they are frustrated by the complexity of the bizarre ISA of the most popular platforms. For example, why the Intel x86 created 95 different instructions for branches? [10]

Not many professionals realized that the big vendors covered their shortcomings by saying there is no need for using assembly language as the C language is efficient enough. But in education, learning a processor architecture has to include the ISA, which is accessible only through the assembly language. Bob Colwell the chief architect for Intel Pentium P6: Pentium II/III, and Celeron, said in his interview *"..it was only when I took an assembly language course that I began to understand what a computer architecture was about.."* [11].

Case in point; the Intel IA-32 or the x86 ISA consists of some 95 assembly instructions for branches alone which present one strong argument why the Intel architecture discouraged assembly language education.

3.1. Intimidating repertoire of arcane and obtuse instructions and mnemonics just for branches

The IA32/x86 ISA has an excess of branch instructions which are more than a complete set of all the instructions in many microprocessors. Many are seldom used, hard to learn and remember.

Table 1. Listing of all 95 branch instructions in left and right tables.

No	*Opcode*	*Instruction*	*Description*		*No*	*Opcode*	*Instruction*	*Description*
1	77	JA rel8	Jump short if Above		49	OF 8F	JG rel32	Jump near if greater
2	73	JAE rel8	Jump short if Above or Equal		50	OF 8D	JGE rel16	Jump near if greater or Equal
3	72	JB rel8	Jump short if Below		51	OF 8D	JGE rel32	Jump near if greater or Equal
4	76	JBE rel8	Jump short if Below or Equal		52	OF 8C	JL rel16	Jump near if Less
5	72	JC rel8	Jump short if Carry		53	OF 8C	JL rel32	Jump near if Less
6	E3	JCXZ rel8	Jump short if CX =0		54	OF 8E	JLE rel16	Jump near if Less or Equal
7	E3	JECXZ rel8	Jump short if ECX =0		55	OF 8E	JLE rel32	Jump near if Less or Equal
8	E3	JRECXZ rel8	Jump short if RCX =0		56	OF 86	JNA rel16	Jump near if Not Above
9	74	JE rel8	Jump short if Equal		57	OF 86	JNA rel32	Jump near if Not Above
10	7F	JG rel8	Jump short if Greater		58	OF 82	JNAE rel16	Jump near if Not Above or Equal
11	7D	JGE rel8	Jump short if Greater or Equal		59	OF 82	JNAE rel32	Jump near if Not Above or Equal
12	7C	JL rel8	Jump short if Less		60	OF 83	JNB rel16	Jump near if Not Below
13	7E	JLE rel8	Jump short if Less or Equal		61	OF 83	JNB rel32	Jump near if Not Below
14	76	JNA rel8	Jump short if Not Above		62	OF 87	JNBE rel16	Jump near if Not Below or Equal
15	72	JNAE rel8	Jump short if Not Above or Equal		63	OF 87	JNBE rel32	Jump near if Not Below or Equal
16	73	JNB rel8	Jump short if Not Below		64	OF 83	JNC rel16	Jump near if Not Carry
17	77	JNBE rel8	Jump short if Not Below or Equal		65	OF 83	JNC rel32	Jump near if Not Carry
18	73	JNC rel8	Jump short if Not Carry		66	OF 85	JNE rel16	Jump near if Not Equal
19	75	JNE rel8	Jump short if Not Equal		67	OF 85	JNE rel32	Jump near if Not Equal
20	7E	JNG rel8	Jump short if Not Greater		68	OF 8E	JNG rel16	Jump near if Not Greater
21	7C	JNGE rel8	Jump short if Not Greater or Equal		69	OF 8E	JNG rel32	Jump near if Not Greater
22	7D	JNL rel8	Jump short if Not Less		70	OF 8C	JNGE rel16	Jump near if Not Greater or Equal
23	7F	JNLE rel8	Jump short if Not Less or Equal		71	OF 8C	JNGE rel32	Jump near if Not Greater or Equal
24	71	JNO rel8	Jump short if Not Overflow		72	OF 8D	JNL rel16	Jump near if Not Less
25	7B	JNP rel8	Jump short if Not Parity		73	OF 8D	JNL rel32	Jump near if Not Less
26	79	JNS rel8	Jump short if Not Sign		74	OF 8F	JNLE rel16	Jump near if Not Less or equal
27	75	JNZ rel8	Jump short if Not Zero		75	OF 8F	JNLE rel32	Jump near if Not Less or equal
28	70	JO rel8	Jump short if Overflow		76	OF 81	JNO rel16	Jump near if Not Overflow
29	7A	JP rel8	Jump short if Parity		77	OF 81	JNO rel32	Jump near if Not Overflow
30	7A	JPE rel8	Jump short if Parity Even		78	OF 88	JNP rel16	Jump near if Not Parity
31	7B	JPO rel8	Jump short if Parity Odd		79	OF 88	JNP rel32	Jump near if Not Parity
32	78	JS rel8	Jump short if Sign		80	OF 89	JNS rel16	Jump near if Not Sign
33	74	JZ rel8	Jump short if Zero		81	OF 89	JNS rel32	Jump near if Not Sign
34	OF 87	JA rel16	Jump near if Above		82	OF 85	JNZ rel16	Jump near if Not Zero
35	OF 87	JA rel32	Jump near if Above		83	OF 85	JNZ rel32	Jump near if Not Zero
36	OF 83	JAE rel16	Jump near if Above or Equal		84	OF 80	JO rel16	Jump near if Overflow
37	OF 83	JA rel32	Jump near if Above or Equal		85	OF 80	JO rel32	Jump near if Overflow
38	OF 82	JB rel16	Jump near if Below		86	OF 8A	JP rel16	Jump near if Parity
39	OF 82	JB rel32	Jump near if Below		87	OF 8A	JP rel32	Jump near if Parity
40	OF 86	JBE rel16	Jump near if Below or Equal		88	OF 8A	JPE rel16	Jump near if Parity Even
41	OF 86	JBE rel32	Jump near if Below or equal		89	OF 8A	JPE rel32	Jump near if Parity Even
42	OF 82	JC rel16	Jump near if Carry		90	OF 8B	JPO rel16	Jump near if Parity Odd
43	OF 82	JC rel32	Jump near if Carry		91	OF 8B	JPO rel32	Jump near if Parity Odd
44	OF 84	JE rel16	Jump near if Equal		92	OF 88	JS rel16	Jump near if Sign
45	OF 84	JE rel32	Jump near if equal		93	OF 88	JS rel32	Jump near if Sign
46	OF 84	JZ rel16	Jump near if Zero		94	OF 84	JZ rel16	Jump near if Zero
47	OF 84	JZ rel32	Jump near if Zero		95	OF 84	JZ rel32	Jump near if Zero
48	OF 8F	JG rel16	Jump near if Greater					

The above 95 instructions are derived from the basic 32 different codes from the mnemonics below:

JO, JNO, JS, JNS, JE, JZ, JNE, JNZ, JB, JNAE, JC, JNB, JAE, JNC, JBE, JNA, JA, JNBE, JL, JNGE, JGE, JNL, JLE, JNG, JG, JNLE, JP, JPE, JNP, JPO, JCXZ, JECXZ.

Additionally there are also the LOOP (LZ, LNZ) and the JUMP (JP) instructions. Compare these with just five for integer operations for the AMIR CPU, using the eXecute instructions:

Table 2. Branch instructions for the AMIR.

No	*Opcode*	*Instruction*	*Description*
1	1B	**X** =	Execute if Equal
2	1C	**X** <	Execute if Less
3	1D	**X** >	Execute if Greater
4	1E	**X** <=	Execute if Less or Equal
5	1F	**X** >=	Execute if Greater or Equal

There is no ambiguity as to what the above do as they center around ZERO result. If condition is true, execute the next instructions, if false jump to the 24-bit offset value. There is no near/far to contend with. There are legacy reasons for having the "short", and "near" mostly centering around the mindset of having variable-byte instructions with 8-bit immediate field, and byte-size registers beginning from the progenitor of x86 architecture, the 16-bit 8086. The only good reason being that they help pipelines when they minimize branches taken.

3.1.1. CPU builds a platform
A platform creates Intellectual-Property that benefits a nation by supporting education with freeware standardized components for freedom to the developing world. Platform also creates an ecosystem of hardware and software industry.

3.1.2 There is no processor from our part of the world
All popular processors whether hard or softcore came from the developed world notably the US with the Intel x86, Europe with the ARM, and Japan from several companies such as Fujitsu, Hitachi, NEC, Mitsubishi, Toshiba, and others. None came from the developing nations.

3.1.3. Softcore/Reconfigurable/FPGA CPU is the future in education and industry
The time has come to lead the way in education of the future in response to Intel's USD16.7 billion gamble in 2015 [12], and Microsoft's server farms use of specialized FPGA processors beginning 2014 codenamed "*Catapult*" making it mainstream cloud equipment [13]. The gist of the purchase is about a whole way of learning and building electronics which today centers around a microprocessor.
Learning digital electronics starts from logic gates; the basic building block of all complex digital circuits in all smart devices. From logic gates, future scientists and engineers learn combinational logic, then FSM (Finite-State-Machines), then functional logic blocks, then on to complex functions interconnecting them, finally leading to microprocessors. These steps can now be integrated onto just one small motherboard consisting of a softcore CPU, the ultimate digital electronics learning and designing environment in school right up to professional level with the same development tools. Even without a physical board, the free software development tools can simulate the design so as the environment can be made available to every computer desk in which a student normally works as freeware laboratory. With freeware CPU which is multi-targeted, our educators in higher learning and industry have a complete set of components on a uniform platform to impart know-how as well as target for commercialization, joining the promising R&D community working their way to the digital future of IoT, internet hardware, and cloud applications.

4. Notable Features of AMIR CPU
The AMIR CPU is not just another 32-bit processor as it has a few innovative schemes of its own. We present here a simple instruction format depicting its consistency in which a cursory programmer's model could be inferred.

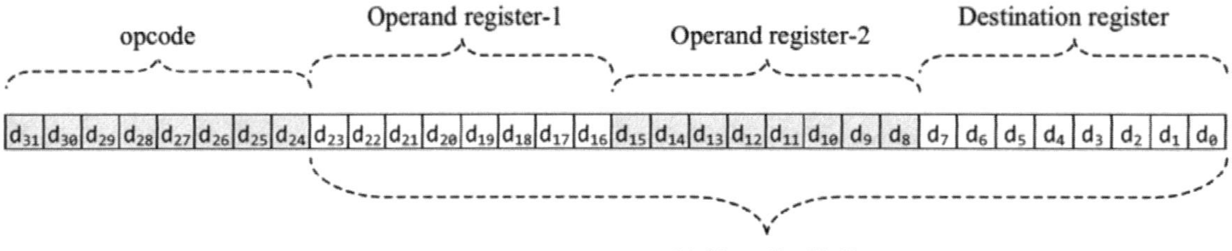

Figure 1. Instruction format bit-field of the AMIR

The opcode is always 8-bit. If no registers are specified, the open 24-bit field will be an ordinal of an offset value. Listed in the following are some of the notable differentiating features of the AMIR CPU architecture that maps to the simplified block diagram below in Figure 2 .

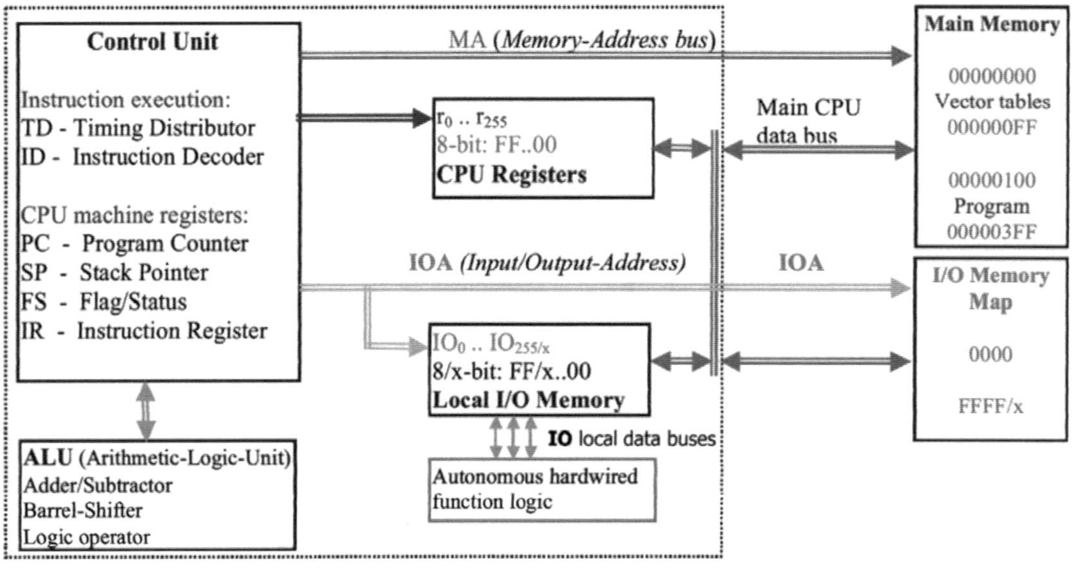

Figure 2. Simplified AMIR CPU architecture block diagram

4.1. Orthogonal instruction set makes easy to learn which is ideal for education

Orthogonality is synonymous to consistency in a processor ISA (Instruction-Set-Architecture). The main criteria for instruction set orthogonality is in the register set, in which all can be used for both addressing and data without specialized or implied use except for rare occasions. This is possible when there are many general-purpose registers available typically sixteen or more. Another criteria is in instruction use where the same instructions can be reused for other operations without creating new specialized instructions and mnemonics, thus making them easier to learn. This is possible with three-register instruction format.

Thus the AMIR CPU was designed with 256 registers using a three-register instruction format, in one stroke lowers the learning curve.

4.2. Better and simple data transfers with separate banks for I/O

Data transfers from *Load-Store* occur some 30% [14,15] which could be more when considering transfers between I/O and memory. Direct data transfers are possible only when the source and destination are in different banks, otherwise either temporary buffer or a DMA engine is required. Ideally data-transfer scheme in different banks should look like this:

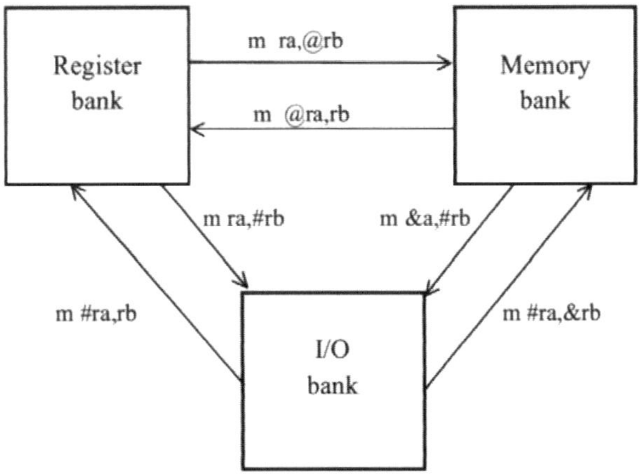

Figure 3. The three-bank direct data transfer architecture of CISC

The Intel x86 provided specialized I/O transfers with the IN and OUT instructions involving registers. The INS and OUTS involving "strings" and memory use implied registers DX, ES, DI which falls in the *non-orthogonal* instruction set category synonymous with inconsistency and hard to learn.
The ARM has no explicit I/O instructions at all as it uses memory-mapped I/O. All processor architectures by default are memory-mapped I/O including Intel x86. Memory-mapped I/O is a lazy man's scheme or "do nothing" I/O design effort. This decision gave rise to unnecessary problems later on, such as cache-coherency [16], and Gigabytes memory module-I/O partitioning. ARM was careless not to foresee this in 1983 when it was designed around memory modules of Kilobytes which was easy to partition among I/O peripherals. To see this problem it created, give a graduate student a project how to access I/O when all 32-bit memory addresses are used by one 4 GB DIMM which is now a common commodity PC part.

The AMIR CPU I/O scheme is revolutionary in that it encourages a change in mindset such that there is no need to think in terms of I/O transfer. All are data transfers from one bank to another bank. The instructions all use the same **M** (*Move*) mnemonic. Transfers between banks are differentiated by the characters @,# producing better *orthogonality* as tabled below.

Table 3. Data transfer instructions for the AMIR.

Transfer operation	Instruction
Register to Memory	m ra, @rb
Memory to Register	m @ra, rb
Register to I/O	m ra, #rb
I/O to Register	m #ra, rb
Memory to I/O	m @ra, #rb
I/O to Memory	m #ra, @rb

4.3. Neither RISC nor CISC – Takes the best features, leave the worst

The AMIR CPU is neither RISC nor CISC. We took the best features of both and leave the worst. Though the AMIR CPU is not microcoded, there is nothing inherently wrong with microcoding technology apart from its inefficient use of transistors for large inputs. The future has to include some adaptation of microcoding because of its speed advantage due to one-level of logic, and adaptibility to correct some errors after manufacturing [11]. LUT (*Look-Up-Table*) uses the same technology as microcoding and is the basic technology used in FPGAs.

The main philosophy of RISC which translates into a few main rules is to create a simple set of instructions that execute fast with consistent instruction format which facilitates pipelining. Thus one of the main rules, in which RISC is often called a *Load-Store* processor is a questionable tenet which we believe to be an error. From this *Load-Store* rule, RISC has to adopt the memory-mapped I/O data transfer when its rule states no direct memory-I/O transfer is allowed. Thus the ideal scheme of Figure 3 cannot be realized, and instead reduced to this:

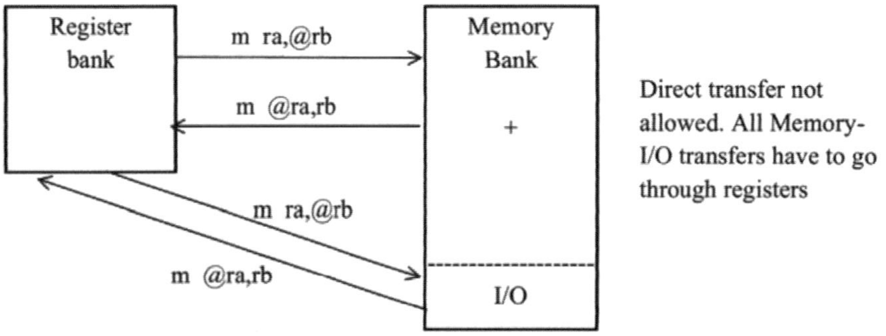

Figure 4. The two-bank indirect data transfer between memory and I/O architecture of RISC

The wisdom for memory-mapped is based on two: 1) All the powerful instruction set used on memory can be used for I/O devices as well, and 2) No complicated hardware needed for isolated I/O operations.

We contend both reasons are now incorrect, though they made sense in the eighties when memory modules were in the kilobytes or megabytes, and software directly controls I/O devices.
Firstly, present memory modules came in Gigabytes that it is difficult if not impossible to partition memory space into I/O addresses. Secondly, current portable OS software discourages direct control of devices other than just transfer of data block to maintain device driver independence.
In another important area where RISC breaks down is when SIMD is required in packet processing, multimedia, and cryptography which the internet heavily relies on. RISC and SIMD are two opposing ideas in that RISC leans on single-cycle instruction whilst SIMD is multi-cycle.
RISC simple *load-store* and *pipelining* concept could be traced to Seymour Cray in 1964 used in the CDC6600, later John Cocke from IBM continued on it in mid 70's and implemented in the IBM801 RISC CPU in 1981 [17]. David Patterson of Berkeley gave the concept its name "RISC" and popularized it with the RISC-I processor in the same year [18]. IBM has now returned to CISC with the latest generation of its mainframe 64-bit processor chip, created in 2000 known as the IBM z/Architecture based on CISC microcode technology which enabled inherent parallel processing on I/O channels. In 2015 IBM announced the z13 CISC microprocessor for their z13 mainframe computers which at the time was the world's fastest microprocessor [19].

The best contribution of RISC are the three-operand instructions, and pipelining though both did not come from RISC which only adopted and popularized it. Both originated from old mainframe implementations. The three-operand instructions came from Honeywell 1800 circa 1974 [20], and pipelining from IBM 7030 *Stretch* circa 1960 [21].
The three-operand instructions made many two-operand instructions obsolete like; Compare, Bit-masking, Test, and Translation/Decoding.

4.4. Flat and linear 256-register bank

Register oriented architecture best maps higher level language to machine implementation, such that all major processors adopted it. The only argument is in the optimum number of registers. Parameters are passed via registers if they fit within the register bank. If not, then via the stack, and if a block of parameters is needed, then via memory pointer. The pioneering processor chips use accumulators due to fabrication constraints that it is now discarded. The Intel x86 still carried this legacy albeit with more spare registers totaling eight. The ARM started with sixteen per mode totaling 6 modes of bank-switched registers. Later processors settled with thirty two general purpose registers. Once selected, this is fixed into the architecture and cannot be easily changed without affecting compatibility.
The AMIR CPU implemented 256 registers using a 24-bit register field specifier in the opcode. This leaves an 8-bit instruction field. The math coprocessor and other hardware accelerators use this same 8-bit opcode specifier to access the private coprocessors' register banks. Therefore theoretically the AMIR CPU places an absolute limit of registers using the 24-bit offset in the 32-bit instruction for coprocessors.

We believe this flat and linear 256-register bank is the best choice for microcontroller applications as it operates in supervisor mode in which during an interrupt, the processor sees all registers thus eliminating bank-switching. For user mode, the local register set can be chosen to be 16 or 32. Bank-switch register sets is now considered a mistake and new processors no longer implement it.

The desire for more registers have been validated by new (2010) ARM 64-bit architecture, the ARMv8 or A64 with 32 registers instead of 16. The A64 also eliminated the bank-switching modes altogether, admitting it as a mistake in its document [22].

4.5. Interrupt scheme that is fast and simple

In almost all processors there are two default interrupt inputs, usually the maskable interrupts (INT) and the non-maskable interrupts (NMI) as in the Intel x86. ARM has the normal maskable interrupt (IRQ) and the fast interrupt (FIQ). If there are more input sources than these, then an interrupt controller must be devised which normally features an interrupt pending stack and a priority scheme.
The AMIR CPU implements a simple single interrupt scheme with an expandable eight input sources with default interrupt vectors that correspond to addresses of interrupt handlers in low memory. A priority encoder with the LSB (Least-Significant-Bit) having the highest priority, gives an output to the *INT vector-port* with a default I/O address =1 which is read by the CPU to match the vector table of interrupt handlers. The interrupt lines are also ORed together into a single INT FF (flip-flop) input which enables the next instruction fetch to be the hardware interrupt opcode. If the INT vector=1 which corresponds to memory address=1, the jump is made to the 32-bit address value found at this location which is the interrupt handler for this interrupt source. Once the interrupt vector is determined, the CPU outputs an INTA (Interrupt-Acknowledge) signal which is combined with the interrupt line address to disable the INT FF. The interrupt source can be a combination of both level-triggered or edge-triggered. Similar to the Intel scheme, there is only one jump required to start the interrupt handler.

IOP Conf. Series: Journal of Physics: Conf. Series **1090** (2018) 012003 doi:10.1088/1742-6596/1090/1/012003

4.6. DMA eliminated with block-transfer instructions

The AMIR CPU is about the only known processor architecture that made the elimination of the DMA possible. DMA is a block transfer between a memory bank and a block of data in another bank usually an I/O in a synchronous-burst data transfer. The AMIR CPU with the isolated memory and I/O banks can execute a synchronous-burst data transfer using the normal moves, M @Ra,#Rb,Rc. If Rc is not specified, the same opcode executes a data transfer of only one 32-bit word. If Rc is specified, the 32-bit data transfer is made synchronously to an I/O port until Rc=0. Rc is recommended to have an upper limit of 0xFFFF or 64 kilowords when the CPU clock runs at 100 Mhz to keep execution time within 20 milliseconds or under 100 hz to be within the recommended multitasking latency period or a video refresh period.

5. Mistakes in the Two Most Successful Architectures: Intel X86, Arm

We selectively compare the notable features of the two most successful processor architectures today; Intel and ARM which happen to represent CISC and RISC respectively. The Intel x86 which AMD also produced as the only competitor accounts for virtually 100% of all PCs in the world including the Apple Macs. On the other front which is the smartphone, with the exit of Intel Atom in 2016, ARM controls virtually 100% of mobile processors. Hence it makes most sense to bring out the notable mistakes on these two architectures. Both are not new architectures as they were designed before 1985 when PCs were just being accepted as office and productivity tools, and the OS for the period was single-tasking floppy-disk based. Chip density was in the 1-micron and above that put a limit into the innovative processor ideas possible. That was the mindset that created the Intel x86 architecture which was 16-bit at birth, and the ARM which was a simple 32-bit at birth. The ARM which came out in 1985 was seven years later than the Intel 8086 and was much superior with more registers and advanced software support of protected mode. This same year too Intel came out with the 80386 with all the desirable support of software OS such as Unix from the upgrade of the 80286 in 1982. So the first ARM appeared in an already better software requirement of the period. Since both represented the pioneering evolution of CISC and RISC designs respectively, naturally we came to expect mistakes in the trial-and-error implementations. We bring out here the most notable mistakes in both architectures that we have learned.

5.1. Mistakes in the Intel x86 32-bit architecture

The progenitor of the venerable Intel x86 architecture was the 8086 16-bit processor which came out in 1978. Though it had evolved into the improved 16-bit 80286 in 1982, the 32-bit 80386 in 1985, the 80486 with built-in math coprocessor in 1989, the superscalar improved 32-bit Pentium P5 in 1993, and the first 64-bit x86 created by AMD came out in 2003. Despite their increased data sizes they remained downward binary compatible. We selected here the most notable mistakes.

5.1.1. The variable bytes instruction set

The x86 came from the CISC mindset with many instruction cycles from the variable bytes instruction set. Intel had to make this classic CISC into RISC and had to retrofit the instructions into a uniform instruction size which RISC had already adopted it from its first inception. The mistakes must be patched with extra logic and silicon to bring them to the next performance level.

It was estimated that 30% of transistors in the Pentium were spent for x86 legacy support and 40% in P6 [23]. Pentium front end logic had to compensate for non-uniform instruction size which affects cache lines, and segmented memory support by a translation stage needed for efficient decoding. This is a significant portion of the 5.5 million transistors in the first P6 version of the Pentium Pro in the mid nineties [23].

5.1.2. Segmentation

This may rank as the top complaint against the x86 processor which caused unnecessary brain twisting and arguably many software bugs in addressing calculations so much so that we bring to light one quote:
"*Computer historians will very likely conclude that the IBM/Microsoft choice of the segmented-memory Intel x86 processor family over the linear address-space Motorola 680x0 family for the IBM PC is one of the greatest mistakes ever made in computer system design.*" (1994) Nelson H. F. Beebe, Center for Scientific Computing, University of Utah.
This memory segmentation is redundant when memory paging in virtual memory does the same thing.

5.1.3. Two-operand instructions

A three-operand instruction typical of RISC makes many two-operand instructions redundant.
With AMIR CPU's three-operand instructions such as the SUB R0,R1,R2 the Intels' CMP (compare), BT (bit-test), TEST, XLAT are unnecessary. Examples given below.
A *Compare* (CMP) instruction requires a corresponding *Jump conditional* (Jcc) instruction.

CMP EAX, EBX ; subtract value in EBX register from EAX register
JE (*32-bit offset*) ; jump if equal an offset 32-bit value from PC (*Program-Counter*)

There are many different compare instructions and mnemonics apart from the jump instructions.
A three-operand instruction uses the same SUB R0,R1,R2 for integer arithmetic for all zero comparisons. The specialized compare instructions are not needed.

Testing a register with a value which is equivalent to masking operation.
Intel created a specialized instruction, TEST EAX, *32-bit data.* A three-operand instruction uses the same AND R0,R1,R2 for Logical operation.

Decoding or translation is a common operation in which a value is required to be converted to another code. Intel created a specialized XLAT (*Translation*) instruction just for this purpose. For example a binary value is needed to be displayed to a Common-Anode 7-segment display. This requires a *Look-Up-Table* (LUT) to be created.

Intel's method is complex and for most occasions hard to remember, that is why Intel seems to encourage to visualize its ISA as a black hole to be abstracted by the C language.

The following example shows how to convert binary 9 to 7-segment *Common-Anode* in a table at location 0x1000 using Intel x86 ISA.

Table 4. Address locations of common-anode codes.

Address	Common-Anode Code
1000	C0
1001	F9
1002	A4
1003	B0
1004	99
1005	92
1006	82
1007	F8
1008	80
1009	90
100A	08

100B	83
100C	C6
100D	A1
100E	86
100F	0E

Intel's XLAT implied instruction which takes no operand uses the 8-bit AL register as the binary value or index register, to index into the DS:BX memory to retrieve the code, and put it back into the AL register. For 16-bit or 32-bit values, such as may be used in internet routing tables, Intel says nothing.

Table 5. Registers used in the LUT conversion for the Intel x86 ISA.

Registers	Value	Comments
EBX	0	32-bit segment address for DS register (Intel's peculiar addressing)
DS	0x1000	starting address of Common-Anode codes
AL	9	binary value for conversion/index-pointer to code in table
AL	0x90	code in table, moved back into AL register

After executing **XLAT**, the code value 0x90 pointed by **DS:EBX** is moved to AL.

By comparison, the AMIR CPU uses the same 3-register **ADD** instruction [24] which can be used for 8-bit, 16-bit, and 32-bit values.

ADD R11,R20,R21

Table 6. Registers used in the LUT conversion for the AMIR ISA.

Registers	Value	Comments
R20	0x1000	starting address of Common-Anode codes
R11	9	binary value for conversion
R21	0x1009	index-pointer to code in table, resulting from the add operation

5.1.4. Too few general purpose registers

The eight general purpose registers have proven to be inadequate as they came from the early popular minimalist implementation of legacy accumulator mindset. They are not strictly general purpose as each are defined for certain operations in implied instructions. The register set is a good example of non-orthogonal design approach which is difficult to learn. This is the reason why we see a lot of *push-pop* stack operations in an x86 (IA-32) assembly listing.

Realizing this shortcoming, the 64-bit Intel x86 (Intel 64) increased the number of general purpose registers from eight to sixteen.

In comparison, the new 64-bit ARM had increased its 16 registers to 32 and eliminating *push-pop* stack operations altogether indicating studies that register numbers have an effect on its necessity.

5.2. Mistakes in the ARM 32-bit architecture

Many of the mistakes of the ARM architecture were already brought out under section 4 "Notable Features of AMIR CPU". We reiterate here from another perspective.

The migration of the ARM architecture from 32-bit to 64-bit provides a good reference for CPU design because of its extensive past implementation studies of what was proven to be good and bad ideas or mistakes. What was discarded implied to be mistakes as the ARMv8 or A64 is a brand new architecture and not an upgrade of the 32-bit ARM or A32. Unlike the Intel 64 which is an extension of the 32-bit x86 (IA-32) and downward binary compatible, the A64 is not binary compatible as the two ISAs are completely different. The A64 can run the old A32 codes by switching states. There are two different processors [22] within a single chip which is not the same as a dual-core which is a network of two clones.

5.2.1 Load-Store only data transfer

RISC employs only *load-store* data transfer that goes through registers as the standardized 1-cycle instruction with pipelining. This enforces data transfers between a register and a memory location with an immediate value or a register pointer. This prevents other innovative ways for direct transfer of data without having to go through registers.

5.2.2 Pure RISC fails in SIMD, packet data transfer

RISC tried so hard to keep everything executing within the same period of time, essentially one instruction or more per cycle to maintain the pipeline in the mindset and period of the early 1980's when single-tasking was the norm. When the internet exploded circa 1994 and with it multimedia, the constraint of single-cycle instruction must be broken at many places. Multimedia requires SIMD which is multicycle instruction, and internet with packet communication lines is efficient with block data transfers.

5.2.3 I/O same bank as memory

Since RISC allows data transfers only between register and memory, it follows that I/O is conveniently lumped together as memory locations at the same time achieving what was thought as an advantage to be able to control devices directly. Today data transfers occur most frequently and in blocks between memory and graphics adapter, and SATA hard disks. RISC of course does not use the *load-store* instructions for these but use DMA. RISC places another constraint in block devices, that is, it must use DMA and the need for complex DMA controller. But by just allowing direct memory-I/O transfer in different banks allowed DMA engines to be removed.

Also as brought in section 4.3, under "4.0 Notable Features of AMIR CPU", having I/O in the same bank as memory can be considered a mistake.

5.2.4. Program-Counter (PC) part of register

Many RISC architectures other than ARM exposed the PC and *Stack-Pointer* (SP) by dedicating registers to them. This is done for flexibility when only the basic leaf procedure call instruction BL (*Branch with Link*) is needed to synthesize other complex recursive subroutine calls. In the ARM, for every recursive function call, requires a few instructions for synthesis. In the end, it converged to only one favored method that it is no longer an advantage to expose the PC. This is why the PC is put back to where it should belong; inaccessible to the programmer in the A64 architecture [22]. It was noted in the ARM A64 documentation that it introduced complications for the design of compilers and complex pipelines.

5.2.5 Bank-switched register file

Bank-switched register file was popularized in RISC I design in which the 32-bit SUN Sparc architecture borrowed in its register windows. From David Patterson's seminal 1982 RISC paper in which he attributed part of RISC I superior performance came from procedure call mechanism using what he called "overlapping register banks". His benchmarks used selected C routines which did not consider multitasking. In fact nowhere in his 25-page long paper multitasking was mentioned. Currently, its not only multitasking that any processor design has to take into consideration as design requirement, but preemptive multitasking. It is this preemptive multitasking that killed Intel's much touted 1989 RISC processor the i80860. All the multi-execution units, separate caches, branch-prediction and other fancy logic could not help the Intel i860 when it performed poorly in context switching or exception handling when caches flushed easily with every interrupt [25].

The SUN Sparc has a complex scalable register files from a minimum of 40 to 520 with each register set partitioned into a 4-byte set of *in, local, out, and global* registers. The SUN Sparc v8 implemented eight sets of register windows for a total of 264 registers [26].

Patterson removed register windowing in his latest 2010 RISC V [27] in favor of a single 32-register file. The 32-bit ARM followed a similar course with six banks of register files, for a total of 96 registers.

The ARM 64-bit A64 removed these register banks in favor of just one 64-bit general purpose register file numbering 32. The reason given in the ARM documentation for removing register banks is that more than half of them were under utilized; implying an architectural mistake [22]. Current OS such as Ubuntu and Windows use preemptive multitasking such that there are many more daemons and background processes that could fit in the limited number of register windows and bank-switched register files. The scheme then became a new bottleneck when the processor spent much time doing housekeeping chores in proportion to useful work.

5.2.6. Interrupts using software table instead of direct vector

The external hardware interrupts IRQ and FIQ use the vectors 0x18 and 0x1C respectively no matter how many interrupt sources are tied. Another branch instructions are usually found at the two addresses typically ldr pc,pc,0x100 creating double jumps. Only then the actual jump addresses are retrieved from an index value. The software counterpart, the SWI instruction has a 24-bit ordinal which further requires instructions to find the actual jump address in the interrupt table.

This drawback was noted and corrected by the Atmel AT91 series ARM CPU by introducing its own proprietary Advanced-Interrupt-Controller (AIC) which routed the single vector into a hardware table of addresses. The ARM has its own interrupt controller, the ARM *PrimeCell VIC* (Vectored-Interrupt-Controller) that does essentially the same thing as the Atmel AIC.

The Intel is quite similar to the Atmel's AIC in that it defaults to an interrupt vector table in memory from addresses 0 to 0xFF.

5.2.7 Conditional codes for branch instructions

Condition codes sometimes called *predication* used in the ARM instruction (and later added to the Intel Pentium) wastes 4 bits in the instruction field with little gain, a novelty in the 80's but not now implemented in most new 32-bit CPU architectures. It is only useful in not stalling pipeline, and reduced code size but gives diminishing returns if the block of codes within a set flag condition needs instructions which modify the flag.

Later on it was made redundant with branch prediction hardware, and complicated the implementation of out-of-order execution. Thus it can be considered a mistake when the ARM A64 dropped most of them.

6. Why Schematic

Professional designers may find HDL (*Hardware-Description-Language*) the ideal method for complex designs needing collaborative partners. It certainly has its merits as demonstrated by the fact that all softcore processors except one were created by it. The main advantage after overcoming the learning curve is the ability to design at a functional level rather than from logic gates, once a desciption is completed and verified, it could be ported to all FPGA platforms.

The steep learning curve required of them may well be justified, but for these three categories of people; 1) beginner, 2) STEM (*Science-Technology-Engineering-Mathematics*) professionals such as teachers, scientists, and mathematicians, and 3) hobbyists, are more suited to schematic design entry as it is symbolic in nature and the same in any platforms with virtually zero learning curve.

There are two most popular HDL design synthesis: VHDL (*Very-High-Speed-Integrated-Circuit*) HDL, and Verilog. We limit our representation of HDL to Verilog because it resembles C and less verbose than VHDL, therefore easier to learn.

Three main grievances against Verilog; steep learning curve, design not self-documenting, cannot import into schematic or suitable block diagram. The free RTL viewer is not a substitute as the designer has no control over its output which cannot convey comprehensible graphic information. The RTL viewer is automatically drawn; take it or leave it, as there is no way to edit. Even if it could be edited, there is no guarantee it will look the same with different compilation. A graphical representation of any functioning work not just circuit design gives the designer a model for any creative mind to work on improvements. A similar analogy can be given in building a 3D model of a machine or a house both of which could be described in terms of (x,y,z) coordinates. Nobody could visualize coordinates, but a 3D solid rendition gives a whole picture in which the creative minds understand.

```
module dff(q,d,c);
    output q;
    input d,c;
    reg q;
    initial
     q=1'b1;
    always @ (posedge c)
       q=d;
    endmodule

module dff1(q,d,clk);
    output q;
    input d,clk;
    reg q;
    initial
    q=1'b0;
    always @ (posedge clk)
    q=d;
endmodule

module ring(q,clk);
    inout [2:0]q;
    input clk;
    dff u1(q[0],q[2],clk);
    dff1 u2(q[1],q[0],clk);
    dff1 u3(q[2],q[1],clk);
endmodule
```

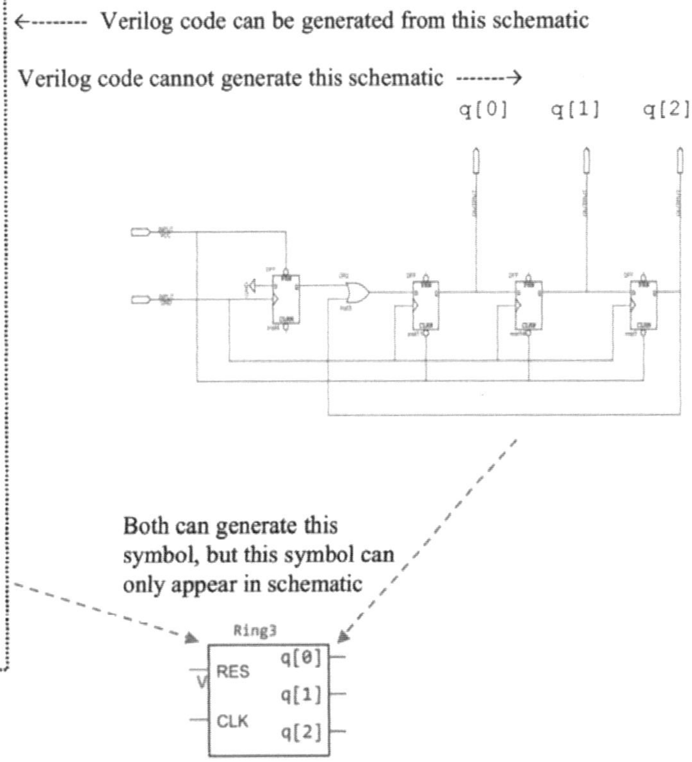

Figure 5. Comparison of Ring-Counter design with Verilog on the left and schematic on the right

An example shown below in Figure 5 is to compare the 3-stage ring-counter in Verilog on the left and its schematic on the right. One glance at the schematic and anyone who understands logic gates will immediately know its function. As for the Verilog code one has to read all the text from top to bottom like reading a story. Both can generate the same symbol which can only appear in schematic. It is this symbol capability of schematic which makes it self-documenting. This symbol is also "alive" in that it is a hierarchical embodiment of schematic design as it may contain other symbols and another complete and complex digital electronic designs in either HDL such as Verilog or schematic. The symbol could also be simulated.

A schematic design space or "block design" as it is aptly named by Altera Quartus II is a hierarchical graphical representation of a block of digital functional ideas consisting of both Verilog and schematic synthesis which can contain schematic design within schematic design or block within block. Whereas a Verilog text space cannot contain any schematic design.

A final plus for schematic is that it can also be converted to Verilog text codes in Altera Quartus II by just clicking the menu "*Create HDL Design File for Current File*" which gives a choice of "*VHDL or Verilog*". The converse is not possible; Verilog codes cannot create a schematic.

7. AMIR CPU as Electronic Components

The AMIR CPU consists of the AMIR CPU core and a separate schematic description for I/O ports which can create different components with different symbols for different applications. Shown below are three AMIR CPU manifestations as components with different number of I/O ports. These processors can operate in parallel within a single FPGA chip. The program memories reside internally in each device.

The simplest small AMIR CPU with a single input and output pair below in Figure 6, the AMIR-1 embodied in CPU4 can be used as a door-bell tone generator, or a remote-control infra-red switch. Remember that each component is a powerful 32-bit processor running at 100 Mhz. The more complex embodiments in CPU1, CPU2, and CPU3 can include separate I/O or FIFO buffers. All these processors have built-in program memory so each can run in parallel in an asymmetric multiprocessing.

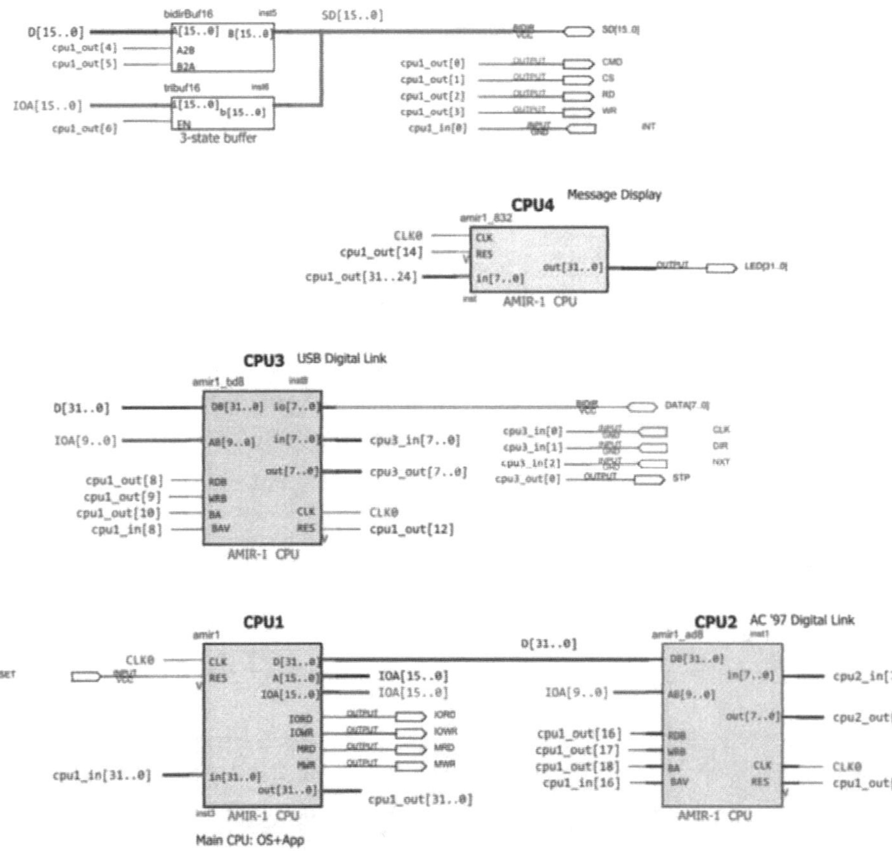

Figure 6. The four variations of AMIR CPU with different I/O port configurations connected in a single schematic for parallel processing in an asymmetric multiprocessing

8. Conclusion

We have presented some of the highlights of a new softcore processor from the developing world represented by an ASEAN country. Though the region may be late in producing a local processor as an answer to the monopolies of the Intel and ARM platforms, the advantages of benefitting from the mistakes of trial-and-error of earlier implementations to create a new design from scratch without having to declare whether it is a RISC or CISC but simply a better architecture is capitalized. The AMIR CPU is not just another processor but presents key improvements that promise better design schemes facilitating innovative creative ideas with elegant simplicity. The weaknesses of the Intel x86 and ARM architectures are motivation enough to create an alternative ICT hardware solution on a freeware platform for the benefit of the developing world without investing in a physical electronic laboratory except for the ubiquitous PCs which are now part of a normal education and work production tool.

Acknowledgements

We are indebted to the support of UTM (*Universiti Teknologi Malaysia*) ICC (*Innovation and Commercialization Center*) at Skudai, Johor Baru, Johor state of Malaysia, in the form of grants to produce a working prototype of the AMIR CPU

Gratitude also goes to KTD (*Kolej Teknologi Darul-Naim Kelantan state of Malaysia*) for facilitating the AMIR Lab in the technological college.

References

[1] Free-Software-Foundation 2017 *Overview of the GNU System*, https://www.gnu.org/gnu/gnu-history.html
[2] Stephen R. Walli, Software Systems, Inc. *POSIX: A Case Study in a Successful Standard*
[3] Linus Torvalds 1991 Newsgroups: *comp.os.minix;* Subject: *Gcc-1.40 and a posix-question*
[4] Xilinx datasheet for XC2064 1985
[5] Bradly Fawcett 1994 Xilinx Xcell15 Q4 *Examining the process*
[6] Wim Roelandts 1999 Xilinx Xcell32 Q2; *15 years of innovation, Major Xilinx milestones*
[7] Jan Gray 1991 ReConFig 2014 Keynote; RISC4005 *The first monolithic general purpose FPGA CPU*
[8] Wikipedia 2018 *Soft microprocessor,*
[9] Wikipedia and OpenCore 2018 *Softcore processors*
[10] Intel-64 and IA-32, 2011 *Architectures Software Developer's Manual Volume 2A: Instruction Set Reference, A-M*
[11] Bob Colwell Aug 2009 *Oral history of Robert P. Colwell* pp 8,26, Intel Fellow, Chief architect of Intel Pentium P6 (*Pentium Pro, Pentium II, Pentium III, Celeron*), Interviewed by Paul N. Edwards, Assoc. Prof., U of Michigan School of Information, Oregon
[12] Eugene Kim Dec 29 2015 Business Insider *3 reasons why Intel pulled the trigger on a massive $16.7 billion deal, its biggest ever*
[13] Clayton Cotterell for Wired Sep 25 2016 *Microsoft bets its future on a reprogrammable computer chip*
[14] Jan Gray Circuit Cellar March 2000, Part-1 pp 27 *Building a RISC System in an FPGA*
[15] James R. Goodman, Wei-Chung Hsu 1986 U of Wisconsin-Madison US pp 375 IEEE *On the use of registers vs cache to minimize memory traffic*
[16] Andrew Tanenbaum 2009 *Modern Operating System* 3rd edition 5.1.3 Memory-Mapped I/O pp 333
[17]W. Warner Dec 22 2004 *Great moments in microprocessor history- The history of the micro from the vacuum tube to today's dual-core multithreaded madness*
[18] C.H Sequin and D.A. Patterson 1982 U of California, Berkeley, US *Design and Implementation of RISC I;*

[19] IBM News Room/News Release Jan 13 2015 Armonk, NY *IBM Launches z13 Mainframe -- Most Powerful and Secure System Ever Built*

[20] Honeywell 1800 product description 1974 pp 15 Honeywell Electronic Data Processing, Wellesley Hills, Massachusetts, US.

[21] G.A. Blaauw and F.P. Brooks Jr. 1997 *Computer Architecture: Concepts and Evolution* section 13.3 *IBM Stretch*. Reading, MA: Addison Wesley

[22] ARM Cortex-A Series version 1.0 2015 *Programmer's Guide for ARMv8-A* Chap-4 *ARMv8 Registers* pp 4-14

[23] Jon Stokes 2007 *Inside the Machine: An Illustrated Introducton to Microprocessors and Computer Architecture* William Pollock (Pub.), Ch 5 *The Intel Pentium and Pentium Pro* pp 92, pp107, No Starch Press, Inc, San Francisco, USA.

[24] Md Nasir, N Azhari, A Baharum 2017 Pahlawan Mikropemproses Sdn Bhd, Kota Bharu Kelantan, Malaysia *AMIR-1 32-bit Microcontroller Instruction Set Architecture*

[25] Paul DeMone Jan 25 2000 real world technologies *Intel's History Lesson* pp 2 *https://www.realworldtech.com/intel-history-lesson/*

[26] SPARC International Inc. Menlo Park California US 1992 *The SPARC Architecture Manual Version 8* pp 9, 24

[27] David A. Patterson et al 2014 EE & Computer Sciences University of California Berkeley US *The RISC-V Instruction Set Manual, Volume I: User-Level ISA Version 2.0*

IOP Conf. Series: Journal of Physics: Conf. Series **1090** (2018) 012004 doi:10.1088/1742-6596/1090/1/012004

Effect of isotope enrichment on performance of lead-lithium blanket of inertial fusion reactor

Masatoshi Kondo[1], Yuu Nakajima[2], Teruya Tanaka[3], Takayoshi Norimatsu[4]

[1] Institute of Innovative Research, Tokyo Institute of Technology, 2-12-1-N1-18 Ookayama, Meguro-ku, Tokyo 152-8550, Japan
Type the author addresses here
[2] Department of Nuclear Engineering, Tokai university, 4-1-1 Kitakaname, Hiratsuka-shi, Kanagawa 259-1292, Japan
[3] Fusion System Research Division, National Institute for Fusion Science, 322-6 Oroshi-cho, Toki-city, Gifu 509-5292, Japan
[4] Institute of Laser Engineering, Osaka university, 2-6 Yamadaoka, Suita, Osaka 565-0871, Japan

E-mail: kondo.masatoshi@lane.iir.titech.ac.jp

Abstract. Simulation studies on neutron transport and material activation for lead-lithium (Pb-17Li) blanket of the inertial fusion reactor "KOYO-FAST" were performed with the PHITS code and the EASY 2005 code. Tritium breeding ratio (TBR) and productions of harmful radioactive isotopes ^{210}Po and ^{203}Hg in the blanket are made clear. The blanket performances improved by enrichment of stable isotope (^{6}Li, ^{204}Pb, ^{206}Pb or ^{207}Pb) are also made clear. The TBR is 1.28 for the blanket with a natural Pb-17Li alloy. The large TBR is allowed because of large blanket space in the reactor chamber. The TBR reaches 1.64 when ^{6}Li is enriched to 90%. In this case, intermediate and thermal neutrons, that are necessary to produce ^{210}Po and ^{203}Hg, are preferentially captured by the ^{6}Li (n, α) T reaction. Then, the productions of ^{210}Po and ^{203}Hg are mitigated. The enrichment of ^{204}Pb leads to a small TBR, because neutrons are captured by ^{204}Pb due to its large cross section for (n, γ) reaction. The TBR of ^{206}Pb or ^{207}Pb enriched Pb-17Li is more or less the same with that of the natural Pb-17Li. In these cases, the productions of ^{210}Po and ^{203}Hg are slightly mitigated because of the removal of ^{208}Pb and ^{204}Pb from the blanket.

1. Introduction

Conceptual design of a fast-ignition laser fusion reactor has been performed in Japan. According to the staged development of the reactor [1], the laser inertial fusion test experimental reactor (LIFT) [1] and the fast ignition laser fusion power plant (KOYO-FAST) [2] are being designed. In the design of the KOYO-FAST, lead lithium alloy (Pb-17Li) is selected as tritium (T) breeder in the liquid blanket system. The first wall of the chamber is protected by the film flow of Pb-17Li from high-energy neutrons, α particles and high-temperature debris. These liquid metal technologies prolong the life time of the reactor chamber and mitigate the difficulties on the maintenance scenario. Some key issues such as film flow stability [3], material compatibility [4], T monitoring [5], T recovery technology [6, 7] and other safety related phenomena [8] are being studied.

Some harmful radioactive nuclides are produced by the activation of Pb-17Li under neutron irradiation. The activation affects on the reactor safety during the reactor operation and maintenance as well as the

Content from this work may be used under the terms of the Creative Commons Attribution 3.0 licence. Any further distribution of this work must maintain attribution to the author(s) and the title of the work, journal citation and DOI.
Published under licence by IOP Publishing Ltd

processing of the activated alloy for re-use or final disposal. Some articles (see, for example, L. Petrizzi et al. [9] and G. Casini et al. [10]) analysed the activation of a Pb-Li blanket in magnet confinement fusion reactors, and reported the associated safety effects. Polonium-210 (^{210}Po) and mercury-203 (^{203}Hg) are widely recognized as the major radioactive nuclides produced in Pb-17Li and are of special concern. ^{210}Po is a radiotoxic alpha-ray-emitting nuclide with a half-life of 138.4 days and is highly mobile due to its high vapour pressure. ^{203}Hg is a radiotoxic gamma-ray-emitting nuclide with a half-life of 46.6 days. Their production must be quantified to plan the maintenance scenario and to prepare contingency plans for the accidental release of radioactive nuclides from the blanket system. Simultaneously, ways to suppress the production of harmful radioactive nuclides must also be investigated.

Natural lithium (Li) consists of two stable isotopes: Li-6 (^{6}Li) and Li-7 (^{7}Li). The T breeding ratio (TBR) is known to increase for high concentrations of ^{6}Li in Li. The technology of ^{6}Li enrichment has been investigated [11]. Lead (Pb) has a large cross section for the (n, 2n) reaction for fast neutrons. Natural Pb consists of the four isotopes ^{204}Pb, ^{206}Pb, ^{207}Pb, and ^{208}Pb. Enriching the Pb isotopes by laser separation and gas centrifugation is possible. It is known that Pb-isotope-enriched alloys have distinctive nuclear characteristics. For example, ^{208}Pb-enriched liquid Pb has been proposed as coolant for fast reactors and acceleration driven systems instead of natural Pb because of its small neutron absorption and moderation [12]. The beneficial effect of the Pb isotope enrichment on the performance of fusion blanket systems has rarely been discussed, and the information is limited.

Simulation studies on neutron transport and material activation of the Pb-17Li blanket of the "KOYO-FAST" were performed. The purpose of the present study is to make clear the tritium breeding performance and the productivity of ^{210}Po and ^{203}Hg of the blanket. The effect of the isotope enrichment on these performances is clarified.

2. Method and model for neutron transport analysis and activation calculation

The KOYO- FAST has four reactor chambers, and the fusion output of each chamber is 800 MW. The output is corresponding to the generation of 2.84×10^{20} neutrons/s with the energy of 14.1MeV. The desired lifetime of the reactor chamber is 20 years. However, the final optical systems have a life time of approximately 2 years, and must be replaced during the maintenance procedure while the reactor is stopped. The availability of the reactor is assumed to be 50% due to the required high frequency of maintenance. In this case, the total operation period of the blanket under neutron irradiation is roughly estimated to be 10 years.

Figure 1 (a) shows the geometry of the reactor chamber. The structure of the reactor chamber is simplified by the removal of the final optics etc. due to its small volume ratio. A point neutron source is situated at the position of the fast ignition. The vessel is made of JIS-STBA26. The structural material of the blanket is the reduced-activation ferritic martensitic steel JLF-1 and SiC/SiC. In the neutron transport calculation, the uniformly mixed breeder layer is simulated. The compositions of the blanket is 90vol.% Pb-17Li, 5vol.% JLF-1 and 5vol% SiC/SiC. The chemical composition of JIS- STBA26 and JLF-1 steels are Fe- 9Cr- 1Mo- 0.45Mn- 0.15C- 0.07Ta -0.5Si -0.05N (typical contents, wt%) and Fe- 9Cr- 2W- 0.5Mn- 0.1C- 0.07Ta- 0.05Si- 0.05N (typical contents, wt%), respectively. The average neutron-wall loading is 4.2 MW/m^{2}.

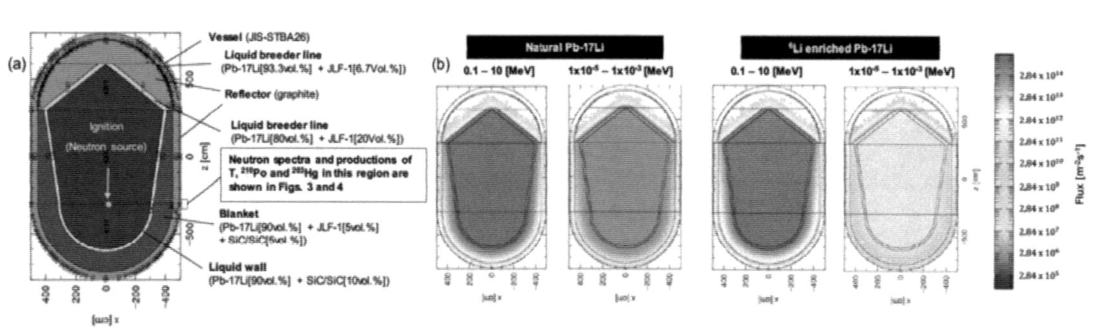

Fig.1 (a) Geometry of reactor chamber
(b) neutron flux distributions in reactor chamber by natural Pb-17Li and ^{6}Li-enriched Pb-17Li

The neutron spectra at various positions in the reactor chamber are obtained by the PHITS Monte Carlo code [13] with the JENDL 4.0 nuclear data library [14] for neutrons with energy between 1.0×10^{-5} MeV and 14.918

MeV. The neutron spectrum for each region of the reactor chamber is used for the following activation analysis. The activation analysis is performed with the EASY2005 code package containing the FISPACT-2005 code [15]. The group structure is GAMII. The period of the neutron irradiation is the total operation period of the blanket, i.e. 10 years. Then, the activation conditions are clarified for the various periods as 1 second, 1 min, 1 hour, 1 day, 30 days, and 1−10 years (with one-year increments). The dilution of the radioactive nuclides in the blanket loop by the circulation of the liquid breeder is not taken into account.

The natural isotopic abundance ratios of ^{6}Li and ^{7}Li are 7.5% and 92.5%, respectively. Those of ^{204}Pb, ^{206}Pb, ^{207}Pb, and ^{208}Pb are 1.4 atom%, 24.1 atom%, 22.1 atom%, and 52.4 atom%, respectively. These isotopes have their own microscopic cross sections for nuclear reactions. Fig. 2 shows the microscopic cross section of these isotopes by the JENDLE 4.0 nuclear library. Pb has larger cross section for the (n, 2n) reaction for fast neutrons than Be as shown in Fig. 2 (a). The cross section for the (n, 2n) reaction of ^{204}Pb is the steepest and that of ^{207}Pb is the widest in the isotopes. ^{204}Pb has the largest cross section for (n, γ) reaction. The ^{204}Pb cross section for the (n, γ) reaction is three orders magnitude larger than ^{208}Pb cross section for thermal neutrons as shown in Fig. 2 (c). The resonance region of the ^{204}Pb cross section is located in the lower energy region than other isotopes.

Table 1 shows the conditions of isotope concentrations in the Pb-17Li in the current work. Each Pb isotope is enriched to 90% in the Pb isotopes to clarify the effect of the each isotope on the blanket performance. Then, the ^{208}Pb, which is the heaviest and initially included at the highest concentration in the natural Pb, is 8.3 mol%. These compositions of Pb-17Li are held fixed during the reactor operation and do not change by burn up or nuclear transformation and. The concentrations of impurities in the Pb-17Li are not taken into account in the current work to feature the effect of the isotope enrichment on the blanket performance.

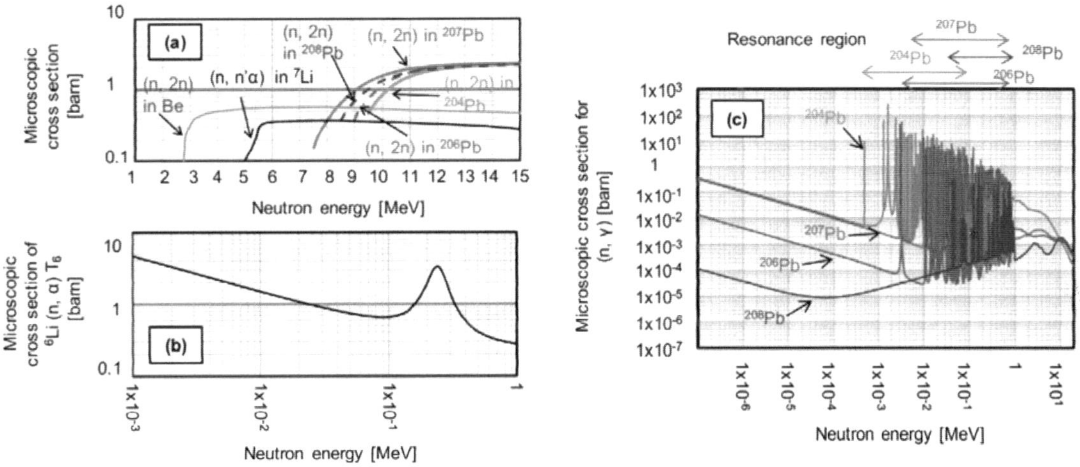

Fig. 2 Microscopic cross sections of ^{6}Li, ^{204}Pb, ^{206}Pb, ^{207}Pb and ^{208}Pb

Table 1 Isotope conditions of Pb-17Li (unit: %)

	Natural	^{6}Li enrich	^{204}Pb enrich	^{206}Pb enrich	^{207}Pb enrich
^{6}Li	1.29	15.3	1.29	1.29	1.29
^{7}Li	15.7	1.7	15.7	15.7	15.7
^{204}Pb	1.16	1.16	74.7	0	0
^{206}Pb	20.0	20.0	0	74.7	0
^{207}Pb	18.3	18.3	0	0	74.7
^{208}Pb	43.5	43.5	8.3	8.3	8.3
Total	100	100	100	100	100

IOP Conf. Series: Journal of Physics: Conf. Series **1090** (2018) 012004 doi:10.1088/1742-6596/1090/1/012004

3. Results and discussions

3.1. Performance of natural Pb-17Li blanket

The TBR is 1.28 as presented in Table 2. This value is greater than that obtained for magnetic- confinement demonstration reactors [15, 16], because the blanket space is not encroached upon by a magnet and the covering ratio of the plasma is nearly 100 %. T is generated mainly in the blanket by the ^{6}Li (n, α) T reaction with the neutron- multiplication $(n, 2n)$ reaction by Pb (Table 2). When the initial T inventory and the T recovery ratio are assumed as 20 kg and 90%, respectively, the doubling time of tritium is estimated approximately 1 year.

The neutron spectra in the liquid wall and the blanket in the region indicated in Fig. 1 (a) are shown with solid blue lines in Fig. 3 (a). The drop in the neutron flux near the neutron energy of 0.24 MeV is due to the large peak in the cross section of ^{6}Li (n, α) T as shown in Fig. 2 (b). Fig. 1 (b) shows neutron flux distributions for high and low energy neutrons in reactor chamber. These figures indicates that neutrons are uniformly diffused in the blanket. The radial profiles of T, ^{210}Po and ^{203}Hg productions in this region are shown in Figs. 4 (a), (b) and (c), respectively. The typical processes to produce ^{210}Po and ^{203}Hg from Pb are given as follows;

$$^{208}\mathrm{Pb}(n,\gamma)\,^{209}\mathrm{Pb} \xrightarrow{\beta} {}^{209}\mathrm{Bi} \rightarrow {}^{209}\mathrm{Bi}(n,\gamma)\,^{210}\mathrm{Bi} \rightarrow {}^{210}\mathrm{Bi} \xrightarrow{\beta} {}^{210}\mathrm{Po}, \quad (1),$$

$$^{204}\mathrm{Pb}(n,2n)^{203}\mathrm{Pb} \rightarrow {}^{203}\mathrm{Pb} \xrightarrow{\beta-} {}^{203}\mathrm{Tl} \rightarrow {}^{203}\mathrm{Tl}(n,p)^{203}\mathrm{Hg}, \quad (2),$$

$$^{206}\mathrm{Pb}(n,\alpha)^{203}\mathrm{Hg}. \quad (3).$$

Table 2 TBR and production of ^{210}Po and ^{203}Hg in reactor chamber after continuous operation for 10 years

		Natural Pb-17Li	^{6}Li enriched Pb-17Li	^{204}Pb enriched Pb-17Li	^{206}Pb enriched Pb-17Li	^{207}Pb enriched Pb-17Li
TBR (T_6, T_7)		1.28 (1.259, 0.024)	1.63 (1.635, 0.03)	0.521 (0.498, 0.023)	1.31 (1.285, 0.024)	1.33 (1.306, 0.024)
Production	^{210}Po	9.17×10^{-6}	7.54×10^{-7}	5.19×10^{-7}	1.72×10^{-7}	1.81×10^{-6}
[g per Pb-17Li kg]	^{203}Hg	5.02×10^{-6}	3.88×10^{-6}	6.41×10^{-5}	1.00×10^{-5}	3.01×10^{-6}

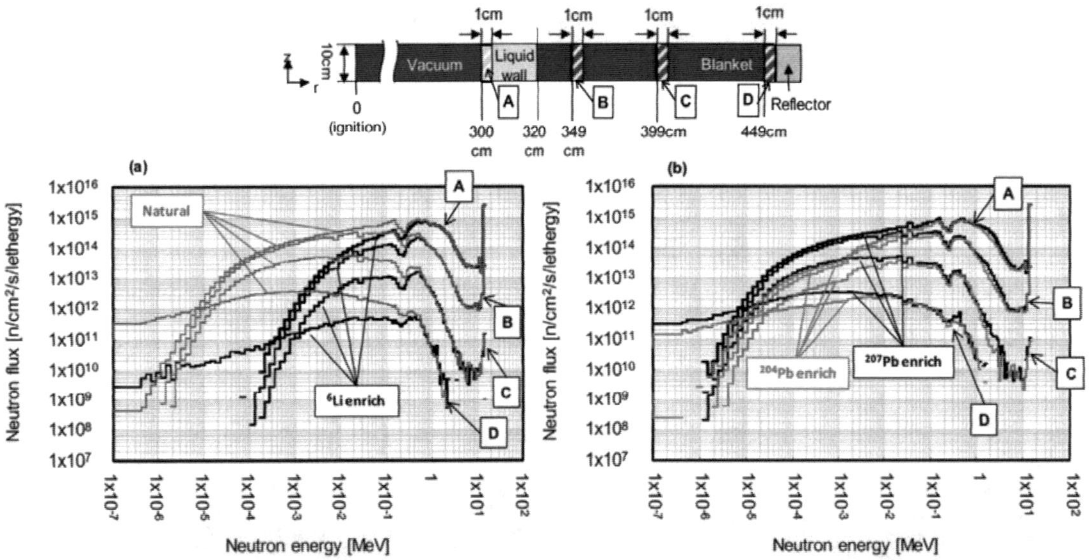

Fig. 3 Blanket neutron spectra in the horizontal cross section at level of ignition point in chamber (a) Natural Pb-17Li and ^{6}Li enriched case, (b) ^{204}Pb- or ^{207}Pb- enriched case

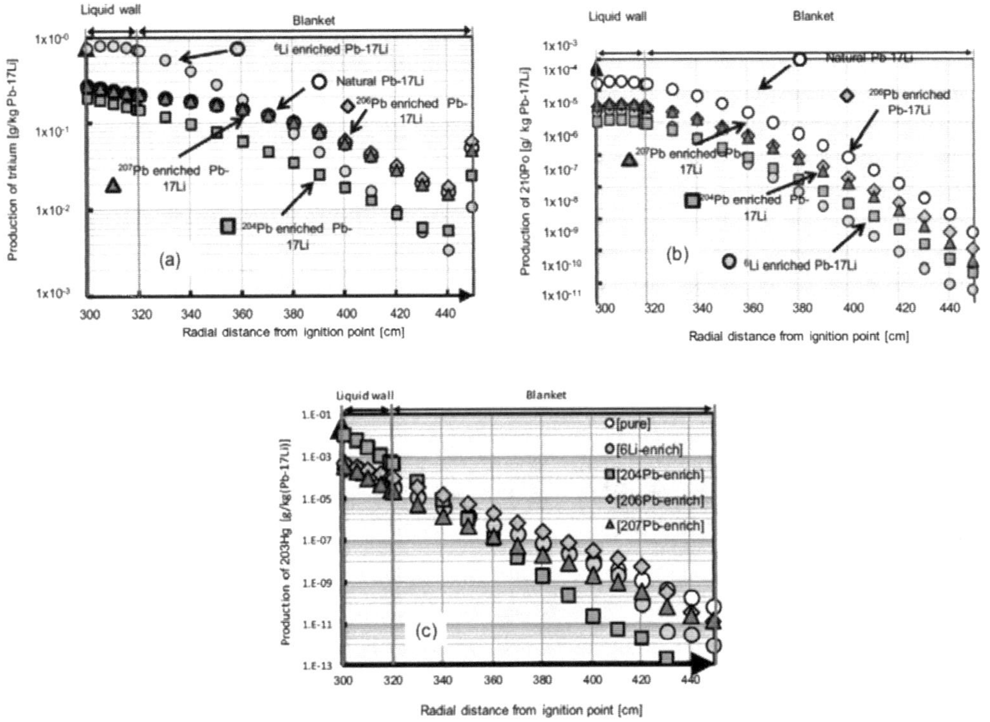

Fig. 4 Radial profiles of nuclide productions, (a) T, (b) ^{210}Po and (c) ^{203}Hg

3.2. Effect of ^{6}Li enrichment

The TBR is 1.64. The neutron spectra in the ^{6}Li-enriched blanket are harder than those in the blanket without enrichment as shown in Fig. 3 (a). The radial profile of tritium production in liquid wall and blanket region (Fig. 4 (a)) indicates that its slope is steeper than that with natural Pb-17Li. The thermal neutron flux in this blanket is much smaller than that without isotope enrichment as shown in Fig. 1 (b). These features are attributed to the reaction rate of the ^{6}Li (n, α) T reaction is dominant in the blanket.

The productions of ^{210}Po and ^{203}Hg are reduced by 91.8% and 22.7%, respectively rather than that without enrichment as presented in Table 2. The reduction attributes to that the reaction rate for the (n, γ) reaction of the Pb isotopes in ^{6}Li-enriched Pb-17Li blanket becomes smaller than that in natural Pb-17Li blanket because the reaction rates of ^{6}Li (n, α) T becomes larger by its enrichment.

3.3. ^{204}Pb-enriched case

The TBR of ^{204}Pb enriched blanket is 0.52 and is much less than that without the enrichment (Table 2). The reason is the large cross section of the (n, γ) reaction. The ^{204}Pb-enriched blanket has lower neutron flux for the energy less than 0.01 MeV as shown in Fig. 3 (b) and the neutron spectrum had a noticeable drop in the neutron flux around 10^{-3} MeV. The drop is attributed to the cross section for the (n, γ) reaction, which has the resonance region in the energy range between 10^{-1} and 10^{-3} MeV. The resonance region overlaps the cross section of ^{6}Li (n, α) T. The production of ^{210}Po was reduced by 94.3%. However, the production of ^{203}Hg is 12 times larger than the natural Pb-17Li case. Fig. 5 shows the trend of isotope productions based on the results of FISPACT calculation. The trend indicates that ^{203}Hg is produced from ^{204}Pb via the formation of ^{203}Pb and ^{203}Tl as presented in Eq. (2).

3.4. ^{206}Pb- or ^{207}Pb-enriched case

^{206}Pb- or ^{207}Pb- enrichment leads to more or less the same TBR vis-à-vis natural Pb-17Li as presented in Table 2. The neutron spectra in these blankets are close to those in natural Pb-17Li blanket (Fig. 2 (b)). Fig. 4 (a)

indicated that the radial profile of the T production in the ^{206}Pb- or ^{207}Pb- enriched blanket is close to that of natural Pb-17Li blanket because the ^{6}Li (*n*, *α*) T reaction is dominant for T production in these cases.

The enrichment of ^{206}Pb or ^{207}Pb leads to the suppression of ^{210}Po production. The production of ^{210}Po is reduced by 98.1% and 80.3% in ^{206}Pb and ^{207}Pb enriched cases, respectively, rather than that without the isotope enrichment. The suppression is due to the removal ^{208}Pb, which produces ^{210}Po by Eq. (1), and the concomitant decrease in relative content of ^{208}Pb. In the ^{206}Pb enriched case, the production of ^{203}Hg becomes 2 times larger than the natural Pb-17Li case due to the reaction given by Eq. (3). The production of ^{203}Hg is reduced by 40.0% in the ^{207}Pb enriched case due to the removal of ^{204}Pb and ^{206}Pb, which produces ^{203}Hg as expressed by Eq. (2) and (3).

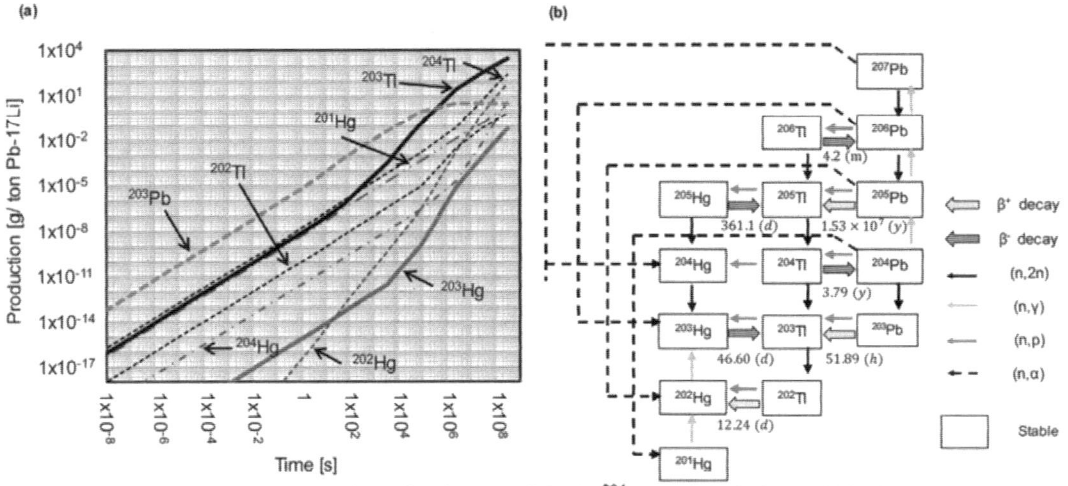

Fig. 5 (a) Production of various nuclides in ^{204}Pb-enriched Pb-17Li blanket
(b) Production pathways of Hg isotopes from Pb isotopes

4. Conclusion

Simulation studies on neutron transport and material activation for lead-lithium (Pb-17Li) blanket of the inertial fusion reactor "KOYO-FAST" were performed. The blanket performances improved by enrichment of some stable isotopes (^{6}Li, ^{204}Pb, ^{206}Pb and ^{207}Pb) were investigated. Major conclusions are follows;

(1) The TBR is 1.28 for natural Pb-17Li blanket. The large TBR is allowed because the blanket space is not limited in the design of inertial fusion reactor, in which superconducting magnet coils are not necessary for plasma confinement.

(2) The TBR is 1.64 for ^{6}Li-enriched blanket. The productions of ^{210}Po and ^{203}Hg are reduced by 91.8% and 22%, respectively. The productions of these isotopes are mitigated because intermediate and thermal neutrons, that are necessary to produce ^{210}Po and ^{203}Hg, are preferentially captured by the ^{6}Li (n, α) T reaction.

(3) TBR is 0.521 for ^{204}Pb-enriched blanket. The small TBR is caused because neutrons are captured by ^{204}Pb due to its large cross section for (n. γ) reaction. The production of ^{203}Hg is promoted by 128% because of the production of ^{203}Ti, which transmutes to ^{203}Hg by (n. γ) reaction.

(4) The enrichment of ^{206}Pb or ^{207}Pb leads to more or less the same TBR with natural Pb-17Li. The production of ^{210}Po is reduced by 98.1% and 80.3% for ^{206}Pb- and ^{207}Pb-enriched cases, respectively. The production of ^{203}Hg is reduced by 40% in the ^{207}Pb-enriched case because of the removal of ^{204}Pb and ^{203}Pb from the blanket.

References

[1] Norimatsu T, Okano K, Kozaki Y, 2016 *J. Plasma Fusion Res* **92** pp 304-10
[2] Kozaki Y, Norimatsu T, Furukawa H, Souman Y, Nishikawa M and Tomabechi K 2007 *J. Plasma Fusion Res* **83** pp 19–27
[3] Kawara Z, Yamamoto K, Kunugi T and Norimatsu T 2010 *Fusion Eng. Des.* **85** pp 2181-86
[4] Kondo M, Takahashi M, Tanaka T, Tsisar V and Muroga T 2012 *Fusion Eng. Des.* **87** pp 1777–87
[5] Kondo M, Muroga T, Katahira K and Oshima T 2008 *J. Power Energy Syst.* **2** pp 590–97
[6] Kondo M, Oshima T, Tanaka T and Muroga T 2012 *Fusion Eng. Des.* **87** pp 1788–93
[7] Veredas G, Fradera J, Fernandez I, Batet L, Penalva I, Mesquida L, Abella J, Sempere J, Martinez I, Herrazti B and Sedano I 2011 *Fusion Eng. Des.* **86**, pp 2365–69
[8] Kondo M and Nakajima Y, 2013 *Fusion Eng. Des.*, **88**, 9–10, pp 2556–59
[9] Petrizzi L, Auditore L, Cambi G, Cepraga D. G and Villari R, 2008 *Fusion Eng. Des.* **83**, pp 1244–48
[10] Casini G, Rocco P, Zucchetti M, 1997 *Fusion Eng. Des.* **17**, pp 351–57
[11] Hoshino T and Terai T 2011 *J. Nucl. Materials* **417**, 1–3, pp 696–99
[12] Khorasanov G. L., *Application of stable lead isotope Pb-208 in nuclear power engineering and its acquisition techniques*, Nova Science Publishers, Inc. ISBN: 978-1-62417-653-1 (2013).
[13] Sato T, Niita K, Matsuda N, Hashimoto S, Iwamoto Y, Noda S, Ogawa T, Iwase H, Nakashima H, Fukahori T, Okumura K, Kai T, Chiba S, Furuta T and Sihver L 2013 *J. Nucl. Sci. Technol.* **50** pp 913–23
[14] Shibata K, Iwamoto O, Nakagawa T, Iwamoto N, Ichihara A, Kunieda S, Chiba S, Furutaka K, Otuka N, Ohsawa T, Murata T, Matsunobu H, Zukeran A, Kamada S and Katakura 2011 J, *J. Nucl. Sci. Technol.* **48** pp 1–30
[15] Forrest R. A and Kopecky J, 2007 *Fusion Eng. Des.*, **82** 15–24 pp 2471–77
[16] Palermo I, Gomez-Ros J. M., Veredas G, Sanz J, Sedano L, 2012 *Fusion Eng. Des.* **87** pp 1019–24
[17] Fischer U, Pereslavtsev P, Grosse D, Weber V, Puma A, Gabriel F, 2010 *Fusion Eng. Des.* **85** pp 1133–38
[18] Ambrose D and Sprake C. H. S. 1972 *J. Chem. Thermodynaics* **4** pp 603–20
[19] Obara T, Miura T and Sekimoto H, *Prog. Nucl. Energy* **47** pp 577-85

The Band Structure of a Rare Earth Element of Promethium (III) Oxide (Pm_2O_3) Calculated Using Density Functional Theory

Nur Afifah Mohd Rafi and Ahmad Nazrul Rosli

Faculty of Sciences and Technology, Universiti Sains Islam Malaysia (USIM) Nilai, 71800 Negeri Sembilan, MALAYSIA.

anazrul84@yahoo.com

Abstract. Luminescence is emission of light by a substance not resulting from heat. It is form of cold body radiation. It can be caused by chemical reactions of the substance with other substances around it, electrical energy, subatomic motions or stress on a crystal. There are many types of luminescence. One of it is phosphorescent. Phosphorescent is made from phosphors such as doped strontium aluminate. Phosphorescent emission light persists long after it has been exposed to light and will fading over time. Promethium (Pm) is low energy beta emitter. It does not degrade the phosphor lattice and the luminosity of the material does not degrade so fast. Promethium based paints are safer than Radium (Ra) and have half-life about more than two years. Promethium usually found in form oxidation called promethium (III) oxide (Pm2O3) and usually applied on watch and phones dial. The study of a band structure of Pm2O3 has not been covered yet. By using density functional theory, we calculated the band gap of promethium (III) oxide from it crystalline structure. The calculated band structure show a clear band gap between valence and conduction band at Fermi level. However the value shows very small gap at 0.286 eV. The density of state and partial density of states has been produced. The calculation has been done using different type of approximation from general gradient approximation (GGA).

1. Introduction

Luminescence is emission of light by a substance that not resulting from heat. This is form of cold body radiation. Luminescence usually caused by chemical reactions, electrical reactions, subatomic motions, or stress on a crystal. This distinguishes luminescence from incandescence [1]. As we know, incandescence is light emitted by a substance that resulted from heating. Nowadays, the dials, hands and scales and navigational instruments and markings are often coated with luminescent materials. There are many types of luminescence such as fluorescence, phosphorescence, and radioluminescence.

Fluorescence is the emission of light by a substance that has absorbed light or electromagnetic radiation. While, phosphorescence is mechanism for producing light similar to fluorescence, but the emission of visible light persists long after it has been exposed to the light. One of materials emitting this type of luminescence is phosphors. Phosphor can absorb energy from the incident radiation that fall onto it and emit photons [1]. Radioluminescence is the phenomenon by which light is produced by materials by bombardment with ionizing radiation, such as beta particles. It contains a radioactive isotope that combined with radioluminescence substance. Radioactive materials that usually used in radioluminescence type of emission are radium (Ra), tritium (3H) and promethium (Pm) [3].

Content from this work may be used under the terms of the Creative Commons Attribution 3.0 licence. Any further distribution of this work must maintain attribution to the author(s) and the title of the work, journal citation and DOI.
Published under licence by IOP Publishing Ltd

Promethium element is relatively low-energy beta-emitter, which unlike alpha emitters, does not degrade the phosphor lattice and the luminosity and does not degrade so fast. Furthermore, promethium is quite safer than radium.

Usually, rare earth elements are ideal elements for spectral conversion because of their high luminescence characteristics. Rare earth elements rich with energy level structure that allows for great flexibility on conversion of photons in wide spectral region [1]. Promethium is one of the rare earth materials, but usually it is form by bombardment of uranium-235 and thermal neutrons.

Promethium is one of the radioactive substances in the lanthanide group [3,4]. Promethium can be prepared artificially in particle accelerators or known as "atom smashers". All of promethium isotopes elements are known are radioactive. The common form of promethium is promethium (III) oxide. Recently, lanthanide group of elements have been doped into wide band gap semiconductor to produce light in different range of wavelength especially at visible light range and UV spectrum at different temperature [5].

Density functional theory (DFT) has been used to simulate the structure of material at the ground state energy to calculate the density of state, energy gap and produce a band structure. The exchange-correlation potential in the Kohn-Sham equation can be solved using local density approximation (LDA), generalized gradient approximation (GGA) and hybrid [6,7,8]. No calculation has been publish on promethium using DFT or other calculation method. In this paper, we calculate the density of state, band gap structure and measure the band gap of promethium (III) oxide with different type of GGA approximation. Previously, we have done a calculation on several types of oxide material and agreed with the experimental result. We also measured the band gap of abundant material using GGA, however it shows small shift when comparing with an experimental result.

2. Material and method

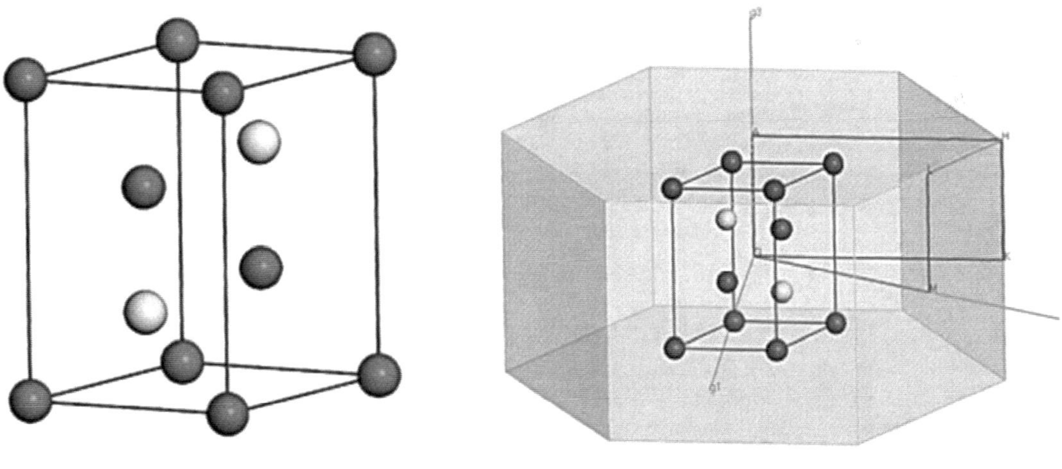

(a) **(b)**

Figure 1: (a) Structure of promethium (III) oxide and (b) Brillouin zone of promethium (III) oxide using CASTEP module.

We are using the density functional theory (DFT) calculation in CASTEP software. There are many approximation can be used in CASTEP software such as local density approximation (LDA), generalized gradient approximation (GGA) and Heyd-Scuseria-Ernzerhof (HSE). The scope of this analysis focused on the band gap and density of states (DOS) of the structure promethium (III) oxide using GGA with different exchange-correlation functional such as Perdew-Burke-Ernzel (PBE), Perdew-Wang 91 (PW91), revised Perdew-Burke-Ernzerhof (RPBE) and Perdew-Burke-Ernzerhof revised for solids (PBESOL). DFT is a predictive calculation and very accurate almost for material with low percentage error [6,7,8].

3. Result and discussion

The calculation has been done using GGA PBESOL, PW91,PBE and RPBE. It show that RPBE has a higher band gap for Pm_2O_3 as shown at table 1. However, this calculation is far from the hybrid method as well as Hubbard-U model.

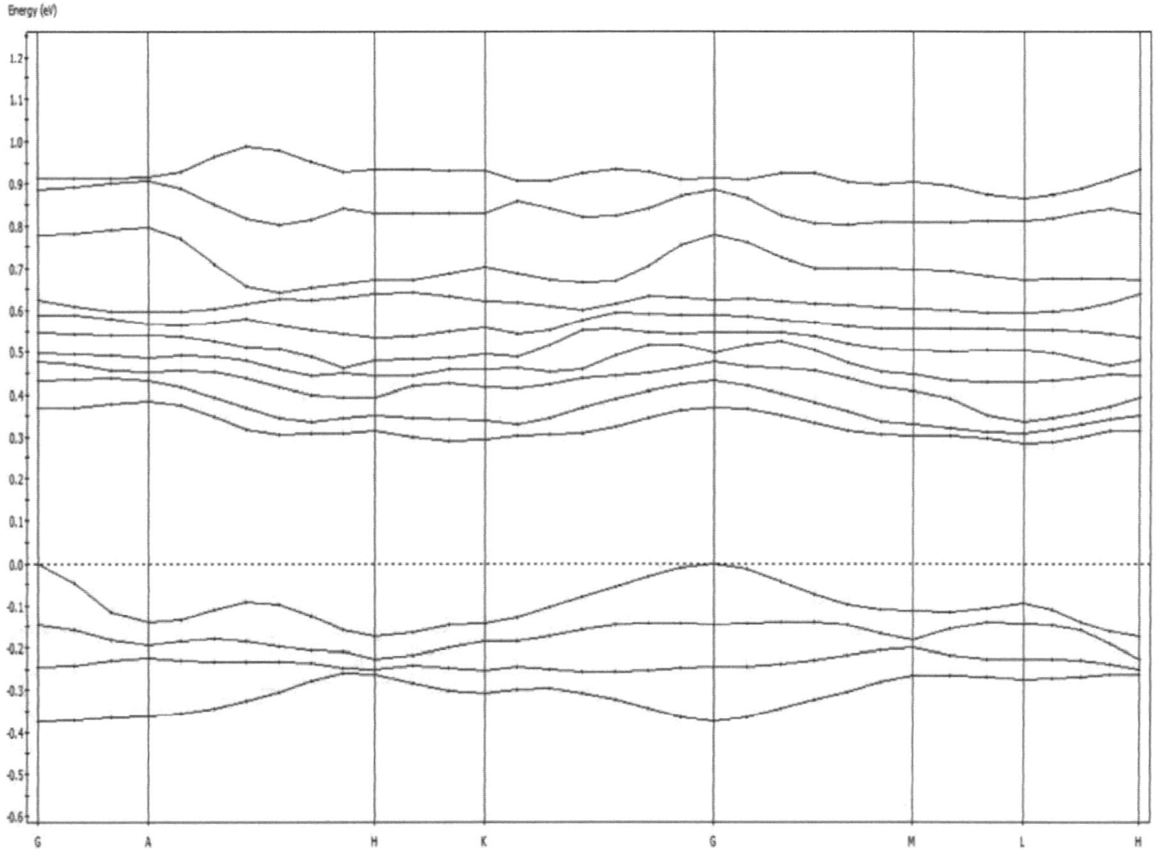

Figure 2: The band structure of Pm_2O_3 calculated using PBESOL approximation of GGA.

Table 1: The value of band gap energy in promethium (III) oxide with different exchange correlation functional of GGA

Exchange correlation functional of GGA	PBESOL	PW91	PBE	RPBE
Band gap (eV)	0.286	0.707	0.790	0.991

4. Conclusion
We able to obtained the band gap of promethium (III) oxide using different approximation of exchange-correlation, but it is very small compared to other research that uses hybrid functional which is within 4.8 to 5.6 eV [2]. The band gap energies obtained is very small in range 0.2~1.0 eV.

References
[1] Kabongo, G. L., Luminescence investigation of zinc oxide nanoparticles doped with rare earth ions. (2013) MSc dissertation. University of South Africa

[2] Gillen, R., Clark, S. J., & Robertson, J., Nature of the electronic band gap in lanthanide oxides. *Physical Review B*, *87*(12), (2013). 125116.

[3] Voncken, J. H. L., The Rare Earth Elements—A Special Group of Metals. in The Rare Earth Element*s*, Springer International Publishing, (2016) 1-13.

[4] Azimi, G., Dhiman, R., Kwon, H. M., Paxson, A. T., & Varanasi, K. K., Hydrophobicity of rare-earth oxide ceramics. *Nature materials*, *12*(4), (2013) 315-320.

[5] Birkhahn, R. H., Chao, L. C., Garter, M. J., Scofield, J. D., & Steckl, A. J., *U.S. Patent No. 6,255,669* (2001).

[6] W. Kohn, Nobel Lecture: Electronic structure of matter-wave functions and density functionals. *Rev. Mod. Phys.* 71(5), (1999) 1253-1266.

[7] W. Kohn & L. J. Sham, Self-Consistent Equations Including Exchange and Correlation Effects. *Phys. Rev.* 140, (1965) A 1133 – A 1138.

[8] P. Hohenberg & W. Kohn, Inhomogeneous Electron Gas. *Phys. Rev.* 136, (1964) B864 – B871.

[9] Basri, Katrul Nadia; Zabidi, Noriza Ahmad; Abu Kassim, Hasan; Rosli, Ahmad Nazrul (2015). Density Functional Theory (DFT) Calculation of Band Structure of Kesterite, Advanced Materials Research, 1107, 491-495.

Modified CANDLE Burnup Calculation System, Its Evolution, and Future Development

Zaki Su'ud[1], M. Ilham[1], Nina Widiawati[1] and Hiroshi Sekimoto[2].
[1]Nuclear and Biophysics Research Division, Faculty of Mathematics and Natural Sciences, Bandung Institute of Technology
[2]Tokyo Inst. of Technology (Emeritus Profesor), Japan
E-mail: szaki@fi.itb.ac.id

Abstract. CANDLE Burn-up has been widely investigated during the last 15 years and give excellent possibility for natural Uranum utilization without the necessity of fuel enrchment plant and fuel reprocessing plant. Modified CANDLE burn-up scheme is a slight modification of CANDLE burn-up scheme by adopting discreet regions in the core. Modified CANDLE burn-up calculation is a complex calculations because it need accurate treatment of nuclear group constants, burn-up calculations, and multigroup diffusion calculations. Another problem come from the fact that the burn-up level of fuel is very high. Therefore the fission produc treatment have to be carried out in a highly accurate apporach and alsolarge number of energy groups is applied. Large number of Fission product is considered durng simulations. The calculations are performed under iterative scheme. For 10 regions and 10 years of burn-up period, first initial power density distributions are assumed and then multi-group diffusion and burn-up calculations are conducted. After 10 years of burn-up process the fuel in the first region is moved to the second region, the fuel in the second region was moved to the third region, the fuel in the third region is moved to the forth region, and so on. The fuel in the 10th region is then taken out. The first region is filled with the fresh fuel of natural uranium. The process is repeated till equilibrium condition achieved. Modified CANDLE burnup also has evolved to radial shuffling and combined axial-radial shuffling scheme.

1. Introduction[1-9]

Nuclear fuel enrichment and nuclear fuel reprocessing are two very sensitive issues related to the nuclear non-proliferation in the world especially when it is carried out in the developing countries . However without these two processes (at least one of them) the optimal nuclear energy utilization will difficult to be achieved. In this study, conceptual design of long life Pb-Bi cooled fast reactors which can be continuously operated by only supplying natural uranium without fuel enrichment plant or fuel reprocessing plant is performed based on the CANDLE concept. Therefore using these type of nuclear power plants optimum nuclear energy utilization including in developing countries can be easily conducted without the problem of nuclear proliferation. In order to enhance calculation accuracy, in the present study large number of energy groups is applied. Large number of Fission product is considered during simulations. The calculations are performed under iterative scheme. For 10 regions and 10 years of burn-up period, first initial power density distributions are assumed and then multi-group diffusion and burn-up calculations are conducted.

Content from this work may be used under the terms of the Creative Commons Attribution 3.0 licence. Any further distribution of this work must maintain attribution to the author(s) and the title of the work, journal citation and DOI.
Published under licence by IOP Publishing Ltd

IOP Conf. Series: Journal of Physics: Conf. Series **1090** (2018) 012006 doi:10.1088/1742-6596/1090/1/012006

2. Design Concept and Methodology

CANDLE type nuclear reactor has an axial movement of active core area. Modified CANDLE type reactor is initially developed in the axial direction moving. In the eraly stage of Modified CANDLE (MCANDLE) the core is divided axially into several equal regions and the fresh fuel is put near the st active regions to accelelrate the breeding process, see Figure 1. [1-7]

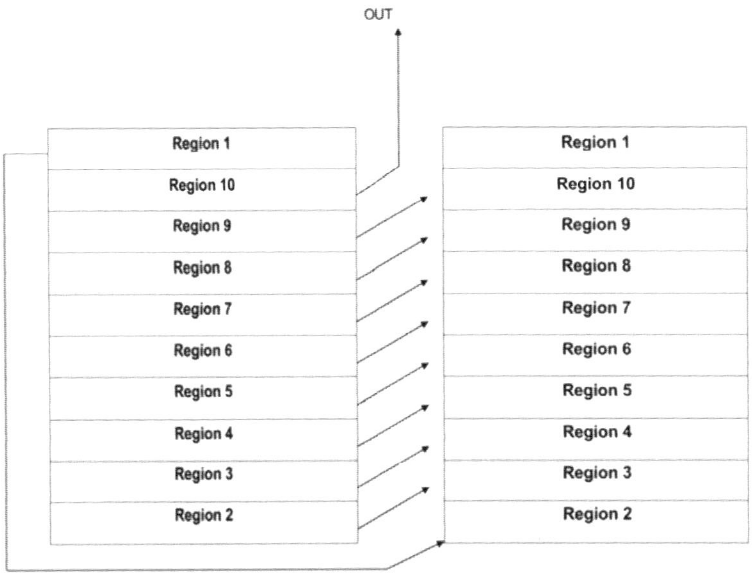

Figure 1 Illustration of Modified CANDLE burn-up scheme with axial shuffling

Consider the above Modified CANDLE reactor with 10 regions and 10 years of burn-up cycle. After 10 years of burn-up process the fuel in the first region is moved to the second region, the fuel in the second region was moved to the third region, the fuel in the third region is moved to the forth region, and so on. The fuel in the 10th region is then taken out. The first region is filled with the fresh fuel of natural uranium. The process is repeated till equilibrium condition achieved. This type of NPP need high conversion/breeding fuel, so a high fuel volume fraction based on nitride fuel is proposed (fuel volume fraction is usually at least 50%). The core is then optimized to fulfill the criteria that the core can maintain criticality during 10 years of burn-up using the proposed refueling scheme.

The calculation scheme of these Modified CANDLE reactors is iterative. First the power density distribution in each region is assumed and used to perform burn-up calculation. Then multigroup diffusion calculation is performed to get power density distribution after the latest burn-up calculation. The results is a new power density distributions which then used to update the burn-up calculations. The process is repeated till rech convergence. The calculation is performed using SRAC code system (SRAC-CITATION system)[10-11] or SLAROM code system[12], and FI-ITB CH1 Code system[13].

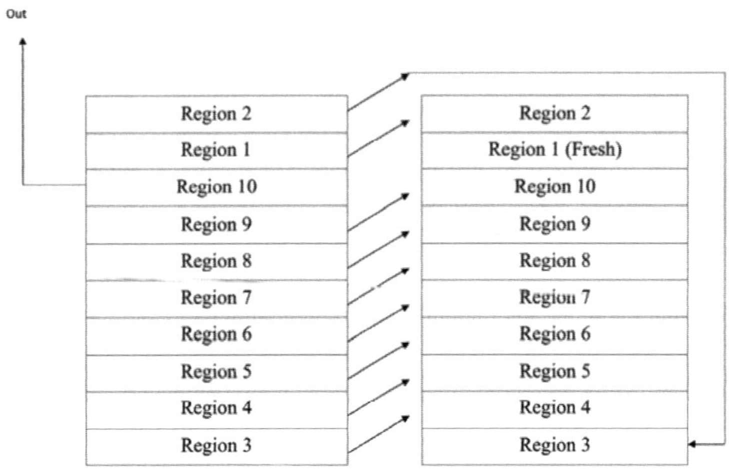

Figure 2 Illustration of Modified CANDLE burn-up scheme with two regions near most active area (type A)

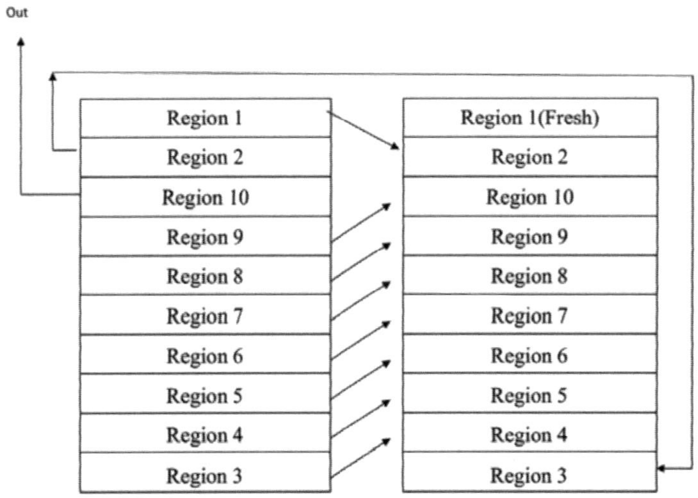

Figure 3 Illustration of Modified CANDLE burn-up scheme with two regions near most active area (type B)

Figures 2-4 shows various configurations of axial shuffling modified CANDLE burnup with more than one regions located the most active reagions. The previous studies have shown that putting one region near the most active regions enhanching criticallity of the system by about 5%. Putting 2 or more regions near the most active regions can increase further the criticallity but not so significant compared to the putting the first region near the most active regions.

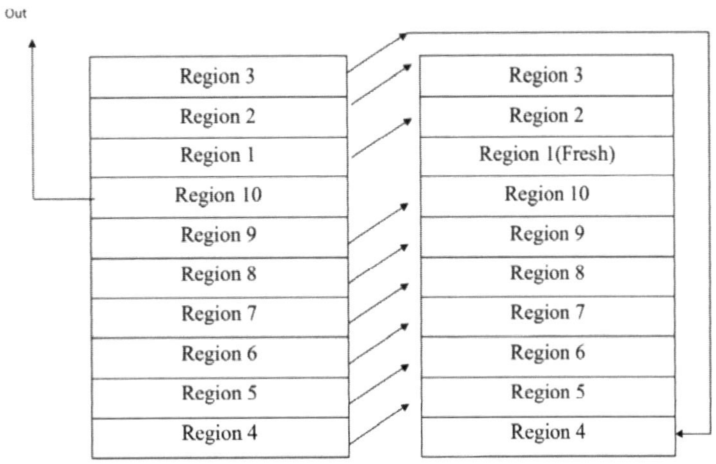

Figure 4 Illustration of Modified CANDLE burn-up scheme with tree regions near most active area (type C)

In the further development the radial shuffling schemes have been investigated. The basic mechanism is still similar that is by dividing the active core into several regions in the radial directions. An example of the Modified CANDLE in the radial direcction is shown in Figure 5.

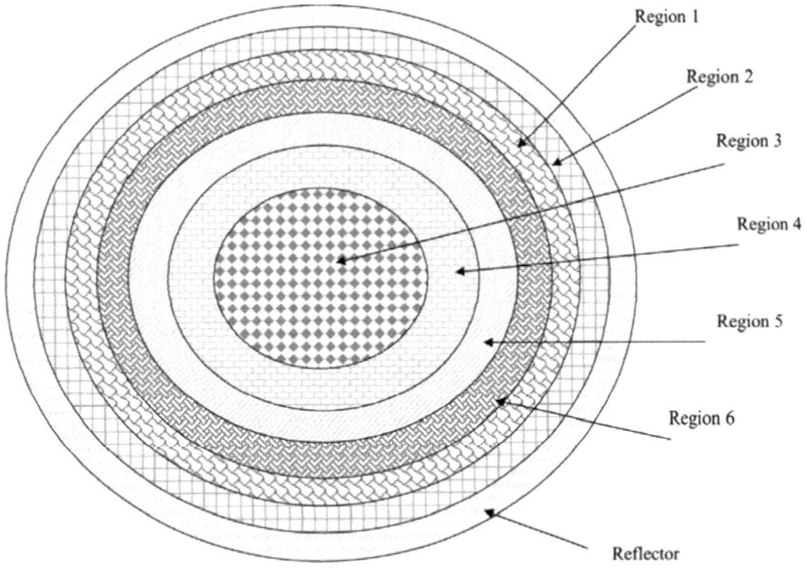

Figure 5 Illustration of Modified CANDLE burn-up scheme of 6 regions with radial direction

The previous study has shown that the concept of Modified CANDLE can work both in axial and radial directions and in both systems the reactors just need natural Uranium as fuel cycle input. Some nvestigations have also been conducted for different number of regions. Figure 6 shows illustrataion of Modified CANDLE burnup scheme with 6 regions in axial directions.

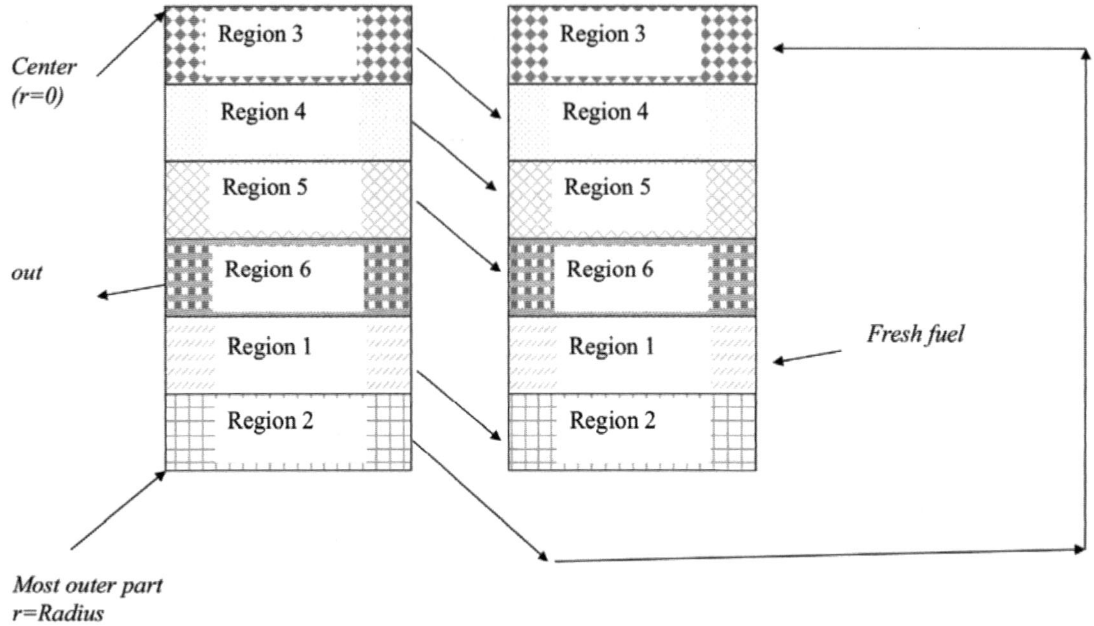

Figure 6 Illustration of Modified CANDLE burn-up scheme of 6 regions with axial direction

In the latest development the combination of axial radial shuffling scheme has also investigated (See figure 7 for example.). Basically axial-radial shuffling scheme give the best criticallity in general and can work with average discharge burn-up of about 20% HM.

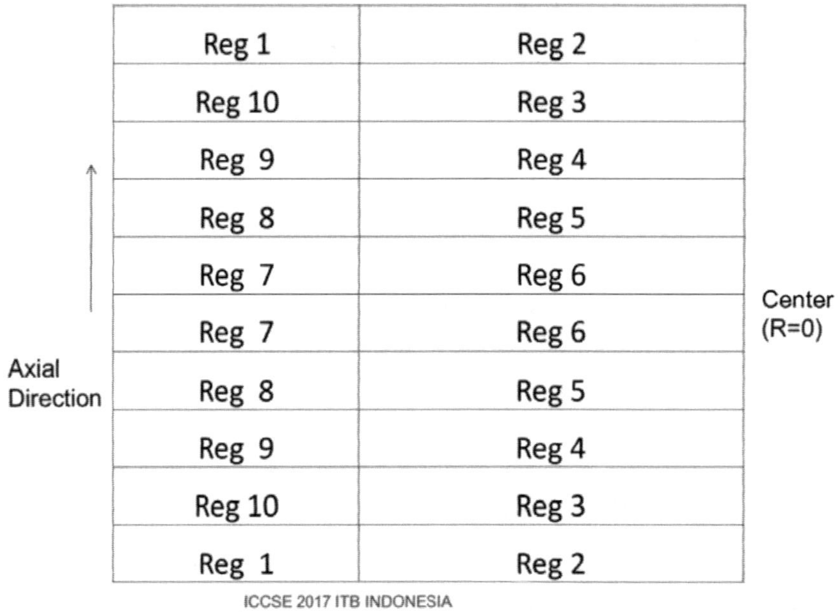

Figure 7 Illustration of Modified CANDLE burn-up scheme of 10 regions with axial-radial combined shuffling

3. Results and discussions

Table 1 shows main paramters for sample parameters for the current simulations. The reactor power is 400 MWt and this is relatively small for Modified CANDLE burn-up system.

Table 1 Main parameter used in the present study

Parameter	Value/description
Power (MWth)	800
Number of equal volume region in core	10
Axial width of each region	17.5 cm
Active core radial width	130 cm
Reflector radial width	50 cm
Reflector axial width	50 cm

Figure 8 shows k-inf change with time for standard axial shuffling Modified CANDLE burnup reactor. It is shown that the fuel experienced accelerated breeding process during first 10 years of burnup history due to its position near the most active regions. The k-inf continuously increases till reach the peak in the 32 years and then decreases.

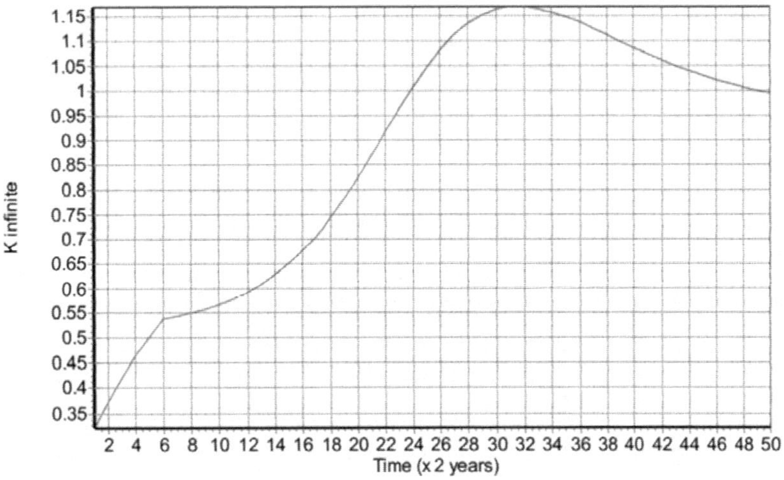

Figure 8 K-inf change with burn-up history for Standard Axial Shuffling

Figure 9 shows k-eff change with time during 10 years of operation. The k-eff value increases during 10 years of operation time. The excess reactivity is about 2% dk/k in average.

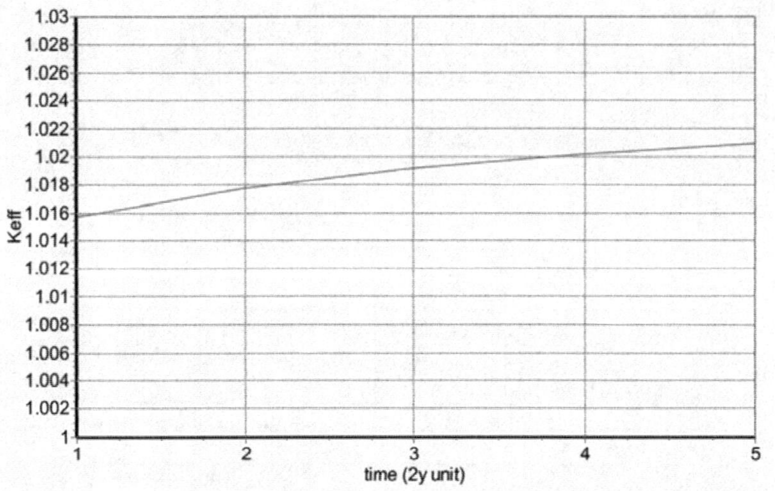

Figure 9 K-eff change with burn-up during 10 years of operation time

Figure 10 shows burnup level change with time. Its is shown that the burn-up level increases significantly after the fuel entering the burning zone which is roughly second half part of the core (Regions 6-10)

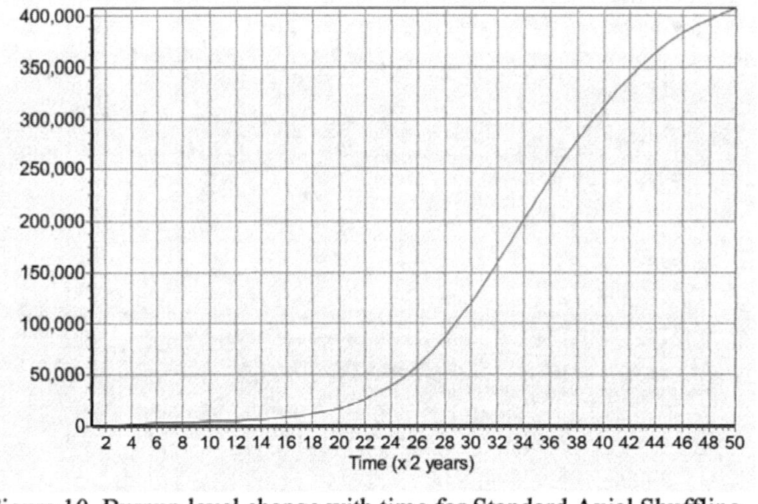

Figure 10 Burnup level change with time for Standard Axial Shuffling

Figure 11 shows Pu-239 atomic density change with burn-up for standard 10 regions axial shuffling Modified CANDLE reactor. It is shown that in the first 10 years the accumulation of Plutonium is accelerated compared to that in regions 2-4. Pu-239 atomic density reaches a peak in about 32 years and then decreases till the end of burn-up history. Next, Figure 12 shows U-238 atomic density change with burnup. It is shown that the U-238 monotonously decreases but the decrease rate is becoming quite high when the fuel entering burning zone (regions 6-10).

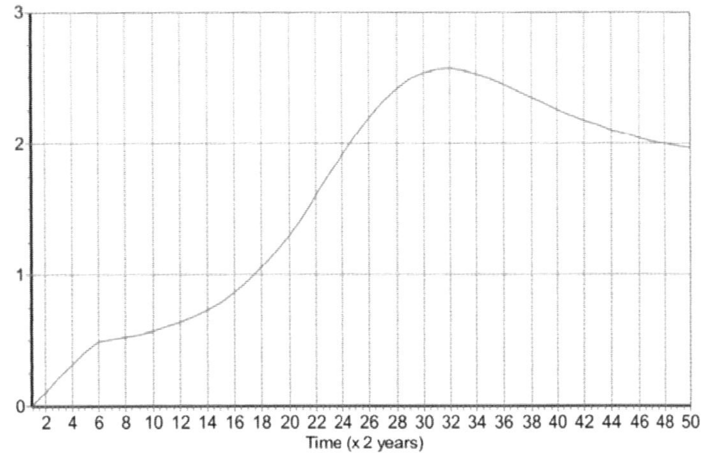

Figure 11 Pu-239 atomic density change with time for Standard Axial Shuffling

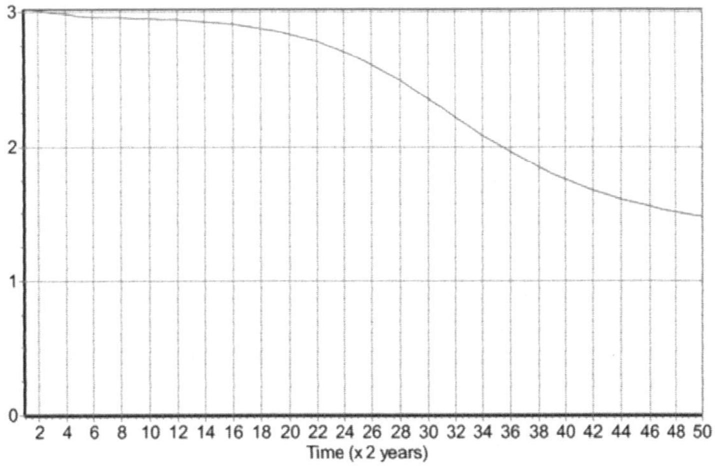

Figure 12 U-238 atomic density change with time for Standard Axial Shuffling

Table 2 shows k-eff initial value (at BOL) during 10 years of operation time for 10 regions with various shuffling scheme with 600 MWt power level. It is shown that the shuffling scheme B Cand D give better criticallity than the standard scheme but the difference is not so significant.(less than 1%dk.k)

Table 2 Comparison of several shuffling type in axial mode MCANDLE

Shuffling type	Reactor Power (MWt)	K-eff (initial)
Standard scheme	600	1.00514
A	600	1.01144
B	600	1.01056
C	600	1.01107

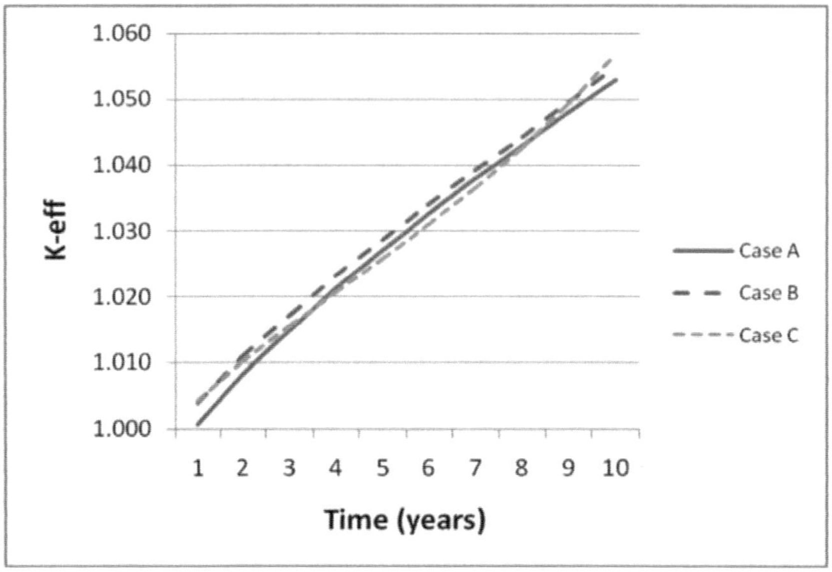

Figure 13 K-eff change with burnup for the radial shuffling as shown in Fiigure 5 with 800MWt (Case A), 850 MWt (Case B) and 900 MWt (Case C)

It is shown in the figure 13 that the radial shuffling Modified CANDLE can be operated 10 years with monotonously increasing k-eff value. Figure 14 shows Pu-239 change with burnup history. It is shown that during the breeding period the Pu-239 atomic density is increasing and reach the peak at about 32 years then decreases due to the shorten of U-238.

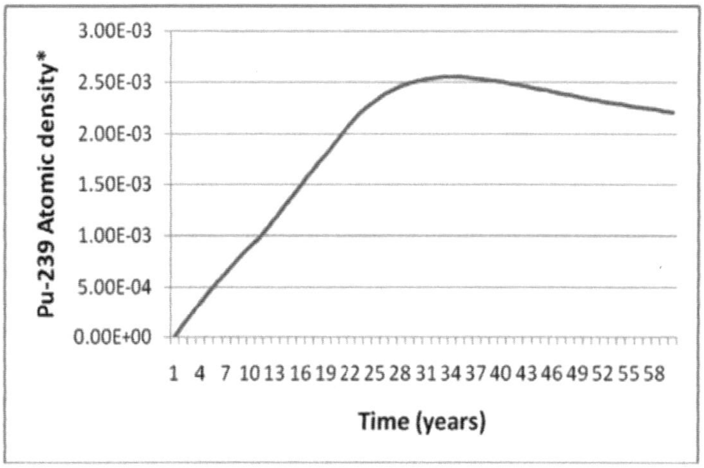

Figure 14 Pu-239 atomic density change with time for radial shuffling

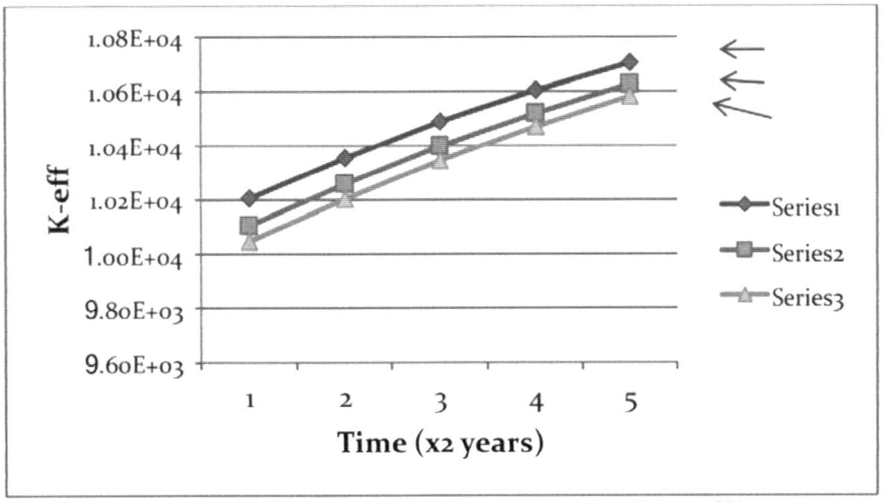

Figure 15 K-eff change with burnup for axial-radial shuffling

Figure 15 shows k-eff change with operation period for axial-radial combined shuffling for 525,550 and 600 MWt power level (series 1, 2, and 3 respectively). The burnup level is about 20% and the system is still be able for operations. Such relatively low burnup level for Modified CANDLE burnup reactor give possibility for the usage of advanced fast reactor material for the oepration of the reactor without the necessity of recladding process during its operation.

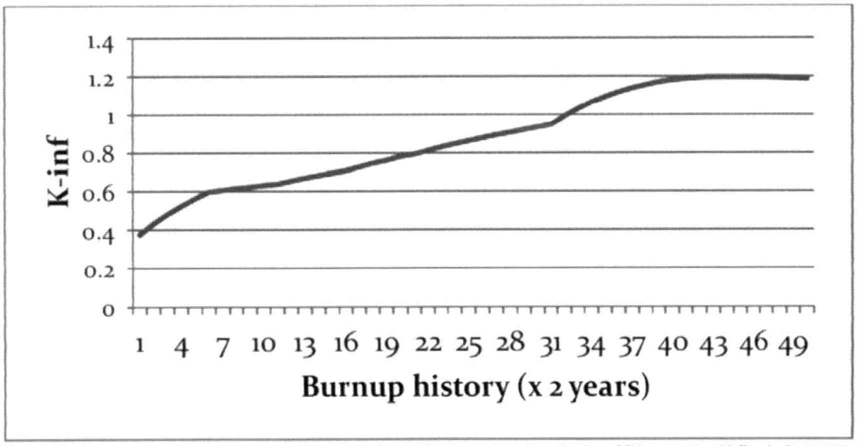

Figure 16 K-enf change with burn-up for axial-radial combined shuffling Modified CANDLE reactor

Figure 16 shows k-inf change with burnup history for 550 MWt case. It is shown that the peak value is reached near the end of burnup history.

4. Conclusion

Modified CANDLE burnup strategy has been successfully used to create long life Pb-Bi cooled fast reactors core which fuel cycle input is natural uranium. Modified CANDLE reactors has evolved from original axial shuffling scheme to radial and combined axial-radial shuffling scheme. The system has been applied to many power level which results relatively flexible dischage burnup level up to around 200 MWd/ton HM by using axial-radial combined shuffling scheme. Axial-radial shuffling have advantages in term of better criticallity so that it can be used to design long life GCFR with burn-up level about 19%HM

References

[1] Zaki Su'ud and H Sekimoto, 2010, IJNEST, Design study of long-life Pb-Bi cooled fast reactor with natural uranium as fuel cycle input using modified CANDLE burn-up scheme, Vol 5, No. 4 , p.347-368

[2] Su'ud, Z., and Sekimoto, H., 2012, IJNEST, Design study of medium-sized Pb-Bi cooled fast reactors with natural uranium as fuel cycle input using modified CANDLE burn-up scheme, Vol. 7 (1) , pp. 23-44.

[3] Zaki Su'ud and H Sekimoto, 2013, Annals of Nuclear Energy, The prospect of gas cooled fast reactors for long life reactors with natural uranium as fuel cycle input, Vol. 54 , pp. 58-66.

[4] Zaki Su'ud and H Sekimoto, 2013, Annals of Nuclear Energy, The prospect of gas cooled fast reactors for long life reactors with natural uranium as fuel cycle input, Vol. 54 , pp. 58-66.

[5] M. Ariani, Z. Su'ud, A. Waris, Khairurrijal, N. Asiah, and M. A. Shafii, Effect of Fuel Fraction on Small Modified Candle Burn-Up Based Gas Cooled Fast Reactors, AIP Conf. Proc. 1325, p. 249

[6] Irka, F.H., Aryani, M., Suud, Z., Sekimoto, H., Desain study of gas cooled fast reactor with natural uranium as fuel cycle input using radial shuffling strategy, AIP Conference Proceedings 1448 , pp. 74-81, 2012

[7] Ariani, M., Su'ud, Z., Monado, F., Waris, A., Khairurrijal, Arif, I., Ferhat, A., Sekimoto, H., Optimization of small long life gas cooled fast reactors with natural Uranium as fuel cycle input, (2013), Applied Mechanics and Materials, 261-262, pp. 307-311

[8] Fiber Monado, Zaki Su'ud, Abdul Waris, Khairul Basar, Menik Ariani, Hiroshi Sekimoto, Application of Modified CANDLE Burnup to Very Small Long Life Gas-cooled Fast Reactor, Advanced Material Research(AMR), Vol. 772,2013, pp. 501-506

[9] Fiber Monado, Menik Ariani , Zaki Su'ud, Abdul Waris, Fiber Monado, Hiroshi Sekimoto, Power Flattening on Modified CANDLE Very Small Long Life Gas-cooled Fast Reactor, AIP Conf. Proc. Vol 1615, 2014, pp. 47-50

[10] K. Okumura et al., 2002, The Comprehensive Nuclear Calculation Code System (SRAC), JAERI Report.

[11] K. Okumura, K. Kaneko, and K. Tsuchihashi , 2005, SRAC95: Generap Purpose Neutronics Code System, JAERI Data/Code 96-015, JAERI, Japan

[12] N. Nakagawa and K.Tsuchihashi, 1984, JAERI Memo, SLAROM: a code for cell homogenization calculation of fast reactor, (Japan Atomic Energy Research Institute, Ibaraki), JAERI M-1294

[13] Zaki Su'ud et al. ,2000, FI-ITB CHI Neutronic code manuals, internal report

Effects of tangential inlet shape and orientation angle on the fluid dynamics characteristics in a biomass burner

Pasymi[1,2], Y W Budhi[1] and Y Bindar[1†]

[1] Energy and Processing System Research Group, Study Programmes of Chemical, Bioenergy and Chemurgy Engineerings, Faculty of Industrial Technology, Institut Teknologi Bandung, Indonesia

[2] Department of Chemical Engineering, Faculty of Industrial Technology, Bung Hatta University, Padang, Indonesia

† *Corresponding author e-mail: yazid@che.itb.ac.id*

Abstract. The perennial crops are potentially used as renewable fuels in the boiler furnace. Due to its specific characteristics, the burner design for this biomass needs to be properly developed. The burner design proposed here is a cylindrical burner having an axial inlet and a pair of tangential injection. This study is aimed to investigate the effect of tangential inlet geometry on the burner performance, through numerical evaluation of fluid dynamics characteristics. The numerical evaluations are conducted using *k-ε* turbulence model under Ansys-Fluent software. The simulation results showed that, at certain orientation angles of tangential inlet, there are back flows from furnace to the internal burner. This phenomenon is responsible to the flame stability in the burner. The turbulence intensity and convective heat transfer coefficient are also influenced by the tangential inlet orientation angle. For same cross sectional area, the rectangular tangential inlet shape generated deeper backflow penetration and higher turbulence intensity than that was done by the circular ones. One of the rectangular shape shortcomings to the circular ones is to produce the higher static pressure, which correlate to the higher burner operating cost. This investigation study concluded that the burner with rectangular tangential inlet shape and orientation angle of 20^0 potentially produces the best burner performance.

Keywords: backflow penetration, biomass burner, heat transfer coefficient, orientation angle, turbulence intensity, static pressure.

1. Introduction

In recent years, along with the depletion of fossil fuel stocks, researchers have begun to develop biomass utilization for electricity generation, particularly for small to medium scale plant [1, 2, 3, 4, 5]. In addition to the renewable objective, biomass is also environmentally friendly (CO_2 neutral) and has a relatively cheap power generation cost. The perennial crops, one of the non-woody biomass, are proven to be much potential biomass types to be used as fuel in the steam power plant. As an example, *miscanthus x giganteus* has an energy potential of ± 442.5 GJ/hectare/year and can substitute 20 tons of sub bituminous coal [1].

Generally, burner design is determined by the process occured in the burner. For solid fuel burners, there are several processes that are expected to take place in a burner, such as the body heating, the devolatilization and the combustion. Some conditions are needed to make processes to take place well in the burner, among others: high temperature, intensive mixing, and sufficient oxygen and residence time [6]. The embodiment of the conditions can be evaluated through the fluid dynamics characteristics in the burner such as turbulence intensity, velocity profile and pressure drop. Furthermore, the velocity profile can illustrate heat transfer rate, residence time and back flow phenomenon.

Some researcher reported that the performance of burner can be increased by involving the tangential flow [7, 8]. The tangential flow has a potential to create the backflow from the furnace to the internal burner. The backflow will carry the hot flue gas from the furnace into the internal burner. This causes the burner temperature to remain high, thus, the process of the particle heating, the devolatilization and

Content from this work may be used under the terms of the Creative Commons Attribution 3.0 licence. Any further distribution of this work must maintain attribution to the author(s) and the title of the work, journal citation and DOI.
Published under licence by IOP Publishing Ltd

the combustion can take place in the burner, continuously. In addition, the presence of the tangential flow may also increase the mixing intensity. One of the challenges of the tangential flow addition in a burner is to minimize the static pressure. The fluid static pressure is a fluid flow resistant that must be resisted by air blower. The greater the fluid static pressure along the burner, the greater the blower power demand and the greater the burner operational cost becomes.

Currently, the use of the tangential flow in the burner is intensified because of its superiorities. The tangential flow in the burner can be generated through the swirler or through the use of the mechanical devices or through the installation of the tangential injection inlets. The biomass burner proposed here consists of a horizontal cylinder having an internally-extended axial inlet and a pair of tangential injection, as shown in Figure 1.

This research is aimed to study the effects of tangential inlet geometry on the burner performance through the fluid dynamics characteristics evaluation. The tangential inlet geometries studied here are the shape and the orientation angle, while the fluid dynamics characteristics evaluated are turbulence intensity, heat transfer rate, pressure drop and back flow penetration.

The fluid turbulence intensity is represented by the equation (1). Variable $\bar{u}_i$ is the mean velocity for each direction component and u_i' is the fluctuation velocity for each direction component [9].

$$I = \frac{\sqrt{((u'^2_x + u'^2_y + u'^2_z)/3)}}{(\bar{u}_x + \bar{u}_y + \bar{u}_z)/3} \quad (1)$$

The heat transfer rate are calculated as convective heat transfer coefficient, as given by equation (2).

$$h = 0.023\left(\frac{k}{D_c}\right)\left(\frac{\rho D_c \bar{u}}{\mu}\right)^{0.8}\left(\frac{c\mu}{k}\right)^{0.4} \quad (2)$$

Variable h is convective heat transfer coefficient, k is thermal conductivity, D_c is burner diameter, c is specific heat, ρ is density and $\bar{u}$ is magnitude velocity. From equation (2) shown that, for isothermal conditions, the convective heat transfer coefficient is directly proportional to the fluid velocity.

2. Investigation methodology

The method used in this research was the Computational Fluid Dynamic (CFD) technique with the Ansys Fluent as the CFD engine. This method has many advantages than experiment, such as low cost, time saving and able to give comprehensive data and information with acceptable results.

2.1 Experimental set-up for simulation

The proposed burner design in this study consisted of a horizontal cylindrical body and an inlet section. The inlet section was a certain length cylinder embedded by the internally- extended axial inlet and a pair of tangential injection, as shown in Figure 1. The mixing between the axial and the tangential flows, including their interaction impacts, occured in the cylindrical body of the burner. The burner fluid dynamics performances were largely determined by the fluid dynamics characteristics in this cylinder. The cylindrical body had a diameter (D_c) of 0.3 m and the cylinder aspect ratio (L_c/D_c) of 1.5. The axial inlet was a cylindrical tube with diameter (D_{ai}) of 0.15 m.

The tangential inlet has two cross-sectional shape variations those were circular and rectangular shape. The circular shape has a diameter (D_{ti}) of 0.062 m, while the rectangular shaped has a width (l) of 0.1 m and thickness (t) of 0.03 m. The cross-sectional area of the two tangential inlet shape was made equal along the study. The best tangential inlet shape would be chosen to study the effect of orientation angle on the burner fluid dynamics performances. The variations of tangential inlet orientation angle used were 15^0, 20^0, 25^0, 30^0, 35^0, 40^0, 45^0, 60^0, 75^0 and 90^0.

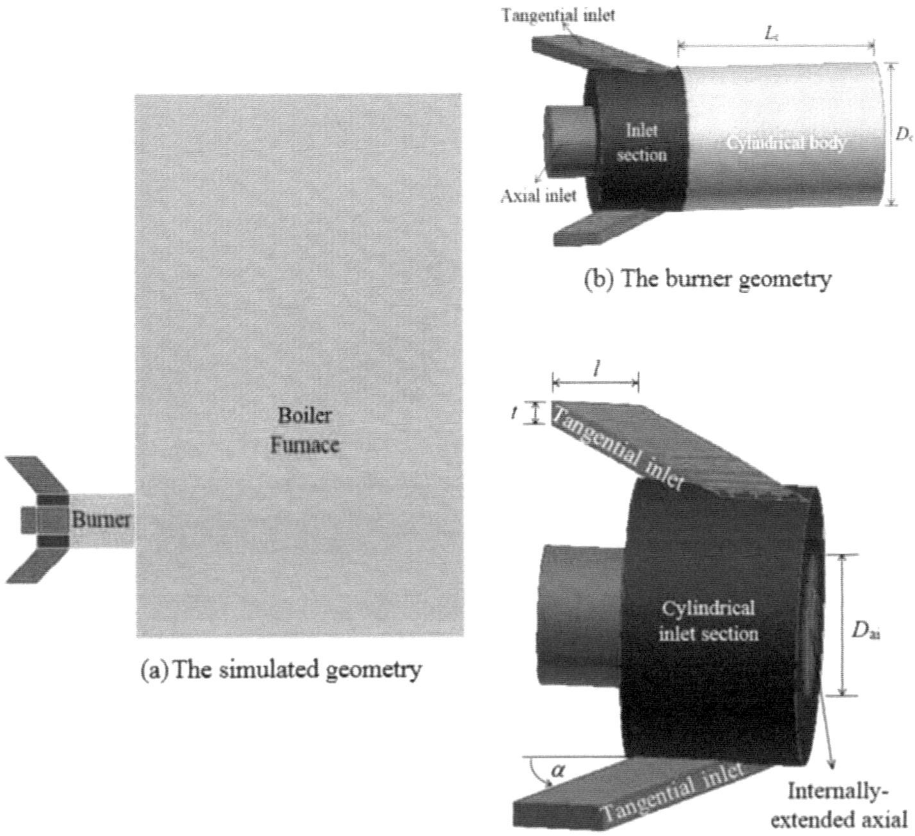

(a) The simulated geometry

(b) The burner geometry

(c) The geometry of inlet section

Figure 1. The geometry of the proposed biomass burner

The simulation was conducted under the constant burner Reynolds number (N_{Re}) and initial tangential intensity (I_{it}). All the simulations in this study were conducted on the N_{Re} of 69,000 and I_{it} of 3,5. The burner Reynolds number (N_{Re}) was calculated using the following formula.

$$N_{\mathrm{Re}} = (\rho D_c \bar{u})/\mu \tag{3}$$

where ρ is air density, D_c is burner diameter, $\bar{u}$ is average magnitude velocity and μ is air viscosity.

While, the initial tangential intensity (I_{ti}) was defined as the ratio of the mass flux entering the tangential inlet to the mass flux of the burner cylinder. Mathematically, I_{it} could be stated by the following equation

$$I_{\mathrm{it}} = (A_c/A_t)\dot{m}_t^2/(\dot{m}_t + \dot{m}_c)^2 \tag{4}$$

where, A_c and A_t are the surface area of burner cylinder and tangential inlet, respectively. Meanwhile $\dot{m}_t$ and $\dot{m}_c$ are the mass flowrate entering the tangential inlet and the mass flowrate entering the burner cylinder.

2.2 Fluid flow model

Theoretically, fluid dynamics characteristics in a chamber are governed by mass and momentum conservation equations. For constant fluid density (ρ) and viscosity (μ), the conservation equations of mass and momentum for turbulent flow were given by the equations (5) to (8). The equations are often known as the Reynolds Average Navier-Stokes (RANS) equations.

$$\frac{\partial \rho}{\partial t} = \frac{\partial \rho \bar{u}_x}{\partial x} + \frac{\partial \rho \bar{u}_y}{\partial y} + \frac{\partial \rho \bar{u}_z}{\partial z} \tag{5}$$

$$\rho \frac{\partial \bar{u}_x}{\partial t} + \sum_{i=x}^{y,z} \rho \bar{u}_i \frac{\partial \bar{u}_x}{\partial i} = -\frac{\partial \bar{p}}{\partial x} + \sum_{i=x}^{y,z} \frac{\partial}{\partial i}\left(\mu_{\text{eff}} \frac{\partial \bar{u}_x}{\partial i}\right) + \rho g_x \tag{6}$$

$$\rho \frac{\partial \bar{u}_y}{\partial t} + \sum_{i=x}^{y,z} \rho \bar{u}_i \frac{\partial \bar{u}_y}{\partial i} = -\frac{\partial \bar{p}}{\partial y} + \sum_{i=x}^{y,z} \frac{\partial}{\partial i}\left(\mu_{\text{eff}} \frac{\partial \bar{u}_y}{\partial i}\right) + \rho g_y \tag{7}$$

$$\rho \frac{\partial \bar{u}_z}{\partial t} + \sum_{i=x}^{y,z} \rho \bar{u}_i \frac{\partial \bar{u}_z}{\partial i} = -\frac{\partial \bar{p}}{\partial z} + \sum_{i=x}^{y,z} \frac{\partial}{\partial i}\left(\mu_{\text{eff}} \frac{\partial \bar{u}_z}{\partial i}\right) + \rho g_z \tag{8}$$

$$\mu_{eff} = \mu + \mu_t \tag{9}$$

The fluid dynamics equations system above had 5 dependent variables, namely the mean pressure ($\bar{p}$), average velocity of x direction ($\bar{u}_x$), the average velocity of y direction ($\bar{u}_y$), the average velocity of z direction ($\bar{u}_z$) and the turbulent viscosity (μ_t), while the available equations were only 4. There are three methods to solve the above RANS equations system, namely Direct Numerical Solution (DNS) method, modeling method (RANS based model) and Large Eddy Simulation (LES) method. In this research, the solution method used was the modeling with a k-ε standard turbulent model. The model was reported by some researchers, has sufficient capability in modeling turbulent swirl flows, especially in low swirl numbers, and has low computational efforts [10, 11].

In the k-ε turbulent model, turbulent viscosity (μ_t) is expressed by the equation

$$\mu_t = C_\mu \rho (k^2/\varepsilon) \tag{10}$$

which c_μ is the empirical constant, ρ is the fluid density, k is the specific turbulent kinetic energy and ε is the dissipation rate of specific turbulent kinetic energy. The occurrence of variables k and ε, resulting in the number of dependent variable on the system of equation is increased to 6 variables, namely ($\bar{p}$), ($\bar{u}_x$), ($\bar{u}_y$), ($\bar{u}_z$), k and ε, while the number of conservation equations available are still 4 equations.

In order to solve simultaneously the above equations system, two additional conservation equations are needed, those are conservation equations for variables k and ε. The formulations of both conservation equations are given by equation (11) and (12).

$$\rho \frac{\partial k}{\partial t} + \sum_{i=x}^{y,z} \rho \bar{u}_i \frac{\partial k}{\partial i} = \sum_{i=x}^{y,z} \frac{\partial}{\partial i}\left(\frac{\mu_{\text{eff}}}{\sigma_k} \frac{\partial k}{\partial i}\right) + G_k - \rho\varepsilon \tag{11}$$

$$\rho \frac{\partial \varepsilon}{\partial t} + \sum_{i=x}^{y,z} \rho \bar{u}_i \frac{\partial \varepsilon}{\partial i} = \sum_{i=x}^{y,z} \frac{\partial}{\partial i}\left(\frac{\mu_{\text{eff}}}{\sigma_\varepsilon} \frac{\partial \varepsilon}{\partial i}\right) + C_{\varepsilon 1} \frac{\varepsilon}{k} G_k - C_{\varepsilon 2} \rho \frac{\varepsilon^2}{k} \tag{12}$$

Thus for k-ε turbulent models, fluid dynamics characteristics are simulated by the equations (5) to (8) and equations (11) and (12). Variables G_k in the above equations is given by the equations (13), while the constant values of each parameter are as follow $C_\mu = 0.09$, $\sigma_k = 1.0$, $\sigma_\varepsilon = 1.3$, $C_{\varepsilon 1} = 1.44$ and $C_{\varepsilon 2} = 1.92$ [12].

$$G_k = 2\mu_t \left[\sum_{i=x}^{y,z} \left(\frac{\partial \bar{u}_i}{\partial i}\right)^2\right] + \mu_t \left[\left(\frac{\partial \bar{u}_x}{\partial y} + \frac{\partial \bar{u}_y}{\partial x}\right)^2 + \left(\frac{\partial \bar{u}_x}{\partial z} + \frac{\partial \bar{u}_z}{\partial x}\right)^2 + \left(\frac{\partial \bar{u}_y}{\partial z} + \frac{\partial \bar{u}_z}{\partial y}\right)^2\right] \tag{13}$$

3. Results and discussion

3.1 Effect of tangential inlet cross-sectional shape

The rectangular tangential inlet shape provided a higher turbulence intensity than the circular shape, that was 238.7 versus 208.4%. The predicted heat transfer coefficient generated from the rectangular shape was also larger than the circular ones. The average values of heat transfer coefficient for both of tangential inlet shapes, using the Diffus-Boelter equation, were 32.22 and 29.29 W/m^2K, respectively. The heat transfer process in the burner has positive correlation to the devolatilization and the combustion process which occurred in the burner.

The rectangular tangential inlet shape resulted in a longer backflow penetration than the circular ones, those were 0.145 and 0.11 m. The comparison of the generated backflow penetration by the two tangential inlet shapes was given in Figure 2. Backflow penetration in this figure was indicated by a negative axial velocity contour. From the picture it was seen that the backflow penetration depth generated by the rectangular tangential inlet shape was longer than the circular shape.

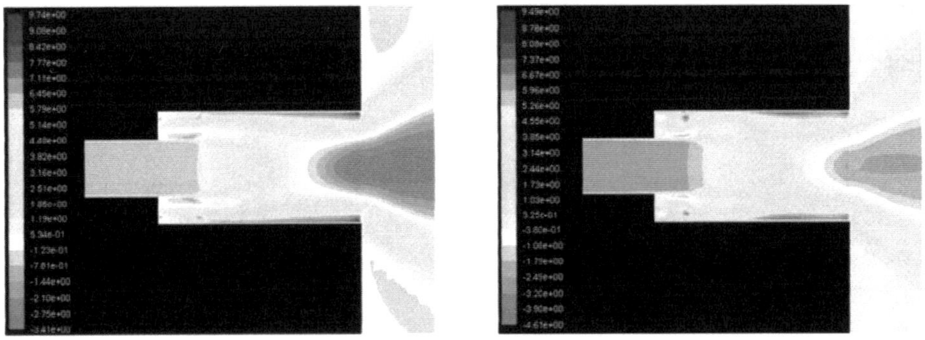

a. Rectangular shape b. Circular shap

Figure 2. Axial velocity contour for both tangential inlet cross section shapes

In the case of the flow resistance, the rectangular tangential inlet shape produced a higher static pressure than the circular ones. The average static pressure in the burner cylinder for both tangential inlet shapes are 51.6 and 42.2 Pascal. The higher the static pressure of the fluid flow, the higher the fluid flow resistance was. It leads to the higher power consumption of the air blower. The comparison of the fluid dynamics performances for both of the tangential inlet shapes were summarized in Table 1.

Table 1. Comparison of fluid dynamics performances of rectangular and circular tangensial inlet shapes.

Performance parameter	Tangential inlet shape	
	Circular	Recta
Turbulence intensity, %	208.4	238.7
Heat transfer coef., W/m^2K	29.29	32.22
Backflow penetration, m	0.11	0.145
Static pressure, Pascal	42.2	51.6

3.2 Effect of tangential inlet orientation angle

The tangential inlet angle also affects the fluid dynamics characteristics in the burner cylinder. When the orientation of tangential inlet configurates a certain angle with the axial axis, then the stream that is formed from the inlet will be tangentially oriented. The larger the tangential inlet angle, the greater the tangential flow frequency and vice versa.

Research on the influence of tangential inlet orientation angle to the burner performance was reported by Ziqiang et al (2016). Of the 3 tangential inlet orientation angles used, ie, 15^0, 30^0 and 45^0, Ziqiang stated that the best tangential inlet orientation angle is an angle of 30^0. At that angle, the largest recirculation zones and uniformly dispersed fuels in the furnace were produced [13].

IOP Conf. Series: Journal of Physics: Conf. Series **1090** (2018) 012007 doi:10.1088/1742-6596/1090/1/012007

To see the effect of the other orientation angle in the range of 15^0 to 45^0, the tangential inlet angle was varied every 5^0. In addition, the orientation angles greater than 45^0, namely 60^0, 75^0 and 90^0, were also simulated. Besides being evaluated based on the recirculation flow pattern, the burner fluid dynamics performances were also evaluated based on other variables such as turbulence intensity, heat transfer and static pressure.

From all tangential inlet orientation angles simulated, the backflow from the furnace to the internal burner always exists, except for the orientation angle of 15^0. The tendency of penetration depth could be stated as the greater the orientation angle, the shorter the backflow penetration depth and vice versa. There is no backflow from the furnace to the internal burner, at the orientation angle of 15^0, is due to the low frequency of the tangential flow that is formed at this angle.

At the orientation angle of 20^0, the backflow penetration approaches the axial inlet output. The mixture of the incoming flow and backflow recirculate to the furnace through the area near the wall. At orientation angles of 25^0 and 30^0, although the depth of backflow penetration is close to the axial inlet output, but the mixture of the two streams tends to lead to the tangential inlet section. For the larger orientation angles (35^0, 40^0, 45^0, 60^0, 75^0 and 90^0), the depth of backflow penetration is shorter. A comparison of the backflow pattern in the burner cylinder for some tangential inlet orientation angles is given in the figure 3.

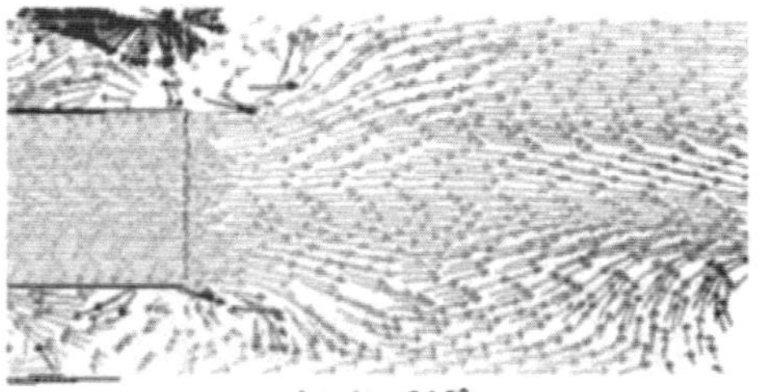
Angle of 15^0

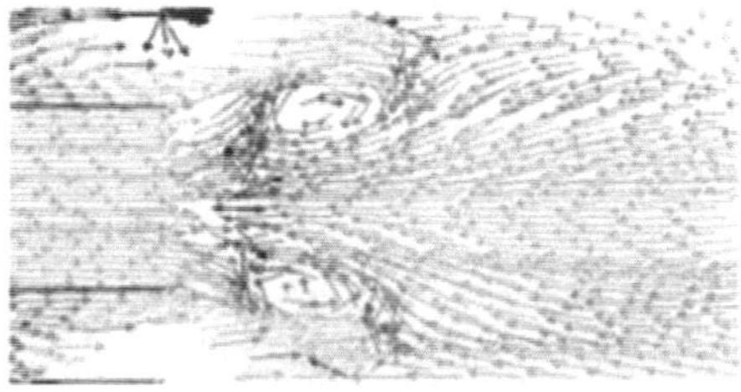
Angle of 20^0

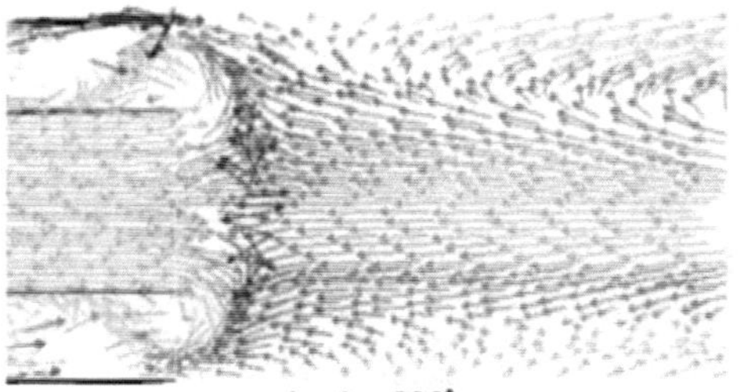
Angle of 25^0

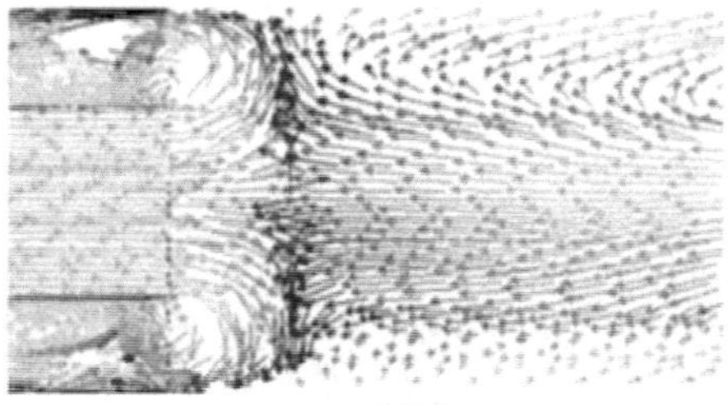
Angle of 30^0

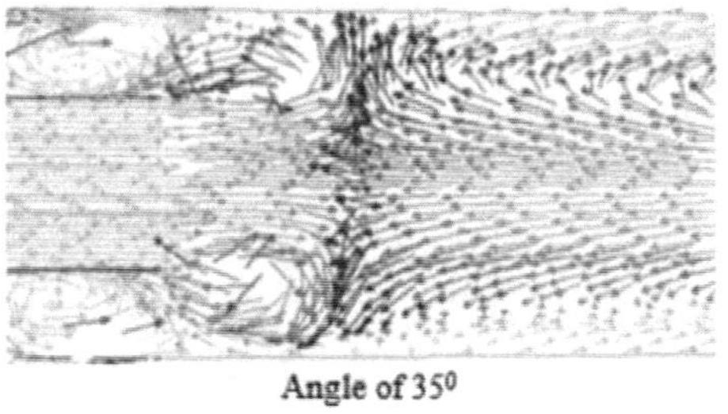
Angle of 35^0

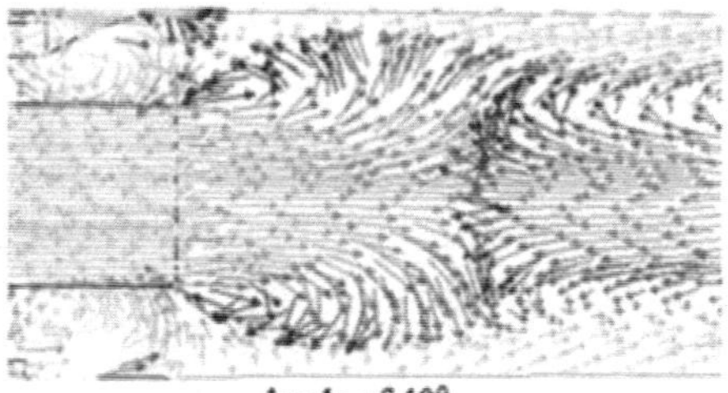
Angle of 40^0

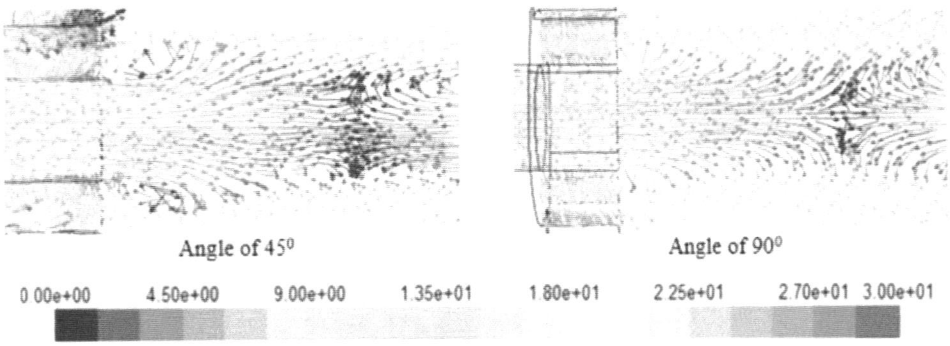

Figure 3. Recirculation flow pattern in burner cylinder for various tangential inlet orientation angles

In term of mixing intensity, the tangential inlet with orientation angle 20^0 yielded the highest turbulence intensity value. This condition is related to the formation of internal recirculation at the mixing zone of the two flow streams. In addition, on this orientation angle, the static pressure was relatively low and the heat transfer rate was relatively high, as shown in figure 4. This indicates that the burner with the tangential inlet angle of 20^0 has the potential to produce the best fluid dynamics performance in the burner. It will allow the devolatilization process and a stable ignition occurs in the burner.

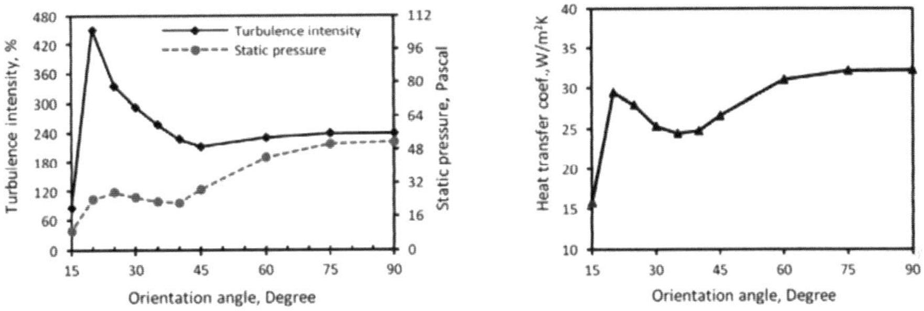

Figure 4. Profile of turbulence intensity, static pressure and heat transfer coefficient in burner cylinder for several tangential inlet orientation angles.

4. Conclusion

The fluid dynamics characteristics in the proposed biomass burner design have been simulated using a standard k-ε turbulent model. The simulation results showed that orientation angle and shape of the tangential inlet influence significanly the fluid dynamic characteristics in the burner cylinder.

At certain orientation angle, the tangential inlet tends to produce back flow from furnace to the internal burner. The tendency of back flow penetration depth could be stated as the greater the orientation angle, the shorter the backflow penetration depth. The turbulence intensity is also inversely proporsionl to the tangential inlet orientation angle, while the static pressure and heat transfer coefficient are directly proportional to it. The rectangular tangential inlet shape produces the deeper backflow penetration depth, the higher turbulence intensity and the greater heat transfer rate, but it requires a higher blower power than the circular shape to overcome the greater static pressure.

The above results have revealed that the best tangential inlet orientation angle of the proposed biomass burner is 20^0 and the appropriate tangential inlet shape is rectangular. Under these conditions, the fluid dynamics characteristics in the burner cylinder potentially produce the best burner performance.

Acknowledgments
We would like to thank to the Ministry of Research, Technology and Higher Education of Indonesia for the awarded research fund to this research on the decentralization research scheme.

References

[1] Dahl J dan Obernberger I 2004 *2nd World Conference on Biomass for Energy, Industry and Climate Protection* Rome Italy

[2] Kops S M B and Malte P C 2004 *Final Technical Report, Energy and Environmental Combustion Laboratory* Department of Mechanical Engineering University of Washington

[3] Siyi L, Bo X, Zhiquan H, Liu S and Maoyun H 2010 *Energy Conversion and Management* **51** 2098-2102

[4] Momeni M, Yin C, Kær S K, Hansen T B, Jensen P A and Glarborg P 2013 *Energy & Fuels* **27** 507−14

[5] Arnao J H S, Ferreira D J O, Santos C G, Alvarez J E, Rangel L P and Park S W 2015 *International Journal of Mechanical, Aerospace, Industrial, Mechatronic and Manufacturing Engineering* **9** (5) 798-801

[6] Nag P K 2002 *Power Plant Engineering* 2nd edition (Singapore: McGraw Hill Company)

[7] Paulrud S dan Nilsson C 2004 *Fuel* **83** 813-21

[8] Al-Abdeli Y M and Masri A R 2015 *Experimental Thermal and Fluid Science* (Accepted Manuscript)

[9] Ansys Incorporation. 2013 *Solver Theory* (United States: Canonsburg)

[10] Nemoda S, Bakic V, Oka S, Zivkovic G and Crnomarkavic N 2005 *Int. J. Heat Mass Transfer* **48** 4623–32

[11] Vazquez J A R 2012 *Thesis* University of Zaragoza Spain

[12] Yazid B 2017 *Rekayasa Komputasi Aliran Turbulen Multidimensi* First edition (Bandung: ITB Press) pp 18-24

[13] Ziqiang L V, Guangqiang L and Yingjie L 2016 *International Journal of Smart Home* **10** 171-80

IOP Conf. Series: Journal of Physics: Conf. Series **1090** (2018) 012008 doi:10.1088/1742-6596/1090/1/012008

Experimental Investigation and Numerical analysis of SO_2 Removal Using Polypropylene Membrane Contactor

D. Ariono, A.N. Hakim, and I.G. Wenten*

Chemical Engineering Department, Institut Teknologi Bandung, Jalan Ganesha 10, Bandung 40132, Indonesia

Corresponding author's e-mail: igw@che.itb.ac.id

Abstract. In this study, polypropylene membrane contactor is designed for SO_2 absorption of simulated flue gas. Water and the aqueous solution of Na_2SO_3 are used as an absorbent. Absorption performance of the longitudinal module is compared to novel transversal membrane contactor module. The result shows that the transversal configuration of hollow fiber module significantly enhances absorption performance over conventional longitudinal configuration. Absorption flux of SO_2 in aqueous Na_2SO_2 is 1.8 times higher than water at optimum Na_2SO_3 concentration and gas flow rate. According to performance stability test, polypropylene hollow fiber membrane is successfully operated for 15 hours maintaining average absorption flux of 20 x 10^{-5} mol.m^{-2}.s^{-1}. Numerical simulation shows good agreement with physical absorption data.

1. Introduction

Nowadays, most of the energy production is made up from the combustion of fossil fuels such as oil, gas, and coal. It is the main source for environmental problems of major concern such as acid rain, photochemical smog, and ozone depletion. To cope with these problems, more stringent emission standards on the acid gasses have been introduced. Generally, well-known packed absorption techniques used for flue gas cleaning are characterized by huge space requirements and high capital cost. Furthermore, conventional wet scrubbing techniques suffer from a number of disadvantages such as loading, flooding, and entrainment [1].

Membrane-based processes can be a suitable alternative to traditional processes. Membrane-based acid gas removal processes have become a feasible alternative to conventional gas separation technologies because of its simplicity and effectiveness. A complete review of membrane-based CO_2 separation process from the natural gas stream is also available in the literature [2]. More specific review of the application and development of superhydrophobic membrane contactor for acid gas removal has been conducted by Himma and Wenten[3]. Polymeric membranes are widely used in CO_2 removal from the natural gas, but few commercially available polymeric membranes which have both high permeability and high selectivity. The effort in enhancing membrane effectiveness has been done in several ways, such as crosslinking, copolymerization, and phase inversion control [4-8]. Molecular engineering of PIM-1/Matrimid blend was carried out by Yong et al.[5] to produce gas separation membranes. Mixed and pure gas permeation tests were performed on these membranes for CH_4 and CO_2. The obtained ideal selectivity and CO_2 permeability were in the range of 4–36 and 12–3355 respectively.

Content from this work may be used under the terms of the Creative Commons Attribution 3.0 licence. Any further distribution of this work must maintain attribution to the author(s) and the title of the work, journal citation and DOI.
Published under licence by IOP Publishing Ltd

Recently, hybrid process of solvent absorption and membrane filtration process has gained more attention as an alternative to conventional acid gas recovery. The substitution of the contacting equipment by membrane device intensifies the process while increasing process efficiency and reducing solvent loses. Process efficiency is increased due to membrane compactness, higher interfacial area, and membrane modularity. When hollow fiber membrane is used, fluids are contacted on the opposite side of the fiber (lumen and shell) and the gas-liquid interface forms at the entrance of membrane pore. Due to this contacting mechanism the available contact area is independent of flow rate. This type of membrane can be an attractive alternative to conventional packet tower because of the following reasons: flooding and unloading free, no emulsions, and the differen in density between fluids is not required. Some application of membrane contactor for mass transfer operation including CO_2 absorption, SO_2 absorption, palm oil deacidification, and iodide oxidation [9-11]. Initial investigation on the applicability of membrane-based acid gas absorption has been completed by Wickramasingh et al., [12]. Since then, several investigation results have been published with mostly in lab scale. Further, a related study on factor determining mass transfer resistance has been conducted. These factors are module configuration [13], membrane wetting [13-16], and mass transfer enhancement by chemical reaction [17-18].

Studies on SO_2 absorption using hollow fiber membrane have been reported in literature. Park et al. [19] investigated the absorption of SO_2 using synthesized PVDF hollow fiber membrane. The effect of various operating variables on membrane performance was evaluated and the different absorbent was also compared. The PVDF hollow fiber membrane contactors exhibited a quite high removal efficiency (up to 85 %) with 2 M NaOH solution, while the Na_2CO_3 solution is the most promising absorbent. Applying seawater as the adsorbent for SO_2 removal from the simulated flue gas through polypropylene hollow fiber membrane contactor was conducted by Sun et al. [20]. Seawater showed superiority compared with aqueous NaOH solution in SO_2 removal especially for high feed gas concentration and large feed gas flux which indicated that seawater has a large capacity on SO_2 absorption. Modeling of SO_2 absorption using hollow fiber ceramic membrane was simulated by Luis et al. [21]. The negative effect of membrane wetting on mass transfer was studied and confirmed in mathematical modeling. In addition, the model also allowed to obtain the best operation condition and solvent selection by incorporating solvent affinity. More recently, experimental study on simultaneous removal of CO_2 and SO_2 was conducted by applying polypropylene hollow fiber membrane as contactor and monoethanolamine (MEA) as absorbent [22]. The experiment results indicated that the membrane could eliminate both components from flue gas. Membrane wetting was also observed during prolonged operation which resulted in the reduction of mass transfer rate. Moreover, increasing gas phase pressure temporarily was found to be effective method to retrieve mass transfer rate. In membrane-based SO_2 absorption, turbulence affects significantly on absorption performance since it controls mass transfer resistance on gas and liquid film. Therefore, flow configuration is an important parameter for gas absorption. Investigation in transversal flow configuration of hollow fiber membrane contactor in CO_2 absorption process was previously reported by Rahmawati et al. [13].

The aim of this study is to investigate the potential of polypropylene membrane contactor in SO_2 removal from synthetic flue gas. In order to get a better understanding of the effect of flow configuration on absorption performance, conventional longitudinal flow membrane contactor is compared to novel transversal flow membrane contactor. In addition, physical and chemical absorption of SO_2 is performed using water and aqueous Na_2SO_3 solution, respectively. Numerical simulation is also conducted to study mass transport inside of membrane module.

2. Numerical Simulation

2.1. Governing equation on gas phase

Differential mass balance equation on gas phase is generated by considering a small cylindrical tube with the porous wall. A mass transfer mechanism is the combination of the convective and diffusive mechanism. Fig. 1. depict 2-dimensional domain of numerical calculation used in this study.

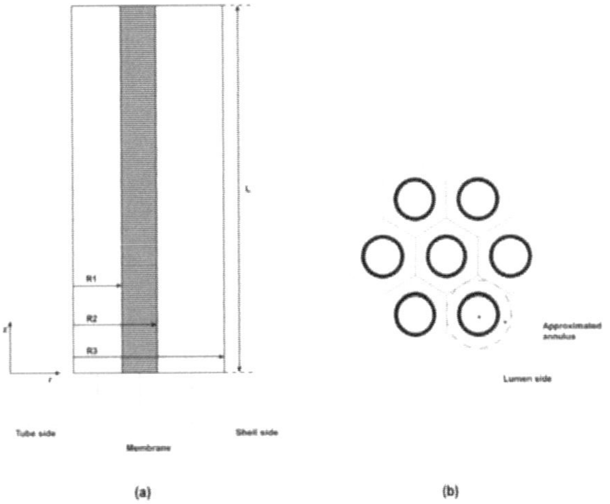

Fig. 1. The axial and radial coordinate of hollow fiber (a) and free surface approximation (b)

Steady-state mass balance for SO_2 on gas phase which consists of the diffusive term and the convective term is described by the following equation:

$$D_{AB}\left[\frac{\partial C_G}{\partial r^2}+\frac{1}{r}\frac{\partial C_G}{\partial r}\right]-V_Z\frac{\partial C_G}{\partial z}-S_R=0 \tag{1}$$

Where D_{AB} is binary diffusion coefficient of $SO_2 - N_2$ in gas phase, $C_G$is SO_2 concentration in the gas phase, V_z is the linear velocity of gas and S_R is source term for reaction. Mass balance equation should be solved simultaneously with momentum equation for convective effect. With the assumption of fully developed laminar flow, velocity profile on tube side is :

$$V_Z=2V_{\max}\left[1-\left(\frac{r}{R_1}\right)^2\right] \tag{2}$$

Where V_m is maximum linear velocity on the tube.
The boundary condition for the system is given by,

at $z=0$, $r=r$, $C_G=C_o$ (3)

at $z=z$, $r=R_1$, $C_G=C_M$ (4)

at $z=z$, $r=0$, $\frac{\partial C_G}{\partial r}=0$ (no flux) (5)

Continuity of flux at the interface between phase and membrane is described by :

at $r=R_1$

$$-D_{AB}\frac{\partial C_G}{\partial r}=-D_M\frac{\partial C_M}{\partial r} \tag{7}$$

at $r=R_2$

$$-D_M\frac{\partial C_M}{\partial r}=-D_{12}\frac{\partial C_L}{\partial r} \tag{8}$$

IOP Conf. Series: Journal of Physics: Conf. Series **1090** (2018) 012008 doi:10.1088/1742-6596/1090/1/012008

Where D_M is diffusion coefficient of SO_2 in membrane pore which is the combination of the binary diffusion coefficient and Knudsen diffusivity and D_{AB} is binary diffusion coefficient for SO_2–N_2 system.

2.2. Governing equation on the membrane

Naturally, mass transport of SO_2 from the gas phase to liquid phase inside membrane pore is mainly due to molecular diffusion. Then the differential mass balance of SO_2 is generated by removing convective term and written as:

$$D_M\left[\frac{\partial C_M}{\partial r^2}+\frac{1}{r}\frac{\partial C_M}{\partial r}\right]=0 \quad (9)$$

The boundary condition for the system on mass transfer domain is given by :

at $z=z$ and $r=R_1$, $C_M=C_G$ (10)

at $z=z$ and $r=R_2$, $C_M=\frac{C_L}{m}$ (11)

Where m is solubility of SO_2 in the aqueous solution of Na_2SO_3

2.3. Governing equation on liquid phase

Steady-state mass balance for the differential element in shell side where liquid absorbent flows is given by :

$$D_{12}\left[\frac{\partial C_L}{\partial r^2}+\frac{1}{r}\frac{\partial C_L}{\partial r}\right]-V_Z\frac{\partial C_L}{\partial z}-SR=0 \quad (12)$$

By adopting Happel free surface approximation, the governing velocity profile on shell side is written as:

$$V_Z=2V_{\max}\left[1-\left(\frac{R_2}{R_1}\right)^2\right]\left[\frac{\left(r/R_3\right)^2-\left(R_2/R_3\right)^2+2\ln\left(R_2/r\right)}{3+\left(R_2/R_3\right)^4-4\left(R_2/R_3\right)^2+4\ln\left(R_2/R_3\right)^2}\right] \quad (13)$$

With boundary condition is given by:

at $z=0$ and $r=r$, $C_L=C_{Lo}$ (14)

at $z=L$ and $r=r$, $\frac{\partial C_L}{\partial r}=0$ (no flux) (15)

at $z=z$ and $r=R_2$, $C_L=mC_M$ (16)

at $z=z$ and $r=R_3$, $\frac{\partial C_L}{\partial r}=0$ (no flux)

Where m is solubility of SO_2 in the aqueous solution of water

In this study, two-dimensional analysis of membrane-based SO_2 absorption is conducted. Set of differential mass balance equation on liquid phase, membrane and gas phase is solved simultaneously by COMSOL MULTIPHYSICS solver with appropriate boundary condition. COMSOL MULTIPHYSICS solves the partial differential equation by finite element approach.

3. Experimental setup

In this study, hydrophobic polypropylene (GDP Filter Indonesia) membrane is used as contactor. The properties of hydrophobic polypropylene membrane are listed in Table 1. SEM micrograph of hydrophobic polypropylene membrane used in this experiment can be seen in Fig.2. The configuration of novel transversal membrane contactor is described in Fig.3. An aqueous solution of Na_2CO_3 (0,18N and 0,25N) and pure water is used as a solvent. Gas-filled pore operation is applied in this experiment since it offers the lowest mass transfer resistance. During experimental test, gas is always flow in lumen side and gas in lumen side since it provides maximum surface area.

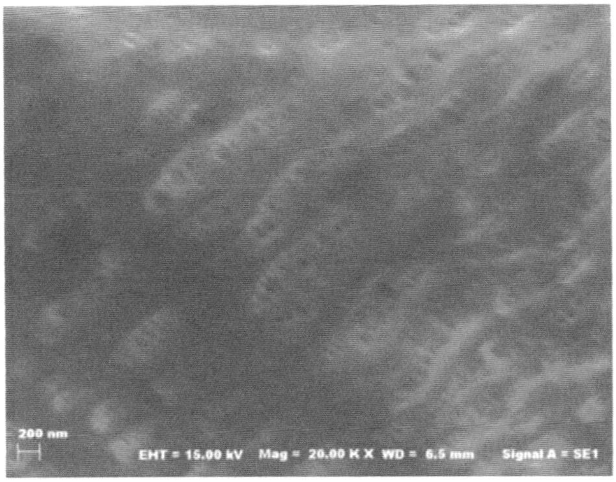

Fig. 2. SEM micrograph of hydrophobic polypropylene membrane surface

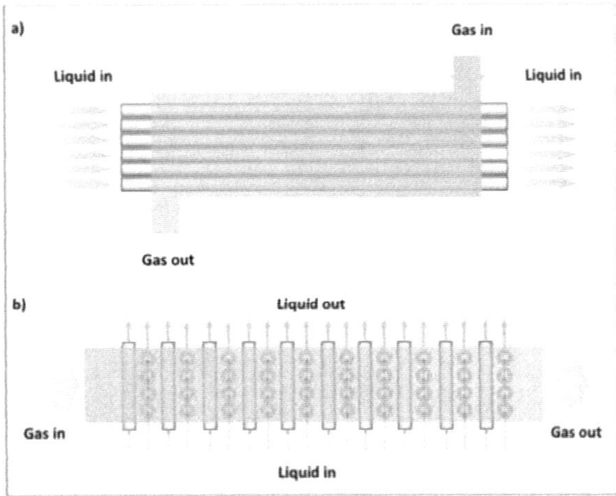

Fig. 3. The configuration of a) longitudinal and b) transversal membrane

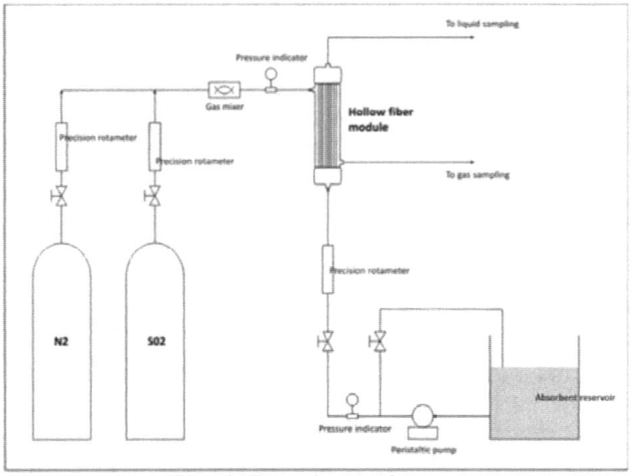

Fig. 4. Experimental setup

Table 1. Properties of membrane contactor used in this study

Membrane material	Polypropylene
Pore diameter (μm)	0.2
Membrane porosity (%)	70
Fiber O.D. (μm)	541
Fiber I.D. (μm)	235
Membrane wall thickness (μm)	153
Membrane effective length (mm)	180
Number of fiber	145
Module I.D. (mm)	12

The experimental set up is shown in Fig. 4. A feed gas containing SO_2 diluted with N_2 is introduced to the system from the pressurized gas cylinder. The concentration of SO_2 on gas stream is maintained < 2%. Feed gas flow is controlled and measured with a regulator valve and precision rotameter. Pressure gauges are installed at the contactor inlet and outlet indicates the gas pressure. Liquid absorbent (pure water or aqueous Na_2SO_3 solution) is pumped from the solvent reservoir by using the peristaltic pump. The instrumentation in the liquid line for the control and measurement of flow and pressure is similar to that in the gas line. Liquid pressure is maintained slightly higher than the gas pressure to prevent bubble formation. The system is first operated for at least half an hour to ensure that a steady state has been reached before collecting data. All the experiments are conducted at room temperature.

Iodimetric reversed titration is used to quantify the amount of captured SO_2. In this method, the sample containing SO_2 is reacted with I_2 where I^- will be liberated during the reaction. The amount of I^- is equivalent to SO_2 in the liquid phase. Determination of I^- is conducted by titration of the solution with a standardized $Na_2S_2O_3$ solution until the equivalent point is reached. Two replication is applied for all of the samples.

4. Result and Discussion

4.1. Effect of operating condition

Turbulence is the most important operating variable in the membrane-based acid gas absorption process since it has an obvious effect on SO_2 absorption flux. Fig. 5 shows the experimental results of SO_2 absorption flux as a function of gas velocity and liquid velocity on chemical absorption where turbulence parameter is characterized by Reynold number (N_{Re}). It is observed that absorption rate of SO_2 increases significantly with the increase of Reynold number. This behavior is expected because the boundary layer thickness of the gas phase in lumen side decreases with increase in turbulence, which leads to the decrease in mass transfer resistance on gas boundary layer. Higher gas flowrate also provides higher concentration difference between phase. The same behavior is also observed on the liquid side. This condition indicates that mass transfer process is controlled by both of gas and liquid mass transfer resistance.

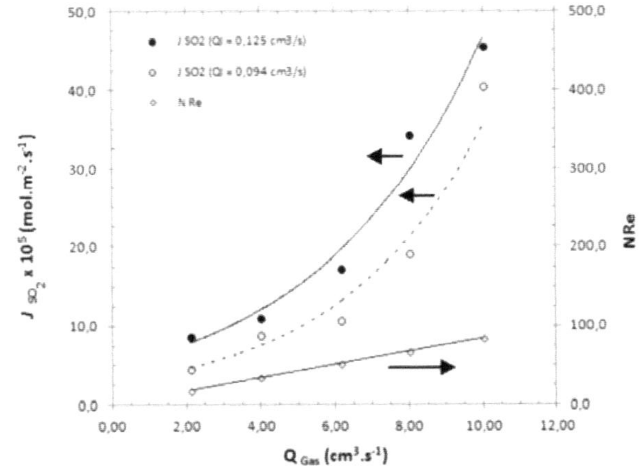

Fig. 5. Effect of gas flow rate on SO_2 absorption flux in various liquid flowrate (Na_2SO_3 concentration = 0.25 N; longitudinal module)

4.2. *Effect of flow configuration*

Comparison between two different flow configuration on membrane-based SO_2 absorption can be seen in Fig. 6. SO_2 Absorption flux in the transversal module and the longitudinal module is compared. The result shows that SO_2 absorption flux in the transversal module is higher than the longitudinal module in all absorbent velocity tested indicated by $J_{trans}/J_{long} > 1$. This result is expected since in the transversal module configuration the flow is perpendicular to the fibers cause greater turbulence than the longitudinal module. The perpendicular flow causes a higher degree of circulation eddy. Moreover, the boundary layer is shorter and equivalent to the circumference of membrane fiber resulting lower mass transfer resistance and higher absorption flux. Whereas in the longitudinal module, the flow is parallel to the fiber and hence the boundary layer equivalent to the fiber length.

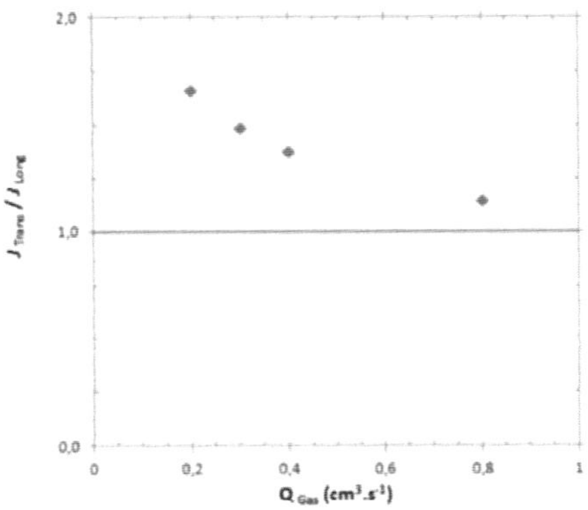

Fig. 6. Performance comparison of longitudinal membrane contactor to novel transversal membrane contactor on physical absorption

4.3. Chemical vs physical absorption

The comparative analysis of absorption performance between chemical and physical absorption is described in Fig. 7. The use of reactive absorbent enhances mass transfer of SO_2 and increases its scrubbing capacity. As the result, for equivalent SO_2 removal, the required absorbent flowrate can be significantly reduced. The reaction involved in SO_2 removal using N_2SO_3 aqueous solution is shown in equation 16 (Park et al., 2008). According to Fig. 7, absorption flux of SO_2 in aqueous Na_2SO_3 is 1,8 times higher than water at optimum Na_2SO_3 concentration and gas velocity. It is also observed that SO_2 removal of chemically enhanced absorption is initially increased then decrease further after passing optimum point. At high velocity, increase in gas flow rate will reduce gas residence time leading to decrease in removal efficiency. In case of physical absorption, SO_2 removal is decreased along with gas flow rate due to its small loading capacity.

$$SO_2 + Na_2SO_3 + H_2O \rightarrow 2NaHSO_3 \qquad (17)$$

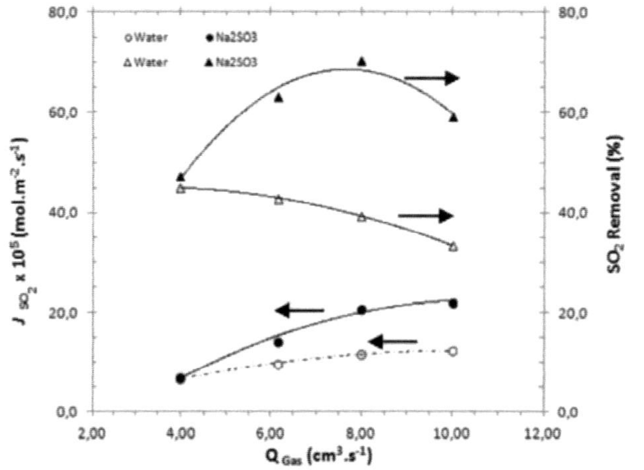

Fig. 7. Comparison of SO_2 absorption flux and % SO_2 removal in physical absorption and chemical absorption (liquid flowrate = 0.125 $cm^3.s^{-1}$; longitudinal module)

According to Fig. 8 it is observed that SO_2 absorption flux increase significantly with the increase of Na_2SO_3 concentration. This condition is expected since reaction rate between Na_2SO_3 and SO_2 is enhanced by the increase of Na_2SO_3 concentration lead to higher SO_2 flux. In addition, a higher concentration of Na_2SO_3 will increase the absorption capacity of the absorbent. However, it should be noted that the higher concentration of solvent employed resulting in higher cost of solvent and energy required for regeneration.

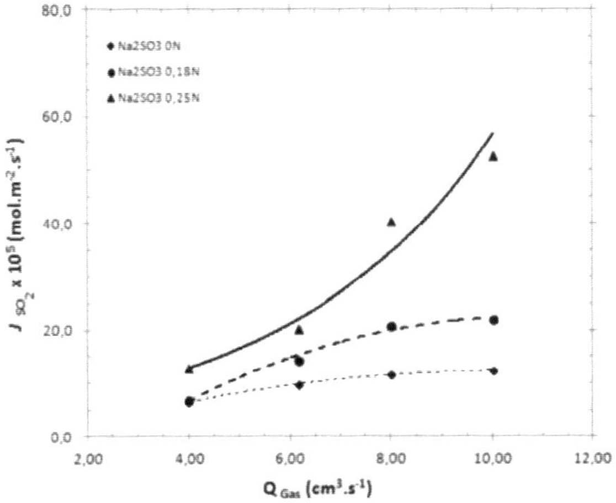

Fig. 8. Effect of Na_2SO_3 concentration on SO_2 absorption flux (liquid flowrate = 0.125 $cm^3.s^{-1}$; longitudinal module)

4.4. Simulation result

Film theory has been widely used to described mass transfer on membrane contactor process. In chemically enhanced gas absorption, mass transfer from the gas phase to liquid phase is driven by concentration difference of chemically active component between phase. The film theory assumes that mass transfer resistances are in series and are represented by the sum of the reciprocals product of the individual mass transfer coefficients. The overall mass transfer coefficient may then be written as:

$$\frac{1}{K_{ov}} = \frac{1}{Hk_G}\frac{d_i}{d_o} + \frac{1}{Hk_M}\frac{d_i}{d_m} + \frac{1}{Ek_L} \tag{15}$$

Where H represents Henry's constant, K_{ov} is overall mass transfer coefficient, k_G, k_M, k_L is individual mass transfer coefficient on gas phase, membrane and liquid phase respectively, d_i is inside diameter of the fiber, d_o is outside diameter of the fiber, d_m is log mean diameter of the fiber and E is enhancement factor due to chemical reaction. Comparison between experimental and calculated value of *K* is shown in Fig 10. The calculated value is in good agreement with experimental data on physical absorption. On the other side, the calculated value of *K* is far lower than the experimental value at chemical absorption.

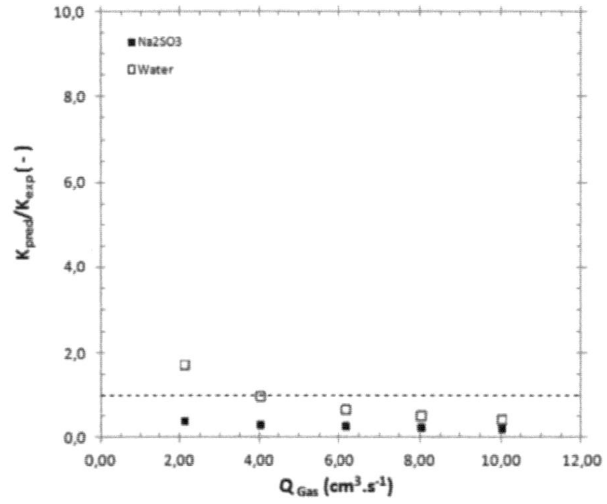

Fig. 9. Comparison overall mass transfer coefficient between experimental and calculated (liquid flow rate = 0.125 $cm^3.s^{-1}$; longitudinal module)

The contour of SO_2 concentration on the liquid phase is depicted on Fig. 10 with arrow shape indicates total flux. Liquid absorbent (water) flows in the shell side of hollow fiber membrane from (*z*/*L*) = 0 in the parallel direction to gas phase. SO_2 is transported from gas phase to liquid phase due to concentration difference between gas and liquid. The highest concentration of SO_2 is found on gas-liquid interface where initial dissolution takes place.

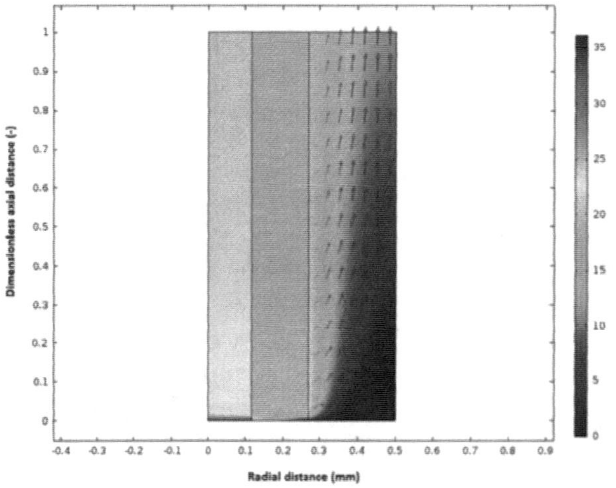

Fig. 10. Concentration of SO_2 on liquid phase (liquid flowrate = 0.125 $cm^3.s^{-1}$; longitudinal module)

Contour of SO_2 concentration on liquid phase at various gas velocity is also presented in Fig. 10. The amount of SO_2 captured by water increases along with gas velocity which is expected since it's provide higher concentration difference. This behavior is also validated by experimental data. Effect of gas inlet velocity and gas initial concentration is shown in Fig.11. SO_2 absorption flux increase along with inlet velocity which is agree with experimental data (Fig.11.a). Expected behavior is also found for variation of initial SO_2 concentration where absorption flux increases due to larger concentration difference between phases (Fig. 11.b).

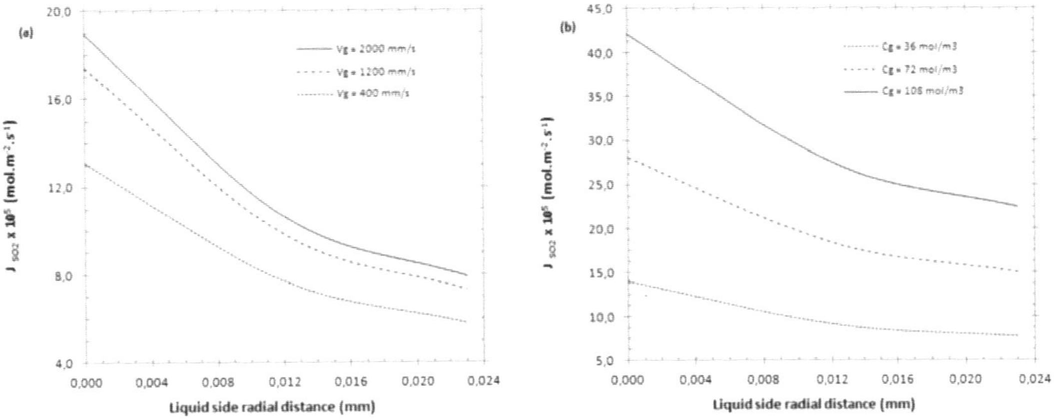

Fig. 11. SO_2 absorption flux at various gas inlet velocity (a) and SO_2 absorption flux at various inlet concentration (b) for physical absorption.

4.5. *Performance stability test*

Fig. 12 shows the experimental results of SO_2 absorption in membrane contactor over 15 h. The liquid absorbents flowed in the shell side and SO_2 in the lumen side of the polypropylene hollow fiber membrane at ambient temperature. During the tests, absorption flux of SO_2 is maintained steadily over 20 x 10^{-5} mol.m^{-2}.s^{-1}. Hydrophobic nature of polypropylene hollow fibers membrane prevents the penetration of liquid absorbent into membrane pores. It prevents additional mass transfer resistance due to entrapment of liquid inside membrane pores.

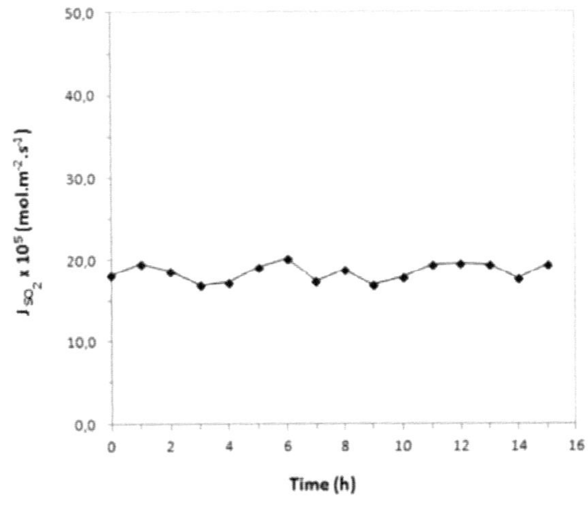

Fig. 12. Performance stability test of membrane contactor in chemical absorption (gas flowrate = 6.15 $cm^3.s^{-1}$; liquid flowrate = 0.125 $cm^3.s^{-1}$; Na_2SO_3 concentration = 0.25 N; longitudinal module)

5. Conclusion

Polypropylene hollow fiber membrane contactor is designed for absorption of SO_2 from synthetic flue gas. The effect of mass transfer enhancement by chemical reaction is also investigated by comparing water and the aqueous solution of Na_2SO_3. The result shows that the absorption performance is controlled by both of gas and liquid flow rate. In comparison with the longitudinal module, transversal module shows potential with higher absorption performance. Significant enhancement of SO_2

IOP Conf. Series: Journal of Physics: Conf. Series **1090** (2018) 012008 doi:10.1088/1742-6596/1090/1/012008

absorption flux is achieved by chemical absorbent with absorption flux 1.8 times higher than water in optimum gas velocity. The membrane contactor is successfully operated for 15 h maintaining average SO_2 absorption flux of 20 x 10^{-5} mol.m^{-2}.s^{-1}. It is also observed that simulation result on absorption flux is in a good agreement with experimental data

Nomenclature

C_G	SO_2 concentration in liquid phase (mol.l^{-1})
C_L	SO_2 concentration in liquid phase (mol.l^{-1})
C_M	SO_2 concentration in liquid phase (mol.l^{-1})
D_{12}	diffusion coefficient of SO_2 in Na_2SO_3 aqueous solution (m^2.s^{-1})
D_M	diffusion coefficient of SO_2 in membrane pore (m^2.s^{-1})
D_{AB}	binary diffusion coefficient for $SO_2 - N_2$ system (m^2.s^{-1})
K_{ov}	overall mass transfer coefficient (m.s^{-1})
k_G	individual mass transfer coefficient on gas phase (m.s^{-1})
k_M	individual mass transfer coefficient on membrane phase (m.s^{-1})
k_L	individual mass transfer coefficient on liquid phase (m.s^{-1})
m	solubility of SO_2 in Na_2SO_3 aqueous solution
N_{Re}	Reynold number
r	radius of fiber (m)
S_R	source term for reaction
V_z	linear velocity of liquid (m.s^{-1})

References

[1] Dindore V Y, Brilman D W F and Versteeg G E 2005 Modeling of cross-flow membrane contactors: mass transfer with chemical reactions J. Membr. Sci. **255** 275–289

[2] Adewole J K, Ahmad A L, Ismail S, and Leo C P 2013 Current challenges in membrane separation of CO_2 from natural gas: A review International Journal of Green House Gas Control **17** 46 – 65

[3] Himma N F and Wenten I G 2017 Superhydrophobic Membrane Contactor for Acid Gas Removal J. Phys.: Conf. Ser. **877** 012010

[4] Kanehashi S, Sato S and Nagai K 2010 Synthesis and gas permeability of hyperbranched and cross-linked polyimide membranes. In: Yampolskii, Y., Freeman, B. (Eds.), Membrane Gas Separation. John Wiley & Sons, Ltd., West Sussex, pp. 3–27

[5] Yong W, Li F, Xiao Y, Li P, Pramoda K, Tong Y W and Chung T S 2012 Molecular engineering of PIM-1/Matrimid blend membranes for gas separation J. Membr. Sci. **407–408** 47–57

[6] Pinnau I and Koros W J 1991 Structures and gas separation properties of asymmetric polysulfone membranes made by dry, wet, and dry/wet phase inversion J. Appl. Polym. Sci. **43** 1491

[7] Song S –W and Torkelson J M 1995 Coarsening effects on the formation of microporous membranes produced via thermally induced phase separation of polystyrene-cyclohexanol solutions J. Membr. Sci. **98** 209

[8] Mikawa M, Nagaoka S and Kawakami H 2002 Gas permeation stability of asymmetric polyimide membrane with thin skin layer: effect of molecular weight of polyimide J. Membr. Sci. **208** 405–414

[9] Ariono D, Chandranegara A S, Widodo S, Khoiruddin and Wenten I G 2018 Flue gas carbon capture using hollow fiber membrane diffuser-separator IOP Conf. Ser.: Mater. Sci. Eng. **285** 012010

[10] Purwasasmita M, Nabu E B P, Khoiruddin and Wenten, I.G. 2015 Non Dispersive Chemical Deacidification of Crude Palm Oil in Hollow Fiber Membrane Contactor J. Eng. Technol. Sci. **47** 426–46

[11] Wenten I G, Julian H and Panjaitan N T 2012 Ozonation through ceramic membrane contactor for iodide oxidation during iodine recovery from brine water Desalination **306** 29–34

[12] Wickramasinghe S R, Semmens M J and Cussler E L 1992 Mass-transfer in various hollow fiber geometries J. Membr. Sci. **69** 235–250

[13] Rahmawati Y and Wenten I G 2010 Wetting Phenomena on The Polypropylene Hollow Fiber Membrane Contactor for CO_2 Absorption Process, 8th International Conference on Membrane Science and Technology ITB Bandung, Indonesia, 29th Nov - 1st Dec, 2010

[14] Keshavarz P, Fathikalajahi J and Ayatollahi S 2008 Analysis of CO_2 separation and simulation of a partially wetted hollow fiber membrane contactor J. Hazard. Mater. **152** 1237 – 1247.

[15] Zhang H -Y, Wang R, Liang D T and Tay J H 2008 Theoretical and experimental studies of membrane wetting in the membrane gas–liquid contacting process for CO_2 absorption J. Membr. Sci. **308** 162–170

[16] Lu J -G, Zheng Y –F and Cheng M –D 2008 Wetting mechanism in mass transfer process of hydrophobic membrane gas absorption, J. Membr. Sci. **308** 180–190

[17] Wang R, Li D F and Liang D T 2004 Modeling of CO_2 capture by three typical amine solutions in hollow fiber membrane contactors Chem. Eng. Proc. **43** 849–856

[18] Korikov A P and Sirkar K K 2005 Membrane gas permeance in gas–liquid membrane contactor systems for solutions containing a highly reactive absorbent J. Membr. Sci. **246** 27–37

[19] Park H Y, Deshwal B R, Kim I W and Lee H K 2008 Absorption of SO_2 from flue gas using PVDF hollow fiber membranes in a gas–liquid contactor J. Membr. Sci. **319** 29–37

[20] Sun X, Meng F and Yang F 2008 Application of seawater to enhance SO_2 removal from simulated flue gas through hollow fiber membrane contactor J. Membr. Sci. **312** 6–14

[21] Luis P, Garea A and Irabien A 2010 Modelling of a hollow fibre ceramic contactor for SO_2 absorption Separ. Purif. Technol. **72** 174–179

[22] Lv Y, Yu X, Tu S-T, Yan J and Dahlquist E 2012 Experimental studies on simultaneous removal of CO_2 and SO_2 in a polypropylene hollow fiber membrane contactor Appl. Ener. **97** 283–288

IOP Conf. Series: Journal of Physics: Conf. Series **1090** (2018) 012009 doi:10.1088/1742-6596/1090/1/012009

Simulation of Semi-Autogenous Grinding (SAG) Mill using Circular-Disks-based Model

R. Sari[1], P. M. Widartiningsih[2], M. A. Martoprawiro[3], L. Hendrajaya[4], S. Viridi[4*]

[1]Master Program in Computational Science, Faculty of Mathematics and Natural Sciences, Institut Teknologi Bandung, Jalan Ganesha 10, Bandung 40132, Indonesia
[2]Master Program in Physics, Faculty of Mathematics and Natural Sciences, Institut Teknologi Bandung, Jalan Ganesha 10, Bandung 40132, Indonesia
[3]Chemistry Department, Faculty of Mathematics and Natural Sciences, Institut Teknologi Bandung, Jalan Ganesha 10, Bandung 40132, Indonesia
[4]Physics Department, Faculty of Mathematics and Natural Sciences, Institut Teknologi Bandung, Jalan Ganesha 10, Bandung 40132, Indonesia

[*]dudung@fi.itb.ac.id

Abstract. System of semi-autogenous grinding (SAG) mill is modeled in this work based on circular-disk objects, which represents not only the ground materials but also the mill liners. Interaction between these objects is through linear spring dash-pot force, which is solved numerically in soft-sphere scheme using molecular dynamics (MD) method. There is also force due to earth gravity. For ground materials a grain consists of several smaller grains, which are attached through some adhesive force. As the ground materials are smashed to the mill liners, when they reach maximum compression, they will be shattered into the smaller grains. Shattering process happens in a special materials motion known as cataracting mode. Friction is not considered in this work due to its complexity.

1. Introduction

The application of semi-autogeneous grinding (SAG) mill is depend on the type of motion. Cascading and cataracting mode are the optimal type of motion to crush the particles into smaller ones.

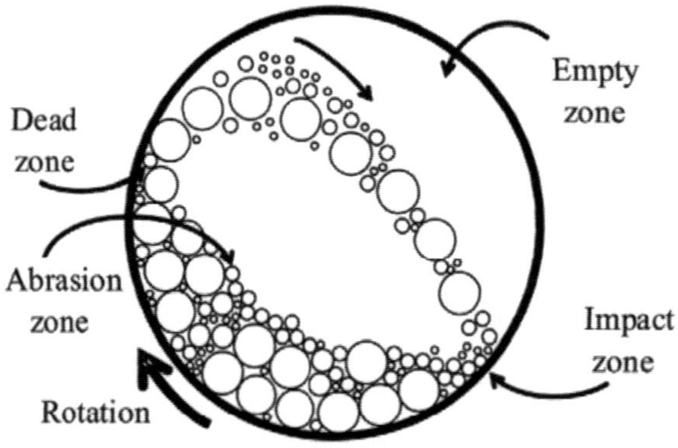

Figure 1. Motion of particles and zones in cataracting mode.

Content from this work may be used under the terms of the Creative Commons Attribution 3.0 licence. Any further distribution of this work must maintain attribution to the author(s) and the title of the work, journal citation and DOI.
Published under licence by IOP Publishing Ltd

There are 4 zones in cataracting mode, consist of dead zone, abrasion zone, impact zone, and empty zone (Wills, 2015). As the mill was rotated in clockwise direction, the particle will move up on the left-bottom side of the mill and free fall in parabolic path once they reach the certain height. The smaller grains will be generated in impact zone.

The experimental research regarding two-dimensional SAG mill had been done using serrated cylindrical wall in order to increase milling efficiency, by optimizing the cataracting mode in grinding process (Faizin *et.al.*, 2016). In this paper, we conducted a simulation and analysis of SAG with circular liner using event-driven simulation.

2. Model

In this work a SAG (semi-autogenous grinding) mill is simplified into a two-dimensional system and modeled as a compound of circular disks. Then, the whole system consists only of two-dimensional circular objects, i.e. the ground materials and also the mill liners.

2.1. Mill model

The mill with radius R will have a set of liners represented by N_L circular disks as given in the following Figure 1.

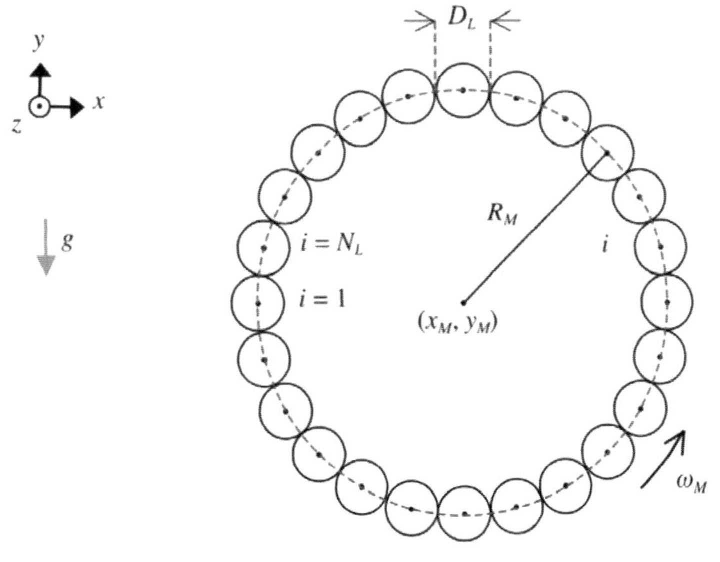

Figure 2. Two-dimensional model of SAG mill with radius R and liners in the form of N circular disks is rotated with angular velocity *ωM*.

Consider two adjacent liner of diameter D which is separated in angular position by $\Delta\theta$ as shown in Figure 2 (left). From Figure 2 (right) it can be obtained that

$$\sin\left(\frac{1}{2}\Delta\theta\right) = \frac{1}{2}\frac{D_L}{R_M}, \tag{1}$$

where

$$N_L = \frac{2\pi}{\Delta\theta}. \tag{2}$$

Substitution of Equation (2) into Equation (1) will give

$$D_L = 2R_M \sin\left(\frac{\pi}{N_L}\right). \tag{3}$$

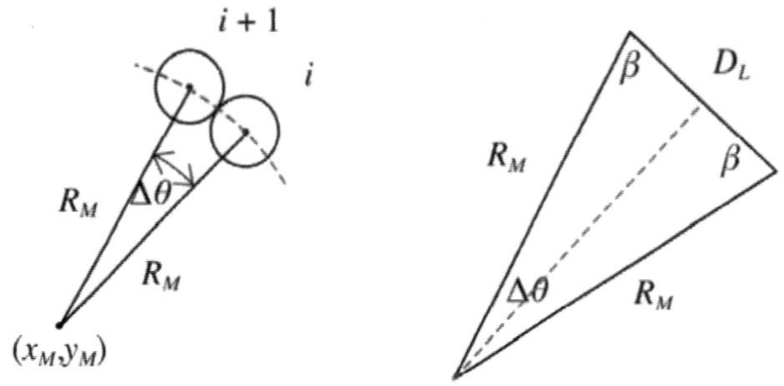

Figure 3. Geometry relation among R_M, D_L, and $\Delta\theta$.

Equation (3) can be easily proven for N_L = 2, 3, 4, and 6 as shown in Figure 3, which are

$$N_L = 2 \Rightarrow D_L = 2R_M \sin\left(\frac{\pi}{2}\right) = 2R_M, \tag{4a}$$

$$N_L = 3 \Rightarrow D_L = 2R_M \sin\left(\frac{\pi}{3}\right) = \sqrt{3}R_M, \tag{4b}$$

$$N_L = 4 \Rightarrow D_L = 2R_M \sin\left(\frac{\pi}{4}\right) = \sqrt{2}R_M, \tag{4c}$$

$$N_L = 6 \Rightarrow D_L = 2R_M \sin\left(\frac{\pi}{6}\right) = R_M. \tag{4d}$$

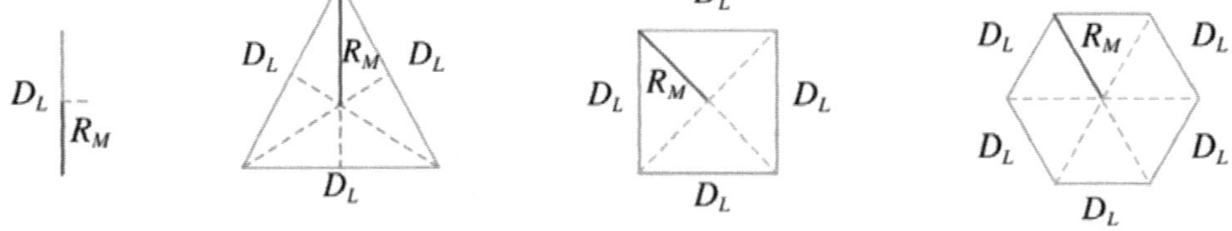

Figure 4. Geometry relation between R_M and D_L for N_L = 2, 3, 4, and 6.

Angular position of liner i relative to first liner is

$$\theta_i = \theta_1 + (i-1)\Delta\theta. \tag{5}$$

Then, it is defined that

$$\theta_1 = \theta_0 + \omega_M t, \tag{6}$$

where θ_0 is initial angular position of the first liner. The angular position is measured to the x axis in counter-clockwise (CCW) direction the same as direction of ω_M, as it is shown previously in Figure 1. With Equations (2), (5), and (6) following relation can be obtained

$$\theta_i = \theta_0 + (i-1)\frac{2\pi}{N_L} + \omega_M t, \quad i = 1, 2, 3, .., N_L, \tag{7}$$

which gives angular position of liner i at every time t. Position of every liner is then

$$x_i = x_M + R_M \cos\theta_i \tag{8}$$

and

$$y_i = y_M + R_M \sin\theta_i. \tag{9}$$

The kinematics of every liner is defined sufficiently using Equation (7)-(9) with input parameters N_L, R_M, and ω_M.

2.2. Interaction forces

Each of the disk-formed ground particles i could have specific value of diameter D_i and density ϱ_i (or mass m_i). These particles interact with each other and also with particles representing mill liners through normal force $\vec{N}$. Then, the two types of particles are also suffering from earth gravitation force $\vec{G}$ and air drag force $\vec{D}$. Only these three types of forces are considered in this work. According to coordinate system used in Figure 1 then gravitation force on particle i is simply

$$\vec{G}_i = -m_i g \hat{e}_y, \tag{10}$$

where g is earth gravity with value about 9.807 m/s^2. Assuming that particles are moving in still air, then

$$\vec{D}_i = -c_D \vec{v}_i. \tag{11}$$

is the formulation for drag force. Relative position of particle i related to particle j is

$$\vec{r}_{ij} = \vec{r}_i - \vec{r}_j \tag{12}$$

and its relative velocity is

$$\vec{v}_{ij} = \vec{v}_i - \vec{v}_j. \tag{13}$$

From Equation (12) unit vector of relative position can be obtained

$$\hat{r}_{ij} = \frac{\vec{r}_{ij}}{r_{ij}} = \frac{\vec{r}_{ij}}{|\vec{r}_{ij}|} = \frac{\vec{r}_{ij}}{\sqrt{\vec{r}_{ij} \cdot \vec{r}_{ij}}}. \tag{14}$$

Overlap between particles i and j is defined as

$$\xi_{ij} = \max\left(0, \tfrac{1}{2}D_i + \tfrac{1}{2}D_i - r_{ij}\right) \tag{15}$$

and it derivative with respect to time t is

$$\frac{d\xi_{ij}}{dt} = -v_{ij}\,\mathrm{sign}(\xi_{ij}). \tag{16}$$

Normal force on particle i due to interaction with particle j is formulated as (Schäfer *et al.*, 1996)

$$\vec{N}_{ij} = \left(k_N \xi_{ij} - \gamma_N \frac{d\xi_{ij}}{dt} \right) \hat{r}_{ij} . \quad (17)$$

2.3. Newton's second law of motion
Acceleration of particle i can be obtained from

$$\vec{a}_i = \frac{1}{m_i} \sum \vec{F} , \quad (18)$$

where $\sum \vec{F}$ is sum of all forces worked on the particle. Supposed that if there are N_L liner particles and N_G ground material particles then from Equations (10), (11), and (17)

$$\vec{a}_i = \frac{1}{m_i} \left(\vec{G}_i + \vec{D}_i - \sum_{j=1}^{N_L} \vec{N}_{ij} + \sum_{\substack{j=1 \\ j \neq i}}^{N_G} \vec{N}_{ij} \right) \quad (19)$$

or

$$\frac{d^2\vec{r}_i}{dt^2} - \frac{1}{m_i} \left[\sum_{j=1}^{N_L} \left(k_N \xi_{ij} - \gamma_N \frac{d\xi_{ij}}{dt} \right) \frac{\vec{r}_i - \vec{r}_j}{|\vec{r}_i - \vec{r}_j|} + \sum_{\substack{j=1 \\ j \neq i}}^{N_G} \left(k_N \xi_{ij} - \gamma_N \frac{d\xi_{ij}}{dt} \right) \frac{\vec{r}_i - \vec{r}_j}{|\vec{r}_i - \vec{r}_j|} \right] + \frac{c_D}{m_i} \frac{d\vec{r}_i}{dt} + g\hat{e}_y = 0 \quad (20)$$

is the equation of motion of particle i, that should be solved. It is actually $N_G \times (N_G - 1 + N_L)$ coupled differential equations. Due to the complexity they cannot be so solved simultaneously, but it can be approached using numerical method, which will be described in Simulation section.

3. Simulation

Equation (20) will be solved using Euler method (Seinfeld *et al.* 1970), that relates motion variable at time $t + \Delta t$ with its value at time t and also it time derivative

$$\left. \frac{d^n f}{dt^n} \right|_{t+\Delta t} = \left. \frac{d^n f}{dt^n} \right|_t + \left. \frac{d^{n+1} f}{dt^{n+1}} \right|_{t+\Delta t} \Delta t , \quad (21)$$

where f stands for $\vec{r}$ if $n = 0$ and for $\vec{v}$ if $n = 1$. Acceleration $\vec{a}$ is obtained from Newton's second law of motion. After one time step Δt of calculation all values of motion variables from all particles have been changed, then Equations (19) and (21) are repeated until the desired simulation repetition. Normally a final stopping criterion is set, e.g. $t = t_{\mathrm{end}}$, where initial time is set to 0 or t_{beg}. In this research, the simulation run at 10^{-3} *s* time step interval, Δt.

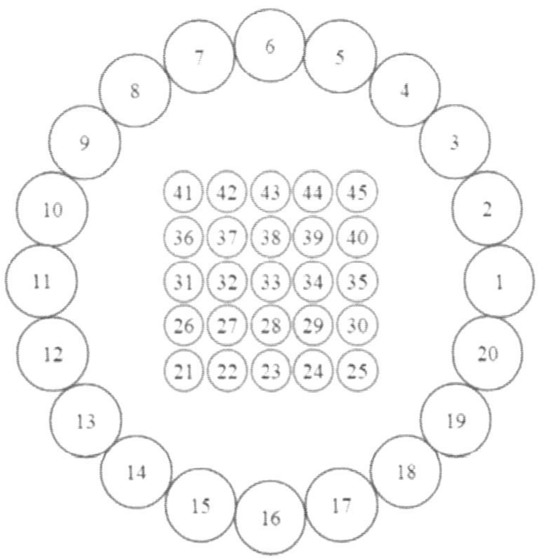

Figure 5. Initial condition of simulation with 20 circular disks as liner and 25 grains inside the mill. The liner was rotated in clockwise direction with angular velocity of 0.8 rad/s.

The number of circular disks and radius of mill are required to be set before the simulation begins. Circular disks and grains are numbered in order to analyse the behaviour of each grains. Numbering on grains is depend on the number of circular disks and the number of the whole particles in the system. The configuration of simulation which was shown in Fig.4 used 45 particles with 20 of them were circular disks. In initial condition, the grains were positioned at the centre of the mill.

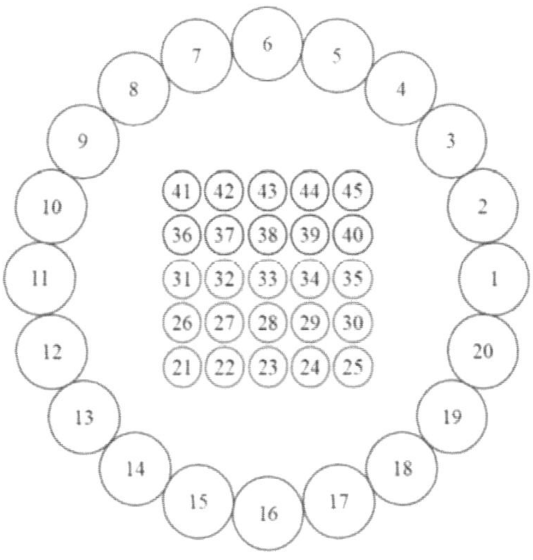

Figure 6. Initial condition of SAG simulation with 10 ball mills at the top of grains configuration.

The ball mills, which were marked with black line, are placed at the top two rows of grains configuration in order to push the grains to reach the bottom of liner wall as the mass was 5 times greater than the particles and speed up the grinding process. The number of ball mills was set lesser than the particles to be milled. In this simulation, we did the same characteristic of rotation with the previous one.

4. Results and discussion

Simulation was conducted using JavaScript program (Viridi and Kurniadi, 2017).

Figure 7. Display of simulation with $N_L = 20$, $N = 45$, $R_M = 38$ at (a) $t = 0$ *s*, (b) $t = 2$ *s*, (c) $t = 3$ *s*, (d) $t = 4$ *s*, (e) $t = 6$ *s*, (f) $t = 7$ *s*, (g) $t = 8$ *s*, (h) $t = 10$ and (i) $t = 12$ *s*.

As the consequence of the gravitational attraction, the grains move in negative direction of *y*-axis, which was shown in Fig.7b. The grains started to reach the bottom of the mill at $t = 2$ *s* and interacted with the rotated liner. Interaction between grains and the circular disks generates the normal force where the direction of movement is affected by the properties of interacting particles such as vector and magnitude of velocity. When the value of normal force takes greater contribution than gravitational force, the movement of particle will obey the normal force direction as shown particle numbered 42 in Fig.7c.

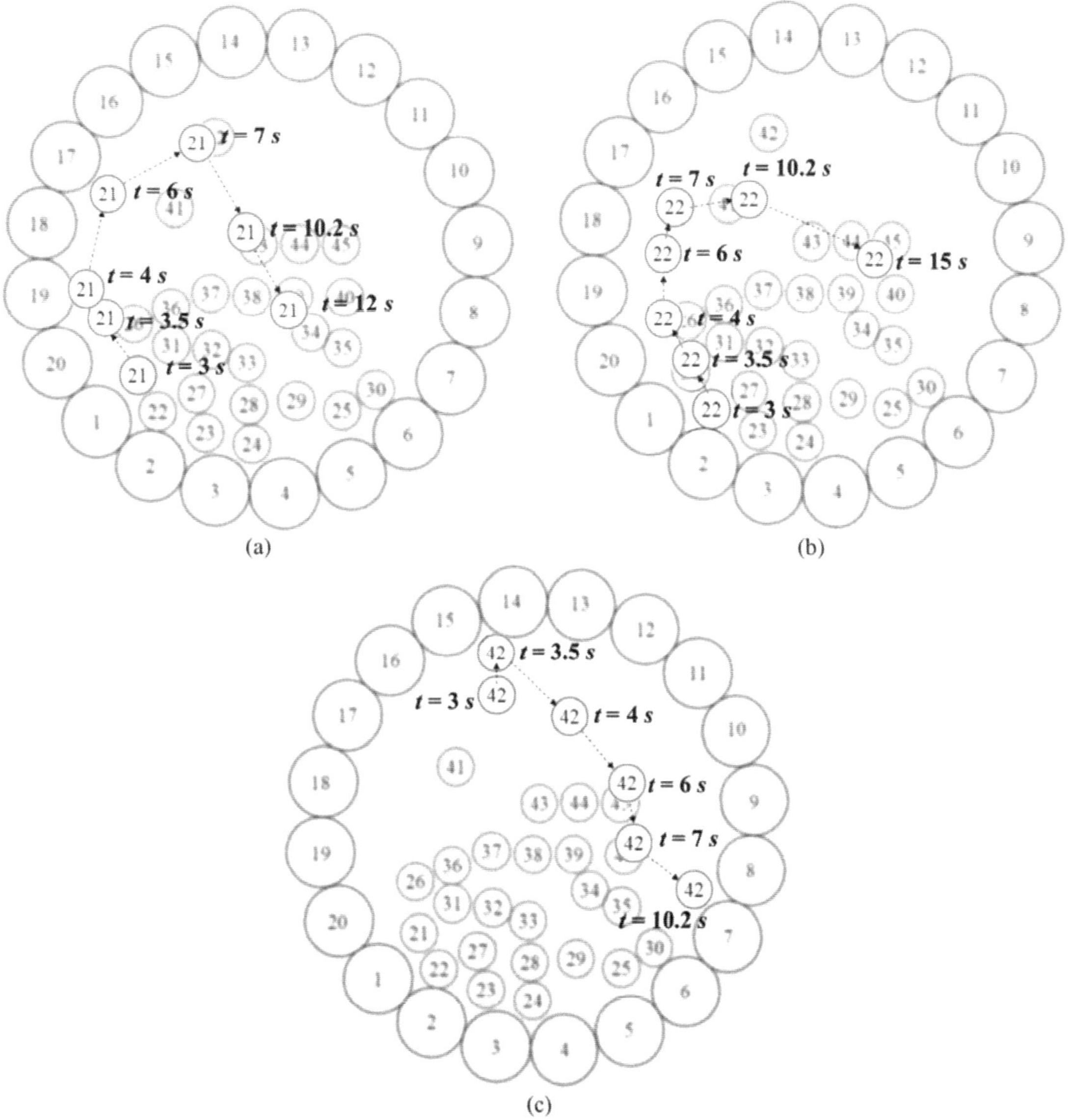

Figure 8. Movement of particle numbered (a) 21, (b) 22, and (c) 42.

Particle numbered 21 and 22 were moving in cascading mode, then become cataracting mode at $t = 7$ *s* forward which were described in Fig.8a and 8b. While particle numbered 42 was moving in cataracting mode by $t = 3$ *s*.

The particles were moving in cascading and cataracting mode according to the rotation of mill which was in clockwise direction. This condition was described in Fig.7e where the particles assembled at the left-bottom side which is convenient with the characteristic of cataracting mode described in Fig.1.

IOP Conf. Series: Journal of Physics: Conf. Series **1090** (2018) 012009 doi:10.1088/1742-6596/1090/1/012009

Figure 9. Display of simulation with $N_L = 20, N = 45, R_M = 38$ at (a) $t = 0$ s, (b) $t = 2$ s, (c) $t = 3$ s, (d) $t = 4$ s, (e) $t = 6$ s, (f) $t = 7$ s, (g) $t = 8$ s, (h) $t = 10$ s, and (i) $t = 12$ s.

In SAG simulation, the particles and the grinding balls were moving down due to gravitational force in first two seconds. The motion of particles which affected by normal force were retained by the falling ball mills, therefore the particles distributed at the lower region of mill if compare to the distribution in Fig.7.

Grinder ball numbered 40 hit the particle numbered 35 in impact zone, allows the particle to be smaller parts. The cataracting mode in this simulation are shown in Fig.9d until Fig.9i. Particle numbered 21, 22, and 26 were moving up due to the rotation of the wall at $t = 6$ s, afterwards they jump to the impact zone. This condition is similar to Fig.8.

5. Conclusion

The simulation program used in this paper offer the adjustable configuration of mill, such as size and number of circular liner and also the radius of mill. The simulation run results an acceptable distribution of grains when the grinding occurs, especially for cascading and cataracting mode. The existence of ball mills can accelerate the grinding process.

Acknowledgements

Author SV would like to thank Program Riset Inovasi ITB in year 2017 with contract number 107x/I1.C01/PL/2017 for supporting this work and also to PT Freeport Indonesia for introducing author to SAG mill system and related problem while visiting Tembagapura site on 7-11 November 2016.

References

[1] Faizin N, Hendrajaya L, Viridi S and Alfisyah M 2016 *Spektra: J. Fis. Aplikasi.* **1** 157
[2] Wills B A, Finch J A 2015 Wills' Mineral Processing Technology: An Introduction to the Practical Aspects of Ore Treatment and Mineral Recovery Eighth Edition, Department of Mining and Materials Engineering, McGill University, Montréal, Canada pp. 147 – 154 ISBN: 978-0-08-097053-0
[3] Schäfer J, Dippel S and Wolf D E 1996 *J. Phys. I France* **6** 5
[4] Seinfeld J H, Lapidus L and Hwang M 1970 *Ind. Eng. Chem. Fundamen.* **9** 266
[5] Viridi S and Kurniadi R 2017 Komputasi Fisika 3: Simulasi dan Visualisasi Fisika dengan JavaScript, Departemen Fisika, FMIPA, Institut Teknologi Bandung, pp. 289 – 294 doi: 10.13140/RG.2.2.35065.67680

IOP Conf. Series: Journal of Physics: Conf. Series **1090** (2018) 012010 doi:10.1088/1742-6596/1090/1/012010

Analysis, Control and Circuit Design of a Novel Chaotic System with Line Equilibrium

A Sambas[1,2], S Vaidyanathan[3], M Mamat[2], M Sanjaya WS[4], S H Yuningsih[4], and K Zakaria[4]

[1]Department of Mechanical Engineering, Universitas Muhammadiyah Tasikmalaya, Indonesia (Corresponding author)
[2]Faculty of Informatics and Computing, Universiti Sultan Zainal Abidin, Kuala Terengganu, Malaysia
[3]Research and Development Centre, Vel Tech University, Chennai, Tamil Nadu, India
[4]Department of Physics, Universitas Islam Negeri Sunan Gunung Djati Bandung, Indonesia

Corresponding author's e-mail address: acengs@umtas.ac.id

Abstract. In this paper, we describe a new 3-D chaotic system with line equilibrium. A detailed analysis of the proposed chaotic system is provided with MATLAB phase portraits, equilibrium points, symmetry, invariance, Lyapunov exponents, Kaplan-Yorke dimension, bifurcation diagram and a Poincarè map. For specific values of the parameters, the proposed system displays chaotic behaviour. We apply adaptive control method for the global chaos synchronization of the new chaotic systems with unknown system parameters. In addition, a new circuit implementation of the new chaotic system is reported and examined in MultiSIM. A good qualitative agreement is shown between the simulations and the MultiSIM results.

1. Introduction

The first experimental model of a chaotic system was obtained by Lorenz [1], when he was modelling the atmospheric convection with a 3-D model in 1963. Lorenz system [1] is a classical chaotic system with a two-scroll chaotic attractor. Since this monumental work by Lorenz, chaos has been studied extensively and many chaotic systems have been developed in the literature. Both theoretical and experimental chaotic systems have been reported and their applications have been discussed as well.

In [2], Rössler constructed a seven-term chaotic system with just one quadratic nonlinearity. In [3-4], Rikitake discovered a new chaotic system to explain the irregular polarity switching of the earth's geomagnetic field. A model to explain the irregular variability in the luminosity of stars was presented in 1976 [5] and this system is known as the Moore-Spiegel system. In [6], Rucklidge discovered a double convection chaotic model in a horizontal layer of Boussinesq fluid with lateral constraints.

In 1994, Sprott described 19 algebraically simple chaotic systems [7]. In 2000, Malasoma proposed the simplest dissipative jerk equation that is parity invariant [8]. In 2009, Sun and Sprott constructed a piecewise exponential jerk system [9]. In the last few years, several new chaotic systems with an infinite number of equilibrium points have been proposed [10-14]. Specifically, the works [10-14] propose chaotic systems with a line or curve of equilibrium points.

Chaos has been widely applied to many scientific disciplines such as physics [15], biology [16], ecology [17], economy [18], random bit generators [19], psychology [20], lasers [21], astronomy [22],

Content from this work may be used under the terms of the Creative Commons Attribution 3.0 licence. Any further distribution of this work must maintain attribution to the author(s) and the title of the work, journal citation and DOI.
Published under licence by IOP Publishing Ltd

chemical reactions [23], memristors [24], neural networks [25], robotics [26], encryption [27], secure communication systems [28-31], DC motor [32], chaotic masking [33], etc.

Synchronization of chaotic systems [34] deals with a pair of chaotic systems called the master and slave systems and the design goal is to synchronize the respective state trajectories of the two systems asymptotically with the help of a controller attached to the slave system.

The synchronization of chaotic systems is a challenging research problem as the state trajectories of chaotic systems are very sensitive to small changes in initial conditions [35]. Several methods have been proposed in the literature for the synchronization of chaotic systems [36-38].

In this paper, we describe a new 3-D chaotic system with line equilibrium. Our novel chaotic system has six terms with four quadratic nonlinearities. In Section 2, we first provide the dynamics and phase portraits of the new chaotic system. We also present a qualitative analysis of the new chaotic system with properties such as dissipativity, equilibrium points, symmetry, invariance, Lyapunov exponents, Kaplan-Yorke dimension, bifurcation diagram and a Poincarè map. In Section 3, we apply adaptive control method to derive new results for the global chaos synchronization of the new chaotic systems with unknown system parameters. In Section 4, we, a new circuit implementation of the new chaotic system is reported and examined in MultiSIM. A good qualitative agreement is shown between the simulations and the MultiSIM results. In Section 5, we give final conclusions of this work.

2. A new 3-D chaotic system

In this work, inspired by the method and structure proposed in [39-40], we announce a new 3-D chaotic system described by the nonlinear dynamics

$$\begin{cases} x = y - xz - yz \\ y = axz \\ z = y^2 - bz \end{cases} \tag{1}$$

where x, y, z are state variables and a, b are positive system parameters.

We shall show that the system (1) is *chaotic* when the system parameters take the values

$$a = 2.8, \quad b = 0.6 \tag{2}$$

For numerical simulations, we take the initial conditions as

$$x(0) = 0.1, \quad y(0) = 0.1, \quad z(0) = 0.1 \tag{3}$$

With the parameter values as in (2) and the initial conditions as in (3), the Lyapunov exponents of the new system (1) are calculated using Wolf algorithm [41] as

$$L_1 = 0.1292, \quad L_2 = 0, \quad L_3 = -1.6192 \tag{4}$$

This shows that the new 3-D system (1) is *chaotic*.

Also, the sum of the Lyapunov exponents of the system (1) is found as

$$L_1 + L_2 + L_3 = -1.49 < 0 \tag{5}$$

Hence, the system (1) is a dissipative chaotic system with a strange chaotic attractor.

The phase portraits of the new chaotic system (1) are shown in Figure 1. The Lyapunov exponents of the new chaotic system (1) are shown in Figure 2.

The Kaplan-Yorke dimension of the new chaotic system (1) is calculated as

$$D_{KY} = 2 + \frac{L_1 + L_2}{|L_3|} = 2.0798 \tag{6}$$

Thus, the new chaotic system (1) has fractional dimension of D_{KY}.

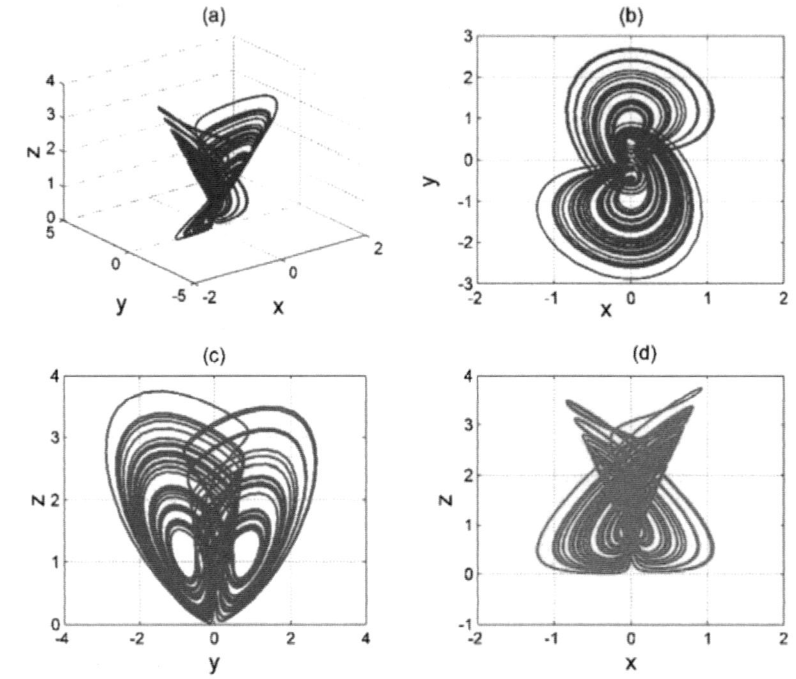

Figure 1. Phase portraits of the new chaotic system

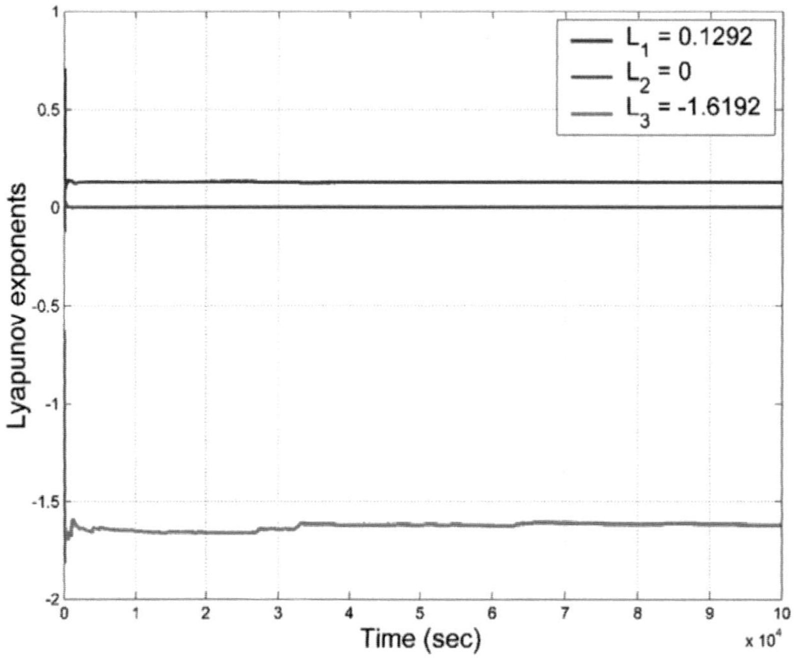

Figure 2. Lyapunov exponents of the new chaotic system

The equilibrium points of the new chaotic system (1) are obtained by solving the following system

$$\begin{cases} y - xz - yz = 0 \\ axz = 0 \\ y^2 - bz = 0 \end{cases} \tag{7}$$

We take the parameter values as in the chaotic case (2).

A simple calculation shows that the new chaotic system (1) has a line of equilibrium points

$$S = \{(x, y, z) \in R^3 : y = 0, z = 0\} \tag{8}$$

which is the x − axis in R^3.

It is easy to see that the new chaotic system (1) is invariant under the change of coordinates

$$(x, y, z) \quad (-x, -y, z) \tag{9}$$

This shows that the system (1) has rotation symmetry about the z − axis.

Also, we find that z − axis is invariant under the flow of the new chaotic system (1). This invariant motion is characterized by the 1-D dynamics

$$z = -bz \quad (b > 0) \tag{10}$$

which is exponentially stable.

Next, to have a detailed view of the new chaotic system (1), the behaviour of the system with respect to the bifurcation parameter b is studied. The new chaotic system (1) displays the expected chaotic behavior for $0.5 \le b \le 0.84$ and for $0.85 \le b \le 0.86$ periodic behavior is noted in the system. Also, the new chaotic system has chaotic behavior for $b > 0.86$ as presented in Figure 3. Poincarè map of the system (1) is also illustrated in Figure 4, which exhibits that the new chaotic system (1) has complex dynamics.

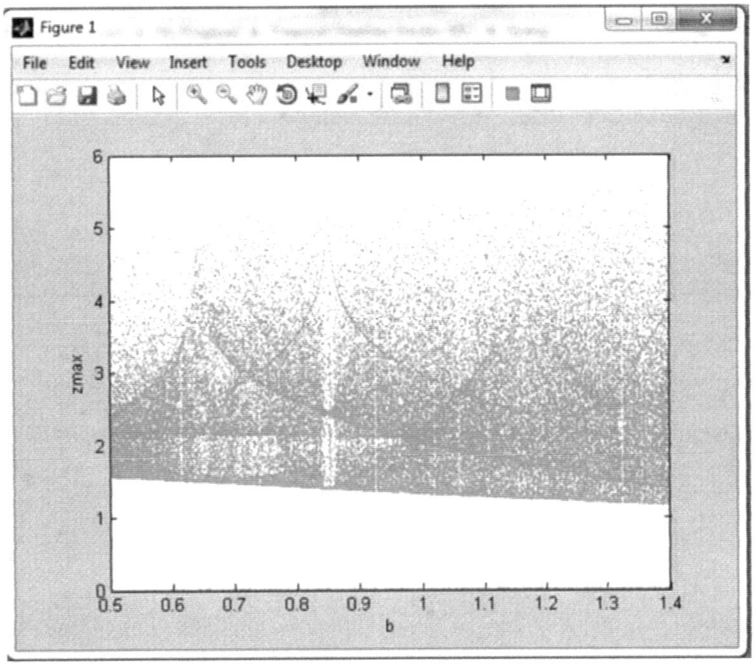

Figure 3. Bifurcation diagram of the new chaotic system (1) with b as varying parameter

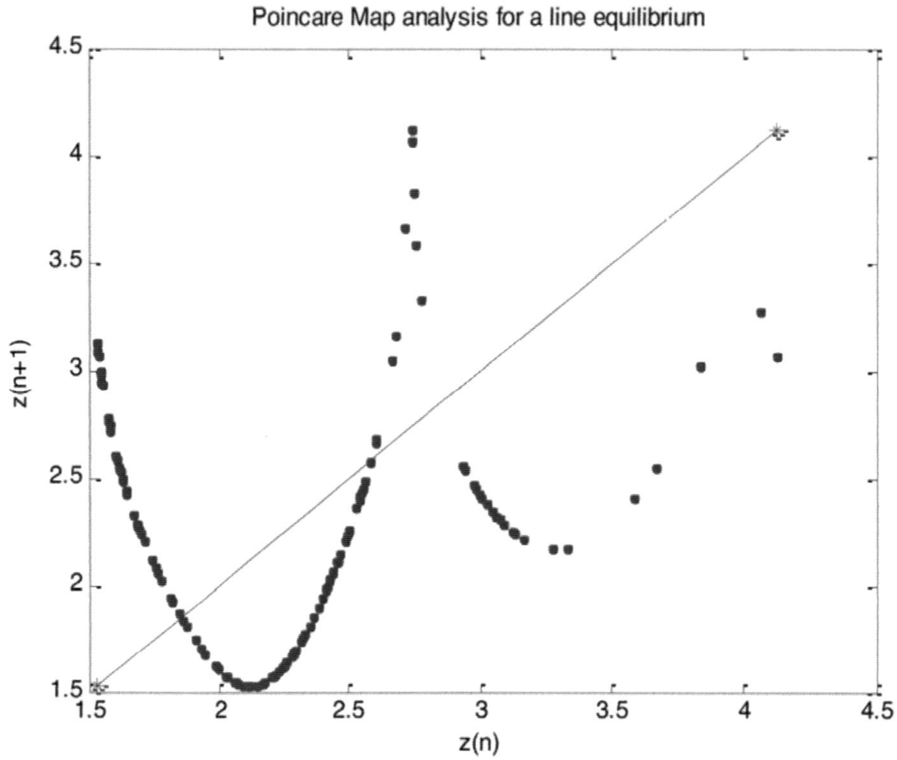

Figure 4. Poincarè map for the new chaotic system (1) when $a = 2.8$ and $b = 0.6$

3. Adaptive synchronization of the new chaotic system

In this section, we apply adaptive control method [42-45] for the global chaos synchronization of identical new chaotic systems with unknown system parameters.

As the master system, we consider the new chaotic system dynamics

$$\begin{cases} \dot{x}_1 = y_1 - x_1 z_1 - y_1 z_1 \\ \dot{y}_1 = a x_1 z_1 \\ \dot{z}_1 = y_1^2 - b z_1 \end{cases} \tag{11}$$

where x_1, y_1, z_1 are the states and a, b are unknown system parameters.

As the slave system, we consider the controlled new chaotic system dynamics

$$\begin{cases} \dot{x}_2 = y_2 - x_2 z_2 - y_2 z_2 + u_x \\ \dot{y}_2 = a x_2 z_2 + u_y \\ \dot{z}_2 = y_2^2 - b z_2 + u_z \end{cases} \tag{12}$$

where x_2, y_2, z_2 are the states and u_x, u_y, u_z are the adaptive controls to be designed.

We define the synchronization error as follows:

$$\begin{cases} e_x = x_2 - x_1 \\ e_y = y_2 - y_1 \\ e_z = z_2 - z_1 \end{cases} \tag{13}$$

Then the error dynamics is obtained as

$$\begin{cases} e_x = e_y - x_2z_2 - y_2z_2 + x_1z_1 + y_1z_1 + u_x \\ e_y = a\left(x_2z_2 - x_1z_1\right) + u_y \\ e_z = -be_z + y_2^2 - y_1^2 + u_z \end{cases} \tag{14}$$

We consider the adaptive control defined by

$$\begin{cases} u_x = -e_y + x_2z_2 + y_2z_2 - x_1z_1 - y_1z_1 - k_xe_x \\ u_y = -\hat{a}(t)\left(x_2z_2 - x_1z_1\right) - k_ye_y \\ u_z = \hat{b}(t)e_z - y_2^2 + y_1^2 - k_ze_z \end{cases} \tag{15}$$

where k_x, k_y, k_z are positive constants.

By substitution of (15) into (14), the following closed-loop system is obtained:

$$\begin{cases} e_x = -k_xe_x \\ e_y = [a - \hat{a}(t)]\left(x_2z_2 - x_1z_1\right) - k_ye_y \\ e_z = -[b - \hat{b}(t)]e_z - k_ze_z \end{cases} \tag{16}$$

To simplify the notations, we define the parameter estimation errors as

$$\begin{cases} e_a = a - \hat{a}(t) \\ e_b = b - \hat{b}(t) \end{cases} \tag{17}$$

Substituting (17) into (16), we obtain the closed-loop system as

$$\begin{cases} e_x = -k_xe_x \\ e_y = e_a\left(x_2z_2 - x_1z_1\right) - k_ye_y \\ e_z = -e_be_z - k_ze_z \end{cases} \tag{18}$$

Time-differentiation of (17) yields the differential equations

$$\begin{cases} e_a = -\hat{a}(t) \\ e_b = -\hat{b}(t) \end{cases} \tag{19}$$

Next, we define a candidate Lyapunov function given by

$$V(e_x, e_y, e_z, e_a, e_b) = \frac{1}{2}\left(e_x^2 + e_y^2 + e_z^2 + e_a^2 + e_b^2\right) \tag{20}$$

Differentiating V along the trajectories of (18) and (19), we obtain

$$V = -k_xe_x^2 - k_ye_y^2 - k_ze_z^2 + e_a\left[e_y(x_2z_2 - x_1z_1) - \hat{a}\right] + e_b\left[-e_z^2 - \hat{b}\right] \tag{21}$$

In view of (21), we consider the following update law for parameter estimates:

$$\begin{cases} \hat{a} = e_y\left(x_2z_2 - x_1z_1\right) \\ \hat{b} = -e_z^2 \end{cases} \tag{22}$$

Next, we state and prove the main result of this section.

Theorem 1. The identical new chaotic systems (11) and (12) are globally and exponentially synchronized by the adaptive control law (15) and the parameter update law (22) for all initial conditions, where k_x, k_y, k_z are positive constants.

Proof. We establish this result by using Lyapunov stability theory [46].

First, we note that the quadratic Lyapunov function V defined by (20) is positive definite in R^5.

After substituting the parameter update law (22) into (21), we obtain the time-derivative of V as

$$\dot{V} = -k_x e_x^2 - k_y e_y^2 - k_z e_z^2 \tag{23}$$

which is negative semi-definite in R^5.

Hence, by Barbarat's lemma [46], we conclude that the closed-loop synchronization error dynamics (18) is globally exponentially stable.

This completes the proof. ■

For numerical simulations, we take the parameters as in the chaotic case (2), i.e. $a = 2.8$ and $b = 0.6$. We take the initial values of the parameter estimates as $\hat{a}(0) = 6.4$ and $\hat{b}(0) = 9.5$.

Also, we take the gain constants as $k_x = k_y = k_z = 10$.

As the initial state of the master system (11), we take

$$x_1(0) = 8.4, \quad y_1(0) = 3.6, \quad z_1(0) = 12.4 \tag{24}$$

As the initial state of the slave system (12), we take

$$x_2(0) = 17.5, \quad y_2(0) = 6.3, \quad z_2(0) = 4.2 \tag{25}$$

Figures 5-7 show the complete synchronization of the new chaotic systems (10) and (11), while Figure 8 shows the time-history of the synchronization errors e_x, e_y, e_z.

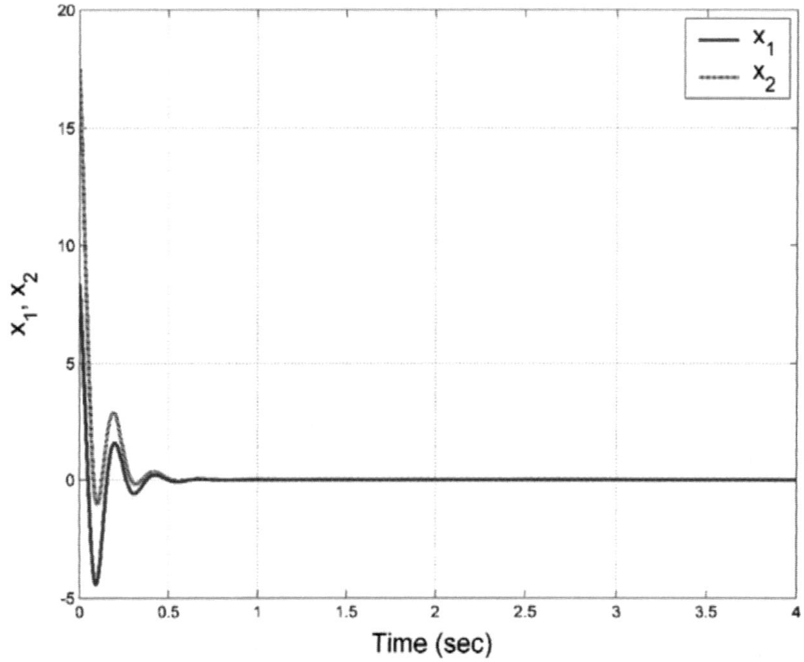

Figure 5. Complete synchronization of the states x_1 and x_2 of the new chaotic systems

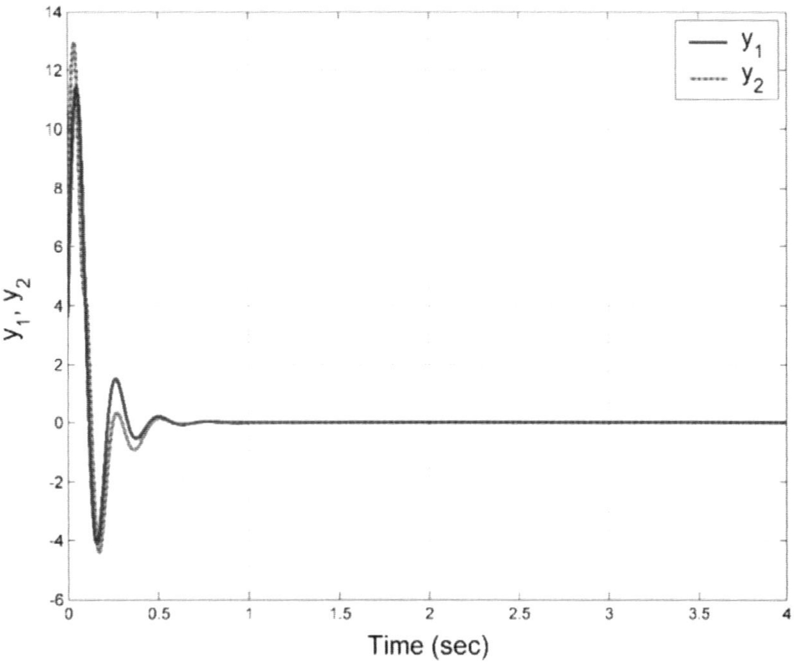

Figure 6. Complete synchronization of the states y_1 and y_2 of the new chaotic systems

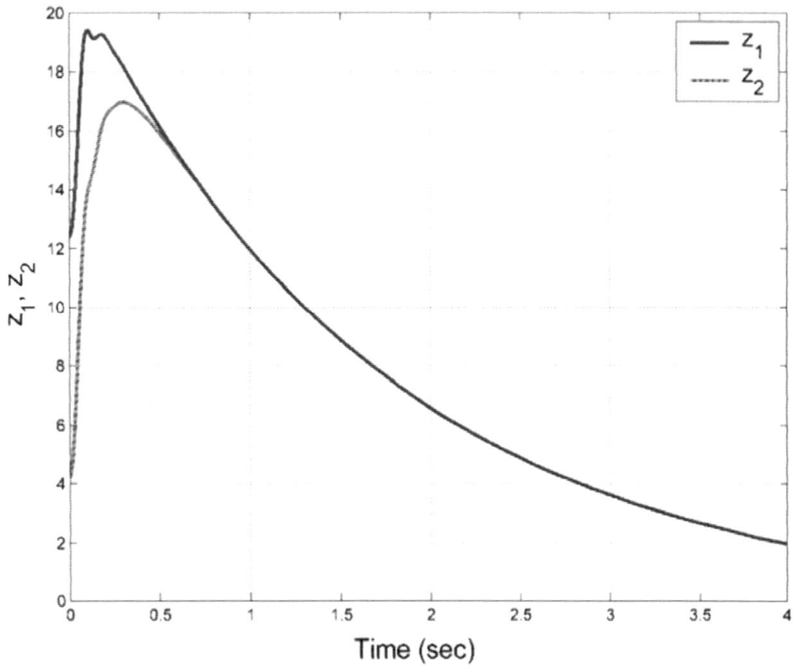

Figure 7. Complete synchronization of the states z_1 and z_2 of the new chaotic systems

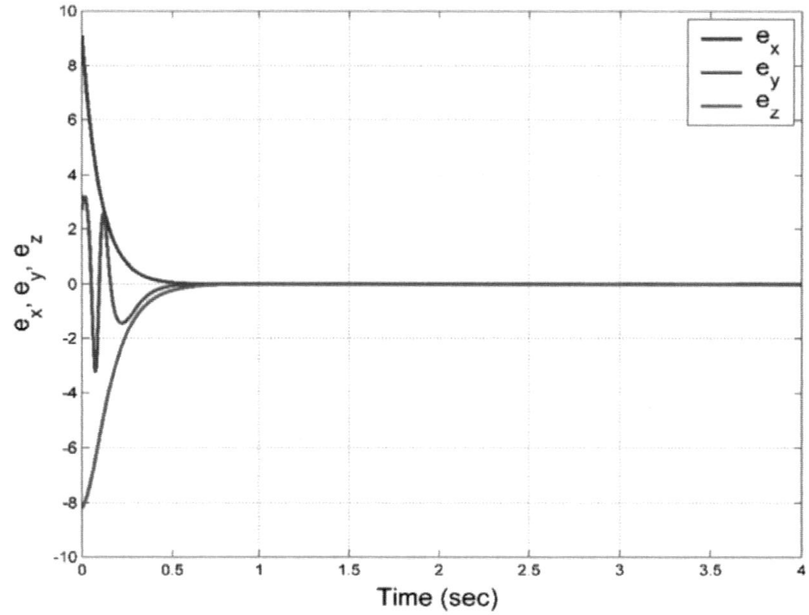

Figure 8. Time-history of the synchronization errors e_x, e_y, e_z

4. Circuit realization of the new chaotic system

In this section, an electronic circuit which emulates the new chaotic system (1) is described to show its feasibility. The circuit design in Figure 9 has been described following an approach based on operational amplifiers [47-52], where the state variables x, y, z of the new system (1) are associated with the voltages across the capacitors C_1, C_2 and C_3 respectively.

By applying Kirchhoff's circuit laws, the corresponding circuital equations of the designed circuit can be written as

$$\begin{cases} \dot{x} = \dfrac{1}{C_1R_1}y - \dfrac{1}{C_1R_2}yz - \dfrac{1}{10C_1R_3}xz \\ \dot{y} = \dfrac{1}{10C_2R_4}xz \\ \dot{z} = \dfrac{1}{10C_3R_6}y^2 - \dfrac{1}{10C_3R_5}z \end{cases} \tag{26}$$

We choose the values of the circuital elements as

$$\begin{cases} R_4 = 14.285K\Omega, R_5 = 666.66K\Omega, R_2 = R_3 = R_6 = 40K\Omega \\ R_1 = R_7 = R_8 = R_9 = R_{10} = 400K\Omega \\ C_1 = C_2 = C_3 = 1nF \end{cases} \tag{27}$$

The circuit has three integrators by using Op-amp TL082CD in a feedback loop and four multipliers IC AD633. The supplies of all active devices are ± 15 volt. With MultiSIM 10.0, we obtain the experimental observations of new system (1) as shown in Figures 10-12. The agreement between the experimental phase portraits with MultiSIM simulation (Figs. 6-8) and the numerical simulation with MATLAB (Fig. 1) confirms the feasibility of our new chaotic system (1).

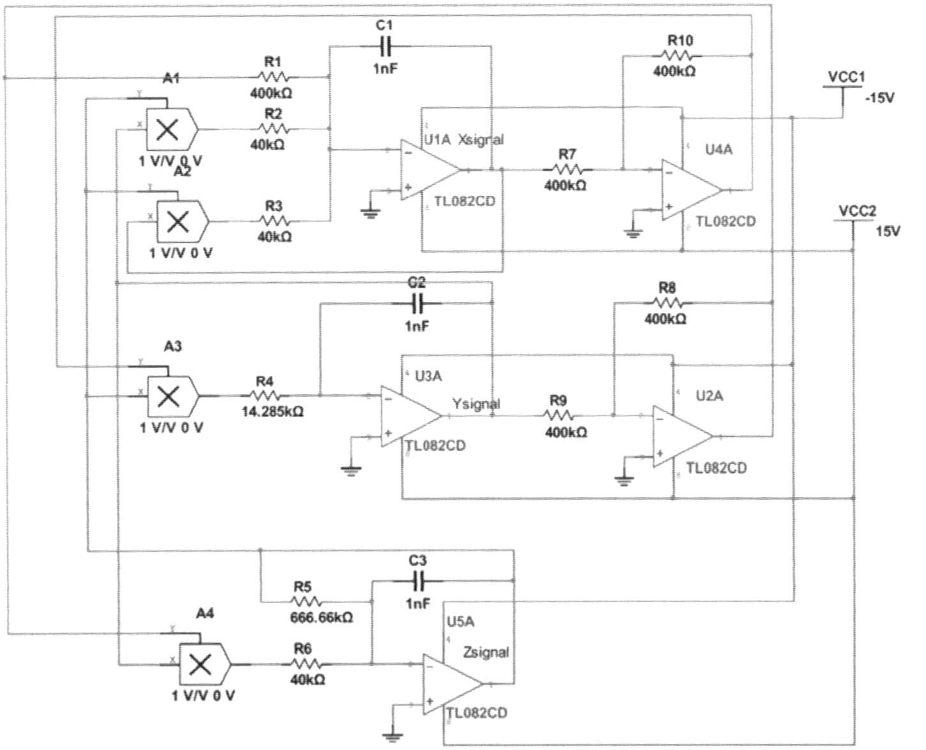

Figure 9 Schematic of the proposed new chaotic system by using MultiSIM

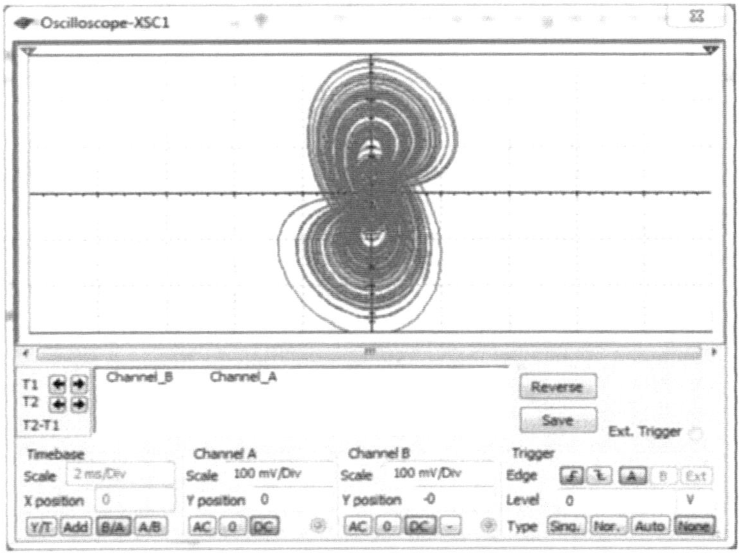

Figure 10 2-D projection of the new chaotic system on the (x, y) plane

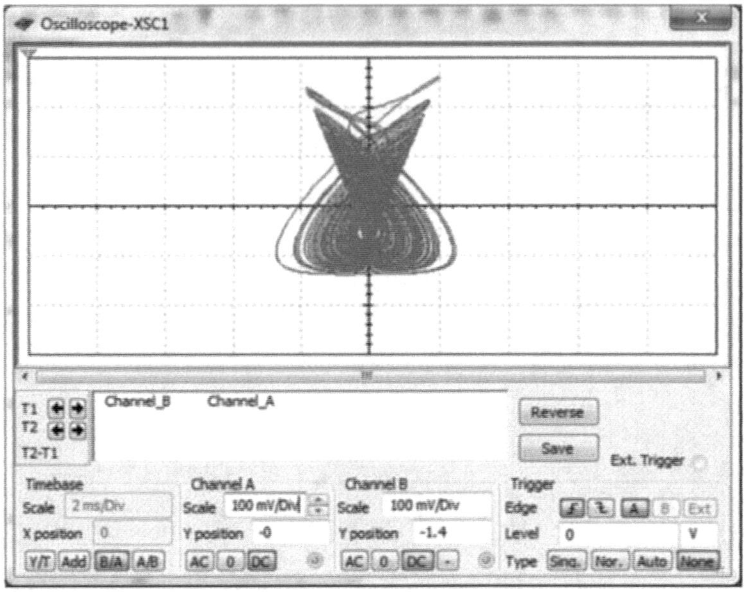

Figure 11 2-D projection of the new chaotic system on the (x, z) plane

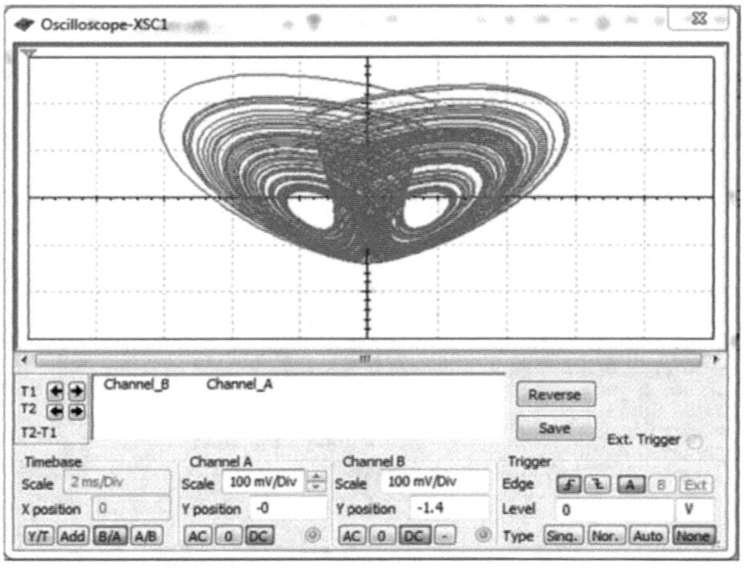

Figure 12 2-D projection of the new chaotic system on the (y, z) plane

5. Conclusions

In this paper, we derived a new chaotic system with a line of equilibrium points. The dynamical behaviors of the modified Rucklidge chaotic system are analyzed, both analytically and numerically, including some basic dynamical properties, symmetry, invariance, phase portraits, Lyapunov exponents, Kaplan-Yorke dimension, bifurcation diagram and Poincarè map. The designed circuit has been implemented and examined using the MultiSIM and numerical results using MATLAB. Comparison of the MultiSIM result and MATLAB simulations showed good qualitative agreement between the chaotic system and its circuitry realization.

References

[1] Lorenz E N 1963 *J. Atmospheric Sciences* **20** 130-41
[2] Rössler O 1976 *Physics Letters* A **57** 397-8
[3] Ito K 1980 *Sci. Lett.* **51** 451-6
[4] Rikitake T 1998 *Proc. Cambridge Philos. Soc.* **54** 89-105
[5] Moore D W and Spiegel E A 1986 *Astrophys. J.* **143** 871-87
[6] Rucklidge A M 1992 J. Fluid Mechanics 237 209-29
[7] Sprott J 1994 *Phys. Lett. E* **50** 647-50
[8] Malasoma J. M. 2000 *Phys. Lett. A* **264** 383-9
[9] Sun K H and Sprott J H 2009 *Internat. J. Nonlinear Sci. Num. Simulat.* **10** 1443-50
[10] Gotthans T and Petrzela J 2015 *Nonlinear Dynamics* **73** 429-36
[11] Zhou P and Yang F 2014 *Nonlinear Dynamics* **76** 473-80
[12] Pham V T, Jafari S, Volos C and Kapitaniak T 2016 *Chaos Solit. Fract.* **93** 58-63
[13] Pham V T, Volos C, Vaidyanathan S and Wang X 2016 *Advances Math Phys.* **2016** 4024836
[14] Pham V T, Jafari S, Volos C, Vaidyanathan S and Kapitaniak T 2016 *Optik* **127** 9111-7
[15] Awrejcewicz J, Supel B, Lamarque C H, Kudra G, Wasilewski G and Olejnik P 2008 *Int. J. Bifur. Chaos* **18** 2883-915
[16] Hannon B and Ruth M 2014 Modeling Dynamic Biological Systems (Berlin: Springer)
[17] Mamat M, Sanjaya W M, Salleh Z and Ahmad M F *2011 J. Sustainability Science and Management* **6** 44-50
[18] Bouali S, Buscarino A, Fortuna L, Fresca M and Gambuzza L V 2012 *Nonlinear Analysis: RWA* **13** 2459-65
[19] Akgul A, Moroz, I, Pehlivan I and Vaidyanathan S 2016 *Optik* **127** 5491-99
[20] Jafari S, Sprott J C and Golpayegani S M R H 2016 *Nonlinear Dynamics* **83** 615-22
[21] Sciamanna M and Shore, K A 2015 *Nature Photonics* **9** 151-62
[22] Daquin J, Rosengren A J, Alessi E M, Feleflie F, Valsecchi G B and Rossi A 2016 *Celestical Mechanics and Dynamical Astronomy* **124** 335-66
[23] Bringer M R, Gerdts C J, Song H, Tice J D and Ismagilov R F 2004 *Phil. Trans. Royal Society of London A* **362** 1087-104
[24] Pham V T, Vaidyanathan S, Volos C K, Jafari S, Kuznetsov N V and Hoang T M 2016 *European Physical Journal: Special Topics* **225** 127-36
[25] Lu, H 2002 *Physics Letters A* **298** 109-116
[26] Sambas A, Vaidyanathan S, Mamat M, Sanjaya W M and Rahayu D S 2016 *Studies in Computational Intelligence* **636** 283-310
[27] Volos C K, Kyprianidis I M and Stouboulos I N 2013 *J. Engineering Science and Technology Review* 6 9-14
[28] Sambas A, Sanjaya W S, Mamat M and Prastio R P 2016 *Studies in Fuzziness and Soft Computing* **337** 133-53
[29] Sambas A, Sanjaya W S and Mamat M, 2015 *J. Engineering Science and Technology Review* **8** 89-95
[30] Sambas A, Sanjaya W S and Halimatussadiyah 2012 *WSEAS Transactions on Systems* **9** 506-15
[31] Xu G, Xu J, Xiu C, Liu F and Zang Y 2017 *Neurocomputing* **227** 108-12
[32] Rajagopal K, Vaidyanathan S, Karthikeyan A and Duraisamy P 2017 *Electrical Engineering* **99** 721-33
[33] Vaidyanathan S and Rajagopal K 2017 *International Journal of Simulation and Process Modelling* **12** 165-78.
[34] Vaidyanathan S and Sampath S 2017 *International Journal of Modelling, Identification and Control* **27** 3-13.
[35] Vaidyanathan S 2016 *Archives of Control Sciences* **26** 471-95.
[36] Azar A T and Vaidyanathan S Advances in Chaos Theory and Intelligent Control (Berlin: Springer)

[37] Vaidyanathan S and Volos 2016 Advances and Applications in Chaotic Systems (Berlin: Springer)
[38] Vaidyanathan S and Lien C H 2017 Applications of Sliding Mode Control in Science and Engineering (Berlin: Springer)
[39] Li C, Sprott J C, and Thio W 2014 *Journal of Experimental and Theoretical Physics* **118** 494-500
[40] Li C and Sprott J C 2014 *Physics Letters A* **378** 178-183
[41] Wolf A, Swift J B, Swinney H L and Vastano J 1985 *Physica D* **16** 285-317
[42] Vaidyanathan S and Rajagopal K 2016 *International Journal of Control Theory and Applications* **9** 175-97
[43] Sampath S, Vaidyanathan V and Pham V T 2016 *International Journal of Control Theory and Applications* **9** 85-100
[44] Nasr S, Bouallegue K, Mekki H and Vaidyanathan S 2016 *International Journal of Control Theory and Applications* **9** 453—66
[45] Vaidyanathan S 2016 *International Journal of Control Theory and Applications* **9** 1-20
[46] Khalil H 2002 Nonlinear Systems (New York: Prentice Hall)
[47] Li X F, Chu Y D, Zhang J G and Chang Y X 2009 *Chaos Solitons & Fractals* **41** 2360-70
[48] Sundarapandian V and Pehlivan I 2012 *Mathematical and Computer Modelling* **55** 1904-15.
[49] Pehlivan I, Moroz I M and Vaidyanathan S 2014 Journal of Sound and Vibration **333** 5077-96.
[50] Li C, Sprott J C and Thio W 2014 *IEEE Trans. Circuits Sys.-II: Exp. Briefs* **61** 977-81
[51] Li C, Pehlivan I, Sprott J C and Akgul A 2015 *IEICE Electronics Express* **12** 1-12
[52] Akgul A, Moroz I, Pehlivan I and Vaidyanathan S 2016 *Optik* **127** 5491-9.

Comparative Evaluation on Several Reactor Type of Actinide Closed-Cycle Schemes

Sidik Permana[1*], Asril Pramutadi[1], Sycilendra Pramuditya[1], Dwi Irwanto[1]

[1]*Nuclear Physics and Biophysics Research Division, Department of Physics, Institut Teknologi Bandung, Jl. Ganesha 10, Bandung, 40132, Indonesia*
**E-mail: psidik@fi.itb.ac.id*

Abstract. Naturally, nuclear fuel resources are coming from uranium and thorium fuels for nuclear energy utilization. As by product reactor operation, some spent nuclear nuclear are produce which can be optimized by optimum recycled fuel option, as well as new fuel resource to increases fuel sustainabililty aspect of nuclear fuel. Several nuclear reactor types and the utilization of some recycled fuel options from their spent nuclear fuel as well as fuel breeding cabalitiy of the reactor based on those uranium and thorium fuel cycles. Some important parameters have been analyzed such as reactor performance of criticality and fuel conversion as well as some actinide productions in the present evaluation based on several fuel cycle schemes. As a basic recycling scheme, an open and a closed-fuel cycle of some actinides was used based on water-cooled reactor system, including ligh water and heavy water coolant. Basic analysis of the system is based on an equilibrium burnup calculation for the present computational approach analysis. Once through case requires more fissile materials than MOX fuel scheme to maintain criticality condition and heavy water coolant requires more fissile material than light water coolant for UOX fuel case, however, it shows the opposite trend for in case of MOX fuel. Thorium cycle requires more fissile material of U-233 for heavy water coolant than light water coolant. In term of light water case, the required fissile material of U-233 is less for thorium case than fissile material of U-235 in MOX fuel case. Fuel conversion ratios are obtained less than unity except for heavy water coolant of MOX fuel case, which obtains slightly more than unity. Thorium fuel shows better fuel breeding capability as well as heavy water coolant, which gives more fuel conversion capability than light water coolant. Actinide compositions of plutonium gives more than 1% for light water coolant and more than 10% for heavy water coolant case. For different fuel scheme cases, it shows that plutonium recycled scheme as MOX fuel case gives higher production of actinide or trans-uranium actinides from neptunium up to Curium than UOX fuel case for H2O coolant. Total even mass plutonium isotope produces higher than total odd mass plutonium isotope for both coolants and heavy water coolant produces more than its production for light water coolant. Better composition of even mass plutonium has shown better proliferation resistant level which is shown by plutonium recycled scheme and heavy water coolant case.

1. Introduction

Nuclear energy sustainability can be improved, which has a similar trend with the renewable energies as a sustainable energy source [1]. Nuclear fuel cycle option and fuel sustainability aspect can be used to improve nuclear fuel sustainability of the reactor, which can be maintained by improving reactor design capability for breeder reactor as well as optimization of fule usage and new fuel exploration and fuel mixing program between uranium and thorium as well as by product actinide such as plutonium and minor actinides. Some sensitive issues for nuclear utilizaton especially in term of recycle program, that are an issue which has corresponding to the nuclear proliferation resistance issues. To reduce the proliferation resistance level, some barriers were used such as matrial barrier of nuclear fuel based on isotopic composition. Some reactor designs are also contributed to the fuel

Content from this work may be used under the terms of the Creative Commons Attribution 3.0 licence. Any further distribution of this work must maintain attribution to the author(s) and the title of the work, journal citation and DOI.
Published under licence by IOP Publishing Ltd

sustainability as well as nuclar proliferation resistant issues espcially when the reactor are using some recyling products such as plutonium from uranium cycle and U-233 production from thorium fuel cycle. For decades some thorium fuel technologies have being developed in conjunction with the uranium fuel technology for commercial purposes, for both conventional and advanced nuclear reactor. One of the excellent thorium fuel programs was a thorium breeder reactor program for breeding program as well as in uranium fuel technology. As an excellent fuel, thorium fuel cycle gives some good features such as higher fuel breeding potential, better fuel stability and better level of proliferation resistance [2-6] in comparison with others fuel cycles. In term of proliferation resistance feature, there are some feature, which can be used as material barrier such as a barrier of isotopic composition. It affects to critical mass, heat-generation rate, spontaneous neutron generation and radiation. In term of isotopic barrier feature such as in isotopic barrier of plutonium, it is mainly comes from some compositions of even mass number of plutonium isotope such as ^{238}Pu, ^{240}Pu and ^{242}Pu for uranium-plutonium cycle which is also in similar dependency for thorium-uranium cycle which is depending on some composition of isotope ^{232}U.

Several nuclear reactor types and the utilization of some recycled fuel options from their spent nuclear fuel as well as fuel breeding cabalitiy of the reactor based on those uranium and thorium fuel cycles. Some important parameters have been analyzed such as reactor performance of criticality and fuel conversion as well as some actinide productions in the present evaluation based on several fuel cycle schemes. As a basic recycling scheme, an open and a closed-fuel cycle of some actinides was used based on water-cooled reactor system, including ligh water and heavy water coolant. Basic analysis of the system is based on an equilibrium burnup calculation for the present computational approach analysis.

Table 1. Reactor parameters

Parameters	Values/Type
Thermal Power Output [MWt]	3000
Coolant	H_2O and D_2O
Fuel Cycle Option	No and some Actinides closed
Supply Fuel Composition	U and Th-U233
Cladding	Zircaloy-4
Fuel pellet diameter [cm]	1.31
Fuel pin outer diameter [cm]	1.45
Moderator to Fuel Ratio (MFR) [-]	1
Burn-up [GWd/t]	36

2. Parameter and Method

As mentioned above, several key parameters has been analyzed on required fissile material for maintaining a criticality condition, nuclear fuel sustainability based on conversion ratio and fuel breeding ratio and actinide production composition for several fuel cycle schemes and water coolant types. Fuel cycle schemes are based on once through case or UOX case and plutonium (Pu) recycled case or MOX fuel case. Water coolants system are based on light water and heavy water coolants which basically adopted well established reactor technology of current LWR plant technology as general design parameter. As shown in **Table 1,** it shows some basic parameter of the reactor systems, which is based on a larger reactor of 3000 MWt power output. It was adopted a moderator to fuel ratio (MFR) of 1 which is a half value compare to a standard MFR of pressurize water reactor (PWR) (MFR =2) that gives the neutron spectra are relatively harder than standard PWR for a fix discharged fuel burnup of 36 GWd/t and a 1.452 cm for a larger fuel pin than standard PWR type [2,6-7]. Fuel cycle scheme of closed cycle fuel that no or some actinides are returned to the reactor and only FPs are sent to final disposal stream. To evaluate criticality condition, fuel sustainability and actinide composition or fuel composition of the reactors, an approach has been used for equilibrium burn-up calculation code (Equilibrium Cell Iterative Calculation System: ECICS). The calculation is coupled with SRAC code and JENDL3.2 as nuclear data library [2,7-18].

3. Results and Discussions

3.1. Required Fissile Material of U-235 and U-233

Criticality condition of the reactors, which affects to reactor operation, can be maintained by loading some fissile materials to have enough fission reaction in the reactor core to have chain reaction for neutron balance to run the reactor in criticality condition. Those fissile materials are mainly from enrichment of U-235 or recycled material of fissile plutonium such as Pu-239 and fissile uranium of U-233 fro thorium fuel cycle basis. Besides adopting those several fuel cycle schemes, a water coolant technology bases are used such as light water and heavy water coolants for comparative purposes. Obtained results of required fissile materials are shown in **Figures 1** and **2,** which is based on required fissile material of U-235 and U-233 for different water coolants. Figure 1 shows fuel cycle scheme of UOX, which means once through case, requires higher fissile materials in comparing with MOX fuel scheme, which means all plutonium materials are recycled. Less required fissile material of U-235 for MOX fuel is estimated from the contribution of fissile Pu-239 and Pu-241 after those materials are recycled into the reactor. Therefore, contribution of U-235 to maintain criticality condition becomes less because of the fission reaction can be maintained also by fissile material of Pu-239 and Pu-241. In term of UOX fuel case or once through fuel case, those plutonium materials are not recycled into the reactor, therefore for maintaining the reactor in critical condition, some fissile material of U-235 are required more. Different coolant type is estimated to give some effect to the reactor criticality as well as to the required fissile materials. It shows that UOX case require more fissile material for heavy water coolant case (D2O) than light water coolant (H2O). Heavy water coolant gives less thermal condition of neutron in the reactor, which causes the reactor becomes harder, therefore to maintain some criticality condition, it requires more fissile materials. In case of MOX fuel, it shows that the opposite trend for heavy water coolant than light water coolant have been obtained for requires fissile materials. Less fissile materials are required for heavy water coolant than light water coolant for MOX fuel case. This condition can be estimated from the recycled plutonium content into the reactor that recycled plutonium from heavy water is higher than light water, which contributes to help the reactor to maintain the reactor in critical condition. Therefore, required fissile material of U-235 becomes less because of the higher contribution of recycled plutonium.

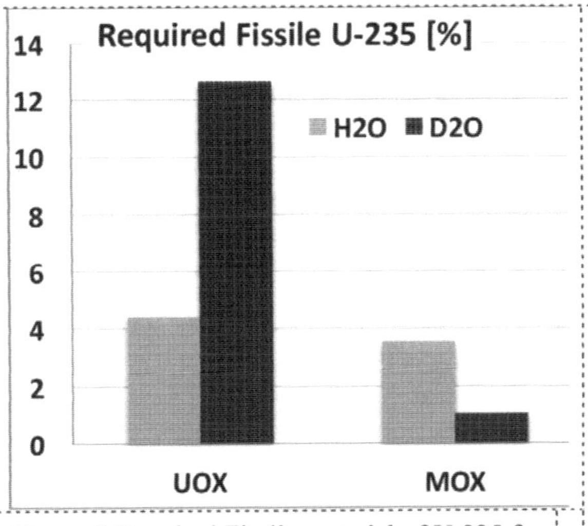

Figure 1 Required Fissile material of U-235 for Different Coolants and Uranium Fuel Cycles

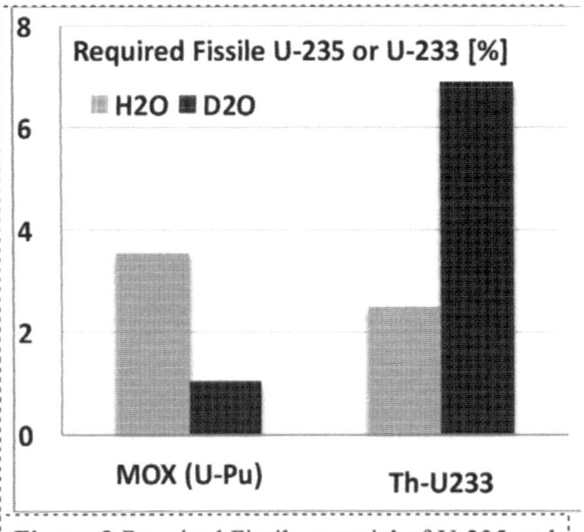

Figure 2 Required Fissile material of U-235 and U-233 for Different Coolants and Fuel Cycles

IOP Conf. Series: Journal of Physics: Conf. Series **1090** (2018) 012011 doi:10.1088/1742-6596/1090/1/012011

In addition, Obtained results of the required fissile material based on uranium and thorium fuel cycles are shown in Figure 2. It shows require fissile material of U-233 of thorium fuel cycle for heavy water coolant is higher than light water coolant, which is similar to UOX fuel case. In term of light water case, the required fissile material of U-233 is less for thorium case in comparing with the required fissile material of U-235 in MOX fuel case. However, It requires more fissile material of heavy water coolant in thorium fuel case than MOX fuel case. Light water coolant in both cases of fuel cycle gives some effect to make the neutron in thermal condition or softened the neutron spectrum. In thermal condition as well as in epithet condition, more fission reaction is obtained by U-233 than U-235 as well as with other fissile materials. Therefore, fissile U-233 is easier to make the reactor critical, which requires less fissile material for criticality condition. In case of heavy water coolant, in this condition of MOX, it uses plutonium to support the U-235 fissile to make critical, therefore supply U-235 becomes less. In thorium case, recycled plutonium does not used, therefore more U-233 materials are needed to maintain the reactor in critical condition.

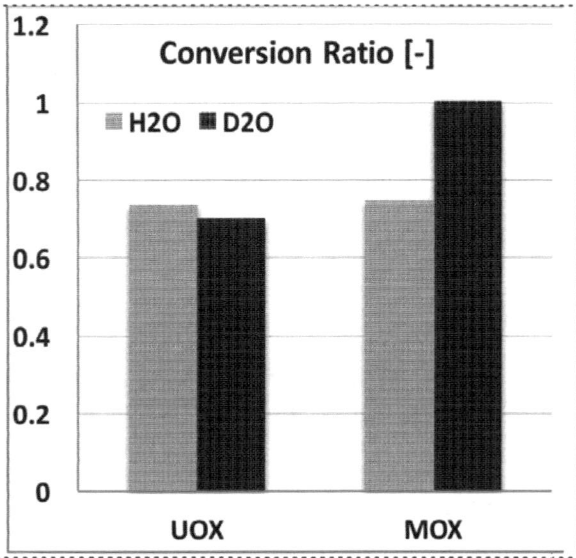

Figure 3 Fuel Conversion Ratio for Different Coolants and Uranium Fuel Cycles

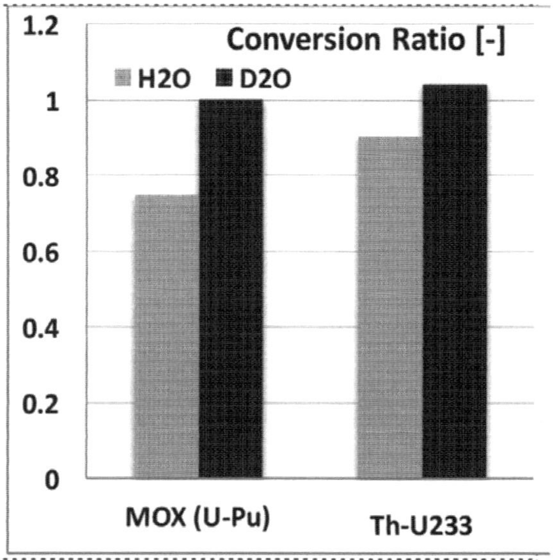

Figure 4 Fuel Conversion Ratio for Different Coolants and Uranium and Thorium Cycles

3.2. Fuel Conversion Ratio For Different Coolant and Fuel Cycles

Fuel breeding ratio can be obtained from fuel conversion ratio when it is higher than unity. In same cases, the fuel conversion is near to unity, which can be estimated as near breeder condition or high fuel conversion ratio. Fuel conversion ration is also can be shown as the fuel capability to convert fuel fertile material into fuel fissile material, which can be used as fissionable material for maintaining reactor criticality. Some obtained results of fuel conversion ratios are shown in **Figures 3** and **4** for different coolants and fuel cycle Schemes. It shows that fuel conversion ratio of light water coolant for UOX and MOX fuels are less than unity as well as fuel conversion ratio for heavy water coolant of UOX fuel case. However for heavy water coolant of MOX fuel case, it obtains slightly more than unity than light water case. In this condition only heavy water coolant can obtain a fuel breeding condition. Thorium fuel case, it shows better fuel breeding capability than MOX and UOX fuel cases. Both heavy water and light water coolants for thorium fuel, which give a fuel conversion ratio level to near breeding and higher than unity to have breeder condition. Heavy water coolant gives more fuel conversion capability and it obtains breeding condition than light water coolant. Higher fuel conversion capability and obtained fuel breeder condition can be estimated from more contribution of recycled plutonium in case of comparison of UOX and MOX fuel cases. These plutonium condition affects to have a better fuel conversion or have a breeder condition in addition to fissile U-235. In addition, thorium fuel, which is mainly the contribution from U-233, obtains better fuel conversion capability than fuel conversion capability of U-235 and Plutonium. These conditions can be estimated from the eta-value condition that U-233 is superior than others in thermal and epi-thermal energy regions, which is also effective to make easier for criticality condition as mentioned in the previous section of criticality aspect.

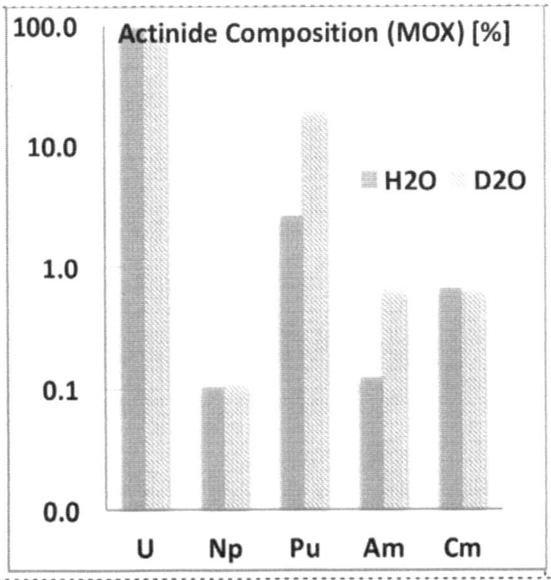

Figure 5 Actinide Composition of Fuel Element for Different Coolants of MOX Fuel

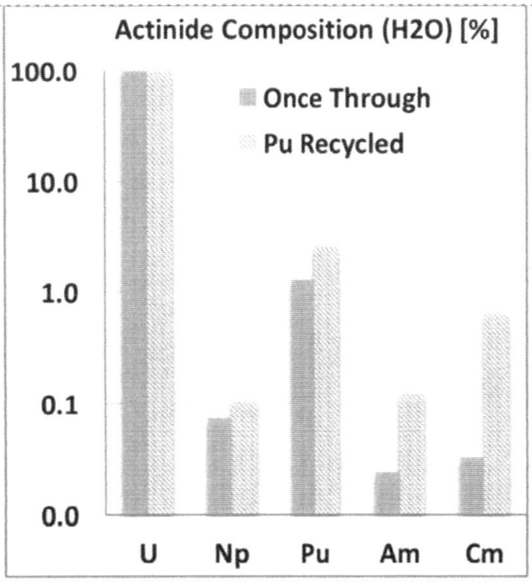

Figure 6 Actinide Composition of Fuel Element for Different Fuel Cycles of H2O Coolant

3.3. Actinide element compositions For Different Coolant and Fuel Cycles

Fuel behaviour of the reactor are effective also to make the reactor condition become critical condition or becomes breeding condition or high conversion condition. Contribution of different fresh fuel as supply fuel as well as different recycled fuels are effective to change the fuel composition trend in the reactor especially for trans-uranium and trans-thorium fuel compositions. Heavy metal nuclide as main nuclide can be represented by these nuclide such as uranium (U), Neptunium (Np), Plutonium (Pu), Americium (Am) and Curium (Cm). These fuel compositions as obtained results are shown in **Figures 5** and **6** for different coolants and fuel cycled schemes. As main fuel composition Uranium is obtained as highest fuel composition and it is followed by plutonium and others actinide compositions. Actinide compositions other than uranium are obtained less, which gives less than 1 % of fuel composition, except for plutonium which gives more than1% for light water coolant and more than 10% for heavy water coolant case. Different coolants are effective to produce more actinides such as plutonium and Americium, however its very limited change for Neptunium and Curium as shown in Figure 5 for MOX fuel case. For different fuel scheme cases, it shows that plutonium recycled scheme as MOX fuel case gives higher production of actinide or trans-uranium actinides from neptunium up to Curium than UOX fuel case, as shown in Figure 6 for H2O coolant. These increasing actinide materials can be estimated from the contribution of recycled Pu material, which leads to increase plutonium production as well as trans-plutonium productions.

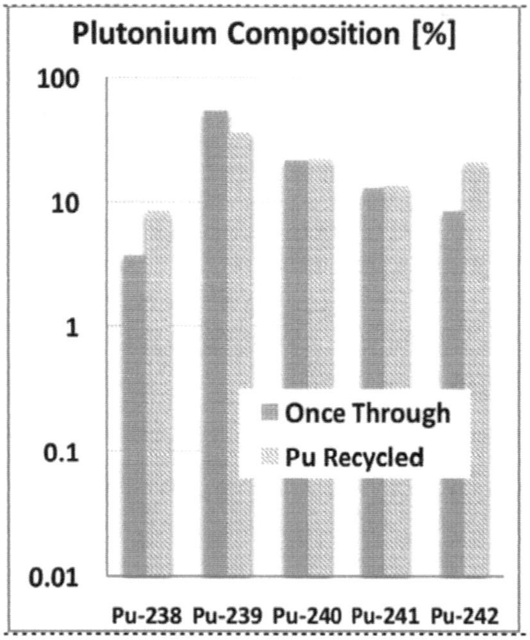

Figure 7 Plutonium Composition for Different Uranium Fuel Cycles of H2O Coolant

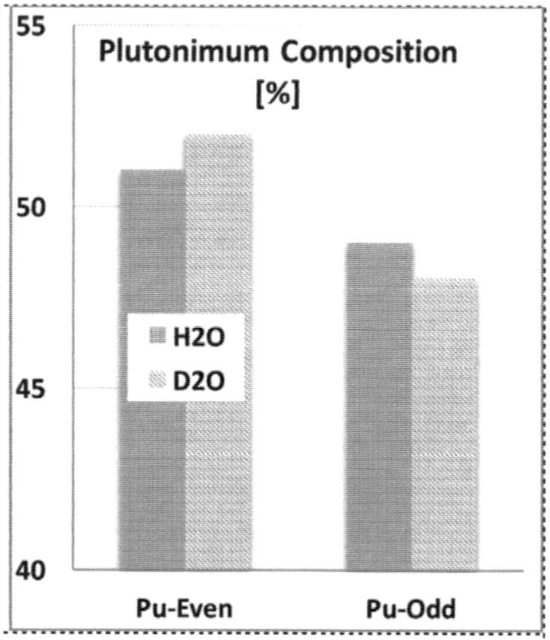

Figure 8 Plutonium Composition for Different Coolants of Plutonium recycled scheme

3.4. Plutonium isotopes compositions For Different Coolant and Fuel Cycles

As mentioned, contribution of elemental composition of actinide will be based on some isotopic composition. Isotopic composition will be composed by even mass and odd mass actinides. In term of plutonium element is based on isotopic plutonium composition of even and odd mass isotopes such as Pu-238, Pu-239, Pu-240, Pu-241 and Pu-242. Obtained results of plutonium isotopes are shown in **Figure** 7 and **8** for different fuel cycle schemes and coolants. Each plutonium isotope has its own trend composition for different fuel cycles and coolants. It shows Pu-239 is produced as a main production, and its followed by Pu-240 and others. Production of Pu-239 is estimated from the conversion process of U-238 after capturing some neutrons and it converts into Pu-239. Pu-240 production is also from capturing neutrons of Pu-239 and Pu-240 converts to Pu-241 after capturing neutron as well as Pu-242. In case of Pu-238 production is estimated from (n,2n) reaction of Pu-239 and it converts to Pu-238. In cane of plutonium recycled or MOX fuel case, it shows better production than once through case for all plutonium isotopes except for Pu-239, which obtains less production. More plutonium productions are coming from the contribution of recycled plutonium loading into the reactor. Some decreasing Pu-239 is estimated from the fission process of Pu-239 is more to maintain the reactor in critical condition. This fission reaction of Pu-239 gives some decreasing composition of Pu-239 for plutonium-recycled scheme in comparing with once through case. Total even mass plutonium isotope, which is a total combination of Pu-238, Pu-240 and Pu-242, produces more composition of more 50% than total odd mass plutonium isotope of less than 50%, which is combination of Pu-239 and Pu-241. Total even mass plutonium production is obtained more for heavy water coolant than its production for light water coolant as shown in Figure 8. While it's total odd mass production becomes less. Higher composition of even mass plutonium gives an indication of the composition of plutonium has better proliferation resistant level for the system, which is shown by plutonium recycled scheme and heavy water coolant case. In case a specific composition of Pu-238, it shows better production for recycled Pu, which is more than 8% as well as less production of Pu-239 of less than 40%. This composition of Pu-238 can be adopted to protect the plutonium to be used for explosive devices based on some even number mass of plutonium isotope because of some internal material barrier of plutonium such as decay heat and spontaneous neutron fission. These plutonium barriers, which is based on even mass plutonium composition was used for some plutonium criteria such as IAEA, Pellaud and Kessler criteria (IAEA, 1972; Pellaud, 2002; and Kessler, 2004) [19-21].

4. Conclusions

Some important key performances have been evaluated such as reactor performance of criticality and fuel conversion as well as some actinide productions based on several fuel cycle schemes open close or once through cycle and a closed-fuel cycle of some actinide was used based on water-cooled reactor system, including ligh water and heavy water coolant. Once through case or UOX case requires higher fissile materials than MOX fuel scheme to maintain criticality condition. In UOX fuel case, heavy water coolant case requires more fissile material than light water coolant (H2O), however, in case of MOX fuel, it shows the opposite trend for heavy water coolant than light water coolant. In case of thorium cycle, it requires more fissile material of U-233 for heavy water coolant than light water coolant, which is similar to UOX fuel case. In term of light water case, the required fissile material of U-233 is less for thorium case in comparing with the required fissile material of U-235 in MOX fuel case. Fuel conversion ratio of light water and heavy water coolants for UOX and MOX fuel cycles are less than unity except for heavy water coolant of MOX fuel case, which obtains slightly more than unity or becomes a fuel breeding condition. Thorium fuel shows better fuel breeding capability than MOX and UOX fuel cases. Heavy water coolant gives more fuel conversion capability and it obtains breeding condition than light water coolant. Actinide compositions other than uranium are obtained less, which gives less than 1 % of fuel composition, except for plutonium which gives more than 1% for light water coolant and more than 10% for heavy water coolant case. Different coolants are effective to produce more actinides such as plutonium and Americium, however its very limited change for Neptunium and Curium for MOX fuel case. For different fuel scheme cases, it shows that plutonium recycled scheme as MOX fuel case gives higher production of actinide or trans-uranium actinides from neptunium up to Curium than UOX fuel case for H2O coolant. Total even mass plutonium isotope produces more composition than total odd mass plutonium isotope for both coolants and it is obtained more for heavy water coolant than its production for light water coolant. Higher composition of even mass plutonium gives an indication of the composition of plutonium has better proliferation resistant level for the system, which is shown by plutonium recycled scheme and heavy water coolant case.

Acknowledgments

We would like acknowledge and extend our gratitude to desentralisasi research program of ministry of research, technology and higher education and research innovation program of ITB for the grant and international conference publications

References

[1] President's meeting of G8 countries, Brazil, China, India and South Africa, 2006 Moscow April 19-20, Contribution of Coal and Nuclear to Sustainable Energy Supply: Perspectives and Problems, Paper presented at *the Russian Academy of Sciences, Academies of Sciences*
[2] P. Sidik, N. Takaki, H. Sekimoto 2007 "Feasible region of design parameters for water cooled thorium breeder reactor," *J. Nucl. Sci. Technol*, **44**, 7, 946-957
[3] T. K. Kim and T. J. Downar 2002 *Nuclear Technology,* **138**,[1] 17-29
[4] L. Michael, G. Otto 1998 *Nucl. Eng. Des.* **180**, 2, 133
[5] U. Turan 2000 *Prog. Nucl. Energy*, **37**, 137
[6] L. B. Freeman et al., 1989 *Nucl. Sci. Eng.*, **102**, 341
[7] P. Sidik, N. Takaki, H. Sekimoto, 2008 *J. Nucl. Sci. Technol*.,. **45**, 7, 589-600
[8] P. Sidik, N. Takaki, H. Sekimoto, 2006 *Ann. Nucl. Energy*, **33**, 561
[9] H. Sekimoto, N. Takaki, 1991 *J. Nucl. Sci. Technol.*, **28**, 941
[10] A. Waris, H. Sekimoto 2001 *J. Nucl. Sci. Technol.*, **38**, 7, 517
[11] A. Waris, H. Sekimoto 2001 *Ann. Nucl. Energy*, **28**, 153

[12] A. Mizutani, H. Sekimoto 1998 *Ann. Nucl. Energy*, **25**, 9, 623
[13] A. Mizutani, H. Sekimoto 1998 *Ann. Nucl. Energy*, **25**, 13, 1011
[14] A. Mizutani, H. Sekimoto 1998 *Prog. Nucl. Energy*, **32**, 713
[15] A. Mizutani, H. Sekimoto 1997 *J. Nucl. Sci. Technol.*, **34**, 6, 596 (1997).
[16] T. Seino, H. Sekimoto 1998 *Ann. Nucl. Energy*, **25**, 4–5, 223
[17] K. Okumura et al., 1996 *Japan Atomic Energy Research Institute (JAERI)*
[18] T. Nakagawa et al., 1995 *J. Nucl. Sci. Technol.*, **32**, 1259
[19] International Atomic Energy Agency, I972, Information Circular, INFCIRC/153
[20] B. Pellaud, 2002 *J. Nucl. Mater. Managements*, XXXI, No. 1
[21] G. Kessler, 2004 The First Int. Sci. Technol. Forum on Protected Plutonium Utilization for Peace and Sustainable Prosperity, March 1–3, Tokyo, Japan

Experimental and numerical investigation of cooling performance of a cold storage in a pharmaceutical industry

A Sularno[1], FXN Soelami[1], and Y Bindar[2]

[1]Department of Engineering Physics, Faculty of Industrial Technology, Institut Teknologi Bandung, Bandung 40132, Indonesia
[2] Department of Chemical Engineering, Faculty of Industrial Technology, Institut Teknologi Bandung, Bandung 40132, Indonesia.

E-mail : agussularno313@gmail.com

Abstract. This paper describes the study of cooling performance of cold storage in a pharmaceutical industry. It was intended to investigate the temperature distribution inside the storage that is an important performance factor in pharmaceutical industry cold storage. Cold storage that used is a ceiling type with the liquid bottle loading.
Temperature distribution and the storage cooling performance were studied using experimental measurement and numerical simulation. Some variation of bottle arrangement and rack arrangement have been observed to show the impact of distribution temperature and cooling performance of cold storage. Surface temperatures of the bottles were measured with different bottles and rack arrangement. The temperature of cold storage was set to 5°C. In numerical simulation, a transient three dimensional Computational Fluid Dynamic (CFD) model was developed to investigate the cooling performance and temperature distribution inside bottle.
At this stage, the results showed that rack arrangement that parallel with the cold room fan and V shape bottle layout has given a good cooling performance (it takes 1480 minutes to reach a stable temperature at the setpoint) and an optimum temperature distribution (with temperature difference of 0.58 °C). For the measurement of the distribution of temperature in the bottle, the mean deviation value between the simulation and the experiment on the measurement of 2 coordinate *points* (X = 0,1 m, Y = 0,3 m, Z = 0 m and X = -0,1 m, Y = 0,3 m, Z = 0 m) were 5,5 % and 7,6 %.

1. Introduction

In a pharmaceutical industry, especially in a biotechnology industry, product quality begins from raw materials until final products. Monitoring for each step process should be done includes storage processing. Those were done to guarantee that the product still in a good quality.

A cold storage was used to store the product that need cold temperature to guarantee its quality. To achieve the desired temperature, a refrigeration system is employed. Using refrigeration system, high energy consumption of the cold storage could happen. Therefore the cold storage should be designed as efficiently as possible with considering of its performance. Most cold storage are designed by experiences which may cause increasing operating cost. Moreover, a uniform airflow field can also improve the refrigeration quality of the products in the cold storage.

Some studies to improve performance of cold storage have been done both experimentally and by numerical simulation. Several studies modeling of temperature distribution in cold storages have demonstrated the effectiveness of various methods (Nahor et al. 2005; Konishi et al. 2009, Tanaka et al.

Content from this work may be used under the terms of the Creative Commons Attribution 3.0 licence. Any further distribution of this work must maintain attribution to the author(s) and the title of the work, journal citation and DOI.
Published under licence by IOP Publishing Ltd

2011, Akdemir et al. 2013, Mahdi et al. 2014). Rajan et al. (2015) investigated performance of cold storage for the different stacking arrangements experimentally. The temperature is the only parameter at which the performance of cold storage depends on.

The objective of this study is to investigate the effect of loading arrangements of vaccine bottles to the cooling performance of cold storage by experiment. It also investigates temperature distribution and cooling rate inside the bottle both experimentally and by numerical simulation.

2. Research Methodology

This research was conducted using two kinds method, by experimental and by numerical simulation using Ansys Fluent.

2.1 Experimental Setup

Thermocouple sensors were used to measure air temperature distribution and the cooling rate of the cold storage

2.1.1 Operation description. Cold storage that used is a ceiling type with dimension 290 cm (L) x 260 cm (W) x 250 cm (H). The refrigeration system works with a vapor compression mechanism with R22 refrigerant. The schematic diagram of the cold storage is shown in figure 1.

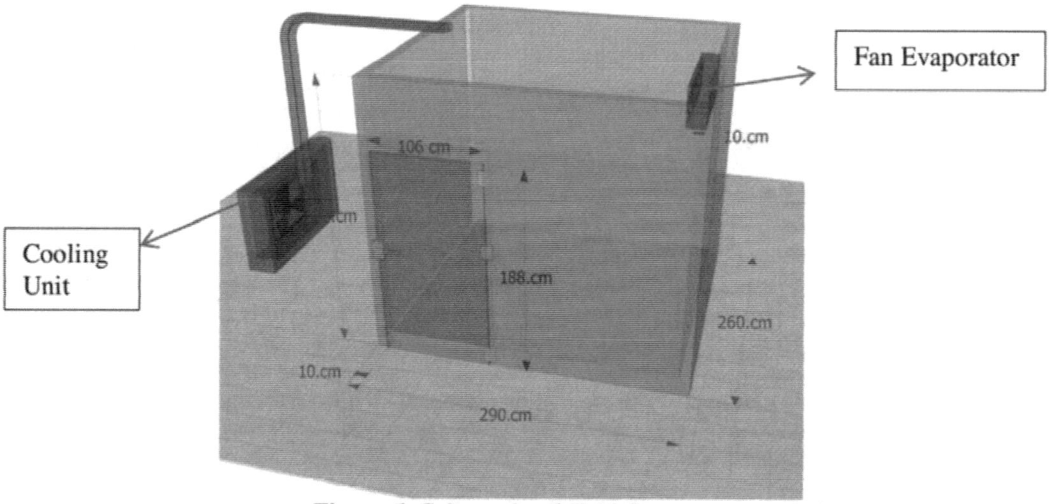

Figure 1. Schematic of Cold Storage

The cold storage was designed for 2 – 8 °C store temperature with set point temperature 5 °C. The cold storage is used to store 24 liquid vaccine bottles with 20 litres volume each bottle. These bottles were arranged in 2 racks and cooled to achieve 5 °C. Distribution temperature measurement was done with thermocouple sensors that put at the outside bottle surface.

2.1.2 Bottle & Rack Arrangement In this research, some variation of bottle and rack arrangements were observed to see optimum condition of the cold storage that can be achieved. Bottle and rack arrangements were designed with considering of cold air flowing through the bottles and personnel accessibility.

a. Arrangement I
In this arrangement, the rack was arranged parallel with fan evaporator and 4 bottles were arranged straightly in each step of the rack. The schematic of arrangement is shown in figure 2.

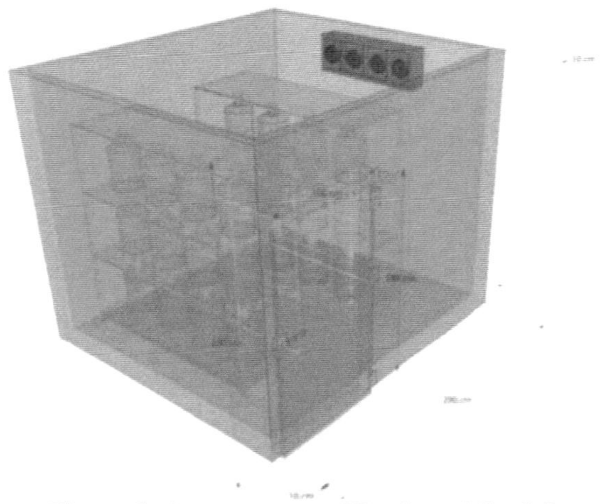

Figure 2. Arrangement of Bottle and Rack I

b. Arrangement II
In this arrangement, the rack was arranged parallel with fan evaporator and 4 bottles were arranged with V shape layout in each step of the rack. The schematic of arrangement is shown in figure 3.

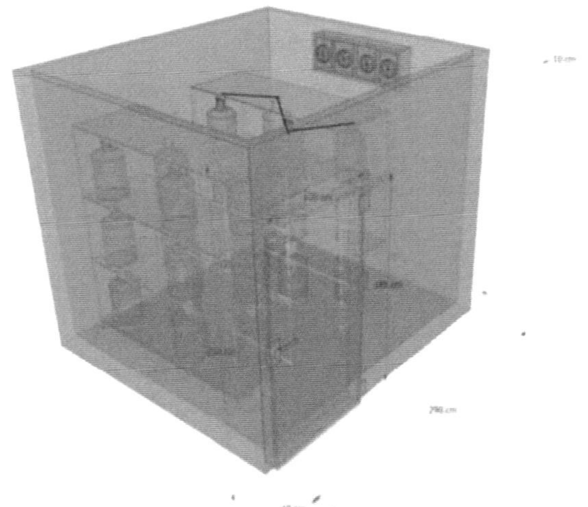

Figure 3. Arrangement of Bottle and Rack II

c. Arrangement III
In this arrangement, the rack was arranged parallel with the door and 4 bottles were arranged straightly in each step of the rack. The schematic of arrangement is shown in figure 4.

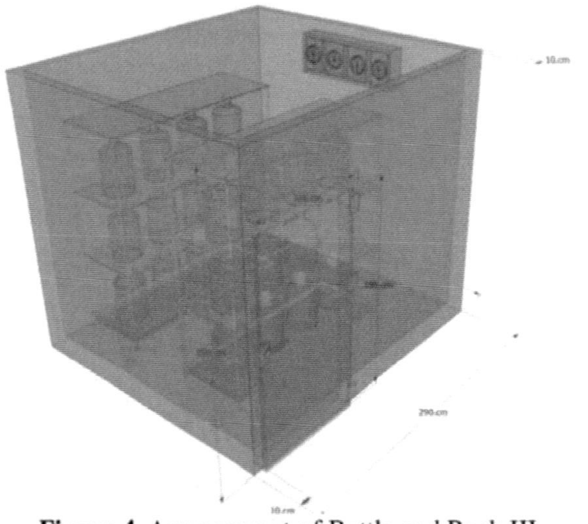

Figure 4. Arrangement of Bottle and Rack III

d. Arrangement IV

In this arrangement, the rack was arranged parallel with the door and 4 bottles were arranged with V shape layout in each step of the rack. The schematic of arrangement is shown in figure 5.

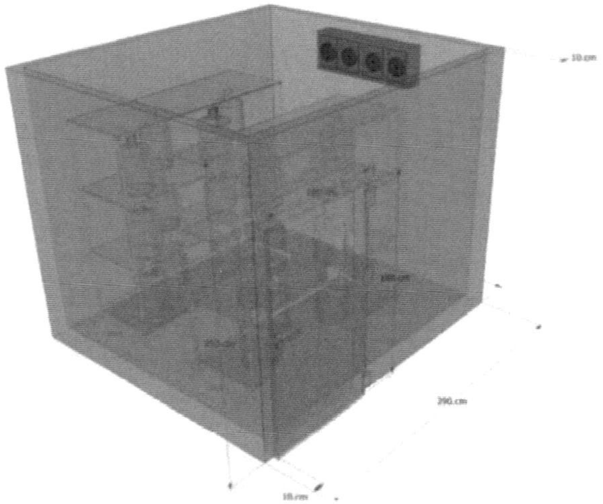

Figure 5. Arrangement of Bottle and Rack IV

2.2 Numerical Simulation Method

In numerical method, simulation of cooling process inside the bottle was carried out. This research is started by drawing geometry of bottle that shown in figure 6. The bottle is drawn according to its original dimension (280 mm in diameter and 520 mm in height). The material of bottle is high densinty polypropilene (hdpp) and the material of liquid is water.

Inside the bottle, it is assumed that heat conduction is the only mode of heat transfer. The equation for the conservation of energy for the bottle with constant properties at transient state can be written as

$$\frac{\partial(\rho C_p\ T)}{\partial t} = \frac{\partial}{\partial x}\left(k\frac{\partial T}{\partial x}\right) + \frac{\partial}{\partial y}\left(k\frac{\partial T}{\partial y}\right) + \frac{\partial}{\partial z}\left(k\frac{\partial T}{\partial z}\right) + q_{gen} \quad (1)$$

Because it was assumed that no heat generation inside the bottle, the equation become

$$\frac{\partial(\rho C_p\ T)}{\partial t} = \frac{\partial}{\partial x}\left(k\frac{\partial T}{\partial x}\right) + \frac{\partial}{\partial y}\left(k\frac{\partial T}{\partial y}\right) + \frac{\partial}{\partial z}\left(k\frac{\partial T}{\partial z}\right) \quad (2)$$

The variables ρ and C_p are fluid density and fluid spesific heat.

The boundary conditions on temperature can be written as

On wall of bottle : $-k\frac{\partial T}{\partial n} = h(T - T_{ext})$ .... (3)

The variable k is thermal conductivity of bottle with liquid and the variable h is heat transfer coefficient of fluid. External temperature was set constant at 5 °C and heat transfer coefficient , h = 15 W m^{-2} K^{-1}.

Figure 6. Geometry of Bottle

3. Result and Discussion

It was intended to see the effect of bottles and rack arrangements on the performance of cold storage. In this section, the average product temperature distribution and cooling rate for various bottles and rack arrangements has been presented. The graphs of average surface bottle temperature distribution inside the cold storage for four different arrangements are plotted as follows.

a. Arrangement I

The average temperature distribution, maximum temperature and minimum temperature of cold storage are plotted as following:

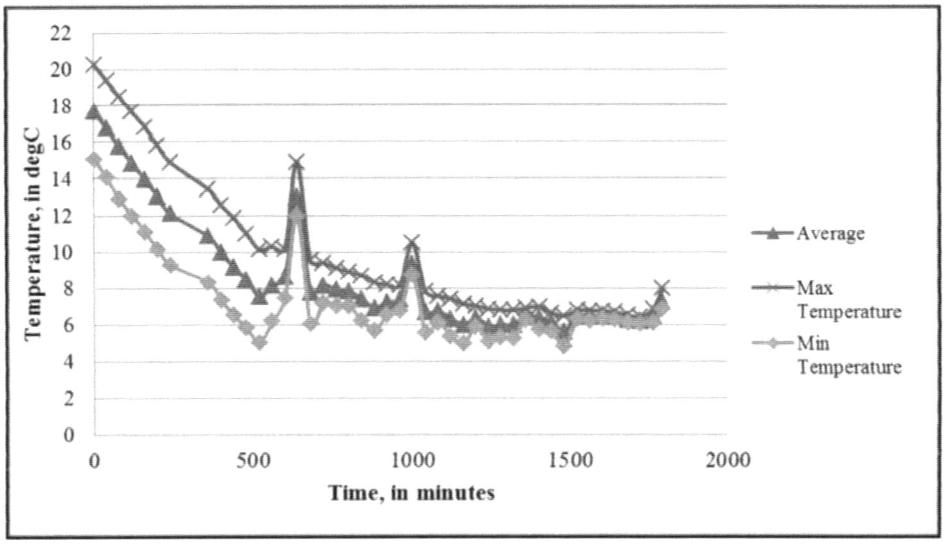

Figure 7. Average temperature distribution in arrangement I

b. Arrangement II

The average temperature distribution, maximum temperature and minimum temperature of cold storage are plotted as following:

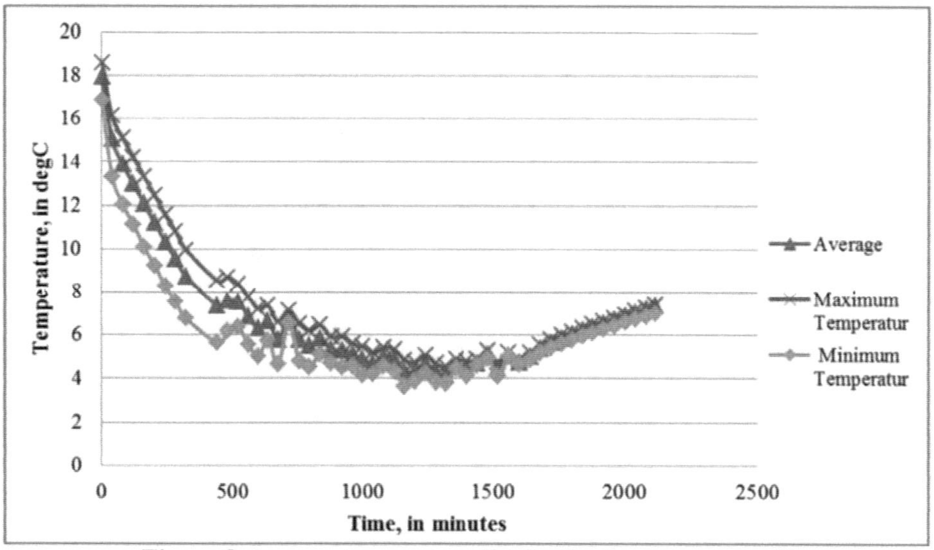

Figure 8. Average temperature distribution in arrangement II

c. Arrangement III

The average temperature distribution, maximum temperature and minimum temperature of cold storage are plotted as following:

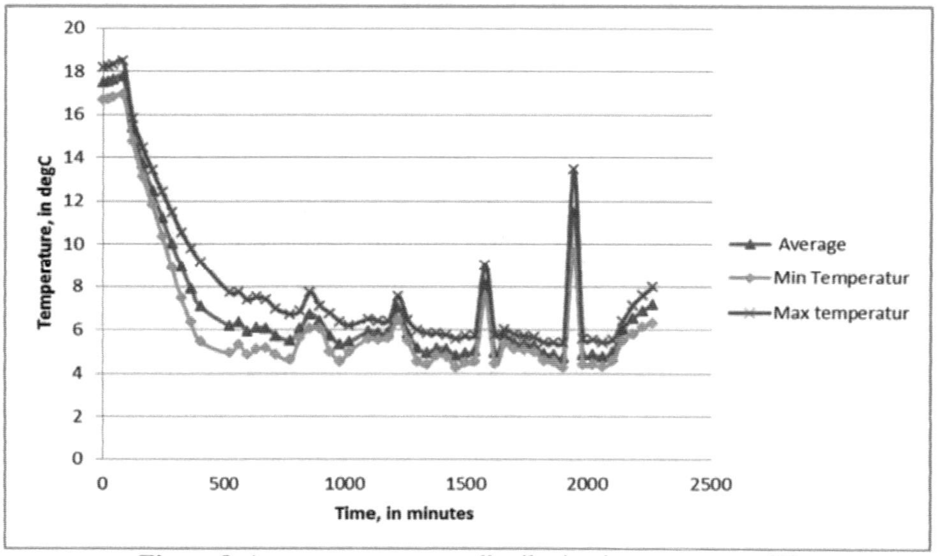

Figure 9. Average temperature distribution in arrangement III

d. Arrangement IV

The average temperature distribution, maximum temperature and minimum temperature of cold storage are plotted as following:

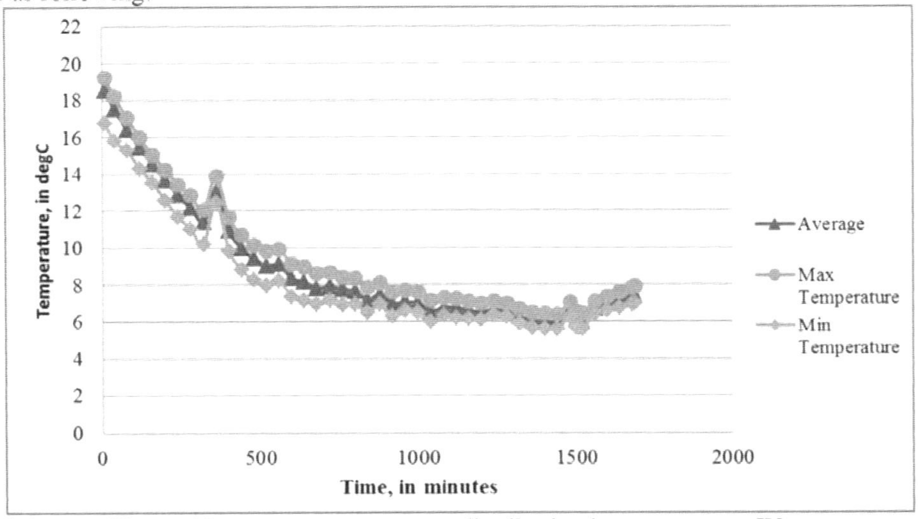

Figure 10. Average temperature distribution in arrangement IV

The performance of cold storage with different arrangement can be summarized as shown in table 1.

Table 1. Cooling performance of cold storage in different loading arrangement

	Temperature Difference, in °C	Cooling time to the stable setpoint temperature, in minutes
Arrangement I	0,79	1510
Arrangement II	0,58	1480
Arrangement III	0,93	1596
Arrangement IV	0,84	1500

Based on experimental result, arrangement II has the best cooling performance. It has more uniform distribution temperature than the other arrangements. It also has less cooling time than the others. Cold air can easily flow through the outer surface of each bottle in this arrangement.

The graphs of distribution temperature around coordinate X = 0 which resulted by numerical simulation method are shown in the following figures.

IOP Conf. Series: Journal of Physics: Conf. Series **1090** (2018) 012012 doi:10.1088/1742-6596/1090/1/012012

Figure 11. Distribution temperature inside the bottle at t = 3000 second

Figure 12. Distribution temperature inside the bottle at t = 30000 second

Figure 13. Distribution temperature inside the bottle at t = 60000 second

Figure 14. Distribution temperature inside the bottle at t = 90000 second

Figure 11 shows cooling process of bottle in the early time. The top of bottle was cooled faster than the other part of bottle. Figure 12-13 show profile of temperature distribution inside the botle during cooling process time. It shows that cooling process begin from the bottle wall. The liquid temperature that close to wall of botlle is colder than the liquid temperature in the center of bottle. Figure 14 shows that temperature around the bottle close to the temperature of environment (5 °C).

Comparison result between simulation and experimental method is shown in figure 15.

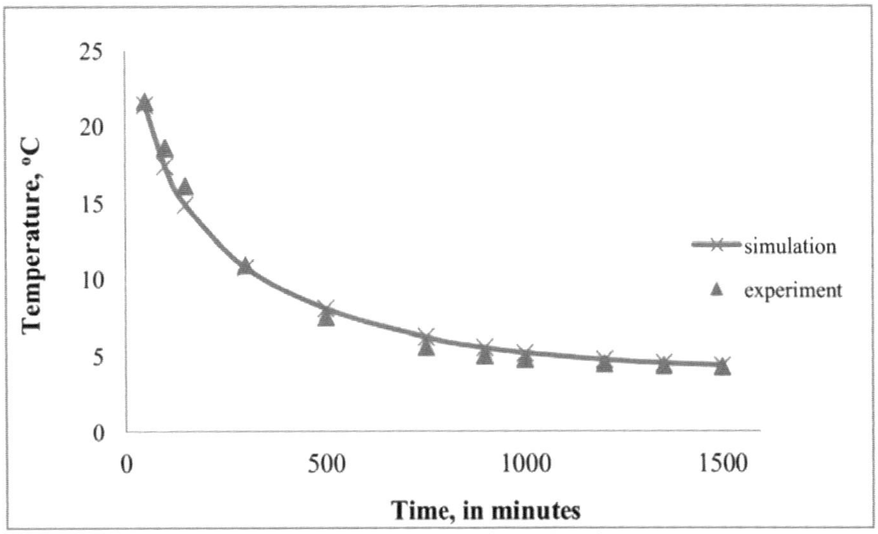

(a) X = 0,1 m Y = 0.3 m Z = 0 m

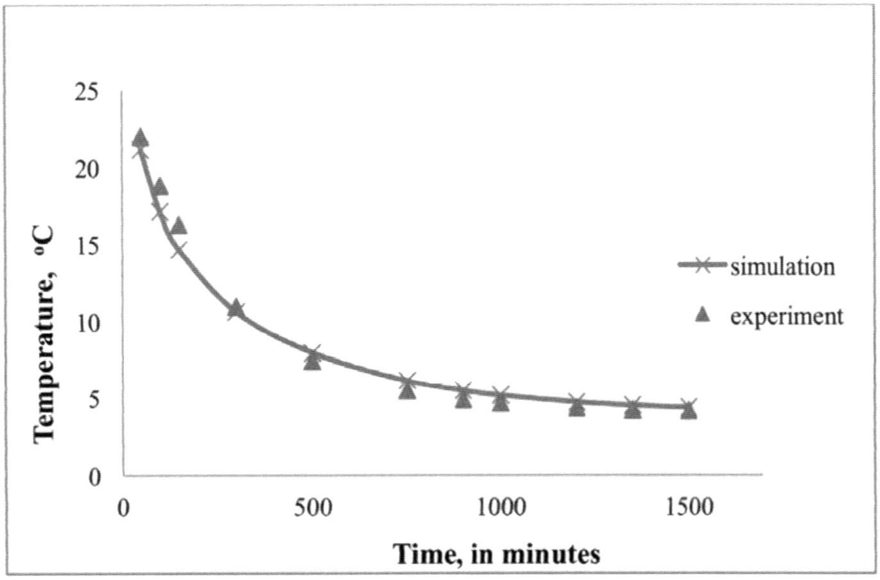

(b) X = - 0,1 m Y = 0.3 m Z = 0 m

Figure 15. Comparison result between simulation method and experimental method.

Figure 15 (a) and 15 (b) show comparison result between simulation and experimental methods inside the bottle. The average difference temperature is about 0,45 °C and 0,64 °C. It results average error about 5,5 % and 7,6 %. In the experiment, temperature is not constant at 5 °C so it can contribute error between simulation and experiment.

4. Conclusion

Based on result of experimental and numerical simulation, it can be concluded that :

- Loading arrangement of the bottles influences cooling performance of the cold storage.
- Numerical simulation can be used to predict heat transfer and to asses cooling performance in the cold storage..
- The model has an error about 5,5 % - 7,6 %

References

[1] Rajan A, Aharwal K R, and Bhadu R. Performance analysis of cold storage for the different stacking arrangements 2015 *India. IJERGS* **3** 455.

[2] Tanaka F, Konishi Y, Kuroki Y, Hamanaka D and Uchino T. The use of CFD to improve the performance of a partially loaded cold store 2010 *Japan. J. Food.Proc.Eng* 1

[3] Nahor H B, Hoang M L, Verbovena P, Baelmans M, and Nicolai B M. CFD model of the airflow, heat and mass transfer in cool stores 2005 *Belgium. Int. J. Refrig.***28** 363.

[4] Akdemir S, Ozturk S, Edis F O,and Bal E. CFD modelling of two different cold stores ambient factors 2013 *Turkey. IERI Procedia* **5** 28.

[5] Mahdi Q S, and Hadi H M. Experimental and numerical investigation of airflow and temperature distribution in a prototype cold storage 2014 *Iraq. IJMET* **5** 123

[6] Ho S H, Rosario L, and Rahman MM**.** Numerical simulation of temperature and velocity in a refrigerated warehouse 2010 *USA. IntJ. Refrig.* **33** 1015.

[7] Chourasia, M K, and Goswami, T K 2007. Simulation of effect of stack dimensions and stacking arrangement on cool-down characteristics of potato in a coldstore by computational fluid dynamics. Biosyst. Eng. **96 (4)** 503.

Mo-99 Isotope Production Calculation of SAMOP Reactor Experimental Facility

M. Iqbal Farezza W[1,3], Syarip[2,4]

[1]Engineering Physics Department, University of Gadjah Mada, Yogyakarta.
[2]Centre for Science & Accelerator Technology, JL. Babarsari, POB 6101 ykbb, Yogyakarta

[3]Email: ryakuzza@gmail.com, [4]E-mail: syarip@batan.go.id

Abstract. The calculation of ^{99}Mo isotope production of SAMOP (Subcritical Assembly for ^{99}Mo Production) experimental test facility using SRAC computer code has been done. The SAMOP experimental test facility is a reactor in subcritical condition fueled with low-enriched uranium in the form of uranyl nitrate solution, with uranium concentration of 300 g U/L, and driven by an external neutron source. The calculation method is using Standard Reactor Analysis Code (SRAC) which is a computer code package system applicable to neutronic analysis for various types and forms of the nuclear reactor core. The calculation results show that the neutron multiplication factor of 0.981 to 0.992 can be achieved with 30.2989 L to 31.4587 L uranyl nitrate in the certain core configuration of SAMOP test reactor. The ^{99}Mo isotope production calculation for 6-day operation (batch) shows that the SAMOP test facility can produce approximately 437.03 mCi (16170.11 MBq) per batch, and the ^{99}Mo production will reach a saturated condition at twelve day operation.

Keywords: ^{99}Mo, isotope production, SAMOP, SRAC, neutron multiplication, uranyl nitrate.

1. Introduction

The need of radioisotopes for nuclear medicine in Indonesia has grown rapidly in recent years. One of the most common aspects of nuclear medicine is diagnostics is using the ^{99m}Tc radioisotope. The increased need for the ^{99m}Tc isotope cannot be fulfilled because ^{99}Mo commonly used as a ^{99m}Tc generator, can only be produced in a nuclear reactor and the half-life of ^{99}Mo is only 66 hours making it difficult to store and mobilize its spread. Solving this problem can be done by shortening the ^{99}Mo production process as well as shortening the mobilization of radioisotopes spreading distance.

Over 10000 hospitals worldwide use radioisotopes in medicine, and about 90 % of the procedures are for diagnosis. The most common radioisotope used in diagnosis is ^{99m}Tc which is daughter isotope of ^{99}Mo, accounting for 80 % of all nuclear medicine procedures worldwide [1, 2]. Commonly, most ^{99}Mo is produced using highly-enriched uranium placed in high-power nuclear research reactors. Recently there are eight medical isotopes producing reactors that provide over 90 % of the world's ^{99}Mo needs. These reactors are government-owned and -subsidized, and the respective governments control the funding, priorities, and operational schedule of these reactors. The unscheduled shutdown of two of these reactors in 2009 and 2010 (Canadian and Dutch reactors)

Content from this work may be used under the terms of the Creative Commons Attribution 3.0 licence. Any further distribution of this work must maintain attribution to the author(s) and the title of the work, journal citation and DOI.
Published under licence by IOP Publishing Ltd

caused worldwide shortages of ^{99}Mo, leading to the delay or cancellation of many medical procedures [3, 4].

The medical community has been plagued by ^{99}Mo shortages due to aging reactors, such as the NRU (National Research Universal) reactor in Canada. There are currently no US producers of ^{99}Mo, and NRU was being shut-down in 2016, which means that another ^{99}Mo shortage is imminent unless a potential domestic ^{99}Mo producer fills the void [5]. The ^{99}Mo producing system without a nuclear reactor and without using highly-enriched uranium is being implemented at National Nuclear Energy Agency (BATAN). Its production process will use a subcritical assembly driven by an external neutron source from neutron generator or a particle accelerator and target that generate neutron in the core of subcritical assembly. Compact neutron generator (CNG) is a particle accelerator, as predicted that the accelerators will probably the best-known uses are for cancer therapy, medical isotope production, and food irradiation [6]. The similar method is being developed by SHINE which plans to produce at least one-half of the U.S. need for ^{99}Mo by 2016 [7].

In general, the production of ^{99}Mo is by splitting the ^{235}U atom with neutrons inside a nuclear reactor so that a critical reactor is required. Licensing issues for critical reactors are also complex because they have to fulfill the very strict technical and safety requirements, and it needs huge funds as well. To solve this problem BATAN or the National Nuclear Energy Agency designed a reactor that could produce ^{99}Mo but did not categorized as the critical reactor and did not use high-enriched uranium. A non-critical reactor or subcritical assembly fueled with uranyl nitrate, called subcritical assembly for ^{99}Mo production (SAMOP) has been designed and will be developed further at the Center for Accelerator Science and Technology (CAST) BATAN [8]. The SAMOP experimental facility as a test facility which use an external neutron source from the beam-port of Kartini TRIGA reactor, which has been identified suitable for this purpose [9]. It is expected that the SAMOP system will reduce much less waste than current ^{99}Mo production methods.

The aim of the research is the calculation and determination of ^{99}Mo specific activities, and determination of optimal irradiation conditions, as well as the effect of some variables of parameters will affect the ^{99}Mo production of the SAMOP reactor. The ^{239}Pu byproducts will also be analyzed considering ^{239}Pu is a fissile material so that the SAMOP reactor can later meet the safeguard requirements. The isotopes calculation will be approximated by the deterministic method applied in Cell Burn of the SRAC computer code.

The use of subcritical aqueous homogenous reactors driven by accelerators presents an attractive alternative for producing ^{99}Mo. In this method, the medical isotope production system itself is used to extract ^{99}Mo or other radioisotopes so that there is no need to irradiate common targets. In addition, it can operate at much lower power compared to a traditional reactor to produce the same amount of ^{99}Mo by irradiating targets [10]. Study to produce ^{99}Mo with activation method by using TRIGA reactor have also been done to fulfill the need of ^{99}Mo as ^{99}m Tc generator for use in nuclear medicine [11,12,13]

2. Description of SAMOP experimental facility

The SAMOP experimental facility as a test facility which use an external neutron source emerge from the radial beam-port of Kartini reactor. The external neutron source has been identified as thermal neutron in order of 10^8 n/cm^2 s [9]. The SAMOP core consists of annular cylindrical tube containing uranyl nitrate [UO_2 $(NO_3)_2$] as fuels and target, surounded by ring of UO_2 $(NO_3)_2$ tubes or TRIGA fuel elements. The TRIGA fuel elements can be loaded in the ring together with UO_2 $(NO_3)_2$ tubes to increase neutron multiplication factor.

The dimension of the SAMOP core, reflector, boral rod neutron absorber, and coolant tank is described in Figure 1. The SAMOP experimental facility is provided by an instrumentation and control system in such that there is a criticality indication, the boral control rod neutron absorber will dropped automatically inserted to the SAMOP reactor core.

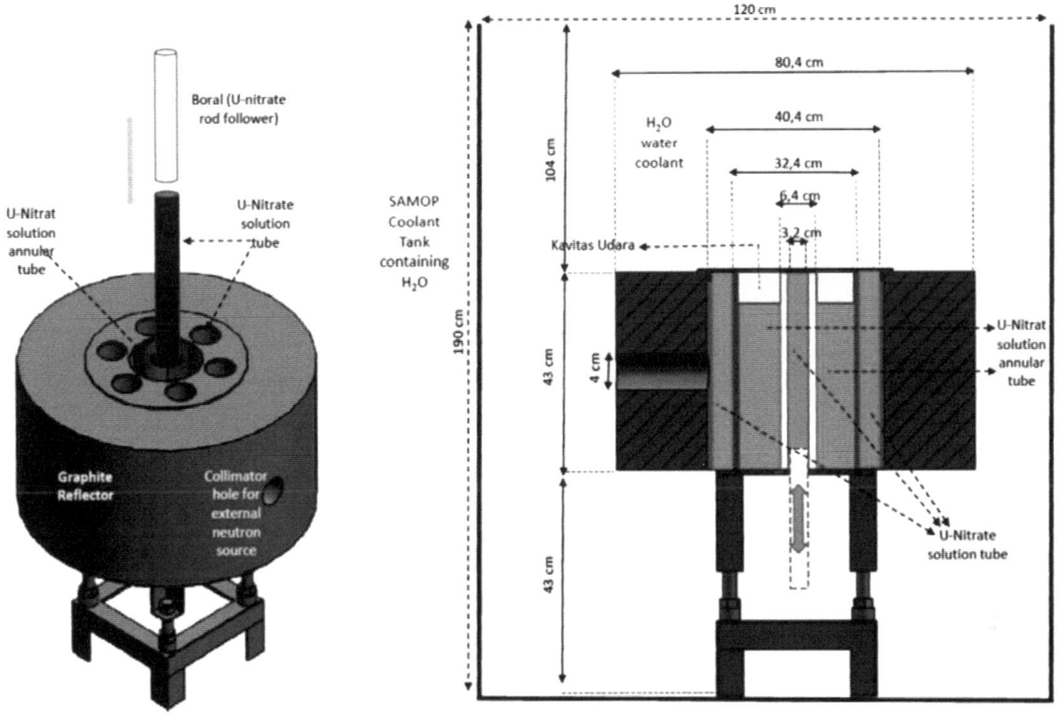

Figure 1. Core and reflector configuration of SAMOP experimental facility

3. Materials and methods

3.1. ^{99}Mo Production methods

In general, the production of ^{99}Mo is by splitting the ^{235}U atom with neutrons inside a nuclear reactor so that a critical reactor is required. There are two primary approaches for producing the medical isotope ^{99}Mo i.e.: fission of ^{235}U or $^{235}U(n,f)^{99}Mo$, which produces ^{99}Mo and other medically important isotopes such as ^{131}I and ^{133}Xe, and neutron capture by ^{98}Mo or $^{98}Mo(n,\gamma)^{99}Mo$ [14]. The fission of ^{235}U produces a large number of fission products, including ^{99}Mo. The global producers of ^{99}Mo use the fission of ^{235}U method instead of neutron capture method to produce ^{99}Mo because of its inefficiencies. However, the neutron capture process can be used to make smaller quantities of ^{99}Mo. Table 1 compares the two methods of ^{99}Mo production [15].

Table 1. Comparison of fission and neutron capture in ^{99}Mo production [15]

$^{235}U(n,f)^{99}Mo$	$^{98}Mo(n,\gamma)^{99}Mo$
Produces high specific activity ^{99}Mo	Produces low specific activity ^{99}Mo
Requires enriched ^{235}U target	Requires highly enriched ^{98}Mo target
Complex chemical processing	Simple chemical processing
Requires dedicated processing facility	Requires high flux neutron source
Generates high-level radioactive waste	Generates minimal waste

The isotope production rate, which is of interest here, is proportional with several conditions as illustrated in the equation (1) below:

$$R \approx n\sigma\phi \qquad (1)$$

where: R is a rate of reaction (i.e., number of reactions per unit time and volume), which is related to the amount of the new substance that can be produced, n is the number of target nuclei present (i.e., the target nuclei density in atoms per unit volume), ϕ is the flux of particles causing the reaction (neutrons per cm^2 per second), and σ is the probability that the reaction will occur, expressed as an area. The calculation steps by using SRAC computer code is described in Figure 2. The analysis of ^{99}Mo production is done as a function of uranyl nitrate enrichment and SAMOP reactor power, at a subcritical conditions or effective neutron multiplication factor (k_{ef}) small than 1 (0.9 to 0.99).

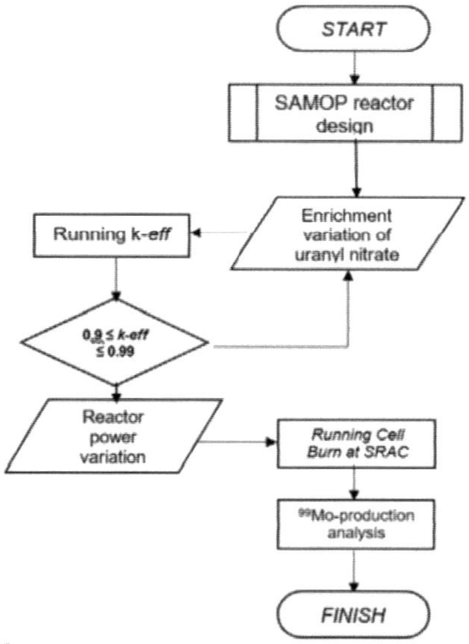

Figure 2. The ^{99}Mo calculation steps by using SRAC computer code.

3.2. SRAC computer code

The SRAC (Standard Reactor Analysis Code) is a set of FORTRAN computer codes created by Japan Atomic Energy Research Institute (JAERI) to analyze and solve the neutronic calculations of various models of thermal reactors [16]. This code can solve the microscopic and macroscopic cross-sections effectively to complete burnup calculations. In core reactor operations, the magnitudes of fission such as neutron flux, nuclide concentration, and temperature will vary as spatial functions. Burnup calculations must be performed to see the correlation of these parameters with time. In the SRAC system, the reactor core burn-up calculations are divided into two stages. First, calculate cell-level burn-up using SRAC to obtain a homogeneous macroscopic cross-section on each burn-up step. The latitude is then tabulated with discrete values of fuel temperature, coolant, and degree of burn-up to model the actual fuel assembly in the reactor core. The next stage is to interpolate the existing data to be distributed in the reactor core for further processing.

Cell Burn Up is an additional code SRAC system that serves to complete the calculation of reactor core burn-up with the theory of neutron diffusion. In the calculation of burn-up cell, the burnup changes of the atomic number density of the nuclides in the burn-up chain reaction model are obtained from solving the depletion equation with the reaction rate of fission, capture, and reaction (n, 2n) [16]. Here is an optional treatment that can be used in cell burn-up calculations: option to calculate the cooling time when burn-up, option to calculate the conversion of the instant or integration ratio defined by the user, burn-up calculations with constant flux levels, and burn-up calculations with fixed density numbers for specific nuclides. Several models of burn-up chains are available to choose for use in calculations [16].

In the cell burnup calculations, there are two types of time-step units adopted in the SRAC code. First is the burnup step unit with relatively common time intervals, and the second is the sub-step unit in each burnup step. When the burn-up step interval is described in the input, the sub-step interval is specified in the code. At each burn-up step, the flux calculations are performed by the selected code. As a result, the distribution of one group collapsed flux and collapsed microscopic cross-sectional for burnable nuclides is obtained. The depletion equation for the interval from the burnup step to n (t_{n-1}, t_n) can be written by equation (2), on the assumption that the relative distribution of microscopic reaction rates for capture, fission, and reaction (n, 2n) does not change in time interval step burn-up.

$$\frac{dN_i^M(t)}{dt} = \sum_{j\neq 1} f_{j\to 1}\lambda_j N_i^M(t) + \sum_{k\neq 1}\{g_{k\to 1}\sigma_{c,k}^M + \gamma_{k\to 1}\sigma_{f,k}^M\}\varphi^M Fact(t)N_k^M(t) - \left[\lambda_i + \left(\sigma_{a,i}^M + \sigma_{n2n,i}^M\right)\varphi^M Fact(t)\right]N_i^M(t) \quad (2)$$

i, j, k : the depleted nuclides, M: depletion zone, associated with M-region in SRAC code, N: atomic number density, λ, f: decay constant and branching ratio, γ, g, h: the yield fraction of each transmutation, φ: the relative flux obtained from the eigenvalue calculation, and Fact (t): the normalization factor for converting the relative flux to absolute flux.

4. Result and discussiuon

The first analysis was to look at how the influence of uranium enrichment variation of the SAMOP reactor to the sub-criticality value. From the results of the first analysis will get the relationship between the criticality levels with the uranium enrichment, then selected the value of enrichment that the criticality level is closest to 1 (subcritical). Once the enrichment value is known, *Cell Burn* is done to determine the amount of ^{99}Mo produced on the SAMOP reactor. The result of cell burn program code will show the relationship between enrichment with ^{99}Mo result. The running result of this program code can also be obtained plutonium production value which is expected still within the safe limits. The second analysis is to vary the power on the SAMOP reactor with variations of 1 kW, 10 kW, 50 kW and 100 kW. The results of this simulation will be found reactor power relationship with ^{99}Mo results generated each increase of 1 kW of power. The third analysis was to run a SAMOP reactor for 24 days to get ^{99}Mo of growth and the optimal operating cycle. The effective multiplication factor (k_{ef}) of the SAMOP reactor was calculated using the SRAC-CITATION code for variation of uranium enrichment, the result is obtained as shown in Figure 3.

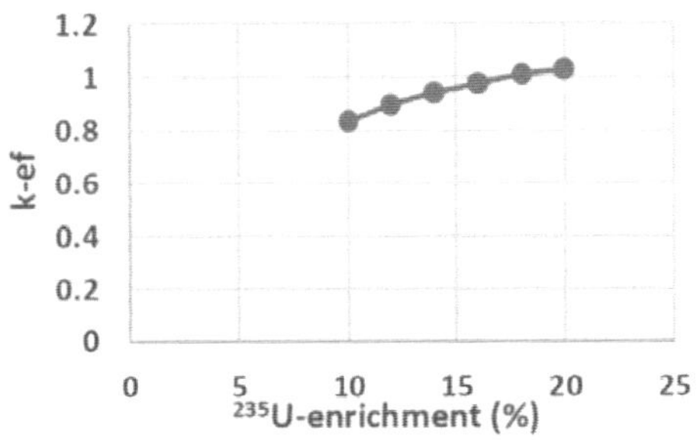

Figure 3. The k_{ef} of the SAMOP reactor experimental facility with variation of fuel enrichment

By using 19.5 % of ^{235}U fuel enrichment, the criticality or neutron multiplication factor was also done for two configurations of SAMOP reactor core i.e. for the core with annular tube surrounded by 8 tubes of $UO_2\ (NO_3)_2$ and 12 tubes in the ring respectively. The criticality calculation was done to investigate the change of sub-criticality levels per day of reactor operation, the result is depicted in Figure 4. It is shown that within ten days operation (for ^{99}Mo production), the sub-criticality level of both SAMOP core configurations is still stable. The calculation results show that the neutron multiplication factor of 0.981 to 0.992 can be achieved with 30.2989 L to 31.4587 L uranyl nitrate in that core configurations. The criticality analysis result was in accordance with refer to similar work [17], and similar work previously done for SAMOP criticality analysis [18,19].

The ^{99}Mo production calculation for 6 days operation (one batch) was done by *Cell Burn* as function of ^{235}U enrichment, and the result is described in Figure 5. It can be seen that the ^{99}Mo production value is in the range of 437.03 mCi, and increasing as uranium enrichment increases. This increase is due to the fact that ^{99}Mo is a fission product of ^{235}U, so that if the value ^{235}U in fuel increases, the ^{99}Mo will also increase. This can be seen from the ^{99}Mo formation scheme in the SRAC manual as seen in Figure 10. The average ^{99}Mo production calculation result is in line with the ^{99}Mo production forecast target for the first 300-480 mCi model for each production cycle of 6 days.

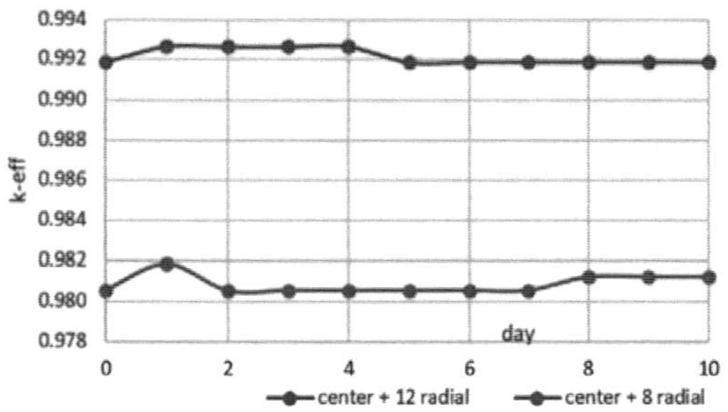

Figure 4. The sub-criticality levels per day of SAMOP experimental reactor operation

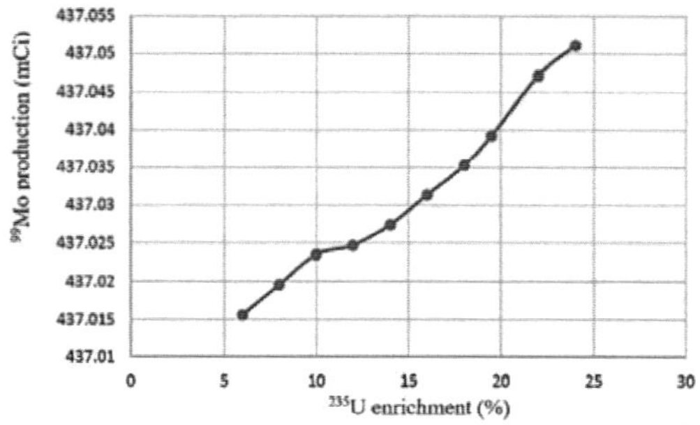

Figure 5. ^{99}Mo production for 6 days operation (per batch) as function of ^{235}U enrichment

The output from the *Cell Burn* is also obtained the ^{239}Pu isotope produced by SAMOP reactor which is very small such as described in Figure 6. It is shown in Figure 6 that ^{239}Pu production in a batch around 0.0207 – 0.031 mg, and the curve shows a decreasing trend in plutonium yields for the increase in ^{235}U enrichment. This is occurred because the increasing in enrichment means decreasing the amount of ^{238}U so that ^{239}Pu which is the result of transmutation of ^{238}U will also decrease.

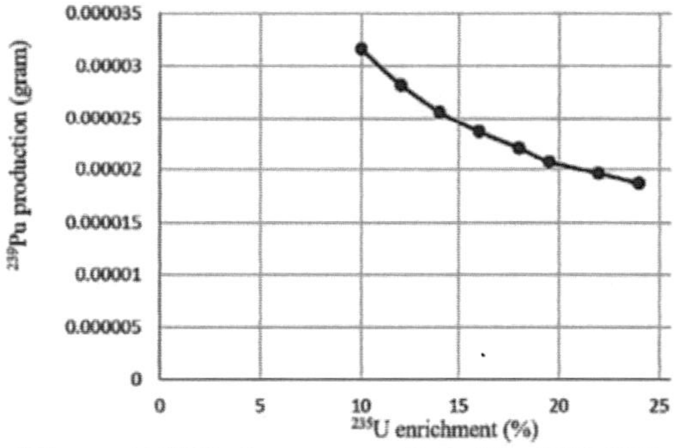

Figure 6. ^{239}Pu isotope produced by SAMOP reactor as function of ^{235}U enrichment

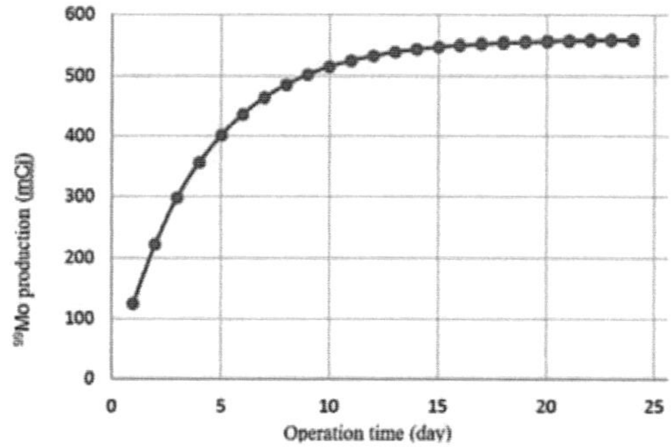

Figure 7. Production of ^{99}Mo as function of SAMOP operation time

The calculation of ^{99}Mo production is also conducted for operating time until 24 days, the result shows that the ^{99}Mo yield reaches saturation value on the 12th day such as depicted in Figure 7. It can be seen that the growth of ^{99}Mo on the first day to day 6 amounted to 437.03 mCi or 16170.11 MBq. While from day 6 to day 12 only amounted to 102.23 mCi so it can be concluded that 6-day operating cycle time is more effective. Therefore, the operation of the SAMOP reactor is determined for 6 days in consideration that 6-day operating time was considered more economical, this is in line with the commonly ^{99}Mo production [14,15].

5. Conclusion

The ^{99}Mo average production of SAMOP experimental facility is 437.03 mCi or 16170.11 MBq per batch (6 days operation). The amount of ^{99}Mo produced continues to increase until the 12th day of SAMOP operation, where the saturation of ^{99}Mo production is achieved. The SAMOP system operate at subcritical condition with the neutron multiplication factor of 0.981 to 0.992 with 30.2989 L to 31.4587 L uranyl nitrate in the reactor core.

Acknowledgement
The authors would like to thank the Director of Center for Accelerator Science and Technology Yogyakarta to Head of Reactor Division and all staffs for their support, and to Secretariat of INSINAS RISTEKDIKTI for budget support to this project (Project Code: InSinas RT-2016-0151).

References

[1] Maroor, R.A.P, A. Dash, & F.F. (Russ) Knapp Jr. Diversification of ^{99}Mo/^{99m}Tc Separation: non–fission reactor production of ^{99}Mo as a strategy for enhancing ^{99m}Tc availability. *Journal of Nuclear Medicine* 56:159-161 (2015).

[2] Mallinckrodt Pharmaceuticals. *Global ^{99}Mo supply chain.* 2017 http://www2.mallinckrodt.com/Nuclear_Imaging/Global_Mo99_Supply_Chain_Reactors.aspx

[3] Pillai, M.R.A., A. Dash & F.F. (Russ) Knapp Jr. 2013 *Sustained availability of ^{99m}Tc: Possible paths forward.* Journal of Nuclear Medicine 54: 313–323.

[4] Pitas, K. Piefer, G. R. *Next Generation Mo-99 Production: SHINE Update*, 2017 Mo-99 Topical Meeting on Molybdenum-99 Production Technology Development, Montreal, QC Canada, September 10-13 (2017). https://mo99.ne.anl.gov/2017/pdfs/

[5] Youker, A.J., S.D. Chemerisov, M. Kalensky, P. Tkac, D.L. Bowers & G.F. Vandegrift. 2013 *A solution-based approach for Mo-99 production: Considerations for nitrate versus sulfate media*. Science and Technology of Nuclear Installations 2013:1–10.

[6] Feder, T. 2015 *DOE: Let's move accelerator technologies to commercial markets*. Physics Today 68(7): 18.

[7] Pitas, K.M., G.R. Piefer, R.V. Bynum, E.N. Van Abel & J. Driscoll. 2013 *SHINE: technology and progress*. In: Proceedings of the ^{99}Mo Topical Meeting on Molybdenum-99 Technological Development. Mackie T.R. & R.F. Radel (Ed.), Chicago, III, USA.

[8] Syarip, Tegas Sutondo, Edi Trijono BS, Endang Susiantini. 2016 *Design & Development of Subcritical Assembly for ^{99}Mo Production (SAMOP)*. Paper presented at the 2nd Intl. Conf. on Science and Technology, Yogyakarta, Indonesia, ID PA-P-02.

[9] Tegas Sutondo and Syarip, 2014. *Beam Characteristics of Pierching and Tangential Beamports of Kartini Reactor*, "Ganendra" Journal of Nuclear Science and Technology, 17 2 83-90.

[10] Z. Gholamzadeh, S.A.H. Feghhi, S.M. Mirvakili, A. Joze-Vaziri, and M. Alizadeh. 2015 *Computational investigation of ^{99}Mo, ^{89}Sr, and ^{131}I production rates in a subcritical $UO_2(NO_3)_2$ aqueous solution reactor driven by a 30-MeV proton accelerator,* Nuclear Engineering Technology 47 875-883.

[11] ET Budisantoso, Syarip. 2002 *Studi Produksi Radioisotop Mo-99 Dengan Bahan Target Larutan Uranil Nitrat Pada Reaktor Kartini*, "Ganendra" Journal of Nuclear Science and Technology, 5 1 1-8.

[12] Artem V. Matyskin, Danas Ridikas, V.S.Skuridin, Johannes Sterba, Georg Steinhauser. 2012 *Feasibility study for production of 99mTc by neutron irradiation of MoO3 in a 250 kW TRIGA Mark II reactor*, J Radioanal Nucl Chem DOI 10.1007/s10967-012-2381-y

[13] Hummel, Andrew J. 2013 *Molybdenum-99 production in the Oregon State TRIGA reactor : analysis of the reactor design using a new LEU target as fuel*, PhD Dissertation, Oregon State University

[14] National Research Council (US) Committee on Medical Isotope Production Without Highly Enriched Uranium. 2009 *Medical Isotope Production without Highly Enriched Uranium*. National Academies Press, Washington DC.

[15] Ottinger, C. L., and E. D. Collins. 1996. Assessment of Potential ORNL Contributions to Supply of Mo-99. Oak Ridge Nat. Lab. Report No. ORNL/TM-13184. Oak Ridge, TN: ORNL.

[16] Okumura Keisuke dan Kugo Teruhiko. 2007 *SRAC 2006: A Comprehensive Neutronics Calculation Code System*. JAEA Technical Document, Ibaraki-ken.

[17] Syarip, E Togatorop and Yassar. 2018 *Molybdenum-99 production calculation analysis of*

SAMOP reactor based on thorium nitrate fuel. Journal of Physics: Conf. Series **978** 012072 doi :10.1088/1742-6596/978/1/012072

[18] Tegas Sutondo, Syarip, Slamet Santoso. 2009 *Safety design limits of main components of the proposed SAMOP system*, Proc. of the 3rd Asian Physics Symposium, Bandung, Indonesia.

[19] Setiadipura, T. & E. Saragi. 2007 *Neutronic aspect of subcritical assembly for Mo ^{99} production (SAMOP) reactor.* Proceedings of International Conference on Advances in Nuclear Science and Engineering in Conjunction with LKSTN, p. 23−26.

On The Markov Chain Monte Carlo Convergence Diagnostic of Bayesian Finite Mixture Model for Income Distribution

I Susanto[1,2], N Iriawan[3*], H Kuswanto[3], Suhartono[3], K Fithriasari[3], B S S Ulama[4], W Suryaningtyas[1] and A A Pravitasari[1]

[1] Doctoral Candidate at Department of Statistics, Institut Teknologi Sepuluh Nopember, Surabaya, Indonesia.

[2] Statistics, Faculty of Mathematics and Natural Sciences, Sebelas Maret University, Surakarta, Indonesia.

[3] Department of Statistics, Institut Teknologi Sepuluh Nopember, Surabaya,Indonesia.

[4] Department of Business Statistics, Institut Teknologi Sepuluh Nopember, Surabaya, Indonesia.

E-mail: nur_i@statistika.its.ac.id

Abstract. The assessment and comparison of income inequality and poverty can be supported by estimating the probability distribution of income. Income distributions which are typically heavy-tailed and positively skewed have been estimated both parametric and nonparametric approach. In parametric approach, finite mixtures distributions have been usefully implemented in the modelling of income distributions which has the multimodal characteristic. The Markov Chain Monte Carlo (MCMC) approach is one of the estimation methods which has a good performance in estimating the parameter of Bayesian finite mixture model. The convergence of the MCMC sampler to the posterior distribution is typically assessed using standard diagnostics methods, i.e., Gelman-Rubin method, Geweke method, Raftery-Lewis method and Heidelberger-Welch method. Those methods can give different results to conclude MCMC convergence condition. In this paper, a real sample income data from the Indonesian Family Life Survey (IFLS) 2015 and BidikMisi 2015 are employed to demonstrate the performance of diagnostics tools that assess convergence of the MCMC algorithm in estimating the parameter of Bayesian finite mixture models.

1. Introduction

Statistical modelling based on a fixed number distributions mixed together that are also known as finite mixtures modelling is an interesting research field with the considerable range of applications. Finite mixture model would able to capture some specific properties of real data such as its heterogeneity, multimodality, kurtosis, and skewness [1]. It means that a mixture distribution can represent as data pattern in data-driven analysis perspective [2]. Presently, finite mixture models are implemented in such varied areas as techno-biometrics, genetics, medicine, finance and economics including income distribution modelling.

In parametric approach, there are many probability distributions alternatives which can be used to estimate the income distribution model in such population. This approach is implemented as part of the analyzing process of the income distribution in a region related to the economic and poverty imbalances which occur in that population. The mixture distribution is relatively flexible for

Content from this work may be used under the terms of the Creative Commons Attribution 3.0 licence. Any further distribution of this work must maintain attribution to the author(s) and the title of the work, journal citation and DOI.
Published under licence by IOP Publishing Ltd

approaching the distribution of income with different sub-populations. Each sub-populations may reflect groups with economic homogeneity [3]. Finite mixture models has been assumed as a new additional flexible extension tools of classical statistical models, though the inference for these models poses a particular challenge on computational aspects [1].

Bayesian inference approach can overcome that computational problem, since it permits the complicated structure of a mixture model to be separated into a simpler structures [4]. However, their analytical parameter estimation of posterior distribution which can contain large integral dimension equation is usually difficult to be found. Thus a simulation method called Markov Chain Monte Carlo (MCMC) is suggested to solve numerically this problem. MCMC approach involves the simulation process which is performed iteratively Markov chains using Monte Carlo method to obtain convergence condition on posterior distribution. Implementation of MCMC in Bayesian analysis requires a proper sampling algorithm to obtain a sample of a distribution. Gibbs sampler is one of the algorithms that are frequently used as a generator of random variables in MCMC [5].

Although MCMC algorithms allow an advantageous computation, it can also encounter a conceivably serious weakness that relates with the convergence of parameter estimation process. Therefore, in Bayesian finite mixture modelling, it should use standard diagnostics tools that verify the convergence of estimated parameter in order to reach the target posterior inference for parameters. The assessment methods of convergence such as Gelman-Rubin method [6], Geweke method [7], Raftery-Lewis method [8] and Heidelberger-Welch method [9] can produce diverse results about the achievement of MCMC convergence condition.

2. Bayesian Finite Mixture Model

We explain finite mixture model in section 2.1 and Bayesian approach for finite mixture model is described in section 2.2.

2.1. Finite Mixture Model

A random variable vector **y** which has discrete or continuous type is said to be derived from a finite mixture distribution, if the probability density function $g(\mathbf{y})$ has a mixture density form that applies to all **y**,

$$g(\mathbf{y}) = w_1 f_1(\mathbf{y}) + \ldots + w_K f_K(\mathbf{y}) \tag{1}$$

where $f_k(\mathbf{y})$ is probability density function for $k = 1, 2, \ldots, K$. $f_k(\mathbf{y})$ is distribution function of k-th mixture component and K is the number of mixture components. $w_1, \ldots, w_K$ are weighting parameters and vector $\mathbf{w} = (w_1, \ldots, w_K)'$ is weighting vector of mixture distribution. The value of $\mathbf{w}$ must fulfil $0 \le w_k \le 1$ and $w_1 + \ldots + w_K = 1$.

In some cases of application, if it is assumed that all components of the finite mixture distribution come from a probability distribution having parameters $\boldsymbol{\theta}$, then equation (1) can be redefined as

$$g(\mathbf{y} \mid \mathbf{w}, \boldsymbol{\theta}) = w_1 f(\mathbf{y} \mid \boldsymbol{\theta}_1) + \ldots + w_K f(\mathbf{y} \mid \boldsymbol{\theta}_K) \tag{2}$$

with $\boldsymbol{\theta} = (\boldsymbol{\theta}_1, \ldots, \boldsymbol{\theta}_K)'$ [1]. A statistical model that implements the concept of finite mixture distribution in the modeling is usually called the finite mixture model.

2.2. Bayesian Approach

Let $\mathbf{y} = \{\mathbf{y}_1, \mathbf{y}_2, \ldots, \mathbf{y}_N\}$ are uncategorized observations from the mixture distribution that are randomly selected. Let $\boldsymbol{\Theta} = (\Theta_1, \Theta_2, \ldots, \Theta_d)' = (\boldsymbol{\theta}_1, \ldots, \boldsymbol{\theta}_K, \mathbf{w})'$ stand for all unknown parameters appearing in the mixture model. By using Bayes theorem, the posterior probability distribution $\pi(\boldsymbol{\Theta} \mid \mathbf{y})$ is proportional to the multiplication between the prior distribution of $\boldsymbol{\Theta}$, $p(\boldsymbol{\Theta})$, and the mixture likelihood, $f_L(\mathbf{y} \mid \boldsymbol{\Theta})$. It can be represented mathematically as follows,

$$\pi(\boldsymbol{\Theta}\mid\mathbf{y})=\frac{f_L(\mathbf{y}\mid\boldsymbol{\Theta})p(\boldsymbol{\Theta})}{f(\mathbf{y})} \propto f_L(\mathbf{y}\mid\boldsymbol{\Theta})p(\boldsymbol{\Theta}) \tag{3}$$

and the mixture likelihood function $f_L(\mathbf{y}\mid\boldsymbol{\Theta})$ takes the form,

$$f_L(\mathbf{y}\mid\boldsymbol{\Theta})=\prod_{i=1}^{N} g(\mathbf{y}_i\mid\boldsymbol{\Theta})=\prod_{i=1}^{N}\left(\sum_{k=1}^{K} w_k f(\mathbf{y}_i\mid\boldsymbol{\theta}_k)\right). \tag{4}$$

After determining prior distribution, the Bayesian approach will perform parameter estimation by integrating the posterior distribution. The integration process is approached by simulation procedure which is commonly known as MCMC method. When the Bayesian approach is implemented on finite mixture modelling, parameter estimation result does not fulfil the asymptotic normality analytically, and guarantee the valid inference in cases where regularity conditions are desecrated, such as when the data is limited and the proportion of mixtures component is too small [10].

2.2.1. Gibbs Sampler. As a method of generating samples, the Gibbs sampler has some advantages. Each step in generating sample, the random values are only considers to be generated from a univariate conditional distributions as a full conditional form of its joint posterior distribution of parameters. Based on [5], the algorithm can be summarized by the following steps:

1. Set an initial values for $\boldsymbol{\Theta}^{(0)}$.
2. For $t=1,2,...,T$ run the following steps iteratively:
   (i) $\boldsymbol{\Theta}=\boldsymbol{\Theta}^{(t-1)}$
   (ii) For $j=1,2,...,d$ update Θ_j using the full conditional distribution of $\Theta_j \sim \pi(\Theta_j\mid\boldsymbol{\Theta}_{\setminus j},\mathbf{y})$
   $\pi(\Theta_j\mid\boldsymbol{\Theta}_{\setminus j},\mathbf{y})$ is full conditional posterior distribution with $\boldsymbol{\Theta}_{\setminus j}=(\Theta_1,...,\Theta_{j-1},\Theta_{j+1},...,\Theta_d)'$.
   (iii) Assign $\boldsymbol{\Theta}^{(t)}=\boldsymbol{\Theta}$ and save it as the generated random samples $\boldsymbol{\Theta}^{(t-1)}$ as a result of $t+1$ step of iterations.

Therefore, by giving the chain $\boldsymbol{\Theta}^{(t)}$, the algorithm will generate the new parameter values as follows

$$\begin{array}{l}
\Theta_1^{(t)} \text{ from } \pi(\Theta_1\mid\Theta_2^{(t-1)},\Theta_3^{(t-1)},...,\Theta_d^{(t-1)},\mathbf{y})\\
\Theta_2^{(t)} \text{ from } \pi(\Theta_2\mid\Theta_1^{(t)},\Theta_3^{(t-1)},...,\Theta_d^{(t-1)},\mathbf{y})\\
\vdots\\
\Theta_j^{(t)} \text{ from } \pi(\Theta_j\mid\Theta_1^{(t)},\Theta_2^{(t)},...,\Theta_{j-1}^{(t)},\Theta_{j+1}^{(t-1)},...,\Theta_d^{(t-1)},\mathbf{y})\\
\vdots\\
\Theta_d^{(t)} \text{ from } \pi(\Theta_d\mid\Theta_1^{(t)},\Theta_2^{(t)},...,\Theta_{d-1}^{(t)},\mathbf{y})
\end{array}$$

The generating process, $\pi(\Theta_j\mid\boldsymbol{\Theta}_{\setminus j},\mathbf{y})=\pi(\Theta_j\mid\Theta_1^{(t)},\Theta_2^{(t)},...,\Theta_{j-1}^{(t)},\Theta_{j+1}^{(t-1)},...,\Theta_d^{(t-1)},\mathbf{y})$, is quite simple and it has a proportional relation as $\pi(\Theta_j\mid\boldsymbol{\Theta}_{\setminus j},\mathbf{y}) \propto \pi(\boldsymbol{\Theta}\mid\mathbf{y})$, where all the variables except Θ_j have constant values.

3. Markov Chain Monte Carlo Convergence Diagnostics

3.1. Gelman-Rubin

The Gelman-Rubin method approach is done by defining several Markov Chains with different initial values and comparing the variance between some Markov Chains with the variance in each Markov Chain. If there is m *Markov Chains* which are mutually independent and if each Markov chain has been taken a number of T iterations, $t = 1, 2, \ldots, T$, then MCMC convergence can be monitored through estimation of potential scale reduction factor (PSRF) formulated as follows

$$\hat{R} = \frac{\hat{V}(\Theta)}{avg} \tag{5}$$

where $\hat{V}(\Theta) = \frac{T-1}{T} avg + \frac{1}{T} B$ is the variance estimate of Θ, $avg = \frac{1}{m}\sum_{i=1}^{m} s_i^2$ with $s_i^2 = \frac{1}{T-1}\sum_{t=1}^{T}\left(\Theta_{it} - \overline{\Theta}_i\right)^2$ and $B = \frac{T}{m-1}\sum_{i=1}^{m}\left(\overline{\Theta}_i - \overline{\overline{\Theta}}\right)^2$ with $\overline{\overline{\Theta}} = \frac{1}{m}\sum_{i=1}^{m}\overline{\Theta}_i$. If the PSRF value is close to 1, then every m Markov Chains converge to the target distribution. Conversely, if the value of PSRF is large, it is necessary to take an extended stage of simulation that allows to reduce the value of $\hat{V}(\Theta)$ or to increase the value of *avg* such that m Markov Chains converge to target distribution [11]. The Gelman-Rubin method for one parameter is further developed by Brooks and Gelman for vector of $\boldsymbol{\Theta}$ and also known as the Brooks-Gelman-Rubin (BGR) method [12].

3.2. Geweke

Suppose that Θ_1 is a parameter of interest and $\Theta_1^{(1)}, \ldots, \Theta_1^{(T)}$ are related simulated values with subsamples A and B as the beginning and the end respectively. The diagnostic test computes

$$Z = \frac{\overline{\Theta}^B - \overline{\Theta}^A}{\sigma_{(\overline{\Theta}^B - \overline{\Theta}^A)}} \tag{6}$$

where $\overline{\Theta}^A$ and $\overline{\Theta}^B$ are means of subsamples and $\sigma_{(\overline{\Theta}^B - \overline{\Theta}^A)}$ is an estimated standard deviation of difference $\overline{\Theta}^B - \overline{\Theta}^A$. Considering Z asymptotically follows the standardized normal distribution, $Z \sim N(0,1)$, if $|Z| > 2$ then the chain is not convergent [5].

3.3. Raftery-Lewis

Suppose $N_{\min}$ as the minimum number of iterations that is needed to achieve the required estimation precision for some function of parameter Θ and the quantile of interest is q and s, then $N_{\min}$ is given by

$$N_{\min} = \left\{F^{-1}\left(\frac{s+1}{2}\right)\right\}^2 \frac{q(1-q)}{r^2} \tag{7}$$

where $F(\cdot)$ is the standard normal cumulative distribution function. The value of dependence factors, $I = N / N_{\min}$, can be used to indicate convergence condition of the chain. If the value is greater than 5, then it implies a convergence failure.

3.4. Heidelberger-Welch

The Heidelberger and Welch method which is applied for the analysis of single chains from univariate observations is developed based on Brownian bridge theory [9]. Suppose $\Theta^{(j)}$ is the j-th iterate in the chain, $f(0)$ is the spectral density of the chain evaluated at zero and $[nt]$ is the greatest integer less than or equal to nt. Let

$$B_n(t) = \frac{S_{[nt]} - [nt]\,\overline{\Theta}}{\left(n f(0)\right)^{1/2}}, \quad 0 \le t \le 1 \tag{8}$$

where $S_0 = 0$, $S_k = \sum_{j=1}^{k} \Theta^{(j)}$ $k \geq 1$, $\bar{\Theta} = \left(\sum_{j=1}^{n} \Theta^{(j)}\right) n^{-1}$, then $B_n = \{B_n(t),\ 0 \leq t \leq 1\}$ converges in distribution to the Brownian bridge as $n \to \infty$.

4. Application

In this section, real sample income data which determines from the Indonesian Family Life Survey (IFLS) 2015 and BidikMisi 2015 are employed to demonstrate the performance of diagnostics tools that assess convergence of the MCMC algorithm in finite mixture models. Computation of MCMC convergence with those diagnostic methods has been integrated in BUGS software: WinBUGS [13] or OpenBUGS [14]. Whereas computing through R software is done with Convergence Diagnostic and Output Analysis (CODA) package [15].

4.1. Indonesian Family Life Survey (IFLS) 2015.

Source of household income response data is processed based on survey data from Indonesian Family Life Survey (IFLS) 2015 which examines the life of household in Indonesia [16]. In this paper, we use household income data in Province of Daerah Istimewa Yogyakarta (DIY) with a sample of 690 households in five districts/cities: Kulonprogo Regency, Bantul Regency, Gunungkidul Regency, Sleman Regency and Yogyakarta City. The histogram kernel plot of the distribution is shown in Figure 1 below

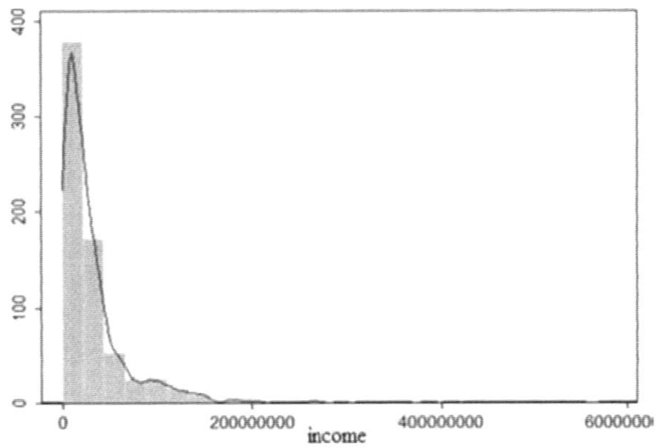

Figure 1 Kernel Histogram Plot Distribution of Household Income Data in Province of DIY.

Based on figure 1, it can be shown that income distribution indicates a multimodal characteristic that can be approached by finite mixture distribution. Using *Mathematica* 11 software, we preliminary presume that income distribution can be modelled by mixture model with three components of the two-parameter Gamma distribution, $Gamma(\alpha, \beta)$,which can be called as three components Gamma mixture,

$$g(\mathbf{y} \mid \mathbf{w}, \boldsymbol{\theta}) = \sum_{k=1}^{3} w_k f(\mathbf{y} \mid \boldsymbol{\theta}_k)$$

where $f(\mathbf{y} \mid \boldsymbol{\theta}_k) = f(\mathbf{y} \mid \alpha_k, \beta_k) = \beta_k^{\alpha_k} \left(\Gamma(\alpha_k)\right)^{-1} y^{\alpha_k - 1} e^{-\beta_k y}$ is two parameter Gamma distribution.

In that model, there are three parameters α, β, w which have to be estimated. Based on [17], the prior distributions which are implemented for each of parameters are

$$p(\mathbf{w}) \propto Dirichlet(\vartheta)$$
$$p(\alpha_k) \propto Exp(\theta)$$
$$p(\beta_k) \propto InverseGamma(\alpha_k, \beta_k)$$

In our research, we focus on estimation convergence of weighting parameters w_1, w_2, and w_3. In the first process, we generated 8,000 iterations which produced indicator values of CODA diagnostic that indicated some convergent problems. It was necessary to take an extended stage of iterations as stated on [11]. Therefore we performed further simulations,i.e.,100,000 iterations so that the effect of raised number iterations could be verified on MCMC convergence. The results of CODA diagnostic are described in Table 1 for case 1 with 8,000 iterations and for case 2 with 100,000 iterations.

Table 1. Indicator values of CODA diagnostics for weighting parameters w_1, w_2, w_3 .

Diagnostics	w_1		w_2		w_3	
	case 1	case 2	case 1	case 2	case 1	case 2
BGR	5.74	1.01	3.05	1.01	1.43	1.04
Geweke	-4.555	-1.260	5.653	1.56	0.116	-0.249
Raftery-Lewis	41.20	60.3	9.37	17.8	22.20	50.8
Heidelberger-Welch	failed	failed	failed	failed	passed	passed

Referring to Table 1, for BGR and Geweke methods, the increasing number of iterations can improve the convergence of MCMC. However, it does not occur for Raftery-Lewis and Heidelberger-Welch methods. In all cases and all parameters, Raftery-Lewis method gives unconvergent indicator values. If we regard from three diagnostics methods: BGR, Geweke and Heidelberger-Welch, it seems that only estimation parameter of w_3 which tends to converge. This problem is caused by a computation process that is trapped on one of the mixture components and the process cannot get out from that mixture component. Historical simulations with two different chains on Figure 2 presents that estimation process of w_3 tend to reach a convergent condition which does not be achieved by estimation processes of w_1 and w_2.

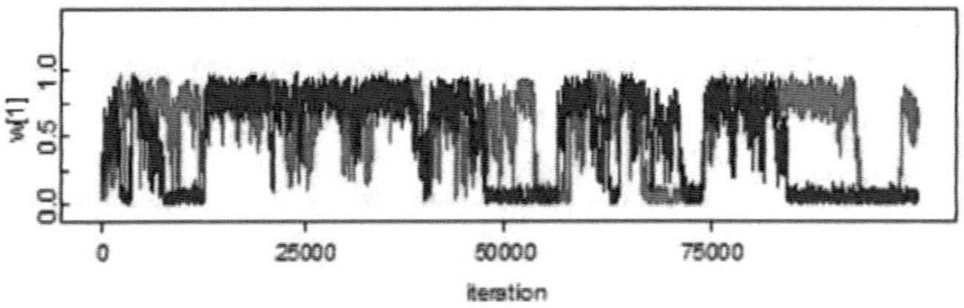

(a) w_1

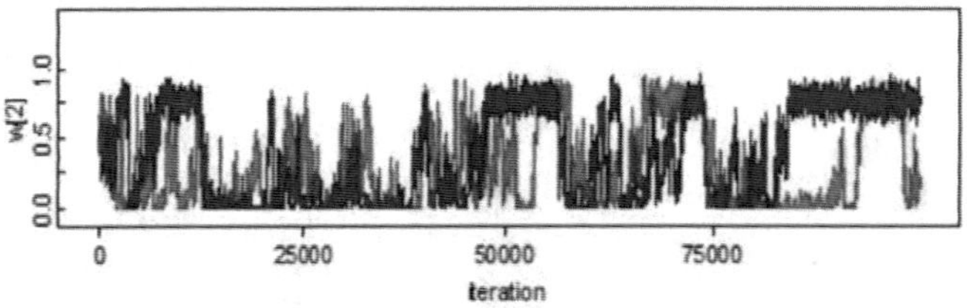

(b) w_2

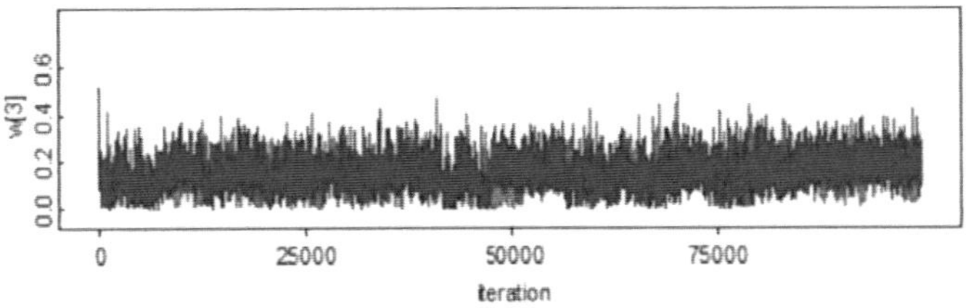

(c) w_3

Figure 2. Historical simulation of w_1, w_2, w_3 on 100,000 iterations.

Those convergence problems can happen, because there is no restriction condition on the mean of each component Gamma mixture, i.e., $\mu_1 < \mu_2 < \mu_3$, which can assure identifiability of component mixture. So re-parameterization of Gamma mixture model that implements restriction condition is done through BUGS program which is given in Appendix. The results of CODA diagnostics after re-parameterization are shown in Table 2.

Table 2. Indicator values of CODA diagnostics for weighting parameters w_1, w_2, and w_3

Diagnostics	w_1		w_2		w_3	
	case 1	case 2	case 1	case 2	case 1	case 2
BGR	1.02	1.00	1.00	1.01	1.02	1.00
Geweke	-0.9699	-0.7212	-0.5645	-0.5383	0.7292	0.2442
Raftery-Lewis	98.4	63	7.78	17	18.7	35.7
Heidelberger-Welch	passed	passed	passed	passed	passed	passed

table 2 shows that indicator values for BGR, Geweke and Heidelberger-Welch methods indicate the Markov chain reaches convergence condition for both cases. However, the convergence conditions are not fulfilled by the Raftery-Lewis method.

Another possible solution to avoid a trapped estimation process on one of the mixture components is to reduce the number of mixture components [10]. So, we simplified the mixture model from three component mixtures to two-component mixtures. The CODA results for two-component Gamma mixture are exposed on Table 3.

Table 3. Indicator values of CODA for two components Gamma mixtures.

Diagnostics	w_1		w_2	
	case 1	case 2	case 1	case 2
BGR	1.03	1	1.03	1
Geweke	0.764	1.115	-0.764	-1.115
Raftery-Lewis	16.8	26.5	9.78	18.5
Heidelberger-Welch	passed	passed	passed	passed

In regarding table 3, it can be seen that BGR, Geweke and Heidelberger-Welch methods give a similar conclusion about convergence parameters achieved. Nevertheless, Raftery-Lewis method still gives an opposite convergence result. For this reason, it has to be required further exploration which concerns about the implementation of Raftery-Lewis on finite mixture models.

4.2. BidikMisi 2015
Source of income response data is processed based on survey data from BidikMisi in 2015. In this paper, we used income data in Province of Daerah Istimewa Yogyakarta (DIY) with a sample of 1,149 households. In this paper, the income distribution is modeled by mixture model with three components of Normal distribution, $N(\mu,\sigma^2)$,

$$g(\mathbf{y}\mid\mathbf{w},\boldsymbol{\theta})=\sum_{k=1}^{3} w_k f(\mathbf{y}\mid\boldsymbol{\theta}_k)$$

where $f(\mathbf{y}\mid\boldsymbol{\theta}_k)=f(\mathbf{y}\mid\mu_k,\sigma_k^2)=\left(2\pi\sigma^2\right)^{-1/2} e^{-\frac{(x-\mu)^2}{2\sigma^2}}$ is Normal distribution. The prior distribution for weighting parameter is $p(\mathbf{w})\propto Dirichlet(\vartheta)$. Similar simulations as IFLS data are conducted for assessment process of MCMC convergence. The results of CODA diagnostic are presented in table 4.

Table 4. Indicator values of CODA diagnostics for weighting parameters w_1, w_2, and w_3 of Normal mixture.

Diagnostics	w_1		w_2		w_3	
	case 1	case 2	case 1	case 2	case 1	case 2
BGR	1.04	1	1.01	1	1.07	1
Geweke	-0.8569	-0.6666	-0.8146	0.4141	1.0277	0.0883
Raftery-Lewis	12.00	23.3	7.93	12.0	55.5	99.0
Heidelberger-Welch	failed	passed	failed	passed	passed	passed

Indicator values which are shown in table 4 give an affirmation that the increasing number of iterations can provide better indicator values of MCMC convergence. Even though, that condition only happens on BGR, Geweke and Heidelberger-Welch methods, while indicator value of convergence base on Raftery-Lewis method has not changed.

Historical simulations with two different chains on Figure 3 describe that estimation process of w_1, w_2, and w_3 tends to reach a convergent condition.

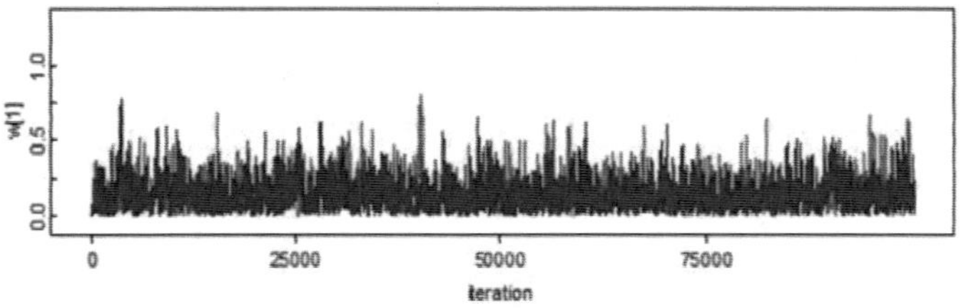

(a) w_1

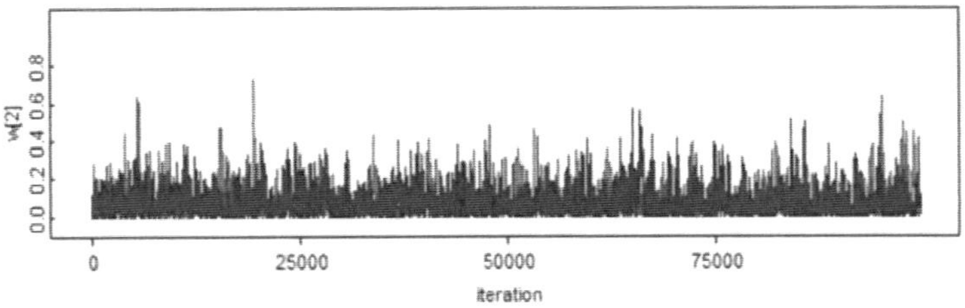

(b) w_2

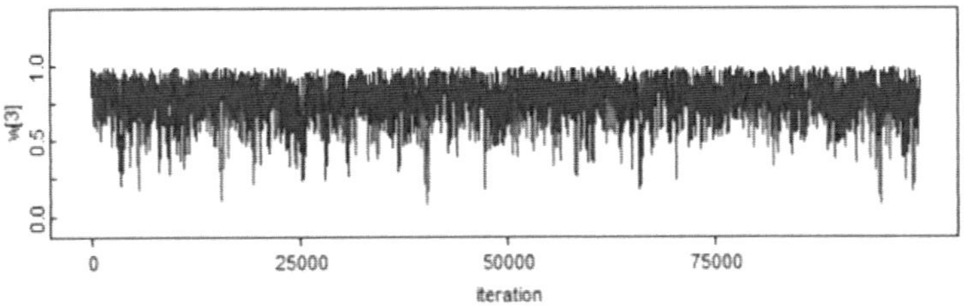

(c) w_3

Figure 3. Historical simulation of w_1, w_2, and w_3 on 100,000 iterations for Normal mixture model.

5. Conclusion

The increasing number of iterations can provide a better indicator value of MCMC convergence on diagnostics tools such as Gelman-Rubin method, Geweke method, and Heidelberger-Welch method. Nonetheless, indicator value of Raftery-Lewis method is not affected by the increment of iteration. Therefore, further researches that concern on Raftery-Lewis method for assessment convergence of Bayesian finite mixture modelling are required.

Acknowledgments

The Authors are grateful to Directorate for Research and Community Service (DRPM) Ministry of Research, Technology and Higher Education Indonesia which support this research under PUPT research grant no. 608/PKS/ITS/2017.

Appendix

```
model{
        for( i in 1 : N ) {
                y[i] ~ dgamma(alpha, beta[i])
                beta[i] <- alpha/mu[T[i]]
                T[i] ~ dcat(w[])
        }
        w[1:3] ~ ddirich(alp[])
        theta ~ dunif(0.0, 10000000)
        theta2 ~ dunif(0.0,10000000)
        mu[3] <- mu[2] + theta2
        mu[2] <- mu[1] + theta
        mu[1] <- 1/lamb
```

```
    lamb ~ dgamma(1.5, 1)
    alpha ~ dexp(1)
}
```

References

[1] Frühwirth-Schnatter S. Finite mixture and Markov switching models New York: Springer Science+Business Media; 2006.

[2] Iriawan N. Modelling and Data-Driven Analysis Surabaya: ITS Press; 2012.

[3] Pittau MG, Zelli R. Empirical evidence of income dynamics across EU regions. Journal of Applied Econometrics. 2006; 21(5): p. 605–628. DOI : 10.1002/jae.855.

[4] Marin JM, Mengersen K, Robert CP. Bayesian Modelling and Inference on Mixtures of Distributions. In Dey DK, Rao CR, editors. Handbook of Statistics Vol.25 : Bayesian Thinking Modeling and Computation. Amsterdam: Elsevier; 2005. p. 459-507.

[5] Ntzoufras I. Bayesian modeling using WinBUGS New york. DOI: 10.1002/9780470434567: John Wiley & Sons; 2009.

[6] Gelman A, Rubin D. Inference from Iterative Simulation using Multiple Sequences. Statistical Science. 1992; 7(4): p. 457 - 511. DOI : 10.1214/ss/1177011136.

[7] Geweke J. Evaluating the Accuracy of Sampling-Based Approaches to the Calculation of Posterior Moments. In Bernardo JM, Smith AFM, Dawid AP, Berger JO, editors. Bayesian Statistics 4. New York: Oxford University Press; 1992. p. 169 - 193. DOI: 10.1.1.27.2952.

[8] Raftery AE, Lewis SM. How Many Iterations in the Gibbs Sampler ? In Bernardo JM, Smith AFM, Dawid AP, Berger JO, editors. Bayesian Statistics 4. New York: Oxford University Press; 1992. p. 762 - 773.DOI : 10.1.1.41.6474.

[9] Heidelberger P, Welch PD. Simulation Run Length Control in the Presence of an Initial Transient. Operations Research. 1983; 31(6): p. 1109 - 1144. DOI: 10.1287/opre.31.6.1109.

[10] Tatarinova T, Schumitzky A. NonLinear Mixture Models: A Bayesian Approach London: Imperial College Press; 2015.

[11] Brooks SP, Roberts GO. Convergence assessment techniques for Markov Chain Monte Carlo. Statistics and Computing. 1998; 8(4): p. 319-335. DOI: 10.1023/A:1008820505350.

[12] Brooks SP, Gelman A. General Methods for Monitoring Convergence of Iterative Simulations. Journal of Computational and Graphical Statistics. 1998; 7(4): p. 434 - 455. DOI: 10.1080/10618600.1998.10474787.

[13] Lunn DJ, Thomas A, Best N, Spiegelhalter D. WinBUGS — a Bayesian modelling framework: concepts, structure, and extensibility. Statistics and Computing. 2000; 10: p. 325-337. DOI : 10.1023/A:1008929526011.

[14] Lunn D, Spiegelhalter D, Thomas A, Best N. The BUGS project: evolution, critique and future directions (with discussion). Statistics in Medicine. 2009; 28: p. 3049 - 3082. DOI: 10.1002/sim.3680.

[15] Plummer M, Best N, Cowles K, Vines K. CODA: Convergence diagnosis and output analysis for MCMC. R News. 2006; 6(1): p. 7-11.

[16] Strauss J, Witoelar F, Sikoki B. The Fifth Wave of the Indonesia Family Life Survey (IFLS5): Overview and Field Report. Working Paper. Santa Monica: RAND Corporation; 2016. Report No.: WR-1143/1-NIA/NICHD.

[17] Wiper M, Insua DR, Ruggeri F. Mixtures of Gamma Distributions with Applications. Journal of Computational and Graphical Statistics. 2001; 10: p. 440 - 454.

Numerical Method and Laboratory Experiment of RC Circuit using Raspberry Pi Microprocessor and Python Interface

W. S. Mada Sanjaya[12*], Dyah Anggraeni[12], Aceng Sambas[2,3] and Rena Denya[1]
[1]Department of Physics, Faculty of Science and Technology, Universitas Islam Negeri Sunan Gunung Djati Bandung, Indonesia
[2]Bolabot Techno Robotic Institute, CV. Sanjaya Star Group, Bandung, Indonesia
[3]Department of Mechanical Engineering, Universitas Muhammadiyah Tasikmalaya, Indonesia

*madasws@gmail.com

Abstract. RC circuit is the basic physics experiment show the charging and discharging on capacitor electronic component. In this paper, we have investigated the numerical simulation from the mathematical model of charging and discharging of capacitor, also we developed experimentally RC circuit using Raspberry Pi single board microprocessor and make interfaces using Python software. The benefit of this system that all of them was open Source and can help the physics experiment easily at the laboratory. To get data acquisition from RC circuit we used IC MCP3008 which have converted an analog to digital data. The experimental result showed the best agreement with the numerical simulation result.

1. Introduction

Physics experiment is an interesting field that always pushing advance in instrumentation and acquisition system. One of the physics experiment is investigating about charging and discharging of the capacitor through a resistor for the study of electricity [1].

RC Circuit can be modeled by Runge-Kutta Ordinary Differential Equations (ODEs). This numerical method can be used to obtain the approximation solution of RC Circuit. Some numerical method can be used to show ODEs system such as; Euler's method [2] and Runge-Kutta's method [3] of different orders.

The classical procedure of experiment is done using stopwatch and voltmeter, the procedure probably the human error occurs. Recently, Calin Galeriu, et al [4] and Pareria [5] have used Arduino microcontroller to investigate the RC Circuit. Other microcontroller systems which same work like Arduino is Raspberry Pi. Different from Arduino, the developer of Raspberry Pi build a mini computer processor (microprocessor/microcomputer) integrate with the control for the electronic project [6]. Some researches using Raspberry Pi to build an apparatus experiment laboratory, such as; robot (e.g., [7], [8]), motor dc [9], thermocouple [10], hydrostatic [11], pulse generator [12], astronomy [13], special relativity [14], and other.

In this research will be described a numerical simulation of RC Circuit using Runge-Kutta 4 Order (RK4) method compared with the exact solution. We will validate RC Circuit experimentally. The system in this research will use Raspberry Pi to investigate the charging and discharging process of RC Circuit. The purpose of this research is to build an effective low cost experimental physics of RC Circuit.

Content from this work may be used under the terms of the Creative Commons Attribution 3.0 licence. Any further distribution of this work must maintain attribution to the author(s) and the title of the work, journal citation and DOI.
Published under licence by IOP Publishing Ltd

IOP Conf. Series: Journal of Physics: Conf. Series **1090** (2018) 012015 doi:10.1088/1742-6596/1090/1/012015

The paper is organized as follows. In section 2, described a basic theory of RC Circuit. In section 3, describe a numerical simulation using Python software. In section 4, hardware design of RC Circuit described the works of the system in detail. In section 5, described the experimental results. Finally, in Section 6, the concluding remarks are given.

2. Basic Theory of RC Circuit

RC is a circuit which contains a series combination by a resistor (R) and a capacitor (C) component and connected to DC voltage [15]. RC circuit contains two processes they are "Charging" and "Discharging", will describe below:

2.1 Charging

In this case, assume that capacitor initially not have a charge. When the switch turns on at $t = 0$, the charge will flow into the resistor then fill the capacitor. In Kirchhoff's law we get;

$$\frac{dq}{dt} = \frac{1}{R}\left(v_0 - \frac{q}{C}\right), \tag{1}$$

where $\frac{dq}{dt}$ is the time rate of change of the charge, v_0 is the initial voltage, C is the capacitance, R is the resistance, and q is the charge. To determine the time rate of change of the voltage on the capacitor we can substitute $q = Cv$ to equation (1) can be written as;

$$\frac{dv}{dt} = \frac{1}{RC}(v_0 - v), \tag{2}$$

with $\frac{dv}{dt}$ the time rate of change of the voltage, v_0 is the initial voltage and v is the voltage. If analyze equation (2) we found the exact solution as;

$$v(t) = v_0\left(1 - e^{\frac{-t}{RC}}\right), \tag{3}$$

when $v(t)$ is the time rate of change of the voltage and t is the time of charging with range $a \le t \le b$. The process of charging capacitor shown as Figure 1.

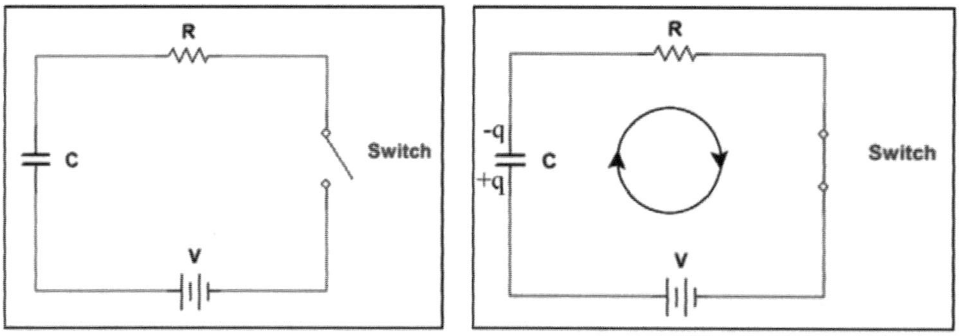

(a) (b)

Figure 1. RC Circuit Charging Process (a) when $t < 0$, (b) when $t > 0$.

2.2 Discharging

In this case, the capacitor charge initially $v_0 = q / C$, while the potential on the resistor is "zero". After $t = 0$, discharge began by the capacitor. Base on Kirchhoff's law is given by (4),

$$\frac{q}{C} + R\frac{dq}{dt} = 0, \tag{4}$$

where q is the charge, C is the capacitance, R is the resistance, and $\frac{dq}{dt}$ is the time rate of change of the charge. In discharging case, to determine the time rate of change of the voltage on the capacitor can substitute $q = Cv$ from equation (4) can be written as (5),

$$\frac{dv}{dt} = -\frac{v}{RC}, \quad (5)$$

with $\frac{dv}{dt}$ is the time rate of change of the voltage and v is the voltage. If analyze equation (5), will get the exact solution as;

$$v(t) = v_0 e^{\frac{-t}{RC}}, \quad (6)$$

where $v(t)$ is the time rate of change of the voltage, v_0 is the initial voltage and t is the time of discharging with range $a \leq t \leq b$. Discharging capacitor process shown in Figure 2.

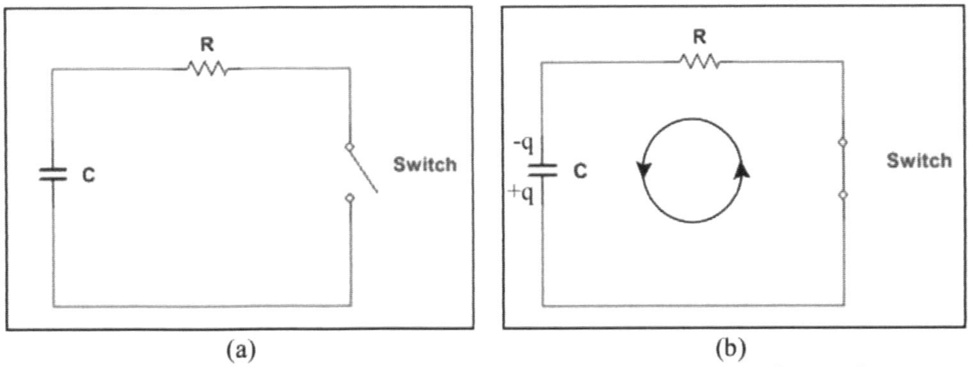

Figure 2. RC Circuit Discharging Process (a) when $t < 0$, (b) when $t > 0$.

3. Numerical Simulation

In this study, we use Runge-Kutta 4 Order Method (RK4) to simulate a Charging or Discharging of RC Circuit system, because of Runge-Kutta allow the differential solution with minimize the truncation error [3]. To calculate the numerical simulation of RC Circuit used Python software. With evaluation of equation (2), the RK4 solution function of charging process obtained;

$$\left.\begin{aligned} v_{i+1} &= v_i + \left(\frac{k_1+2k_2+2k_3+k_4}{6}\right) \\ k_1 &= h\left(\frac{1}{RC}(v_0 - v_i)\right) \\ k_2 &= h\left(\frac{1}{RC}\left(v_0 - \left(v_i + \frac{k_1}{2}\right)\right)\right) \\ k_3 &= h\left(\frac{1}{RC}\left(v_0 - \left(v_i + \frac{k_2}{2}\right)\right)\right) \\ k_4 &= h\left(\frac{1}{RC}\left(v_0 - \left(v_i + \frac{k_3}{2}\right)\right)\right) \end{aligned}\right\}, \quad (7)$$

with $h = \frac{b-a}{N}$ called step size (N is a positive integer). Thus, from equation (5) the discharging process have RK4 solution function show in (8).

$$\left.\begin{aligned} v_{i+1} &= v_i + \left(\frac{k_1+2k_2+2k_3+k_4}{6}\right) \\ k_1 &= h\left(\frac{v_i}{RC}\right) \\ k_2 &= h\left(\frac{\left(v_i+\frac{k_1}{2}\right)}{RC}\right) \\ k_3 &= h\left(\frac{\left(v_i+\frac{k_1}{2}\right)}{RC}\right) \\ k_4 &= h\left(\frac{\left(v_i+\frac{k_1}{2}\right)}{RC}\right) \end{aligned}\right\}, \qquad (8)$$

v_i+1 used to find the value of voltage v increased. The process will be iterated for N - 1 times. From equation (7) and (8) we set h = 0.1, R = 1 kΩ, C = 1 mF and v_0 = 3.3 volt. Last, we will obtain the approximation value for the solution of RC Circuit system.

After calculating the numerical simulation, the solution will be plotted using Matplot library for Python software. Then, we can investigate the graphic form of charging and discharging process of RC Circuit. From the algorithm obtain graphics as Figure 3.

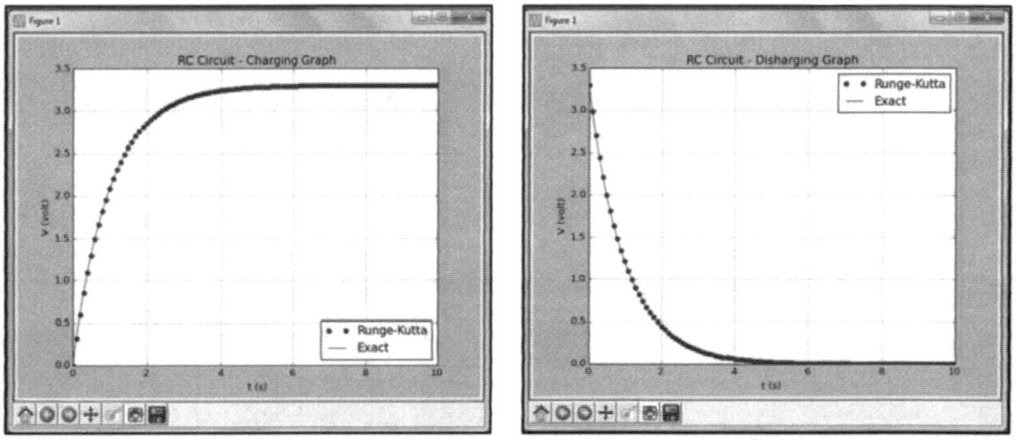

(a) (b)

Figure 3. Numerical Simulation of RC Circuit (a) Charging, (b) Discharging.

Figure 3 shows the plot of capacitor voltage when the RC Circuit charging and discharging. On Figure 3 (a) shows the voltage v across the capacitor voltage (initially not have a charge) will rise exponentially. Thus, Figure 3 (b) show initially capacitor fully charged, then the capacitor discharge through the resistor and the capacitor voltage will fall exponentially with time. The rate of increasing and decreasing of the voltage with time will be dependent upon the product of capacitance and resistance called time constant. Figure 3 shows the best agreement between theoretical solution (3) and (6) with the numerical solution (7) and (8).

4. Hardware Design of RC Circuit

In this section will describe the hardware design of RC Circuit shown in Figure 4. The basic RC Circuit hardware design in this research consist of basic RC Circuit, ADC, and Raspberry Pi model type B+. When the switch turns on (charging process), the charge will flow into the resistor then fill the capacitor.

And when switch turn off (discharging process), the charge on the capacitor will decrease. Both of the processes will obtain the voltages data. To read the real-time of voltage data, Raspberry Pi need ADC (Analog to Digital Conversion) using IC MCP3008. The voltage data will be processed on Raspberry Pi using Python software. Then, the algorithm will show the result as real-time plotting interface.

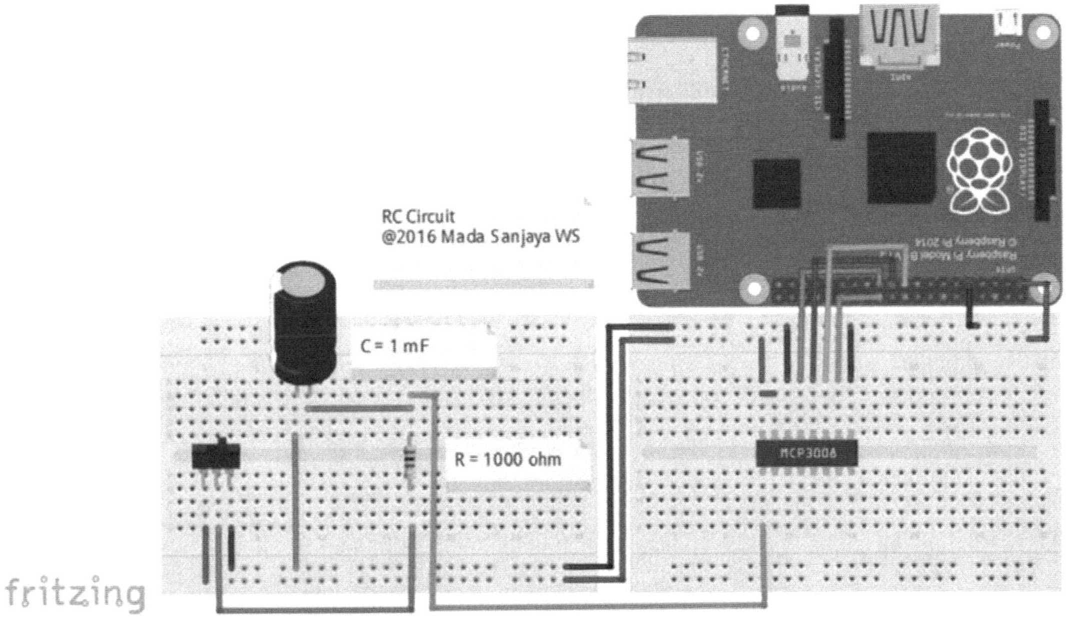

Figure 4. RC Circuit Scheme using Fritzing.

5. Experimental Result

From Figure 5 shown the hardware realization of the RC Circuit system based Raspberry Pi. This hardware realization consists by basic RC Circuit, ADC using IC-MCP3008, Raspberry Pi model type B+, 7-inch mini LCD (for display real-time plotting graphic of charging and discharging RC Circuit) and mini keyboard (for control). In this research, we have set a fully integrated Raspberry Pi based system for measuring capacitance with data visualization and processing.

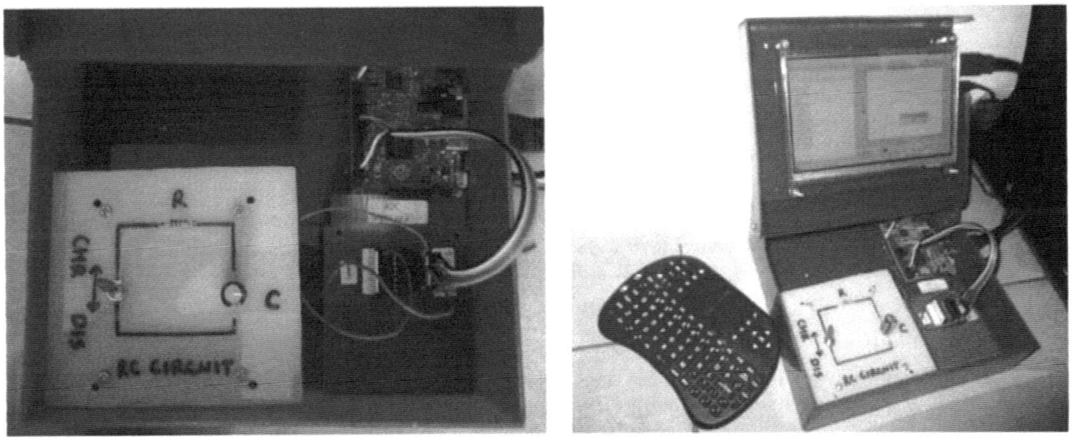

(a) (b)

Figure 5. Realization of RC Circuit (a) RC Circuit and Raspberry Pi System, (b) System Design.

We obtain the experimental observation of charging and discharging RC Circuit system (the RC component configuration same as the simulation) shown in Figure 6. As compared to Figure 3 and

Figure 6 shows the best agreement between theoretical solution, numerical simulation and the realization circuit is confirmed.

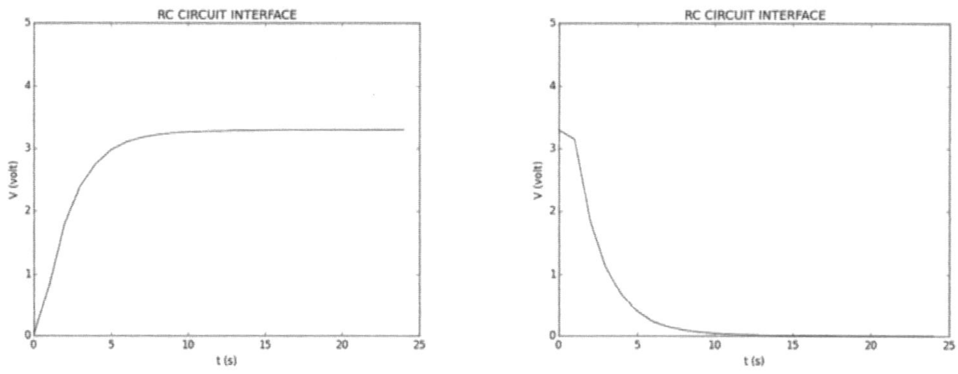

(a) (b)

Figure 6. Real time RC Circuit Graphic Result (a) Charging, (b) Discharging.

6. Conclusions

In this research, we have studied the RC circuit system. This research shows that the RK4 method successfully can solve RC system, and showing the best agreement if compared with the theoretical solution. Using a Raspberry Pi, we have verified experimentally, with an excellent quantitative agreement, the mathematical model and the numerical simulation that describes the charging or discharging of the capacitor. Raspberry Pi allows us to build a real-time data acquisition and can be used for data visualization and processing. All system can be integrated become mini experiment kit of RC Circuit. Finally, we trust that Raspberry Pi platform is an excellent hardware for low-cost open-source laboratory instrumentation of physics education and another field.

Acknowledgment

The authors would like gratefully acknowledge the financial support from LPM2M UIN Sunan Gunung Djati Bandung, Indonesia.

References

[1] Serway R A and Jewett J W 2004 *Physics for Scientists and Engineers Sixth Edition* (Thomson) p 60-63

[2] Roslan U A M, Salleh Z and Kihcman A 2010 Solving Zhou's Chaotic System Using Euler's Method *Thai Journal of Mathematic* **8 2** 299-309

[3] Roslan U A M, Shalleh Z and Kilicman A 2013 Solving Zhou Chaotic System Using Fourth-Order Runge-Kutta Method *World Applied Sciences Journal* **21 6** 939-944

[4] Galeriu C, Letson C and Esper G 2015 An Arduino Investigation of the RC Circuit *The Physic Teacher* **53** 285-288

[5] Pereira N S A 2016 Measuring The RC Time Constant with Arduino *IOP Publishing* 1-10

[6] Tooley M 2015 *Electronic Circuits: Fundamentals and Applications Fourth Edition* (New York: Taylor & Francis) p 337

[7] Vanitha M, Selvalakshmi M and Selvarasu R 2016 Monitoring and Controlling of Mobile Robot Via Internet Through Raspberry PI Board *Second International Conference on Science Technology Engineering and Management* 462-466

[8] Siagian P and Shinoda K 2015 Web Based Monitoring and Control of Robotic Arm Using Raspberry Pi *International Conference on Science in Information Technology* 192-196

IOP Conf. Series: Journal of Physics: Conf. Series **1090** (2018) 012015 doi:10.1088/1742-6596/1090/1/012015

[9] Reck R M and Sreenivas R S 2016 Developing a New Affordable DC Motor Laboratory Kit for an Existing Undergraduate Controls Course *American Control Conference* 2801-2808

[10] Arnold F, DeMallie I, Florence L and Kas D O 2015 Method for Collecting Thermocouple Data Via Secured Shell Over A Wireless Local Area Network In Real Time *Review Of Scientific Instruments* **86** 96-100

[11] Balula S, Henriques R, Fortunato J, Pireira T, Borges H, Amarante-Segundo G and Fernandes H 2015 Distributed E-Lab Setup Based On The Raspberry Pi: The Hydrostatic Experiment Case Study *IEEE* 282-285

[12] Fedrizzi M and Soria J 2015 Application of A Single-Board Computer As A Low-Cost Pulse Generator *IOP Publishing* **25** 1-4

[13] Maulana F, Soegijoko W and Yamani A 2016 Utilising Raspberry Pi as A Cheap And Easy Do It Yourself Streaming Device for Astronomy *IOP Publishing* 1-4

[14] Singh P and Hedgeland H 2015 Special Relativity In The School Laboratory: A Simple Apparatus For Cosmic-Ray Muon Detection *IOP Publishing* 317-323

[15] Sanjaya W S M 2015 *Metode Numerik Berbasis Python* (Yogyakarta, Indonesia: Penerbit Gava Media) p 209-214

IOP Conf. Series: Journal of Physics: Conf. Series **1090** (2018) 012016 doi:10.1088/1742-6596/1090/1/012016

Finite Difference Methods for Simulation of Water Waves Generated by Moving Topography

Sudi Mungkasi[1] and Nikenasih Binatari[2]

[1] Department of Mathematics, Faculty of Science and Technology, Sanata Dharma University, Yogyakarta, Indonesia
[2] Department of Mathematics Education, Faculty of Mathematics and Natural Science, Yogyakarta State University, Yogyakarta, Indonesia

E-mail: sudi@usd.ac.id, nikenasih@uny.ac.id

Abstract. In this paper, we consider water waves excited by moving topography. This problem usually occurs in the tsunami generation and propagation. We use the shallow water wave model governing this wave generation and propagation problem. Two finite difference methods are tested to solve the problem. The first is the forward in time centred in space finite difference method. The second is the Adams-Bashforth--Adams-Moulton method. We obtain that the second gives more flexibility in terms of the choice of time step value.

1. Introduction

Water waves in the ocean can be generated by a number of sources, such as wind, lunar influence, ship movement, bottom motion (moving topography), etc. This paper considers water waves that are generated by moving topography. The moving topography can initiate water to move, which is called the wave generation, and it is then followed by the wave propagation.

Water surface waves generated by moving topography have been an interest of a number of authors, as these can be used to simulate tsunami generation and propagation. Tinti, et al. [1] studied tsunami generation by submarine slides using the shallow-water theory. Didenkulova, et al. [2] derived some analytical solutions for a basin of variable depth relating to tsunami waves generated by submarine landslides. Jamin et al. [3] conducted experiments on the generation of surface waves by a moving topography. Case studies on this topic have also been conducted, such as by Grilli, et al. [4] regarding the coastal tsunami hazard model from submarine mass failures. These previous studies indicate that a reliable numerical method is needed in the simulation of surface wave generation and propagation.

In order to fulfil the mentioned need, this paper investigates the performance of popular finite difference methods used to solve the generation and propagation problems of surface waves due to moving topography. Two numerical methods are presented and their performances are discussed. We take the forward in time centred in space (FTCS) and the Adams-Bashforth--Adams-Moulton (ABAM) methods.

The remainder of this paper is organised simply as follows. Section 2 provides the mathematical model governing water waves. Section 3 contains the numerical methods under consideration for solving the mathematical model. Numerical results are given in Section 4. Finally, we draw some concluding remarks in Section 5.

Content from this work may be used under the terms of the Creative Commons Attribution 3.0 licence. Any further distribution of this work must maintain attribution to the author(s) and the title of the work, journal citation and DOI.
Published under licence by IOP Publishing Ltd

2. Mathematical model

The mathematical model for water waves generated by moving topography is [1]:

$$\eta_t + [uH]_x = -h_t \,, \tag{1}$$

$$u_t + g\eta_x = 0\,. \tag{2}$$

Here t is the time variable, x is the space variable, $\eta = \eta(x,t)$ is the level of water surface, $u = u(x,t)$ is water velocity, $h = h(x,t)$ is the bed topography, and g denotes the acceleration due to gravity. In addition, $H(x,t) = \eta(x,t) + h(x,t)$. The bed topography is assumed to be impermeable, but it depends on variables x and t. An illustration of water waves and their variables is given in Figure 1.

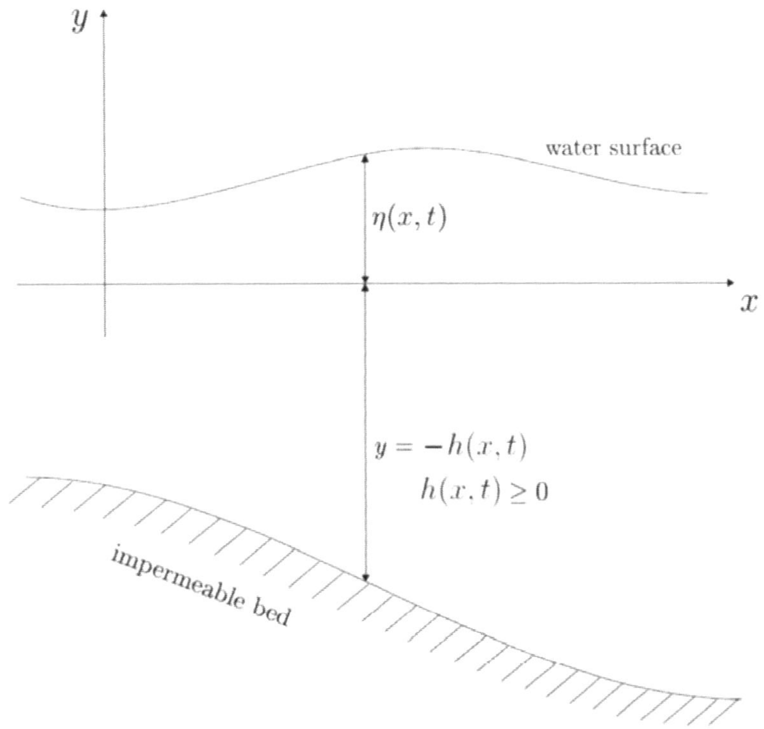

Figure 1. Illustration of water waves and their variables.

For smooth solutions, the model (1) and (2) is actually a rewriting of using the potential function relation to the velocity $u = \partial\phi/\partial x$, where ϕ is the potential function. This simplifies equation (2), so equations (1) and (2) can also be written as

$$\eta_t + [H\phi_x]_x = -h_t \,, \tag{3}$$

$$\phi_t + g\eta = 0\,. \tag{4}$$

In this paper, we solve equations (3) and (4) using finite difference methods. The methods are presented in the next section.

3. Finite difference methods

We present two numerical finite difference methods. The first is the forward in time centred in space (FTCS) finite difference method. The second is the Adams-Bashforth--Adams-Moulton (ABAM) finite difference method.

Let us assume that the space domain $a \le x \le b$ is discretised using a uniform spatial step size Δx. Then we have spatial points $x_i = a + i\Delta x$, where $i = 0, 1, 2, \ldots$. In addition, we assume that the time domain $t \ge 0$ is discretised using a uniform temporal step size Δt. Then we have temporal points $t^n = n\Delta t$, where $n = 0, 1, 2, \ldots$. Using the finite difference framework, we use $\eta_i^n \approx \eta(x_i, t^n)$, $H_i^n \approx H(x_i, t^n)$, $h_i^n \approx h(x_i, t^n)$, $\phi_i^n \approx \phi(x_i, t^n)$, etc.

The FTCS scheme being considered is a two-step method. The first step is

$$\eta_i^* = \eta_i^n - \frac{\Delta t}{4\Delta x^2}[H_{i+1}^n(\phi_{i+2}^n - \phi_i^n) - H_{i-1}^n(\phi_i^n - \phi_{i-2}^n)] - \frac{1}{2}(h_i^{n+1} - h_i^{n-1}), \tag{5}$$

$$\phi_i^* = \phi_i^n - \Delta t\, g\, \eta_i^n. \tag{6}$$

The second step is

$$\eta_i^{**} = \eta_i^* - \frac{\Delta t}{4\Delta x^2}[H_{i+1}^*(\phi_{i+2}^* - \phi_i^*) - H_{i-1}^*(\phi_i^* - \phi_{i-2}^*)] - \frac{1}{2}(h_i^{n+1} - h_i^{n-1}), \tag{7}$$

$$\phi_i^{**} = \phi_i^* - \Delta t\, g\, \eta_i^*. \tag{8}$$

After those two steps, we take the average for the values at time t^{n+1}, as follows

$$\eta_i^{n+1} = \frac{1}{2}(\eta_i^{**} + \eta_i^n), \tag{9}$$

$$\phi_i^{n+1} = \frac{1}{2}(\phi_i^{**} + \phi_i^n). \tag{10}$$

The ABAM method consists of two steps too [5-6]. The Adams-Bashforth step is

$$\eta_i^* = \eta_i^n + \frac{\Delta t}{12}(23F_i^n - 16F_i^{n-1} + 5F_i^{n-2}), \tag{11}$$

$$\phi_i^* = \phi_i^n + \frac{\Delta t}{12}(23G_i^n - 16G_i^{n-1} + 5G_i^{n-2}). \tag{12}$$

Here

$$F_i^n = -\frac{1}{4\Delta x^2}[H_{i+1}^n(\phi_{i+2}^n - \phi_i^n) - H_{i-1}^n(\phi_i^n - \phi_{i-2}^n)] - \frac{1}{2\Delta t}(h_i^{n+1} - h_i^{n-1}), \tag{13}$$

$$G_i^n = -g\eta_i^n. \tag{14}$$

Using the predicted values η_i^*, ϕ_i^*, we evaluate F_i^{n+1} and G_i^{n+1}. Then, we use the Adams-Moulton step, as follows

$$\eta_i^{**} = \eta_i^* + \frac{\Delta t}{12}(5F_i^{n+1} + 8F_i^n - F_i^{n-1}), \tag{15}$$

$$\phi_i^{**} = \phi_i^* + \frac{\Delta t}{12}(5G_i^{n+1} + 8G_i^n - G_i^{n-1}). \tag{16}$$

After those two steps, we take the average for the values at time t^{n+1}, as follows

$$\eta_i^{n+1} = \frac{1}{2}(\eta_i^{**} + \eta_i^n), \tag{17}$$

$$\phi_i^{n+1} = \frac{1}{2}(\phi_i^{**} + \phi_i^n). \tag{18}$$

We use these FTCS and ABAM methods to simulate water waves generated by the motion of water bed topography. We present our simulations in the next section.

4. Numerical results

We present our numerical results and discussion. When quantities are not written with their units, we assume that they have SI units with the MKS system.

In our simulations, we assume that initially water surface and its bed topography are horizontal. At time $t = 0$, water is still and its depth is $H_0 = 4000$ everywhere. We consider the space domain $|x| \leq$ 80,000. The acceleration due to gravity is $g = 9.81$ ms^{-2}. Suppose that during the first T =10 seconds, the bottom topography is perturbed, so the water depth is defined as

$$H(x,t) = H_0 - f(x)S(t) \tag{19}$$

where

$$f(x) = \begin{cases} 1 & \text{if} \quad |x| \leq 10{,}000 \\ 0 & \text{if} \quad |x| > 10{,}000 \end{cases} \tag{20}$$

and

$$S(t) = \begin{cases} 0.5(1-\cos(\pi t/T)) & \text{if} \quad t \leq T \\ 1 & \text{if} \quad t > T. \end{cases} \tag{21}$$

The space domain is discretised using the spatial step size $\Delta x = 25$. The time domain is discretised using the temporal step sizes $\Delta t = \Delta x/1000, \Delta x/500$, and $\Delta x/250$. We will investigate which numerical method is more flexible with respect to the temporal step size.

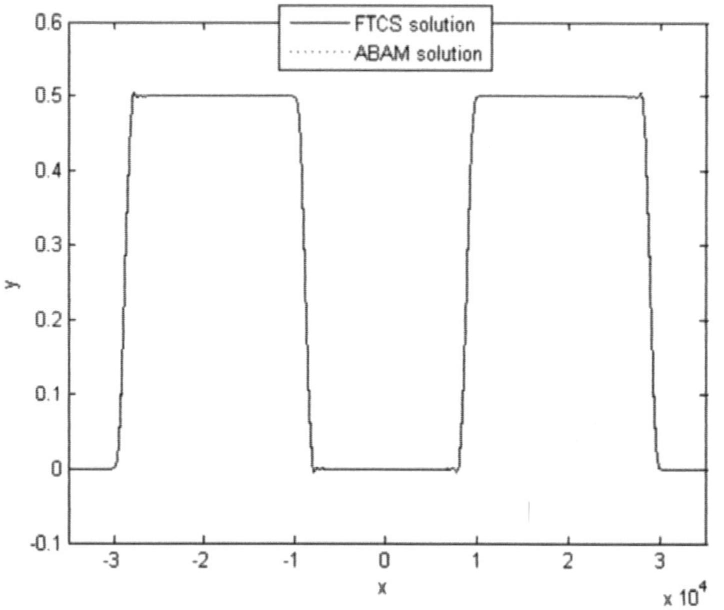

Figure 2. Numerical solution at time $t = 100$ using $\Delta t = \Delta x / 1000$.

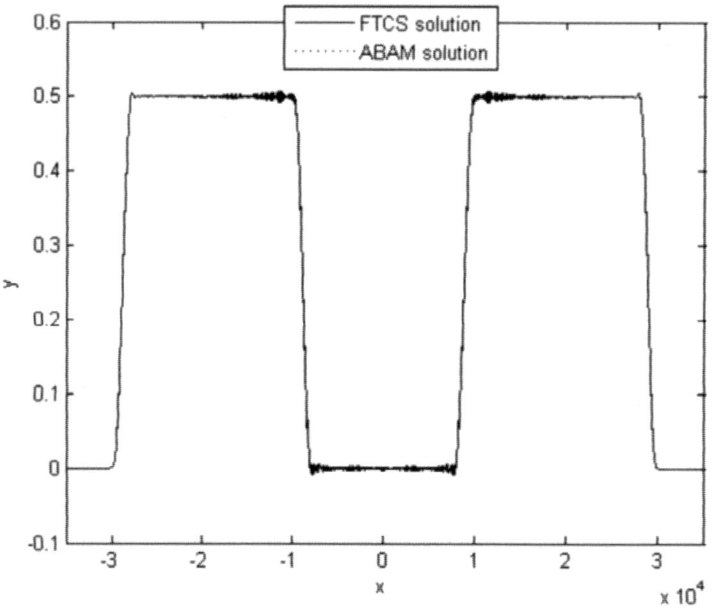

Figure 3. Numerical solution at time $t = 100$ using $\Delta t = \Delta x / 500$.

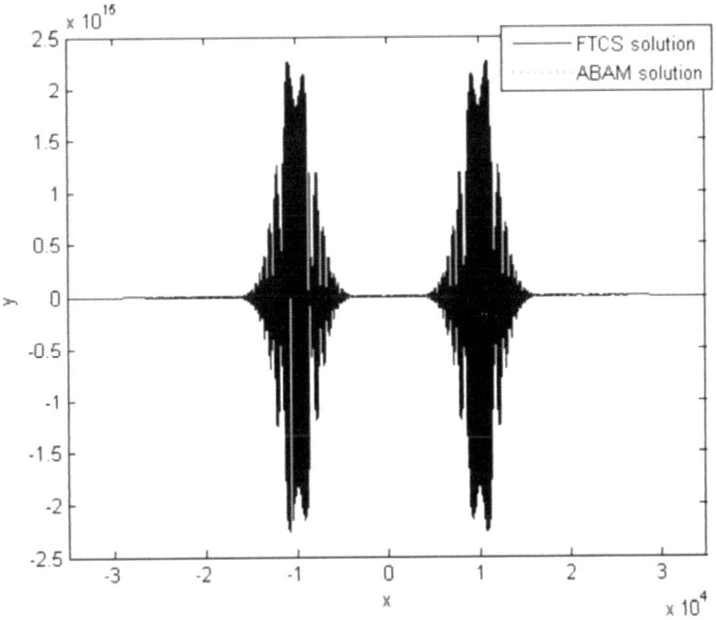

Figure 4. Numerical solution at time $t = 100$ using $\Delta t = \Delta x / 250$.

Numerical simulations are stopped for time $t = 100$. Figures 2-4 show the numerical solution for the water surface at time $t = 100$ using the temporal step sizes $\Delta t = \Delta x / 1000, \Delta x / 500$, and $\Delta x / 250$ respectively. In Figure 2, both the FTCS and ABAM methods produce the same results (with negligible discrepancies). Figure 3 shows that while the ABAM method gives a very much similar to the results in Figure 2, the FTCS method starts producing artificial oscillations. Completely incorrect results are generated by the FTCS method when the time step (the temporal step size) is taken smaller, as shown in Figure 4. In Figure 4, we observe that when the time step is $\Delta t = \Delta x / 250$, the FCTS is unstable, whereas the ABAM method is still able to solve the problem.

5. Conclusion

We have investigated the performance of the forward in time centred in space (FTCS) and Adams-Bashforth--Adams-Moulton (ABAM) methods in solving problems of water waves generated by moving topography. Both methods are in the finite difference family. We obtain that the ABAM method gives more flexibility in terms of the time step choice. The FTCS method is still able to solve the problems, but the time step should be taken smaller than that for the ABAM method.

Acknowledgements
This work of Sudi Mungkasi was financially supported by Sanata Dharma University. Sudi Mungkasi thanks Sanata Dharma University for the financial support.

References

[1] Tinti S, Bortolucci E and Chiavettieri C 2001 Tsunami excitation by submarine slides in shallow-water approximation *Pure and Applied Geophysics* **158** 759

[2] Didenkulova I, Nikolkina I, Pelinovsky E and Zahibo N 2010 Tsunami waves generated by submarine landslides of variable volume: analytical solutions for a basin of variable depth *Natural Hazards and Earth System Sciences* **10** 2407

[3] Jamin T, Gordillo L, Ruiz-Chavarría G, Berhanu M, and Falcon E 2015 Experiments on generation of surface waves by an underwater moving bottom *Proceedings of the Royal*

Society A **471** 20150069

[4] Grilli S T, Shelby M, Kimmoun O, Dupont G, Nicolsky D, Ma G, Kirby J T and Shi F 2017 Modeling coastal tsunami hazard from submarine mass failures: effect of slide rheology, experimental validation, and case studies off the US East Coast *Natural Hazards* **86** 353

[5] Wiryanto L H 2005 Numerical solution of Boussinesq equations as a model of interfacial-wave propagation *Bulletin of the Malaysian Mathematical Sciences Society* **28** 163

[6] Mungkasi S 2015 Numerical solutions to unsteady wave problems over porous media *IndoMS Journal on Industrial and Applied Mathematics* **2** 33

IOP Conf. Series: Journal of Physics: Conf. Series **1090** (2018) 012017 doi:10.1088/1742-6596/1090/1/012017

Approximate Solution of Riccati Differential Equations and DNA Repair Model with Adomian Decomposition Method

Rahmat Al Kafi[1], Bariqi Abdillah[2], Sri Mardiyati[1]

[1]Department of Mathematics, Universitas Indonesia, Depok, Indonesia
[2]Faculty of Computer Science, Universitas Indonesia, Depok, Indonesia

E-mail: rahmat.alkafi@sci.ui.ac.id

Abstract. This paper investigates the approximate solution of Riccati differential equations, and DNA repair model which is formed by Riccati equation. The existence result of the solution is obtained by Adomian decomposition method (ADM). Some initial value problems are solved by the method to demonstrate the main results. The obtained results are compared with the Runge-Kutta method to verify the validity of the ADM. It shows that the ADM has better results than the Runge-Kutta method. This results confirm that the ADM is a suitable method for approaching the exact solution of Riccati differential equations and the proposed model.

1. Introduction

Riccati differential equation, formulated by the Italian mathematician Jacopo Francesco Riccati, is a first-order nonlinear ordinary differential equations that arise in different areas of mathematics and physics. Moreover Riccati differential equation appears on the DNA repair problem, which the solution of this equation interpreted the average number of DNA double-strand breaks (DSBs) per cell at a certain time [1]. Analytically, some forms of this differential equation are difficult to solve exactly and explicitly by elementary methods. However there is a good numerical method that can be used to obtain the approximate solution of this equation, i.e. Adomian Decomposition Method (ADM), presented by George Adomian in 1988 [2]. The method has been examined to solve some class of linear and nonlinear equations which the obtained solution rapidly closed to the exact solution, as in [3-5]. In this paper, we will discuss the use of ADM to approximate the exact solution of Riccati differential equation and the DNA repair model, and perform some numerical simulations.

2. Analysis of Method

2.1. Adomian Decomposition Method

ADM is a method to resolve various types of nonlinear differential equations without requiring a linearization or prerequisite nonlinearity [2]. ADM generates a solution in the form of series whose terms are determined by a recursive relation of the Adomian polynomial. The ADM will be described by following the procedures [2]. To introduce this method, we consider the operator equation $F[u(t)] = g(t)$, where F represents a general nonlinear ordinary differential operator and g is a given function. The linear part of F is decomposed into $L + R$, where L is an invertible operator and R is the reminder of F. Thus the equation can be written as

Content from this work may be used under the terms of the Creative Commons Attribution 3.0 licence. Any further distribution of this work must maintain attribution to the author(s) and the title of the work, journal citation and DOI.
Published under licence by IOP Publishing Ltd

$$(L+R+N)u = g$$

or

$$Lu = g - Ru - Nu \tag{1}$$

where N is a nonlinear part of F.

Applying the operator L^{-1} formally to the Eq. (1), we obtain $u = L^{-1}g - L^{-1}Ru - L^{-1}Nu$. Suppose h is the solution of the homogeneous equation $Lu = 0$, with the given initial/boundary conditions. Then the general solution of Eq. (1) is

$$u = h + L^{-1}g - L^{-1}Ru - L^{-1}Nu \tag{2}$$

The next problem is the decomposition of the nonlinear term Nu. Adomian developed a very elegant technique, such that the approximate solution of Eq. (2) can be represented as

$$u = u_0 + \lambda u_1 + \lambda^2 u_2 + \ldots = \sum_{n=0}^{\infty} \lambda^n u_n \tag{3}$$

with λ is constant whereas $u_0, u_1, u_2, \ldots, u_n$ are sought. If the nonlinear operator N is attempted to Eq. (3) then

$$N(u) = N(u_0 + \lambda u_1 + \lambda^2 u_2 + \ldots)$$

Expanding $\mathrm{N(u)}$ to Maclaurin series with respect to λ we obtain

$$N(u) = A_0 + \lambda A_1 + \lambda^2 A_2 + \ldots = \sum_{n=0}^{\infty} \lambda^n A_n \tag{4}$$

with $A_n = \frac{1}{n!}\frac{d^n}{d\lambda^n} N(\sum_{k=0}^{n} \lambda^k u_k)$, where the components of A_n are called Adomian polynomials that are generated for each non-linearity Nu [1].

As an example, if $N(u) = f(u)$, the Adomian polynomials are given as:

$$A_0 = f(u_0)$$
$$A_1 = u_1 f'(u_0)$$
$$A_2 = \frac{u_1^{\ 2}}{2!} f''(u_0) + u_2 f'(u_0)$$
$$A_3 = \frac{u_1^{\ 3}}{3!} f'''(u_0) + u_1 u_2 f''(u_0) + u_3 f'(u_0)$$
$$\vdots$$

Now suppose $\mathrm{L^{-1}Ru}$ and $\mathrm{L^{-1}Nu}$ have an λ order, then Eq. (2) can be written as

$$u = h + L^{-1}g - \lambda L^{-1}Ru - \lambda L^{-1}Nu \tag{5}$$

If Eq. (3) and (4) are substituted for Eq. (5), then we obtain

$$\sum_{n=0}^{\infty} \lambda^n u_n = h + L^{-1}g - \lambda L^{-1}(R(\sum_{n=0}^{\infty} \lambda^n u_n)) - \lambda L^{-1}(\sum_{n=0}^{\infty} \lambda^n A_n) \tag{6}$$

Equating the coefficients of equal powers λ on both sides of Eq. (6), we obtain

$$u_0 = h + L^{-1}g$$
$$u_1 = -L^{-1}R(u_0) - L^{-1}(A_0)$$
$$u_2 = -L^{-1}R(u_1) - L^{-1}(A_1)$$
$$\vdots$$

In general can be expressed by the recursive relations

$$u_n = -L^{-1}R(u_{n-1}) - L^{-1}(A_{n-1}), n \geq 1$$

Therefore, the approximate solution is given by

$$u = u_0 + u_1 + u_2 + \dots$$

3. Application and Numerical Simulations

3.1. ADM for Solving Riccati Differential Equations

With reference to [6], Riccati differential equations form is given by the equation below

$$\frac{du}{dx} = A(x)u^2 + B(x)u + C(x) \tag{7}$$

where A, B, and C are the real functions of the real argument x.

Now, let consider the Riccati differential Eq. (7), with the initial condition

$$u(0) = D(x)$$

The ADM requires that Eq. (7) be expressed in terms of operator form as

$$Lu - B(x)u - A(x)N_1(u) = C(x) \tag{8}$$

where $L = \frac{d}{dx}$ and $N_1(u) = u^2$.

By ADM, the approximate solution of Eq. (7) is the infinite series $u(x) = \sum_{n=0}^{\infty} u_n(x)$ and the nonlinear term is $N_1(u) = u^2 = \sum_{n=0}^{\infty} A_n$, where A_n is the Adomian polynomial of u^2. The first component of Adomian polynomial is given by

$$\begin{aligned} A_0 &= N_1(u_0) = u_0^2 \\ A_1 &= u_1 N_1'(u_0) = 2u_0 u_1 \\ A_2 &= \frac{u_1^2}{2!} N''(u_0) + u_2 N'(u_0) = u_1^2 + 2u_0 u_2 \\ A_3 &= \frac{u_1^3}{3!} N'''(u_0) + u_1 u_2 f''(u_0) + u_3 f'(u_0) = 2u_0 u_3 + 2u_1 u_2 \\ &\vdots \end{aligned} \tag{9}$$

Now if both side of Eq. (8) are attempted by the operator L^{-1}, we obtain

$$u - L^{-1}[B(x)u + A(x)N_1(u)] = h + L^{-1}C(x) \tag{10}$$

If $u(x) = \sum_{n=0}^{\infty} u_n(x)$ and $N_1(u) = \sum_{n=0}^{\infty} A_n$ are substituted for Eq. (10), we get

$$\sum_{n=0}^{\infty} u_n(x) - L^{-1}[B(x)\sum_{n=0}^{\infty} u_n(x) + A(x)\sum_{n=0}^{\infty} A_n] = h + L^{-1}C(x) \tag{11}$$

If equating the coefficients of equal powers of λ on both side of Eq. (11), yields the recursive relation

$$\begin{aligned} u_0 &= h + L^{-1}C(x) \\ u_n &= L^{-1}[B(x)u_{n-1} + A(x)A_{n-1}], n \geq 1 \end{aligned} \tag{12}$$

Hence, the approximate solution of Eq. (7) is

$$u(x) = u_0(x) + u_1(x) + u_2(x) + \dots$$

with $u_0, u_1, u_2, \ldots$ is given by (12).

To give a better understanding of the methodology, the following two examples will be discussed.

Example 1: Consider the following Riccati differential equation

$$\frac{du}{dx} = u^2 - 2xu + x^2 + 1 \tag{13}$$

subject to the initial condition

$$u(0) = \frac{1}{2}$$

The exact solution of the initial condition problem (13) is given by

$$u(x) = \frac{1}{2-x} + x$$

Substituting the initial condition problem (13) into (12) and using (9) to calculate the Adomian polynomials, yields the following recursive relation

$$\begin{aligned} u_0 &= \frac{1}{2} + L^{-1}(x^2+1) \\ u_n &= L^{-1}[-2xu_{n-1} + A_{n-1}], n \geq 1 \end{aligned} \tag{14}$$

By using (14), the first three terms of the approximate solution $u(x)$ is given by

$$u_0 = \frac{1}{2} + \frac{1}{3}x^3 + x$$

$$u_1 = \frac{x^7}{63} + \frac{x^4}{12} - \frac{x^3}{3} + \frac{x}{4}$$

$$u_2 = \frac{2x^{11}}{2079} + \frac{x^8}{112} - \frac{2x^7}{63} + \frac{x^5}{20} - \frac{x^4}{12} + \frac{x^2}{8}$$

Hence, the third-term approximate solution for the initial condition problem (13) is given by

$$u(x) = u_0 + u_1 + u_2 = \frac{2x^{11}}{2079} + \frac{x^8}{112} - \frac{x^7}{63} + \frac{x^5}{20} + \frac{x^2}{8} + \frac{5x}{4} + \frac{1}{2}$$

The graphs of the approximate solution of (13) by ADM and Runge-Kutta method, and the error comparison between both methods are given in Fig. 1 and Fig. 2 respectively.

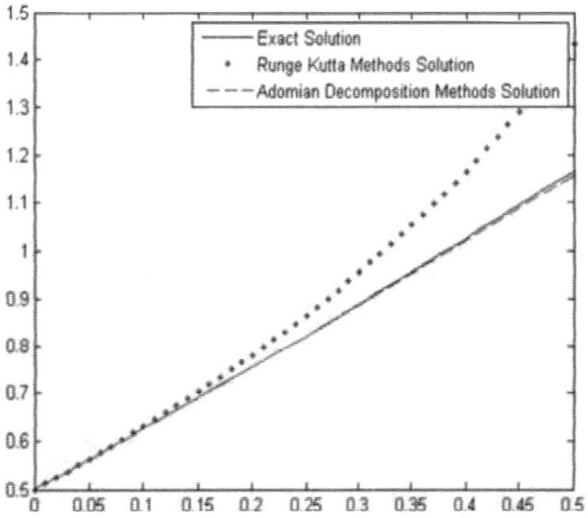

Figure 1. Comparison of the exact solution, Runge-Kutta, and ADM.

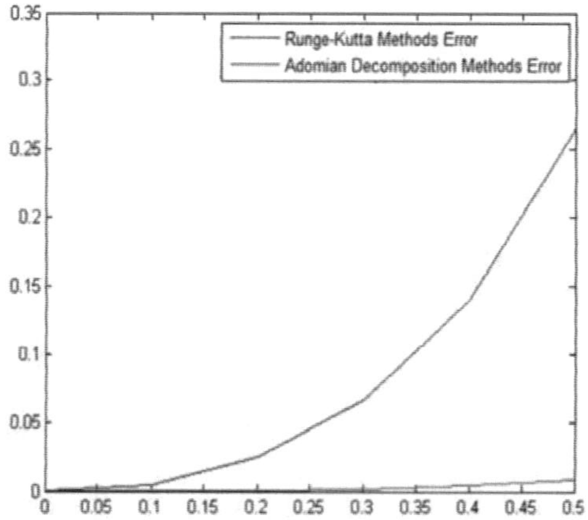

Figure 2. Error comparison between ADM and Runge-Kutta.

It can be seen in Fig. 1, that the ADM has a better approximation than the Runge-Kutta method. Moreover Fig. 2 shows clearly that the accuracy gained by ADM is much better. The error value of both methods shows an enhancement on every iteration. However, the error values of the ADM are smaller than the Runge-Kutta method.

The following table is given to see the accuracy of both methods on the initial value problem (13).

Table 1. The absolute error of the approximate solution with ADM and Runge-Kutta

n	x_n	Exact Solution	Runge-Kutta Solution u	Runge-Kutta Error	ADM Solution u	ADM Error
0	0.0	0.5000	0.5000	0.0000	0.5000	0.0000
1	0.1	0.6263	0.6314	0.0051	0.6263	0.0001
2	0.2	0.7556	0.7805	0.0250	0.7550	0.0005
3	0.3	0.8882	0.9548	0.0665	0.8864	0.0019
4	0.4	1.0250	1.1657	0.1407	1.0205	0.0045
5	0.5	1.1667	1.4323	0.2656	1.1577	0.0089

Example 2: Consider the following Riccati differential equation

$$\frac{du}{dx} = xu^2 + (1-2x)u + x - 1 \tag{15}$$

subject to the initial condition

$$u(0) = -1$$

The exact solution of the initial condition problem (15) is

$$u(x) = 1 + \frac{1}{-\frac{3}{2}e^{-x} - x + 1}$$

Substituting the initial condition problem (15) into (12) and using (9) to calculate the Adomian polynomials, yields the following recursive relation

$$\begin{aligned} u_0 &= -1 + L^{-1}(x-1) \\ u_n &= L^{-1}[(1-2x)u_{n-1} + xA_{n-1}], n \geq 1 \end{aligned} \tag{16}$$

By using (16), the first three terms of the approximate solution $u(x)$ is given by

$$u_0 = \frac{x^2}{2} - x - 1$$

$$u_1 = \frac{x^6}{24} - \frac{x^5}{5} - \frac{x^4}{4} + \frac{3x^3}{2} + x^2 - x$$

$$u_2 = \frac{x^{10}}{240} - \frac{17x^9}{540} - \frac{x^8}{480} + \frac{341x^7}{840} - \frac{x^6}{5} - \frac{37x^5}{20} - \frac{x^4}{8} + \frac{5x^3}{3} - \frac{x^2}{2}$$

Hence, the third-term approximate solution for the initial condition problem (15) is given by

$$u(x) = u_0 + u_1 + u_2 = \frac{x^{10}}{240} - \frac{17x^9}{540} - \frac{x^8}{480} + \frac{341x^7}{840} - \frac{19x^6}{120} - \frac{41x^5}{20} - \frac{3x^4}{8} + \frac{19x^3}{6} + x^2 - 2x - 1$$

The graphs of the approximate solution of (15) by ADM and Runge-Kutta method, and the error comparison between both methods are given in Fig. 3 and Fig. 4 respectively.

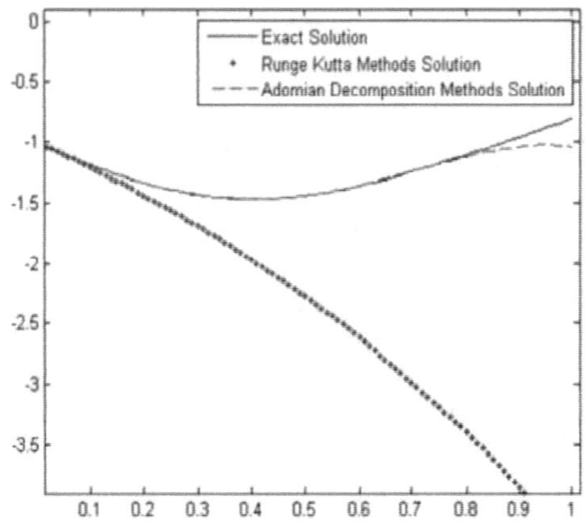

Figure 3. Comparison of the exact solution, Runge-Kutta, and ADM.

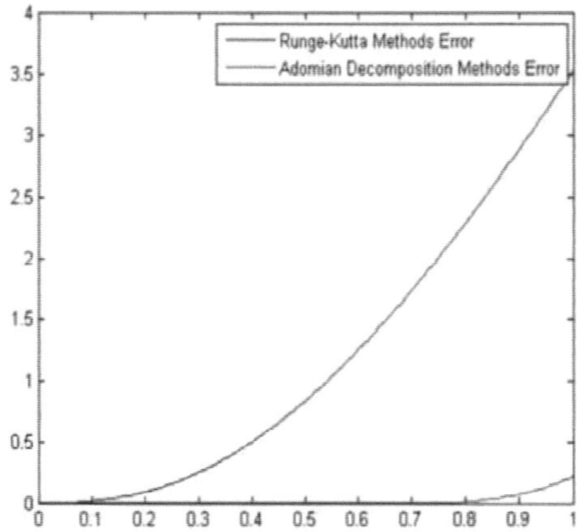

Figure 4. Error comparison between ADM and Runge-Kutta.

It is clear, as in the previous case, that the ADM has a better approximation than the Runge-Kutta method (See Fig. 3). The error value of both methods shows an enhancement on every iteration, however the error values of the ADM are smaller than the Runge-Kutta method (See Fig. 4).

The following table is given to see the accuracy of both methods on the initial value problem (15).

Table 2. The absolute error of the approximate solution with ADM and Runge-Kutta

n	x_n	Exact Solution	Runge-Kutta		ADM	
			Solution u	Error	Solution u	Error
0	0	-1.0000	-1.0000	0.0000	-1.0000	0.0000
1	0.2	-1.3359	-1.4374	0.1015	-1.3359	0.0000
2	0.4	-1.4662	-1.9690	0.5028	-1.4679	0.0017
3	0.6	-1.3629	-2.6145	1.2516	-1.3604	0.0025
4	0.8	-1.1097	-3.3971	2.2874	-1.1245	0.0148
5	1	-0.8122	-4.3446	3.5324	-1.0401	0.2279

3.2. ADM for Solving DNA Repair Model

Riccati differential equation has been used very heavily, one of the applications is in radiotherapy modeling and in studying damage to cells by ionizing radiation. The most important radiation damage is to chromatin, e.g. DNA double-strand breaks (DSBs). DNA DSBs are one of the most toxic of these lesions and must be repaired to preserve chromosomal integrity. Most DBSs are repaired during the next half hour or so, and a few are not repaired. Suppose U be the number of DSBs, then the average number of DSB per cell at time t is given by

$$\frac{dU}{dt} = \delta R - \frac{1}{\tau}U - \gamma U^2 \tag{17}$$
$$\delta, \tau, \gamma = const. \geq 0$$

where δ is the average number of DSBs induced per unit dose, R is radiation dose rate, τ is a repair time constant, and γ is a binary reaction rate constant in the sense of mass-action chemical kinetics [1].

The following illustration of using ADM for solving Eq. (17) is given. Consider the following DSBs repair model:

$$\frac{dU}{dt} = 150 - 5U - U^2 \tag{18}$$

If there are 50 number of DSBs repair at time $t = 0$, then yields the initial condition

$$U(0) = 50$$

The exact solution of the initial condition problem (18) is given by

$$U(t) = \frac{130e^{25t} + 120}{13e^{25t} - 8}$$

The recursive relation of the initial condition problem (18) is

$$\begin{aligned} U_0 &= 50 + L^{-1}(150) \\ U_n &= L^{-1}[-5u_{n-1} - A_{n-1}], n \geq 1 \end{aligned} \tag{19}$$

By using (19), the first three terms of the approximate solution U(t) is given by

$$U_0 = 50 + 150\text{t}$$

$$U_1 = \text{-}7500\text{t}^3 - 7875\text{t}^2 - 2750\text{t}$$

$$U_2 = \text{-}450000\text{t}^5 + 393750\text{t}^4 + 550625\text{t}^3 + 144375\text{t}^2$$

Hence, the third-term approximate solution for the initial condition problem (18) is given by

$$U(t) = U_0 + U_1 + U_2 = \text{-}450000\text{t}^5 + 393750\text{t}^4 + 543125\text{t}^3 + 136500\text{t}^2 - 2600\text{t} + 50$$

The other components can be determined similarly by using the recursive relation (19). The following graph shows the approximate solution of (18) with ADM for $n = 2$, $n = 3$, and $n = 4$. The validity of this method shown by comparing the approximate solutions of the ADM, and the exact solution.

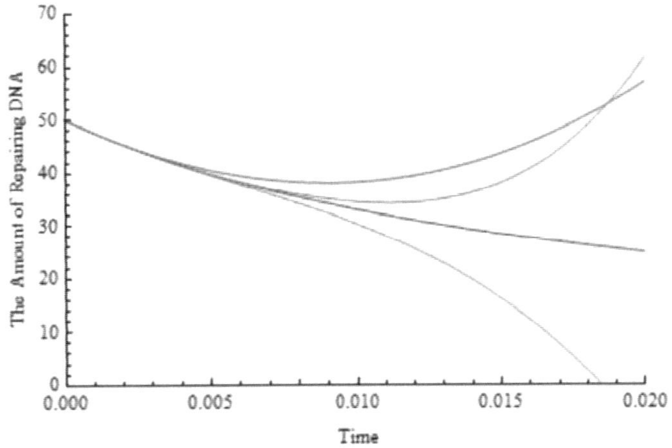

Figure 5. Comparison between the exact solution and the approximate solution with ADM. Blue line: the exact soluton ; purple line: n=2 ; yellow line: n=3 ; green line: n=4

It can be seen, the solution of the ADM is closer to the exact solution for greater order. Mathematically, the Eq. (17) can be written as

$$\frac{dU}{dt} = (\delta - U)(\gamma U + R)$$

where $\tau = \dfrac{1}{R-\delta\gamma}$. It means, Eq. (18) can be expressed as

$$\frac{dU}{dt} = (10-U)(U+15)$$

The following figure shows the phase plane of the DNA repair model (18).

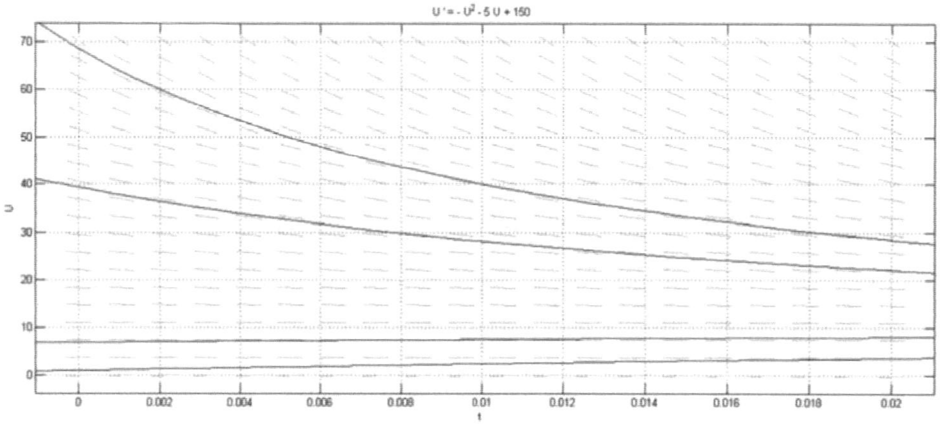

Figure 6. Phase plane of DNA repair problem (19) with $\delta = 10$, $R = 15$, and $\gamma = 1$

Based on Fig. 6, it can be analyzed that the number of repairing DNA will increase and approach to ten, if the number of induced dose DNA is below ten. If the number of induced dose DNA are in between forty and seventy, then the amount of repairing DNA will decrease and approach to twenty.

4. Conclusion

In this paper, the Adomian decomposition method has been applied to finding the approximate solution of the Riccati differential equations and the DNA repair model. All of the examples show that the numerical results of the proposed method are better than the Runge-Kutta method. The method was clearly very efficient in approaching the exact solution of the proposed equation. The flexibility of the method and reduction in the size of computational components give this method wider applicability.

References

[1] Hlatky L R and Hahnfeldt D 2001 *J. Math. Comput. Model.* **33** 1297-1305
[2] Adomian G 1988 *J. Math. Anal. Appl.* **85** 501–544
[3] Adomian G 1994 *Solving Frontier Problems of Physics: The Decomposition Method 1st edition* (Boston: Kluwer Academic Publishers)
[4] Evans D J and Raslan K R 2004 *Int. J. Comput. Math.* **00** 1-6
[5] Sweilem N H and Kadar M M 2010 *J. Appl. Math. Comput* **217** 495-505
[6] Edward C H and Penney D E 2007 *Elementary Differential Equation 6th edition* (New Jersey: Pearson)

IOP Conf. Series: Journal of Physics: Conf. Series **1090** (2018) 012018 doi:10.1088/1742-6596/1090/1/012018

An Evaluation of A Solid Biomass Cook Stove in Small Household Industry

Suhartono[1], Fitria Gasela[1], Anis Khoirunnisa[1], Suharto[2]
[1]Department of Chemical Engineering, Faculty of Engineering, Universitas Jenderal Achmad Yani, Cimahi 40533, Indonesia
E-mail: suhartono@lecture.unjani.ac.id, gaselatria@gmail.com, anis.kh09@gmail.com
[2]Research and Development Division for Mineral Technology, Indonesia Institute of Sciences, Jln. Ir. Sutami KM. 15, South Lampung 35361, Indonesia
E-mail: harto_berg@yahoo.com

Abstract. This paper presents the design of solid biomass natural convection stove test with that solid biomass as fuel in one of the small household industries. Corncob, coconut shell and wood chips waste were utilized using the stove with combustion chamber diameter of 15 cm and the height of 45 cm. The performance test of the stove includes combustion flame temperature, visual flame color and water boiling test as well as economic evaluation.The results of the test showed the start-up time of 10.55-16.34 min., the specific fuel consumption (S_c) of 0.51 to 0.76 kg/h, with the thermal efficiency (η_{th}) of 20-21% has a maximum fire power of 22.97 kW and 8.76 kW using coconut shells and wood chips, respectively. The biomass consumption cost for boiling water was about IDR 130-300/kg of water. The biomass utilization for frying fish gave a higher profit margin of IDR 1,500/kg fish compared to LPG. From this study, the performance test is met to the Indonesian National Standard of biomass stove (SNI 7926:2013) and suitable for the households, especially small household industries in rural areas. Further study related to CO and particulates emissions will be conducted in an attempt to improve this biomass stove design. Life cycle assessment of the used of solid biomass for heat production will be studied as well.

1. Introduction

Biomass a renewable energy resource used for cooking in small household industries in West Bandung District, West Java, Indonesia. In this rural area recorded at least 23 small household industries. Each industry uses approximately 34.2 kg/day of liquid petroleum gas (LPG) to fulfill the daily energy needs. Recently, LPG is no longer easy to come in term of high fuel cost and availability. LPG is utilized in combination with solid biomass as their main fuel for cooking. About 70% of the LPG consumption is replaced by the solid biomass as fuel using conventional biomass stove in these industries. Corncob, coconut shell and wood chips as the solid biomass residues of agriculture crops are used as main fuel for the production process and meanwhile LPG as a secondary fuel for smaller tasks such as frying fish crackers as a result of their products. In 2015, the biomass energy potential from agriculture waste of corncob, wood chips, and coconut shell are 48,629 GWh, 205,581 GWh, and 6,758 GWh, respectively [1, 2] and will continue to increase every year. As an illustration of corn production in West Bandung district is about 11,960 tons/year with about 50% of it as corncob waste (13.03 GWh/year). Despite its enormous potential, the utilization of solid biomass waste such as corncobs has not yet been fully utilized as fuel. Some of the solid biomass of corncobs, wood, coconut shell is used as heating by burning on the ground, while the rest is disposed of as waste.

Most people in these household industries, cooking is performed on the open fire, usually enclosed between the three or more stones fire in U-shape which act as pot supports as a traditional solid biomass stove. Poor combustion will negatively affect thermal performance of this traditional

Content from this work may be used under the terms of the Creative Commons Attribution 3.0 licence. Any further distribution of this work must maintain attribution to the author(s) and the title of the work, journal citation and DOI.
Published under licence by IOP Publishing Ltd

stove. Incomplete combustion and low heat transfer efficiency (ratios of energy entering the pot to that liberated by combustion) will decrease the stove efficiency. In general, such combustions fire have a thermal efficiency of no more than 10% compared to the above 20% obtained by using a properly designed solid biomass cooking stove [3, 4]. Low heat transfer efficiencies is a significant factor affect stove's efficiency, while energy loss due to incomplete combustion is about 8% of total energy input [3, 5]. However, the incomplete combustion of solid biomass fuel will release harmful emissions and affect high levels of indoor air pollution. The emissions are associated with serious adverse health effects that causing the acute respiratory problem. The harmful emission includes CO, unburned hydrocarbons, and NOx.

Many designs, construction, and testing of the improved solid biomass stove are reported by researchers. Although, the development of solid biomass cooking stove is not the earliest technologies. Unfortunately, there today exist no clear-cut internationally accepted design standards for solid biomass cooking stove [3, 5, 6, 7]. This is due to a lack of knowledge and understanding in combustion science and stove design. The currently available local solid biomass stoves do not represent the best in modern engineering design.This local solid biomass stove has a low efficiency and consequently high emissions from unburned fuel.

Commonly, the aimed of the improvements are to increase the stove thermal efficiency, such as improved air/fuel mixing and secondary air to combust the biomass fuel more completely which resulted in lower pollutant emissions. Therefore, the challenge of designing a solid biomass cooking stove is not only a technical issue but a social issue as well.The improvement of a solid biomass stove in this work is undertaken on the economics, environmental and health considerations. Another goal is to be flexible to different types of solid biomass and ease of operation. The LPG needs of the all small household industries in the rural area can be replaced with solid biomass waste. In present work, the modification on the existing solid biomass stove designs was conducted by providing sufficient air to enhance the complete combustion, it means to increase combustion flame temperature. The high combustion flame temperature can reduce the level harmful air pollution and increase combustion efficiency. The solid biomass stove performance was evaluated according to some of the parameters required by the National Standardization Agency of Indonesia (BNSi) for biomass stove (SNI 7926: 2013) [8].

2. Biomass Resource in Indonesia

Most of the biomass in Indonesia is crop residues from non-edible plant parts as remain of by-products after crop processing, such as pilling, milling, and extraction. These biomass residues and can potentially be used as energy source in the household and small industry sector for cooking in the rural area [9]. These residues contribute significantly up to 5% energy consumption in Indonesia in 2025 as one of new and renewable energy resources [2, 9]. Table 1 shows the total biomass residue production in the year 2015 from some provinces in Indonesia [2]. The estimation of biomass base new and renewable energy potential derived of agricultural and plantation from solid residues are approximately 399,536 GWh.

Table 1. Solid biomass energy potential in Indonesia [1, 2].

Region	Energy (GWh)			
	Rice husk	Corncob	Wood chips	Coconut shell
Jawa	65,652	26,534	5,053	1,532
Sumatera	36,292	8,103	41,785	2,243
Kalimantan	13,298	715	68,649	381
Bali, Nusa Tenggara	7,486	5,039	3,309	399
Sulawesi	15,304	7,705	19,275	1,613
Maluku	323	474	7,789	558
Papua	213	59	59,721	32
Total Potential (GWh)	138,568	48,629	205,581	6,758

In Indonesia, agricultural residues vastly meet the household energy demands in rural areas, for example, the potential energy of corncob, coconut shell and wood chips from agricultural plantations in West Bandung District were estimated 1.413 GWh, 0.196 GWh and 5.973 GWh in a year respectively [9]. In this region, around 80 percent of the economic activities are based on agriculture about 84% of those solid biomass residues are used as a renewable energy source. These residues are utilized as fuel using a conventional simple stove in the household and small industry sector for cooking [9] and the remains are used for several others purposes. The typical agricultural residues fuel has the ultimate analysis by mass as carbon content is 30 to 50%, hydrogen is around 6%, oxygen is 30 to 45%, nitrogen is around 2% and HHV is about 17 MJ/kg [4, 6, 7], [9-12]. According to the report, the energy in about 4 kg of the solid biomass of agricultural residues approximately substitutes about ±1 kg of LPG [9,10].

3. Solid Biomass Combustion

In order to convert solid biomass into useful heat energy, it has to undergo combustion. This process is used every day in households and small household industries for heating and/or cooking. Solid biomass combustion is a process in which the fuel is burnt with the oxygen from the air at sustainable elevated temperatures to release heat in burners. This combustion process is a complex interaction of physical and chemical processes and affected by the physicochemical properties of the fuel (size, shape, density, moisture content, fixed carbon content, volatile matter, etc.), quantity and mode of air supply (primary and secondary air), and the conditions of the surroundings (temperature, wind, humidity, etc.) [3, 13].

Solid biomass fuels are primarily composed of carbon, hydrogen, and oxygen. Ideally, all hydrogen and carbon would split off and combine with the oxygen in the air to create water vapor, carbon dioxide, and heat. Flame temperatures can exceed 2000°C, depending on the heating value and moisture content of the fuel (the fuel characteristics), the amount of air used to burn the fuel and the design, configuration and construction of the solid biomass stove. The high-flame temperature is closely related to thermal efficiency and tends to increase it. This condition can be achieved when complete combustion occurs, which lowers losses and extracts all of the energy from the biomass as fuel. In practice, to obtain perfect combustion is very difficult due to various factors. The technical factors that may affect perfect combustion are fuel/air ratio, mode of fuel supply, and primary and secondary air supplies.The perfect fuel to air ratio is needed for the stoichiometric combustion takes place. In practice, in order to achieve complete combustion, it is necessary to increase the amount of air to the combustion process to ensure the burning of all of the biomass fuel. The amount of air that must be added to make certain all energy will be retrieved is known as excess air. Typical excess air required for various stove design is in the range of 5% to 50% [4], depending on the fuel properties and the stove configuration.

4. Biomass Stove Design

The design of this solid biomass stove was developed to be implemented in a small household industry by consideration to enhance the combustion quality. The natural convection mode is a basic principle design of this stove [14, 15] with the aims to increase efficiency and reduce emissions. Providing the sufficient air for combustion by the adequate draft for enhancing the combustion process. A proper air-fuel ratio can increase combustion efficiency. The primary and secondary combustion air ports on the stove will affect to fuel consumption rate, power fire, combustion flame temperature as well as thermal efficiency. Consequently, the stove geometry, stove size, loading fuel door, combustion air inlet ports/holes and grate (fuel bed) in stove design must be able to ensure in accommodation of combustion air properly.

Based on the considerations of a solid biomass stove as described above, solid biomass stove was designed by using selected parameters as follows; a stove was constructed using mild steel sheet with 8 mm thickness. The total stove height was 45 cm with a height of 35 cm combustion chamber. The external diameter and internal diameter stove were 15 cm and 13.5 cm, respectively. The loading fuel door size of 7cm x 13 cm, equipped with 12 and 40 holes of primary and secondary air inlet with a diameter of each hole was 2 cm and 8 mm, respectively, Figure 1.This improved biomass cooking stove and its implementation have the potential impacts to decrease exposure indoor air pollution, reduce the dependence of the LPG fuel and household industrial cost (fuel substitution and economic aspect).

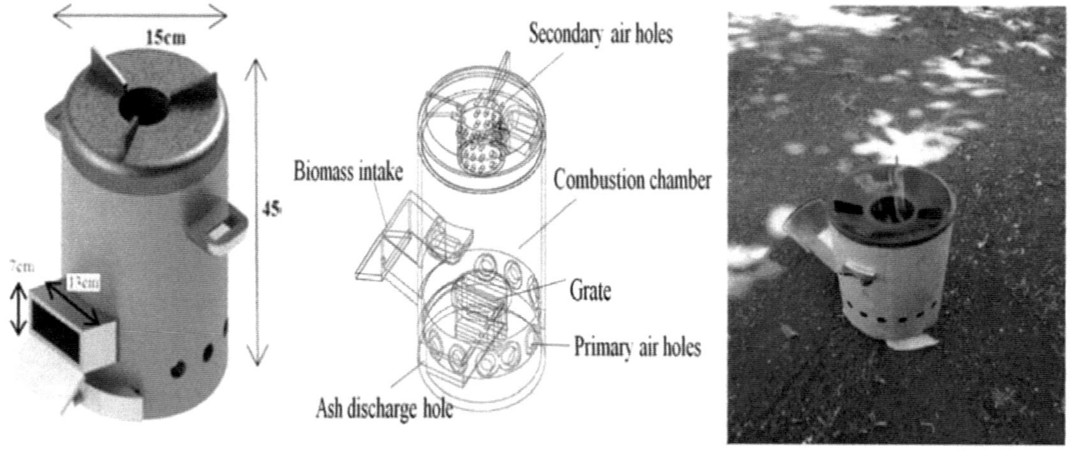

(a) Sketch of biomass stove (b) Visualized flame

Figure 1. Sketch of improved biomass stove and visualized flame

5. Experiment

Corncobs, coconut shells and wood chips waste were used as fuel for evaluating of a solid biomass cook stove in a small household industry in West Bandung District. Such biomass is very abundant in this area, but it is very less in its use as an energy resource. The modified of a conventional natural draft biomass stove with the specification as described in the above subsection was used in this study.This stove design was not equipped with a blower, therefore the need of combustion air was supplied to the chamber due to the natural draft. This air flows due to the difference in density, the low-temperature zones on the stove will have greater air density than in higher ones. The combustion air will move naturally from low temperatures to high temperatures. The adequacy of secondary air for complete combustion of biomass pyrolysis gases due to the presence of air intake through secondary air holes.

The three kinds of biomass waste as feedstock for testing biomass stove were cut into small pieces and store in a dry place for further use. Elements and proximate analysis were carried out before testing by using the method suggested by ASTM. The physical and thermal properties of biomass feedstock used for evaluating the stove performance as well as economic evaluation. The measured parameters during the study were boiling time, start-up time, flame temperature, thermal efficiency, air flow rate, and biomass consumption. From the results of this stove, performance test study is expected to fulfill some of the parameters set by the Indonesian National Standard of biomass stove (SNI 7926:2013). While other parameters such as CO and particulates emissions will be conducted in an attempt to improve this biomass stove design in further study.

The biomass weight of about 0.5-1.5 kg was used in the test to measure the startup time and the fuel consumption. Specific fuel consumption, S_c was calculated using the equations below [8]:

$$S_c = \frac{\Delta m_k}{\Delta t} \tag{1}$$

where Δm_k is the fuel mass that has been burned during the test (kg) and Δt is the duration of the test. The fuel (biomass) consumption was determined by evaporating a known quantity of water which is recommended as standard water-boiling test [8, 9, 16]. The consumption of fuel was also evaluated by frying a quantity of food until cooked. In the current work, to compute the thermal efficiency, experiments were conducted to study energy consumed for boiling amount of water at atmospheric pressure (at about 100 °C) and for frying amount of food until cooked (at the boiling point of cooking oil, about 250 °C). The weight measurement was done using a digital strain gauge base weighing balance and the temperature, while the temperature of water and frying oil, as well as surface flame temperature of flue gas combustion, were measured using a digital thermocouple Krisbow-KW06-283. The secondary air flow rate on secondary holes inlet for fuel combustion air was measured using a digital anemometer Krisbow-KW06-653. The experiments were repeated three time and thermal efficiency was determined by measuring the amount of heat absorbed by the fuel, and dividing by the total amount of fuel used. The ratio of energy used for cooking with heat energy available in the fuel was defined as [17]:

$$\text{Thermal efficiency} = \frac{\text{Heat utilized}}{\text{Heat supplied}} \tag{2}$$

The experiment results of the above parameters have to meet some of the following standards (SNI 7926: 2013); the maximum of specific fuel consumption, S_c is 1kg/hr. and the minimum thermal efficiency, η_{th} is 20%. Another additional requirement parameter is power output of about 2.5 to 5 kW or equivalent to 0.5 to 1.0 kg/hr of biomass fuel consumption with a heating value of 18 MJ/kg.

The output power of biomass cooking stove output is a measured of the total energy produced by the fuel during one hour. The output power was calculated by using equation [18]:

$$P_o = \frac{F_{Sc} \times H_{vf} \times \eta_{th}}{3600} \tag{3}$$

where P_o is the output power (kW), F_{Sc} is consumption rate of biomass (kg), η_{th} is the thermal efficiency and H_{vf} is the fuel calorific value (kJ/kg). In this experiment, the power was also evaluated as a ratio of the biomass energy consumed by the stove per unit time (in kW) as fire power (F_P) and mathematically expressed as [5].

$$F_P = \frac{(m_f \times H_{vf})}{(60 \times (t_f - t_i))} \tag{4}$$

where (t_f - t_i) is the duration of the specific experiment phase. The experiment was conducted with both cold and warm start conditions. The cold start begins with the stove at room temperature, while the warm start, a fire was reset immediately after cold start phase.

6. Results and discussion

6.1 Biomass properties as fuel

Table 2 shows the proximate analysis, the ultimate analysis and the calorific value of some selected agricultural biomass waste. The structural components and chemical elements of biomass are influence factors on combustion characteristics and thermal utilization. These components and elements depending on the origin and type of biomass. C, H, and O are the main elements of these solid biomass waste. In this work, the calorific values were estimated using Dulong's formula based on elements composition. The calorific values of corncob, coconut shell, and wood chip were obtained about 17275 kJ/kg, 20890 kJ/kg and 18321 kJ/kg, respectively. The higher content of C and H increase to the calorific values, while the O content decreases the calorific value. The C content of coconut shell is higher than those of corncob, it means the slightly higher calorific value of coconut shell [19]. The lower calorific value of wood chip was influenced by H content due to the formation of water (moisture) [19]. The ash content also decreased the net calorific value, moisture and ash absorb some of the heat.

Since the value of this biomass calorific values is still within the range of calorific values of biomass in general. It is reasonable to be used as the basis for calculating the parameters for evaluating the performance of the natural draft stove.

Table 2 The proximate analysis, the ultimate analysis and the calorific value of biomass

Characteristic	Corncob		Coconut shell		Wood chip	
Ultimate Analysis (%):						
C	45,01		47,89		51,28	
H	6,45		6,09		7,23	
O	46,3		45,75		46,78	
N	0,26		0,22		0,28	
S	0,11		0,05		0,34	
Ash	1,87		7,56		18,65	
Proximate Analysis :						
Moisture Content (% wb)	10,6	-	6,51	-	7,30	8,69**
Volatile Matter (% db)	75,52	74,58**	68,82	13,90**	76,75	77,33**
Fix Carbon (% db)	22,61	23,90**	17,11	83,70**	14,82	11,70**
Calorific Value (kJ/kg)	17275*	-	20890*	-	18321*	-
Bulk density(g/l)	-	332,06**	-	562,06**		362,06**

* Calculated by Dulong's formula
** Experiment

6.2 Stove performance

Tabel 3 shows the values of some stove performance during hot start phase and cold start phase using a different kind of biomass fuel. Various fuel properties were found to affect the stove performance. Slightly lesser time (about 1 min.) was required to start up and to boil the water during hot start phase than in cold start. Coconut shell waste and wood chips waste provide longer start-up time of about 1 minute compared to corncob. This can be understood by the fact that the biomass fuel in the stove would be at slightly higher temperature than room temperature. This increases the amount of volatile vapor that would burn more easily. The higher volatile matter allows the fuel to ignite quickly into gas. This is due to heat transfer to the fuel and the faster combustion process occurs in the hot start phase than the cold start phase and due to the percentage of volatile matter content.

Table 3. Specific start-up, operation time and boiling time

Type of biomass	Weight (kg)	*Start-up time* (min.)		Operation time (min.)		Boiling time (min)	
		Hot start	**Cold start**	**Hot start**	**Cold start**	**Hot start**	**Cold start**
Corncob	0.50	10.55	9.25	12.78	13.56	4.02	4.16
	0.75	11.49	11.25	15.11	15.61	3.54	3.67
	1.00	12.14	12.70	16.37	16.48	3.66	3.51
	1.50	12.61	-	18.14	-	3.33	-
Coconut shell	0.50	12.11	11.03	14.09	13.78	4.37	4.09
	0.75	14.47	13.25	16.21	15.67	3.60	3.82
	1.00	14.79	14.68	18.19	16.21	3.28	3.54
	1.50	15.09	-	19.55	-	2.98	-
Wood chip	0.50	11.93	12.30	13.88	13.95	4.61	4.12
	0.75	14.07	14.75	16.69	16.78	3.79	3.89
	1.00	15.75	15.98	17.46	17.56	3.31	3.67
	1.50	16.34	-	20.21	-	3.06	-

In addition, corncobs have a large number of void fractions to facilitate incoming combustion airflow and improve air and combustible elements mixing that enable good combustion. In this experiment, there was no significant difference of operation time for the various amount of biomass in both hot and cold phase. This can be explained by the experiments that average of burning rate was found about 0.2 kg/min for all kinds biomass fuels.

The maximum flame temperature on the natural draft stove using three kinds of biomass fuel was about 1126 °C. The flame temperature of about 795-869 °C with reddish orange flame color was observed when corncob utilized as fuel. Whereas the utilization of coconut shells tends to produce orange bluish flame at the temperature of 706-963 °C and wood waste fuels also tend to produce bluish orange flame at the temperature of 912-1126 °C, as shown in Table 4. The flame temperature is affected by the type of fuel, the amount of fuel and air supplied. The high flame temperature determines the amount of released heat, as the flame temperature gets higher, the heat released by the fire increases. Different flame colors were influenced by some gas content in biomass fuel. The orange flame color was caused by a lot of CO and unburned of carbon particles. When more CO in the flue combustible gases closer to the red flame color. While the blue color on the flame indicates that a lot of combustible gases of CH_4 and H_2 [5]. In this work, the height and shape of the resulting flame tend to be unstable due to turbulent airflow.

Table 4. Flame profile

Type of biomass	Flame Temperature (°C)	Flame color
Corncob	795 – 869	Reddish orange
Coconut shell	706 – 963	Orange
Wood chip	912 – 1126	Bluish orange

The other stove performance indicators were conducted by boiling 1 kg of water through calculations based on modified of water boiling test. The time required for boiling 1 kg of water about 3-4 minutes, with the consumption of biomass fuel 105-121 g (specific fuel consumption, S_c about 0.51 - 0.76 kg/hour). The use of coconut shell as fuel in this stove gave a shorter boiling time than corncobs and wood chips. This was influenced by the combustion process because of the greater the burning rate the greater the firepower and the amount of flammable gas produced. The higher of combustion quantity would cause the greater of heat combustion. It could be understood that utilized coconut shell as fuel would cause faster boiling of water. But in this case, the higher of the calorific value of fuel resulted in a little bit more energy lost to the environment, as well as the rate of evaporation of boiling water becomes larger, however no significant effect on thermal transfer efficiency.

Tabel 5. Comparison of SNI standard parameters and experiments

Parameters performance	SNI standard	Experimental data
Thermal efficiency, η_{th} (%)	Minimum 20	20 – 21
Combustion efficiency, η_c (%)	Minimum 0.96 or 0.04	–
Specific fuel consumption, Sc (kg/jam)	Maximum 1	0.51 – 0.76
Emission of CO (g/kg)	Maximum 67	–
Particulate emissions, PM 2,5 (mg/kg)	Maximum 1500	–

The performance of biomass stoves that represented as thermal efficiency is influenced by a number of factors. The first factor is insulation or coating along the combustion chamber. Insulation will minimize the rate of heat loss in the combustion chamber wall by conduction. There is also an adequate air supply factor, which will ensure the complete combustion of biomass [6]. The biomass utilization using this stove gave the thermal efficiency, η_{th} of about 20 - 21%. And a maximum fire power of 22.97 kW and 8.76 kW using coconut shells and wood chips, respectively.

From several evaluation parameters that have been accomplished, some of them have met the SNI standard requirements of SNI 7926: 2013. While the combustion efficiency (η_c), CO emissions and particulate emissions have not been measured yet. The more details of parameters are presented in Table 5.

6.3 Economic Studies

In the current work, the economical was evaluated by the implementation of the biomass stove in PT. Tiga Oscar Besaudara, in the area of West Bandung District. The economical was conducted on the basis of the used of biomass and excluding the investment of the stove. The economic calculation was based on the ratio of energy used for cooking/frying, as well as for boiling of water with heat energy available in the fuel. By the thermal efficiency of about 20 %, the cost for boiling 1 kg of water about IDR 130-300/kg water. The price of this economical was still much cheaper than using kerosene stove, IDR 188/kg of water and LPG stove, IDR 250/kg of water. While the economic calculation of the biomass utilization for frying fish using this stove gave the difference between the benefit of the use of LPG and the use of biomass was IDR 1,500/kg fish.

7. Conclusions

The specific fuel consumption, S_c of solid biomass waste utilized using this stove was in the range of 0.51-0.76 kg/h. The coconut shells and wood chips utilization using this stove produced a maximum fire power of 22.97 kW and 8.76 kW, respectively and gave the maximum flame temperature was 1126 °C with a bluish orange flame. About 3-4 minutes was needed to boil 1 kg of water with thermal efficiency in the range of 20-21%. The average of burning rate was found about 0.2 kg/min for all kinds biomass fuels. The biomass waste utilization gave the most economical cost than the used of LPG. This stove performance have met some parameters of the SNI standard requirements of SNI 7926: 2013. By the results of this study, small industries in West Bandung District are suggested to utilize the biomass waste using this stove to substitute LPG needed.

8. Further study

Since the finding of the experiments is encouraging, work is in progress to measure CO and particulates emissions. The experiment is also being planned to improve this biomass stove design and scale up for the medium industrial needs. The estimation of the life cycle environmental impacts of the solid biomass combustion and to identify its benefits will also be studied. The data will be submitted in subsequent paper.

Acknowledgement

We wish to acknowledge the financial support from DRPM-RISTEKDIKTI through the PUPT program for the year 2016-2018. We also wish to thank Universitas Jenderal Achmad Yani for supporting this work through the provision of laboratory space and equipment. We also wish to make special reference to Mr. Aidil and Mr. Sukeni for their technical support and cooperation during the experiment. All those who supported this work, directly and indirectly, are greatly acknowledged.

References

[1] Republic of Indonesia (RI) 2006. *Blueprint for national energy management 2006-2025*, Jakarta.

[2] Ministry of energy and mineral resources (ESDM)2014. *Handbook of Energy & Economic Statistics of Indonesia 11st edn.*, Jakarta.

[3] Saturday A, Sule E P, Ogbona E F, and Anslem N E 2016. Design and thermal analysis of an energy efficient solid biomass stove, *International Journal of Scientific Development and Research (IJSDR)* 1 **(8)** 166-174.

[4] Jenkins B M, Baxter L L, Miles Jr T R and Miles TR 1998. Combustion properties of biomass, *Fuel Processing Technology*, **54** 17–46.

[5] Berrueta V M, Edwards R D, and Masera O R 2008. Energy performance of wood-burning cookstoves in Mexico, *Renewable Energy*, **33** 859–70.

[6] Ayo S A 2009. *Design, Construction and Testing of an Improved Wood Stove*, AU J.T, 12-18.

[7] Tryner J, Wilson B D, and Marchese A J 2014. The effect of fuel type and stove design on emissions and efficiency of natural-draft semi gasifier biomass cookstoves," *Energy for Suatainable Development*, **23** 99-109.

[8] Indonesian standardization bureau 2013. *Indonesia's national standard-performance biomass stove*. BSN,Jakarta, Indonesia.

[9] Suhartono, Prasetyo B D, and Azizah I N 2016. Synthetic gas (syngas) production in downdraft corncob gasifier and its application as fuel using conventional domestic (LPG) stove, *Journal Engineering and Applied Science*, 11 (8) 5238-43.

[10] Suhartono, Putri TA, and Fauziah L 2017. Performance evaluation of a pressurized cooking stove," *International Journal on Advanced Science, Enginering and Information Technology*, 7 (4) 1255-1261.

[11] Khalil R A, Bach Q V, Skreiberg Ø, and Tran K Q 2013. Performance of a Residential Pellet Combustor Operating on Raw, *Energy & Fuels*, 4760−69.

[12] Das B K and Hoque S M N 2014. Assessment of the Potential of Biomass Gasification for Electricity Generation in Bangladesh, *Journal of Renewable Energy* 1-10.

[13] Overend R P 2009. Direct combustion of biomass, in *Renewable Energy Sources Charged With Energy from the Sun and Originated from Earth-Moon Interaction*, Evald E. Shpilrainaytex, Ed. Evald E. Shpilrainaytex, USA: Encyclopedia of Life Support Systems (EOLSS), vol. 2.

[14] Febriansyah H, Setiawan A A, Suryopranoto K, and Setiawan A 2014. *Gama stove: Biomass stove for palm kernel shells in Indonesia*, Indonesia EBTKE CONEX 2013 **47** 123-132.

[15] Agenbroad J, DeFoort M, Kirkpatrick A, and Kreutzer C 2011. A simplified model for understanding natural convection driven biomass cooking, *Energy for Sustainable Development* **15** 169-175.

[16] Visser P. *The Nederlands: Biomass technology group B.V.*, AE Enschede.

[17] Kartikeyan N S, Agarwaal A, Santhiyanarayanan K, Natarajan R 2008.Use of vegetable oil as fuel to improve the efficiency of cooking stove, *Renewable energy* **33** 2423-27.

[18] Rathod P, Waghmare, Wasewar, Kalyani L 2015. *Performance evaluation and heat transfer studies on biomass gasifier cook-stove*. Nagpur 440010, India: Chemical Engineering Department, Visvesvaraya National Institute of Technology (VNIT).

[19] Brunner T, Barnthaler G, Ombernberger I 2006. Chemical properties of solid bio fuels-significance and impact, *Biomass and bioenergy* **30** 973-982.

IOP Conf. Series: Journal of Physics: Conf. Series **1090** (2018) 012019 doi:10.1088/1742-6596/1090/1/012019

Numerical Study of Radiation Heat Transfer on Supplementary Firing with Gas Turbine Load Variations

Muhammad Haidar Fakhri*, Teguh Hady A., Arrad Ghani S.
Power Plant System Engineering, Electronics Engineering Polytechnic Institute of Surabaya, Surabaya, Indonesia
*fakhrialex@gmail.com

Abstract
In a combined cycle, it is rare to find the use of supplementary firing in Heat Recovery Steam Generator (HRSG). However, supplementary firing is actually responsible for 200%-400% steam production increase due to additional combustion process inside HRSG. This process evidently produced radiation heat transfer so it is appealing to conduct a further research on it.This study attempts to evaluate 3 dimensional radiation heat transfer in the industrial supplementary firing. Discrete Ordinates (DO) method and Weighted Sum-Gray Gases (WSGGM) model are combined to outline the radiation process, whilst non-premixed with Probability Density Function (PDF) Chemical Equlibrium model is used for combustion analysis. Evaluation is conducted using variations of load 50%, 75%, and 100% of Gas Turbine (GT) industrial datasheet. Validation is build upon load 100% of GT datasheet. Based on the evaluation, datasheet outcome shown that simulation possessed 2.90% and 5.28 % deviation for temperature and pressure respectively. From wall heat flux radiation, this study found that load 50% of gas turbine variation had the best trend, are emit radiation (22413.50 W/m^2) and absorb radiation (-11847.5 W/m^2) compared to the other options available.
Keyword: Supplementary firing, Radiation, Load Gas Turbine,

1. Introduction

Combined cycle system applied to HRSG aims to improve the efficiency by utilizing exhaust gas from gas turbine. Which still having heat that can be used to heat the tube. There are some supplementary firing added in HRSG to increase the steam production to 200-400% [1]. Supplementary firing utilizes exhaust gas from gas turbine as oxidizer. The combustion types that occurred in supplementary firing is non-premixed. Then, the effect of this combustion is the radiation heat transfer.

Several previous studies already discuss about the importance and factors that increase the radiation heat transfer in combustion cases. Combustion cases [2] use some radiation model to know the experimental result of radiation heat transfer. In the industrial combustion, thermal radiation has important role when the temperature is very high. In addition, sometimes radiation transfer is often overlooked in the calculation process because it is considered too mathematically complex. Ignoring the radiation transfer especially in combustion system may lead to computation error [3]. In paper [2] shows the radiation heat transfer predominates on the furnace walls. The concentrations of combustion air will affect the radiation heat transfer. This is delivered on study [4] which analyzed the characteristic of combustion by varying the combustion air. Turbulence will also affect the combustion process. This is because of the influence of chemical reaction between air and fuel [5].

This study aims to know the radiation heat transfer on HRSG furnace wall using Computational Fluid Dynamic (CFD). There are 7 supplementary firing arranged vertically. Combustion air is obtained from gas turbine exhaust gas. There are 3 variations of load from GT that are, 50%, 75%, and 100%. Each load has different concentrations of air and inlet pressure. This will be discussed to see the effect of radiation on the furnace wall.

2. CFD model

2.1 Pressure Based and Solver

Simulation is conducted using equation Reynold-Averanged Navier Stokes (RANS) for the continuity (equation 1) and momentum (equation 2). Solver using double precisions pressure based

Content from this work may be used under the terms of the Creative Commons Attribution 3.0 licence. Any further distribution of this work must maintain attribution to the author(s) and the title of the work, journal citation and DOI.
Published under licence by IOP Publishing Ltd

(segregated) for flowing liquid. Special discretization from scalar is made with second order upwind scheme and Pressure Staggering Option (PRESTO) for determine the pressure. These are selected because the fluid inside is assumed flowing with incompressible flow. Semi-Implicit Method for Pressure-Linked Equations (SIMPLE) algorithm is used to pressure-velocity coupling [2,6,7] Equation:

$$\frac{\partial}{\partial xi}(\rho u_i) = 0 \tag{1}$$

$$\frac{\partial}{\partial x_j}(\rho u_i u_j) = -\frac{\partial \rho}{\partial x_i} + \frac{\partial}{\partial x_j}\left[\mu\left(\frac{\partial x_i}{\partial x_j} + \frac{\partial x_j}{\partial x_i} - \frac{2}{3}\delta_{ij}\frac{\partial u_l}{\partial x_l}\right)\right] + \frac{\partial}{\partial x_j}\left(-\rho\overline{u_i' u_j'}\right) \tag{2}$$

2.2 Viscous k-ε Realizable

The value of k-ε realizable is better for non-premixed combustion cases. This is shown in experiments comparing k-ε standarts and k-ε realizable [8] in oxy-fuel combustion. In some combustion studies this tubulence [6,8]. also used. Here is the k-ε Realizable equation:

$$\frac{\partial(\rho\varepsilon)}{\partial t} + \frac{\partial(\rho\varepsilon u_j)}{\partial x_j} = \frac{\partial y}{\partial x_j}\left[\left(\mu + \frac{\mu_t}{\sigma_\varepsilon}\right)\frac{\partial \varepsilon}{\partial x_j}\right] + \rho C_1 S_\varepsilon - \rho C_2 \frac{\varepsilon^2}{k+\sqrt{v\varepsilon}} + C_{1\varepsilon}\frac{\varepsilon}{k}C_{3\varepsilon}G_b \tag{3}$$

$$C_1 = max\left[0.43, \frac{\eta}{\eta+5}\right] \tag{4}$$

$$\eta = S_\varepsilon^k$$

$$S = \sqrt{2S_{ij}S_{ij}} \tag{5}$$

Value σ_k= 1, σ_ε=1,2, C_2=1,9, $C_{1\varepsilon}$=1,44, and $C_{3\varepsilon}$ described with equation:

$$C_{3\varepsilon = tanh\left|\frac{v_i}{u_\perp}\right|} \tag{6}$$

2.3 Radiative Transfer Equation

This model is used with specification from opaque wall inside the interior domain. The equation from this transfer radiation is for finite number from discrete solid angles [9]. The contribution from phase particle is shown in equation:

$$\nabla\overrightarrow{I s} + (a + a_p + \sigma_s)I(\vec{r}.\vec{s}) = an^2\frac{\sigma T^4}{\pi} + E_p + \frac{\sigma_p}{4\pi} + \int_0^{4\pi} I(\vec{r}.\vec{s})\Phi(\vec{s}.\vec{s})d\Omega' \tag{7}$$

Where the value of α, α_p are gas and particle from absorption coefficient, σ_s and σ_p are gas and particle from scattering coefficient, Ep is the equivalent from particle emission σ is Boltzmann constant, I is the intensity of radiation, T is the local temperature, Φ is the scattering phase function and Ω is solid angle. DO is used well at oxy fuel combustion in optical thickness[9,10.11]

2.4 Species non-premixed combustion

Non-premixed combustion is considered efficient because include some solution from transport equation for one or two scalar equation to be converted from equation for each species. Scalar conversion is mixture fraction that defined as thermo-chemical reaction of the fluid. Mixture fraction can be expressed with atomic mass fraction [9], as shown in equation

$$f = \frac{Z_i - Z_{i,ox}}{Z_{i,fuel} - Z_{i,ox}} \tag{8}$$

This is the equation that shows the oxidizing agent from inlet flow and equation from fuel value of inlet flow. If the value of diffusion assumed to be same, the equation can be reduce/simplify to single equation for mixture fraction. (f). This equation is also suitable used in turbulent flow [9,12]. In this model, all thermo-chemical scalar (species fraction, density and temperature) is mixed to mixture fraction. The form of this is PDF. This is used as a sub grid local fluctuations consideration among species and described by beta-function and PDF experiment process. [8,9]. Below is The Favre mean equation (density-averaged) mixture fraction:

$$\frac{\partial}{\partial}(\rho\bar{f}) + \nabla.(\rho\bar{v}\bar{f}) = \nabla.\left[\frac{\mu_t}{\sigma_t}\nabla\bar{f}\right] + S_m + S_{user} \quad (9)$$

The value of S_m is mass transfer that happen from liquid phase or reaction from another phase to gas phase. If simplified, the equation become like this:

$$\frac{\partial}{\partial}(\rho\overline{f'^2}) + \nabla(\rho\bar{v}\,\overline{f'^2}) = \nabla.\left[\frac{\mu_t}{\sigma_t}\nabla\overline{f'^2}\right] + C_g\mu_t(\nabla\,\bar{f})^2 - C_d\rho\frac{\varepsilon}{k}\overline{f'^2} + S_{user} \quad (10)$$

2.5WSGGM (Weight Sum-Gray Gases Model)

WSGGM based is used to determine wall heat flux and radiative source terms to be expected [8,13]. An approach foo absorption from exhaust gas is need to be done to see spread inside the furnace. Here is the equation:

$$\varepsilon = \sum_{i=0}^{I} a_{\varepsilon,i}(T)\left(1 - e^{-k_i p \vec{s}}\right) \quad (11)$$

$a_{\varepsilon,i}$ is the emissivity of the weighted factor for gray gases and quantity, k_i is the absorption coefficient of gray gases, p is the partial pressure of all the absorbed gases, $\vec{s}$ is *path length* [8].

3. Geometri dan Boundary Conditions

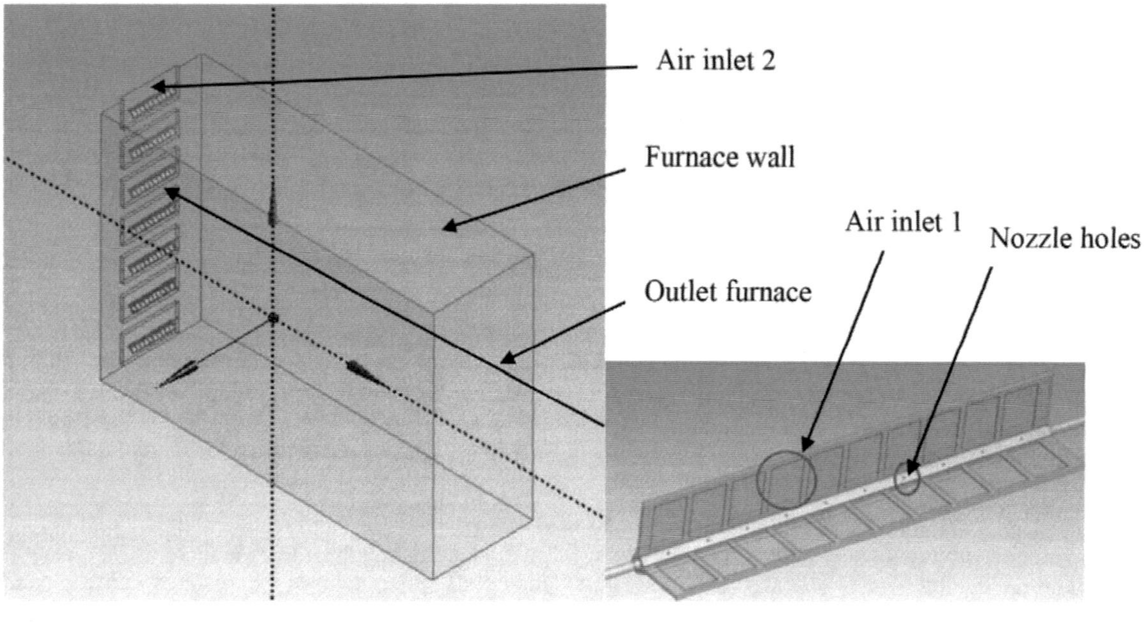

Figure 1 Furnace and burner are simplified in simulation.

The geometry of the simulation is a representation of the image in the data sheet. There are several forms in the furnace and a simplified burner. This is done to save computation time. However, in general it does not cause large error between the data sheet and the simulation

There are 5 boundaries conditions that must be set to run the simulation. The inlet air has 2 boundarys with air inlet 2 is an additional of air inlet 1 form combustion air. The concentration of combustion air and the inlet pressure of each load are different. Air concentration and inlet pressure can be seen in table 2. The hole nozzle is the way out of the fuel in table 2 and each variation load is set to have the same fuel concentration and inlet pressure. Wall is set with steel material. The outlet side has the same backflow temperature but different backflow pressure in table 2. This simulation has a tetrahedral mesh form. This mesh has 1334359 nodes and 7321245 elements.

Table 1. Boundary conditions input at simulation

	Nozzle burner	**Air inlet 1**	**Air inlet 2**	**Wall furnac e**	**Outlet**
Boundary conditions	Pressure inlet	Pressure inlet	Pressure inlet	Wall	Pressure outlet
Total pressure (Pa)	150.000	Load 50%= 1716 Load 75%= 1814 Load 100%= 2451	Load 50%= 1716 Load 75%= 1814 Load 100%=2451	-	Load 50%= 1500 Load 75%= 1500 Load 100%= 2000
Temperature (K)	300	Load 50%= 634 Load75%= 682 Load 100%= 685	Load 50%= 634 Load75%= 682 Load 100%=685	1000	Load 50%= 1000 Load75%= 1000 Load 100%= 1000
Tubulence model	Intensity and hydraulic diameter	Intensity and hydraulic diameter	Intensity and hydraulic diameter	-	Intensity and hydraulic diameter
Hydarulic diameter (m)	0,01	1,67	0,185	-	6,49
Material	Mixture	Mixture	Mixture	steel	Mixture

Table 2. Flue gas and fuel concentration inlet

	50%	**75%**	**100%**
Flue gas concentration	O_2: 15,17% N_2: 73,46% Ar: 0,88% CO_2: 2,42% H_2O:8,07%	O_2: 13,61 % N_2: 72,92 % Ar: 0,87 % CO_2: 3,17% H_2O:9,43%	O_2: 13,45% N_2: 72,56% Ar: 0,87% CO_2: 3,24% H_2O: 9,56%
Fuel concentration	CO_2:4,984% N_2: 0,065% CH_4: 86,039% C_2H_6:3,974% C_3H_8: 2,808% iC_4H_{10}: 0,596% nC_4H_{10}: 0,710% iC_5H_{12}: 0,285% nC_5H_{12}: 0,186% C_6^+: 0,353%	CO_2:4,984% N_2: 0,065% CH_4: 86,039% C_2H_6:3,974% C_3H_8: 2,808% iC_4H_{10}: 0,596% nC_4H_{10}: 0,710% iC_5H_{12}: 0,285% nC_5H_{12}: 0,186% C_6^+: 0,353%	CO_2:4,984% N_2: 0,065% CH_4: 86,039% C_2H_6:3,974% C_3H_8: 2,808% iC_4H_{10}: 0,596% nC_4H_{10}: 0,710% iC_5H_{12}: 0,285% nC_5H_{12}: 0,186% C_6^+: 0,353%
***Suppl. Firing air inlet temp.* (K)**	634	682	685
***Gas pressure* (pascal)**	150000	150000	150000

4. Result and Analysis

4.1 Validation model

Validation data use 2 parameters, are temperature and pressure. Those validation parameter use load 100%. From simulation result, point x=-3.75, y= 4.60, z=-1.26 used for temperature parameter.

Temperature read from the simulation is 1156.31 K while the value from data sheet is 1123 K, so the deviation values of temperature is 2.90%. The second validation use pressure at point x=-3.76,y=-3.83, z=-1.26, at that point the pressure value is 1894.39 Pascal, while the value from data sheet is 2000 Pascal, so the deviation value is 5.28%.

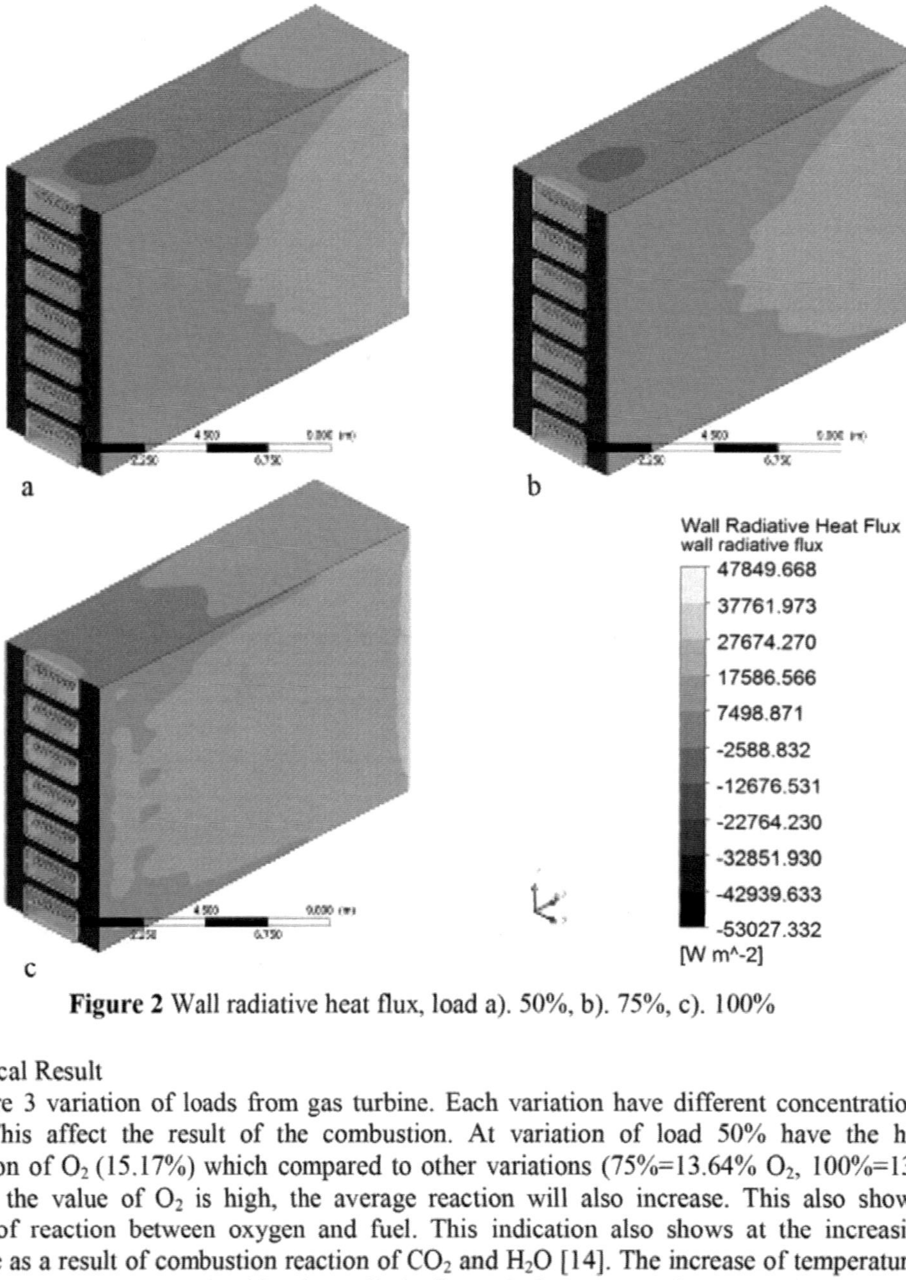

Figure 2 Wall radiative heat flux, load a). 50%, b). 75%, c). 100%

4.2 Numerical Result

There are 3 variation of loads from gas turbine. Each variation have different concentration and pressure. This affect the result of the combustion. At variation of load 50% have the highest concentration of O_2 (15.17%) which compared to other variations (75%=13.64% O_2, 100%=13.45% O_2). When the value of O_2 is high, the average reaction will also increase. This also shows the perfection of reaction between oxygen and fuel. This indication also shows at the increasing of temperature as a result of combustion reaction of CO_2 and H_2O [14]. The increase of temperature will affect to the radiation that received by the wall. At figure 2 shows at ranging are emitted (47849.558 W/m^2) and absorbed (-53027.332 W/m^2) radiation heat flux from the combustion. Dark color shows that it is closer to heat source, so the value will become smaller or negative.

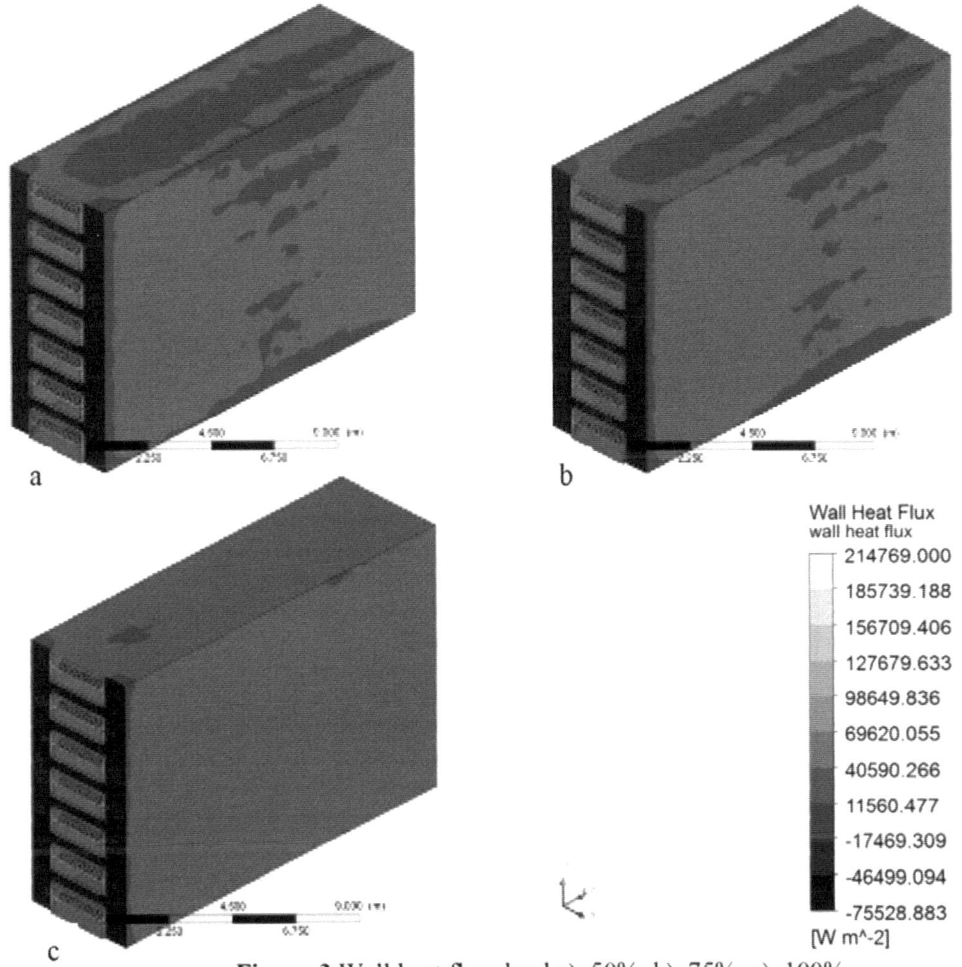

Figure 3 Wall heat flux, load a). 50%, b). 75%, c). 100%

The value of heat flux is the total from convection and radiation. From fgure 3, load 50% has darkest color compared to load 75% and 100%. If we see from top side of the furnace, load increase makes the color become brighter. Load 100% has the biggest flue gas input pressure that is 2451 Pascal and highest input temperature (685 K). If we see from the data sheet, load 100% has the highest mass flow flue gas that is 129.7 kg/s. Load 75% and 50 % each has 107,5 kg/s and 107.3 kg/s. Those show the convection value from load 100% is very high. Because the value of convection come from the heat flux minus the radiative heat flux. The value of radiatve heat flux in load 100 % is smallest than others. On the other side, those could be more faster for cooling process inside the furnace. This is strengthened with low O_2 and high value of CO_2 and H_2O. Molecule CO_2 and H_2O will hamper the perfection of the combustion.

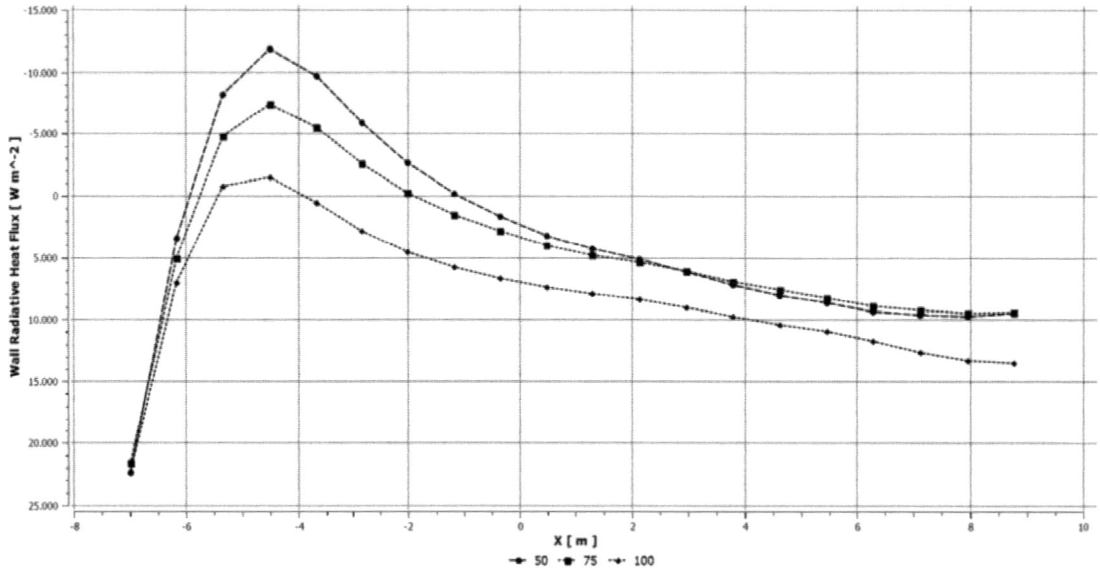

a). Wall radiative heat flux

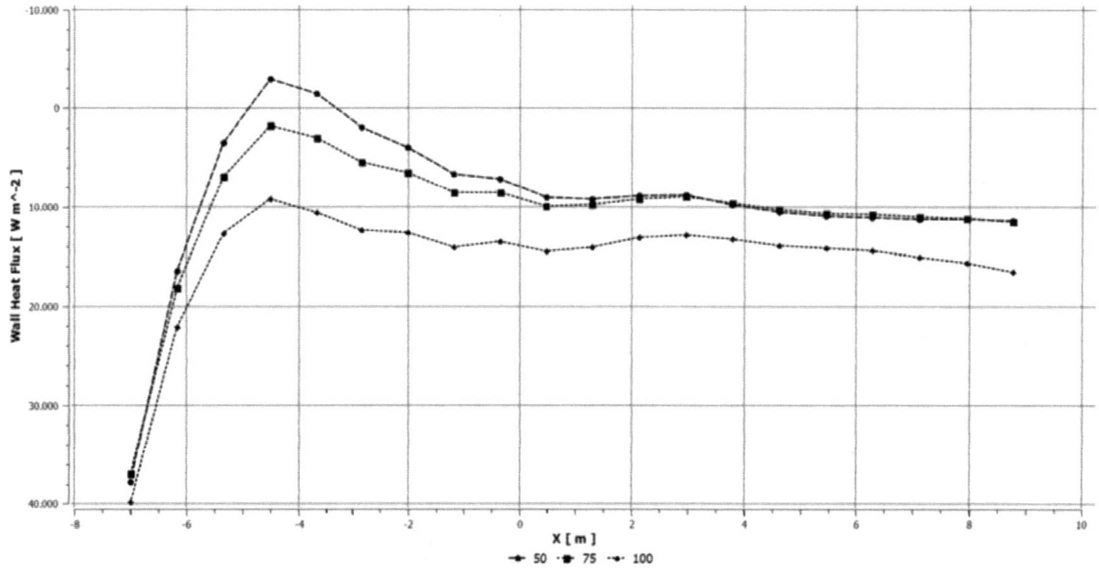

b). Wall heat flux

Figure 4 (a). Wall radiative heat flux and (b). Wall heat flux at x=8.77, y=5.11, z=-1.18

The first graphic shows the value of radiation wall that absorbed and emitted come back to another surface The graphic above is taken at x=8.77, y=5.11, z=-1.18 as long as x which is at the center from above the furnace. Load 50% shows the highest number of emit in (22413.5 W/m^2) and absorb radiation (-11847.50 W/m^2) the radiation of combustion in furnace. Even for the heat flux, load 50 % has the highest ranges, it means that total of the convection and radiation-emitting (37696 W/m^2) - absorbing (-2896.6 W/m^2) respectively.

5. Conclusion

This study aims to see the influence from variation of exhaust gas GT to radiation in the furnace. There are 3 loads, they are 50%, 75% and 100%. From the result of the analysis, this study conclude that load 50% emit radiation (22413.5 W/m^2) and absorb (-11847.5 W/m^2) higher than the other variations.

Acknowledgement

Thank you to PT Kaltim Daya Mandiri for helping this study by providing the data in this study.

References

[1] V. Ganapathy, *Steam Generators and Waste Heat Boilers: For Process and Plant Engineers*. 2014.

[2] M. A. Rajhi, R. Ben-Mansour, M. A. Habib, M. A. Nemitallah, and K. Andersson, "Evaluation of gas radiation models in CFD modeling of oxy-combustion," *Energy Convers. Manag.*, vol. **81**, pp. 83–97, 2014.

[3] E. P. Keramida, H. H. Liakos, M. A. Founti, A. G. Boudouvis, and N. C. Markatos, "Radiative heat transfer in natural gas-fired furnaces," *Int. J. Heat Mass Transf.*, vol. **43**, no. 10, pp. 1801–1809, 2000.

[4] K. Andersson and F. Johnsson, "Flame and radiation characteristics of gas-fired O2/CO2 combustion," *Fuel*, vol. **86**, no. 5–6, pp. 656–668, 2007.

[5] G. P. Merker, C. Schwarz, G. Stiesch, and F. Otto, *Günter P. Merker · Christian Schwarz · Gunnar Stiesch · Frank Otto Simulating Combustion*. .

[6] R. Prieler, B. Mayr, M. Demuth, D. Spoljaric, and C. Hochenauer, "Application of the steady flamelet model on a lab-scale and an industrial furnace for different oxygen concentrations," *Energy*, vol. **91**, pp. 451–464, 2015.

[7] R. Prieler, B. Mayr, M. Demuth, B. Holleis, and C. Hochenauer, "Prediction of the heating characteristic of billets in a walking hearth type reheating furnace using CFD," *Int. J. Heat Mass Transf.*, vol. **92**, pp. 675–688, 2016.

[8] M. Ghadamgahi, P. Ölund, T. Ekman, N. Andersson, and P. Jönsson, "A Comparative CFD Study on Simulating Flameless Oxy-Fuel Combustion in a Pilot-Scale Furnace," *J. Combust.*, vol. 2016, pp. 1–11, 2016.

[9] D. A. Granados, F. Chejne, J. M. Mejía, C. A. Gómez, A. Berrío, and W. J. Jurado, "Effect of flue gas recirculation during oxy-fuel combustion in a rotary cement kiln," *Energy*, vol. **64**, pp. 615–625, 2014.

[10] E. Karampinis, N. Nikolopoulos, A. Nikolopoulos, P. Grammelis, and E. Kakaras, "Numerical investigation Greek lignite/cardoon co-firing in a tangentially fired furnace," *Appl. Energy*, vol. **97**, pp. 514–524, 2012.

[11] H. B. Vuthaluru and R. Vuthaluru, "Control of ash related problems in a large scale tangentially fired boiler using CFD modelling," *Appl. Energy*, vol. **87**, no. 4, pp. 1418–1426, 2010.

[12] S. Septiawan, "Studi Numerik Karakteristik Pembakaran Natural Gas di dalam Boiler Furnace dengan Variasi Sudut Swirl Vanes pada Radially Stratified Flame Core Burners," vol. **2**, no. 1, 2013.

[13] R. Johansson, K. Andersson, B. Leckner, and H. Thunman, "Models for gaseous radiative heat transfer applied to oxy-fuel conditions in boilers," *Int. J. Heat Mass Transf.*, vol. **53**, no. 1–3, pp. 220–230, 2010.

[14] M. A. Nemitallah and M. A. Habib, "Experimental and numerical investigations of an atmospheric diffusion oxy-combustion flame in a gas turbine model combustor," *Appl. Energy*, vol. **111**, pp. 401–415, 2013.

IOP Conf. Series: Journal of Physics: Conf. Series **1090** (2018) 012020 doi:10.1088/1742-6596/1090/1/012020

Analysis Service Engineering Development Lifecycle: an Object-Oriented Approach

Ade Hodijah

Computer Engineering and Informatics, Bandung State of Polytechnic (POLBAN)
Jl. Geger Kalong Hilir, Ds. Ciwaruga, Parongpong, Kab. Bandung Barat, Jawa Barat 40012
e-mail: adehodijah@jtk.polban.ac.id

Abstract. Service Engineering (SE) is defined as a set of activities introducing a new business service. An object-oriented methodology is defined as principles and procedures to develop object-oriented software. This paper proposes an object-oriented methodology embedded SE framework in re-defining business service as a service-oriented application with detail implementation by using Activities and Artifacts. The proposed model for SE lifecycle consists of Process, Notation, and Tools modeled by utilizing Unified Process (UP), Business Process Model and Notation (BPMN), a programming language that supports the Business Process Management System (BPMS). In this research, the explanation is organized into three parts: (1) the brief review of the existing SE framework, (2) the elaboration of the proposed model, and (3) the detail of the SE lifecycle model with survey results from case studies. The framework is proposed to help practitioners and researchers to conduct service-oriented application by employing principles and methodology offered by the object-oriented approach.

1. Introduction

SE represents software engineering process at a higher level of abstraction. If the main purpose of software engineering is a software system, the main purpose of SE is business service system [1]. Business service in this research refers to a common definition of a service as business activity provided by a service provider to service consumer in order to create value for the customer [2]. There are three main factors to make an organization get more value than other competitors, which are a business organization, product or services produced, and ongoing business process organization [3].

There is a method to evaluate and improve business process, called Business Process Management (BPM) [3]. The use of BPM method in this research defines an SE framework into two workflows process of UP disciplines. Those workflows are Business Modeling of a current business process (As-Is) and implemented business process (To-Be). Based on the result of Requirement analysis for a solution of information system or application. BPM is applied on the stage of the SE framework.

The purpose of the SE framework is to provide service in a layer of information and technology. Particularly, in the form of web services can automate the interaction between services. Hence, the enterprise architecture has to be redefined using the approach of Service Oriented Architecture (SOA). The SOA approach can integrate business process and information technology infrastructure. Analysis and design of service are derived from the organization's business process [4].

In SOA development requires a methodology, such as Service Oriented Unified Process (SOUP) from [5]. SOUP includes the complete stage of software engineering lifecycle for SOA development side (SOUP for Initial SOA Deployment) until the maintenance side (SOUP for Ongoing SOA

Content from this work may be used under the terms of the Creative Commons Attribution 3.0 licence. Any further distribution of this work must maintain attribution to the author(s) and the title of the work, journal citation and DOI.
Published under licence by IOP Publishing Ltd

Management). However, SOUP already explained the documentation, but it still leaves room for adaptation. Hence, this research gives a practical approach for SOA development by using a framework of SE [1].

A comprehensive SE framework for delivering business service as service-oriented application based on SOA methodology has been proposed by Suhardi et al. [6]. The SE framework consists of 8 activities grouped into three stages: (1) identification, (2) design, and (3) prototyping.

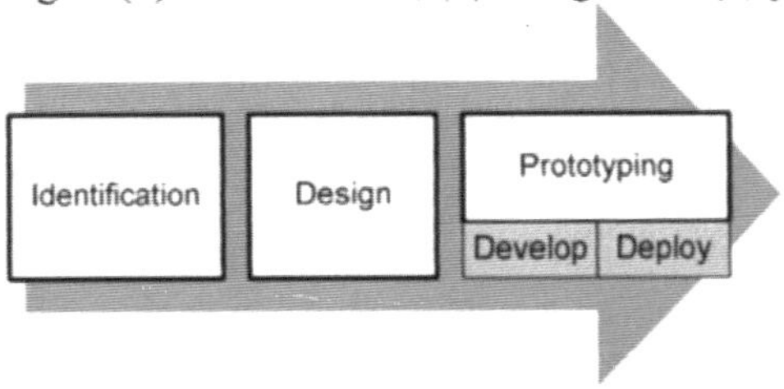

Figure 1. SE framework (adapted from [6])

The main analysis tool proposed in Identification stage that is the Business Model Canvas (BMC), then in Design stage uses Service Blueprinting. The use of BMC and Service Blueprinting aim to replace the process of service analysis and service design from SOA methodologies.

There are several technical platforms introduced in Prototyping stage along with proposed technology that is Microsoft's Windows Communication Foundation (WCF) [6]. To achieve the goal within the proposed lifecycle model in detailing Activities and Artifacts, Bizagi technology is recommended due to its rapid process automation. Bizagi support BPM and SOA with real-time analytics [7]. The BPM component consists of process modeling, process execution, and process monitoring [8]. Specifically, the SOA component is in the form of web service technology [1].

2. The Proposed Model

The idea of this research is to use True Requirement Investigation from Identification phase of SE framework [1] by using Business Modeling and Requirements phase of UP [9] as a prior workflows process for the design process. The Business Modeling tool uses flowchart as business process of As-Is. The Requirements uses basic analysis activities [10] to analyze the priority components which consists of: (1) list of problems, (2) list of directives, and (3) list of opportunity.

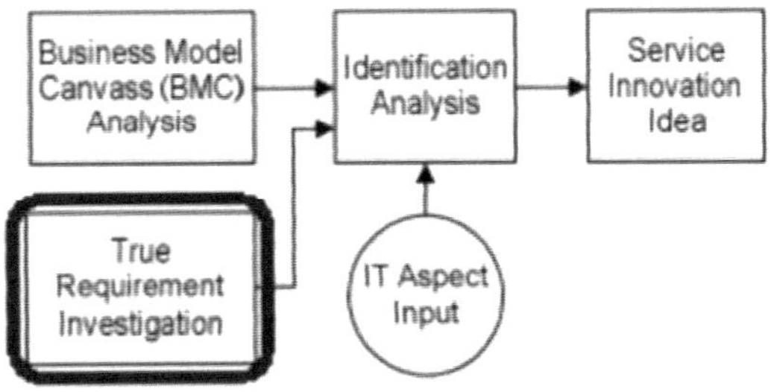

Figure 2. Identification stage: Business Modeling and Requirements as business process As-Is (adapted from [1], [9])

Furthermore, BPM is used in Identification Analysis by applying analysis value added to business service as service-oriented Idea [3].

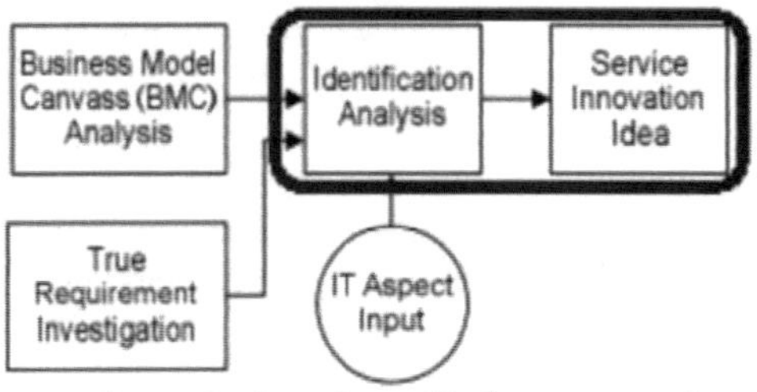

Figure 3. Identification stage: apply analysis value added to get service-oriented Idea (adapted from [1], [3])

Service Blueprinting can be used to describe service-oriented as business process of To-Be designed in SOA approached by using BPM notation (BPMN) model [8].

Figure 4. Design stage: modeling Service Blueprinting as business process of To-Be use BPMN (adapted from [1], [3], [8])

Bizagi as BPM system (BPMS) makes the implementation of immediate business requirements by providing monitor and continuously improve the business process [7]. In this research, for the Prototyping stage is only a Develop process which is performed.

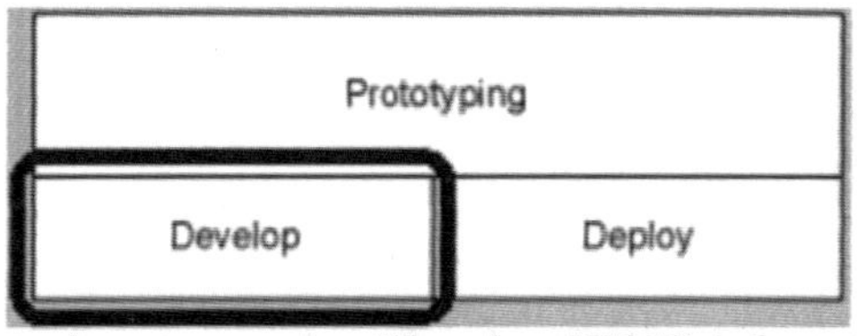

Figure 5. Prototyping stage: using Bizagi as BPMS (adapted from [1], [7])

Finally, an object-oriented methodology from UP [9] is used in detailing Activities and Artifacts with BPMN as a modeling language.

3. Results and Discussion

There are three phases of UP that are used in this research which are (1) inception, (2) elaboration, and (3) construction. Each phase consists of five workflows process: (1) business modeling, (2) requirements, (3) analysis & design, (4) implementation, (5) test.

Table 1. Activity and Artifact of UP (adapted from [9], [11])

UP Phase	Activity	Artifact
Inception [U1]	Analysis problem domain	- Domain model (class diagram without method)
	Analysis software	- Use case diagram
Elaboration [U2]	Analysis problem domain	- Revision of domain model (class diagram without method)
	Analysis software	- Revision of use case diagram - Use case scenario - Sequence diagram - Class diagram (analysis)
	Design software	- User interface / UI - Subsystem - Deployment diagram

Elaboration [U2]	Design software	- Throw-away prototype - Class diagram (design)
	Implementation software	- Source code
Construction [U3]	Design software	- Revision of UI - Revision of subsystem - Revision of deployment diagram - Revision of class diagram design
	Implementation software	- Revision of source code
	Testing integration	- Test plan - Test evaluation

Details of activities and artifacts from the table 1 show that the stages of software development have not been prepared for web service composition [6]. The software based on SOA consists of two parts: (1) a set of services as service provider and (2) client applications as service consumer [12]. Hence, the additional activities required for defining the service that is service analysis and service design with reference to SOA lifecycle [4]. In this research, the SOA lifecycle consists of: (1) service-oriented analysis, (2) service-oriented design, (3) service development, (4) service testing.

Table 2. SOA lifecycle (adapted from [1], [4])

SOA Lifecycle	Activities	Artifacts
Service-oriented analysis [S1]	Determine scope (problem domain)	- Business process of legacy system - Business requirement automation
Service-oriented design [S2]	Determine technology implementation	- Service candidate model (entity-centric, task-centric) - Service design for SOA layer (application, business, orchestration)
Service development [S3]	Implementation of service	- Service source code for each SOA layer - Web service catalog
Service testing [S4]	Test / evaluate the service	- Test plan for each web service in client side based on a design of the business process. Tools to test are: (1) unit test, (2) integration test, (3) system/service test, and (4) user acceptance test [1]. - Test evaluation

Notation for modeling Artifacts from table 2 uses BPMN with BPEL as a standard language for defining and executing BPM [7]. Technically, BPEL as a language that can be compiled by an application supporting orchestration of SOA layer [4]. BPMS role in this research is to support the technology needs of BPMN and BPEL. The following figure is the proposed lifecycle model for SE based on UP methodology.

Table 3. The proposed lifecycle model in detail

UP	SOA	Activities	Artifacts
U1	S1	Analysis problem domain [A1]	- Domain model (class diagram without method)
		Analysis business process (As-Is) [A2]	- Legacy system with glossary - Business process As-Is - Business specification: scope, software, hardware - Service automation candidate
		Analysis application requirements [A3]	- Use case diagram - Use case scenario / supplementary specification - Vision
U2	S2	Analysis problem domain [A1]	- Revision of domain model (class diagram without method) - Revision of legacy system with glossary - Revision of business specification: scope, software, hardware - Business process To-Be with value added
		Analysis service model (To-Be) [A4]	- Revision of service automation candidate - Service candidate model (entity centric, task centric)
		Analysis client software [A5]	- Revision of use case diagram - Revision of use case scenario - Revision of vision - Sequence diagram - Class diagram (Process Data ini Bizagi)
		Design service [A6]	- Service candidate and technology - Service operation - Deployment diagram for service - Business process To-Be in BPMN for service (Model the Process in Bizagi) - User interface for service (Forms Creation in Bizagi)
		Design client software [A7]	- Subsystem - Deploymen diagram for client software - Business process To-Be in BPMN for client software (Model the Process in Bizagi) - Revision of class diagram (Process Data ini Bizagi) - User interface for client software (Forms Creation in Bizagi)
		Design service architecture [A8]	- Infrastructure design for desktop/web/mobile application
U2	S3	Implementation [A9]	- Technology specification: software, hardware - Source code: model, controller, UI (Business Rules in Bizagi)
U2	S4	Testing integration [A10]	- Test plan: unit test, integration test, system/service test.
U3	S2	Analysis service model (To-Be) [A4]	- Revision of service candidate model (entity centric, task centric)

U3	S2	Design service [A6]	- Revision Class diagram (Process Data ini Bizagi) - Revision of Service candidate and technology - Revision of service operation - Revision of deployment diagram for service - Revision of business process To-Be in BPMN for service (Model the Process in Bizagi) - Revision of class diagram (Process Data ini Bizagi) - Revision of User interface for service (Forms Creation in Bizagi)
U3	S3	Design client software [A7]	- Revision of subsystem - Revision of deploymen diagram for client software - Revision of business process To-Be in BPMN for client software (Model the Process in Bizagi) - Revision of Class diagram (Process Data ini Bizagi) - Revision of User interface for client software (Forms Creation in Bizagi)
		Implementation [A9]	- Revision of infrastructure design for desktop/web/mobile application Revision of technology specification: software, hardware - Revision of source code: model, controller, UI (Business Rules in Bizagi)
U3	S4	Testing integration [A10]	- Revision of test plan: unit test, integration test, system/service test.

From table 3, the proposed lifecycle model is below (I-Inception; E-Elaboration; C-Construction; s-start; r-refine):

Table 4. Iteration of the proposed lifecycle model

Activity	**Artifact**	**I**	**E**	**C**
A1	Domain model (class diagram without method)	s	r	
	Business process To-Be with value added		s	
A2	Legacy system with glossary	s	r	
	Business process As-Is	s		
	Business specification: scope, software, hardware	s	r	
	Service automation candidate	s	r	
A3, A5	Use case diagram	s	r	
	Use case scenario / supplementary specification	s	r	
	Vision	s	r	
A4	Service candidate model (entity centric, task centric)		s	r
A5	Sequence diagram		s	r
A5, A6,A7	Class diagram (Process Data ini Bizagi)		s	r
A6	Service candidate and technology		s	r
	Service operation		s	r
	Deployment diagram for service		s	r

A6	Business process To-Be in BPMN for service (Model the Process in Bizagi)		s	r
	User interface for service (Forms Creation in Bizagi)		s	r
A7	Subsystem		s	r
	Deploymen diagram for client software		s	r
	Business process To-Be in BPMN for client software (Model the Process in Bizagi)		s	r
	User interface for client software (Forms Creation in Bizagi)		s	r
A8	Infrastructure design for desktop/web/mobile application		s	r
A9	Technology specification: software, hardware		s	r
	Source code: model, controller, UI (Business Rules in Bizagi)		s	r
A10	Test plan: unit, system		s	r

4. Case Studies

The first case study was performed before a definitive formal framework was devised. The goal was to provide an integrated e-government implementation [13]. The case study was started with a strategic analysis of the EKPPD (Evaluasi Kinerja Penyelenggaraan Pemerintahan Daerah) with BMC. Service blueprinting tool was only used lightly to specify the data dissemination service based on a request from SKPD (Satuan Kinerja Penyelenggaraan Pemerintahan Daerah). SOA method was employed to produce three layers of service: task-centric, entity-centric, and application utility. The service interface is described by using WSDL.

The second case study was done after the first draft of the framework was formalized. The case study employed SE based on an object-oriented methodology of room usage permit in POLBAN [14]. The work produced two BPM diagrams: as-is business process to describe existing situation and to-be business process to propose the new service based on Porter's value chain analysis in performing true requirement investigation. Modeling notation was complemented with basic UML diagrams: Use Case, Class Diagram, and Sequence Diagram.

The third case study also select the process of room usage permit in POLBAN, but focusing on the design of proposed external data service for the agency as an external user. The work was also done by following the completed and formalized framework. BPMN diagram and Bizagi were used as BPMS in supporting the simplicity of service analysis as SE Identification, service design as SE Design, and service implementation as Prototyping of SE.

The results from the proposed framework utilization in the third case study were taught in a Proyek 3 Diploma degree class for a semester. At the end of the semester, there are three questions of a survey that was conducted to 32 students of the class to assess the simplicity of using the proposed framework. The response are given under 5-scale likert respond, from "strongly agree" to "strongly disagree". The questions and their results are presented in the following charts.

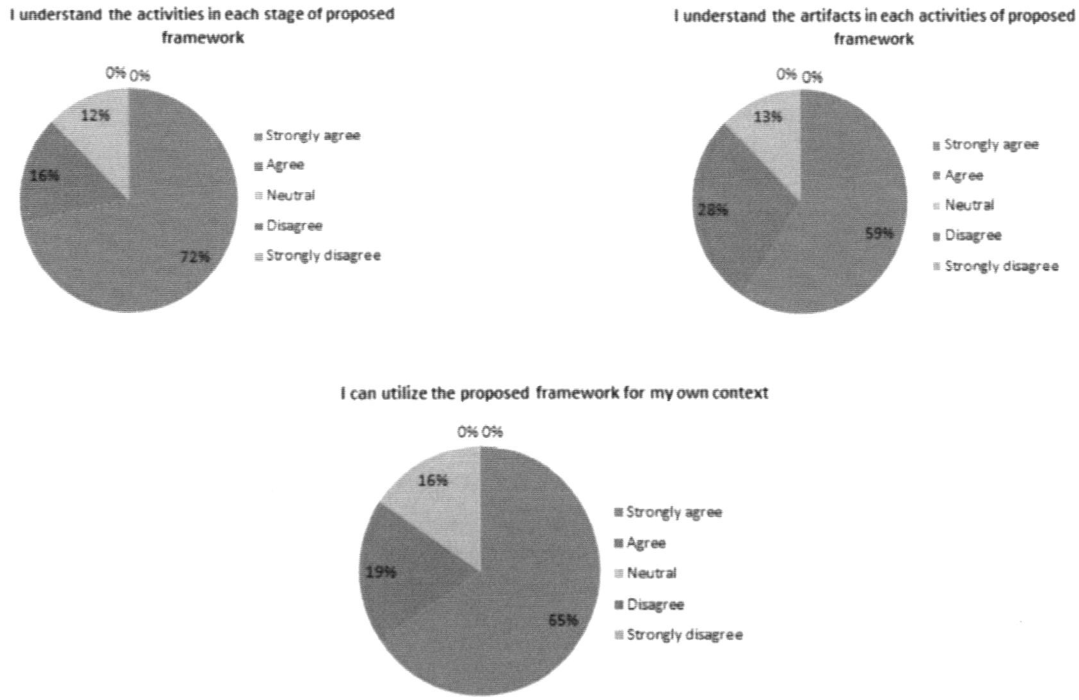

Figure 6. Survey results on comprehension to the proposed framework

The results from figure 6 show that more than 83% of the respondents agree that they able to adopt the proposed framework regarding the explicitness of activities and artifacts in implementing service-orientation application.

5. Conclusion

Alternative software development for adopting SOA lifecycle within SE framework is proposed here along with several trial case studies. The proposed framework is emphasized on the simplicity of the object-oriented approach by using specific tools, notation, and process for each activity and artifact of introducing a new business service.

There is technology to support business service as a service-oriented application, Bizagi as Tools. The technology is used to ensure the proposed service analysis is implementable by using BPMN as Notation for service design process. The analysis and design of service will continue to be improved implementation in each UP phase as Process.

Acknowledgments

Alhamdulillah, acknowledgment is delivered to all lecturers majoring in computer engineering and to all staff of UPPM Polytechnic State of Bandung who has given the opportunity to carry out "Penelitian Dosen Pemula 2016".

Reference

[1] Suhardi, Robin Doss, and Purnomo Yustianto, *Service Engineering Based on Service Oriented Architecture Methodology*, TELKOMNIKA, vol. 13, no. 4, December 2015.

[2] J Cardoso, K Voigt, and M Winkler, *Service engineering for the internet of services*, Enterprise Information System, pp. 15-27, 2009.

[3] Harmon, Paul. *Business Process Change, A Guide for Business Managers and BPM and Six Sigma Professional. 2nd Edition.* Burlington, USA. Morgan Kaufmann Publishers is an imprint of Elsevier. 2007. [Online].

[4] Erl, Thomas. *Service-Oriented Architecture (SOA): Concepts, Technology, and Design*. Prentice Hall, 2005.

[5] Mittal, Kunal. *Service Oriented Unified Process (SOUP)*. 2016. [Online].

[6] Suhardi, Pathya Budhiputra, and Purnomo Yustianto, *Service Engineering Framework: A Simple Approach*, in International Conference of Information Technology System and Innovation (ICITSI), Bandung-Bali. 2014.

[7] Bizagi Web Services (SOA Layer). [Online].

[8] Bajwa, Imran S. *SOA Embedded in BPM: A High-Level View of Object Oriented Paradigm*. 2011.

[9] Larman, Craig. *Applying UML And Patterns: An Introduction to Object-Oriented Analysis and Design and the Unified Process, 2nd Edition*. 2004.

[10] Jeffrey Whitten and Lonnie Bentley. *System Analysis and Design Methods*. New York, USA: McGraw-Hill, Inc. 2007.

[11] Desenta, Salma. *Development of a Configuration and Monitoring Tools for Software Project with Unified Process Methodology*. Undergraduate thesis. STEI-ITB. 2006.

[12] MATEI, Gheorghe. *SOA and BPM, a Partnership for Successful Organizations*. Informatica Economica. 2011.

[13] Hodijah, Ade. *Perancangan Sistem Portal Evaluasi Kinerja Penyelenggaraan Pemerintahan Daerah (EKPPD), Studi Kasus: Pemerintahan Kota Cimahi*. Thesis STEI-ITB. 2012.

[14] Hodijah, Ade, Joe Lian Min, dan Nurjannah Syakrani. *Rancang Bangun Aplikasi REST Web Service Berbasis Service Oriented Architecture Menggunakan Metodologi Unified Process*. Prosiding 7th IRWNS Politeknik Negeri Bandung. 2016.

IOP Conf. Series: Journal of Physics: Conf. Series **1090** (2018) 012021 doi:10.1088/1742-6596/1090/1/012021

Design Study of 600 MWt Long Life Modular Gas Cooled Fast Reactors

Fareha M.A., Ratna D. Syarifah, Zaki Su'ud, Neny Kurniasih.
[1]Nuclear and Biophysics Research Division, Faculty of Mathematics and Natural Sciences, Bandung Institute of Technology
E-mail: szaki@fi.itb.ac.id

Abstract. Design study of long life 600 MWt gas cooled fast reactors which can be operated for 25 years without refueling or fuel shuffling has been performed. A small Gas-cooled Fast Reactors (GFRs) core loaded with uranium nitride fuel (UN) and mixed nitride fuel (UN-PuN) have been calculation by SRAC code system and FI-ITB CH1 code. The results shows that it has been obtained some designs which can be operated for 25 years with relatively excess reactivity. The maximum excess reactivity is 2% dk/k. The reactor core is divided into three regions, the first regon is filled with highest content of plutonium and located in the most outer region. The second part of the core is filed with second highest plutonium content and located in the inner part of first region. The third region is filled wiith the lowest content of plutonium and located in the central part of the core. At the begnning of life the first reactor is the most dominant, but as time proceeds the second and the third regions of the core become more important. The average burn-up is around 10% HM

1. Introduction

As a lesson learned from the TMI II, Chernobyl, Fukushima Nuclear Accidents, inherent safety capability becomes necessary for next generation Nuclear Power Plants (NPPs) to avoid severe accidents causing core melting. Gas cooled fast reactors is one of the Generation IV NPP which has inherent safety capability based on optimal reactivity feedback[1-8]. Small and medium long-life modular reactors are very prospective for remote area with small-medium power consumption level such as many small islands in Indonesia[7-13].

In this study small/medium sized (600 MWt) gas cooled fast reactor designs which can be operated more than 20 years without refueling or fuel shuffling has been investigated. Considering helium gas limitation in transporting heat from the reactor cores, we limit average power density to be less than 75 W/cc. Therefore relatively large core designs were taken and pancake type cores are chosen to reduce severe heat transport burden in axial direction. Such system need high breeding/convertion ratio fuel to produce enough plutonium during its operation. To obtain NPP design with inherent safety capability, reactivity swing minimization is important to limit unprotected rod-run-out transient over power (UTOP) accident [8-13].

Content from this work may be used under the terms of the Creative Commons Attribution 3.0 licence. Any further distribution of this work must maintain attribution to the author(s) and the title of the work, journal citation and DOI.
Published under licence by IOP Publishing Ltd

2. Design Concept and Methodology

The reactor core configuration adopt 3 region active core model as shown in the Figure 1 as follows.[10-13]

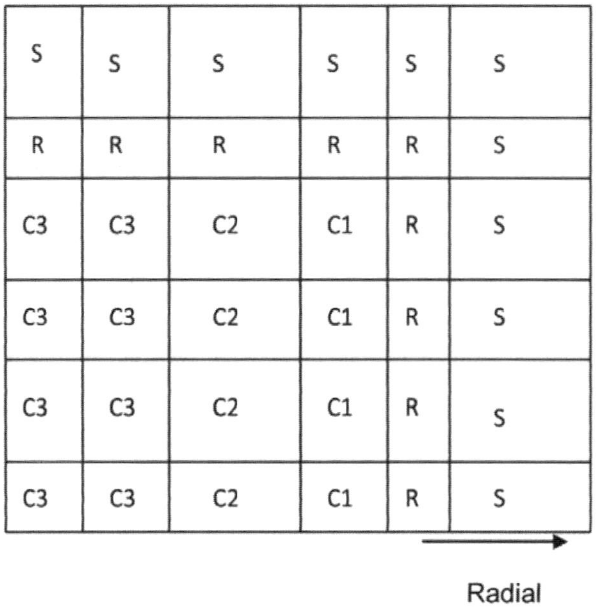

Figure 1 Reactor core configuration used in this study. C1,C2,and C3 is core region, R is reflector, and S is shielding

The configuration is adopted from previous study shown in the references. The Core I (C1) region is given highest plutonium content, the core II (C2) is given lower Plutonium content than C1 and the Core III (C3) is given relatively low Plutonium content. Therefore at the Beginning of life (BOL) the core I plays most important role as the power peak is located in this region. However as burn-up proceeds, the role of C2 and C3 increases and the power peak gradually shifted toward central direction. The lower content of Plutonium in the regions C2 and C3 give higher convertion ratio to enhance life time of the core.

The calculation method adopts Multi groups diffusion model with 60-70 energy groups using two dimensional R-Z geometry diffusion and Burnup Calculations. The burnup Calculation include 30 nucleus Uranium, Plutonium, Americium, Curium, and other minor actinides. The group constant is calculated using SRAC code system especially PIJ module[14-15]. The multigroup diffusion and burn-up calculations are conducted using FI-ITB CH1 code[16].

3. Results and Discussions

Table 1 shows the main parameters used in this study. The reactor power is 600 MWt of gas cooled fast reactor type. At the beginning we perform parametric survey calculations and then optimization process is conducted.

Table 1 Main parameters used in the study

No	Parameter	Value
1	Power(MWt)	600
2	Operation time (years)	25
3.	Fuel material	UN/PuN
4.	Number of main core regions	3
5.	Region radial width(cm) Reg 1/2/3	55/30/20
6.	Axial height (cm)	175 cm
7.	Reflector wit(cm)	30
8.	Pu content (%)	14/12.5/7
9.	Fuel, structural dan coolant volume fractions of Region 1	35%/20%/45%
10	Fuel, structural dan coolant volume fractions of Region 2	35%/20%/45%
11	Fuel, structural dan coolant volume fractions of Region 3	35%/20%/45%

Figure 2 shows k-inf parametric survey results for various Plutonium percentations. The calculations were performed for plutonium percentations from 8% to 15%

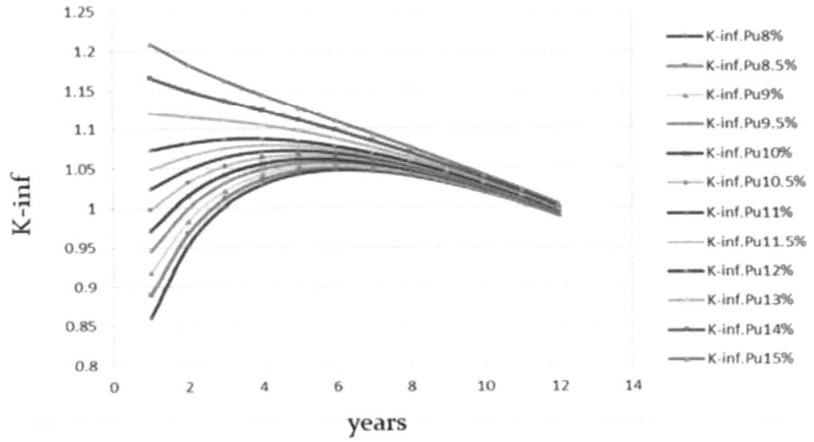

Figure 2 K-inf as a function of burnup time for various Plutonium percentations.

It is shown in figure 2 that for low plutonium content the k-inf initially less than 1.0 (sub critical) but the k-inf value tend to increase till 6 years and then decreases. On the other hands, fuel with the high plutonium percentage tend to continuously decreases with initial value much higher than 1.0 in order to achieve 10 years life time.

Figure 3 shows k-eff as a function of burnup time for optimized design. It is shown that the reactor can be operated for 25 years without refuelling or fuel shuffling with excess reactivity less than 2% dk/k. The K-eff pattern initially decreases, then around 6 years start to increase till about 23 years and then start to decrease. The k-eff pattern change is basically influenced by composition in each core regions in which the outer core has the highest content of Plutonium while the central core has the lowest content of Plutonium at the BOL.Therefore the conversion ratio of the central core is the highest. At the end of life (EOL) the central region becomes dominant.

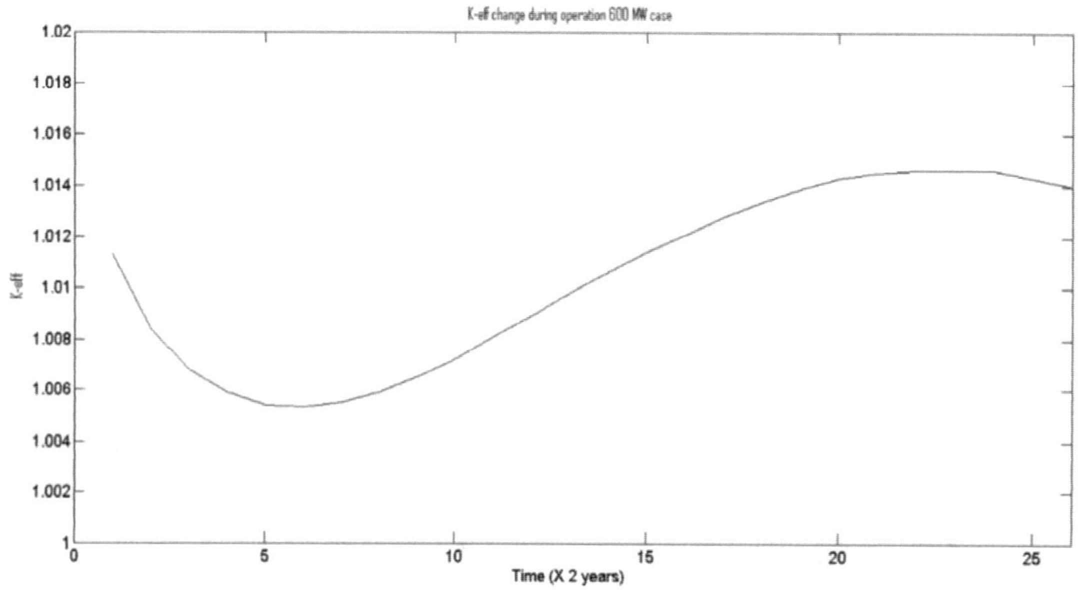

Figure 3 K-eff change with burn-up for optimized design

Figure 4 shows the k-inf pattern change with burn-up for the most outer core (Core I). It is shown that the K-inf pattern is monotonously decreases due to relatively high plutonium content. Figure 5 shows conversionratio change with burn-up for the Core I. During the operation conversion ratio is always above 1.0 but its value continuously decreases. At the BOL its value is about 1.1 and at the EOL its value decreases to about 1.01.

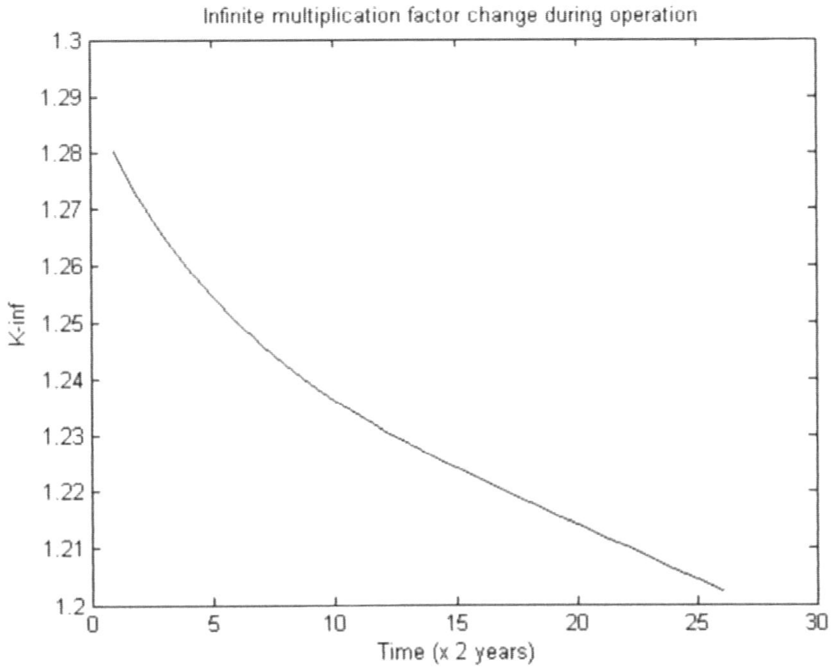

Figure 4 K-inf pattern change with burnup time for the Core I

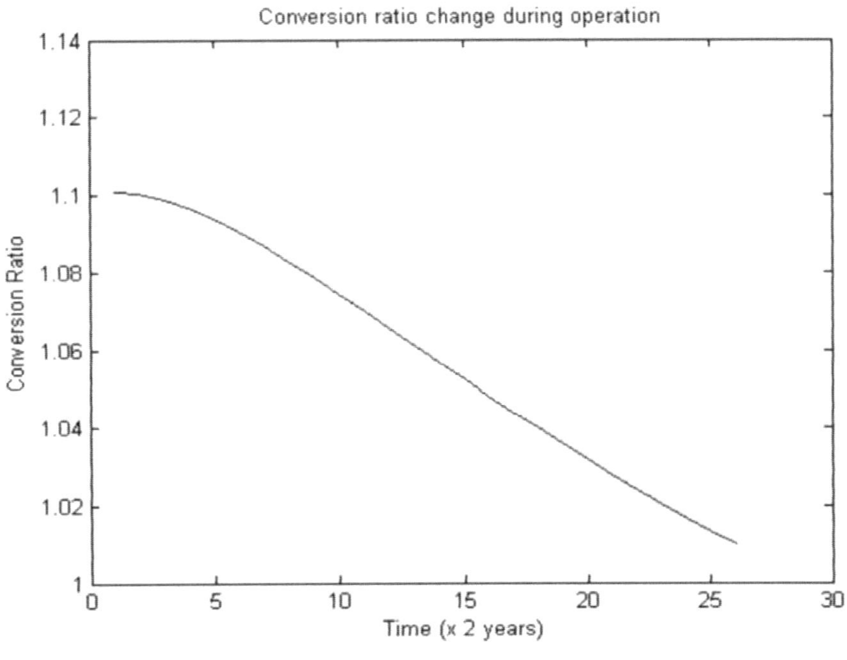

Figure 5 Conversion ratio pattern change with burnup time for the Core I

Figure 6 shows Pu-239 atomic density change for the Core I region. It is shown that the BOL the Pu-239 atomic density increases with monotonously reduced rate of change. At the EOL its value is relatively flat. At the EOL the U-238 to Pu-239 atomic density ratio is much smaller compared to that of BOL due to reduction of U-238 atomic density, as shown in Figure 7, and the increase of Pu-239 atomic density.

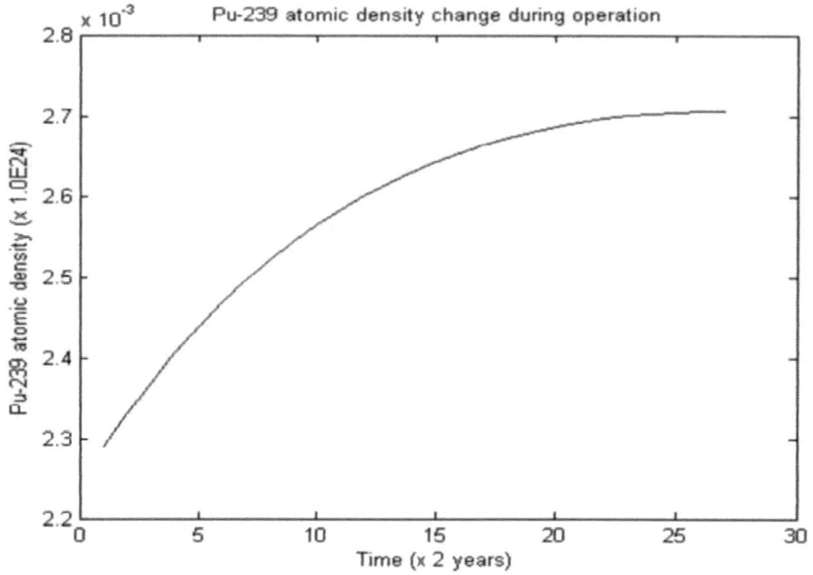

Figure 6 Pu-239 atomic density pattern change with burnup time for the Core I

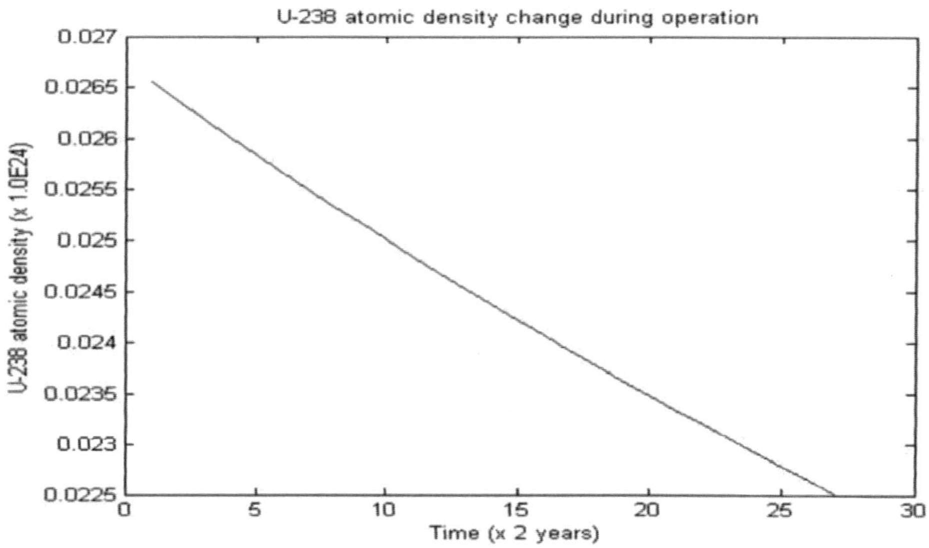

Figure 7 U-238 atomic density pattern change with burnup time for the Core I

Figure 8 shows K-inf pattern change with burn-up for the Core II. Its pattern is similar to that of the Core I but with lower value of K-inf due to lower Plutonium content. Figure 9 shows conversion ratio pattern of the Core II. It is shown that initially the conversion ratio increases till about medium of life (MOL) but then decreases. This pattern is different from that of The Core I. The Plutonium content in the core II is smaller than that of the Core I. The shift of the power peak from the outer part of the core (the Core I) to the center direction has also influence on the conversion ratio pattern.

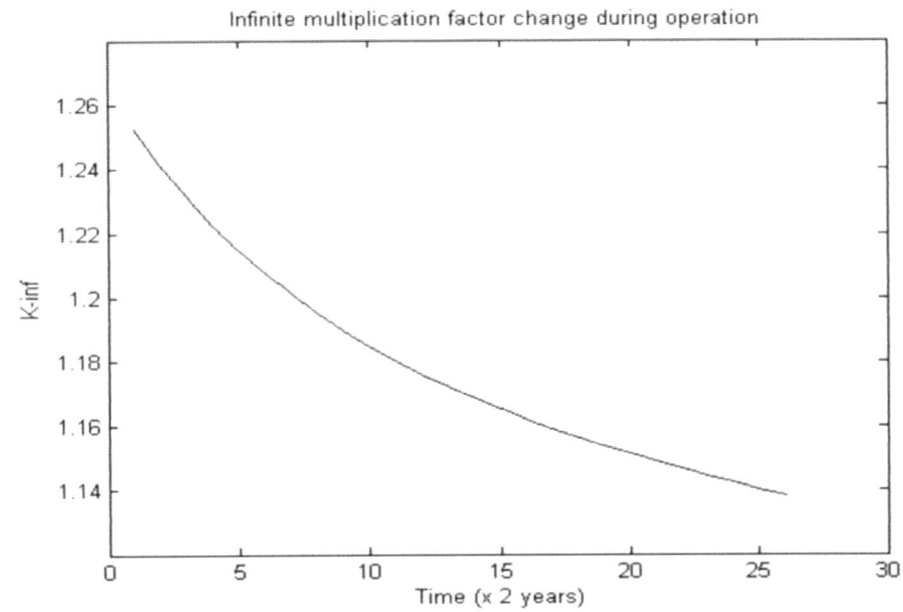

Figure 8 K-inf pattern change with burnup time for the Core II

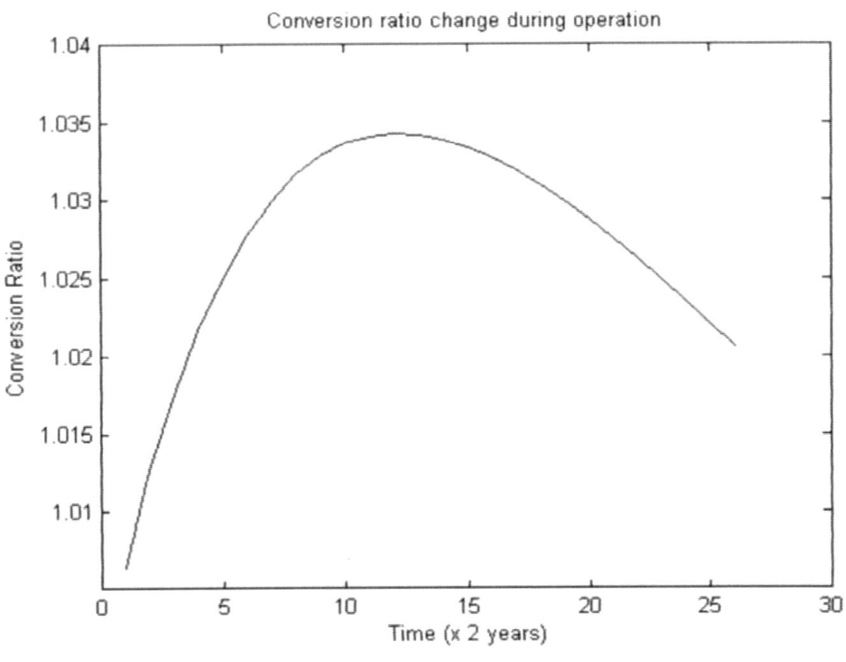

Figure 9 Conversion ratio pattern change with burnup time for the Core II

Figure 10 shows Pu-239 atomic density pattern change during burn-up. It is monotonously increases and even though the rate of increase decreases till the end of life there is still no reversal pattern.

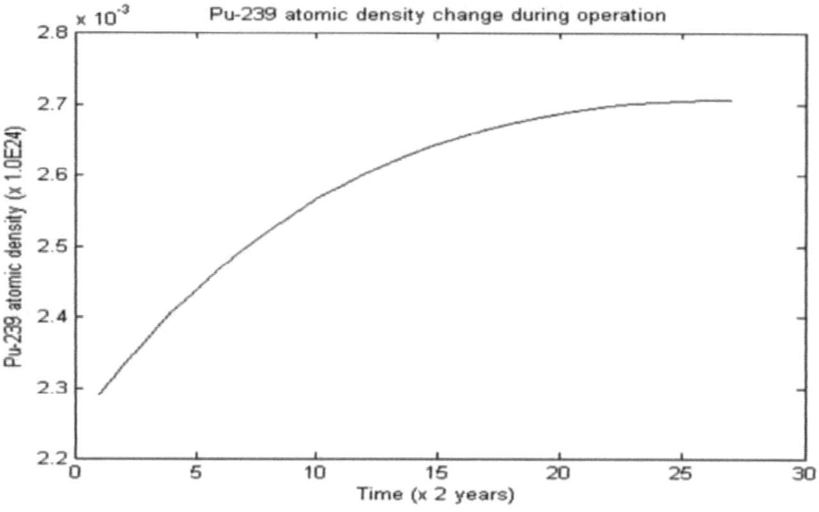

Figure 10 Pu-239 atomic density pattern change with burnup time for the Core II

4. Conclusion

Design study of long life 600 MWt gas cooled fast reactors which can be operated for 25 years without refueling or fuel shuffling has been performed. The results shows that it has been obtained some designs which can be operated for 25 years with relatively low excess reactivity. The maximum excess reactivity is smaller than 2% dk/k. The reactor core is divided into three regions, the first regon is filled with highest content of plutonium and located in the most outer region. The second part of the core is filed with second highest plutonium content and located in the inner part of first region. The third region is filled wiith the lowest content of plutonium and located in the central part of the core. At the beginning of life the first reactor is the most dominant, but as time proceeds the second and the third regions of the core become more important. The average burn-up is around 10% HM

References

[1] Zaki Su'ud and H Sekimoto, 2013, Annals of Nuclear Energy, The prospect of gas cooled fast reactors for long life reactors with natural uranium as fuel cycle input, Vol. 54 , pp. 58-66.

[2] M. Ariani, Z. Su'ud, A. Waris, Khairurrijal, N. Asiah, and M. A. Shafii, Effect of Fuel Fraction on Small Modified Candle Burn-Up Based Gas Cooled Fast Reactors, AIP Conf. Proc. 1325, p. 249

[3] Irka, F.H., Aryani, M., Suud, Z., Sekimoto, H., Desain study of gas cooled fast reactor with natural uranium as fuel cycle input using radial shuffling strategy, AIP Conference Proceedings 1448 , pp. 74-81, 2012

[4] Ariani, M., Su'ud, Z., Monado, F., Waris, A., Khairurrijal, Arif, I., Ferhat, A., Sekimoto, H., Optimization of small long life gas cooled fast reactors with natural Uranium as fuel cycle input, (2013), Applied Mechanics and Materials, 261-262, pp. 307-311

[5] Fiber Monado, Zaki Su'ud, Abdul Waris, Khairul Basar, Menik Ariani, Hiroshi Sekimoto, Application of Modified CANDLE Burnup to Very Small Long Life Gas-cooled Fast Reactor, Advanced Material Research(AMR), Vol. 772,2013, pp. 501-506

[6] Fiber Monado, Menik Ariani , Zaki Su'ud, Abdul Waris, Fiber Monado, Hiroshi Sekimoto, Power Flattening on Modified CANDLE Very Small Long Life Gas-cooled Fast Reactor, AIP Conf. Proc. Vol 1615, 2014, pp. 47-50

[7] RD Syarifah, Z Su'ud, K Basar, D Irwanto, SC Pattipawaej, M Ilham: Comparison of uranium plutonium nitride (U Pu N) and thorium nitride (Th N) fuel for 500 MWth Gas Cooled Fast Reactor (GFR) longlife without refueling, International Journal of Energy Research, Article in Press

[8] RD Syarifah, Y Yulianto, Z Su'ud, K Basar, D Irwanto: Neutronic Analysis of Thorium Nitride (Th, U233) N Fuel for 500MWth Gas Cooled Fast Reactor (GFR) Long Life without Refueling, Key Engineering Materials 733, 2017, p. 47-50

[9] DN Krisna, Z Su'ud: Study Neutronic of Small Pb-Bi Cooled Non-Refuelling Nuclear Power Plant Reactor (SPINNOR) with Hexagonal Geometry Calculation, Journal of Physics: Conference Series 799 (1), 2017, 012021

[10] RD Syarifah, Z Su'ud, K Basar, D Irwanto: Fuel Fraction Analysis of 500 MWth Gas Cooled Fast Reactor with Nitride (UN-PuN) Fuel without Refueling, Journal of Physics: Conference Series, Vol. 799 (1), 2017, 012022

[11] SC Pattipawaej, Z Su'ud: Preliminary Study of Gas Cooled Fast Breeder Reactor with Heterogen Percentage of Uranium–Plutonium Carbide based fuel and 300 MWt Power, Journal of Physics: Conference Series 799 (1), 2017, 012029

[12] Su'ud, Z, Design study of Small Pb-Bi cooled Non-refueling Nuclear Power Reactors (SPINNORs), (2013), Applied Mechanics and Materials, 261-262, pp. 296-301

[13] H.Sekimoto and S. Zaki, 1995, Nucl. Technol. , Design study of lead and lead-bismuth cooled small long-life nuclear power reactors using metallic and nitride fuel, 109,p. 307-313.

[14] K. Okumura et al., 2002, The Comprehensive Nuclear Calculation Code System (SRAC), JAERI Report.

[15] K. Okumura, K. Kaneko, and K. Tsuchihashi , 2005, SRAC95: Generap Purpose Neutronics Code System, JAERI Data/Code 96-015, JAERI, Japan

[16] Zaki Su'ud et al. ,2000, FI-ITB CHI Neutronic code manuals, internal report

Optimization of texture of the multiple textured lubricated contact with slip

M Muchammad*, N Sinaga, B Yunianto, Arijanto, M F Noorkarim, M Tauviqirrahman

Mechanical Engineering Department, Engineering Faculty, Diponegoro University, Semarang, Central Java, Indonesia

*m_mad5373@yahoo.com

Abstract. Development in surface modifications including texturing and boundary slip has shown promising outcomes in enhancing the hydrodynamic lubrication performance. In this work, the modified Reynolds equation was considered in order to evaluate the tribo-characteristics of partially textured contact with boundary slip. Finite volume method coupled with tridiagonal matrix algorithm was used to solve non-linear Reynolds theory. For maximizing the load support, the optimization procedure was carried out using the exact optimization method. Results showed encouraging improvements in load support behaviour by shifting the multiple-texture to exit zone of the contact. It was also confirmed that the improvement of the load support of around 300% using the optimized textured lubricated contact could be achieved.

1. Introduction

Recently, the research related to textured contacts was paid much attentions by researchers. This is because texturing method is able to enhance lubrication characteristics. From the perspective of hydrodynamic lubrication theory, some significant efforts have been directed toward studying the mechanisms of textured features in more detail. Yagi et al. [1] investigated the magnitude of the load-carrying capacity enabled by textured feature in hydrodynamic lubrication. Two boundary conditions were compared, i.e. periodic boundary condition and an atmospheric boundary condition. Zhang et al. [2] used a rectangular array of circle dimples to modify the film thickness profile. They found that the load support of lubricated contact may be enhanced through appropriate arrangement of textures partially covering its sleeve. In recent lubrication, Shinde and Pawar [3] based on the multi-objective optimization approach determined the optimal surface texturing parameters to improve performance of journal bearing.

In addition to surface texturing, the boundary slip became a popular method to enhance the lubrication performance. Lin et al. [4] studied the influence of large-area texture/slip surface, on the performance of lubricated sliding contact. It was found that texture/slip surface would not affect the pressure and load support when it locates at cavitation zone. Tauviqirrahman et al. and Muchammad et.al [5,6] evaluated the location of the boundary slip on the textured surface based on modified Reynolds equation considering cavitation. Susilowati et al. [7] studied the influence of gap ratio of the parallel sliding contact on the performance of slip-textured bearing. The authors found that the gap ratio affect significantly the friction force and the volume flow rate. In general, combined method of texturing with the boundary slip gives more advantages compared to pure texturing with respect to the load support as

Content from this work may be used under the terms of the Creative Commons Attribution 3.0 licence. Any further distribution of this work must maintain attribution to the author(s) and the title of the work, journal citation and DOI.
Published under licence by IOP Publishing Ltd

well as the friction. Thus, the study of improvement of tribological performance via boundary slip and surface texturing and simultaneously have received an explosion of interest recently.

Starting from this main frame, the main objective of the present paper is to determine the pattern of multiple partially textured surface which is able to maximize the resulted load support. The position of the multiple texturing on the lubricated contact is of particular interest. The exact optimization method is used to optimize such pattern.

2. Methodology

In classical mechanic, the derivation of classical Reynolds equation with a Newtonian lubricant is based on assumption of no-slip between the lubricant and the contacting surfaces. In the present work, the model of lubrication is based on the fact that slip in the lubricant exists in the interface. The proposed wall-slip model leads to a modified Reynolds equation. Such lubrication can be described as follows [5]:

$$\frac{\partial}{\partial x}\left(\frac{h^3}{\eta}\frac{\partial p}{\partial x}\left(1+\frac{3\alpha\eta}{h+\alpha\eta}\right)\right)=\frac{\partial}{\partial x}\left(6Uh\left(1+\frac{\alpha\eta}{h+\alpha\eta}\right)\right) \tag{1}$$

The physical meanings of the symbols in Equation 1 are as follows: h the film thickness (gap) at location, p the lubrication film pressure, α the slip coefficient, and η the lubricant viscosity. The modified Reynolds equation (Eq. (1)) is discretized over the flow using the finite volume method, and is solved using tridiagonal matrix algorithm (TDMA). By employing the discretization scheme, the computed domain is divided into a number of control volumes using a grid with uniform mesh size. The grid independency is validated by various numbers of mesh sizes.

For all following computations, the parameters used with respect to the slip-textured characteristics as well as their operating conditions are described in detail in Table 1. The bearing configuration studied here as reflected in Fig. 1 is multiple textured contact which consists of number of texture cells with slip introduced at the leading edge of the contact for improved load support [5]. In this work, it is assumed that the exit length is constant. In the present study, the Reynolds cavitation model is adopted.

Table 1. Main parameters of the bearing.

Parameters	Symbol	Value
Bearing length	B	0.02 m
Inlet length	a	0.0004 m (a/B = 0.02)
Pocketed zone length	b	0.0006 m (b/B = 0.03)
Exit land breadth	c	0.010 m (c/B = 0.5)
Sliding velocity	U	1 m/s
Viscosity	η	0.01 Pa s
Atmosphere pressure	P_{atm}	100 kPa
Minimum film thickness	h_o	1 μm
Pocket depth	h_d	2.5 μm
Total film thickness	h_p	3.5 μm
Slip coefficient	α	0.02 m^2s/kg

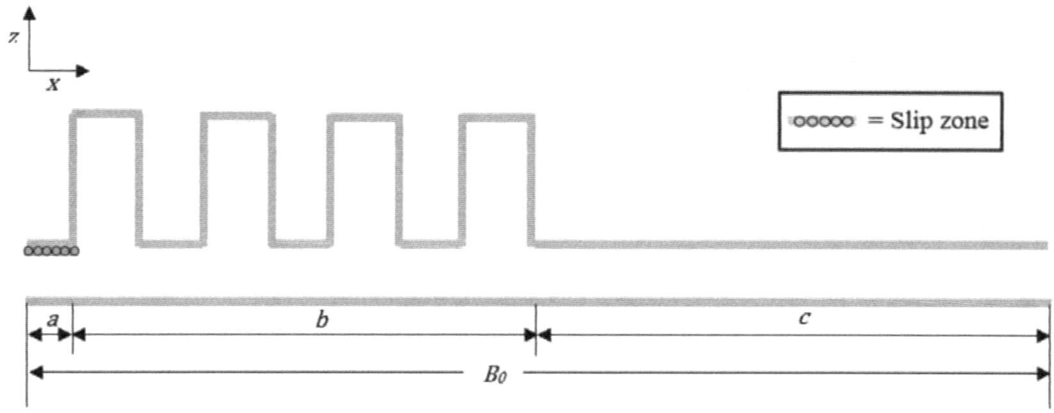

Figure 1. Schematic representation of multiple textured bearing with slip.

In order to maximize the performance of lubrication, the texturing parameters (position of texture cells, pocket length and inlet length) are optimized. The object of optimization is to maximize the hydrodynamic load support. The optimization analysis is performed using developed computer code. The design variables, and objective function are referred to as the optimization variables. The design variables are slip zones a and texture zone b as indicated in Fig. 1. Design variables are independent quantities that area varied in order to achieve the optimum design. The objective function is the dependent variable that we are attempting to maximize and in this case the objective function is the maximum dimensionless load support W. The simulation results will be presented in dimensionless form, i.e. P for dimensionless pressure ($P = ph^2 / \eta BU$), W for dimensionless load support ($W = wh^2 / (U\eta B_0^2)$) in which w is the load per unit length.

3. Results and Discussion

Figure 2 presents the optimization results of the hydrodynamic pressure. The initial design of the slip-textured pattern as a starting design for the optimization simulation is defined as Table 1. It can be seen from Figure 2 the hydrodynamic pressure induced by the initial design is much lower than by the optimal design. From the physical point of view, the longer slip inlet length of the optimal design gives more positive effect in altering the flow characteristic, i.e. enhancing the hydrodynamic pressure. As is known from published works [5-8], the presence of the slip length at the leading edge of the contact can retard the cavitation phenomena. This prevents the rupture of the film leading to the reduced load support.

Figure 2 also reflects that the improvement of the hydrodynamic pressure and thus the enhanced load support can be achieved by shifting the multiple-texture to exit zone of the contact. This result is clearly shown in Figure 3. It can be seen that for the case of optimal texture, the distance between texture cells becomes closer. This may be the second reason why the load support becomes larger for optimal case. More lubricant may trap in texture cells, so the gradient pressure induced by the dynamic of the lubricant is faster and larger.

With respect to the texture cell profile, it can be observed from Figure 3 that pocket length for optimal design is larger than the initial design. This is consistent with the previous result mentioned earlier which states that more reservoir (i.e. texture) for lubricant may alter the flow and lead to the enhanced hydrodynamic pressure.

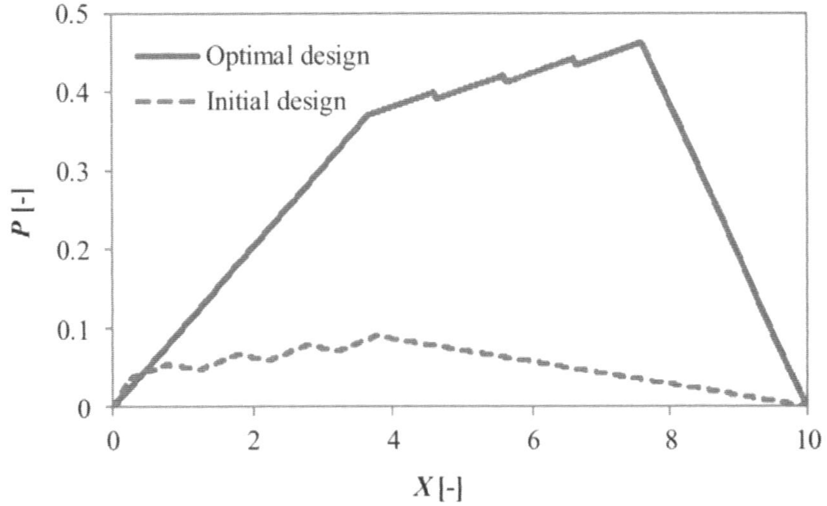

Figure 2. Comparison of the hydrodynamic pressure obtained between initial design and optimal design of multiple slip-textured pattern.

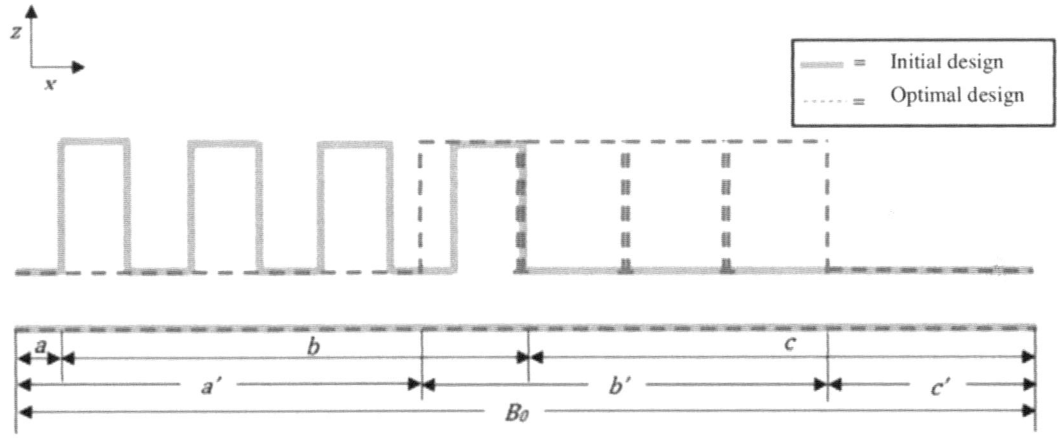

Figure 3. Comparison of the scheme of the multiple slip-textured pattern hydrodynamic pressure between initial design and optimal design.

Figure 4 shows the the global solution of dimensionless load support W for each textured pattern during optimization process. It can be seen that the load support increases with increasing the number of combination data (i.e. slip and texture pattern) and finally reaches the steady condition. While Table 2 presents the optimization result (i.e. the optimal design) in comparison with the initial design. From this table, the improvement of the load support of around 300% using the optimized textured lubricated contact can be achieved.

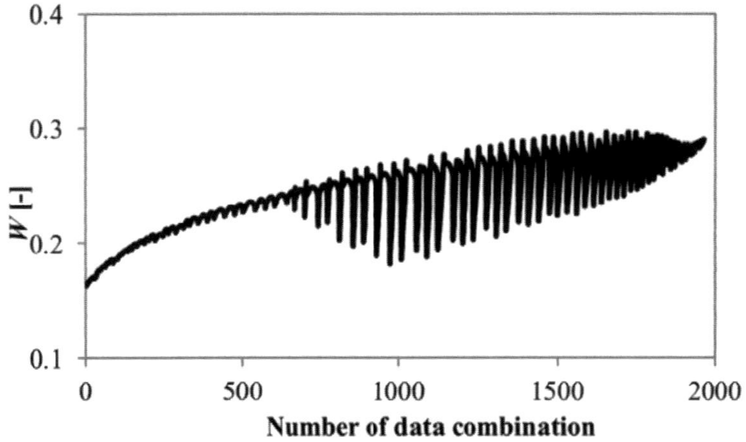

Figure 4. Prediction of dimensionless load support W during optimization process.

Table 2. Comparison of the main parameters between initial design and optimal design of multiple slip-textured pattern.

Parameter	Initial pattern	Optimal pattern
Inlet length [mm]	0.5	7.3
Pocket length [mm]	1.5	9.2
Outlet length [mm]	10	3.5
Load carrying capacity [N]	3.01×10^5	1.18×10^6

4. Conclusion

In the present work, the texture parameters of the slip-textured contact are optimized. The modified Reynolds equation with slip was used to solve the lubrication combined with optimization library based on the exact optimization method. Based on the explanation above, the main conclusion can be drawn as follows:

1. The load support behaviour can be improved by shifting the multiple-texture to exit zone of the contact.
2. Based on the optimization results, the improvement of the load support of around 300% using the optimized textured lubricated contact can be achieved.

References

[1] Yagi K, Sato H and Sugimura J 2015 *Tribol. Online* **10** 232
[2] Zhang H, Dong G, Hua M, Guo F and Chin KS 2015 *Ind. Lubr. Tribol.* **67(4)** 359
[3] Shinde AB and Pawar PM 2017 *Tribol. Int.* **In press** http://dx.doi.org/10.1016/j.triboint.2017.04.041.
[4] Lin Q, Wei Z, Wang N and Chen W 2015 *Ind. Lubr. Tribol.* **67(3)** 216
[5] Tauviqirrahman M, Ajie WK, Yohana E, Muchammad M and Jamari J 2016 *Int. J. Eng. Tech.* **8(2)** 913
[6] Muchammad, Tauviqirrahman M, Jamari and Schipper DJ 2017 *Tribol. Online* **12(2)** 67
[7] Susilowati S, Tauviqirrahman M, Jamari J and Bayuseno AP 2016 *Tribol. – Mater. Surf. Interfaces* **10(2)** 85
[8] Muchammad M, Tauviqirrahman M, Jamari J and Schipper DJ 2017 *Lubr. Sci.* **29(3)** 133

Solving Unconstrained Global Optimization Problem Using Parameter Free Filled Function in Cooperation with Jameson Gradient Based Method

Herlina Napitupulu[1], Ismail Bin Mohd[2], Endang Soeryana Hasbullah[3], Sukono[4], Sudradjat Supian[5]

[1)]School of Informatics and Applied Mathematics
Universiti Malaysia Terengganu
Kuala Terengganu, Malaysia
[2)]Institute for Mathematical Research (INSPEM)
Universiti Putera Malaysia
Selangor, Malaysia
[1,3,4,5)]Department of Mathematics
FMIPA Universitas Padjadjaran
Sumedang, Indonesia

napitupuluherlina@gmail.com

Abstract. Global optimization problem still becomes a challenges due to the problem on locating the global optimum of multimodal function. How to reach the better minimizer from the current minimizer and how to decide that the obtained minimizer is the desired one are both major challenges on solving global optimization problem. Filled function method is one of the recent considered deterministic easy applied methods which concerned to the mentioned problems. The basic concept of filled function method is firstly by minimizing the objective function (first phase) then to build such an auxiliary function which to be minimized (second phase) in order to locate a point with lower function value than the current minimizer of the objective function. In the second phase, a local minimization method can be applied. Newton's method is considered to be fast method on finding the zero of gradient of quadratic function, but may be very expensive or infeasible to determine the Hessian matrix in the case of complex problems. The Jameson gradient based method is the search procedures which avoid the need to store an estimate of the Hessian as well as its inverse and do not require exact line searches. In this paper, an algorithm in cooperation of parameter free filled function method and Jameson gradient method are introduced for solving global optimization problem with two variables. The algorithm is implemented to some benchmark test function. The numerical performance of the method on solving two-dimensional global optimization problems is presented.

1. Introduction

In Mathematical programming, global optimization problems has become a high concern topic, since the application spread widely such as in engineering, biology, medicine, and many more. Generally, the difficulty of designing algorithm method for global optimization is how to determine the obtained minimizer is a global one.

Content from this work may be used under the terms of the Creative Commons Attribution 3.0 licence. Any further distribution of this work must maintain attribution to the author(s) and the title of the work, journal citation and DOI.
Published under licence by IOP Publishing Ltd

Flled function method firstly introduced by Ge [1], with the two adjustable parameter. A number of filled functions modifications from [1] were proposed. However, there is no efficient criterion to choose the parameter appropriately. Then a question appears to be; Is it possible to build a free-parameter filled function? Ma et al. [2] proposed a parameter free filled function with n variable, however this function discontinuous in particular point. Mohd et al. [3], have proposed a parameter free filled function which is successfully solving single variable multi modal of general global optimization problems. Referring to this idea of filled function, Napitupulu et al [4] proposed a parameter free filled function, combining together with partial radius of curvature, Newton method, steepest descent for solving two variables of twice continuously differentiable objective function. Napitupulu and Mohd [5] proposed the modification of vector direction in [4] for solving two dimensional global optimization problem. The filled function are more preferred to be used among others deterministic method since more efficient in computationally and several advantages.

2. Filled Function Method

The filled function considered in this paper is a special function as a bridge for one point to jump from a basin to a lower basin until a global minimizer of a considered multimodal function $f: R^2 \to R^1$ can be found. The following three assumptions are applied for this process.

Assumption 1 $f(x)$ is a twice continuously differentiable function

Assumption 2 $f(x)$ has only a finite number of minimizers

Assumption 3 $f(x) \to \infty$ as $\|x\| \to \infty$

The Assumption 3 implies the existence of a closed compact bounded domain $D \subset R^n$ such that D contains all isolated minimizers of $f(x)$ and the value of $f(x)$ when x is on the boundary of D is greater than any values of $f(x)$ when x is in the interior of D.

The special parameter free filled function of $f(x)$at the isolated local minimizer x^* so called IG's function proposed in [3] is defined by

$$F_{IG}\left(x,x^*\right)=\begin{cases} -\int_{x^*}^{x} f(s)-f(x^*)ds & \left(x^* \le x\right) \\ -\int_{x}^{x^*} f(s)-f(x^*)ds & \left(x < x^*\right). \end{cases}$$

The IG's function is slightly modified as follows. Suppose that a point x_0 is chosen from the domain of $f(x)$ to be used as a starting point for minimization of $f(x)$ to obtain a local isolated minimizer x_0^*.

Consider a function $f: X \subset R^2 \Rightarrow R$ where X is a box defined by

$X = \{(x_1,x_2) \mid x_{1I} \le x_1 \le x_{1S} \wedge x_{2I} \le x_2 \le x_{2S}\} = (\underline{x}_1 = [x_{1I}, x_{1S}], \underline{x}_2 = [x_{2I}, x_{2S}])$

where x_{jI} and x_{jS} $(j=1,2)$ are the infimum and supremum of the interval $\underline{x}_j = [x_{jI}, x_{jS}]$ respectively.

Consider a function $F: D \subseteq X \subset IR^2 \Rightarrow IR$ where D is a box defined by

$D = \{(x_1,x_2) \mid x_1^{Z-} \le x_1 \le x_1^{Z+} \wedge x_2^{Z-} \le x_2 \le x_2^{Z+}\} = (\underline{x}_1 = [x_1^{Z-}, x_1^{Z+}], \underline{x}_2 = [x_2^{Z-}, x_2^{Z+}])$

Assume that $\mathbf{x}^* = (x_1^*, x_2^*) \in D$ is a current isolated local minimizer of f. Let x_1^{Z+}, x_1^{Z-} are the first nearest points lied on the right and left hand sides of $\mathbf{x}_1^*$ respectively while x_2^{Z+} and x_2^{Z-} are the first nearest points lied on the right and left hand sides of $\mathbf{x}_2^*$ respectively such that

$$f(x_1^{Z+}, x_2^{Z+}) = f(x_1^{Z+}, x_2^{Z-}) = f(x_1^{Z-}, x_2^{Z+}) = f(x_1^{Z-}, x_2^{Z-}) = f(x^*).$$

The domain D can be divided into four sub domains D_0, D_1, D_2, and D_3 as shown in Fig. 4 such that

$$D = \bigcup_{i=1}^{4} D_i = D_1 \cup D_2 \cup D_3 \cup D_4$$

where

$$D_1 = \left([x_1^*, x_1^{Z+}], [x_2^*, x_2^{Z+}]\right),$$

$$D_2 = \left([x_1^*, x_1^{Z+}], [x_2^{Z-}, x_2^*]\right),$$

$$D_3 = \left([x_1^{Z-}, x_1^*], [x_2^*, x_2^{Z+}]\right),$$

$$D_4 = \left([x_1^{Z-}, x_1^*], [x_2^{Z-}, x_2^*]\right)$$

In this section we will recall about parameter free filled function in two dimensional case [4], that is the extension from filled function [3], with the form F_i $(i = 0,1,2,3)$, so called IHR's Function, which defined on sub domain D_i $(i = 0,1,2,3)$, is given by

$$F_i\left(x, x^*\right) = \begin{cases} F_{i1}\left(x, x^*\right) = -\left(\int_a^b \left(f(s_1, x_2) - f\left(x_1^*, x_2^*\right)\right) ds_1\right) & (a \le b) \\ F_{i2}\left(x, x^*\right) = -\left(\int_c^d \left(f(x_1, s_2) - f\left(x_1^*, x_2^*\right)\right) ds_2\right) & (c \le d) \end{cases}$$

where a, b, c and d are shown in Table 1.

Table 1. Determination for boundary of integration a, b, c and d

i	0	1	2	3
Boundary of integration	$a = x_1^*, b = x_1$ $c = x_2^*, d = x_2$	$a = x_1^*, b = x_1$, $c = x_2, d = x_2^*$	$a = x_1, b = x_1^*$, $c = x_2^*, d = x_2$	$a = x_1, b = x_1^*$, $c = x_2, d = x_2^*$

The definition of IHR's function is given in Definition 1 and its validity as a parameter free filled function is proved by Theorem 1-Theorem 3 as presented in detail in [4].

Definition 1 $F\left(x, x^*\right) = \left(F_1, F_2, F_3, F_4\right)\left(\mathbf{x}, \mathbf{x}^*\right)$ defined previously is so called as IHR's filled function of $f(x)$ at an isolated local minimizer x^* if

(i) x^* is a maximizer of $F_{IHR}(x, x^*)$.

(ii) F_{IHR} has no stationary point in set $H_1 = \left\{x \mid f(x) > f(x^*), x \in D\right\}$

(iii) There is a point $x \in H_2 = \left\{x \mid f(x) \le f(x^*),\ W(x) = 0,\ x \in D \setminus \left\{x^*\right\}\right\}$ such that x is a stationary point of F_{IHR}

where a, b, c and d are as in Table 1 and

$$W(x) = \begin{cases} W_1(x) = \dfrac{\partial}{\partial x_2}\left(\int_a^b f(s_1, x_2) ds_1\right), (j = 1) \\ W_2(x) = \dfrac{\partial}{\partial x_1}\left(\int_c^d f(x_1, s_2) ds_2\right), (j = 2) \end{cases}$$

Theorem 1 *If $x^* = (x_1^*, x_2^*)^T \in R^2$ is an isolated minimizer of f on $D = \bigcup_{i=0}^{3} D_i$, then x^* is an isolated maximizer of F_{IHR}.* □

Theorem 2 *If $(x_1^*, x_2^*)^T$ is a local isolated minimizer of f on $D = \bigcup_{i=0}^{3} D_i$ and*

$$H_1 = \left\{(x_1,x_2)^T \middle| f(x_1,x_2) > f(x_1^*,x_2^*),(x_1,x_2)^T \in D \setminus \left\{(x_1^*,x_2^*)^T\right\}\right\}$$
$$= \left([x_1^{Z-},x_1^*) \cup (x_1^*,x_1^{Z+}],[x_2^{Z-},x_2^*) \cup (x_2^*,x_2^{Z+}]\right),$$

then $F_{IHR}\left((x_1,x_2)^T,(x_1^*,x_2^*)^T\right)$ *has no stationary point in the set* H_1. □

Theorem 3 *If* $H_2 = \left\{(x_1,x_2)^T \mid f(x_1,x_2) \le f(x_1^*,x_2^*), W(x_1,x_2)=0,(x_1,x_2)^T \in D \setminus (x_1^*,x_2^*)^T\right\} \neq \varnothing$ *where*

$$W = \begin{cases} \frac{\partial}{\partial x_2}\left(\int_a^b f(s_1,x_2)ds_1\right) & (j=1) \\ \frac{\partial}{\partial x_1}\left(\int_c^d f(x_1,s_2)ds_2\right) & (j=2) \end{cases}$$

and a,b,c,d *are as in Table 1, then there is a point* $\mathbf{x}=(x_1,x_2)^T \in H_2$ *such that* x *is a stationary point of* $F_{IHR}((x_1,x_2)^T,(x_1^*,x_2^*)^T)$.□

3. Jameson Gradient Based Optimization Method Modification [6]

In [4] and [5] the process of minimizing filled function is using Newton's method. However, even Newton's method is considered to be fast method on finding the zero of gradient of quadratic function, but may be very expensive or infeasible to determine the Hessian matrix in the case of complex problems. The Jameson gradient based method is the search procedures which avoid the need to store an estimate of the Hessian as well as its inverse and do not require exact line searches.

We use the following updates to find the zeros of gradient of filled function $\nabla F_{ij}(\mathbf{x},\mathbf{x}_m^*)=0$, that is the modification of Jameson gradient method by replace the estimates of error, e_n, by 1,

$$\mathbf{x}_{n+1} := \mathbf{x}_n - \frac{\alpha e_n g_n}{\beta + \frac{1}{4} g_n^T g_n},$$

with $\beta \ge 1$. We define the value of $\alpha := 0.1$ and $\beta := 1$ to be applied in the algorithm method.

4. Algorithm method

The following algorithm method is generally includes of two methods of optimization, that are the steepest descent method in phase one and modification of Jameson gradient method in phase two.

Data (initialization): specify *n*, $\mathbf{x}_0$, D_i, d>0, set $i=1$, $j=1$, $m=1$

step 1 specify initial step size of steepest descent $\alpha_0 > 0$, minimize $f(\mathbf{x})$ starting at $\mathbf{x}_0$ to obtain $\mathbf{x}_m^*$

step 2 construct F function $F_{ij}(\mathbf{x},\mathbf{x}_m^*)$ at $\mathbf{x}_m^*$

step 3 if $i \le n$ then choose direction e_i for respected defined dimension and go to step 4 else **stop**

step 4 set $c := 1$

step 5 set initial point $\mathbf{x}_0^F = \mathbf{x}_m^* + cde_i$.
 if $\mathbf{x}_0^F \in D_i$ then go to step 6
 else if $j < n$ then $j := j+1$, go to step 4
 else $j=1; i := i+1$; and go to step 3

step 6 if $f(\mathbf{x}_0^F) < f(\mathbf{x}_m^*)$ then set $\mathbf{x}_0 := \mathbf{x}_0^F$, go to step 1
 else if one of cond_1 true then go to step 7
 else $c := c+1$ and go to step 5

step 7 by Jameson gradient based method, with initial $\mathbf{x}_0^F$ solve $\nabla F_{ij}(\mathbf{x},\mathbf{x}_m^*)=0$ to obtain $\mathbf{x}^F$

step 8 if cond_2 and cond_3 hold then set $\mathbf{x}_0 := \mathbf{x}^F$, $m := m+1$, go to step 1
 else $c := c+1$ go to step 5

Note:
cond₁ : at least there is exists a j $(j=1,2)$ such that the following condition holds

$$\left(\frac{\partial F_{ij}}{\partial x_j}(\mathbf{x}=\mathbf{x}_m^*+cde_i)\right)^2 < \left(\frac{\partial}{\partial x_j}F_{ij}(\mathbf{x}=\mathbf{x}_m^*+(c-1)de_i)\right)^2$$

cond₂ : at least one of following conditions holds for a specified j

$$\text{(i). } \frac{\partial^2}{\partial x_j^2}F_{ij}(\mathbf{x}=\mathbf{x}^F)>0 \text{ and (ii). } \left|\frac{\partial^2}{\partial x_j^2}F_{ij}(\mathbf{x}=\mathbf{x}^F)\right|<10^{-4}$$

cond₃ : $\left|f(\mathbf{x}^F)-f(\mathbf{x}_m^*)\right|<10^{-2}$

5. Numerical Results

Following benchmark tests function are used to test the performance of the algorithm method.

- Six Hump Back Camel Function

$$\min f(x_1,x_2)=4x_1^2-2.1x_1^4+\frac{1}{3}x_1^6-x_1x_2-4x_2^2+4x_2^4$$
$$s.t. -3\le x_1,x_2\le 3$$

- Rastrigin Function

$$\min f(x_1,x_2)=x_1^2+x_2^2-\cos(18x_1)-\cos(18x_2)$$
$$s.t. -2\le x_1,x_2\le 2$$

Numerical experiment is done by visual C++ program, Processor Intel ® Core i3, 2048 MB RAM. The process of minimizing $f(\mathbf{x})$, steepest descent method with elimination line search step size (see [7]) is used, which its step size is made to be absolute. Stopping criteria for steepest descent method is $|\nabla f|\le 10^{-3}$, while for Jameson's method is $|\nabla F|<2\cdot 10^{-4}$.

The numerical results of the given problems are presented in Table 2 and Table 3. The symbols used in the table are as follows:

k : number of iteration

F_{ij} : the evaluated Three-Dimensional Parameter Free Filled Function

α_0 : initial step size of steepest descent method

c : integer number to be multiplied to search direction

$\mathbf{x}_0$: initial point for steepest descent method in $f(\mathbf{x})$

$\mathbf{x}_k^*$: k-th obtained minimizer

$\mathbf{x}_0^F$: initial point for Newton's method in F_{ij}

$\mathbf{x}^F$: point obtained by Newton's method or satisfied $f(\mathbf{x}^F)<f(\mathbf{x}_k^*)$

f_0, f_k^*, f_0^F, f^F: function value of $\mathbf{x}_0, \mathbf{x}_k^*, \mathbf{x}_0^F, \mathbf{x}^F$

Table 2. Numerical Results of Six Hump Back Camel Function

k	α_0, F_{ij}, c, d	Results
0	-	$\mathbf{x}_0=(2,1)^T$, $f_0=5.73333$
	$\alpha_0=0.001$	$\mathbf{x}_1^*=(1.60711,-0.568658)^T$, $f_1^*=2.10425$
1	F_{31}, $c=7$,	$\mathbf{x}_0^F=(1.36211,-0.323658)^T$, $f_0^F=2.38716$
	$d=0.1$	$\mathbf{x}^F=(0.910266,-0.166634)^T$, $f^F=2.1059$

α_0=0.001	**x_2^* = (-0.0898664,-0.712651)^T, f_2^* = -1.03163**

Table 3. Numerical Results of Rastrigin Function

k	α_0, F_{ij}, c, d	**Results**
0	-	x_0=(0.5,-0.5)^T, f_0 = 2.32226
	α_0 = 0.001	x_1^*=(0.346924,-0.346924)^T, f_1^* = -1.7578
1	F_{31}, c=320,	x_0^F=(0.0269238,-0.0269238)^T, f_0^F = -1.76825
	d = 0.1	x^F=(0.0269238,-0.0269238)^T, f^F = -1.76825
	α_0=0.001	**x_2^* = (-1.178x10^{-14},1.178x10^{-14})^T, f_2^* = -2**

From the numerical results of Table 2, from the starting point $\mathbf{x}_0^F$, the point x^F=(0.910266,-0.166634)^T is obtained which has the same value to the minimizer x_1^* that is f^F = 2.1059. Thus this point is used to obtain the next minimizer of the problem by steepest descent routine, that is x_2^* = (-0.0898664,-0.712651)^T, f_2^* = -1.03163. While from Table 3, for Numerical Results of Rastrigin Function, after the first minimizer is obtained, the point that has function value lower than the current minimizer is obtained, that is f_0^F = -1.76825. The numerical results shows that the algorithm success in obtain the global minimizer of the given tests function with the given initial points with value of $\alpha := 0.1$ and $\beta := 1$.

6. Conclusion

In this paper, we introduce the parameter free filled function method in cooperation with Jameson gradient based method modification for solving global optimization of $f : R^2 \to R$. The filled function $F(\mathbf{x}, \mathbf{x}^*)$ at a current local minimizer $\mathbf{x}^*$ of f is used as an auxiliary function for finding the zeros of $\nabla F(\mathbf{x}, \mathbf{x}^*)$ or to meet the function with lower function value, $f(\mathbf{x}) < f(\mathbf{x}^*)$. In the phase of exixtance of filled function, the Jameson gradient based method modification is used. This method has an advantages compared to Newton's method, since it avoid the need to store an estimate of the Hessian as well as its inverse and do not require exact line searches. The algorithm method that we proposed is in cooperation of steepest descent method, Jameson gradient based method modification with $\alpha := 0.1, \beta := 1$, and the two dimensional parameter free filled function. Based on the numerical performance of the given test functions, we conclude that our proposed method is promising in successfully obtain global minimizer of unconstrained global optimization problem with two variables.

Acknowledgments

This work was supported by: 1) Grant research of *Riset Kompetensi Dosen Unpad* (RKDU) under project leader of Dr. Sukono MM, MSi.; 2) Unpad Academic Leadership Grant (ALG) under the project leader of Prof. Dr. Sudradjat Supian, MS.

References

[1] Ge R 1990 A Filled Function Method for Finding A Global Minimizer of A Function of Several Variables *J. Mathematical Programming* **46** (2) pp 191-204
[2] Ma S, Yang Y, and Liu H 2010 A Parameter Free Filled Function For Unconstrained Global Optimization *J. Applied Mathematics and Computation* 215 pp 3610-3619

[3] Mohd I B, Wen G K, and B. D. Yosza 2008 Fungsi Pengisian Sebagai Satu Kaedah Pilihan Dalam Pengoptimuman Sejagat *Jurnal KALAM (Karya Asli Lorekan Ahli Matematik)* **1**(1) pp 6-14.

[4] Napitupulu H, Mohd I B, and Pandiya R 2014 *Proc. of the 3rd int. Conf. on Global Optimization and Its Applications* Kuala Terengganu: Penerbit UMT pp 218-140

[5] Napitupulu H, and Mohd I B 2016 Vector Direction of Filled Function Method on Solving Unconstrained Global Optimization Problem, *AIP Conf. Proc.* 1707, 050011 (doi: 10.1063/1.4940843)

[6] Jameson A 1995 Gradient Based Optimization Methods *MAE Technical Report No. 2057* Princeton University

[7] Antoniou A, and Lu W S 2007 *Practical Optimization, Algorithm and Engineering Application*. New York: Springer

Design improvement of table fan blade in term of its vibration characteristic using finite element method

O Kurdi[1], S A Widyanto[1], A Suprihanto[1], D B Wibowo[1], M A Mat Norman[2], G P Mohana Sundram[2]

[1]Mechanical Engineering Department, Diponegoro University, Indonesia
[2]Faculty of Engineering and Life Sciences, Universiti Selangor, Selangor, Malaysia

ojokurdi@ft.undip.ac.id

Abstract. The main purpose of this paper is to improve vibration characteristic, which is natural frequency, of fan blade to get a good natural frequency. An existing table fan model TF1610 of three speeds with 16 inches blades was used for the study. To complete this study, finite element method has been used to perform the simulation and the analysis of the table fan blade. Components of the table fan are drawn using CATIA software. By using finite element software, the drawn table fan design was run for the analysis and the simulation to get the natural frequencies of the table fan blade and compared with the natural frequencies of the motor. The results show that, for Polycarbonate material, shape 4 is considered as the best shape. Shape 2 is considered as the best shape for Polyethylene Terephthalate and Polypropylene material.

1. Introduction

A poor design of fan blade can cause fatigue and a minor crack or fracture to the blade. Vibration characteristic, which is natural frequency, is one of the important parameters in designing fan blade to prevent resonance. The table fan produce sound as noise from the rotation of the fan blades and motor. Vibration or noise will be produced if the fan is not balanced properly that will further results in resonance. To avoid resonance, the fan blade natural frequency and motor frequency should in large difference. Therefore, it is important to design fan blade with good natural frequencies.

Tandon & Nakra [1] have investigated the high noise generated by table fan. The main sound produced by fan was analysed by grid layer. The highest level of sound intensity was produced at two regions, namely at motor and base of the table fan region. High level of noise was produced by both motor and base of the table fan.

Shinici Noda et.al. [2] have analyzed the spatial resonance frequency and its resonance mode of fan channel by FEM and experiments. They concluded that the resonance appears inside fan cover of the motor. The spatial resonance mode around a fan cover was Helmholtz resonance mode. N. S. Aher et al. [3] have carried out the analysis and extract the natural frequencies of cooling tower fan blades of different sizes, so that, the designer can ensure that the natural frequencies will not be close to the frequency of the main excitation forces in order to avoid resonance. The Rayleigh method has been applied to find the fundamental frequency of blades by considering blade as a Stepped, Taper, continuous cantilever beam. A MATLAB program has been developed to predict the fundamental

Content from this work may be used under the terms of the Creative Commons Attribution 3.0 licence. Any further distribution of this work must maintain attribution to the author(s) and the title of the work, journal citation and DOI.
Published under licence by IOP Publishing Ltd

frequencies. Also, three dimensional models of blades have been developed in Unigraphics NX 5 and modal analysis is carried out by ANSYS 14.5.

M. NagaKiran et al. [4] have investigated the effect of various number of fan blades and various of materials to the efficiency of the fan. By observing the analysis results, for all materials, the analyzed stress values are less than their respective yield stress values, so using all the three materials is safe under given load conditions. The strength of the composite material E Glass is more than that of other 2 materials Mild Steel and Aluminium Alloy. By observing the analysis results, the displacement and stress values are less when 8 blades are used. Therefore, they concluded that composite material E Glass and using 8 blades are recommended as the best design because of its high efficiency.

This research aims to simulate various design/model of blades of table fan in order to get the best design that have good characteristic in vibration. The best model of blade should have natural frequencies that far enough from the natural frequency of motor, therefore the possibility of resonance can be reduced. The result of this research can improve the level of health of human being since the noise from fan very disturb to the human being earnest.

The marking system is a new system that is used for this analysis based from the similar research done by Gasporani et al. [5]. Gasporani et al find the differences between 53 Hz and 55 Hz with 10 different modes by using TCSLDV method. The differences of the frequencies are then converted to percentage (%) form, while in this research we have given marks based on the differences of the frequencies. The reason for giving marks for this study is, three speed of table fan motor is considered and from those three speeds, the difference for lowest and highest frequency for each speed is calculated. The increase of the blade thickness leads to decrease in the stresses and deformations and that is due to the higher reduction in structural stiffness. Therefore, thin blades give greater stresses and deformations than the thick blades [6].

2. Finite Element Simulation

2.1. Model of fan

CATIA V5 (Computer Aided Three-dimensional Interactive Application) works as the developing tool of the table fan design. CATIA software is popular among the 3D model software's. The model created based on existing model which is from TF1610 of three speeds with 16 inches blades. The blade of the model has a length of 399 mm. The hub of the fan blade is created by using a circular shape with the thickness of 40 mm and width of 36 mm.

Five different shape with different thickness were proposed to replace the old design. Each shape of the blade will be analyzed with 2mm, 3 mm and 4 mm thickness. For this analysis, 3 different thicknesses were chosen. In previous research, five different thickness of the curvature blade was selected by Abdullah and Schlattmann [6]. The increase of the blade thickness leads to decrease in the stresses and deformations. Therefore, thin blades give greater stresses and deformations than the thick blades. Figure 1 showed the proposed model to replace the existing model.

2.2. Material properties

The modelling procedure also requires the specification of material properties. These properties may include: Modulus of Elasticity, Poisson's Ratio, Coefficient of Thermal Expansion, Modulus of Rigidity, Density and similar. The specific material properties required will depend upon the type of analysis being performed. Three different types of materials are used for the table fan model analysis. All the materials and its properties are stated in Table 1.

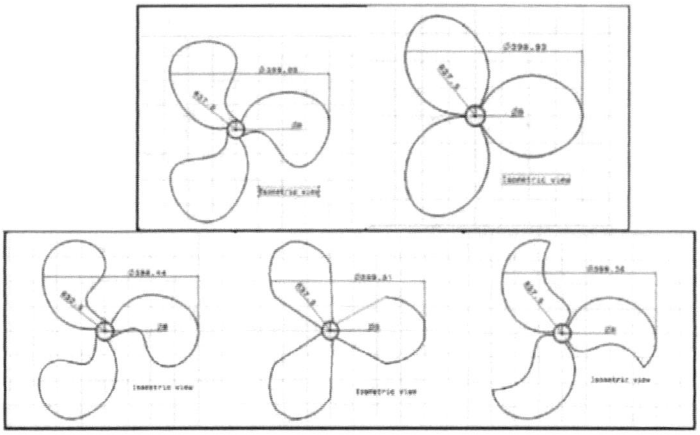

Figure 1: Five different shapes of table fan blade

Table 1. Properties of table fan blade material

Types of Materials	Density (tonne/mm^3)	Young Modulus (MPa)	Poisson Ratio	Tensile Strenght (MPa)
Polycarbonate	1.2×10^{-9}	2.0×10^3	0.37	70
Polyethylene Terephthalate	1.38×10^{-9}	2.8×10^3	0.30	55
Polypropylene	8.55×10^{-10}	1.3×10^3	0.45	40

2.3. Boundary condition

Boundary conditions define constraints on the model. These typically are limitations on displacement in the analysis of mechanical parts. There are no boundary conditions for this analysis since the real condition of blade fan is free free.

3. Result and discussion

Five different shape of fan blade is designed with different thickness of the fan blade. Each shape of the blade will be analyzed with 2mm, 3 mm and 4 mm table fan blade thickness. Polycarbonate, Polyethylene Terephthalate and Polypropylene are the three different materials being used for analysis. Results that are obtained is then tabulated according to the frequency produced by the table fan motor, which is 13.67 Hz at low speed, 16.45 Hz at medium speed and 19.20 Hz at high speed. The lowest and highest value of the frequency will be taken to find the difference between the frequency produced by table fan blade and the frequency produced by the table fan motor. After the differences being calculated, the marks of range 1 to 5 will be given according the highest differences. Total marks for each shape is 30 marks. So, the shape that gain the highest mark will be consider as the best design for each material according to its blade thickness.

3.1. Polycarbonate material

Figure 2 shows the result of marks of 2 mm blade's thickness made of Polycarbonate material. The best shapes for this 2mm blade thickness of Polycarbonate material are shape 2, shape 4 and shape 5. From the graph it can be observed that the highest mark obtained which are 20. For shape 2, the differences of frequency at low speed are 13.367 Hz and 13.367 Hz, at medium speed are 16.45 Hz, and 1.534 Hz while at high speed are 1.152 Hz and 75.173 Hz. For shape 4, the differences of frequency at low speed are 0.208 Hz and 64.855 Hz, at medium speed are 2.988 Hz, and 62.075 Hz while at high speed are 5.738 Hz and 59.325 Hz. For shape 5, the differences of frequency at low speed are 13.67 Hz, 13.67 Hz, at medium speed are 16.45 Hz, and 1.042 Hz while at high speed is 1.168 Hz and 74.884 Hz.

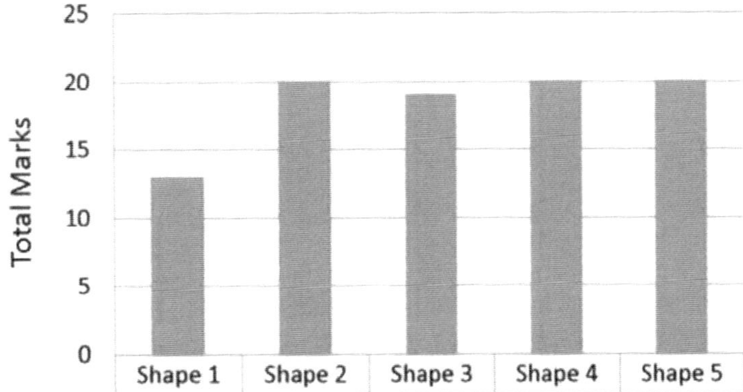

	Shape 1	Shape 2	Shape 3	Shape 4	Shape 5
2mm Thickness of Polycarbonate	13	20	19	20	20

Figure 2. Mark of 2 mm blade's thickness made of Polycarbonate material

3.2. Polyethylene Terephthalate material

The best shape for this 2mm blade thickness of Polyethylene Terephthalate material is shape 2. From the graph, it can be observed that the highest mark obtained is 22. For shape 2, the differences of frequency at low speed are 13.67 Hz, 5.8 Hz, at medium speed are 16.45 Hz, and 3.02 Hz while at high speed are 19.20 Hz and 0.27 Hz. The data for marks of 2 mm blade's thickness made of Polyethylene Terephthalate material was stated in Figure 3.

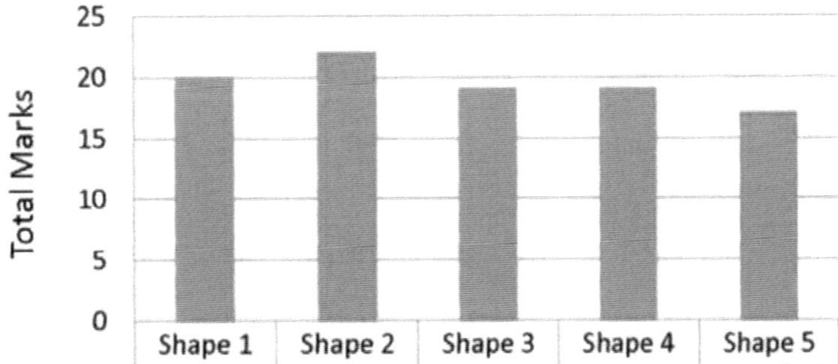

	Shape 1	Shape 2	Shape 3	Shape 4	Shape 5
2mm Thickness of Polyethylene Terephthalate	20	22	19	19	17

Figure 3. Mark of 2 mm blade's thickness made of Polypropylene material.

3.3. Polypropylene material

The best shape for this 2mm blade thickness of Polypropylene material is shape 2. From the graph it can be seen that it has, the highest mark obtained which are 23. For shape 2, the differences of frequency at low speed are 13.67 Hz, 5.253 Hz, at medium speed are 16.45 Hz, and 2.473 Hz while at high speed are 0.263 Hz and 74.413 Hz. Figure 4 shows the result of marks of 2 mm blade's thickness made of Polypropylene material

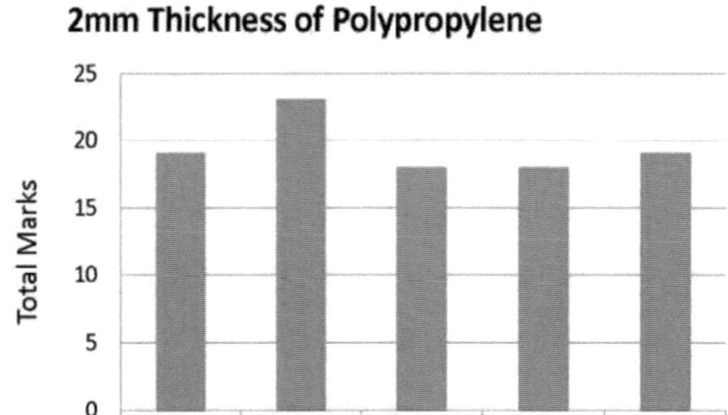

	Shape 1	Shape 2	Shape 3	Shape 4	Shape 5
2mm Thickness of Polypropylene	19	23	18	18	19

Figure 4. Mark of 2 mm blade's thickness made of Polypropylene material.

From the table above, it is shown that the existing model is not at optimum or best level of performance. Some other shapes and improved models are having better natural frequencies result compare to the existing model. It need to replace the existing design with the model/shape that has the highest mark. It gives the lowest probability for resonance phenomena occurred in the fan system.

4. Conclusions

After an observation and analysis have been made, it is proven that the study on the natural frequencies of a table fan blade by using finite element method is accomplished and the objective is achieved successfully. The aim of analysis programmed is to provide information of the important of designing a table fan impeller model that have a huge different in its natural frequencies compared to the natural frequencies of the table fan motor.

The natural frequency of the table fan blade is also have been redesigned so that the best natural frequency can be achieved as an improvement. Based on the research and development planning, a better natural frequency from the output result by simulation of the table fan impeller which varies from the frequencies of the motor is achieved as expected. Once a better natural frequency is achieved, resonances also have been avoided. For Polycarbonate material, shape 4 is considered as the best shape. Shape 2 is considered as the best shape for Polyethylene Terephthalate and Polypropylene material.

References

[1] Tandon, N., Nakra, B.C. 1994, *Eng. Design.* **25**, pp 25-30
[2] Shinici Noda, Sueyoshi Mizuno, Katsumi Suzuki, 2005, *ForumAcostum***,** 779-784
[3] N. S. Aher , S. S. Sarawade,L. S. Dhamande, D. P. Bhaskar, 2016, *IOSR Journal of Mechanical & Civil Engineering (IOSRJMCE)***,** 44-59
[4] M.NagaKiran, T.R.Sydanna, G.Siva Prasad, K.Sagar Kumar, M.Venkateswarlu, 2015, *Int. Journal of Innovative Research in Science, Eng. and Tech.* **4:8**, 7633-7647
[5] Gasparoni, A., Allen, M.S., Yang, S., Sracic, M.W., Castellini, P. and Tomasini, E.P, http://silver.neep.wisc.edu/~msallen/Gasparoni-Allen-etal--TCSLDV-AIVELA-2010-finalver.pdf
[6] Abdullah, O.I. and Schlattmann, J. 2012, *Advanced Theory Applied Mechanical***,** **5**, 263-275

3D simulation of the lubrication film in journal bearing using Fluid-Structure Interaction (FSI)

S Ramdhani, I Haryanto and M Tauviqirrahman

Laboratory for Engineering Design and Tribology, Mechanical Engineering Department, Engineering Faculty, Diponegoro University,Semarang, Central Java, 50275, Indonesia

E-mail: dhannyzee@gmail.com

Abstract. Most of the studies of journal bearing performance are based on a two-dimensional calculation approach which lead to the questionable conclusions. In addition, the deformation of the surface is often ignored in the analysis. In the present study, modern technique for analyzing the lubrication performance is developed based on fluid-structure interaction (FSI) method to get more realistic result. In this way, the deformation which may affect the lubricant thickness will be recalculated based on the film thickness profile in each step. The validation of the developed FSI technique is also carried. The simulation results show that FSI method can be used to analyze the performance of current journal bearing calculation in more accurate compared to conventional method.

1. Introduction

Today the design of journal bearings is performed by solving the Reynolds differential equation.This equation is based on the assumption that the film thickness profile is negligible compared to the other lubrication film dimensions both in axial and circumferential directions. Based on literature survey, numerous workers have paid much attention to calculation programs using Reynolds equation.

Gertzos et al. [1] examined the performance characteristics and the core formation in a hydrodynamic journal bearing lubricated with a Bingham fluid. Their main conclusion was that the load carrying capacity, the film pressure, and the frictional force of a Bingham solid are larger than those of a Newtonian fluid. Rao [2] based on Reynolds equation studied the journal bearing based on the extent of the slip region on the bearing surface. An extended Reynolds equation was proposed based on the slip length model.It was found that an increase in the stability threshold can be achieved with a higher value of the slip length and a smaller extent of the slip region on the bearing surface. Zhang et al [3] presented a mathematical model and a computational methodology to simulate the complicated flow behaviors of the journal microbearing in the slip regime. They concluded that the slip-flow boundary condition enhances the stability of the gas-lubricated journal bearing-rotor system.Wang et al. [4] based on Reynolds approach investigated the two dimensional spiral oil wedge sleeve bearing based on the critical shear stress model. They highlighted that the special structure of spiral oil wedge and the increase of eccentricity ratio make slip more difficult to occur.Solghar et al [5] explored thermohydrodynamic characteristics of journal bearings with two axial grooves by means of computational fluid dynamics technique. Kalavathi et al. [6] numerically investigated the effect of surface roughness on finite porous journal bearing with heterogeneous slip/no-slip surface based on generalized Reynolds approach. It was shown that pressure and load support increases with surface

Content from this work may be used under the terms of the Creative Commons Attribution 3.0 licence. Any further distribution of this work must maintain attribution to the author(s) and the title of the work, journal citation and DOI.
Published under licence by IOP Publishing Ltd

roughness for the bearing with slip/no-slip surface.In recent publication, Tauviqirrahman et al. [7] explored effect of cavitation on the lubrication performance of the journal bearing operating with non-Newtonian lubricant. In their study, however, two-dimensional analysis of bearing was adopted.

Based on literature survey, one can find that journal bearing calculationswere mostly based on a two-dimensional analysis and the analyses was conducted by ignoring the deformation of the journal bearing. Thus, in the present paper, the main contribution is to investigate the performance of journal bearing based on three dimensional approach considering the deformation of the journal bearing. The CFD (computational fluid dynamic) technique based on Navier-Stokes equations coupled with solid dynamic equation also called FSI method (fluid-structure interaction) is applied using commercial software ABAQUS.

2. Methodology

In the present work, the lubrication problem is solved by the Navier-Stokes equation and continuity equations. The Navier–Stokes (N-S) equations are solved over the domain using the commercial software package ABAQUS. The Navier–Stokes and the continuity equations can be expressed, respectively,

$$\rho \frac{Du_i}{Dt} = -\frac{\partial p}{\partial x_i} + \rho G_i + \frac{\partial}{\partial x_j}\left[2\eta e_{ij} - \frac{2}{3}\eta(\nabla . u_i)\delta ij\right] \quad (1)$$

$\nabla \bullet$ u= 0 (2)

The boundary condition used here is the inlet as well as the outlet adopts the pressure boundary. The inlet as well as the outlet pressure is set to zero. The no-slip condition is assumed, while the angular velocity is 48.1 rad/s.

For the solid analysis, a finite element method is applied, and ABAQUS is used to calculate the deformation of the bearing. By concept, the fluid and structure affect each other, the hydrodynamic pressure exerts a force onto the solid structure (i.e. bearing surfaces) and makes it move or deform.In the present work, the number of nodes used is 131,596, while the number of elements is 65,404. The type of mesh used is linear hexahedral elements.

3. Results and Discussion

3.1. Fluid analysis

Figure 1 shows the hydrodynamic pressure of journal bearing predicted by the fluid analysis based on the Navier-Stokes equations coupled with continuity equation. It can be observed that the maximum hydrodynamic pressure occurs at the range of angle 60-110°.

It can also be observed from Figure 1 that the cavitation phenomena exist at the divergent area (blue colour). This is as expected that the rupture of the film occurs due to the lack of lubricant supply at the divergent area.

For detail, the hydrodynamic pressure is shown in Figure 2. It is found that the transition region from convergent to divergent occurs at the angle of 180°. Based on Figure 2, the symmetric profile is highlighted. This condition is valid when the cavitation model is not considered as conducted in the present study.

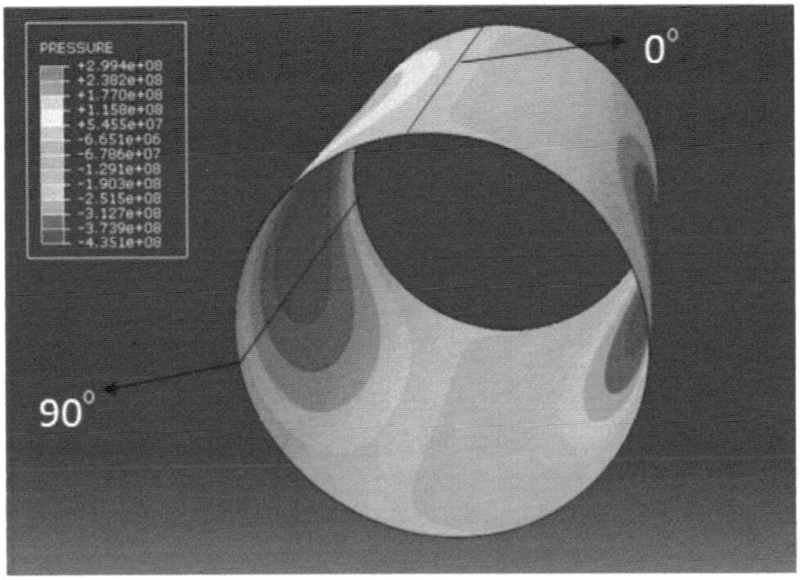

Figure 1. Contour of hydrodynamic pressure of journal bearing

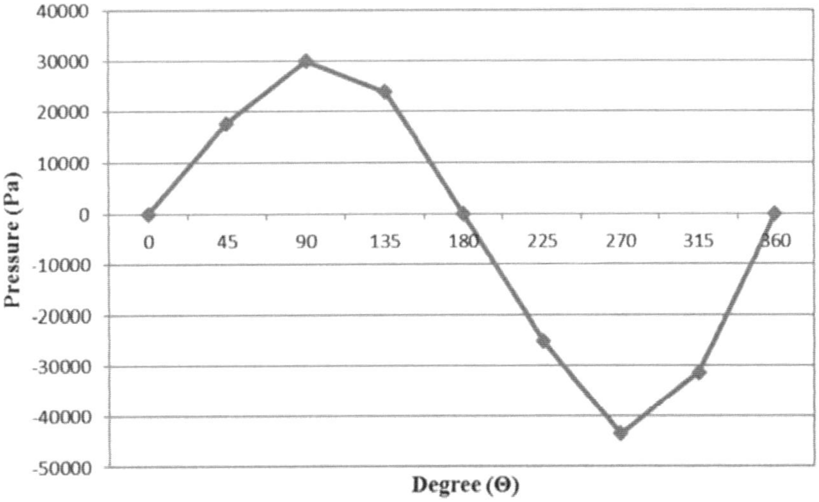

Figure 2. Distribution of hydrodynamic pressure of journal bearing

3.2. Solid analysis

Figure 3 shows the deformation contour of the sleeve as well as the journal. It seems that the deformation occurs on certain area of the bearing. The most possible explanation is that the pressure received by solid structure is local, which indicates that some area of solid is able to handle the possible deformation. This result mean that the deformation of the journal bearing is big enoghr compared to the dimension of the journal. Thus, for analysis of the journal bearing, the deformation which may occur cannot be ignored for obtaining more realistic and accurate result.

Figure 4 depicts the von Mises stress contour of journal bearing. It can be found that the maximum value of the stress is much lower than the yield strength of the material used. As a note, the for all

IOP Conf. Series: Journal of Physics: Conf. Series **1090** (2018) 012025 doi:10.1088/1742-6596/1090/1/012025

following computation, the material of bearing as well as the journal has the yield strength of 2.5 x 10^8 Pa. It means that the operating condition is safe to used, and thus the design of the bearing can be chosen.

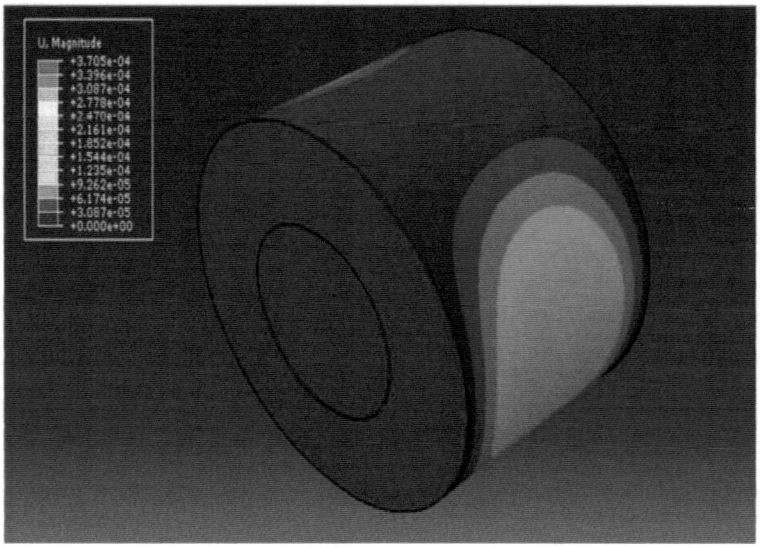

Figure 3. Contour of deformation of journal bearing

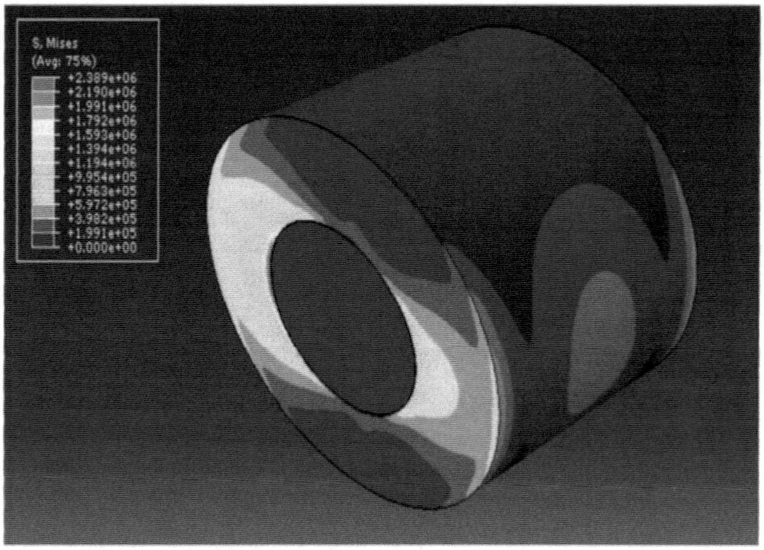

Figure 4. Contour of von Mises stress of journal bearing

4. Conclusions

In the present work, three-dimensional work analysis of journal bearing was conducted based on Navier-Stokes equation. The commercial software ABAQUS was used to investigate its capability in handling the fluid-structure analysis. Based on the explanation above, the conclusions can be drawn that the fluid-structure interaction method developed here is successful to apply in case of journal bearing analysis. In getting more realistic and accurate results, the deformation that occurs can not be ignored.

References

[1] Gertzos KP, Nikolakopoulos PG and Papadopoulos CA 2008 *Tribol. Int.* **41** 1190
[2] Rao TVVLN 2009 *Tribol. Trans*.**52**(6) 750
[3] Zhang WM, Zhou JB and Meng G 2011 *Tribol. Int*.**44** 887
[4] Wang LL, Lu CH, Ge PQ and Chen SJ 2012 *Proc. Inst. Mech part J: J. Eng. Tribol***226**(5) 362
[5] Solghar AA, Brito FP, Abdolzadeh M and Farajpour A 2015 *Lubr. Sci*.**27** 83
[6] Kalavathi GK, Dinesh PA and Gururajan K 2016 *Tribol. Int.* **102** 174
[7] Tauviqirrahman M, Muthik B, Muchammad M, Pratomo AW and J. Jamari 2016 *Int. J. Eng.Tech*.**8(6)** 2541

The application of genetic algorithm optimization on quadratic investment portfolio without a risk-free asset under Value-at-Risk

Sukono[1], S. Supian[1*], H. Napitupulu[1], Yuyun Hidayat[2*], Adam Sukma Putra[3]

[1]Department of Mathematics, Faculty of Mathematics and Natuaral Sciences, Universitas Padjadjaran, INDONESIA
[2]Department of Statistics, Faculty of Mathematics and Natuaral Sciences, Universitas Padjadjaran, INDONESIA
[3]Department of Geophysics, Faculty of Mathematics and Natuaral Sciences, Universitas Padjadjaran, INDONESIA

*Corresponding author's e-mail: sukono@unpad.ac.id; yuyunhidayat@unpad.ac.id; sudradjat@unpad.ac.id

Abstract. In this paper, we performed the Genetic Algorithm within problems of quadratic investment portfolio without a risk-free asset under Value-at-Risk. The limitation of this study is that the risk of an investment portfolio measured by Value-at-Risk, and each investor has the nature of risk aversion. To solve these problems: First, we established the mean vector and covariance matrix. The second step was to define the vector mean and covariance matrices for the formulation of Value-at-Risk of the investment portfolio. Third, using the mean vector and Value-at-Risk established the model. To complete the optimization problem, we performed the Genetic Algorithm. The results show that the trade-off between risk and expected return does not only depend on the type of investor but also on the size of the investment. The Genetic Algorithm certifies us the robust solution in the optimization problem because of its natural ability to locate the global minimal. Moreover, genetic algorithm can be used as an effective way in numerical completion of the optimization of quadratic investment portfolio. In a realistic investment situation, it has likely more constraints. For example, the restriction on short-selling, is need to be considered.

Keywords: Genetic algorithm, mathematical economics, investment portfolio, Value-at-Risk, short-selling

1. Introduction

The capital market takes an important role in economic activity of development countries. It is indicated by the activities of investing in stocks in capital market [5]. An investor who will invest in stock market should be able to understand the risks, which contains uncertainty in the future. This means that it contains an element of risk for investors [3], [11]. By this background, the investors should have an accurate information in making a decision of selecting portfolio stocks in capital market. In this paper, both deterministic and heuristic approach has been tested to the portfolio selection problem based on Markowitz 1952 to address an optimal choice of the portfolio, which has a maximum return and the minimum risk. Markowitz model has an effectiveness to give the steps to

Content from this work may be used under the terms of the Creative Commons Attribution 3.0 licence. Any further distribution of this work must maintain attribution to the author(s) and the title of the work, journal citation and DOI.
Published under licence by IOP Publishing Ltd

investors in preparing its portfolio and also gives the weight of the allocation of funds in the market shares. Markowitz model is often known as the model of Mean-Variance [7], [12]. Mean-Variance is a quadratic portfolio model, because variance is in the form of a quadratic function [4]. In this research, The Value at Risk (VaR) (heuristic) and the Genetic Algorithm (GA) are addressed to set the optimum portfolio [6], [2].

VaR is a method of calculating market risk to determine the risk of loss that can occur in a portfolio, either single instrument or multi instrument [16]. The advantage of VaR are that this method focuses on downside risk, independent of the distribution assumption of return, and its flexibility be applied to all traded financial products. The number obtained from the measurement by this method is the result of aggregate or comprehensive calculation of product risks as a whole [13]. VaR provides an estimation of the likelihood or probability of incurring losses that is greater than the number of losses that have been determined. VaR also takes into account to changes in prices of existing assets and their effects on other assets. This allows the measurement of the reduced risk resulted by the diversification of some investment or portfolio instruments [1], [17].

The Genetic Algorithm (GA) has been widely known as the robust method in solving many scientifical problem which has a non-linierity issue and complex solution [10]. The GA is one of a heuristic method that is a branch of the evolutionary algorithm, which is a technique to solve complex optimization problems by mimicking the process of evolution of living things [9]. The Algorithm can be applied to many studies of sciences. For example in chemical problem to estimate the fittest parameters in kinematic model [15], and the optimization process in chemical reactions and for solving the complex model in accelerator physics by implicating the GA to its multi-objectives optimization [14]. Nowadays, The GA can be also applied in economics studies such as a model of cobweb-type in which has the series of cyclical nature of prices and quantities and the Optimization in market of stocks and the model of portfolio [15].

GA does not involve deterministic calculations, Goldberg argues that GA has characteristics that need to be known, so that it can be distinguished from other searching or optimization procedures: (1).GA's works with encoding of problem solution sets based on predefined parameters and not the parameters themselves. (2).GA's conducting a search for a population of a number of individuals who are the solution to problems not just of an individual. (3) GA's is the information of objective function (*fitness*), as a way to evaluate individual individuals who have the best solution, not the derivatives of a function. (4) GA's use the rules of opportunity transition, not deterministic rules. The following advantages make GA differ compared to other deterministic methods in the problem of optimization. In this study, GA has been applied to the optimization problems to provide optimization results with various possible combinations of $\mathbf{w}$ values as the weight factors for each stock which are being close to the most optimum solution.

The essence of forming a portfolio is to allocate capital to various investment alternatives, so that overall investment risk can be minimized. The optimum portfolios are categorized as efficient if they have the same level of risk, yet able to provide a higher profit rate, or are able to generate the same level of profit, but with a lower risk. The comparative study is presented by simultaneously defining the solution under Value at Risk (VaR) and the GA methods to determine the optimum weight combination of a portfolio with no risk-free assets. As a numerical illustration, some stocks are traded on the Indonesian capital market.

2. Mathematical models

In this part, the basic concepts of the mathematical model are explained which is to be used in the next part. Furthermore, the description moves to discuss the model of VaR investment portfolio and The GA optimization.

2.1. Basic concept

Suppose the length of time of the investment is expressed by $t_0 = 0$, and $t_1 = 1$, (the initial investment and the end investment). Investors set a portfolio with expectation of values $E[V_1]$ at time $t_1 = 1$, is high. Since V_1 fluctuates, it is expected that the risk $\mathrm{Var}[V_1]$ is minimum. Suppose V_0 initial investment, and a simple asset market with $n \geq 2$ risky assets with spot prices S_t^k, where $t = 0,1$ and $k = 1,\dots,n$. Note that S_0^k is known, whereas S_1^k is not. We also allow positions in a risk-free zero-coupon bond that cost B_0 at time 0, and pays one unit of the chosen currency at time 1 [8], [11].

A position in the risky assets is represented by a vector $\mathbf{h} = (h_1,\dots,h_n)^T \in \Re^n$, where h_k is the number of units of asset number k which is held over the time period by the investor. We let h_0 denote the position in the risk-free bond. The prices or market values at time $t = 0$ and $t = 1$ of an affordable portfolio are [8]:

$$h_0 B_0 + \sum_{k=1}^{n} h_k S_0^k \leq V_0 \text{ and } V_1 = h_0 + \sum_{k=1}^{n} h_k S_1^k .$$

If no risk-free bond is available, then we simply set $h_0 = 0$. The next we take the initial monetary value of the position in the kth asset, $w_k = h_k S_0^k$ and $w_0 = h_0 B_0$. With monetary portfolio weights the current and future portfolio values can be expressed as [8]:

$$w_0 + \sum_{k=1}^{n} w_k \leq V_0 \text{ and } w_0 \frac{1}{B_0} + \sum_{k=1}^{n} w_k \frac{S_1^k}{S_0^k}, \tag{1}$$

It is seen that determining the optimal allocation of initial capital V_0 requires the knowledge of the expected value μ and covariance matrix $\boldsymbol{\Sigma}$ of the vector $\mathbf{R}$, where [8]:

$$\mathbf{R}^T = \left(\frac{S_1^k}{S_0^k}, \dots, \frac{S_1^n}{S_0^n} \right).$$

With $R_0 = 1/B_0$ and $\mathbf{w}^T = (w_1,\dots,w_n)^T$ we may write $V_1 = w_0 R_0 + \mathbf{w}^T \mathbf{R}$, and therefore [11]:

$$E[V_1] = w_0 R_0 + \mathbf{w}^T \boldsymbol{\mu}, \tag{2}$$

and

$$Var[V_1] = \mathbf{w}^T \boldsymbol{\Sigma} \mathbf{w}. \tag{3}$$

We assume that the covariance matrix $\boldsymbol{\Sigma} = Cov(\mathbf{R}) = E[(\mathbf{R} - \boldsymbol{\mu})(\mathbf{R} - \boldsymbol{\mu})^T]$ is positive definite: $\mathbf{w}^T \boldsymbol{\Sigma} \mathbf{w} > 0$ for all $\mathbf{w} \neq 0$. By definition, any covariance matrix is symmetric and also positive-semidefinite: for any $\mathbf{w} \neq 0$, $\mathbf{w}^T \boldsymbol{\Sigma} \mathbf{w} = Var(\mathbf{w}^T \mathbf{R}) \geq 0$. Therefore, assuming that $\boldsymbol{\Sigma}$ is positive definite is equivalent to assuming that $\boldsymbol{\Sigma}$ is invertible, or equivalently, that all the eigenvalues of $\boldsymbol{\Sigma} > 0$ [8], [11].

2.2. The Model of portfolio investment based on Value-at-Risk

In this part, we apply the VaR method to the optimization of portfolio without a Risk-Free asset. It is assumed that the return has a certain distribution, and the risk of portfolio is measured by the VaR method according to Down in 2002, model of Risk measurement [16]. VaR assessment model for the portfolio p is defined as $VaR_p = -V_0\{\mu_p + z_\alpha \sigma_p\}$. Since $w_0 R_0$ is risk-free asset and constant, therefore, referring to equation (2) and (3), The VaR of portfolio p can be written as [6]:

$$VaR_p = -V_0\{\mathbf{w}^T\boldsymbol{\mu} + z_\alpha(\mathbf{w}^T\Sigma\mathbf{w})^{1/2}\}. \tag{4}$$

Where negative sign represents a loss, V_0 the initial capital invested, and z_α the percentile of the standard normal distribution when the level of significance is given as $(1-\alpha)\%$.

Thus, the objective function of the investment portfolio model is to maximize $\{\mu_p - \rho VaR_p\}$ or maximize $\left\{\mathbf{w}^T\Sigma + \frac{c}{2V_0}V_0[\mathbf{w}^T\Sigma + z_\alpha(\mathbf{w}^T\Sigma\mathbf{w})^{1/2}]\right\}$. hence the investment portfolio optimization model that needs to be completed is [16]:

$$\text{Maximum}\left\{(1+\frac{c}{2})\mathbf{w}^T\boldsymbol{\mu} + \frac{c}{2}z_\alpha(\mathbf{w}^T\Sigma\mathbf{w})^{1/2}\right\}. \tag{5}$$

Subject to $\mathbf{w}^T\mathbf{I} \le V_0$

Assume that there are two investments with return R and $\widetilde{R}$, investors who invested V_0 have expectations the same result, then the value of the constant c, can be determined through the equality that [8]:

$$E[V_0R] - \frac{c}{2V_0}VaR[V_0R] = E[V_0\widetilde{R}] - \frac{c}{2V_0}VaR[V_0\widetilde{R}],$$

So we get

$$c = \frac{2\{E[V_0R] - E[V_0\widetilde{R}]\}}{\{VaR[V_0R] - VaR[V_0\widetilde{R}]\}}.$$

2.3. The Model of portfolio investment based on genetic algorithm

In order to solve the equation (5), we applied the GA proposed by Holland in 1975 [14]. The GA initially generates a random value of 'gen' $\mathbf{w}^T$ within a certain interval, which is associated in a 'chromosome'. This algorithm allows the competition between each chromosome [15], which brings each potential solution to the optimization problem. A set of chromosome called 'population' hence performed an iteration (generation) to determine the fittest parameters of the 'objective function'. In this case, the equation (5) is the objective function. The generation process consist of evaluating the 'fitness function' adapted to create a new population until the optimum chromosomes has been addressed. This operation is made by setting the genetic operator: number of population and Chromosome are 100 and 5, number of generation is 1000, the cross over and mutation rate are set to be 0.25.

For an ideal state, we assume that $\mathbf{w}^T\mathbf{I}$ is equal to V_0, hence 'controlled' fitness function is

$$\begin{aligned}\mathbf{w}^T\mathbf{I} &= V_0 \\ \mathbf{w}^T\mathbf{I} - V_0 &= 0\end{aligned}. \tag{6}$$

where $\mathbf{I}^T = (1,1,1,1)$. Which refer to the equation (5) as the objective functions.

Since we wanted to maximize the objective function, we adopted the 'Roulette-Wheel' selection, so that the value of e^U are greater than 0. Where U is the objective functions. The selection of the new individual follows this formula [15]

$$w_i = \frac{f_i}{\sum_{i=1}^{n} f_i}. \tag{7}$$

where w_i is the new individual of each $\mathbf{w}$ and f is the fitness value for each individual. n is the size of population. In this case, we applied a random cross over and mutation algorithm which has a probability of 0.25 for each gen in the chromosomes to be crossed over or mutated [15].

3. Result and discusion

In this part, The result of the numerical data are presented included the GA result and the discussion section

3.1. Numerical data

In this part, we intended to show how the application of the model has been formulated, to analyze stock data traded on the capital market. Stock data analyzed can be accessed through the website http://www.finance.go.id//. The data consists of 5 (five) selected shares, for the period of January 2, 2016 up to April 28, 2017. The data includes shares: INDF, DEWA, AALI, LSIP, and ASII.

The average return value of the five stocks is given in a vector respectively $\boldsymbol{\mu}^T$ = (0.0154, 0.0390, 0.0033, 0.0088, 0.0003). While the value of variance of return with the value of covariance between returns of the five shares is given in the form of covariance matrix $\boldsymbol{\Sigma}$ and its inverse $\boldsymbol{\Sigma}^{-1}$,

$$\boldsymbol{\Sigma} = \begin{bmatrix} 0.0026 & 0.0001 & -0.0002 & 0.0002 & 0.0001 \\ 0.0001 & 0.0028 & 0.0003 & 0.0000 & 0.0001 \\ -0.0002 & 0.0003 & 0.0013 & 0.0006 & 0.0004 \\ 0.0002 & 0.0000 & 0.0006 & 0.0019 & 0.0003 \\ 0.0001 & 0.0001 & 0.0004 & 0.0003 & 0.0002 \end{bmatrix} \quad \boldsymbol{\Sigma}^{-1} = 10^4 \begin{bmatrix} 0.0450 & -0.0022 & 0.0365 & -0.0018 & -0.0916 \\ -0.0022 & 0.0370 & -0.0092 & 0.0039 & -0.0048 \\ 0.0365 & -0.0092 & 0.2310 & -0.0022 & -0.4723 \\ -0.0018 & 0.0039 & -0.0022 & 0.0694 & -0.1008 \\ -0.0916 & -0.0048 & -0.4723 & -0.1008 & 1.6441 \end{bmatrix}$$

The $\mathbf{I}^T$ = (1, 1, 1, 1, 1) is the unit vector, where the initial capital is assumed as Misalkan V_0 = 1 currency unit. Next, vectors $\boldsymbol{\mu}^T$, vectors $\mathbf{I}^T$, and inverse covariance matrices $\boldsymbol{\Sigma}^{-1}$, are simultaneously used for the calculation of investment portfolio optimization. Here, it is assumed that the sale and purchase of short selling stock is not permitted. This optimization process is done by using Mean-VaR model. In this optimization process, the values of risk aversion constants $c > 0$ are determined sequentially from the smallest value to the largest value as can be seen in figure 1. The optimization process of the portfolio investment has been done referring to $\alpha = 5\%$ which give $z_{5\%}$ = -1.645. The risk aversion constants $c > 0$ are Which satisfies the assumption that short sales are not allowed is $8 \leq c \leq 20$.

3.2. Genetic Algorithm Result

Based on the numerical data and the information discussed in section 3.1, then the data and all related parameters are used to optimize the investment portfolio. The process of investment portfolio optimization is done by using genetic algorithm to determine the weight composition, for any given risk aversion, which has an interval value of $8 \leq c \leq 20$. After generating 100 random samples of $\mathbf{w}$, we have 13 chromosomes, which have the best fitness functions as shown in Table 1.

Table 1. Genetic algorithm result of portfolio optimization

c	Weght for stocks of					Sum of weight	Mean	Variance	VaR	Ratio
	INDF	DEWA	AALI	LSIP	ASII					
8	**0.0115**	**0.0936**	**0.4092**	**0.0644**	**0.4213**	**1.0000**	**0.0059**	**0.0003**	**0.0222**	**0.2642**
9	0.1487	0.1825	0.2646	0.0570	0.3472	1.0000	0.0109	0.0012	0.0470	0.2316
10	0.2083	0.2191	0.2038	0.0538	0.3150	1.0000	0.0130	0.0022	0.0646	0.2013
11	0.2443	0.2446	0.1638	0.0518	0.2955	1.0000	0.0144	0.0030	0.0755	0.1905
12	0.2690	0.2606	0.1377	0.0505	0.2822	1.0000	0.0153	0.0036	0.0832	0.1837
13	0.2872	0.2723	0.1186	0.0495	0.2724	1.0000	0.0160	0.0041	0.0889	0.1794

14	0.3012	0.2815	0.1038	0.0487	0.2648	1.0000	0.0165	0.0045	0.0933	0.1764
15	0.3124	0.2888	0.0920	0.0481	0.2587	1.0000	0.0169	0.0048	0.0969	0.1743
16	0.3216	0.2947	0.0823	0.0476	0.2538	1.0000	0.0172	0.0051	0.0998	0.1725
17	0.3293	0.2996	0.0742	0.0472	0.2497	1.0000	0.0175	0.0053	0.1022	0.1711
18	0.3357	0.3038	0.0674	0.0469	0.2462	1.0000	0.0177	0.0055	0.1043	0.1700
19	0.3413	0.3074	0.0615	0.0467	0.2431	1.0000	0.0179	0.0057	0.1060	0.1691
20	0.3462	0.3106	0.0564	0.0463	0.2405	1.0000	0.0181	0.0058	0.1076	0.1683

The listed number of the $\mathbf{w}$ chromosomes in Table 1 are hence being evaluated for each given risk aversion c to determine the mean, variance, and the VaR of investment portfolio using the equation (2), (3), and (4). The result are presented in Table 1 as well. The Ratio represents the proportion of the Mean toward VaR relatively.

The efficeint-frontier curve are generated by plotting the Mean versus VaR as shown in Figure 1. This efficient frontier curve is a distribution of coordinate points (Mean, VaR), for each risk aversion value within interval $8 \leq c \leq 20$.

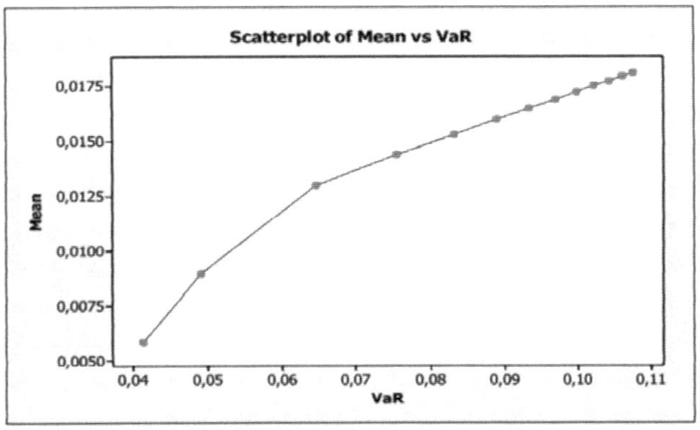

Figure 1. Efficient frontier curve of portfolio investment

A detail explanation of the Figure 1 are presented in the next section.

3.3. Discusion

As it can be seen in Table 1, the different values for each risk aversion are addressed to the shift of composition of Investment portfolio weight. As a result, it appears that the values of the mean and Value-at-Risk portfolio are also changing. The greater values of the investment portfolio mean are followed by the greater Values-at-Risk, which is in this case, it is a measurement of investment portfolio risk. This indicates that the greater the value of investment portfolio expectations, it is always followed by the greater investment risk.

The distribution of 'red-dots' in the efficeint-frontier curve in Figure 1. Represents the appropriate points for investors to make investments, according to the level of risk preferences described as risk aversion. Throughout the efficient-frontier curve, the minimum point occurs when the value of risk aversion $c = 8$, and the composition of the investment portfolio weight is as vector $\mathbf{w}^T = (0.0115, 0.0936, 0.4092, 0.0644, 0.4213)$. This minimum point, also refer to the minimum portfolio. This minimum portfolio gives a mean value of 0.0059, and Value-at-Risk of 0.0222. While the maximum point occurs when the value of risk aversion $c = 20$, and the composition of the investment portfolio weight is as vector $\mathbf{w}^T = (0.3462, 0.3106, 0.0564, 0.0463, 0.2405)$. This maximum point, also refer to the maximum portfolio. This maximum portfolio gives a mean value of 0.0181, and Value-at-Risk of 0.1076. Throughout the efficient-frontier curve, in addition to the minimum portfolio and maximum portfolio, there is also an optimum portfolio. This optimum portfolio is indicated by the ratio between mean and Value-at-Risk where it has a greatest value, which is 0.2642. The optimum portfolio occurs when the value of risk aversion $c = 8$, and the composition of the investment portfolio weight is as vector $\mathbf{w}^T$ = (0.0115, 0.0936, 0.4092, 0.0644, 0.4213). In this case, the optimum portfolio is also a minimum portfolio.

This analysis is expected to provide an overview for investors, who will invest funds especially on investment portfolios in the five stocks discussed in this paper. However, this study can be a good references to make a self-analysis for other stocks that are available in the market.

4. Conclusion

We have done the portfolio optimization modeling without risk-based on Value-at-Risk and the Genetic Algorithm. According to the research, it can be concluded as follows. By applying the algorithm upon the optimization of investment portfolio, the optimum portfolio is located when the value of risk aversion $c = 8$ and the the composition of the investment portfolio weight is as vector $\mathbf{w}^T$ = (0.0115, 0.0936, 0.4092, 0.0644, 0.4213). This optimum portfolio gives the expected return at 0.0059, and the Value-at-Risk at 0.0222. In this study, the optimum portofolio play a role as maksimum global portfolio, which has the greatest ratio of mean to Value-at-Risk at 0.2642. This optimum portfolio is located in the minimum portfolio point which means that the minimum portfolio is the optimum portfolio as well. The maximum portfolio has the value of mean at 0.0181, and the Value-at-Risk at 0.2642 where the value of risk aversion $c = 20$. However, since its ratio is lower than 0.02642 (only 0.1683). this maximum portfolio is not the maximum global. The application of VaR gives a robust result where the solution is unique. In contrary, The GA gives not only one robust solution, but also offers many alternative solutions to the problem. This might be close to the real problem that can be complex and needs many alternative choices since the global stock market has lots of uncertainty. The GA's can generate lots of potential solution according to its ability based on random search of solution. However, in the optimization problem, we expect a unique solution which gives us the optimum robust solution. Therefore, we can obtain the solution by evaluating all of the potential solution by calculating each fitness. The higher the fitness value of the chromosome $\mathbf{w}^T$ then the chromosome election probability will be higher. In this study the result of $\mathbf{w}^T$ has the highest fitness value. This global optimum portfolio has the largest ratio of the mean to the risk value.

5. Acknowledgment

Further thanks to the Rector, Director of DRPMI, and Dean of FMIPA, Universitas Padjadjaran, which has a grant program of the Academic Leadership Grant (ALG) under the coordination of Prof. Dr. Sudradjat, and a grant program of Competence Research of Lecturer of Unpad (*Riset Kompetensi Dosen Unpad*/RKDU) under the coordination of Dr. Sukono, which is a means to increase research activities and publications to researchers at the Universitas Padjadjaran.

6. References

[1] Bansal, Y., Kumar, S., and Verma, P. 2014. Commodity Futures in Portfolio Diversification: Impact on Investor's Utility. *Global Business and Management Research: An International Journal,* Vol. 6, No. 2 (2014).

[2] El hachlou, M., Guennoun, Z. and Hamza, F. 2012. Stocks Portfolio Optimization Using Classification and Genetic Algorithms. Applied Mathematical Sciences, Vol. 6, 2012, no. 94, pp. 4673 – 4684.

[3] Gambrah, P.S.N and T.A. Pirvu, T.A. 2014. Risk Measures and Portfolio Optimization. *Journal Risk Financial Management.* 2014, 7, pp. 113-129; ISSN 1911-8074. doi:10.3390/jrfm7030113.

[4] Ghaemi, R.J., Sun, and Kolmanovsky, I.V. 2009. An Integrated Perturbation Analysis and Sequential Quadratic Programming Approach for Model Predictive Control, *Automatica* 45 (2009) 2412_2418, Journal Homepage: www.elsevier.com/locate/automatica

[5] Ghosh, A. and Mahanti, A. 2014. Investment Portfolio Management: A Review from 2009 to 2014. *Proceedings of 10th Global Business and Social Science Research Conference 23 -24 June 2014,* Radisson Blu Hotel, Beijing, China, ISBN: 978-1-922069-55-9.

[6] Goh, J., Lim, K.G., Sim, M., and Zhang, W. 2011. Portfolio Value-at-Risk Optimization for Asymmetrically Distributed Asset Returns. *NUS Risk Management Institute (RMI).* March 4, 2011.

[7] Golafshani, Z.Y. and Emamipoor, S. 2015. Portfolio optimization using two methods of mean-variance analysis and mean risk in Tehran Stock Exchange. *Technical Journal of Engineering and Applied Sciences*, ISSN 2051-0853 ©2015 TJEAS. ©2015 TJEAS Journal-2015-5-S/128-133. Available online at www.tjeas.com.

[8] Hult, H., Lindskog, F., Hammarlid, O., and Rehn, C.J. 2012. *Risk and Portfolio Analysis. Principles and Method.* Springer Series in Operation Research and Financial Engineering. New York: Springer Science & Business Media.

[9] Kapiamba, J.N., Ulungu, B.E.L., and Mubenga, P.K. 2015. Simulated Annealing vs Genetic Algorithm to Portfolio Selection. *International Journal of Scientific and Innovative Mathematical Research (IJSIMR),* Volume 3, Issue 5, May 2015, PP 18-30. ISSN 2347-307X (Print) & ISSN 2347-3142 (Online). http://www.arcjournals.org.

[10] Lin, C.C. and Liu, Y.T. 2008. Genetic Algorithms for Portfolio Selection Problems with Minimum Transaction Lots. *European Journal of Operational Research* 185 (2008), pp. 393–404.

[11] Pinasthika, N. and Surya, B. 2014. Optimal Portfolio Analysis With Risk-Free Assets Using Index-Tracking And Markowitz Mean-Variance Portfolio Optimization Model. *Journal Of Business And Management*. Vol . 3, No.7, 2014: 737-751.

[12] Plunus1, S., Gillet, R., and Hübner, G. 2015. Equivalent Risky Allocation: The New ERA of Risk Measurement for Heterogeneous Investors. *American Journal of Industrial and Business Management*, 2015, 5, 351-365. Published Online June 2015 in SciRes. http://www.scirp.org/journal/ajibm.

[13] Ogryczak, W., Przyłuski, M., and Sliwinski, T. 2015. Portfolio Optimization with Reward-Risk Ratio Measure based on the Conditional Value-at-Risk. *Proceedings of the World Congress on Engineering and Computer Science* 2015 Vol II. WCECS 2015, October 21-23, 2015, San Francisco, USA.

[14] Sadaf, A. and Ghodrati, H. 2015. An Improved Genetic Algorithm Method for Selection and Optimizing the Share Portfolio. *International Journal of Computer Science and Mobile Computing*, Vol.4 Issue.1, January- 2015, pp. 342-353. Available Online at www.ijcsmc.com.

[15] Sefiane, S. and Benbouziane, M. 2012. Portfolio Selection Using Genetic Algorithm. *MPRA* Paper No. 41783, posted 12. October 2012 08:08 UTC. Online at https://mpra.ub.uni-muenchen.de/41783/

[16] Sukono, Sidi, P., Bon, A.T., and Supian, S. 2017. Modeling of Mean-VaR Portfolio Optimization by Risk Tolerance When the Utility Function is Quadratic. *Statistics and its Applications*. AIP Conf. Proc. 1827, 020035-1–020035-9; doi: 10.1063/1.4979451. Published by AIP Publishing. 978-0-7354-1495-2/. pp. 1-9.

[17] Wang, X. 2013. Theoretical Analysis of Financial Portfolio Model. *iBusiness*, 2013, 5, pp. 69-73. http://dx.doi.org/10.4236/ib.2013.53B015. Published Online September 2013 (http://www.scirp.org/journal/ib).

IOP Conf. Series: Journal of Physics: Conf. Series **1090** (2018) 012027 doi:10.1088/1742-6596/1090/1/012027

Effect of texture shape on the hydrodynamic lubrication performance of the sliding textured contact: a numerical approach

Susilowati[1,2], M Tauviqirrahman[2], J Jamari[2], AP Bayuseno[2]

[1]Laboratory for Research and Unit Operation of Chemical Engineering, Chemical Engineering Department, Industrial Technology Faculty, Universitas Pembangunan Nasional "Veteran", East Java, Indonesia

[2]Laboratory for Engineering Design and Tribology, Mechanical Engineering Department, Engineering Faculty, Diponegoro University, Indonesia

zuzisukasno@gmail.com

Abstract. Currently, there is a great interest in the use of texturing as well as the slip because these surface are able to improve the lubrication performance and thus higher life time of the bearing. In the present work, the modified Reynolds equation considering mass conserving cavitation model is established on the basis of Navier-slip model, in which the slip-textured bearing is considered. The texture shape of the single textured bearing with slip located at the inlet is of particular interest. Various shapes of texture, i.e. triangular, rectangular, and parabola are explored with respect to the hydrodynamic lubrication performance (i.e. the load support and the friction). The inlet length containing slip condition is also varied. The results show that triangular shape of texture gives the higher load support compared to other shape for all value of the inlet length. However, with respect to the friction force, rectangular shape gives the lowest friction force, whilst the triangular shape produces a contrary result.

1. Introduction

Surface texturing as a tools for improving the tribological characteristics of mechanical components has been under intensive exploration over the last two decades. Many techniques have been introduced to create surface texture of various geometries and patterns by several researchers. Ma and Zhu [1] have paid attention to the elliptical-shape dimples with various depths, diameters, area ratios and different operation parameters. The friction coefficient was of particular interest. With the same focus for improving the friction, Uddin and Liu [2] presented the design of a new 'star-like' texture shape. They showed that triangle effect is the most dominant in reducing the friction. They also found that the new optimum star-like texture reduces the friction coefficient by 80%, 64.39%, 19.32% and 16.14%, as compared to ellipse, chevron, triangle and circle, respectively. In recent lubrication, Zhang et al. [3] confirmed that the elliptical-shaped textures have preferable tribological behaviors of low friction coefficient for light load situation.

Recently, it is also well-known that the performance of the lubricated sliding can be improved by combining the surface texturing with the boundary slip. Aurelian et al. [4] presented the results about a partially textured bearing with wall slip conditions. One of their findings was that inadequate locating of the textured-slip pattern can lead to the deterioration of the bearing performance. Rao et al. [5]

Content from this work may be used under the terms of the Creative Commons Attribution 3.0 licence. Any further distribution of this work must maintain attribution to the author(s) and the title of the work, journal citation and DOI.
Published under licence by IOP Publishing Ltd

investigated the effects of texture/slip configuration on the enhancement in load support and the reduction in coefficient of friction for slip-textured pattern both in slider and journal bearing. Extensive work of combined effect of slip and texture has been carried out by Tauviqirrahman et al. [6]. They showed that in comparison with a solely textured surface, the load support of the combined textured/wall slip pattern can be increased by around 300% using the optimized slip parameters. Later, Lin et al. [7] explored the effect of large-area texture/slip surface, especially the area and position of large-area texture surface on journal bearing considering cavitation. They concluded that texture/slip surface would not affect the pressure and load-carrying capacity when it locates at cavitation zone. In recent publication, related to the slip position, it is confirmed that adding slip at the leading edge of the single-textured contact is considered more efficient than extending the slip to the textured region with respect to the load support and friction force [8].

While considering this scientific context, the main aim of this work is to study the effect of texture shape on the performance of the sliding contact with the slip located at the leading edge of the contact varying the slip inlet length. A numerical approach is proposed based on modified Reynolds equation for the choice of texture shape with respect to the load support as well as the friction force. The load support and the friction force are of particular interest.

2. Analysis

The most common shape of texture used in modeling the slider bearing is rectangular. However, in real application, other texture shapes can also be used as seen in Figure 1. Following this insight, in this work the effect of the comparison of the texture variation on the lubrication performance is of particular interest.

In the present study, the shape used is triangular, rectangular, and parabola as seen in Figure 1. For all following computations, the parameters used with respect to the bearing characteristics as well as their operating conditions are described in detail in Table 1. The bearing configuration studied here as reflected in Fig. 1 is single texture bearing with slip located at the leading edge of the contact. It should be noted that introducing slip at the inlet gives the positive effect on improving the load carrying capacity of the bearing [8,9]. In this work, the inlet length containing slip boundary condition is also of particular interest.

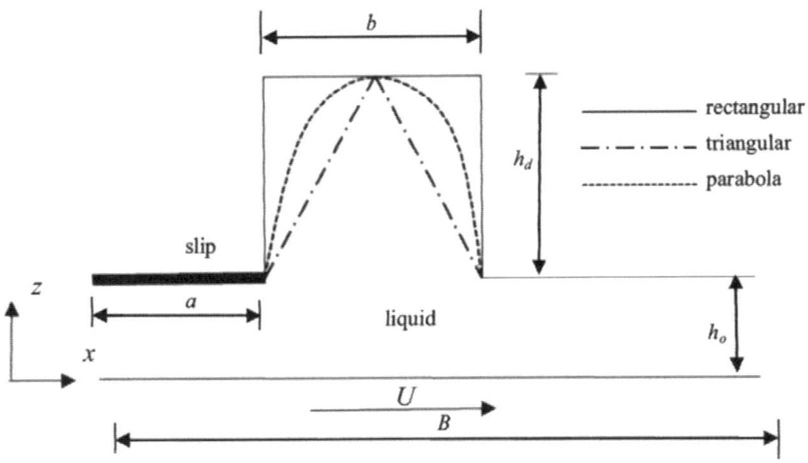

Figure 1. Three texture shapes studied here.

Table 1. Single texture bearing configuration studied here

Bearing length	B	0.02 m
Inlet length	a	0.0004-0.0136 m (a/B = 0.02-0.68)
Pocket length	b	0.006 m (b/B = 0.3)
Sliding velocity	U	1 m/s
Viscosity	η	0.01 Pa s
Atmospheric pressure	P_{atm}	100 kPa
Minimum film thickness	h_o	1 μm
Pocket depth	h_d	2.5 μm
Slip coefficient	α	0.02

In the present study, the modified Reynolds equation is adopted based on the work of Susilowati et al. [10]. The model of lubrication presented here is based on the fact that slip at the interface between lubricant and surface will exist. It is worth noting that the slip length model is used to address the modeling of the boundary slip for the hydrodynamic analysis after the shear stress exceeds the limiting shear stress. Equation 1 represents the lubrication model as follows:

$$\frac{\partial}{\partial x}\left(h^3\frac{h^2+4h\eta(\alpha_a+\alpha_b)+12\eta^2\alpha_a\alpha_b}{h(h+\eta(\alpha_a+\alpha_b))}\frac{\partial p}{\partial x}\right)=6\eta U\frac{\partial}{\partial x}\left(\frac{h^2+2h\alpha_a\eta}{h+\eta(\alpha_a+\alpha_b)}\right)$$

$$-6\eta\tau_{ca}\frac{\partial}{\partial x}\left(\frac{\alpha_a h(h+2\alpha_b\eta)}{h+\eta(\alpha_a+\alpha_b)}\right)+6\eta\tau_{cb}\frac{\partial}{\partial x}\left(\frac{\alpha_b h(h+2\alpha_a\eta)}{h+\eta(\alpha_a+\alpha_b)}\right)-12\eta U\frac{\alpha_a\eta}{h+\eta(\alpha_a+\alpha_b)}\frac{\partial h}{\partial x}$$

$$\frac{\partial}{\partial x}\left(h^3\frac{h^2+4h\eta(\alpha_a+\alpha_b)+12\eta^2\alpha_a\alpha_b}{h(h+\eta(\alpha_a+\alpha_b))}\frac{\partial p}{\partial x}\right)=6\eta U\frac{\partial}{\partial x}\left(\frac{h^2+2h\alpha_a\eta}{h+\eta(\alpha_a+\alpha_b)}\right) \quad (1)$$

The physical meanings of the symbols in Equation 1 are as follows: h the film thickness (gap) at location, p the lubrication film pressure, α the slip coefficient, τ_c the critical shear stress (subscripts a and b denote the stationary and moving surface, respectively), and η the lubricant viscosity. It should be noted that when the slip is located at the stationary surface as indicated in Figure 1, the α_b will be equal zero as well as the τ_{cb}. For all following computation, the slip is assumed to occur on the stationary surface.

In this study, dealing with the derivation of modified Reynolds equation the conditions for cavitation are based on the Reynolds cavitation theory to model the rupture zone for the textured surface. The modified Reynolds equation (Eq. (1)) is discretized over the flow using the finite volume method, and is solved using tridiagonal matrix algorithm (TDMA). By employing the discretization scheme, the computed domain is divided into a number of control volumes using a grid with uniform mesh size. The grid independency is validated by various numbers of mesh sizes.

3. Results and Discussion

3.1. Effect of inlet length on the load support

In this work, the lubrication performance of the slider bearing is represented by the load support. It is obtained by integrating the hydrodynamic pressure over the surface area. Figure 2 shows the effect of the slip inlet length a on the load support W. It is noted that in Figure 2, the inlet length is rationalized by the bearing length B. It can be observed based on Figure 2 that the trend of load support profiles for three texture shapes is similar, i.e. increasing the inlet length to certain value enhances the load carrying capacity. However, after that, increasing the a will reduce the W significantly. When the inlet length is 0.45 times the bearing length, the maximum load support is achieved for all texture shapes studied here. Based on this figure, it can be observed that the shape of triangular texture gives the best performance of the slider bearing with respect to the load support for all range of the inlet length. However, the difference of the predicted load support for three texture shapes is relatively small, especially when a/B = 0.45. The highest difference of the prediction of W occurs when a/B = 0.1. It can be observed that the triangular shape seems superior to other shapes.

Other interesting result is there is a critical value of the inlet length a/B. Based on the simulation results, this point is 0.45. On the other words, for inlet length values which is below critical value (i.e. $a < 0.45\ B$), the difference of the prediction of the W increases by reducing the inlet length with slip. It is also noted that when the dimensionless inlet length exceeds the critical value of the a/B, such difference increases with increasing the inlet length. From the physical point of view, it indicates that in addition to the slip condition, the texture itself has significant effect on altering the flow characteristic.

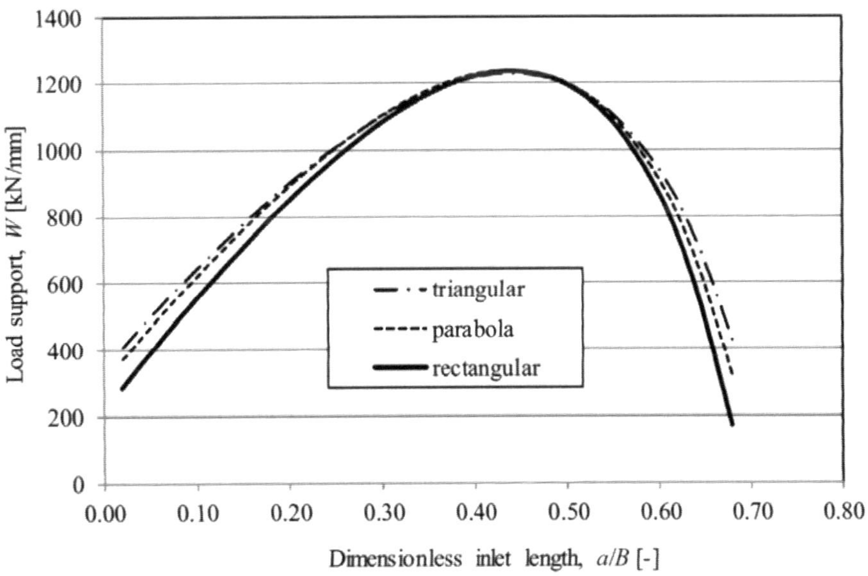

Figure 2. Effect of inlet length with slip a/B on the load support W varying the texture shapes

3.2. Effect of inlet length on the friction force

Figure 3 depicts the correlation between the dimensionless inlet length a/B and the friction force F. It can be observed that the same trend of friction force profile for three textures shapes considered here is highlighted. Increasing the inlet length with slip condition reduces the friction forces significantly. This is as expected due to the nature characteristic of the slip boundary which is hydrophobic. High wetting of the surface containing slip leads to the decrease in the friction reduction.

Regarding to the textures shape, based on Fig. 3, it is also found that the rectangular texture produces the lowest friction force for all value of the inlet length, whilst the triangular gives the highest friction force. This result seems to be contrary with the previous finding which stated that triangular shape leads to positive effect in terms of the load support. The most possible explanation is that based on Fig. 3, the difference of the prediction of the friction force among the texture shapes is relatively small; it is just 5%. The highest difference is calculated when the inlet length is set to maximum, that is, when $a/B = 0.68$.

In real application, the rectangular texture shape seems become a technique that is easier to adopted compared to other shapes. Lowering the friction force by texturing the bearing with rectangular shape will enhance the life time of the bearing. This is because the stiction which may exist due to high friction can be prevented.

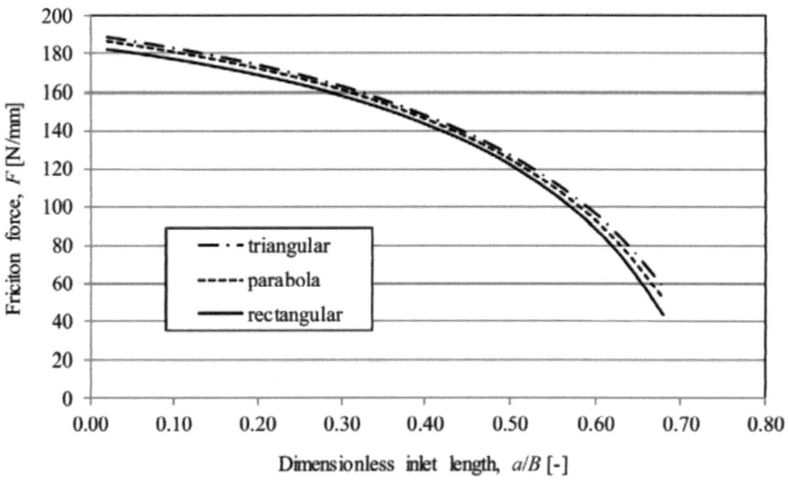

Figure 3. Effect of inlet length with slip a/B on the load support W varying the texture shapes

4. Conclusions

In the present paper, the texture shape of the textured contact with the boundary slip which may alter the flow characteristics of the textured contact was of particular interests varying the inlet length. Three texture shapes, i.e. triangular, parabola and rectangular were considered here. The modified Reynolds equation with slip was used to solve the lubrication. Based on the explanation above, the main conclusion can be drawn as follows:

1. With terms of the load support, triangular shape of textured contact is recommended to use.
2. With terms of the friction force, rectangular shape of textured contact is preferable.
3. The most interesting finding is that when the slip inlet length is set to 0.45 times the bearing length, the highest load support as well as the low friction can be achieved. This result can be regarded as a guideline for designing the textured bearing with slip.

References

[1] Maa C and Zhub H 2011 *Tribol. Int.* **44** 987
[2] Uddin MS and Liu YW 2016 *Biosurf. Biotribol.* **2** 59
[3] Zhang H, Dong GN, Hua M and Chin KS 2017 *J. Tribol.* **139** 011701
[4] Aurelian F, Patrick M and Mohamed H 2011 *Tribol. Int.* **44** 868
[5] Rao TVVLN, Rani AMA, Nagarajan T and Hashim FM 2012 *Tribol. Int.* **56** 121
[6] Tauviqirrahman M, Muchammad, Jamari and Schipper DJ 2014 *Tribol. Trans.* **57** 134
[7] Lin Q, Wei Z, Wang N and Chen W 2015 *Ind. Lubr. Tribol.* **67** 216

[8] Tauviqirrahman M, Ajie WK, Yohana E, Muchammad M and Jamari J 2016 *Int. J. Eng. Tech.* **8(2)** 913
[9] Muchammad M, Tauviqirrahman M, Jamari J and Schipper DJ 2017 *Lubr. Sci.* **29** 133
[10] Susilowati S, Tauviqirrahman M, Jamari J and Bayuseno AP 2016 *Tribol. - Mater. Surf. Interfaces* **10**(2), 85

IOP Conf. Series: Journal of Physics: Conf. Series **1090** (2018) 012028 doi:10.1088/1742-6596/1090/1/012028

Numerical Study of Nuclear Ground State Properties of Fuel Elements of Nuclear Reactor by Using *Skyrme-Hartree-Fock-Bogoliubov* Method

Yacobus Yulianto[1,a], **Santi Hatmanti**[2,b], **Zaki Su'ud**[1,c]

[1]Nuclear Physics and Biophysics Research Division, Physics Department, Mathematics and Natural Science Faculty, Institut Teknologi Bandung, Jl. Ganesha 10 Bandung, 40132, Indonesia.

[2]Computational Science, Mathematics and Natural Science Faculty, Institut Teknologi Bandung, Jl. Ganesha 10 Bandung, 40132, Indonesia.

E-mail: [a]yacyulianto@gmail.com, [b]santi.hatmanti02@gmail.com, [c]zakisuud@gmail.com

Abstract. In this study, it is investigated the total energy, nuclear radii, and neutron skin thickness of nuclear reactor fuel elements by using *Skyrme* interaction collaborated with *Hartree-Fock-Bogoliubov* method. Both proton and neutron densities were calculated by using SIII set parameters. The calculation results of this study are in good agreement with the related experimental results.

1. Introduction

The increase of human population in the world causes the increase of energy demand. Because the sources of non-renewable energies tend to be decreased, it is needed to develop the new alternative energy sources to provide the world energy demands. One of the new alternative energy sources is nuclear power plant, which utilizes the controlled nuclear reaction. In nuclear reactor design, nuclear reaction models are required to provide the estimation of particle-induced reaction cross-section, especially for unavailable experiment data or the difficult cross-section measurement by experiment. In development of reaction systematic, it is useful to understand the nuclear properties, especially the used nuclei on the nuclear reactor.

Nuclear density is one of the basic quantities to describe nuclear structure, which can provide detail information about nuclei internal structure, related to the wave functions of proton and neutron. In addition, nuclear charge radius is also the basic quantity, which represents the most useful observables information in nuclear structure analysis. Nuclear charge radius can provide information about nuclear shape. A method that can be used to study about nuclear structure, especially nuclei far from stability, is Skyrme-Hartree-Fock-Bogoliubov (SHFB) method, which has been used in spherical nuclei treatment extensively [1].

The SHFB method has been performed by Oberacker et al. to investigate the ground-state properties of several neutron-rich nuclei which are close to neutron drip-line [2]. Bassem et al. have used the SHFB method to study the ground-state properties of Mo and Ru [3] isotopes. Even, the SHFB method has been also used by Teran to study about the nuclei far from

Content from this work may be used under the terms of the Creative Commons Attribution 3.0 licence. Any further distribution of this work must maintain attribution to the author(s) and the title of the work, journal citation and DOI.
Published under licence by IOP Publishing Ltd

IOP Conf. Series: Journal of Physics: Conf. Series **1090** (2018) 012028 doi:10.1088/1742-6596/1090/1/012028

stability [4]. In this study, it is investigated the ground state properties of some common fuel materials, especially ^{232,233}Th, ^{233,235,238}U, $^{238-241}$Pu, and ^{241}Am. The total energies of those materials have been calculated by using SHFB method with SIII set parameters. It is also calculated the nuclear root-mean-square (rms) radii of charge, neutron, and proton. This study aims are to calculate numerically the ground-state energy of some common fuel materials (especially ^{232,233}Th, ^{233,235,238}U, $^{238-241}$Pu, and ^{241}Am) by using the SHFB method with SIII set parameters and to simulate the local densities as well as the local potentials of those nuclei.

2. Theory

2.1. The Skyrme-Hartree-Fock-Bogoliubov Method

In this study, interactions among the nucleons in the nucleus are represented by using Skyrme interaction. In Skyrme interaction, it is possible to work in coordinate space. It is very useful for nuclei close to the neutron drip-line. Skyrme interaction was first proposed by T.H.R. Skyrme in 1959 [5] and has been applied to many nuclei with great achievement in nuclear structure investigations [6, 7, 8]. To calculate the total energy of nucleus, it can be performed by using the following formula:

$$E = \int d^3\vec{r}\left[\frac{\hbar^2}{2m}\left(1-\frac{1}{A}\right)\tau + \varepsilon_{Skyrme} + \varepsilon_{pair} + \varepsilon_{Coul}\right] \tag{1}$$

where τ is the total kinetic energy density (the factor in parentheses represents the one-body part of the center-of-mass correction), ε_{Skyrme} is the Skyrme energy density, ε_{pair} is the pairing energy density, and ε_{Coul} is the Coulomb energy density. The detail explanation of the energy derivation can be found in several articles [1, 8, 9]. Next, the Skyrme energy density can be approached as:

$$\begin{aligned}\varepsilon_{Skyrme} =& \tfrac{1}{2}t_0\left(1+\tfrac{1}{2}x_0\right)\rho^2 - \tfrac{1}{2}t_0\left(x_0+\tfrac{1}{2}\right)\sum_q \rho_q^2 + \tfrac{1}{4}t_1\left(1+\tfrac{1}{2}x_1\right)\left[\rho\tau + \tfrac{3}{4}\left(\nabla\rho\right)^2\right] \\ &- \tfrac{1}{4}t_1\left(x_1+\tfrac{1}{2}\right)\sum_q\left[\rho_q\tau_q + \tfrac{3}{4}\left(\nabla\rho_q\right)^2\right] + \tfrac{1}{4}t_2\left(1+\tfrac{1}{2}x_2\right)\left[\rho\tau + \tfrac{1}{4}\left(\nabla\rho\right)^2\right] \\ &- \tfrac{1}{4}t_2\left(x_2+\tfrac{1}{2}\right)\sum_q\left[\rho_q\tau_q + \tfrac{1}{4}\left(\nabla\rho_q\right)^2\right] - \tfrac{1}{16}\left(t_1x_1+t_2x_2\right)J^2 + \tfrac{1}{16}\left(t_1-t_2\right)\sum_q J_q^2 \\ &+ \tfrac{1}{12}t_3\rho^\gamma\left[\left(1+\tfrac{1}{2}x_3\right)\rho^2 - \left(x_3+\tfrac{1}{2}\right)\sum_q\rho_q^2\right] + \tfrac{1}{2}W_0\left[J\nabla\rho + \sum_q J_q\nabla\rho_q\right]\end{aligned} \tag{2}$$

where t_0, t_1, t_2, t_3, x_0, x_1, x_2, x_3, W_0, and γ are Skyrme set parameters, ρ represents the density, and J is the spin-orbit density. The isospin label q represents proton or neutron, where the densities, without an isospin label in all equations, represent total densities.

Table 1. The SIII set parameters for *Skyrme* interaction [13].

Parameter	SIII	Parameter	SIII	Parameter	SIII	Parameter	SIII
t_0	-1128.75	x_0	0.45	t_3	14000	W_0	120
t_1	395	x_1	0	x_3	1	γ	1
t_2	-95	x_2	0				

The pairing energy density can be approached by using:

$$\varepsilon_{pair} = \sum_q \tfrac{1}{4}t_0'\left(1-x_0'\right)\tilde{\rho_q}^2+\tfrac{1}{4}t_1'\left(1-x_1'\right)\left[\tilde{\rho}_q\tilde{\tau}_q+\tfrac{1}{4}\left(\nabla\tilde{\rho}_q\right)^2\right] + \left[\tfrac{1}{8}t_2'\left(1+x_2'\right)+\tfrac{1}{4}W_0'\right]\tilde{J}_q^2+\tfrac{1}{24}t_3'\left(1-x_3'\right)\rho^{\gamma'}\tilde{\rho}_q^2 \tag{3}$$

For Coulomb energy density, the direct and the exchange parts can be respectively expressed as [10]:

$$\varepsilon_{Coul} = \tfrac{1}{2}e^2\int d^3r'\frac{\rho_p(r)\rho_p(r')}{|r-r'|} - \tfrac{3}{4}e^2\left(\tfrac{3}{\pi}\right)^{1/3}\rho_p^{4/3}\left(r'\right) \tag{4}$$

The observed nuclei in this study have been treated as spherical nucleus, which means that the treatments were performed only for radial part. In the SHFB framework, for spherical symmetry, the densities, the kinetic energy densities, and the spin-orbit densities of nucleons (normal and pairing) can be respectively approached by using:

$$\rho\left(r\right) = \frac{1}{4\pi r^2}\sum_{n\ell j}\left(2j+1\right)u_2^2\left(n\ell j,r\right) \tag{5}$$

$$\tilde{\rho}(r) = -\frac{1}{4\pi r^2}\sum_{n\ell j}\left(2j+1\right)u_1\left(n\ell j,r\right)u_2\left(n\ell j,r\right) \tag{6}$$

$$\tau\left(r\right) = \sum_{n\ell j}\frac{\left(2j+1\right)}{4\pi r^2}\left[\left(u_2'\left(n\ell j,r\right)-\frac{u_2\left(n\ell j,r\right)}{r}\right)^2+\left(\frac{\ell\left(\ell+1\right)}{r^2}u_2^2\left(n\ell j,r\right)\right)\right] \tag{7}$$

$$\tilde{\tau}(r) = -\sum_{n\ell j}\frac{\left(2j+1\right)}{4\pi r^2}\left[\left(u_1'\left(n\ell j,r\right)-\frac{u_1\left(n\ell j,r\right)}{r}\right)\left(u_2'\left(n\ell j,r\right)-\frac{u_2\left(n\ell j,r\right)}{r}\right) + \frac{\ell\left(\ell+1\right)}{r^2}u_1\left(n\ell j,r\right)u_2\left(n\ell j,r\right)\right] \tag{8}$$

$$J(r) = \frac{1}{4\pi r^3}\sum_{n\ell j}\left(2j+1\right)\left[j\left(j+1\right)-\ell\left(\ell+1\right)-\tfrac{3}{4}\right]u_2^2\left(n\ell j,r\right) \tag{9}$$

$$\tilde{J}(r) = -\frac{1}{4\pi r^3}\sum_{n\ell j}\left(2j+1\right)\left[j\left(j+1\right)-\ell\left(\ell+1\right)-\tfrac{3}{4}\right]u_1\left(n\ell j,r\right)u_2\left(n\ell j,r\right) \tag{10}$$

where u is the radial wave function, n is pincipal quantum number, ℓ is the orbital angular momentum, and j is the total angular momentum. From Eq. (5), it can be calculated the rms (root-mean-square) radii of proton and neutron by using

$$\left\langle r_q^2\right\rangle = \int_0^{R_{box}} r^2\rho_q(r)d^3r \tag{11}$$

The difference between neutron radius and proton radius, well-known as the neutron skin thickness, can be calculated by using [11]

$$t = r_n - r_p \tag{12}$$

The effective mass (normal and abnormal) due to the dependence of energy density ε on the kinetic densities can be approached by using

$$M_q = \tfrac{\hbar^2}{2m_q^*} = \tfrac{\hbar^2}{2m}+\tfrac{t_1}{4}\left[\left(1+\tfrac{x_1}{2}\right)\rho-\left(x_1+\tfrac{1}{2}\right)\rho_q\right]+\tfrac{1}{4}t_2\left[\left(1+\tfrac{x_2}{2}\right)\rho+\left(x_1+\tfrac{1}{2}\right)\rho_q\right] \tag{13}$$

$$\tilde{M}_q = \tfrac{1}{4}t_1'\left(1 - x_1'\right)\rho_q \tag{14}$$

For particle-hole (Hartree-Fock) field, it is given by

$$\begin{aligned} U_q = & t_0\left(1+\tfrac{1}{2}x_0\right)\rho - t_0\left(x_0+\tfrac{1}{2}\right)\rho_q + \tfrac{1}{4}t_1\left(1+\tfrac{1}{2}x_1\right)\left(\tau - \tfrac{3}{2}\Delta\rho\right) - \tfrac{1}{2}W_0\left[\nabla J + \nabla J_q\right] \\ & - \tfrac{1}{4}t_1\left(x_1+\tfrac{1}{2}\right)\left(\tau_q - \tfrac{3}{2}\Delta\rho_q\right) + \tfrac{1}{4}t_2\left(1+\tfrac{1}{2}x_2\right)\left(\tau + \tfrac{1}{2}\Delta\rho\right) \\ & + \tfrac{1}{4}t_2\left(x_2+\tfrac{1}{2}\right)\left(\tau_q + \tfrac{1}{2}\Delta\rho_q\right) + \tfrac{1}{12}t_3\left(1+\tfrac{1}{2}x_3\right)(2+\gamma)\rho^{\gamma+1} \\ & - \tfrac{1}{12}t_3\left(x_3+\tfrac{1}{2}\right)\left[\gamma\rho^{\gamma-1}\sum_{q'}\rho_{q'}^2 + 2\rho^{\gamma}\rho_q\right] + \tfrac{1}{24}t_3'\left(1-x_3'\right)\gamma'\rho^{\gamma'-1}\sum_{q'}\tilde{\rho}_{q'}^2 \end{aligned} \tag{15}$$

where the pairing field (particle-particle) is given by

$$\tilde{U}_q = \tfrac{1}{2}t_0'(1-x_0')\tilde{\rho}_q^2 + \tfrac{1}{4}t_1'(1-x_1')\left[\tilde{\tau}_q - \tfrac{1}{2}\Delta\tilde{\rho}_q\right] + \tfrac{1}{12}t_3'(1-x_3')\rho^{\gamma'}\tilde{\rho}_q \tag{16}$$

The form factors of spin-orbit fields and its pairing respectively can be approached by using

$$B_q = -\tfrac{1}{8}(t_1x_1 + t_2x_2)J + \tfrac{1}{8}(t_1 - t_2)J_q + W_0\nabla(\rho + \rho_q) \tag{17}$$

$$\tilde{B}_q = \left[\tfrac{1}{2}t_2'(1+x_2') + W_0'\right]\tilde{J}_q \tag{18}$$

where t_0', t_1', t_2', t_3', x_0', x_1', x_2', x_3', W_0', and γ' are the parameter in the particle-particle channel of Skyrme force parameters implemented in the HFBRAD code [9].

2.2. *The Numerov method in HFB equation*

The matrix form of the fields can be defined as

$$\mathcal{M} = \begin{pmatrix} M & \tilde{M} \\ \tilde{M} & M \end{pmatrix}, \qquad \mathcal{U} = \begin{pmatrix} U-\lambda & \tilde{U} \\ \tilde{U} & -U+\lambda \end{pmatrix} \tag{19}$$

$$\mathcal{U}_{so} = \begin{pmatrix} B & \tilde{B} \\ \tilde{B} & -B \end{pmatrix} \frac{j(j+1) - \ell(\ell+1) - \frac{3}{4}}{2r} \tag{20}$$

The HFB equations can be written as

$$\left[-\frac{d}{dr}\mathcal{M}\frac{d}{dr} + \mathcal{U} + \mathcal{M}\frac{\ell(\ell+1)}{r^2} + \frac{\mathcal{M}'}{r} + \mathcal{U}_{so}\right]\begin{pmatrix} u_1 \\ u_2 \end{pmatrix} = E\begin{pmatrix} u_1 \\ u_2 \end{pmatrix} \tag{21}$$

This equation can be written as an equation with no differential operator in the coupling terms and no first order derivative, i.e.

$$\begin{aligned} -M^*\tfrac{d^2}{dr^2}f_1 + Vf_1 + Wf_2 = Ef_1 \\ M^*\tfrac{d^2}{dr^2}f_2 + Vf_2 + Wf_1 = Ef_2 \end{aligned} \tag{22}$$

This last form, with no first order derivative of the functions, is particularly suitable for numerical integration by the Numerov algorithm. In this study, it is used the Numerov (Cowell) method as the standard technique to solve the integration of second-order differential equations [9, 12]. Consider the function $y(r)$ as the solution of the differential equation

$$y'' = F(y, r) \tag{23}$$

Based on the Numerov algorithm, it is used the finite difference formula for three consecutive points on a mesh with step h

$$y_{n+1} - 2y_n + y_{n-1} = \frac{h^2}{12}\left(y''_{n-1} + 10y''_n + y''_{n+1}\right) + O(h^6) \tag{24}$$

where $O(h^6)$ is the error in this relation. Combining eq. (24) with eq. (23), it can be obtained the relation

$$\left(1 - \frac{h^2}{12}F_{n+1}\right)y_{n+1} - 2\left(1 + \frac{5h^2}{12}F_n\right)y_n + \left(1 - \frac{h^2}{12}F_{n-1}\right)y_{n-1} = 0 \tag{25}$$

In this study, to perform numerical calculation, the HFBRAD code [9] has been utilized with SIII set parameters. The SIII set parameters are set up by the adjustment of Skyrme interaction parameters through the binding energy and charge radii fittings of some semi-magic nuclei. The SIII set parameters work well for nuclei close to sphericity with small deformations [13]. In this study, the grid has been set at 0.1 fm with 400 mesh point. The side purpose of this study is to observe the precision of HFBRAD code in calculation of nuclear energy for nuclei far from stability. The calculation results have been compared with the related experiment data [14, 15].

3. Results and Discussion

3.1. Total Energy and Nuclear Radii

The calculation results of total energy for each nucleus investigated in this study are shown in Table 2. From Table 2, it can be seen that the calculation results are in good agreement with the experiment results with 0.687%-0.952% of discrepancies. The total energy of ^{233}Th is almost similar with that of ^{233}U although ^{233}Th and ^{233}U have different numbers of proton and neutron (^{233}Th has 90 protons with 143 neutrons and ^{233}U has 92 protons with 141 neutrons). It is also similar for the cases of ^{238}U with ^{238}Pu and ^{241}Pu with ^{241}Am.

Table 2. The total energies of fuel elements.

Nucleus	Exp. [14] (MeV)	This study (MeV)	discr.* (%)	Nucleus	Exp. [14] (MeV)	This study (MeV)	discr.* (%)
^{232}Th	-1766.686	-1753.916	0.723	^{238}Pu	-1801.268	-1784.689	0.920
^{233}Th	-1771.472	-1759.187	0.693	^{239}Pu	-1806.914	-1790.722	0.896
^{233}U	-1771.721	-1759.541	0.687	^{240}Pu	-1813.449	-1796.708	0.923
^{235}U	-1783.863	-1771.033	0.719	^{241}Pu	-1818.690	-1802.649	0.882
^{238}U	-1801.689	-1787.934	0.763	^{241}Am	-1817.929	-1800.632	0.952

*discrepancy

The charge, mass, proton, and neutron radii for each nucleus have been also calculated. The calculation results are shown in Table 3. As well as the results of total energies, the calculation results of charge radii are also in good agreement with the related experiment results. From Table 3, it can be indicated that the charge, mass, and proton radii increase accordingly with the increase of nucleon number. From those results above, the obtained results of HFBRAD code are in good agreement with the related experiment results. It is indicated that the HFBRAD code can be a useful tool to investigate the ground state properties of nuclei used on nuclear reactor, especially heavy nuclei.

Table 3. The nuclear radii of fuel elements (the units of radii and neutron skin are in fm).

Nucleus	Neutron number	Charge radius Exp. [15]	Radius Charge	Radius Mass	Radius Proton	Radius Neutron	Neutron skin thickness
^{232}Th	142	5.7848±0.0124	5.8099	5.8306	5.7546	5.8783	0.1237
^{233}Th	143	-	5.8160	5.8393	5.7607	5.8882	0.1275
^{233}U	141	5.8203±0.0049	5.8335	5.8420	5.7784	5.8831	0.1047
^{235}U	143	5.8337±0.0041	5.8456	5.8591	5.7906	5.9029	0.1123
^{238}U	146	5.8571±0.0033	5.8630	5.8842	5.8082	5.9315	0.1234
^{238}Pu	144	5.8535±0.0378	5.8771	5.8838	5.8224	5.9235	0.1011
^{239}Pu	145	5.8601±0.0378	5.8829	5.8920	5.8283	5.9330	0.1047
^{240}Pu	146	5.8701±0.0379	5.8886	5.9002	5.8340	5.9424	0.1084
^{241}Pu	147	5.8748±0.0379	5.8943	5.9083	5.8397	5.9517	0.1119
^{241}Am	146	5.8928±0.0042	5.9012	5.9082	5.8468	5.9479	0.1011

3.2. The Profiles of Density and Potential

In this study, it is also investigated the profiles of density and potential for each fuel element to understand its distribution. In Figure 1, it can be seen that neutron densities are larger than proton densities because neutron numbers in each observed nucleus are larger than proton numbers. Proton densities of all observed nuclei in this study are similar to each other and so are the neutron densities. The widths of proton densities for all nuclei are almost similar to each other and so are the widths of neutron densities. The densities of proton and neutron decreased drastically in radius 5-10 fm and approached to zero smoothly at r $\geq$ 10 fm.

In Figure 2, it can be seen that neutron potentials are deeper than that of proton in radius

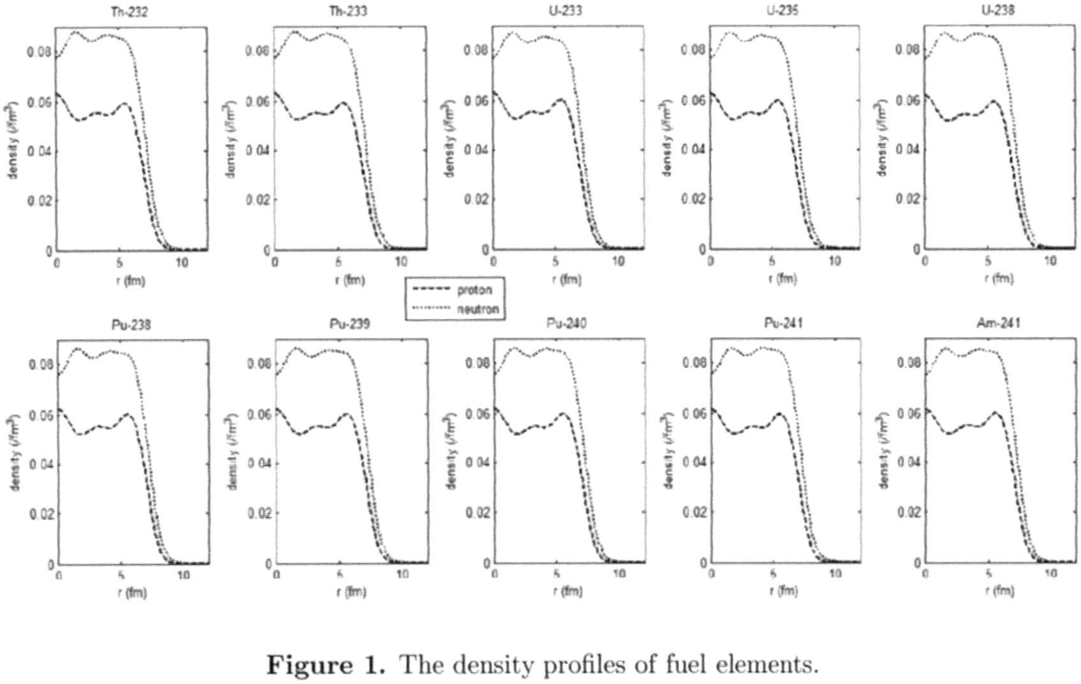

Figure 1. The density profiles of fuel elements.

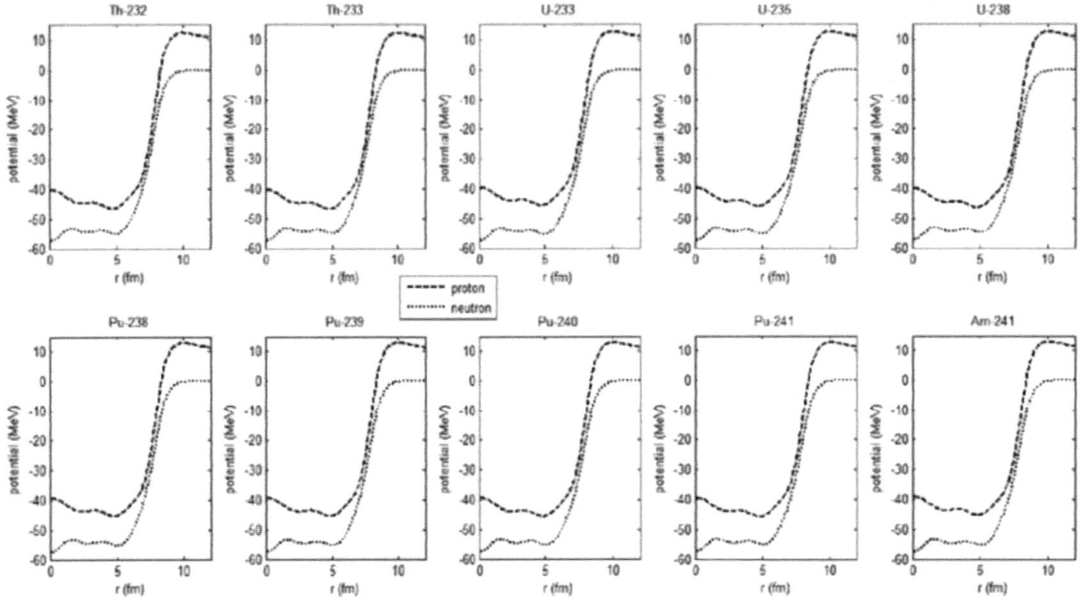

Figure 2. The potential profiles of fuel elements.

0-5 fm. In other hand, proton potentials are larger than neutron potential in r $\geq$ 10 fm. Proton potentials of all observed nuclei in this study are similar to each other and so are the neutron potentials. The widths of proton and neutron potentials are almost similar. From those results above, it can be seen that the results of this study are in good agreement with the results of the related experiment.

4. Summary

In this study, it has been performed the investigation of the total energy, nuclear radii, and neutron skin thickness of nuclear reactor fuel elements, especially ^{232,233}Th, ^{233,235,238}U, $^{238-241}$Pu, and ^{241}Am, by using Skyrme-Hartree-Fock-Bogoliubov method. Both proton and neutron densities were calculated by using SIII set parameters. The obtained results of this study have achieved the good agreement with the related experiment results as well as the results of HFBRAD code. It can be indicated that SHFB method applied in the HFBRAD code can be a useful tool to investigate the ground-state properties of nuclei far from stability.

5. Acknowledgment

The authors are grateful to Prof. K. Bennaceur (France) for guide in using HFBRAD code. This study is fully supported and funded by Advance Nuclear Laboratory of ITB.

References

[1] Stoitsov M V, Dobaczewski J, Nazarewicz W and Ring P 2005 *Comput. Phys. Commun.* **167** 43-63.
[2] Oberacker V E, Umar A S, Teran E, and Blazkiewicz A 2003 *Phys. Rev.* C **68** 064302.
[3] El Bassem Y and Oulne M 2017 *Nucl. Phys.* A **957** 22-32.
[4] Teran E B 2003 Ph.D. Dissertation (Tennessee: Vanderbilt University).
[5] Skyrme T H R 1959 *Nucl. Phys.* **9** 615-634.
[6] Vautherin D and Brink D M 1972 *Phys. Rev.* C **5** 626-647.
[7] Reinhard P G 1991 Skyrme-Hartree-Fock Model of the Nuclear Ground State *Computational Nuclear Physics 1 - Nuclear Structure* ed S E Koonin, (Berlin Heidelberg: Springer-Verlag) chapter 2 pp 28-50.

[8] Dobaczewski J, Flocard H and Treiner J. 1984 *Nucl. Phys.* A **422** 103-139.
[9] Bennaceur K and Dobaczewski J 2005 *Comput. Phys. Commun.* **168** 96-122.
[10] Skalski J 2001 *Phys. Rev.* C **63** 024312.
[11] Myers W D and Swiatecky W J 1969 *Ann. Phys.* **55** 395-505.
[12] Dahlquist G and Björck A 1974 *Numerical Methods* (New Jersey: Prentice-Hall).
[13] Beiner M, Flocard H and Giai N V 1975 *Nucl. Phys.* A **238** 29-69.
[14] Wang M, Audi G, Wapstra A H, Kondev F G, MacCormick M, Xu X and Pfeiffer B 2012 *Chinese Phys.* C **36** 1603-2014.
[15] Angeli I and Marinova K P 2013 *Atomic Data and Nuclear Data Tables* **99** 69-95.

IOP Conf. Series: Journal of Physics: Conf. Series **1090** (2018) 012029 doi:10.1088/1742-6596/1090/1/012029

Frontier Orbitals of Dehydrogenated Tetrahydrocurcumin in Water Solvent: A Theoretical Study

Lusia S. P. Boli[1], Vera Khoirunisa[1,2], Adhitya G. Saputro[1,3], Mohammad K. Agusta[1,3], Febdian Rusydi[4], Heni Rachmawati[3,5], Hermawan K. Dipojono[1,3*]

[1]Engineering Physics Research group, Institut Teknologi Bandung, Indonesia
[2]Engineering Physics Program, Institut Teknologi Sumatera, Indonesia
[3]Research Center for Nanoscience and Nanotechnology, Institut Teknologi Bandung, Indonesia
[4]Research Center for Computational Quantum Engineering, Faculty of Science and Technology, Universitas Airlangga, Surabaya, Indonesia
[5]School of Pharmacy, Institut Teknologi Bandung, Bandung, Indonesia

[*]Corresponding author : dipojono@gmail.com

Abstract. We studied two frontier orbitals - the highest occupied molecular orbital (HOMO) and lowest unoccupied molecular orbital (LUMO)- of tetrahydrocurumin (THC) using density-functional theory (DFT) in water solvent. These orbitals were observed in THC molecule without one hydrogen atom (dehydrogenated THC). The loss of hydrogen atom is due to the transfer of the atom from THC molecule toward reactive oxygen species (ROS) (Hydrogen atom transfer –HAT- mechanism). We began our investigation by optimizing dehydrogenated THC at three X-H sites. Then, water solvent was added by using polarized continuum model (PCM) method. This study observed that dehydrogenated THC at two O-H sites has wider gap of HOMO-LUMO compare to C-H site.

1. Introduction

Studying dehydrogenation of THC is an initial step to understand the role of THC as antioxidant based on hydrogen atom transfer (HAT). In biological system, excessive amount of reactive oxygen species (ROS) promotes oxidative stress [1]. Thus, THC is proposed as one of the antioxidant agents that can prevent oxidative stress in biological system. In playing its role, THC is expected to transfer one of its hydrogen atom toward ROS to neutralize the free radical molecule[2]. This is called hydrogen atom transfer mechanism. Due to HAT mechanism, THC is left with one less hydrogen atom or so called dehydrogenated THC. A good antioxidant agent is expected to maintain the stability of its dehydrogenated form[3].

There are two frontier orbitals used to assess stability of dehydrogenated THC. They are the highest occupied molecular orbital (HOMO) and lowest unoccupied molecular orbital (LUMO). Gap between this two orbitals will decide the ability of electrons to jump from HOMO to LUMO. Wide gap means is correlated to high stability while narrow gap is correlated to low stability[4].

In this research, we intend to study two frontier orbitals of THC produced by THC dehydrogenated using first principles. We choose density functional theory (DFT) as our research framework since it can produce frontier orbitals of dehydrogenated THC by constructing electronic structures only. Thus

Content from this work may be used under the terms of the Creative Commons Attribution 3.0 licence. Any further distribution of this work must maintain attribution to the author(s) and the title of the work, journal citation and DOI.
Published under licence by IOP Publishing Ltd

this is called first principles study. Result of this study will guide us to decide the stability of dehydrogenated THC after transffering a hydrogen atom from an X-H site. We analyze the HOMO and LUMO energy of three X-H sites of THC to lead us to our final conclusion. THC performes better antioxidant reactivity in polar medium [5], so we choose polar water medium as our simulation environment. We are aware of the problematic of PCM in modeling water as solvent in our previous works where the trend predicted by DFT-PCM calculations always break when using water solvent [6,7].

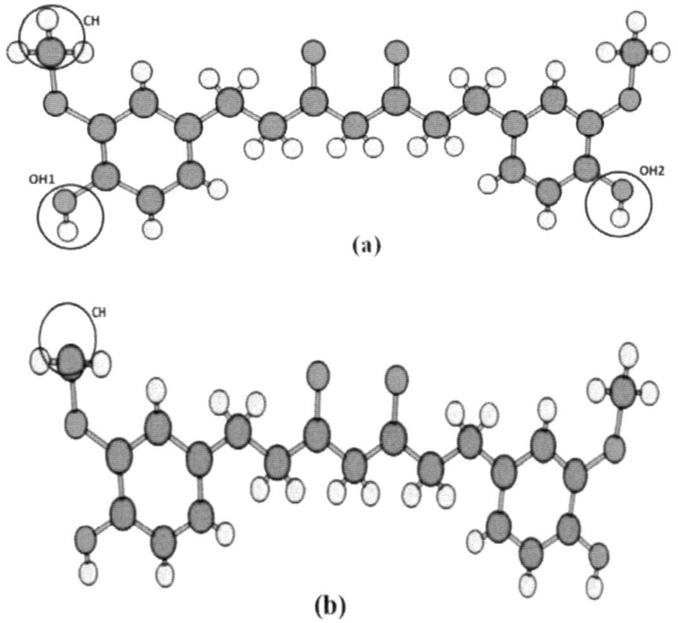

Figure 1. Molecular geometry of (a) THC molecule and (b) dehydrogenated THC produced by abstracting a hydrogen atom from CH site.

2. Computational details

We performed geometry optimization of three THC radicals at site OH1, OH2 and CH using UM05-2X functional and 6-31++G(d,p) basis set. Some authors have successfully used M05-2X functional to study the antioxidant properties of polyphenols [8, 9]. The developer of this functional was also suggesting it for thermochemistry calculations [10]. Since our system was an open shell system of THC dehydrogenated, we performed an unrestricted calculation. Our simulation was done in water environment to represent a polar medium. This solvent effect was included using polarized continuum model (PCM), where water was modelled as cavity that interacts electrostatically with THC radical placed within the cavity [11]. All calculations were performed using Gaussian 09 suite program [12] at 298.15 K.

3. Results and discussion

Figure 1(a) shows model of our molecule of interest which is a THC molecule. Site CH, OH1 and OH2 mark three targeted sites with the ability of transferring a hydrogen atom. We abstract one hydrogen atom from each of this site respectively to create dehydrogenated THC. Figure 1(b) shows dehydrogenated THC at C-H site. Our DFT calculations coupled with PCM confirms that the geometry of our system is in minimum energy with doublet spin states.

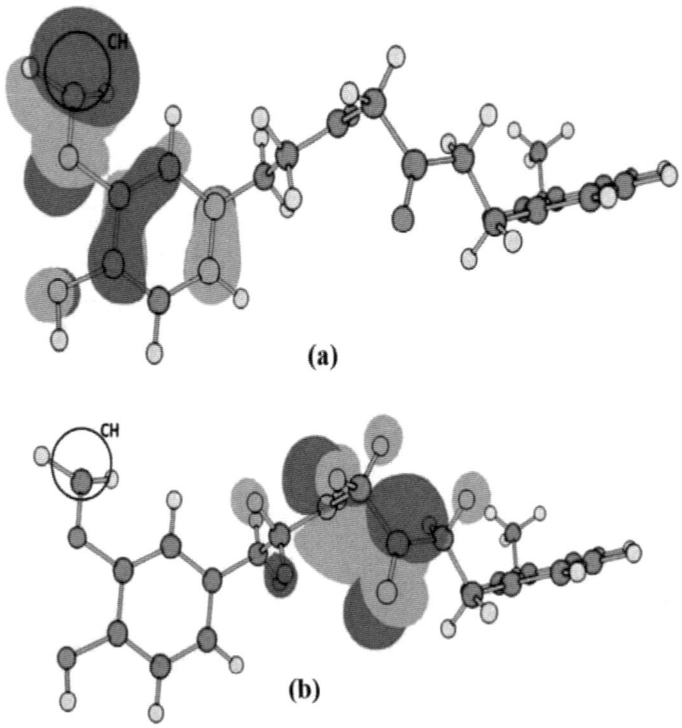

Figure 2. (a) HOMO and (b) LUMO profiles of dehydrogenated THC at CH site.

Figure 2 presents HOMO and LUMO profiles of dehydrogenated THC at C-H site. As shown in 2(a), the HOMO profile is located at dehydrogenated site while in 2(b), the LUMO profile is located in the middle of the molecule. This result is also true for site OH1 and OH2.

Figure 3 summarizes the relative HOMO-LUMO energy of our three dehydrogenated THC systems. In general, O-H sites have higher relative HOMO-LUMO energy than CH site. Higher relative HOMO-LUMO energy means electrons cannot jump easily from HOMO to LUMO. As a result, dehydrogenated THC at O-H sites is more stable than at C-H sites. Site OH1 and OH2 possess very similar value of relative energy HOMO-LUMO. This result predicts that hydrogen atom transfer from both sides will produce the same stability of dehydrogenated THC in water environment. We compare our results to one of our previous work for relative HOMO-LUMO of THC in gas phase [13] and conclude that water solvent does not affect the stability of dehydrogenated THC.

4. Conclusion

The study of frontier orbitals of dehydrogenated THC in water solution showed that when hydrogen atom transfer occured at O-H site, the stability of dehydrogenated THC is higher than when it occured at C-H site. Hydrogen atom transfer at two O-H sites in THC will produce dehydrogenated THC with the same stability.

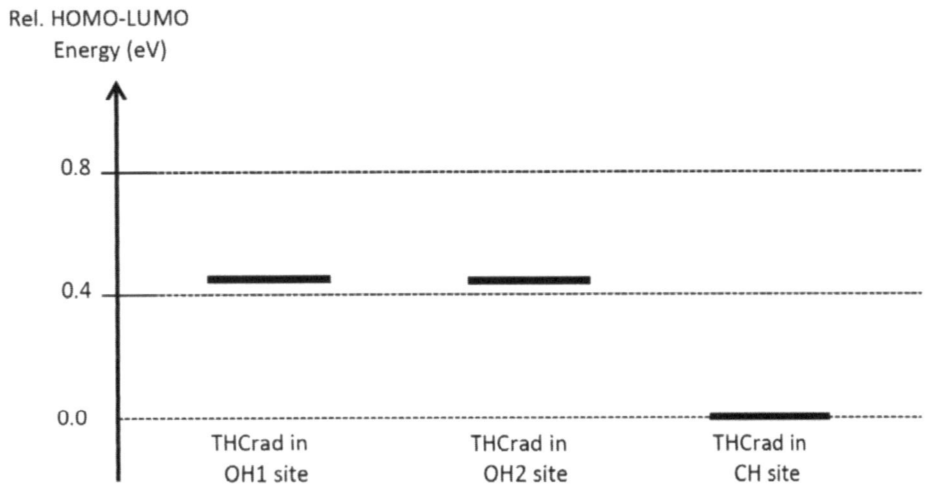

Figure 3. Relative energy diagram of THC radicals in OH1, OH2 and CH sites

Acknowledgement
The authors are grateful to Lembaga Pengelola Dana Pendidikan (LPDP) that has sponsored this research.

References

[1] Halliwell B 1991 Reactive oxygen species in living systems: source, biochemistry, and role in human disease *The American Journal of Medicine* **91** S14-S22

[2] Portes E, Gadrat C and Castellan A 2007 A comparative study on the antioxidant properties of tetrahydrocurcuminoids and curcuminoids *Tetrahedron* **63** 9092-99

[3] Ingold K U and Pratt D A 2014 Advances in radical-trapping antioxidant chemistry in 21st century: a kinetics and mechanisms perspective *Chemical Reviews* **114** 9022-9046

[4] Anh N T 2007 Frontier Orbitals a practical manual *John Wiley & Sons*

[5] Khopde S M, Priyadarsini K I, Guha S N, Satav J G, Venkatesan P and Aswathanarana Rao M N 2000 Inhibition of radiation-induced lipid peroxidation by tetrahydrocurcumin: possible mechanisms by pulse radiolysis *Bioscience, Biotechnology and Biochemistry* **64** 503-509

[6] Fadilla R N, Aisyah N D, Dipojono H K and Rusydi F 2017 The first-principle study on the stability of trans-HCOH in various solvents *IOP Conf. Series:Journal of Physics: Conf.Series* **853** 012031

[7] Rusydi F, Shukri G, Saputro A G, Agusta M K, Dipojono H K, Suprijadi S 2017 Dipole strength calculation based on two-level system approximation to study Q/B-band intensity ratio of ZnTBP in solvent *Journal of the Physical Society of Japan* **86** 044706

[8] Martínez A, Galano A and Vargas R 2011 Free radical scavenger properties of α-mangostin: thermodynamics and kinetics of HAT and RAF mechanisms *Journal of Physical Chemistry B* **115** 12591-8

[9] Iuga C, Alvarez-Idaboy J R and Russo N 2012 Antioxidant activity of trans-resveratrol toward hydroxyl and hydroperoxyl radicals: a quantum chemical and computational kinetic study *Journal of Organic Chemistry* **77** 3868-77

[10] Zhao Y, Schultz N E, and Truhlar D G 2006 Design of density functionals by combining the method of constraint satisfaction with parametrization for thermochemistry, thermochemical kinetics, and noncovalent interactions *J.Chem. Theory Comput* **2** 364-84

[11] Mennucci B , Tomasi J, Cammi R, Cheeseman J R, Frisch M J, Devlin F J, Gabriel S and Stephens P J 2002 Polarizable continuum model (PCM) calculations of solvent effects on

optical rotations of chiral molecules *Journal of Physical Chemistry* **106** 6102-13
[12] Wallingford CT Gaussian 09, revision C.01 *Gaussian Inc* Wallingford CT 2010
[13] Boli L S P, Khoirunisa V, Saputro A G, Agusta M K, Rusydi F, Rachmawati H and Dipojono H K 2017 Theoretical study on frontier orbitals of tetrahydrocurcumin in hydrogen atom transfer mechanism *IOP Conference Proceedings* (submitted, under review)

IOP Conf. Series: Journal of Physics: Conf. Series **1090** (2018) 012030 doi:10.1088/1742-6596/1090/1/012030

Hybrid Cryptosystem Implementation Using IDEA and Knapsack Algorithm for Message Security

Dian Rachmawati[1 *], Maya Silvi Lydia[1 *], Wynda Arianni Siregar[1 *]

[1] Departemen Ilmu Komputer, Fakultas Ilmu Komputer dan Teknologi Informasi, Universitas Sumatera Utara, Jl. Universitas No. 9-A, Medan 20155, Indonesia.

[*]Email: dian.rachmawati@usu.ac.id, maya.silvi@gmail.com, wyndasiregar@gmail.com

Abstract. One way of maintaining the information confidentiality is by using a cryptographic technique such as hybrid cryptosystem. A hybrid cryptosystem is a combination of symmetric and asymmetric algorithms. On symmetric algorithms, the time which it takes to perform encryption and decryption is shorter than an asymmetric algorithm. While the advantages of asymmetric algorithms are only needed to maintain the confidentiality of the private key without the need to frequently change the key as the symmetric algorithm. In this algorithm, there is no need to send the key private key as in the system of symmetry. This research uses the IDEA of symmetry algorithm and algorithm of knapsack asymmetry. The IDEA of the algorithm is to secure a text message, while the knapsack algorithm is to maintain the confidentiality of the message key distribution. Information has acquired in the form of a string. Because the IDEA is a block cipher, then the length of the text characters is out of stock divided by 8. The use of a hybrid cryptosystem with Knapsack algorithm IDEA and managed to keep the integrity of the data, so the data is encoded can return to its original form.

1. Introduction

Cryptography is a technique of securing and ensuring the authenticity of data which is consisting of two processes namely encryption and decryption. Encryption is a process to make a data becomes unreadable by unauthorized parties such as a password or coded language. While decryption is the opposite of encryption, i.e., the process of making the information or data that has been encrypted to be readable by the party entitled.

The use of a symmetric cryptographic algorithm and asymmetric cryptography to secure data, for now, is considered inadequate as cryptanalysis develops. In this research, a cryptographic method used is hybrid cryptosystem. The hybrid cryptosystem is a combination of symmetric algorithms and asymmetric algorithms. Hybrid cryptosystem used in this study is a combination of IDEA and Knapsack algorithms.

The IDEA (International Data Encryption Algorithm) is a symmetric algorithm that operates on an open-width 64-bit message block and a 128-bit key length. While the Knapsack algorithm is an asymmetric cryptographic algorithm that uses a public key to encrypt messages and use private key (private key) to decrypt messages.

Content from this work may be used under the terms of the Creative Commons Attribution 3.0 licence. Any further distribution of this work must maintain attribution to the author(s) and the title of the work, journal citation and DOI.
Published under licence by IOP Publishing Ltd

2. Method

2.1. Hybrid Cryptosystem

Hybrid Cryptosystem is one of the method in cryptography which works by combining both of symmetric and asymmetric algorithm to secure plaintext [7]. In this hybrid system, encryption/decryption of messages using cryptographic key-symmetry, while the symmetry key is encrypted / decrypted with public-key cryptography. A symmetry key (also called a session key) has generated by either party and encrypted a message with that key. The session key has encrypted with the recipient's public key, and the session key will be sent together with an encrypted message. The recipient initially decrypts the session key with its private key, then decrypts the message with the session key [2].

2.2. International Data Encryption Algorithm (IDEA)

IDEA is one of the most secure block cipher algorithms available to the public [4]. First, The IDEA has discovered by Xuejia Lai and James Massey in 1990. Formerly, it was called PES (Proposed Encryption Standard). The following year, after Biham and Shamir have demonstrated their derivative cryptanalysis, the designers of the algorithm strengthened their ciphers to against the attack and named the new algorithm as the IPES (Improved Proposed Encryption Standard). And in 1992, IPES changed its name to IDEA (International Data Encryption Algorithm) [1].

IDEA is a block cipher, which operates on a 64 bits plaintext block. The key length is 128 bits. The same algorithm has used for encryption and decryption process. IDEA applies the following incompatible algebraic operations:

XOR or ⊞.

Added modulo 2^{16} or ⊞.

Multiplication modulo $2^{16} + 1$ or ⊙ [6].

The IDEA algorithm can be divided into three major parts, namely the key forming algorithm, encryption, and decryption.

1. IDEA Key Formation Process

The IDEA method has an input key 128 bits identical to 32 hexadecimal digits or 16 characters processed to produce 52 subkeys, with the details of each of the six subkeys was to be used in 8 rounds and four subkeys for output transformation.

2. IDEA Encryption Process

The IDEA algorithm encryption process is a 64 bits plaintext and divided it into four sub-blocks with 16 bits long, i.e., X1, X2, X3, X4. These four sub-blocks serve as inputs for the first-phase iteration of the algorithm. In total there are eight iterations. At each iteration, four sub-blocks are XOR-aligned, added, multiplied by the other and with six 16-bit subkeys.

Among the iterations of the second and third sub-blocks are interchangeable. Finally, four sub-blocks have combined with four subkeys in the output transformation. At each stage, the following sequence has done:

1. Multiply X1 with K1 mod $(2^{16} + 1)$.
2. Add X2 with K2 mod 2^{16}.
3. Add X3 with K3 mod 2^{16}.
4. Multiply X4 with K4 mod $(2^{16} + 1)$.
5. XOR results from step 1 and 3.
6. XOR results from step 2 and 4.
7. Multiply the result from step 5 with K5 mod $(2^{16} + 1)$.
8. Add the result from step 6 and 7 mod 2^{16}.
9. Multiply the result from step 8 with K6 mod $(2^{16} + 1)$.
10. Add the result from step 7 and 9 mod 2^{16}.
11. XOR results from step 1 and 9.
12. XOR results from step 3 and 9.
13. XOR results from step 2 and 10.

14. XOR results from step 4 and 10.

The output of each round is the four sub-blocks generated in steps 11, 12, 13 and 14. The blocks 12 and 13 are swapped (except for the last round) so that the input of the next round is a combination of steps 11, 12, 13, And 14. After eight rounds, the following output transformation will be performed:

1) Multiply X1 with K1 mod subkey ($2^{16} + 1$).
2) Add X2 with subkey K2 mod 2^{16}.
3) Add X3 with K3 mod 2^{16} subkey.
4) Multiply X4 with K4 mod subkey ($2^{16} + 1$)

3. IDEA Decryption Process

The decryption process is the same as the encryption process. The difference only lies in the rules of the subkey. The order of the subkey was reversed with the encryption process, and its subkey is converted. The subkey on the output transformation step of the encryption process is converted and used as a subkey in round 1 of the decryption process. Subkey of round 8 is converted and used as subkey on round 1 and 2 in the decryption process. Subkeys which has used for encryption are different with subkey which have used for decryption. The first difference is in the order of use of the subkey. The second difference, namely the formation of decryption subkey using multiplexed inverse multiplexing modulo $2^{16} + 1$ and inverse operations of modulo 2^{16} subkey.

2.3. Knapsack Public Key Algorithm

Many conventional knapsack cryptosystems use particular secret sequences for letting the decryption easy. For example, a super-increasing sequence has used in Merkle-Hellman Knapsack scheme [5]. The superincreasing knapsack algorithm is a weak algorithm since the ciphertext can be decrypted easily into O (n). The non-super increasing knapsack or Knapsack algorithm is a problematic problem (computational) in knapsack algorithm group because it takes time in an exponential order to solve it. However, superincreasing knapsack can be modified into non-super increasing knapsack by using the public key (for encryption) and private key (for decryption). The public key is a non-super increasing sequence, while the private key remains a superincreasing sequence. This modification has discovered by Ralph Merkle and Martin Hellman [2].

The basic idea behind Merkle-Hellman's encryption scheme is to create a subset of problems that can be solved quickly. This easy-to-solve problem is by hiding the superincreasing sequence by multiplication modulo and permutation [3].

Algorithm for generating the public key and private key are:

a.Determine the superincreasing sequence.

b.Multiply each element in the sequence with n modulo m. The modulus m should be a number greater than the sum of all elements in the sequence, whereas the multiplier n should have no union factor with m.

c.Multiplication results will be the public key while the superincreasing sequence became the private key.

Encryption Process of Knapsack Algorithm

The encryption process was done with the following stages:

a. Encryption was done in the same way by using previous knapsack algorithm.

b. At first, the plaintext was broken into blocks of bits that are equal in length to the cardinality of the public key row.

c.Multiply each bit in the block with the corresponding element in the public key

Decryption Process of Knapsack Algorithm

The process of decryption has done with the following stages: Decryption has done by using the private key. The receiver of the message counts n^{-1}, i.e., inverses n modulo m, such that $n \cdot n^{-1} = 1$ (mod m). Multiply each cryptogram with n^{-1} mod m, then declare the result as the sum of private key elements to obtain plaintext by using a search algorithm to super increasing Knapsack.

3. Results and Discussions

The illustration of hybrid cryptosystem by using IDEA and Knapsack is shown in figure 1. The recipient shares the Knapsack key to the sender. The sender encrypts the message using IDEA algorithm and encrypts the IDEA key using Knapsack key. After the sending process, the receiver decrypts the IDEA cipher key. The IDEA key is used to open the ciphertext.

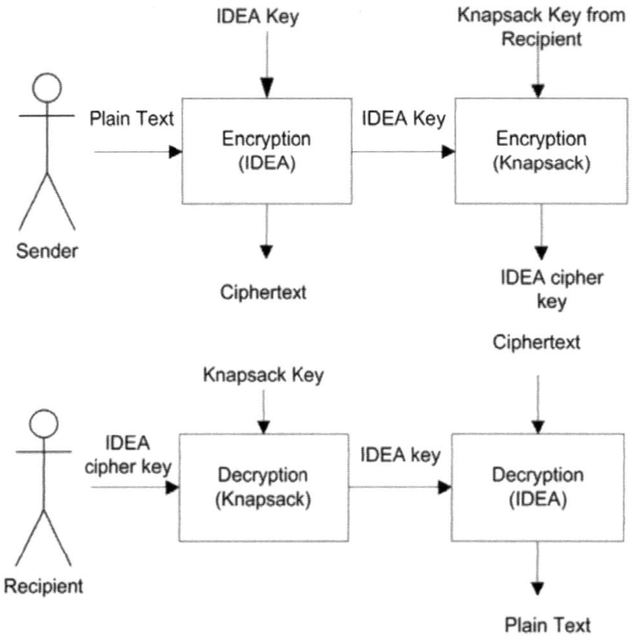

Figure 1. Hybrid cryptosystem IDEA and Knapsack

Data integrity is a parameter used in testing the implementation of hybrid cryptosystem IDEA and knapsack algorithms. There are two tests done. The first is examining the encryption and decryption of messages (files). And the second is IDEA encryption and decryption key testing. Each test has conducted some experiments.

Table 1. Encryption and Decryption of the message

No.	Encryption and Decryption of the message	
	Encryption	**Decryption**
1	*Input :* a. lain text with 16 characters: **Algoritma IDEA64** b. DEA Key: **fasilkom-ti2012C** *Output:* **8C6070BA2AD9CF22 969D404F26D83230**	*Input :* a. DEA Key: **fasilkom-ti2012C** b. ipher text: **8C6070BA2AD9 CF 22969D404F26D8 3230** *Output:* **Algoritma IDEA64**
2	*Input :*	*Input :*

	a. lain text with 24 characters: **Enkripsi & Dekripsi IDEA** b. DEA Key: **UniversitasMedan** *Output:* **2D5817E8B414C023A58DFEF9262C907611661A7698224306**	a. DEA Key: **UniversitasMedan** b. ipher text: **2D5817E8B414C023A58DFEF9262C907611661A7698224306** *Output:* **Enkripsi & Dekripsi IDEA**
3	*Input :* a. lain text with 32 characters:**Sub key, modulo, invers,plainteks** b. DEA Key: GangMelati04psp *Output:* **E10683B68B6F19FB662D590673B685D5A60FE0EDE148A0372E90A78CD5E9523D**	*Input :* a. DEA Key: **GangMelati04psp** b. ipher text: **E10683B68B6F19FB662D590673B685D5A60FE0EDE148A0372E90A78CD5E9523D** *Output:* **Subkey, modulo, invers,plainteks**

In Table 1 it can be seen that the result of ciphertext has built in the form of the hexadecimal number. Each time was encrypting the block cipher (requires eight characters each encrypt one block), it will generate 16 digits of the hexadecimal number. In other words, the number of characters from the encryption test results (in hexadecimal digits) is two times the number of input characters. This test successfully returns the ciphertext to plaintext.

Table 2. Encryption and Decryption of IDEA Key

No.	Encryption and Decryption of IDEA Key	
	Encryption	**Decryption**
1	*Input:* a.Plain Key: **fasilkom-ti2012C** b.Public Key: 14 28 49 105 210 420 360 240 M = 480, N = 7 *Output (Cipher key):* 857 317 782 527 707	*Input:* a.Private Key: 2 4 7 15 30 60 120 240 M = 480, N^{-1} = 343 b.Cipher key: 857 317 782 527 707 887 1307 947 919 602 527 514 154 394 514 628 *Output (Plain key):*

	887 1307 947 919 602 527 514 154 394 514 628	**fasilkom-ti2012C**
2	*Input:* a.Plain Key: **UniversitasMedan** b.Public Key: 39 65 117 247 494 372 154 308 M = 603, N = 13 *Output (Cipher key)*: 992 1202 984 955 862 583 891 984 801 490 891 1239 862 554 490 1202	*Input:* a.Private Key: 3 5 9 19 38 75 151 302 M = 603, $N^{-1} = 232$ b.Cipherkey: 992 1202 984 955 862 583 891 984 801 490 891 1239 862 554 490 1202 *Output (Plain key):* **UniversitasMedan**
3	*Input:* a.Plain Key: GangMelati04 psp b.Public Key: 102 136 272 544 64 128 256 512 M = 1024, N = 17 *Output (Cipher key)*: 1032 920 856 1304 840 1048 600 920 1080 984 816 944 272 952 1720 952	*Input:* a.Private Key: 6 8 16 32 64 128 256 512 M = 1024, $N^{-1} = 241$ b.*Cipher key:* 1032 920 856 1304 840 1048 600 920 1080 984 816 944 272 952 1720 952 *Output(Plain key):* **GangMelati04 psp**

In Table 2 can be seen that the result of the cipher key is a decimal number with the number of cipher key blocks being two times the number of Knapsack public key cards. This test successfully returns the cipher key to plain key (IDEA symmetry key).

From the result of table 1 and table 2, thus, it can be concluded that testing using the hybrid cryptosystem technique of IDEA and Knapsack in this system has met the integrity of the data.

Table 3. Encryption and Decryption time

No.	Length of Message (characters)	Encryption Time (ms)	Decryption Time (ms)
1	8	5	4
2	16	8	6
3	24	10	6
4	32	12	7
5	40	13	7
	Total	48	30
	Average	9.6	6

From table 3, it can be assumed that the longer number of text characters in the message encrypted, the time required to encrypt the message will be longer. Although there is constant time for the greater length of text in the decryption process (e.g., text with characters' length 16 and 24 have the same decryption time of 6 ms). The system takes less time to decrypt the message than it does the message encryption process.

4. Conclusions

Based on the discussion and the results of the research, the following conclusions have obtained:
1. With the message security mechanism that is hybrid cryptosystem technique, message encryption using IDEA can protect the information contained in the message and encryption key using Knapsack algorithm can secure the distribution of key.
2. IDEA and Knapsack algorithms meet the parameters of data integrity. The result of the message decryption test is ciphertext will be the same as the initial plaintext (original message). And the result of the cipher key decryption test is will be the same as the key that has been used for the message encryption process.
3. From testing five files with text and length of different characters, the average time of encryption is 9.6 ms. While the average decryption time is 6 ms. Therefore, it can be concluded that the decryption time of the message is shorter than the time of message encryption.

Acknowledgments
The authors gratefully acknowledge that the present research has supported by Fund Dissemination IPTEKS Research Results for Lecturers / Researchers Universitas Sumatera Utara.

References

[1] Schneier, B. 1996. *Applied Cryptography: Protocols, Algorithms and Source Code in C.* Jhon Wiley & Sons, Inc: New York.
[2] Munir, R. 2006. *Kriptografi.* Informatika Bandung: Bandung.
[3] Agarwal, A. 2011. *Encryption messages using the Merkle-Hellman Knapsack cryptosystem.* International Journal of Computer Science and Network Security (IJCSNS) Vol. 11, No. 5:12-14.
[4] N. Hamdy, K. Shehata, S. Elagooz and M. Helmy, *Design and implementation of fast inverse modulo (2^{16} +1) multiplier used in idea algorithm key schedule on FPGA*, Electrical, Electronic, and Computer Engineering, 2004. ICEEC '04. 2004 International Conference on, Cairo, Egypt, 2004, pp. 507-511.doi: 10.1109/ICEEC.2004.1374513
[5] S. Hamasho, Y. Murakami and M. Kasahara, *A systematic encryption algorithm for knapsack scheme using random sequence*, 2012 7th International Conference on Computing and Convergence Technology (ICCCT), Seoul, 2012, pp. 514-517.
[6] S. L. C. Salomao, V. C. Alves, and E. M. C. Filho, *A two-level pipelined implementation of the IDEA cryptographic algorithm*, Proceedings. XI Brazilian Symposium on Integrated Circuit Design (Cat. No.98EX216), Rio de Janeiro, 1998, pp. 158-161. doi: 10.1109/SBCCI.1998.715431
[7] Dian Rachmawati, Amer Sharif, Jaysilen, and Mohammad Andri Budiman, *Hybrid cryptosystem using tiny encryption algorithm and luc algorithm*, IOP Conference Series: Materials Science and Engineering 300 (2018), no. 1, 012042.

Source Term Analysis of SAMOP Reactor Experimental Facility

Wahid L[1,3], M I Farezza W[1,4], Syarip[2,5]
[1]Department of Nuclear Engineering and Engineering Physics, University of Gadjah Mada, Yogyakarta
[2]Centre for Accelerator Science & Technology, Jl. Babarsari, POB 6101 ykbb, Yogyakarta

Email: [3]luthfi_doank@yahoo.com,[4]ryakuzza@gmail.com, [5]syarip@batan.go.id

Abstract. The Subcritical Assembly for Molybdenum-99 Production (SAMOP) system by using low enriched uranyl nitrate solution as reactor fuel is being developed and designed in the Centre for Accelerator Science & Technology, Yogyakarta. The source term analysis of the SAMOP experimental facility is done in conjunction with the radiation shielding design. The method used is by using MCNPX software for fuel burn-up calculations. The burnup calculation is done both in the reactor operation without ^{99}Mo extraction and with 100% ^{99}Mo extraction per batch (6 day + 1 day reactor off) along 1 month operation (5 batch). The calculation was done for two conditions i.e. without and with ^{99}Mo extraction. The calculation result shows that the radioactivity inside the SAMOP reactor core reach 523 Ci for the condition without ^{99}Mo extraction. The effective dose at a distance of 50 cm from the outer surface of the SAMOP reactor coolant tank is 1600 mSv/h. When the SAMOP reactor is operated then followed by ^{99}Mo extraction, the overall radioactivity of the reactor core is slightly lower than the above value. The calculation results also shows that the highest dose from volatile nuclide is caused by ^{132}Te isotope i.e. 92.12 mSv/h.

Keywords: Subcritical, reactor, source term, ^{99}Mo, extraction, MCNPX, radioactivity.

1. Introduction

The ^{99}Mo radioisotope is indispensable as a ^{99m}Tc generator, of which ^{99m}Tc radioisotope is the most widely used radioisotope for diagnostics in the nuclear medicine [1,2,3]. However, the increased need for the ^{99m}Tc isotope can't be fulfilled because the ^{99}Mo commonly used as a ^{99m}Tc generator can only be produced in a nuclear reactor. The half-life of ^{99}Mo which only 66 hours is also a factor making it difficult to store and mobilize its spread. This problem can be solved by shortening the ^{99}Mo production process as well as shortening the mobilization of radioisotopes spreading distance.

In general, the ^{99}Mo production is done by splitting the ^{235}U nuclide, which ^{235}U is irradiated with thermal neutrons inside a nuclear reactor so that a critical reactor is required. The licensing issues for critical reactors are complex because they have to fulfill strict requirements. This problem requires a very huge funds to realize, excluding the cost of waste management products. To solve this problem BATAN or the National Nuclear Energy Agency designed a reactor that could produce ^{99}Mo but did not categorized as critical reactor and did not use high-enriched uranium. This reactor is simple and safer. The reactor applies subcritical reactor method with an external neutron source or an isotropic neutron source, known as SAMOP or subcritical assembly for ^{99}Mo production [4].

This research is based on previous studies, related to SAMOP design and safety analysis [5,6]. The SAMOP experimental facility as a test facility which use an external neutron source from the beam-port of Kartini TRIGA reactor, which has been identified suitable for this purpose [7]. SAMOP

Content from this work may be used under the terms of the Creative Commons Attribution 3.0 licence. Any further distribution of this work must maintain attribution to the author(s) and the title of the work, journal citation and DOI.
Published under licence by IOP Publishing Ltd

system is similar with an Aqueous Homogenous Reactor (AHR) but its operate a subcritical condition. AHR is a type of reactor in which uranium is dissolved in water. The fuel used is a mixture of coolants such as water which also act as a moderator and uranium salt, often referred to as homogeneous reactors. The high negative temperature reactivity coefficient makes the AHR more stable than conventional reactors. Another positive aspect of AHR is its small size and low total power [8,9]. However AHR cannot be used to generate electrical power because the fuel should not boil, otherwise the fuel concentration will increase due to evaporation of the water and the system is no longer critical. Using AHR, other isotopes can be extracted from solutions other than 99Mo. Until 2016, there are only 5 Aqueous Homogenous Reactors operating in accordance with IAEA databases [9,10].

The purpose of this research is to analyze the source term of SAMOP reactor experimental facility by using simulation method. The simulation modeling is done using MCNPX software [11]. The analysis results will be used for designing the radiation shielding of SAMOP experimental facility. In this source term calculation, the burn-up calculations is done. The burnup calculation is done both in the reactor operation without ^{99}Mo extraction and with 100% ^{99}Mo extraction per batch (6 day + 1 day reactor power off) within 1 month operation or 5 batch.

2. Modelling of SAMOP experimental facility

The SAMOP experimental facility as a test facility which use an external neutron source emerge from the radial beam-port of Kartini reactor. The external neutron source has been identified as thermal neutron in order of 10^8 n/cm^{2}s [7]. The SAMOP core consists of annular cylindrical tube containing uranyl nitrate [UO_2 $(NO_3)_2$] as fuels and target, surrounded by ring of UO_2 $(NO_3)_2$ fuel rods. The TRIGA fuel elements can be loaded in the ring together with UO_2 $(NO_3)_2$ tubes to increase neutron multiplication factor. Uranyl nitrate that is used in this research has enriched to 19.75% ^{235}U. Figure 1 and Figure 2 shows the dimensions of SAMOP experimental facility, and the dimensions of core and reflector system.

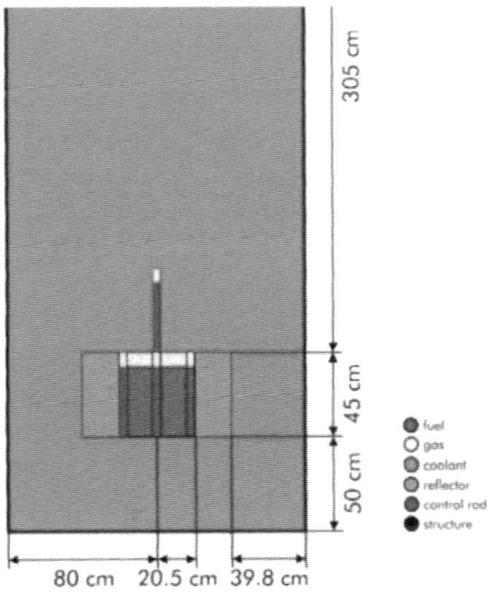

Figure 1. SAMOP experimental facility

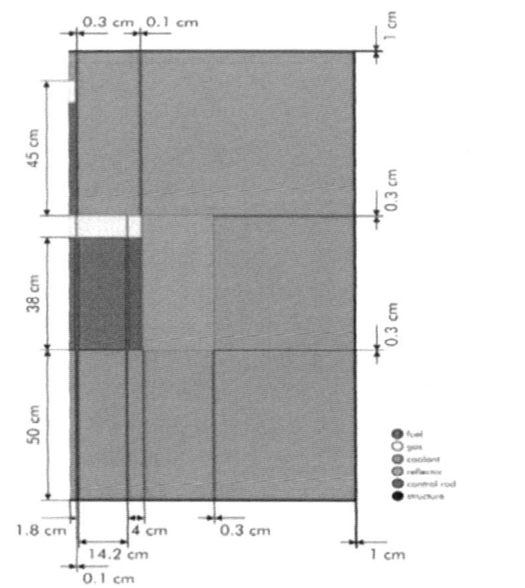

Figure 2. Dimensions of core and reflector system

3. Methods

3.1. Isotope production calculation

Isotope production calculation, or source term calculation is a calculation of isotope production within reactor operation that can be formed in the reactor through nuclear fission reaction, fusion (in case of fusion reactor), and nuclear transmutation reaction. Isotope that produce within reactor could be stable

IOP Conf. Series: Journal of Physics: Conf. Series **1090** (2018) 012031 doi:10.1088/1742-6596/1090/1/012031

or may be unstable that may lead to nuclear decay and emitting radiation, such as alpha, beta or gamma ray radiation. These three types of radiation might be harmful for environment, reactor operator, and people. The calculation of isotope production becomes important to do in order to calculate the effective dose of radiation received by the worker from the operation of the reactor. From this calculation, we will find the number of isotopes present in the reactor core and the resulting radioactive activity.

Isotope production calculation can be done using burnup calculation that provided by neutronic or reactor code system, such as MCNPX, Serpent, SCALE, etc. In this study, MCNPX is used as a code to solve burnup calculations and produces information of number of isotopes produced inside the reactor core. Both of the previously described SAMOP reactor core configurations are then modeled using MCNPX and then burnup codes are run for two conditions, reactor cores operation without extraction with 6-day operating intervals, and reactor cores operation with 100% ^{99}Mo extraction in all fuel rods per batch. One batch of SAMOP operation is 6 days operation in full power 1kW and one day reactor power off to extract ^{99}Mo.

3.2. Effective dose calculation

The effective dose calculation resulting from the radioisotope present in the reactor core is done by multiplying the radioisotope activity with the corresponding specific gamma-ray dose constants. The specific gamma-ray dose constants value of the radioisotopes is derived from the ORNL technical document for dosimetry and radiological assessment [12].

This effective dosage calculation does not include the dose received due to the gamma ray radiation exposure of the neutron capture (n, gamma) when the reactor operates. In other words, the dose calculations reported in this article are limited to doses of radionuclides decaying inside the reactor core. Further consideration is desirable when selecting which radionuclides contribute significantly to the dose [13]. With medium sized power reactors, it usually suffices to consider the following set of radionuclides: whole body noble gases (particularly ^{88}Kr, ^{135}Xe and ^{133}Xe); thyroid (particularly ^{131}I, ^{133}I); lung/internal: volatile nuclides (e.g. ^{131}I, ^{132}Te, ^{106}Ru, ^{134}Cs, ^{137}Cs), and ^{90}Sr for scenarios of high core temperatures (>1000°C).

3.3. MCNPX

MCNPX™ is a general purposes Monte Carlo radiation transport code designed to track many particle types over broad ranges of energies. It is the next generation in the series of Monte Carlo transport codes that began at Los Alamos National Laboratory nearly sixty years ago. MCNPX 2.6.0 is the latest Radiation Safety Information Computational Center (RSICC) release of the code, following the 2005 release of MCNPX 2.5.0 [PEL05]. MCNPX 2.5.0 was a superset of MCNP4C [BRI00] and MCNPX 2.4.0 [WAT02b]. MCNPX 2.6.0 includes many new capabilities, particularly in the areas of transmutation, burnup [FEN06a, FEN06b, FEN08], and delayed particle production. Many new tally source and variance-reduction options have been developed. Physics improvements include a new version of the Cascade-Exciton Model (CEM), the addition of the Los Alamos Quark-Gluon String Model (LAQGSM) option, and a substantial upgrade to muon physics. The code is compatible with MCNP5, and references to MCNPTM in this manual refer to the MCNP5 version [11].

Reactor core modeling using MCNP can be done using vised application provided by the MCNP, or by creating its own input code using a text editor application. The MCNP input code structure consists of cell cards, surface cards and data cards. The cell card contains information about part of the reactor core that is bounded by surface defined on the surface card. Material information such as the type of material used, material density, cell volume (optional), and the type of particles tracked within those cells also defined here. The data card contains information about the details of the material information used in the previous cell card and what type of calculation that user wants MCNP to solve, such as the calculation of reactor core burnup. Flowchart of the calculation is shown in Figure 3.

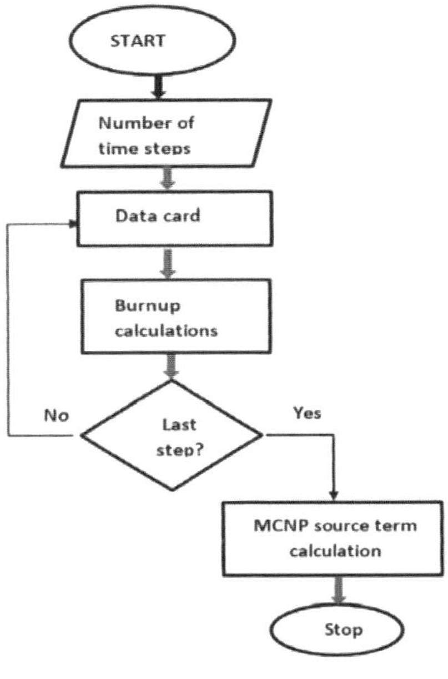

Figure 3. Flowchart of the source term calculation

In the case of calculating core burnup of the SAMOP reactor when operating in ^{99}Mo extraction mode, at each step of the burnup, the material data composition that used within the reactor core was updated due to the extraction process of ^{99}Mo and the change of material composition from previous burnup calculation [14].

4. Result and discussion

4.1. SAMOP operate without ^{99}Mo extraction

The first analysis was to look the quantity of radionuclide on SAMOP reactor core that operate without ^{99}Mo extraction for 3 months using 6-day intervals. The time period of 3 months were selected to see the long-term effects of radioactive material growth during SAMOP operation. The overall radioactivity of the SAMOP reactor core during the first 3 months can be seen in the Figure 4. The effective gamma-radiation dose generated from the SAMOP reactor core at a distance of 50 cm from the outer side of the tank (1 meter from the radial section) can be seen in the Figure 5.

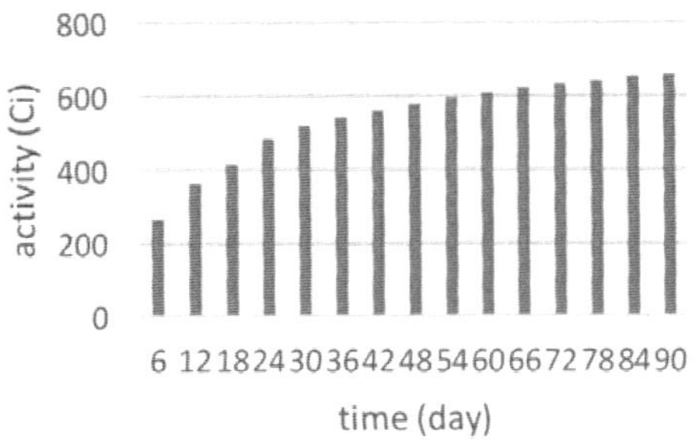

Figure 4. Overall core activity without ^{99}Mo extraction

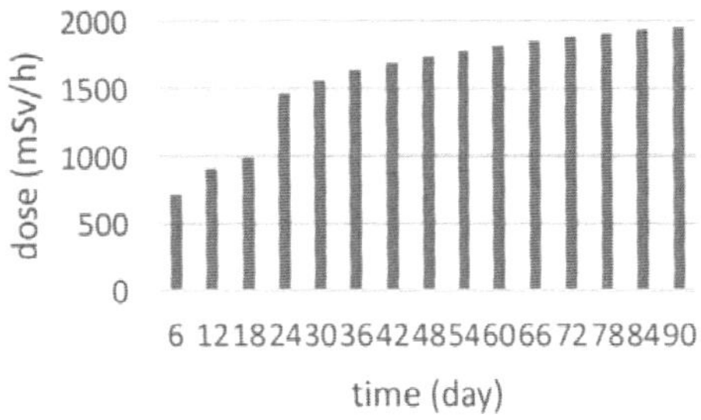

Figure 5. Overall effective gamma dose 50 cm from outer coolant tank without ^{99}Mo extraction

The total activity of radioactive substances in the reactor core until day 30th reaches 516.681 Ci with effective gamma-ray dose of 1557.523 mSv/h. The largest radioactivity is come from of ^{239}Np which is 89.32 Ci, followed by ^{133}Xe 54.68 Ci and then ^{99}Mo that can reach 51.29 Ci. The radioactivity of radioactive substances in the reactor core increases exponentially, and the effective dose of gamma radiation increases significantly after 24 days of operation. This happens because the growth of ^{140}La isotope with half-life of 1.7 d is shown by MCNP after 24 days of operation due nuclides with atom fractions below 10^{-10} for a material are zeroed by MCNP. ^{140}La itself derived from the fission reaction, plus the neutron capture by ^{139}La and the beta decay of ^{140}Ba. Furthermore, the specific gamma-ray dose constant for ^{140}La is greatest among the other radionuclides, 3.424 10^{-4} mSv/MBq or 12.66880 mSv/Ci. It can be seen in Figure 4, the overall effective dose of the reactor core due to the source term (radioactive material) at a distance of 50 cm from the outer surface of the coolant tank can be as high as 1600 mSv/h after 1 month, 1860 mSv/h in the 2nd month, and 1980 mSv/h in the third month.

The amount of radionuclides that important to be taken into account because its activity over 1 month operation of SAMOP can be seen in Table 1. It can be seen from the specific gamma-ray dose information that the largest dose among the nuclides is caused by ^{140}La with 471.91 mSv/h (1 m from point source). The highest dose from volatile nuclide is caused by ^{132}Te (half-life 3.3 d) i.e. 100.66 mSv/h.

Table 1. Amount of source term (radionuclide) of SAMOP core without ^{99}Mo extraction

Nuclide	Specific γ dose (mSv/h/Ci)	Radioactivity (Ci)					Max. effective dose (mSv/h)
		day 6	day 12	day 18	day 24	day 30	
Sr-90		0.0196	0.0392	0.0587	0.0783	0.0978	
Ru-106					0.0022	0.1722	
Te-132	2.7931	24.20	33.23	35.25	35.84	36.03	100.64
I-131	2.8268	8.658	15.32	18.87	21.00	22.27	62.95
Xe-133	1.0297	25.37	41.97	49.59	53.08	54.68	56.30
Cs-137	3.7629	0.0197	0.0394	0.0590	0.0787	0.0984	0.3702
La-140	12.6688				33.06	37.25	471.91
Np-239	5.1282	73.87	86.85	89.7	89.25	89.32	458.05

4.2. SAMOP operate with ^{99}Mo extraction

The next analysis was to look at the quantity of radionuclide on SAMOP reactor core with ^{99}Mo extraction mode for 5 batch or approximately 1 month operation. The overall activity of the SAMOP reactor core during the first month can be seen in the Figure 6. The effective gamma-radiation dose generated from the SAMOP reactor core at a distance of 50 cm from the outer side of the tank (1 meter from the radial section) can be seen in the Figure 7. The amount of radionuclides that important to be taken into account for the radiation safety is the activity over 5 batch operations (approx. 34 day) of SAMOP, can be seen in Table 2. It can be seen that the maximum specific gamma-ray (γ) dose i.e. after day 30th, the largest dose among the nuclides is caused by ^{239}Np with 438.15 mSv/h (1 m from point source). The highest dose from volatile nuclide is caused by ^{132}Te i.e. 88.91 mSv/h.

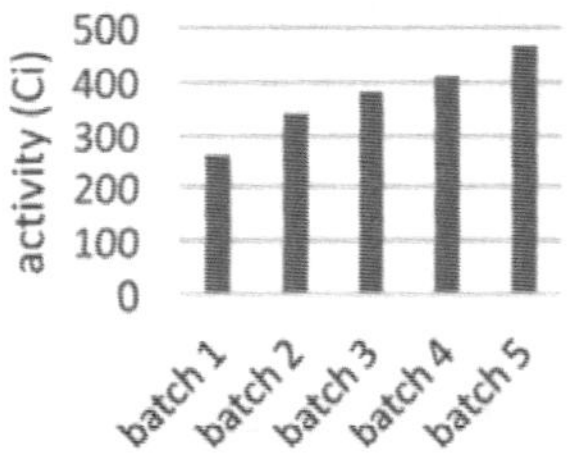

Figure 6. Overall core activity with ^{99}Mo extraction

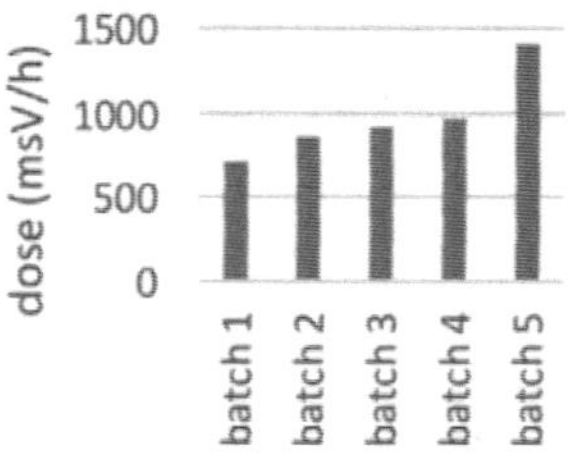

Figure 7. Overall effective gamma dose 50 cm from outer coolant tank with ^{99}Mo extraction

Then to show the maximum dose contained in the fuel rod, the nuclide information in the fuel rod in the center of the reactor is shown in Table 3. The fuel rod is selected because this fuel rod has the highest activity among the other fuel rods on the radial section of the reactor core. It can be seen from Table 3 that the highest activity and effective gamma-ray dose at a distance of 1 m from the fuel rod is still owned by a Type 1 configuration. But when all the radioactive materials inside the reactor are added up, the condition reverses.

Table 2. Amount of source term (radionuclide) of SAMOP core with 100% ^{99}Mo extraction

Nuclide	Specific γ dose (mSv/h/Ci)	Radioactivity (Ci)					Max. effective dose (mSv/h)
		day 6	day 12	day 18	day 24	day 30	
Sr-90		0.0196	0.0392	0.0588	0.0784	0.0979	
Te-132	2.7931	24.20	29.71	30.92	31.66	31.83	88.91
I-131	2.8268	8.66	13.41	16	17.42	18.19	51.42
Xe-133	1.0297	25.37	35.43	39.42	40.99	41.63	42.87
Cs-137	3.7629	0.020	0.039	0.059	0.079	0.099	0.3707
La-140	12.6688				0.4589	32.69	414.14
Np-239	5.1282	73.87	83.77	85.22	85.33	85.44	438.15

Table 3. Nuclide density an effective dose in the fuel rod in the center of the SAMOP reactor

No	Nuclide	mass (g)	Activity (Ci)	Eff. dose (mSv/h)	No	Nuclide	mass (g)	Activity (Ci)	Eff. dose (mSv/h)
1	U-235	2.29E+01	4.95E-05	1.68E-04	16	Te-132	1.59E-06	4.84E-01	1.35E+00
2	U-236	8.68E-05	5.62E-09	4.14E-09	17	I-131	2.29E-06	2.84E-01	8.01E-01
3	U-238	9.31E+01	3.13E-05	2.04E-05	18	Xe-133	3.20E-06	5.99E-01	6.16E-01
4	Np-239	6.14E-06	1.42E+00	7.30E+00	19	Cs-135	1.43E-05	1.64E-08	0.00E+00
5	Pu-239	5.06E-05	3.14E-06	9.46E-06	20	Cs-137	1.62E-05	1.41E-03	5.31E-03
6	Rb-87	4.24E-06	3.64E-13	0.00E+00	21	Ba-140	7.65E-06	5.60E-01	9.14E-01
7	Sr-89	6.43E-06	1.87E-01	1.52E-04	22	La-140	9.59E-07	5.34E-01	6.76E+00
8	Sr-90	9.94E-06	1.40E-03	0.00E+00	23	Ce-141	1.08E-05	3.08E-01	2.25E-01
9	Y-91	7.54E-06	1.85E-01	3.70E-03	24	Ce-142	1.56E-05	7.88E-19	0.00E+00
10	Zr-93	1.01E-05	2.55E-08	0.00E+00	25	Ce-143	1.06E-06	7.07E-01	1.80E+00
11	Zr-95	9.88E-06	2.12E-01	9.85E-01	26	Ce-144	1.46E-05	4.64E-02	1.08E-02
12	Nb-95	6.25E-07	2.46E-02	1.18E-01	27	Pr-143	7.47E-06	5.03E-01	0.00E+00
13	Mo-99	1.20E-06	5.78E-01	6.51E-01	28	Nd-145	1.03E-05	4.22E-19	0.00E+00
14	Tc-99	5.15E-06	8.82E-08	4.05E-13	29	Nd-147	2.63E-06	2.13E-01	2.95E-01
15	Ru-103	4.52E-06	1.46E-01	4.84E-01	30	Pm-147	2.05E-06	1.90E-03	0.00E+00
					31	TOTAL	510.21	7.00	22.33

5. Conclusion

The source term of SAMOP reactor experimental facility has been analyzed, the analysis result shows that the overall radioactivity of the reactor core will increase exponentially during reactor operation period. The radioactivity of the source term is dominated by three radioisotopes i.e. ^{239}Np, ^{133}Xe, and ^{99}Mo. The calculation result shows that the radioactivity inside the SAMOP reactor core reach 523 Ci for the condition without ^{99}Mo extraction, and the effective dose a distance of 50 cm from the outer surface of the SAMOP reactor coolant tank is 1600 mSv/h. When the SAMOP reactor is operated and followed by ^{99}Mo extraction, then the analysis result shows that the overall radioactivity of the reactor core is slightly lower than the above value. It can be also concluded that the highest dose i.e. 92.12 mSv/h is come from volatile nuclide ^{132}Te.

Acknowledgement
The authors would like to thank the National Nuclear Energy Agency of Indonesia for the MCNPX software. Other than that, the author is also thank to the Director of Center for Accelerator Science and Technology Yogyakarta to Head of Reactor Division and all staffs for their support, and to Secretariat of INSINAS RISTEKDIKTI for budget support to this project (Project Code: InSinas RT-2016-0151).

References
[1] Maroor, R.A.P, A. Dash, & F.F. (Russ) Knapp Jr. Diversification of 99Mo/99mTc Separation: non–fission reactor production of 99Mo as a strategy for enhancing 99mTc availability. Journal of Nuclear Medicine 56:159-161 (2015).
[2] Mallinckrodt Pharmaceuticals. *Global ^{99}Mo supply chain*. 2017 http://www2.mallinckrodt.com/Nuclear_Imaging/Global_Mo99_Supply_Chain_Reactors.aspx
[3] Pillai, M.R.A., A. Dash & F.F. (Russ) Knapp Jr. 2013 *Sustained availability of ^{99m}Tc: Possible paths forward*. Journal of Nuclear Medicine 54: 313–323.
[4] BATAN. 2013 *Perangkat Reaktor Subkritik Untuk Memproduksi Mo-99 (Subcritical Assembly for Mo-99 Production)*, Patent No. P00200500760, http://e-statushki.dgip.go.id/index.php/penelusuran/paten/P00200500760
[5] Tegas Sutondo, Syarip, Slamet Santoso. 2009 *Safety design limits of main components of the proposed SAMOP system*, Proceedings of the 3rd Asian Physics Symposium (APS 2009), Bandung, Indonesia.
[6] Syarip, Tegas Sutondo, Edi Trijono, E. Susiantini. 2016 *Design & Development of Subcritical Assembly for ^{99}Mo Production (SAMOP)*. Paper presented at the 2nd International Conference on Science and Technology, Yogyakarta, Indonesia, ID PA-P-02.
[7] Tegas Sutondo and Syarip, 2014. *Beam Characteristics of Pierching and Tangential Beamports of Kartini Reactor,* "Ganendra" Journal of Nuclear Science and Technology, 17 2 83-90.
[8] Ball R M 1998 *Present status of the use of LEU in aqueous reactors to produce Mo-99,* International Meeting on Reduced Enrichment for Research and Test Reactors.
[9] International Atomic Energy Agency (IAEA). 2008 *Homogeneous Aqueous Solution Nuclear Reactors for the Production of Mo-99 and Other Short Lived Radioisotopes*. IAEA TECDOC Report 1601. Vienna, Austria.
[10] OECD-NEA. 2017 *Medical isotope supply review: ^{99}Mo / ^{99m}Tc market demand and production capacity projection 2017-2022*. Nuclear Development NEA/SEN/HLGMR 2.
[11] D. B. Pelowitz, 2008 MCNPX User's Manual Version 2.6.0, LANL, New Mexico
[12] Laurie M. Unger and D. K. Trubey, 1982 *Specific Gamma-Ray Dose Constants for Nuclides Important to Dosimetry and Radiological Assesment*, Oak Ridge National Laboratory, Tennesse.
[13] IAEA, 2008 *Derivation of The Source Term and Analysis of The Radiological Consequences of Research Reactor Accidents*, International Atomic Energy Agency, Vienna.
[14] Dedy Prasetyo Hermawan, Rionaldy and Syarip, 2017 *Neutronic analysis of SAMOP reactor experimental facility using SCALE computer code,* in *the International Conference on Computation in Science and Engineering*, Bandung.

Neutronic Analysis of SAMOP Reactor Experimental Facility Using SCALE Code System

Dedy P Hermawan[1,3], Rionaldy[1] and Syarip[2,3]

[1] Department of Nuclear Engineering and Engineering Physics, Faculty of Engineering, Universitas Gadjah Mada, Jalan Grafika No. 2, Yogyakarta 55284, Indonesia.
[2] Centre for Accelerator Science and Technology, National Nuclear Energy Agency, Yogyakarta 55281, Indonesia

[3] E-mail : dedy.prasetyo.h@gmail.com, syarip@batan.go.id

Abstract. The Subcritical Assembly for ^{99}Mo Production (SAMOP) rector is considered a suitable candidate to produce ^{99m}Tc isotope on a small scale. The design of SAMOP reactor using uranyl nitrate (UN) as fuel has been studied at National Nuclear Energy Agency. The neutronic analysis of the SAMOP Experimental Facility will be reported in this paper. The SAMOP reactor core design consists of an annular tube containing UN surrounded by the ring of UN tubes, and a fuel tube absorber follower is inserted in the center of the core. The work covers the criticality calculation to secure the sub-criticality level and neutron flux distribution analysis. The computational method is using SCALE computer code. The analysis result shows that for the required neutron multiplication factor of 0.98 to 0.99 (sub-criticality level) can be achieved by a combination of fuel (UN) height in the annular tube of 38 cm surrounded by 5 – 8 of UN tubes in the ring. For the UN height in the annular tube of 39 cm, it should be surrounded by 3 – 6 of UN tubes in the ring.

Keywords: ^{99}Mo, SAMOP, uranyl nitrate, SCALE code, reactor, criticality calculation.

1. Introduction

Technetium-99m (^{99m}Tc) is the most commonly used isotope for medical purposes but since ^{99m}Tc has a short half-life of 6.0058 hours, a stock pile cannot be produced. Therefore the parent isotope of ^{99m}Tc, which is ^{99}Mo, is delivered to the medical centers so when ^{99m}Tc isotope is needed, it can be extracted from ^{99}Mo. This extraction process called ^{99m}Tc generator [1,2].

The current conventional method to produce ^{99}Mo is by irradiating low enriched uranium (LEU) or high enriched uranium (HEU) targets in a high neutron flux environment such as in a nuclear reactor for a couple of days. After irradiation or fission process, the targets are brought to the reprocessing facility to extract ^{99}Mo and then prepared to be shipped to the medical centers. In the fission method, the fission yield of ^{99}Mo is about 6.1%, so a larger quantity of ^{99}Mo produced could be reached with this process. However the fission method required considerable capital investment and produced large quantities of radioactive waste [2].

The other method is by irradiating ^{98}Mo with neutron to convert natural ^{98}Mo with neutron capture reaction and became ^{99}Mo, but this process produced much lower ^{99}Mo than fission method using HEU or LEU targets. It's because the cross section of ^{98}Mo (n,γ) ^{99}Mo reaction is small (σ_{th} ~ 0.14 barns),

Content from this work may be used under the terms of the Creative Commons Attribution 3.0 licence. Any further distribution of this work must maintain attribution to the author(s) and the title of the work, journal citation and DOI.
Published under licence by IOP Publishing Ltd

resulting a very small quantity of ^{98}Mo is converted to ^{99}Mo. In order to produce high quality ^{99m}Tc generators from capture reaction method, advanced generator technologies are required [2,14,15].

The demand for ^{99}Mo as medical isotope will continue to increase in the future. Therefore, a lots of research about optimizing the production of ^{99}Mo and medical isotopes had been done. Ball R M describe that an Aqueous Homogeneous Reactor (AHR) fueled with uranyl nitrate could be used to produce medical isotopes [3]. Since then, the initiated program to assess the feasibility of utlizing AHRs for the production of medical isotopes in Russian, United States, Japan, France, Canada, China, Indonesia and other countries, have been done [4,5,6]. In Indonesia, The National Nuclear Energy Agency investigates a Subcritical Assembly for ^{99}Mo Production (SAMOP) reactor based on uranyl nitrate solution [7,8].

An idea to design a facility dedicated to producing ^{99}Mo radioisotope for medical application, named as a Sub-critical Assembly for ^{99}Mo Production (SAMOP) had been launched few years ago [9,10]. The idea came up in conjunction with the rising problems related to the radioisotope production such as the international conversion program that would affect the high enriched uranium supply used in the current's practice of the radioisotope production. Another problem related with the current's method used in the production process of the medical radioisotopes is the inefficiency of the ^{235}U utilization; i.e. around 1 %, and the rest containing high ^{235}U concentration must be wasted and sent back to the original exporting country.

The proposed design of the SAMOP system was intended to solve the above two problems i.e. it was designed to use of low enriched uranyl nitrate (UN) solution, serving for both the reactor fuel and ^{99}Mo producing source as well. Based on this concept, the irradiated uranium solution can be reused again after completing the extraction process and reconditioning process, then transferred back to the core tank for the next irradiation process and hence, improving the efficiency of the ^{235}U utilization.

This paper presents a criticality calculation to secure the sub-criticality level the SAMOP experimental facility, by using SCALE computer code. The SCALE computer code system is a widely used modeling and simulation suite for nuclear safety analysis and design that is developed, maintained, tested, and managed by the Reactor and Nuclear Systems Division (RNSD) of the Oak Ridge National Laboratory (ORNL). SCALE provides a comprehensive, verified and validated, user-friendly tool set for criticality safety, reactor physics, radiation shielding, radioactive source term characterization, and sensitivity and uncertainty analysis [11,12]. The analysis was also intended to determine the allowable number of UN solution tubes or TRIGA fuels inserted into the ring or peripheral holes to keep the core subcritical at any condition. From the previous calculations, the preferable nominal uranium density is 300 g U/L [10].

2. SAMOP experimental facility

The schematic diagram of overall concept of SAMOP system is depicted in Figure 1.

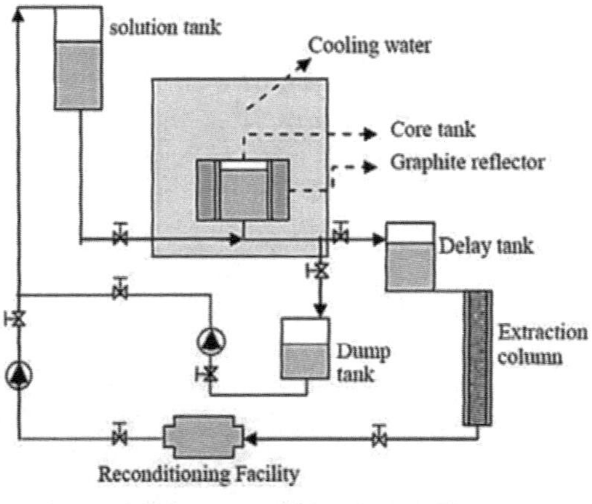

Figure 1. Diagram of The SAMOP system.

The SAMOP core fuelled by uranyl nitrate solution [UO_2 $(NO_3)_2$]. As shown in Figure 1, the UN fuel solution is continuously pumped around. In the delay tank short-lived isotopes will decay, in the extraction column the ^{99}Mo will be recovered. The recondition facility ensures a constant concentration of uranium and the solution tank used as a the buffer tank for the fuel.

The SAMOP experimental facility as a test facility for SAMOP design which use an external neutron source from the radial beam-port of Kartini reactor. The external neutron source has been identified as thermal neutron in order of 10^8 n/cm^2 s [13]. Figure 2 shows the schematic layout of SAMOP experimental facility, and Figure 3 shows the core system. The SAMOP core consists of annular cylindrical tube containing uranyl nitrate [UO_2 $(NO_3)_2$] as fuels and target, surounded by ring of UO_2 $(NO_3)_2$ tubes or TRIGA fuel elements. The TRIGA fuel elements can be loaded in the ring together with UO_2 $(NO_3)_2$ tubes to increase neutron multiplication factor. The dimension of the SAMOP core, reflector, boral rod neutron absorber, and coolant tank is described in Table 1. The cooling system of SAMOP is using H_2O as its coolant and and also as an axial radiation shielding. The SAMOP experimental facility is provided by an instrumentation and control system in such that there is a criticality indication, the boral control rod neutron absorber will dropped automatically inserted to the SAMOP reactor core. The system shall ensure the sub-critical state at any condition with adequate margin.

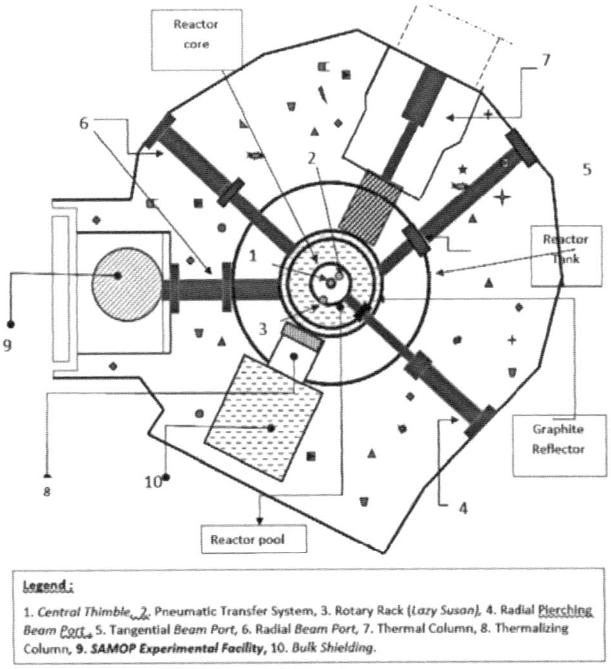

Figure 2. Layout of SAMOP experimental facility [13]

Figure 3. SAMOP reactor core

Table 1. The dimension of SAMOP reactor

Component	Dimension
Annular core:	
- diameter	32.4 cm
- height	43 cm
Ring of UN tubes :	1.70
- diameter of ring	40.4 cm
- tube (fuel) diameter	3.2 cm

Component	Dimension
Coolant tank:	
- diameter	120 cm
- height	400 cm
Neutron absorber tube:	
- diameter	3.2 cm
- height	38 cm

3. Methodology

The neutronic calculation in this work is done using SCALE (Standardized Computer Analyses for Licensing Evaluation) code system which developed and maintained by Oak Ridge National Laboratory (ORNL) under contract with the U.S. Nuclear Regulatory Commission (NRC) [11]. Within SCALE code system, module KENO IV was used to analyse the effective neutron multiplication factor (k_{eff}) of SAMOP reactor. KENO IV calculates k_{eff} of a 3-D system using the Monte Carlo method. In KENO IV module, the neutron source, the material and geometry of SAMOP reactor is modeled.

The material composition of SAMOP reactor is given in the Table 2. Uranyl nitrate with uranium concentration of 300 gU/L and U^{235} enrichment of 20% is used in this work. The specification of uranyl nitrate is given in Table 3.

Table 2. Material composition of SAMOP reactor

Material	Density (g/cm^3)
Uranyl nitrate	2.81
SS304	7.94
Dry air	1.205×10^{-3}
Graphite	1.70
H_2O	9.982×10^{-1}

Table 3. Material specification of uranyl nitrate

Nuclide	Atomic Density (atom/barn cm^3)
U-235	1.537245×10^{-4}
U-238	6.07131×10^{-4}
H-1	5.48779×10^{-2}
N-14	1.521711×10^{-3}
O-16	2.743895×10^{-2}

SAMOP reactor core design consists of an annular tube containing uranyl nitrate fuel surrounded by the ring of uranyl nitrate tubes and a fuel tube absorber follower is inserted in the center of the core. The effective multiplication factor (k_{eff}) that required for this SAMOP design is between 0.98 – 0.99. In order to reach the required value of k_{eff} for SAMOP reactor, uranyl nitrate tubes and fuel tube absorber follower are inserted one by one into the reactor. The fuel tube absorber follower is inserted first into the center of annulus then followed by 12 uranyl nitrate tubes into the ring. There are three variation of uranyl nitrate fuel's height in both annulus and tubes are applied in this work. Those variation are 38 cm, 39 cm and 40 cm. This condition are applied to analyse the effect of fuel's height in either annulus or tubes to the value of k_{eff}.

The safety aspect also investigated in this work. In order to secure the value of k_{eff} did not exceed 1.00, SAMOP reactor has a fuel tube absorber follower in the center of its the core. There are 2 variety of absorber material that investigated in this work. Those absorber material are B_4C (boron carbide) and Boral. To analyse the effect of each absorber material, the fuel tube absorber follower is inserted slowly from 0% to 100% of fuel tube absorber follower is inserted into the core.

4. Result and Discussion

The value of effective multiplication factor (k_{eff}) represented the criticality condition of the SAMOP reactor is calculated with SCALE code system using KENO IV module for different fuel's height in either annulus and tubes. The first analysis was to investigate the influence of fuel's height variation in annulus and tubes around the ring to k_{eff} value.The results of this analysis will show the relationship between the criticality levels with the fuel's height and we can determined how many fuel tubes needed to satisfy the required criticality of SAMOP reactor at specific fuel's height. The geometrical core design model of SAMOP reactor with KENO IV can be seen in Figure 4 and Figure 5. The SCALE running results of the ring tubes fuel's height variations at annulus fuel's height 38 cm is shown in Figure 6.

The first analysis results shows that the variation of the ring tubes fuel's height didn't have significant influence toward the k_{eff} value of the SAMOP reactor. In all of the fuel's height variations in the ring tubes shows that to reach the value of required k_{eff} for SAMOP reactor, the number of fuel tubes needed are between 4 to 10 tubes in the ring and one tube in the center of the core. In total, 5 – 8 uranyl nitrate tubes are needed to reach the required k_{eff} value of SAMOP reactor.

Further SCALE running results for the annulus fuel's height variations with the ring tubes fuel's height at 38 cm is shown in Figure 7. The results shows that the variation of the annulus fuel's height have significant influence toward the k_{eff} value of SAMOP reactor. As the annulus fuel's height increase, the k_{eff} value also increasing significantly. Thus, the number of fuel tubes inserted into the core could be reduced. With the annulus fuel's height 38 cm, fuel tubes that needed to reach the required k_{eff} value for SAMOP reactor is between 5 – 8 tubes. At the height of 39 cm, it need between 3 – 6 tubes and at the height of 40 cm, it need 2 – 5 tubes to reach the required k_{eff} of SAMOP reactor experimental facility. The results show that the increase of the annulus fuel's height will significantly increasing the k_{eff} of SAMOP reactor and also reduced fuel tubes that needed to reach the required k_{eff} of SAMOP reactor experimental facility. The analysis result of criticality values (k_{eff}) is in accordance with the similar system conducted by other researchers [3,4]

The results will be useful for conducting UN fuel loading procedures, known as criticality or sub-criticality experiment, as well for deciding the optimum operation and production scenario of SAMOP reactor experimental facility in the future.

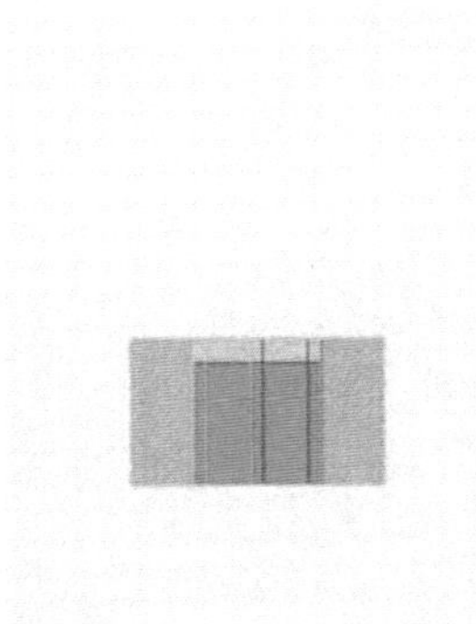

Figure 4. Axial cross section of SAMOP reactor design in KENO IV.

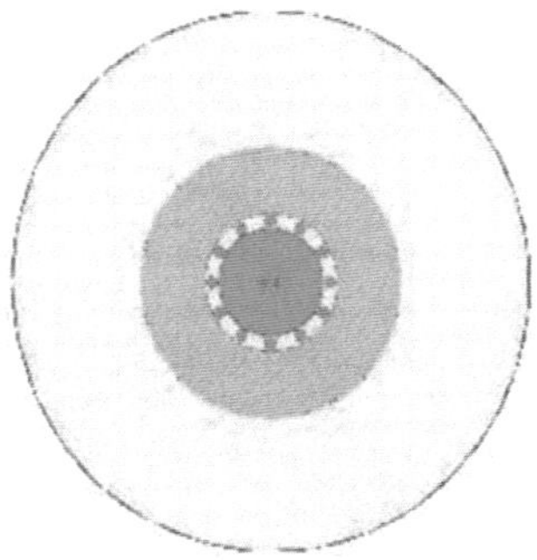

Figure 5. Radial cross section of SAMOP reactor design in KENO IV.

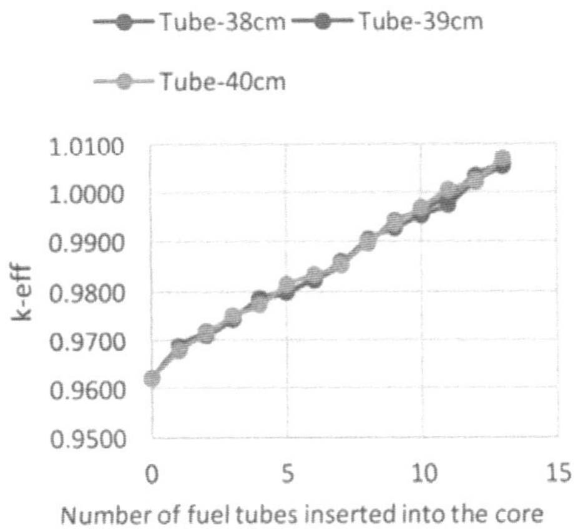

Figure 6. The k_{eff} of the SAMOP reactor with variation of the ring tubes fuel's height and annulus fuel's height at 38 cm.

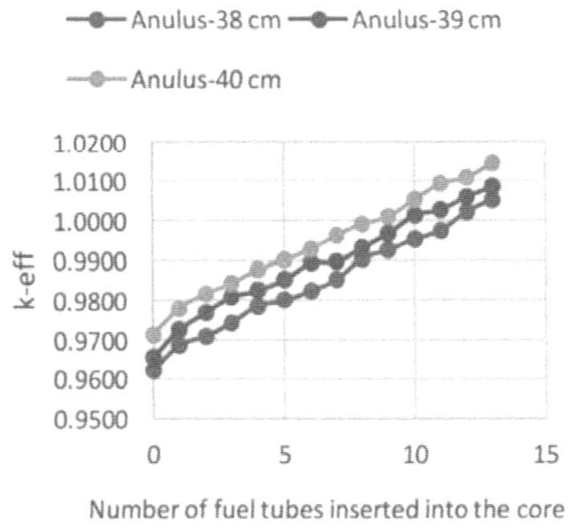

Figure 7. The k_{eff} of the SAMOP reactor with variation of the annulus fuel's height and the ring tubes fuel's height at 38 cm.

The next analysis is to investigate boron carbide (B_4C) and boral as neutron absorber materials in SAMOP reactor experimental facility. The control rod as neutron absorber which is followed by UN tube or known as control rod fuel follower is located in the central of SAMOP reactor core (see Figure 4). The analysis was done to compare the effectivity between both materials to control the fission reaction. The comparison of absorber rod worth between B_4C and boral as the reactor control material is shown in Figure 8. This analysis result is also in a good agreement with the results for the similar system previously conducted by other researchers [7,8,9].

The result show that B_4C has lower reactivity than boral. When B_4C absorber rod is fully inserted into the core, it reactivity worth is around -4.4 %Δk/k, whereas boral's rod reactivity worth is around -3.8 %Δk/k. It means that B_4C rod absorb more neutron than boral rod when it is inserted into the core, thus the value of k_{eff} will be reduced more. This result will be useful for deciding which material is suitable for establishing the safety system of SAMOP reactor experimental facility in the future.

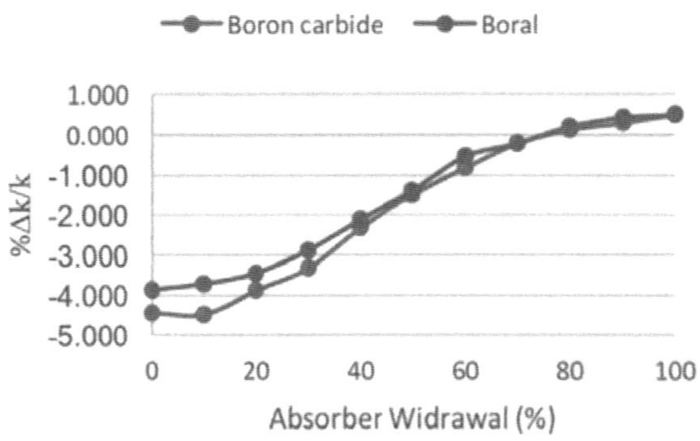

Figure 8. The comparison of B_4C and boral neutron control rod worth.

5. Conclusion

The required sub-criticality level i.e. neutron multiplication factor (k_{eff}) of 0.98 to 0.99 of SAMOP system can be achieved by a combination of uranyl nitrate (UN) fuel height in the annular core of 38 cm surrounded by 5 – 8 of UN tubes in the ring. For the UN height in the annular tube of 39 cm, it should be surrounded by 3 – 6 of UN tubes in the ring, and 2 – 5 UN tubes are needed when the height of UN in annular tube is 40 cm. The variations in the ring UN tubes fuel's height of SAMOP reactor didn't have significant effects toward the k_{eff} value but the variations in the annulus fuel's height did have a significant effect. For the neutron absorber rod, the B_4C material has a lower reactivity worth than boral. The reactivity worth is -4.4 %Δk/k when B_4C rod is fully inserted and -3.8 %Δk/k when boral rod is fully inserted into the core of SAMOP reactor.

Acknowledgment

The authors would like to thank the Director of Center for Accelerator Science and Technology Yogyakarta, to Head of Reactor Division and all staffs for their support, and to Secretariat of INSINAS RISTEKDIKTI for budget support to this project (Project Code: InSinas RT-20116-0151).

References

[1] International Atomic Energy Agency (IAEA). 2008 *Homogeneous Aqueous Solution Nuclear Reactors for the Production of Mo-99 and Other Short Lived Radioisotopes.* IAEA TECDOC Report 1601. Vienna, Austria. Maroor Raghavan Ambikalmajan Pillai et.al. 2013 *Sustained Availability of ^{99m}Tc: Possible Paths Forward*, Journal of Nuclear Medicine 54: 313–323.

[2] Ball R M. 1998 *Present status of the use of LEU in aqueous reactors to produce Mo-99,* International Meeting on Reduced Enrichment for Research and Test Reactors.

[3] Huisman M V. 2013 *Medical Isotope Prodduction Reactor,* Master Thesis Applied Physics, Delft University of Technology.

[4] S. Rijnsdorp BSc. 2014 *Design of a small Aqueous Homogeneous Reactor for production of ^{99}Mo Improving the reliability of the supply*, Thesis Master of Science in Applied Physics, at Delft University of Technology.

[5] OECD-NEA. 2017 *Medical isotope supply review: ^{99}Mo / ^{99m}Tc market demand and production capacity projection 2017-2022*. Nuclear Development NEA/SEN/HLGMR 2.

[6] BATAN. 2013 *Perangkat Reaktor Subkritik Untuk Memproduksi Mo-99 (Subcritical Assembly for Mo-99 Production)*, Patent No. P00200500760 http://e-statushki.dgip.go.id/index.php/penelusuran/paten/P00200500760

[7] Setiadipura, T. & E. Saragi. 2007 *Neutronic aspect of subcritical assembly for Mo 99 production (SAMOP) reactor*. Proceedings of International Conference on Advances in Nuclear Science and Engineering in Conjunction with LKSTN, p. 23–26.

[8] Tegas Sutondo, Syarip, Slamet Santoso. 2009 *Safety design limits of main components of the proposed SAMOP system*, Proceedings of the 3rd Asian Physics Symposium (APS 2009), Bandung, Indonesia.

[9] Syarip, Tegas Sutondo, Edi Trijono BS, Endang Susiantini. 2016 *Design & Development of Subcritical Assembly for ^{99}Mo Production (SAMOP)*. Paper presented at the 2nd International Conference on Science and Technology, Yogyakarta, Indonesia, ID PA-P-02.

[10] Marshall W J and Rearden B T, 2013 *Criticality Safety Validation of SCALE 6.1*, Reactor and Nuclear Systems Division, ORNL/TM-2011/450 Revised.

[11] Fox P B and Hollenbach D F, 2005 *KENO-VI Validation*, ORNL/TM-2004/60, Oak Ridge National Laboratory, Oak Ridge, Tenn.

[12] Tegas Sutondo and Syarip, 2014. *Beam Characteristics of Pierching and Tangential Beamports of Kartini Reactor,* “Ganendra” Journal of Nuclear Science and Technology, 17 2 83-90.

[13] Z. Gholamzadeh, S.A.H. Feghhi, S.M. Mirvakili, A. Joze-Vaziri, and M. Alizadeh. 2015 *Computational investigation of ^{99}Mo, ^{89}Sr, and ^{131}I production rates in a subcritical $UO_2(NO_3)_2$ aqueous solution reactor driven by a 30-MeV proton accelerator*, Nuclear Engineering Technology 47 875-883.

[14] ET Budisantoso, Syarip. 2002 *Studi Produksi Radioisotop Mo-99 Dengan Bahan Target Larutan Uranil Nitrat Pada Reaktor Kartini,* “Ganendra” Journal of Nuclear Science and Technology, 5 1 1-8.

IOP Conf. Series: Journal of Physics: Conf. Series **1090** (2018) 012033 doi:10.1088/1742-6596/1090/1/012033

Neutronic Analysis of UN-PuN Fuel use FI-ITB-CHI Code for 500MWth GFR Long Life Without Refueling

Ratna Dewi Syarifah[1,2], Zaki Su'ud, Khairul Basar and Dwi Irwanto

[1]Nuclear Physics and Biophysics Research Division, Physics Department, Faculty of Mathematics and Natural Science, Bandung Institute of Technology Indonesia, Jalan Ganesha 10 Bandung 40132, INDONESIA
[2]Sekolah Tinggi Teknologi Nuklir, Badan Tenaga Nuklir Nasional (National Nuclear Energy Agency), Jalan Babarsari, Caturtunggal, Kec. Depok, Kabupaten Sleman, Daerah Istimewa Yogyakarta 55281, INDONESIA

E-mail: syarifah.physics@gmail.com

Abstract. Neutronic analysis of UN-PuN fuel use FI-ITB-CHI code for 500MWth GFR long-life without refuelling has been done. The reactor use fast neutron spectrum, helium coolant and stainless steel for the cladding. The FI-ITB-CHI is code to calculate the neutronic analysis both fuel pin and core calculation. The parametric survey of various Plutonium percentages with homogeneous core configuration and variation of fuel volume fraction has been done. The variation is from 8% percentage of Pu up to 15% percentage of Pu. The 11% percentage of Plutonium has k-eff value more stable than the others, or we can say that it has critical condition. And the variation of fuel volume fraction is from 40% up to 60%. When the fuel volume fraction is increase, the reactor can be operated longer than small fuel volume fraction. The 60% of fuel volume fraction has optimum value in this case. For heterogeneous core configuration, there is three type of fuel, i.e. fuel 1 (F1), fuel 2 (F2) and fuel 3 (F3). The percentage plutonium of F3 is varied from 13%, 13.5% and 14%. The optimum value for heterogeneous core configuration is use case 1 which is use 8% percentage of plutonium for F1, 10% for F2 and 14% for F3. The reactor can be operated almost 25 years without refueling.

1. Introduction

Gas Cooled Fast Reactor is one of Generation-IV reactor used high temperature helium coolant and fast neutron spectrum. The fast spectrum affords more sustainable use of uranium resources and waste minimization. GFR can operate with an outlet temperature up to 850°C. In this temperature, it is possible to use as hydrogen production [1].

Neutronic analysis can be calculated with some of neutronic calculation code such as SRAC, SCALE, MCNP, FI-ITB-CHI etc. FI-ITB-CHI code has been patent and has been benchmarking in IAEA-tecdoc-1652 with little small reactors without on-site refuelling: neutronic characteristics, emergency planning and development scenarios [2].

Neutronic analysis of UN-PuN fuel use FI-ITB-CHI code for 500MWth GFR long-life without refuelling has been done. Beforehand, the research of GFR with nitride fuel use SRAC Code has been

Content from this work may be used under the terms of the Creative Commons Attribution 3.0 licence. Any further distribution of this work must maintain attribution to the author(s) and the title of the work, journal citation and DOI.
Published under licence by IOP Publishing Ltd

calculated before [3-7]. The neutronic analysis use FI-ITB-CHI code also has been calculated to Pb-Bi cooled reactor [1,8-10]. In this research, we want to calculate analyse GFR used FI-ITB-CHI code.

2. Design Concept and Calculation Methods

Table 1 show the parameter design of gas cooled fast reactor. The fuel use uranium plutonium nitride fuel. The uranium is uranium natural and the plutonium is spent fuel from light water reactor. The percentage of Plutonium is taken from spent fuel PWR (U-fueled) with burn up 33 MWd/ton. The percentage of Pu-238 is 1.8%, Pu-239 is 58.7%, Pu-240 is 24.2%, Pu-241 is 11.4% and Pu-242 is 3.9% [11].

Nitride fuel has high melting point so it can be operated in high temperature. The cladding use stainless steel, it oftentimes uses to be cladding material. The core reactor dimension use balance core, means the diameter active core as same as height active core.

Table 1. Parameter design of the reactor

Parameter	Specification
Power (MWth)	500
Fuel material	Uranium Plutonium Nitride
Cladding material	Stainless Steel
Coolant material	Helium
Fuel volume fraction (%)	40% - 60%
Cladding volume fraction (%)	10%
Coolant volume fraction (%)	30% - 50%
Active core diameter & height (cm)	250
Reflector radial & axial width (cm)	50
Pin pitch (cm)	1.45
Reactor life (years)	25

The calculation use FI-ITB-CHI code both fuel cell calculation and core calculation. The fuel cell calculation varied the fuel volume fraction from 40% up to 60%. The core calculation, there are two calculation, i.e. homogeneous calculation and heterogeneous calculation. Figure 1 show half heterogeneous core configuration with different width in variation fuel.

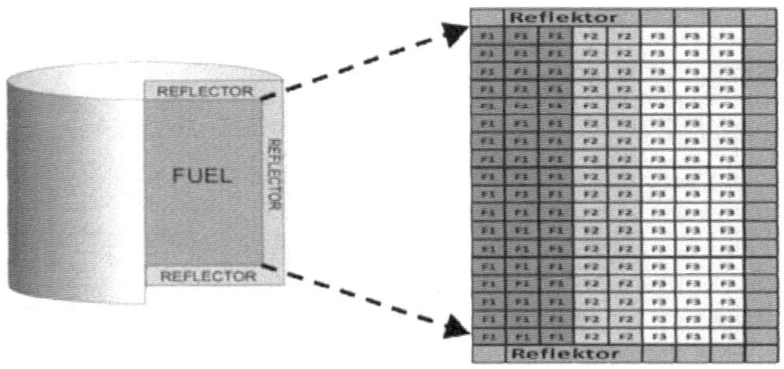

Figure 1. Half heterogeneous core configuration with different width in variation fuel

Fig. 1 show hexagonal geometries for fuel pin calculation (PIJ) by SRAC2006 results calculation. Fig. 2 show half heterogeneous core configuration with different width in variation fuel. There is three variation fuel, i.e. fuel 1 (F1), fuel 2 (F2) and fuel 3 (F3). For each fuel there is a width for each. The F1 is located in the centre of the core, after F2, and last F3.

FI-ITB-CHI has been benchmarking with six participant team in IAEA-tecdoc-1652 [2]. The calculation uses two dimensional R-Z geometry calculations. The energy group neutron is divided by 8 energy group. Figure 2 show FI-ITB-CHI calculation scheme. For iteration, we can take amount of time depend on how many burn up time that we needed in the calculation.

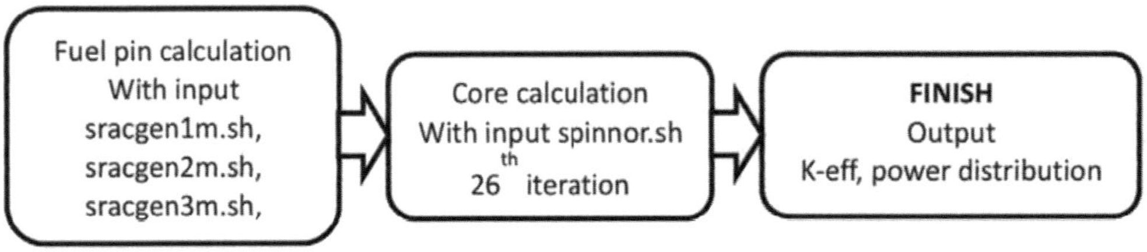

Figure 2. The FI-ITB-CHI calculation scheme

3. Results and Discussion

Figure 3 shows effective multiplication factor (k-eff) value of gas cooled fast reactor with variation percentage of Plutonium. The cell and core calculation use FI-ITB-CHI code. The variation of Plutonium is from 8% up to 15% use homogeneous core configuration, it means in core reactor there is one type of percentage plutonium of fuel. The homogeneous configuration fuel is used to calculate the parametric survey value for the next calculation.

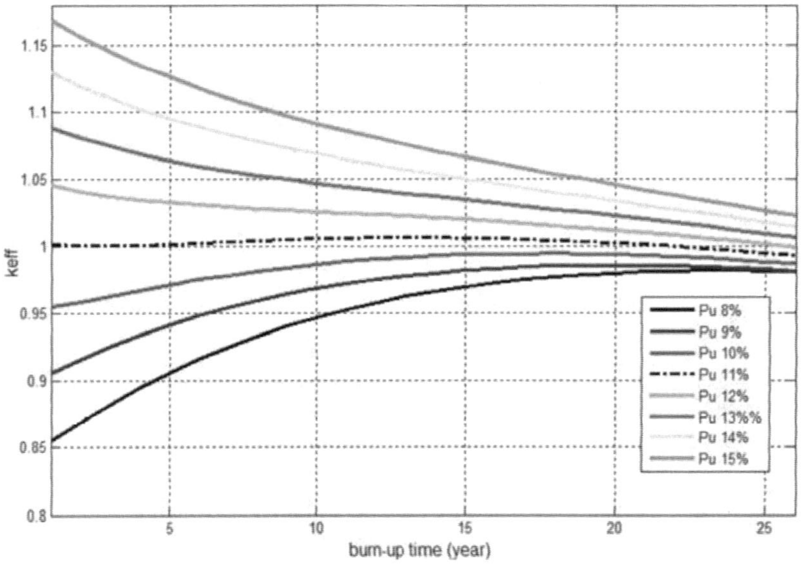

Figure 3. Effective multiplication factor (k-eff) value of gas cooled fast reactor with variation percentage of Pu

When the percentage of Plutonium 8% up to 10%, k-eff value in the subcritic value (k-eff<1). The graphics of percentage 8%, 9% and 10% is increasing. It shows there is breeding, means there is increasing of Plutonium amount in the reactor from fast neutron which absorb by U-238.

Percentage of Plutonium is 11% has k-eff value more stable than the others, or we can say that it has critical condition. Whereas the percentage of Plutonium 12% up to 15% have k-eff value more than one (k-eff>1) or we can say that they have supercritical condition. The k-eff value of percentage plutonium 12% - 15% is decreasing. From that case show that there is burning continuously in the reactor. Burning means that the number of fuel (fissile contain, in here Plutonium) is decrease continuously.

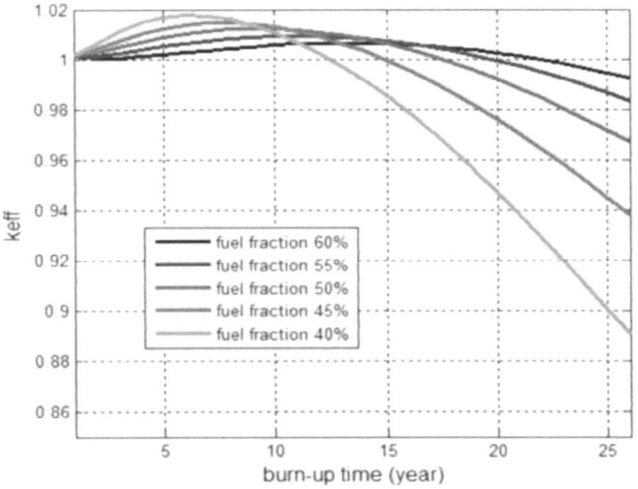

Figure 4. Effective multiplication factor (k-eff) value of gas cooled fast reactor with variation of fuel volume fraction (homogeneous core)

Figure 4 show k-eff value homogeneous core configuration with variation of fuel volume fraction. From Figure 3 or parametric survey, we know that the stable k-eff value is when the percentage of Plutonium 11%. From that case, we take 11% Pu as a standard to calculate fuel fraction variation. In this case, the volume fuel fraction is varied from 40% up to 60%. From Figure 4, the black graph use 60% fuel volume fraction, the blue graph 55% continuously up to the green one 40% fuel volume fraction. When the fuel volume fraction is increase, the reactor can be operated longer than small fuel volume fraction.

Figure 5 shows k-eff value use heterogeneous core configuration. Heterogeneous core configuration means in one core there is more than one type of percentage plutonium in fuel. In here we use three type of fuel, i.e. fuel 1 (F1), fuel 2 (F2) and fuel 3 (F3). F1 located in the central of the core, after that F2, than F3. The radius of F1 is 35 cm from centre, radius F2 is 25 cm from F1 and radius F3 is 35 cm from F2. There are three variation fuels that has been done (see Table 2), we call it case 1, case 2 and case 3. For all variation Plutonium percentage for F1 and F2 are unchanged (constant). The percentage plutonium of F3 is varied from 13%, 13.5% and 14%.

Table 2. Three variation fuel for heterogeneous core configuration

Type	F1	F2	F3
Case 1	8%	10%	14%
Case 2	8%	10%	13.5%
Case 3	8%	10%	13%

Figure 5 shows k-eff value for heterogeneous core configuration. The green line is case 1, the blue one is case 2 and the red one is case 3. For all case use 60% fuel volume fraction. From the graph, we can see that for case 1 and 2 has similar graphic trend, and case 3 different. From figure 3, if we use case 1, the reactor can be operated until almost 25 years without refueling, because the k-eff value is more than one. And for case 2 and case 3 the reactor is subcritical because they have k-eff value less than one. In this case, the optimum value is use case 1 which is use 8% percentage of plutonium for F1, 10% for F2 and 14% for F3.

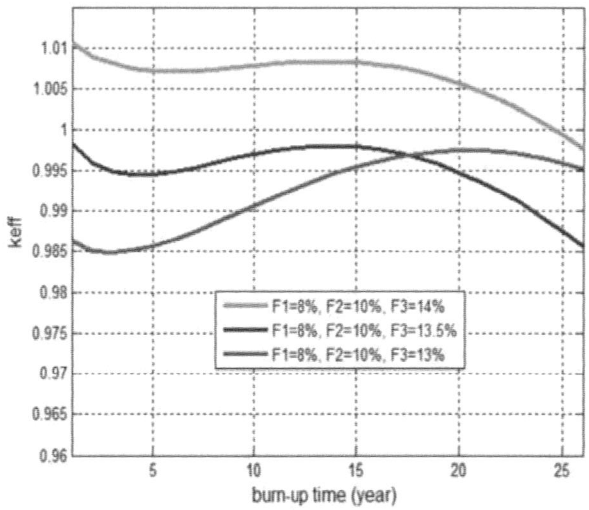

Figure 5. Effective multiplication factor (k-eff) value of gas cooled fast reactor for heterogeneous core configuration

4. Conclusion

Parametric survey with various percentage of plutonium show the critical condition reach when 11% percentage of Pu and 60% fuel volume fraction. When the fuel volume fraction is increase, the reactor can be operated longer. The optimum value of heterogeneous core configuration when use 8% percentage of plutonium for F1, 10% for F2 and 14% for F3. The reactor can be operated almost 25 years without refueling.

Acknowledgment

The authors express their gratitude for the support from The Ministry of Research Technology and Higher Education of The Republic of Indonesia through the scheme of Penelitian Unggulan Perguruan Tinggi (PUPT) RISTEKDIKTI.

References

[1] GIF (The Generation IV International Forum) 2002 A technology Roadmap for Generation IV Nuclear Energy System (U.S DOE Nuclear Energy Research Advisory Committee) p 1-22
[2] IAEA-TECDOC-1652 2010 Small Reactors Without On-site Refuelling: Neutronic Characteristics, Emergency Planning and Development Scenarios (International Atomic Energy Agency) p 62
[3] Ratna D S, Yacobus Y, Zaki S, Khairul B and Dwi Irwanto 2016 Matec Web Conferences **82** 03008
[4] Ratna D S, Zaki S, Khairul B, Dwi Irwanto 2017 IOP Conf. Series: Journal of Physics: Conf. Series **799** 012022
[5] Ratna D S, Zaki S, Khairul B, Dwi Irwanto 2017 Key Engineering Materials **733** 47-50
[6] Ratna D S, Zaki S, Khairul B, Dwi Irwanto 2016 IOP Conf. Series: Journal of Physics: Conf. Series **776** 012103
[7] Ratna D S, Zaki S, Khairul B, Dwi Irwanto 2017 IOP Conf. Series: Journal of Physics: Conf. Series **877** 012064
[8] Zaki S and Hiroshi S 1993 Nuclear Engineering and Design **140** 251-260
[9] Zaki S 1998 Progress in Nuclear Energy **32** 571-577
[10] Zaki S and Hiroshi S 1996 Nuclear Engineering and Design **162** 205-222
[11] Alan E W Albert B R 1939 Fast Breeder Reactors (Pergamon Press) p 243

Small Signal Stability Analysis of Wind Turbine Penetration in Sulselrabar Interconnection System

Mustadir Darusman B[1,2*], Ansar Suyuti[1], Indar Chaerah Gunadin[1]

[1]Department of Electrical Engineering Hasanuddin University, [2]Health Ministry of the Republic of Indonesia.
[1]Jl. Poros Malino KM. 6 Borongloe Gowa, South Sulawesi, 92171, Indonesia
[2]Jl. Perintis Kemerdekaan KM. 11 Makassar, South Sulawesi, 90245, Indonesia

E-mail: [2]mustadir.darusman@gmail.com, [1]ansarsuyuti@gmail.com, [1]indarcg@gmail.com

Abstract. This paper present and explain the analysis of power flow and small-signal stability of the electrical system in the islands of South Sulawesi, East and West (Sulselrabar) located in Indonesia. The research focused on the Eigen values analysis that used to investigate small-signal stability performance of the main electricity grid when the fault conditions with and without control equipment as well as the inclusion of wind turbines large scale on the system. Power Systems Analysis Toolbox (PSAT), which integrated in MATLAB is used to develop a network topology in Sulselrabar. This study allowed us to determine the optimal location of the placement of Power System Stabilizer (PSS) to dampen the oscillation. The simulation result shows the placement of PSS on two generators can add system stability and help the system to resume operation at the balance point. Time domain simulation methods are used to see the frequency response of the rotor speed.

1. Introduction

The growth of electricity consumers lead to increase problems in the stability of the power system and become an important issue in the planning of electric power systems [1-4]. During the system under normal conditions, the balance between the needs of active and reactive power supply can be assured. The stability of the system will has a significant effect when all the isolated electricity networks interconnected as planned in the development of electricity networks [1-5].

Because of its large and complex system [6], the low frequency oscillation phenomenon needs more attention to the potential oscillation power problems that have a direct impact on the stability of the system [5-9]. The stability of the power system components which is related to the system in providing balance and improving the state during disturbances [7], one of the effects of these disorders is the electromechanical oscillation.

The electromechanical oscillation consists of two mode which are the local oscillation mode and inter-area oscillation mode [8]. This study focuses on the phenomenon of small-signal in the electrical network systems in South Sulawesi, East and West (Sulselrabar), one of the islands in eastern Indonesia. With small signal stability analysis, the performance of the power system in a minor disturbances state can be evaluated [10]. Small Signal Stability can be defined the system's ability to stay in synchronization when subjected to small perturbations [11]. The small disturbances phenomenon has frequency range between 0 to 2 Hz [5]. Whereas the local oscillation mode has

Content from this work may be used under the terms of the Creative Commons Attribution 3.0 licence. Any further distribution of this work must maintain attribution to the author(s) and the title of the work, journal citation and DOI.
Published under licence by IOP Publishing Ltd

typical frequency range of 1-2 Hz and the inter-area oscillation mode has a range of frequencies below 1 Hz [8].

Power System Analysis Toolbox (PSAT) software is used to analysis study of power system [12]. This toolbox capable of perform power flow studies [13], the stability of the small-signal and time domain simulation to see the response and rotor angular velocity. To provide damping during the placement of Power System Stabilizer (PSS) isolation is connected to the Automatic Voltage Regulator (AVR) become priority [7]. PSS placement is based on eigenvalue damping which obtained when simulation is done.

2. Research Methodology

These case studies are used as a test case to the electric grid interconnection system Sulselrabar. Furthermore, the electrical system topology Sulselrabar modeled and modified using a Power Systems Analysis Toolbox (PSAT), which has been integrated in Matlab Simulink®. This toolbox is open source software used to analyze and study the power system [14]. The toolbox can run the power flow, continuation power flow, optimal power flow, small-signal stability analysis and time domain simulation. All operations can be assessed by means of Graphical User Interfaces (GUI) and simulink which provides a variety of tools for network design.

2.1. Power System Modeling

All dynamic models of the system are available and have been documented in the manual PSAT. Include synchronous generator model, turbine governor and Automatic Voltage Regulator (AVR) model. Simple version of the synchronous generator contained in PSAT model system, where the dynamic parameter of each of the synchronous generator as the input to d-axis and q-axis. Dynamic parameters are shown in Table 1 All mechanical damping and disregarded of the effect on the overall constellation generator. AVR models used is the model 3 and used for all generators but with different parameters. At the PSAT, Turbine Governor represent in two types of models, namely: models 1 and 3. Where the thermal power plant in the state and represented the first model (1) and the hydro plant model expressed and represented the third model (3). However, in this study, Model 3 [15] is used. In this model, some properties of the system regarded as inelastic penstock where the inertia of the water to be considered, and also the ideal turbine. For simplicity of analysis [18-21], static load models are also used [1]. As for wind turbine used doubly fed induction generator (DFIG).

2.2. Modeling of the Sulselrabar Power Grid

Indonesia is an archipelagic country that has a complex electrical system between the island and other islands. Sulawesi islands comprising South Sulawesi, East and West (Sulselrabar) used for this research study. Interconnection system Sulselrabar strut by several generating units where the supply of Gas Power Generation and Sengkang Steam Power, Bakaru and Poso Hydro, Jeneponto and Tello Steam is supply to the largest generation system interconnect Sulselrabar [16]. Sulselrabar interconnection system consists of 15 generators, 44 bus 47 lines and 34 load spread, whereas the voltage varies from 30, 70, 150 to 275 kV [17-20].

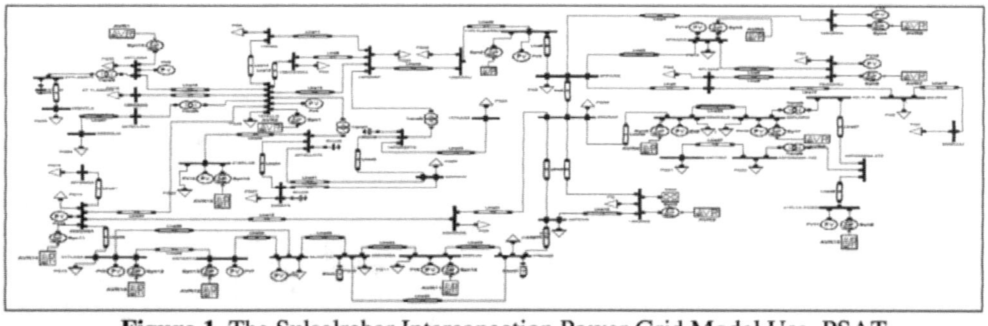

Figure 1. The Sulselrabar Interconection Power Grid Model Use PSAT

3. Case Study

The case studies in this research is to analyze the small signal before and after the inclusion of wind turbine large scale on Jeneponto bus at Sulselrabar interconnection system as shown in Figure 1 Dynamic parameters and excitation of each generator is shown in Table 1 and Table 2 to see the response of the system using power flow studies, eigenvalue and time domain simulation. The first simulation model of the system under normal conditions, whereas system must demonstrate its stability. It is important before doing the next phase simulation.

After condition is stable by eigenvalue analysis, the second stage simulation is by giving disturbance to the system in the form of termination of the channel on bus of Parepare and bus of Barru. The first intervention given time is 1 second and the second intervening time is 200 second.

Table 1. Dynamic Parameter Sulselrabar

No	Pembangkit	X_d (pu)	X'_d (pu)	X''_d (pu)	X_q (pu)	X'_q (pu)	X''_q (pu)	ra	xl
1	PLTA Bakaru	0,924	0,268	0,27	0,553	0,276	0,27	0	0.12
2	PLTA Teppo (Pinrang)	2,08	0,385	0,261	1,12	0,274	0,261	0	0.186
3	PLTD Suppa	2,08	0,385	0,261	1,12	0,267	0,261	0	0.186
4	PLTU Barru	2,363	0,199	0,204	2,182	0,395	0,204	0	0.107
5	PLTU Tello	1,182	0,0995	0,102	1,091	0,1975	0,102	0	0.107
6	PLTD Agrekko (Tello Lama)	2,363	0,199	0,204	2,182	0,395	0,204	0	0.107
7	PLTD Sgmnsa	2,08	0,385	0,261	1,12	0,337	0,261	0	0.186
8	PLTU Arena (jeneponto)	2,08	0,385	0,261	1,12	0.485	0,261	0	0.186
9	PLTA Tangka manipi sinjai	1,924	0,268	0,27	1,553	0,256	0,27	0	0.12
10	PLTGU Sengkang	2,31	0,2	0,12	0,553	0,6	0,12	0	0.6
11	PLTD Malea (Torja)	2,08	0,385	0,261	1,12	0,337	0,261	0	0.186
12	PLTD Palopo	2,08	0,385	0,261	1,12	0,337	0,261	0	0.186
13	PLTA Bili-Bili	2,08	0,385	0,261	1,12	0,330	0,261	0	0.186
14	PLTA Poso	0,924	0,268	0,27	0,553	0,368	0,27	0	0.12
15	PLTD Tallasa	2,08	0,385	0,261	1,12	0,485	0,261	0	0,186

*Data source = PT PLN Persero. AP2B Sulselrabar

Table 2. Data Exitation Sulselrabar

No	Generator	KA (p.u)	TA	VAmax	VAmin
1	PLTA Bakaru	400	0,04	0,71	-0,71
2	PLTA Teppo (Pinrang)	1	0,02	1	-1
3	PLTD Tallasa	1	0,02	1	-1
4	PLTD Suppa	1	0,02	1	-1
5	PLTU Barru	1	0,02	1	-1
6	PLTU Tello	100	0,04	1	-1
7	PLTD Agrekko (T.Lama)	100	0,04	1	-1
8	PLTD Sgmnsa	10	0,02	18,3	-18,3
9	PLTU Arena (Jeneponto)	10	0,02	18,3	-18,3
10	PLTA POSO	400	0,04	0,71	-0,71
11	PLTA Tmatipi (sinjai)	4	0,02	5,99	-5,99
12	PLTGU Sengkang	300	0,04	1	-1
13	PLTD Malea (Toraja)	10	0,02	18,3	-18,3
14	PLTD Palopo	4	0,02	5,99	-5,99
15	PLTA Bili-Bili	4	0,02	5,99	-5,99

*Data source = PT PLN Persero. AP2B Sulselrabar

The third simulation is the modeling system with the addition of PSS control the event of termination of channel. The simulation further analyzes the system when the inclusion of wind turbines on a jeneponto bus with and without PSS control equipment.

PSS placement will be based on the critical eigenvalue or close to zero. There are 44 buses, 47 lines, 6 transformers, 15 generators and 34 loads will become the object of this research.

This study will use conventional thermal generators units, where for Turbine Gevernor (TG) used Model 1, while Hydroelectric unit used Model 3 TG and used to the whole system. As for wind turbine used model doubly fed induction generator (DFIG) with Nominal wind speed 15.00 (m/s).

4. Results and Discussions

In this paper, comparison between the system between with and without using the control damper and damper control is done. The simulation results were observed the deviation rotor angle (δ) or rotor angular velocity (ω) of each generating unit. Then, small signal stability analysis is used with the system. Figure 2 shows the PSAT Sulselrabar eigenvalue models under normal conditions, dynamics condition without PSS and dynamics condition using the PSS.

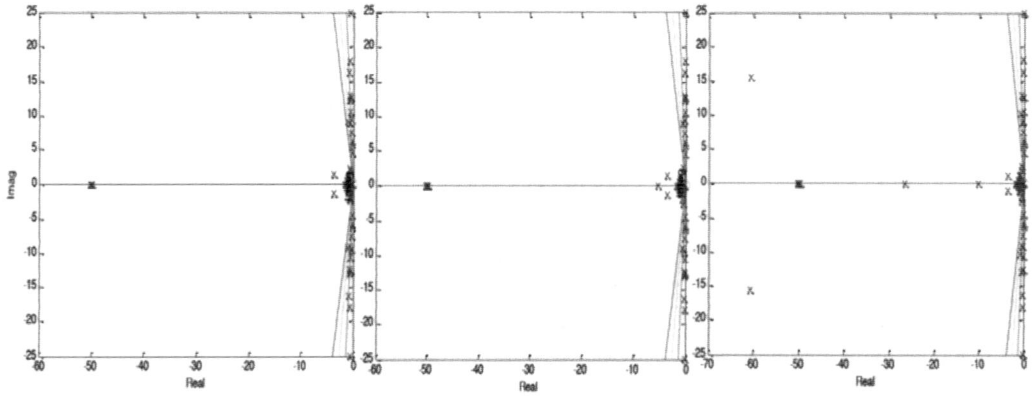

(a) (b) (c)

Figure 2. Eigenvalue use PSAT. (a) Normal conditions, (b) In the dynamical circumstances, (c) In the dynamical circumstances using PSS

Figure 2a shows the normal conditions on the electricity network of Sulselrabar interconnection system. In this condition has not been given any input in the form of interference or load shedding. Figure 2b shows the dynamics system after being given the disruption of the termination of the transmission line between the buses 4 and 11 (bus Barru and Bus Parepare). As for Figure 2c shows the shift of eigenvalue after the addition of control given in the form of PSS. The placement of PSS is randomly based on critical or near-critical value of eigenvalue. From the results of the experiment placement of the maximum PSS is on the two generators in reducing oscillations of the system.

4.1. System Simulation before Entrance wind turbine.

For analysis purposes, the participation factor used to determine the bus and generator weakly damped and most affected by instability. From the analysis found that the generator 2, 6, 7, 12 and 13 represents the weakest bus with eigenvalue levels approaching critical value. There are 111 state and 31 pairs resulting complex to be initialized and observed.

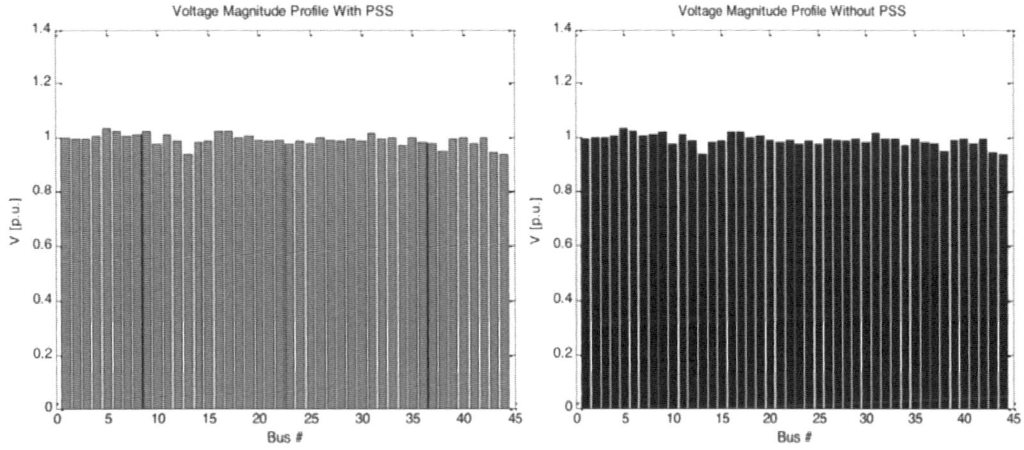

(a) (b)

Figure 3. Voltage magnitude Profile. (a) Without installation of PSS, (b) With PSS installation

Figure 3a and Figure 3b shows the voltage profile of the whole bus is still in the voltage allowed range by the Energy Regulatory Commission from 0.95 to 1.05 p.u. Where in Figure 3a represents the voltage profile using with PSS and Figure 3b represent the voltage profile picture without the PSS.

Table 3. Improvement of Voltage Magnitude Before and After using PSS

Bus	Without PSS	With PSS	Voltage Improvment	Bus	Without PSS	With PSS	Voltage Improvment
10SUPPA	0,99802	0,99909	0,00106	30TBNGA	0,97440	0,97594	0,00154
11PLTUBARRU	0,99933	0,99560	-0,00373	31TLASA	0,98780	0,98989	0,00210
12BARRU	1,00003	0,99719	-0,00284	32MAROS	0,97647	0,97960	0,00313
13PNKEP	1,00813	1,00780	-0,00034	33PGAYA	0,99737	0,99946	0,00208
14PNKEP70	1,03413	1,03436	0,00022	34JNPTO	0,99146	0,99356	0,00210
15TNASA	1,02598	1,02620	0,00022	35BKMBA	0,98821	0,98992	0,00172
16BOSOWA	1,00761	1,00749	-0,00012	36SNJAI	0,99488	0,99618	0,00130
17KIMA	1,01405	1,01429	0,00024	37BONE	0,98516	0,98711	0,00195
18TELLO	1,02315	1,02396	0,00081	38MKALE	1,01720	1,01803	0,00083
19PKANG	0,97703	0,97780	0,00077	39PLOPO	0,99821	0,99879	0,00057
1SKANG	1,01259	1,01505	0,00246	3SPENG	0,99685	0,99916	0,00231
20TELLO70	0,98763	0,98853	0,00090	40LTUPA	0,97298	0,97386	0,00088
21BRLOE	0,93965	0,94066	0,00101	41PLTA POSO	0,99821	0,99885	0,00064
22MNDAI	0,98359	0,98443	0,00083	42POMANA 275	0,98274	0,98345	0,00071
23DAYA	0,98687	0,98775	0,00087	43POMANA 150	0,98040	0,98110	0,00071
24TELLO30	1,02316	1,02397	0,00081	44POSO	0,95173	0,95241	0,00069
25BWAJA	1,02316	1,02397	0,00081	4PPARE	0,99417	0,99537	0,00121
26TLAMA	1,00021	1,00124	0,00103	5PRANG	0,99835	0,99924	0,00089
27TLAMA70	1,00981	1,01085	0,00104	6PLMAS	0,98156	0,98237	0,00081
28BNTLA	0,99311	0,99413	0,00102	7BAKARU	0,99872	0,99928	0,00056
29SGMSA	0,98608	0,98764	0,00156	8MJENE	0,94631	0,95709	0,01078
2SDRAP	0,99239	0,99469	0,00230	9MMUJU	0,94802	0,95880	0,01077

For improvement of voltage magnitude as seen in Table 3, there is a change before and after the installation of PSS. Where there is improvement for each bus voltage. For example at a bus 10 (Suppa bus), an improvement of voltage from 0.99802 p.u becoming 0.99909 p.u or voltage changes 0.00106 p.u after the addition of PSS.

Table 4. Linear Value Analysis of The System Without PSS

Before Aplication of PSS						
Eig.#	**Most associated states**	**Real part**	**Imaginary part**	**Frequency**	**Damping ratio**	**Mode oscillations**
Eig As # 58	omega_Syn_2, delta_Syn_2	-0,0262	4,4241	0,7041	-0,0059	Inter Area
Eig As # 53	omega_Syn_6, delta_Syn_6	-0,0615	6,0832	0,9682	-0,0101	Inter Area
Eig As # 16	omega_Syn_7, delta_Syn_7	-0,0016	24,9226	3,9665	-0,00006	Lokal
Eig As # 41	delta_Syn_12, omega_Syn_12	-0,0084	12,6267	2,0096	-0,00067	Lokal
Eig As # 43	delta_Syn_13, omega_Syn_13	-0,0394	10,7185	1,7059	-0,0037	Lokal

Table 5. Linear Value Analysis of The System With PSS

After Aplication of PSS						
Eig.#	**Most associated states**	**Real part**	**Imaginary part**	**Frequency**	**Damping ratio**	**Mode oscillations**
Eig As # 60	delta_Syn_2, omega_Syn_2	-0,0269	4,4226	0,7039	-0,0061	Inter Area
Eig As # 56	delta_Syn_6, omega_Syn_6	-0,0767	6,1139	0,9731	-0,0125	Inter Area
Eig As # 18	omega_Syn_7, delta_Syn_7	-0,0017	24,9237	3,9667	-0,00007	Lokal
Eig As # 43	omega_Syn_12, delta_Syn_12	-0,0089	12,6211	2,0087	-0,00070	Lokal
Eig As # 48	omega_Syn_13, delta_Syn_13	-0,0586	10,7180	1,7058	-0,0055	Lokal

Table 4 and Table 5 shows participation factor of eigenvalue before and after the addition of PSS. From Table. 4. it can be seen that the value of damping ratio increased for the fifth generator. For the second generator damping ratio increased from -0.0059 to -0.0061. Also with the next generator the damping ratios will increase or will improve the eigenvalue. With the correct input parameters of the PSS, the real and imaginary parts of eigenvalue 59-60 improved from-0.0262 ± j4.4241 become -0.0269 ± j4.4226.

The next thing to observe is the rate speed of the rotor to return to a stability position after interference. From the simulation results using time domain on the PSAT, shows good damping occurs after the addition of PSS. Whereas the rotors speed dampen oscillations and back at the point of equilibrium faster than before using a PSS. This suggests that the placement of PSS and PSS proper tuning values can restore the stability of the system and reduce oscillation well in old or new balance point. Here is a graph of rotor speed (ω) on generators 2, 6, 7, 12 and 13 which are the weakest generators of the simulation results prior to the entry of wind turbines. The black graph is without the PSS and blue graph is using the PSS.

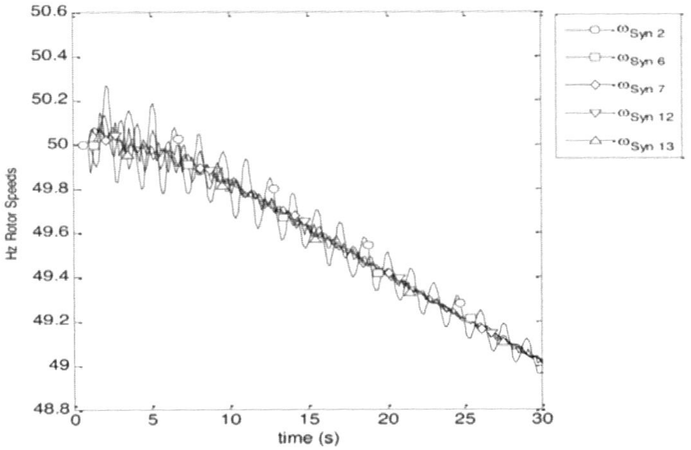

Figure 4. The Rotor Speed of The Generator (ω) during load release between Bus 4 and 11 without using PSS.

From the simulation results above indicate the difference graph generator rotor speed on the state of good dynamics with and without PSS control equipment. Figure 4 shows the rotor speed response to damp the oscillation and return to the equilibrium point more slowly in the appeal using PSS. For example, it can be seen in generator two, in which the rotor speed when its dynamic insulated and downhill. Isolation and degradation the speed at the start of the first second until the seconds to 30 and continued to decline to below the rate of 49.3 Hz. Free time to survive in the value of 49.8 Hz and 49.9 Hz climb back up to speed eventually decreasing.

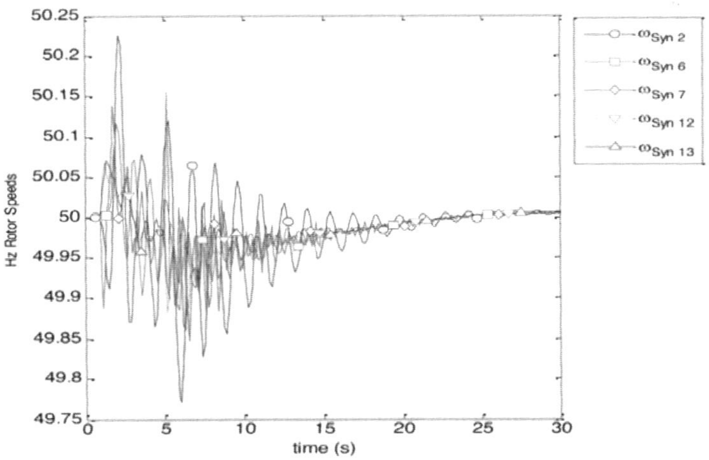

Figure 5. The rotor speed of the generator (ω) during load release between bus 4 and 11 with PSS control.

Figure 5 Indicates otherwise, where the fluctuations in the speed and isolation occurring in the five generators do not exceed the point 49.7 Hz at the lower boundary and 50.3 Hz at the upper limit. Slowly from the 7th to the 30th seconds the speed increases and the oscillation damping is good until it returns to its normal density. This shows the use of PSS control equipment in reducing the occurrence of better and faster oscillations than without the use of control equipment.

4.2. Simulation system after the entry of wind turbine.

The next observation is to insert wind turbines on the Jeneponto bus. The DFIG wind turbine model is used. Tables 6 and Tables 7 show the participation factor of the eigenvalues before and after the addition of PSS after the inclusion of wind turbines. It can be seen that the damping value of the ratio increases for the five generators. For generator six, the damping ratio increased from -0.005488 up to -0.005748. Likewise with the generator further increase or improve the eigenvalue after the addition of PSS. With correct input parameters of PSS, the real and imaginary parts of the eigenvalues of each generator are improved. In the two generators eg. from -0.0426 ± 5.0118i it has improved to -0.0425 ± 5.0189i. Likewise with generators 6,7,12, and 13.

Table 6. Linear Value Analysis of The System Without PSS

Before Aplication of PSS						
Eig.#	**Most associated states**	**Real part**	**Imaginary part**	**Frequency**	**Damping ratio**	**Mode oscillations**
Eig As # 58	omega_Syn_2, delta_Syn_2	-0,0426	5,0118	0,7977	-0,008508	Inter Area
Eig As # 54	omega_Syn_6, delta_Syn_6	-0,0316	5,7527	0,9156	-0,005488	Inter Area
Eig As # 17	delta_Syn_7, omega_Syn_7	-0,0030	24,7455	3,9384	-0,000120	Lokal
Eig As # 40	omega_Syn_12, delta_Syn_12	-0,0052	12,5657	1,9999	-0,000413	Lokal
Eig As # 48	delta_Syn_13, omega_Syn_13	-0,0499	10,6240	1,6909	-0,004699	Lokal

Table 7. Linear Value Analysis of System with PSS

After Aplication of PSS						
Eig.#	**Most associated states**	**Real part**	**Imaginary part**	**Frequency**	**Damping ratio**	**Mode oscillations**
Eig As # 61	delta_Syn_2, omega_Syn_2	-0,0425	5,0189	0,7988	-0,008466	Inter Area
Eig As # 57	delta_Syn_6, omega_Syn_6	-0,0331	5,7546	0,9159	-0,005748	Inter Area
Eig As # 20	delta_Syn_7, omega_Syn_7	-0,0030	24,7376	3,9371	-0,000123	Lokal
Eig As # 42	delta_Syn_12, omega_Syn_12	-0,0053	12,5704	2,0006	-0,000421	Lokal
Eig As # 51	delta_Syn_13, omega_Syn_13	-0,0509	10,6175	1,6898	-0,004789	Lokal

For rotor speed, from the simulation results obtained that the entry of wind turbine in the interconnection system of Sulselrabar very influential on the stability of the system. From Figure 6, it can be seen that the fifth rotor density of the generator fluctuates from the first second to the 30th seconds. At 25 seconds the rotor speed penetrates 50.03 Hz at the upper boundary and 49.97 in lower limit.

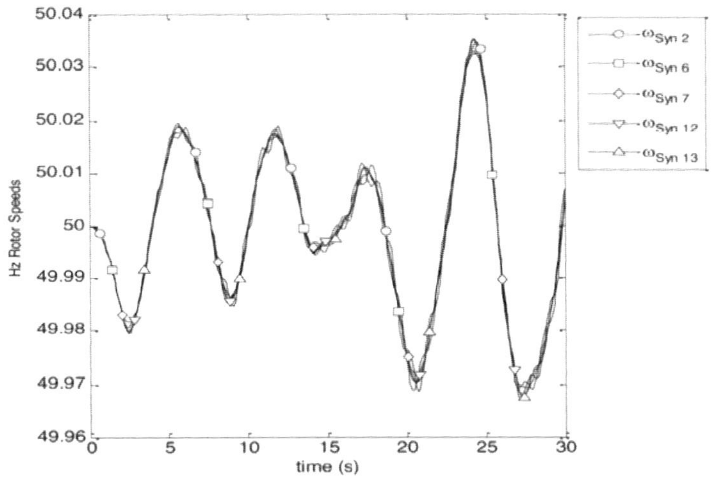

Figure 6. The rotor speed of the generator (ω) upon entry of wind turbines without PSS control

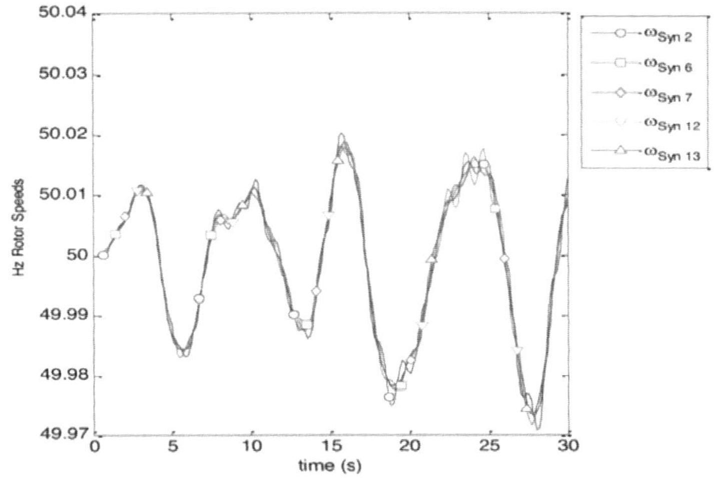

Figure 7. The rotor speed of generator (ω) upon entry of wind turbine with PSS control.

Meanwhile, with the addition of PSS control, the rotor speed on the five generators is relatively more stable in reducing the oscillations that occur. Figure 7 shows where the speed of the five generators is more stable. The rotor speed does not exceed 50.02 on the upper boundary and 49.97 at the lower limit and gradually returns to its normal density.

5. Conclusions

This paper explains the power flow and small-signal stability. Through the PSAT toolbox integrated on Matlab, multi-machine system of interconnected systems modeled Sulselrabar. Conventional thermal generating units, whereas for Turbine Governor (TG) used Model 1, while using the unit Hydroelectric Model 3 TG. For wind turbine using model doubly fed induction generator (DFIG).

Effective implementation and proper placement of the device PSS can cope with the dynamics of multi-machine network stability in the system and increase the critical eigenvalues as well, whether that the oscillation mode between areas and the local oscillation mode. Besides that the implementation of PSS can reduce and minimize overshoot oscillation frequency oscillation. Eigenvalues, voltage, frequency, and the factor observed with either participation in this study.

This research can provide a review for PT PLN electrical organizers to add control equipment on the system to prevent and dampen the occurrence of small signal oscillation in Sulselrabar system. Further research can be developed by adding generators from other renewable resources, which will enter the Sulselrabar electrical system. Optimization of PSS can also be used on systems with transient disturbances. In addition there are several methods that can be done to improve the stability of the system in Sulselrabar, such as using control equipment STATCOM, UPFC, TCSC, SVC or equipment energy storage, SMEs (Superconducting Magnetic Energy Storage), CES (Capacitive Energy Storage) and BES (Battery Energy Storage) that can be researched.

References

[1] Tristan Jr.G. M., 2016 Power FLow and Small Signal Stability Analysis on the Interconnected Philippine Power Grid, *Sciencedirect, Perspectives in Science*. Volume 8. sciencedirect, 2. 589–591. Web.

[2] Pitono, Joko et al, 2013 Power Generation Optimization Based on Steady State Stabilify Using Particle Swarm Optimization (PSO) Limit, *International Review on Modelling and Simulations*, vol. 6, pp. 2–7.

[3] Indar C.G, Soeprijanto, A., and Penangsang, O., 2012 Steady State Stability Assessment Using Extreme Learning Machine Based on Modal Analysis, *International Review of Electrical Engineering* ,vol. xx, no. June.

[4] Indar C.G, Muhammad,A., et al, 2012 Determination of Steady State Stability Margin Using Extreme Learning Machine 2 Research Method, *Wseas Transsactionon Power System*, vol. 7, no. 3, pp. 91–103.

[5] Salahi, Rim B E N et al, 2014 Smal Signal Syability of The Tunisian Interconnected Power System, *IEEE*. N.p., Print.

[6] Indar C.G, Soeprijanto, A., and Penangsang, O., 2010 *Real Power Generation Scheduling to Improve Steady State Stability Limit in the Java-Bali 500 kV Interconnection Power System*, vol. 4, no. 12, pp. 1771–1775.

[7] Ali, I,. Mehdi, K., and OP. M., 2011 Study of a major oscillations event in northeastern area of the Iranian power network, *Electric Power Systems Research*, vol. 81, no. 7, pp. 1292–1298.

[8] Mondal, D., Abhijit, C., and Aparajita, S., 2014 in Power System Small Signal Stability Analysis and Control, 1st ed., *Academic Press*, p. 328.

[9] Lin, Kyaw Myo., 2014 *Small Signal Stability Assessment of MEPE Test System in Free and Open Source Software*, vol. 8, no. 11, pp. 1707–1715,.

[10] Motjaba, K., Mahdi, E.,1996 *Power-system small-signal stability and maximum loadability* 1996, vol. 21, no. 2.

[11] Prabu,K., 1994 *Power System Stability and Control*, McGraw-Hill,Inc.,pp. 699.

[12] Federico, M., 2005 An Open Source Power System Analysis Toolbox, vol. 20, *IEEE Transaction On Power Systems*, pp. 1199–1206, Vol. 20, No. 3

[13] Robert, W., Rakibuzzama, S., and N Mithulananthan, 2014 Expanding Power System Analysis Toolbox (PSAT) Functionalities for Better Result Interpretation, *Australasian Universities Power Engineering Conference*, 1–6. Print.

[14] Federico, M., Luigi, V and Juan C, 2008 An Open Source Power System Virtual Laboratory : The PSAT Case and Experience, *IEEE Transactions on Education*, pp. 17–23, vol. 51, no.1

[15] Yuwa, C., et al, 2012 Development and Implementation of a Nordic Grid Model for Power System Small-Signal and Transient Stability Studies in a Free and Open Source Software, *IEEE*,

pp. 1–8.

[16] Muhammad, D.J., and Yudhi, L. C., 2012 *Studi Aliran daya dan Hubung Singkat Sistem Interkoneksi 150 kV Sulawesi Selatan Dengan ETAP*, Skripsi, FT-Politeknik Negeri Ujung Pandang, MAKASSAR.

[17] Muhammad, B.N., and Ardiaty. A., 2016 Network Losses Reduction Due To New Hydro Power Plant Integration, in *ICITACEE*, no. 1, pp. 181–185.

[18] Muhammad, B.N., and Ardiaty, A., 2016 Network Losses-Based Economic Redispatch for Optimal Energy Pricing in a Congested Power System, *in Energy Procedia*, vol. 100, pp. 311–314.

[19] Ardiaty, A., and Muhammad, B., 2016 An Analytical Method for Optimal Capacitors Placement from the Inversed Reduced Jacobian Matrix, *in Energy Procedia*, vol. 100, pp. 307–310.

[20] Ardiaty, A., and Muhammad, B.N, 2016 Voltage Drop Simulation at Southern Sulawesi Power System Considering Composite Load Model, *in ICITACEE*, 2016, pp. 169–172.

[21] Muhammad, B.N., Ardiaty, A., Ramesh, C.B., 2014 Transmission management for congested power system : A review of concepts , technical challenges and development of a new methodology, *Renew. Sustain. Energy*, Rev., vol. 38, pp. 572–580.

IOP Conf. Series: Journal of Physics: Conf. Series **1090** (2018) 012035 doi:10.1088/1742-6596/1090/1/012035

Expansion of Filter Design from GHz to THz with Metamaterial Hexagonal Split Ring Resonator

M. F. Kurnia[1], R.F. Syahputra[1*], Saktioto[1]

[1]Physics Department, FMIPA, Universitas Riau, Pekanbaru 28293, Indonesia

*Corresponding author's e-mail: romifadlisyahputra@yahoo.com

Abstract. Metamaterial application of hexagonal structures with two hexagonal ring-centered, hexagonal ring resonators (SRR-H) for frequency filter has greatly developed but in microwave frequency range. This paper proposes both higher frequency filter and energy. The SRR-H is designed and computed in the 300GHz - 300THz with outer and inner radii of SRR-H respectively for 0.5μm and 0.36μm to 0.35μm and 0.21μm. The simulation results show that the SRR-H size has an effect on the resonance frequency with the largest refractive index occurring at low frequency with the refractive index peak of -6.0242 corresponding to 9.765THz for SRR-H structure of outer and inner radius 0.4μm and 0.26μm. Resonance frequencies tend to shift at high frequencies when the geometrical size of the structure is reduced. This SRR-H filter can be used in THz frequency because the width of the resonance frequency occurs in SRR-H structure.

1. Introduction

Over the last two decades, electromagnetic metamaterials have grown rapidly with unique electromagnetic effects such as invisibility cloaking [1,2], negative refractive index [3.4] and super resolution [5]. Generally, the material is designed and consists of a resonator ring separator (SRR) and placed on the substrate [6]. The most common SRR modifications made are round and square SRR. Other basic structures have also been investigated such as the structure of the letters 'S' [6,7], omega [8], chiral [9] and helix [10]. The advantages of SRR-H compared to other structures have a high absortivity of up to 99% [11]. Hexagonal SRR is also designed for cloaking purposes, negative refractive index and super resolution. The hexagonal structure places more effective and can be used for wide applications [7]. Negative refractive index investigation was first performed on microscale [3] and now it develops in Terahertz frequency response [12], near-infrared, and visible light [13-15].
Metamaterial in far-infrared or THz region is considered as a slight development of its frequency region in electromagnetic (EM) spectrum, but it is very attractive for a wide range of applications [16,17]. This non-ionizing radiation can easily penetrates materials such as plastics, fabrics and paper [18]. The size of SRR-H can be made small (in order of μm) because the THz radiation has high energy and frequency so it is necessary to expand the SRR-H metamaterial design.

Metamaterials also have extensive applications such as sub-wavelength imaging, electronic components and optical applications, antennas, absorber in solar cells, invisibility technologies, detectors and biosensors [19]. Based on the above studies, this article aims to design and operate the hexagonal resonator ring structure (SRR-H) to increase the higher frequency and energy, as to investigate the SRR and analyze the transmission function and wave propagation reflection. The

Content from this work may be used under the terms of the Creative Commons Attribution 3.0 licence. Any further distribution of this work must maintain attribution to the author(s) and the title of the work, journal citation and DOI.
Published under licence by IOP Publishing Ltd

frequency used is typically at 300GHz - 300THz. These frequencies have taken into consideration to the requirements of optical metamaterials, where SRR-H is observed at various sizes using HFSS™ software.

2. Literature Review

The main characteristics of optical metamaterials, particularly optical metamaterials, are the EM properties of materials derived from their structural forms, and not the compositions of their constituents [20]. The size of the optical metamaterial is less than half the wavelength coming to avoid the diffraction effect [6]. Modified from the Nicolson-Ross-Weir (NRW) method has been done by Ziolkowski [21]. The NRW equation is used extensively to calculate the permittivity and the complex permeability of the waveguide by measuring the S-parameter. As the waves propagate toward the material-air boundary, a part of the waves will be reflected and transmitted to the boundary plane. If the length of the waveguide is tm, then the Equation of the S11 reflection function and the transmission function S21 is as follows [6].

$$S_{11} = \Gamma_{total} = \frac{\Gamma_1(1-z^2)}{1-\Gamma_1^2 z^2} \quad (1)$$

$$S_{21} = T_{total} = \frac{z(1-\Gamma_1^2)}{1-\Gamma_1^2 z^2} \quad (2)$$

where $z = e^{-2j\theta}$ represents the wave impedance and $\theta = kt_m$ is wave phase, so the parameter quantity and difference of S_{21} and S_{11} can be expressed,

$$V_1 = S_{21} + S_{11} \quad (3)$$

$$V_2 = S_{21} - S_{11} \quad (4)$$

Assume $|\Gamma_1| \leq 1$, so that it is obtained the relations as follows,

$$k = \frac{2}{jt_m}\frac{(1-V_1)(1+\Gamma_1)}{1-\Gamma_1 V_1} \quad (5)$$

$$\mu_r = \frac{2}{jkt_m}\frac{1-V_2}{1+V_2} \quad (6)$$

$$\varepsilon_r = \mu_r + j\frac{2S_{11}}{kt_m} \quad (7)$$

$$n = \sqrt{\varepsilon_r}\sqrt{\mu_r} \quad (8)$$

where k is wave propagation (m^{-1}), μ_r is relative permeability ($H.m^{-1}$), ε_r is relative permittivity ($F.m^{-1}$) and n is refractive index.

3. Structure Design of SRR-H

The SRR-H structure is designed and operated using HFSS [6]. The SRR-H cell unit consists of two concentric aluminum rings with inner radius, R_1 and the outer radius, R_2 as shown in Table 1 and Figure 1. SRR-H is set on the FR4-Epoxy substrate and simulated from 300GHz to 300THz to describe frequency filters. The boundary conditions are symmetrically set perfect electrical and magnetic conductors. Its mechanism works by launching the EM source on SRR-H which will be measured by S11 and S21 parameters. The values of S11 and S21 are then calculated using the NRW Equation to obtain refractive index values and resonance frequencies. Table 1 shows the design of SRR-H size in which the substrate has axial dimension with thickness of ts = 0.09μm, ring thickness tc

= 0.03μm and gap c_1 and c_2 are set constant i.e., 0.07μm and 0.04μm, and ring width is constant, l = 0.1μm .

Tabel 1. SRR-H parameter input

Size	R_1 (μm)	R_2 (μm)	a (μm)	b (μm)
r1	0.36	0.5	1	1
r2	0.31	0.45	0.9	0.9
r3	0.26	0.4	0.8	0.8
r4	0.21	0.35	0.7	0.7

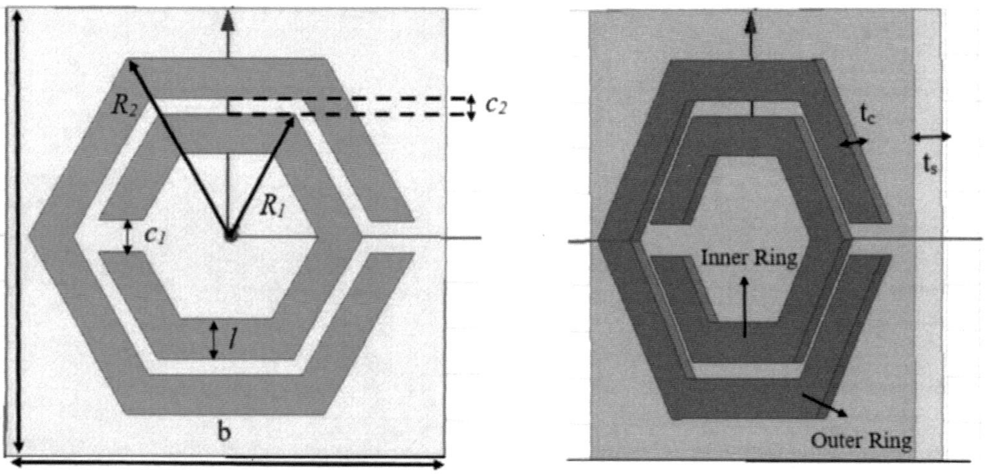

(a) (b)

Figure 1. Structure of SRR-H. (a) front view. (b) side view.

4. Results and Discussion

Figure 2 and 3 depict the effects of frequency corresponding to energy involved to wave propagation. Geometrical structure changes cause resonance frequency shifts as shown in Figure 3. Resonance frequencies tend to shift larger with structure size. At first resonance frequency (f01) all the structural variations have the same resonant frequency, but at second (f02) and third (f03) resonance frequencies have a uniform resonance frequency shift. This indicates that at lower frequency, the permittivity oscillates over the higher energy but it cannot work at much higher frequency since the change of refractive indices is affected by the geometry structure of both radius gaps. Either real or imaginary parts, the period oscillations are nearly similar but, the expansion of attenuation over the frequency is more affected in imaginary part. Real part having the oscillation is more dominant rather than the attenuation intensity.

In Figure 2, the red line, S11 is the reflection line. The purple line, S21 is the transmission line. The S11 value has several peaks where the filter of waves occurs. It is the same as S21 but the peak tends to exist at low frequencies compared to S11. This is due to EM waves are more easily filtered at lower frequencies rather than at high frequencies. As can be seen in Y1 peak, the filter goes down when the frequency increases. Transmission waves can maintain at high value when reflection waves tend to decrease. The sum of S11 and S21 is unstable due to the absorption effect and the medium of the structure [22]. It is not surprised when the peak of transmission is fluctuated and cannot be predicted along the frequencies. EM propagating to the ring structure is not only trapped by the ring gap but also protected by the hexagonal structure which is not resonant to the geometry, so that the filters can operate well at the reflection regimes in S11.

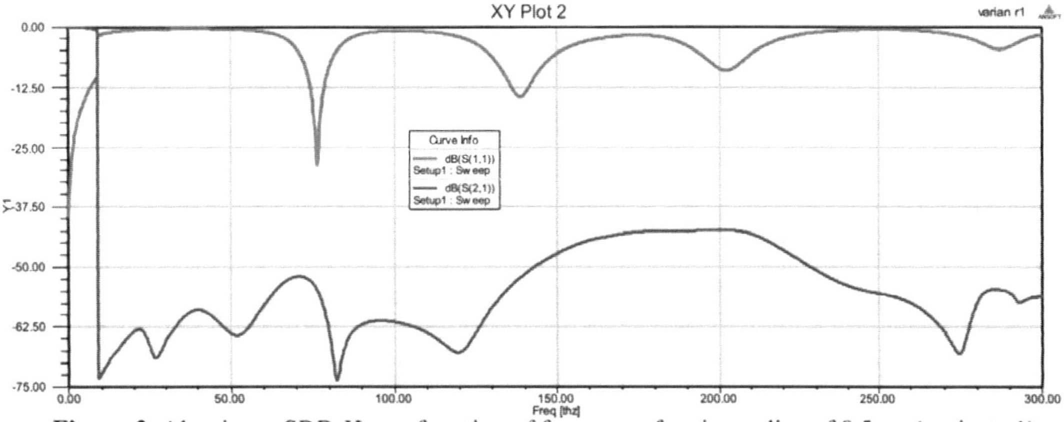

Figure 2. Aluminum SRR-H as a function of frequency for ring radius of 0.5μm (variant r1).

Tabel 2. Resonant frequency of SRR-H

Size	f01 (THz)	Im(n)	f02 (THz)	Im(n)	f03 (THz)	Im(n)
r1	9.255	-5.0204	75.54	-1.3849	135.78	-1.0224
r2	9.765	-5.3335	84.42	-1.4272	150.165	-1.0298
r3	9.765	-6.0242	95.145	-1.4621	168.225	-1.0264
r4	11.775	-5.709	109.395	-1.5313	191.34	-1.0213

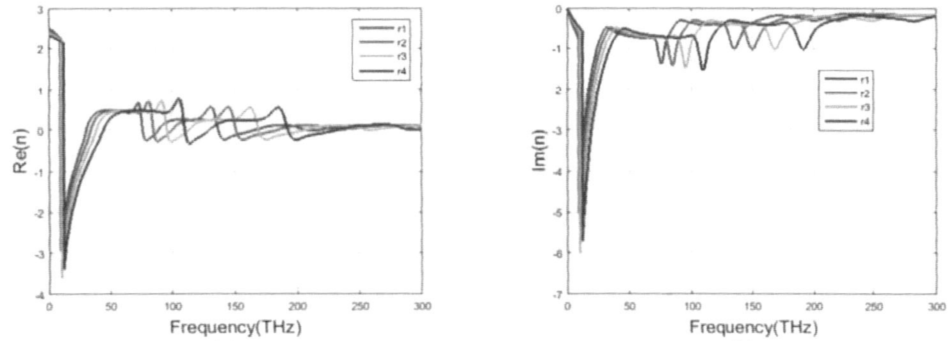

(a) (b)

Figure 3. Effect of geometry size of SRR-H over resonance frequency. (a) Real part of refractive indices (b) Imaginary part of refractive indices.

The SRR-H structure size variation provides a negative refractive index over a wide frequency range which is useful for extensive filter applications, invisibility technologies or for other EM applications. The other results are depicted in Figure 3 and Table 2 where the size variation has a peak trend, but on the peak r4 variation in f02 and f03 is smaller than f01. The peak shift of r1, r2, r3 and r4 is nearly periodic, and by increasing the frequency, the peaks of wave are widened and weak until reach zero.

In Figure 4, the frequency difference from the large to small ring radius moves from $\Delta f = \sim$ 100THz to $\Delta f = \sim$ 60THz which explains that f03 can maintain high frequencies. The largest refractive index values occur at low frequencies with peak refractive index of -6.0242 obtained at a frequency of 9.765THz for r3 structure. The outer ring condition cannot support f01 significantly compared to f02 and f03 corresponding to enhancement of its radius respectively, so that the frequency decay is much dominant for f03.

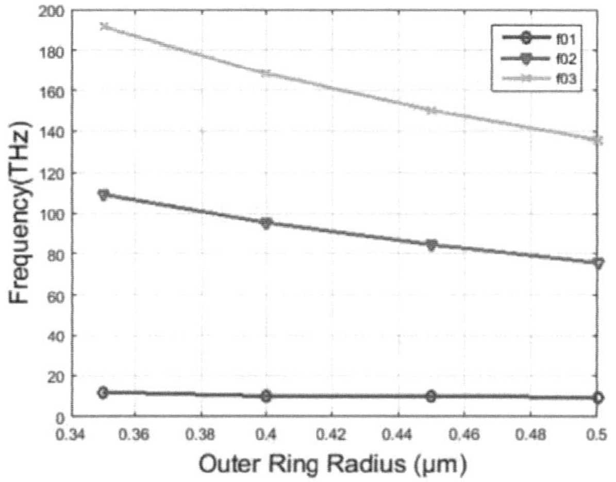

Figure 4. Resonance frequency shift over outer ring radius.

5. Conclusion

The structure of two central hexagonal rings, SRR-H has been designed and operated for high energy and frequency range. There are resonance frequencies that occur at intervals of 300GHz - 50THz, 50THz - 125THz and 125THz - 200THz. The metamaterial size of SRR-H gives the resonant frequency tends to expand when the size of the structure is reduced. The refractive index of SRR-H metamaterials has a resonance response on the Re(n) and imaginary Im(n) structure. This simulation has successfully investigated the hexagonal ring resonator characteristics to filter waves at high frequency.

Acknowledgment

We would like to thank the Universitas Riau, Indonesia for generous support in this research and Ministry of Research,Technology and Directorate General of Higher Education, Indonesia for source of funding through SIMLITABMAS in 2018.

References

[1] Pendry J B, Schurig D and Smith D R 2006 Controlling electromagnetic fields. *Science* **312**, 1780

[2] Schurig D, Mock J J, Justice J B, Cummer S A, Pendry J B, Starr A F and Smith D R 2006 Metamaterial electromagnetic cloak at microwave frequencies *Science* **314** 977

[3] Shelby R A, Smith D R and Schultz S 2001 Experimental verification of a negative index of refraction. Science **292** 77

[4] Smith D R, Padilla W J, Vier D C, Nemat-Nasser S C and Schultz S 2000 Composite medium with simultaneously negative permeability and permittivity *Phys. Rev. Lett.* **84** 4184

[5] Fang N, Lee H, Sun C and Zhang X 2005 Sub-diffraction-limited optical imaging with a silver superlens. Science **308** 534

[6] Patel N 2008 Theory, simulation, fabrication and testing of double negative and near zero metamaterials for microwave application (Thesis of Electrical Engineering Department, California Polytechnic State University)

[7] Benosman H and Hanece N B 2013 Multi-band meta-material structures based on hexagonal shaped *Int. J. Phys. Sci.* **8** 803

[8] Aydin K, Li Z, Bilgie S and Ozbay E 2008. Experimental and Numerical Study of Omega Type Bianisiotropic Metamaterials Combined with a Negative Permitivity Medium. *Phot. Nano. Fund. Appl.* **6** 116

[9] Matra K and Wongkasem 2009 Left-handed chiral isotropic metamaterials: analysis and detailed numerical study *J. Opt. A- Pure Appl. Op.***11** 074011

[10] Huang H-H and Hung Y-C 2013. Numerical analysis and the effective parameter retrieval of helical metamaterials *Photonic and Phononic Properties of Engineered Nanostructures III* (San Francisco, California, USA, February 02, 2013) *Proceeding of SPIE* **8632**: 863222

[11] Lee J, Young Y J and Lim S 2012 Ultra-thin polarization independent absorber using hexagonal interdigital metamaterial. *ENTRI Journal* **31** 126

[12] Yen T J, Padilla w j, fang n, vier d c, smith d c, pendry j b, basov d n and zhang x 2004 terahertz magnetic response from artificial materials *Science* **303** 1491

[13] Dolling G, Enkrich C, Wegener M, Soukoulis C M and Linden S 2006 Simultaneous negative phase and group velocity of light in a metamaterial Science **312** 892

[14] Yuan H K, Chettiar U K, Wenshan C, Kildishev A V, Boltasseva A, Drachev V P and Shalaev V M 2007 A negative permeability material at red light. Opt. Express 2007, 15, 787207877.

[15] Zhang S, Fan W, Panoiu N C, Malloy K J, Osgood R M and Brueck S R 2006 Optical negative-index bulk metamaterials consisting of 2D perforated metal-dielectric stacks. Opt. Express 2006, 14, 6778-6787.

[16] Ferguson B and Zhang X C 2002 Materials for terahertz science and technology *Nat. Mater.* **1**, 26

[17] Williams G P , Filling the THz gap - high power sources and applications *Rep. Prog. Phys.* **69**, 301

[18] Argyros A 2013 Microstructures in polymer fibres for optical fibres, thz waveguides, and fibre-based metamaterials *ISRN Optics* **2013** 785162

[19] Li J and Huang J 2013 *Time Domain Finite Element Methods for Maxwell's Equation in Metamaterials* (Heidelberg: Springer-Verlag) pp 1-18

[20] Salvatore S 2015. Optical metamaterials by block copolymer self-assembly. Doctoral Thesis of University of Cambridge. Published by Springer-Verlag, Switzerland.

[21] Ziolkowski R W 2016 Metamaterials and antenna *Handbooks of Antenna Technologies* ed Z N Chen, D Liu, H Nakano, X Qing and T Zwick (Singapore: Springer) pp 287-320

[22] Saktioto T, Syahputra R F, Punthawanunt S, Ali J and Yupapin P 2017. GHz frequency filtering source using hexagonal metamaterial splitting ring resonators. *Microw. Opt.l Techn. Lett.* **59** 1337

IOP Conf. Series: Journal of Physics: Conf. Series **1090** (2018) 012036 doi:10.1088/1742-6596/1090/1/012036

Analytical Method of Atomic Density Determination of Uranyl Nitrate Solution

Syarip[1,2], Tegas Sutondo[1]

[1]Centre for Science & Accelerator Technology, JL. Babarsari, POB 6101 ykbb, Yogyakarta 55281, Indonesia.

[2]*E-mail: syarip@batan.go.id*

Abstract. A basic theory of an analytical calculation method to determine the atomic density of each individual nuclide contained in a homogeneous system of uranyl nitrate solution is presented. Atomic density or number density is a number of atoms or nuclides per cm^3 or in a unit of atoms per barn-cm. This parameter is one of the important factors in defining the accuracy of neutronic calculations as such used in reactor core design. This work was carried out as part of neutronic design calculation of the SAMOP (subcritical assembly for ^{99}Mo isotope production) project, where uranyl nitrate solution was used as fuel with uranium enrichment of 20%. A homogeneous solution made by dissolving uranyl nitrate salt [UO_2 $(NO_3)_2$] into H_2O solvent has been used as an approach to this model. Two kinds of solution were described i.e. one that was made based on the amount of $UO_2(NO_3)_2$ dissolved in the solution and another one based on the amount of uranium in the solution. Expectedly this method can be used as an alternative way in estimating the atom density, solution density (g/cc) and the corresponding atom and weight fractions which is important in criticality calculations.

Keywords: Analytical method, SAMOP, atomic density, solution, uranyl nitrate.

1. Introduction

Atomic density (N) or commonly termed as number density is number of atoms or nuclides per cm^3 (atoms/cm^3), or in unit of atoms per 10^{24} barn cm (atom / barn cm). This parameter is one of important factors that would contribute to the accuracy of the neutronic calculations result, as such used in reactor core design, core power distributions etc. There are several methods that can be accurately determine the material compositions, which can be performed by means of measurement using special equipment, such as mass spectrometry [1,2,3], or gamma spectrometry [4], and so forth.

In case when there is no such composition measuring equipment available, or lack of data available for neutronic calculations, which often requires many different variations of material composition, an analytical method might be helpful as an approach to perform the simulations. Recently study on a possible use of reactor using solution based fuel, commonly termed as aqueous homogeneous reactor (AHR) is being of great interest for some research institutions. This includes power generation reactor, such as molten salt reactors (MSR) [5,6,7] and those dedicated for producing medical radioisotope ^{99}Mo [8,9,10] which in many cases offers more benefit compared to that using solid fuel or ^{235}U target as used in the present technique .

For the purpose of study related with the neutronic calculations, the atomic density of each nuclide contained in the solution is of important parameter, and for most cases of the data needed are not available. In this respect an analytical approach might help solve the problem, especially when needed simulation of solutions for many different variations of concentrations.

Content from this work may be used under the terms of the Creative Commons Attribution 3.0 licence. Any further distribution of this work must maintain attribution to the author(s) and the title of the work, journal citation and DOI.
Published under licence by IOP Publishing Ltd

This paper presents a basic theory of the analytical method to determine the atomic density of each nuclide contained in uranyl nitrate solution, of different concentrations and ^{235}U enrichments. This method has been used as the basis of calculation in the previous study when dealing with the preliminary study of the proposed SAMOP (Subcritical Assembly for ^{99}Mo Production), [11, 12]. The formula includes two kinds of uranyl nitrate solution i.e. the solution's concentration which was prepared based on the amount (gram) of uranyl nitrate salt dissolved, and the one that based on the amount of uranium in the solution. Further, this method can be used as the basis approach in determining the atom density or the solution density (g/cc) and the associated weight or atom fractions of other homogeneous uranium solution as well.

2. Basic Theory

Atomic density of a nuclide (n) in general can be determined based on the following equation

$$n = \frac{M\,N_A}{V} = \frac{w\,N_A}{AV} = \frac{\rho\,N_A}{A} \tag{1}$$

where M is molarity of the material or nuclide (g/A), N_A is Avogadro's number = 6.0225 x 10^{23} atom/mol, V is volume of the nuclide (cm^3), w is mass of nuclide (g), A is atomic mass or atom's number of the nuclide, and ρ is density of the nuclide (g/cm^3). For a material composed of several nuclides, the atomic density of each individual nuclide (n_i) can be determined based on the molarity of each nuclide contained in the material (M_i). For this case, data on the compositions i.e atomic fraction or weight fraction of each composing nuclide is needed.

Uranyl nitrate or sometimes referred as uranium nitrate is uranium salt, with chemical formula $UO_2(NO_3)_2$ and it is soluble in water[13,14]. Table 1 presents the physical properties and compositions of $UO_2(NO_3)_2$.

Table 1. Physical properties and compositions of uranyl nitrate [$UO_2(NO_3)_2$] [13, 14]

Physical form	Density (g/cc)	Melting point (°C)	Boiling point (°C)	Molecular weight
Solid	2.087	60	118	394.04
		Compositions		
Nulide	Mass number	Number of atom		Molar mass (g/mol)
U-total	238.028	1		238.028
Nitrogen	14.007	2		28.014
Oxygen	16	8		128.000
Total = Molecular weight of uranyl nitrate				394.040

The molecular weight or mass number of uranyl nitrate as in the Table 1, is for natural uranium. This value will be sligthly reduce with the increasing of ^{235}U enrichment.

Uranium, naturally consists of 3 main isotope i.e ^{234}U, ^{235}U and ^{238}U, with the compositions as presented in Table 2. The content of isotope ^{234}U is relatively very small compared to the other 2 isotopes, and in some cases such as used in criticality calculations, the presence can be omitted from the calculation. The compositions of the three uranium isotopes as a function of ^{235}U enrichment has been estimated based on a semi empirical approach [16]. Table 3 Presents the compositions of the 2 main isotope (without ^{234}U) as a function of ^{235}U enrichment.

Tabel 2. Compositions of 3 uranium isotopes of natural uranium [15]

Parameter	^{234}U(A = 234,041)	^{235}U(A = 235,0439)	^{238}U(A = 238,0508)	Total
Atom (%)	0.0054%	0.72%	99.275%	100%

Weight (w %)	0.0053%	0.711%	99.284%	100%
Activity (%)	48.9%	2.2%	48.9%	100%
Activity/g of U	12356 Bq	568 Bq	12356 Bq	25280 Bq

Table 3. Compositions of uranium and atom weight of U (total) as a function of ^{235}U enrichments

^{235}U enrichment	Weight Fraction		Nuclide Fraction		Atom Weight
(w %)	^{235}U	^{238}U	^{235}U	^{238}U	U (total)
0.71	7.110×10^{-3}	9.929×10^{-1}	7.200×10^{-3}	9.928×10^{-1}	238.029
2.50	2.500×10^{-2}	9.750×10^{-1}	2.531×10^{-2}	9.747×10^{-1}	237.976
5.00	5.000×10^{-2}	9.500×10^{-1}	5.061×10^{-2}	9.494×10^{-1}	237.900
7.50	7.500×10^{-2}	9.250×10^{-1}	7.589×10^{-2}	9.241×10^{-1}	237.825
10.00	1.000×10^{-1}	9.000×10^{-1}	1.011×10^{-1}	8.989×10^{-1}	237.750
12.50	1.250×10^{-1}	8.750×10^{-1}	1.264×10^{-1}	8.736×10^{-1}	237.675
15.00	1.500×10^{-1}	8.500×10^{-1}	1.516×10^{-1}	8.484×10^{-1}	237.600
17.50	1.750×10^{-1}	8.250×10^{-1}	1.768×10^{-1}	8.232×10^{-1}	237.525
19.75	1.975×10^{-1}	8.025×10^{-1}	1.995×10^{-1}	8.005×10^{-1}	237.457

3. Methodology

Uranyl nitrate solutions can be prepared through several ways, such as by dissolving the uranyl nitrate [$UO_2(NO_3)_2$] into water solvent or by dissolving U_3O_8 with nitrate acid (HNO_3) solvent. In this work, the first method will be used as the base for evaluation, i.e by dissolving the [$UO_2(NO_3)_2$] salt into water. The concentration of uranyl nitrate in the solution can be determined based on the amount (gram) of the uranyl nitrate salt dissolved, or based on the amount (gram) of uranium in the solution. In this respect, the determination of atomic density of each nuclide for the two cases will be evaluated here:

3.1. Number density based on the mass of uranyl nitrate in the solution

For any weight of uranyl nitrate salt (w_{UN}) The number of $UO_2(NO_3)_2$ molecules can be determined based on the following equation:

$$n_{UN} = \frac{w_{UN}}{A_{UN}} N_A \tag{2}$$

where A_{UN} is the molecular weight of uranyl nitrate which is about 394.040 for natural uranium. This value will be slightly reduce with the increasing of ^{235}U enrichment. Based on the isotopic compositions of uranyl nitrate, the total number of atom for each composing nuclide can then be determined using equation (3). The number of each uranium isotopes can then be found by multiplying with the nuclide fraction of the respective isotopes, which the values depending on the ^{235}U enrichment. When the uranyl nitrate salt is dissolved into water solvent, the number of oxygen atom is a summation of two components i.e which is from the $UO_2(NO_3)_2$ and from water (H_2O), and the amount would depend on the concentration of uranyl nitrate in the solution.

$$\begin{aligned} &\#Uranium: n_U = n_{UN} = \frac{w_{UN}}{A_{UN}} N_A ; \\ &\#Nitrogen: n_{Ni} = 2.n_{UN} = 2\frac{w_{UN}}{A_{UN}} N_A ; \\ &\#Oxygen: n_O = 8.n_{UN} = 8\frac{w_{UN}}{A_{UN}} N_A ; \end{aligned} \tag{3}$$

To determine the number of hydrogen and oxygen atom from the component of water, the volume and hence the weight fraction of water in the solution needs to be determined. For every one litre of solution, the volume fraction of uranyl nitrate (V_{UN}) and water(V_{wat}) can be determined as follows:

$$V_{UN} = \frac{w_{UN}}{\rho_{UN}} = \frac{w_{UN}}{2{,}807} \quad cm^3$$
$$V_{wat} = 1000 - V_{UN} \quad cm^3 \tag{4}$$

where ρ_{UN} = density of uranyl nitrate salt (2,807 g/cm^3). Hence the weight fraction of H_2O (w_{wat}) for each amount of uranyl nitrate to be solved can be determined as follows:

$$w_{wat} = V_{wat} \cdot \rho_{wat}$$
$$= \rho_{wat}(1000 - \frac{w_{UN}}{2{,}807}) \tag{5}$$

where ρ_{wat} = density of water (g/cm^3).

The number of atom hydrogen and oxygen from component of water can then be determined as follows:

$$n_H = 2\frac{w_{wat}}{A_{wat}}N_A = 2\frac{w_{wat}}{18{,}016}N_A$$
$$n_O = \frac{w_{wat}}{18{,}016}N_A \tag{6}$$

Hence the total number of oxygen atoms in the solution for any concentrations of uranyl nitrate solution becomes:

$$n_O(tot) = \left(8\frac{w_{UN}}{A_{UN}} + \frac{w_{wat}}{18{,}016}\right)N_A \tag{7}$$

Based on equation (3), (5), (6) and (7) the number density of each nuclide contained in one litre of solution can then be determined i.e by deviding the calculated total number of atom with 1000 cm^3. For uranium, the number density for the 3 isotopes, must be multipled by their atom fractions, which are depending on the ^{235}U enrichment, as presented in Table 3.

3.2. *Number density based on the amount of uranium in the solution*

In case the uranyl nitrate solution is prepared by dissolving an amount of uranium metal into nitride acid (HNO_3) or other ways instead of dissolving the uranyl nitrate salt into water, the weight fraction or atom density of each nuclide can be estimated based on the following approach. First we need to calculate the weight of each nuclide of the associated $UO_2(NO_3)_2$ salt. The weight of uranium isotopes for any amount (gram) of uranium (w_U) can be determined based on eq. (8)

$$w_{25} = fw_{25} * w_u$$
$$w_{28} = fw_{28} * w_u \tag{8}$$

Where fw_{25} and fw_{28} are the wight fractions of ^{235}U and ^{238}U respectively which correspond to the enrichment. The weight of nitrogen and oxygen can then be determined based on the following equation.

$$w_N = \frac{2*14}{A_U} * w_u$$
$$w_O = \frac{8*16}{A_U} * w_u \tag{9}$$

where A_U is the atom weight of uranium total which is a function of the ^{235}U enrichment, as presented in Table 3. The weight of $UO_2(NO_3)_2$ salt (W_{UN}) now can be found from eq. (8) and (9), from which the associated volume (V_{UN}) and the volume of water can be determined as in eq. (4). The weight of water (w_{wat}) in the solution can then be determined as in eq. (5). The total weight of the solution, the density (g/cc) and the weight fraction of each nuclide can be determined thereafter. Further, the number of atoms for each nuclide can then be determined using the general eqautions (3), (6) and (7). The corresponding number density of each nuclide in the solution can then be determined.

4. Result and discussion

Table 4 summaries the calculated number densities of each nuclide contained in the solutions for several concentrations of $UO_2(NO_3)_2$ and ^{235}U enrihments, and Table 5 presents the corresponding weight fractions. The results show that the number density of uranium isotopes will change depending on both concentration and ^{235}U enrichment. The number density of hydrogen and oxygen are changing very slowly with the changes of concentration for both cases.

Table 4. Atom density for several $UO_2(NO_3)_2$ concentrations and ^{235}U enrichments

Concentration (g / l)	150	200	300	400	500
Density (g / cc)	1.0781	1.1042	1.1563	1.2083	1.2604
Nuclide	Atom density (atom / barn cm) ; Enrichment =10 w%				
U-235	2.3207×10^{-5}	3.0942×10^{-5}	4.6413×10^{-5}	6.1884×10^{-5}	7.7356×10^{-5}
U-238	2.0622×10^{-4}	2.7496×10^{-4}	4.1244×10^{-4}	5.4992×10^{-4}	6.8741×10^{-4}
N	4.5886×10^{-4}	6.1181×10^{-4}	9.1771×10^{-4}	1.2236×10^{-3}	1.5295×10^{-3}
H	6.2107×10^{-2}	6.0504×10^{-2}	5.7298×10^{-2}	5.4091×10^{-2}	5.0885×10^{-2}
O	3.2889×10^{-2}	3.2699×10^{-2}	3.2320×10^{-2}	3.1940×10^{-2}	3.1561×10^{-2}
Nuclide	Atom density (atom / barn cm) ; Enrichment =15 w%				
U-235	3.4801×10^{-5}	4.6401×10^{-5}	6.9602×10^{-5}	9.2803×10^{-5}	1.1600×10^{-4}
U-238	1.9472×10^{-4}	2.5962×10^{-4}	3.8943×10^{-4}	5.1924×10^{-4}	6.4905×10^{-4}
N	4.5903×10^{-4}	6.1204×10^{-4}	9.1806×10^{-4}	1.2241×10^{-3}	1.5301×10^{-3}
H	6.2107×10^{-2}	6.0504×10^{-2}	5.7298×10^{-2}	5.4091×10^{-2}	5.0885×10^{-2}
O	3.2890×10^{-2}	3.2700×10^{-2}	3.2321×10^{-2}	3.1942×10^{-2}	3.1563×10^{-2}
Nuclide	Atom density (atom / barn cm) ; Enrichment =19.75 w%				
U-235	4.5810×10^{-5}	6.1080×10^{-5}	9.1621×10^{-5}	1.2216×10^{-4}	1.5270×10^{-4}
U-238	1.8379×10^{-4}	2.4505×10^{-4}	3.6758×10^{-4}	4.9010×10^{-4}	6.1263×10^{-4}
N	4.5920×10^{-4}	6.1227×10^{-4}	9.1840×10^{-4}	1.2245×10^{-3}	1.5307×10^{-3}
H	6.2107×10^{-2}	6.0504×10^{-2}	5.7298×10^{-2}	5.4091×10^{-2}	5.0885×10^{-2}
O	3.2890×10^{-2}	3.2701×10^{-2}	3.2322×10^{-2}	3.1944×10^{-2}	3.1565×10^{-2}

Table 5. Nuclide weight fraction for several $UO_2(NO_3)_2$ concentrations and ^{235}U enrichments

Concentration (g / l)	150	200	300	400	500
Density (g / cc)	1.0781	1.1042	1.1563	1.2083	1.2604
Nuclide	Atom weight fraction ; Enrichment =10 w%				
U-235	8.3052×10^{-3}	1.0812×10^{-2}	1.5488×10^{-2}	1.9761×10^{-2}	2.3680×10^{-2}
U-238	7.5703×10^{-2}	9.8557×10^{-2}	1.4118×10^{-1}	1.8012×10^{-1}	2.1585×10^{-1}
N	9.8937×10^{-3}	1.2880×10^{-2}	1.8450×10^{-2}	2.3540×10^{-2}	2.8209×10^{-2}
H	9.5652×10^{-2}	9.0985×10^{-2}	8.2282×10^{-2}	7.4330×10^{-2}	6.7034×10^{-2}
O	8.1045×10^{-1}	7.8677×10^{-1}	7.4260×10^{-1}	7.0225×10^{-1}	6.6523×10^{-1}
Nuclide	Atom weight fraction ; Enrichment =15 w%				
U-235	1.2463×10^{-2}	1.6225×10^{-2}	2.3241×10^{-2}	2.9652×10^{-2}	3.5534×10^{-2}
U-238	7.1525×10^{-2}	9.3117×10^{-2}	1.3338×10^{-1}	1.7018×10^{-1}	2.0393×10^{-1}
N	9.8975×10^{-3}	1.2885×10^{-2}	1.8457×10^{-2}	2.3549×10^{-2}	2.8220×10^{-2}
H	9.5652×10^{-2}	9.0985×10^{-2}	8.2282×10^{-2}	7.4330×10^{-2}	6.7034×10^{-2}
O	8.1046×10^{-1}	7.8679×10^{-1}	7.4264×10^{-1}	7.0229×10^{-1}	6.6528×10^{-1}
Nuclide	Atom weight fraction ; Enrichment =19.75 w%				
U-235	1.6415×10^{-2}	2.1370×10^{-2}	3.0612×10^{-2}	3.9056×10^{-2}	4.6803×10^{-2}
U-238	6.7552×10^{-2}	8.7945×10^{-2}	1.2598×10^{-1}	1.6073×10^{-1}	1.9261×10^{-1}
N	9.9011×10^{-3}	1.2890×10^{-2}	1.8464×10^{-2}	2.3558×10^{-2}	2.8230×10^{-2}

H	9.5652×10^{-2}	9.0985×10^{-2}	8.2282×10^{-2}	7.4330×10^{-2}	6.7034×10^{-2}
O	8.1048×10^{-1}	7.8681×10^{-1}	7.4267×10^{-1}	7.0233×10^{-1}	6.6533×10^{-1}

Table 6 and Table 7 present the results when using solution concentration based on the amount of uranium in the solution. The results show that both the number densities and the weight fractions of uranium isotopes will be higher than those calculated based on the uranyl nitrate [$UO_2(NO_3)_2$] concentration quite significantly. This phenomenon is caused by the change of water content in both cases, such as shown in Table 4 to Table 7. Figure 1 depicts comparison of the number density of total uranium atom in the solution for several concentration of both $UO_2(NO_3)_2$ and uranium in the solution for U^{235} enrichment of 19.75 w%. Figure 1 and 2 depict the comparison of both number density and the weight fraction of the uranium (total) as a function of concentration for the enrichment of 19.75 w%, for the two cases.

Table 6. Atom density for several uranium concentrations and ^{235}U enrichments

Concentration (g / l)	150	200	300	400	500
Density (g / cc)	1.1295	1.1726	1.2589	1.3452	1.4315
Nuclide	Atom density (atom / barn cm) ; Enrichment =10 w%				
U-235	3.8441×10^{-5}	5.1255×10^{-5}	7.6883×10^{-5}	1.0251×10^{-4}	1.2814×10^{-4}
U-238	3.4161×10^{-4}	4.5548×10^{-4}	6.8322×10^{-4}	9.1097×10^{-4}	1.1387×10^{-3}
N	7.6011×10^{-4}	1.0135×10^{-3}	1.5202×10^{-3}	2.0270×10^{-3}	2.5337×10^{-3}
H	6.5032×10^{-2}	6.4403×10^{-2}	6.3146×10^{-2}	6.1890×10^{-2}	6.0633×10^{-2}
O	3.2516×10^{-2}	3.2202×10^{-2}	3.1573×10^{-2}	3.0945×10^{-2}	3.0316×10^{-2}
Nuclide	Atom density (atom / barn cm) ; Enrichment =15 w%				
U-235	5.7662×10^{-5}	7.6883×10^{-5}	1.1532×10^{-4}	1.5377×10^{-4}	1.9221×10^{-4}
U-238	3.2263×10^{-4}	4.3018×10^{-4}	6.4527×10^{-4}	8.6036×10^{-4}	1.0754×10^{-3}
N	7.6059×10^{-4}	1.0141×10^{-3}	1.5212×10^{-3}	2.0282×10^{-3}	2.5353×10^{-3}
H	6.5033×10^{-2}	6.4406×10^{-2}	6.3150×10^{-2}	6.1895×10^{-2}	6.0639×10^{-2}
O	3.2517×10^{-2}	3.2203×10^{-2}	3.1575×10^{-2}	3.0947×10^{-2}	3.0319×10^{-2}
Nuclide	Atom density (atom / barn cm) ; Enrichment =19.75 w%				
U-235	7.5922×10^{-5}	1.0123×10^{-4}	1.5184×10^{-4}	2.0246×10^{-4}	2.5307×10^{-4}
U-238	3.0460×10^{-4}	4.0614×10^{-4}	6.0921×10^{-4}	8.1228×10^{-4}	1.0153×10^{-3}
N	7.6105×10^{-4}	1.0147×10^{-3}	1.5221×10^{-3}	2.0295×10^{-3}	2.5368×10^{-3}
H	6.5035×10^{-2}	6.4408×10^{-2}	6.3154×10^{-2}	6.1899×10^{-2}	6.0645×10^{-2}
O	3.2518×10^{-2}	3.2204×10^{-2}	3.1577×10^{-2}	3.0950×10^{-2}	3.0322×10^{-2}

Table 7. Atom weight fraction for several uranium concentrations and ^{235}U enrichments

Concentration (g / l)	150	200	300	400	500
Density (g / cc)	1.1295	1.1726	1.2589	1.3452	1.4315
Nuclide	Atom weight fraction ; Enrichment =10 w%				
U-235	1.3281×10^{-2}	1.7057×10^{-2}	2.3832×10^{-2}	2.9738×10^{-2}	3.4932×10^{-2}
U-238	1.1953×10^{-1}	1.5351×10^{-1}	2.1449×10^{-1}	2.6764×10^{-1}	3.1439×10^{-1}
N	1.5645×10^{-2}	2.0093×10^{-2}	2.8074×10^{-2}	3.5031×10^{-2}	4.1149×10^{-2}
H	8.6669×10^{-2}	7.9720×10^{-2}	6.7252×10^{-2}	5.6383×10^{-2}	4.6824×10^{-2}
O	7.6487×10^{-1}	7.2962×10^{-1}	6.6635×10^{-1}	6.1120×10^{-1}	5.6270×10^{-1}
Nuclide	Atom weight fraction ; Enrichment =15 w%				
U-235	1.9921×10^{-2}	2.5585×10^{-2}	3.5746×10^{-2}	4.4604×10^{-2}	5.2395×10^{-2}
U-238	1.1289×10^{-1}	1.4498×10^{-1}	2.0256×10^{-1}	2.5276×10^{-1}	2.9690×10^{-1}

N	1.5655×10^{-2}	2.0105×10^{-2}	2.8090×10^{-2}	3.5051×10^{-2}	4.1172×10^{-2}
H	8.6663×10^{-2}	7.9714×10^{-2}	6.7243×10^{-2}	5.6373×10^{-2}	4.6813×10^{-2}
O	7.6487×10^{-1}	7.2962×10^{-1}	6.6636×10^{-1}	6.1121×10^{-1}	5.6272×10^{-1}
Nuclide	Atom weight fraction ; Enrichment =19.75 w%				
U-235	2.6229×10^{-2}	3.3685×10^{-2}	4.7064×10^{-2}	5.8726×10^{-2}	6.8981×10^{-2}
U-238	1.0658×10^{-1}	1.3687×10^{-1}	1.9123×10^{-1}	2.3862×10^{-1}	2.8029×10^{-1}
N	1.5664×10^{-2}	2.0116×10^{-2}	2.8106×10^{-2}	3.5070×10^{-2}	4.1194×10^{-2}
H	8.6658×10^{-2}	7.9707×10^{-2}	6.7235×10^{-2}	5.6363×10^{-2}	4.6802×10^{-2}
O	7.6487×10^{-1}	7.2962×10^{-1}	6.6636×10^{-1}	6.1122×10^{-1}	5.6273×10^{-1}

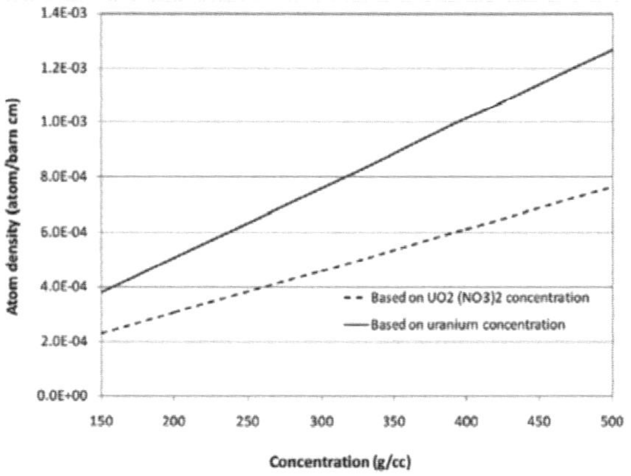

Figure 1. Comparison atom density of uranium in the solution as a function of $UO_2(NO_3)_2$ and uranium concentration.

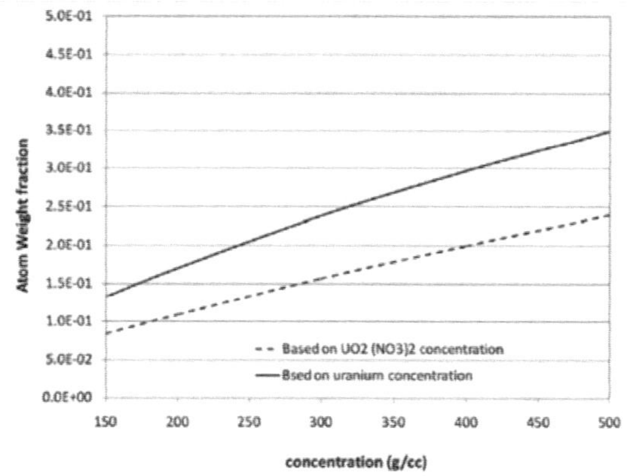

Figure 2. Comparison atom weight fraction of uranium in the solution as a function ofUO_2 $(NO_3)_2$ and uranium concentration.

In a nuclear reactor design calculation, some variations of ^{235}U enrichment are commonly needed. This will affect the isotopic compositions of the main uranium isotopes i.e. ^{235}U and ^{238}U for the respective enrichment. From the neutronic calculations point of view, for a system containing homogeneous uranium solution the results prove that in general, there is a range of uranium concentration giving the smallest of critical dimension, which is as the result of two opposing factors, i.e., the increase of reaction rate on one side, and decreasing the $H/^{235}U$ atomic ratio on the other side, as the uranium concentration increase [7,11]. Such phenomena is shown in Figure 3, where the density of hydrogen in the solution is slightly decreasing with the increasing of UO_2 $(NO_3)_2$ and uranium concentration.

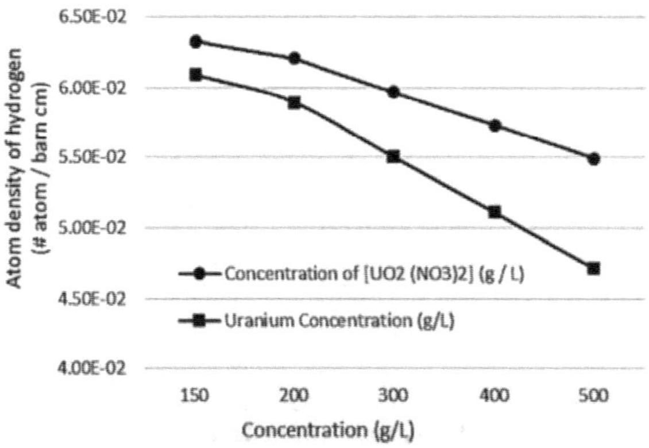

Figure 3. Comparison atom density of hydrogen in the solution as a function of UO_2 $(NO_3)_2$ and uranium concentration

5. Conclusion

A basic theory of an analytical method to determine the atomic or number density or solution density along with the associated atom and weight fraction of each nuclide contained in a homogeneous uranyl nitrate solution has been described. It includes some procedures either the solution concentration defined based on the amount of uranyl nitrate salt [$UO_2(NO_3)_2$] dissolved, or that was made based on the uranium content in the solution. The results conclude that the number density and the corresponding weight fraction of uranium isotopes of the solution defined based on the uranium content, tend to be higher by quite significantly compared with those defined based on the amount of uranyl nitrate salt [$UO_2(NO_3)_2$] dissolved. The difference is caused by the change of water content in both cases. This method can be used as the basic approach for neutronic calculations of a system containing homogeneous uranium solution.

Acknowledgements

This work was part of design assessment activities, related to the neutronic aspect of the SAMOP (Subcritical Assembly for Mo^{99} Production) system, INSINAS Project No.: RT-2016-0151 titled "Production technology development and application of ^{99}Tc for medical diagnoses", funded by Ministry of Research, Technology& Higher Education (Ristekdikti). The author would like to thank to the director of PSTA-BATAN Dr. Susilo Widodo and his staffs for supporting the implementation of the project.

References

[1] ASTM C1474-00. 2011 *Standard Test Method for Analysis of Isotopic Composition of Uranium in Nuclear-Grade Fuel Material by Quadrupole Inductively Coupled Plasma-Mass Spectrometry* STANDARD by ASTM International

[2] Zsolt Varga, Adrian Nicholl, Maria Wallenius, Klaus Mayer. 2012 *Development and validation of a methodology for uranium radiochronometry reference material preparation*, Analytica Chimica Acta 718 25– 31.

[3] Yoko Saito-Kokubu, Daisuke Suzuki, Chi-Gyu Lee, Jun Inagawa, Masaaki Magara, Takaumi Kimura. 2012 *Application of a Continuous Heating Method Using Thermal Ionization Mass Spectrometry to Measure Isotope Ratios of Plutonium and Uranium in Trace Amounts of Uranium–Plutonium Mixture Sample*, International Journal of Mass Spectrometry, 310 52– 56S.

[4] Burger, et al. 2012 *Uranium and plutonium analysis of nuclear material samples by multi-collector thermal ionisation mass spectrometry: Quality control, measurement uncertainty, and metrological traceability*, International Journal of Mass Spectrometry 311 40 – 50

[5] K. Nagy, J.L. Kloosterman, D. Lathouwers, T.H.J.J. van der Hagen, Parametric studies on the fuel salt composition in thermal molten salt breeder reactors. 2008 *International Conference on the Physics of Reactors "Nuclear Power: A Sustainable Resource"*, Casino-Kursaal Conference Center, Interlaken, Switzerland.

[6] C.E.Okon, and T. Abram. 2014 *Comparative Modelling of Molten Salt Reactor (MSR) Passive Cooled Drain Tanks,* American International Journal of Research in Science, Technology, Engineering & Mathematics AIJRSTEM 14-312

[7] W.F.G. van Rooijen∗, Y. Shimazu, N. Yamano. 2015 *Criticality uncertainty dependence on nuclear data library in Fast Molten Salt Reactors*, The Fourth International Symposium on Innovative Nuclear Energy Systems INES-4, Energy Procedia 71 3 – 13

[8] G.R. Piefer, K.M. Pitas, E.N. Van Abel. 2011 *Mo-99 production using a subcritical assembly*, in *Proceedings of the 1st Annual Mo-99 Topical Meeting*, Santa Fe, NM, USA.

[9] S. Rijnsdorp BSc. 2014 *Design of a small Aqueous Homogeneous Reactor for production of ^{99}Mo. Improving the reliability of the supply*, Thesis Master of Science in Applied Physics, at Delft University of Technology.

[10] Amanda J. Youker, Sergey D. Chemerisov, Michael Kalensky, Peter Tkac, Delbert L. Bowers, and George F. Vandegrift, 2013 *A Solution-based approach for Mo-99 production: considerations for nitrate versus sulfate media*, Hindawi Publishing Corporation, Science and Technology of Nuclear Installations, Vol 2013, Article ID 402570.

[11] Tegas Sutondo. 2007 *Affect of reflectors on criticality parameters of homogeneous system of uranyl nitrate solution*, Proceedings of International Conference on Advances in Nuclear Science and Engineering in Conjunction with LKSTN 87-92.
[12] Tegas Sutondo, Syarip, Slamet Santoso. 2009 *Safety design limits of main components of the proposed SAMOP system*, Proceedings of the 3rd Asian Physics Symposium (APS 2009), Bandung, Indonesia.
[13] Roberts, D.E. and Modise, T.S. 2007 *Laser removal of loose uranium compound contamination from metal surfaces.* Applied Surface Science 253, 5258–5267.
[14] Uranyl Nitrate, $UO_2(NO_3)_2$. 2012 http://uranium.atomistry.com/uranyl_nitrate.html
[15] IAEA. 1989 *In Situ Leaching of Uranium: Technical, Environmental and Economic Aspects*, IAEA-TECDOC-492, 172 p.
[16] Tegas Sutondo, 2008 *Metoda semi empiris untuk menentukan komposisi isotop uranium.*", Jurnal Ipetek Nuklir GANENDRA, 8 2 45 – 51
[17] Seung Gy Ro, Duck Kee Min, Jung-Kyoon Chon. 1982 *Atom number densities for uranil nitrate solution*, Journal of Korean Nuclear Society, 14 8 103-109
[18] Satoshi SAKURAI, Masakatsu MIYAUCHI and Shoichi TACHIMORI. 1987 *Density equations of $U0_2(NO_3)_2$-$HN0_3$ and $U0_2(N0_3)_2$-$Gd(NO_3)_3$-HNO_3 solutions*, Journal of Nuclear Science and Technology, 24 5 415-419.

Driver Drowsiness Detection Based on Face Feature and PERCLOS

Suhandi Junaedi and Habibullah Akbar

Master in Computer Science, Bina Nusantara University
Jl. Kebon Jeruk Raya No. 27 Jakarta, Indonesia

Email: habibullah.akbar@binus.ac.id

Abstract. Driving vehicles are complex and require undivided attention to prevent road accidents. Fatigue and distraction are a major risk factor that causes traffic accidents, severe injuries, and a high risk of death. Some progress has been made for driver drowsiness detection using a contact-based method that utilizes vehicle parts (such as steering angle and pressure on the pedal) and physiological signals (electrocardiogram and electromyogram). However, a contactless system is more potential for real-world conditions. In this study, we propose a computer vision based method to detect driver's drowsiness from a video taken by a camera. The method attempts to recognize the face and then detecting the eye in every frame. From the detected eye, iris regions for left and right eyes are used to calculate the PERCLOS measure (the percentage of total time that eye is closed). The proposed method was evaluated based on public YawDD video dataset. The results found that PERCLOS value when the driver is alert is lower than when the driver is drowsy.

1. Introduction

Driving activities require full attention and a large amount of brainpower. Moreover, driving requiring proper coordination of movement of the hands and feet, shifting gears while watching other vehicles. Lack of attention can be fatal for drivers, pedestrian, and economic activities. Therefore, it is important for a country to have solid rules and regulations to maintain safety in land transportation. Besides, the complexity and road characteristics can influence driver's attention. For example, in Indonesia, the current road infrastructure is far from optimal in a sense the number of vehicles produced and sales is not proportional with regards to road infrastructure [1]. This is true especially for big cities such as Jakarta, Jambi, and Surabaya. It is estimated that in 2020, traffic fatality in Indonesia may reach 40,000 per year [1].

Let the government remains responsible to control the vehicle growth and poor infrastructure. Besides these road characteristics, drivers violating traffic law (extreme speeding, drinking, and traffic light offense) also held largely responsible for vehicles accidents. Even with sufficient road infrastructure and excellent driving personalities, accidents are still unavoidable.

According to [2], there is a strong relationship between fatigue and safety risks in driving. Fatigue may be influenced by health and sleep-related problems. Extreme fatigue can lead to driver drowsiness which has been regarded as the culprit of road accidents and may lead to severe injuries, and a high risk of death. In this regard, drowsiness is referred as the decrement or loss of alertness that led the driver to fall asleep while driving.

Content from this work may be used under the terms of the Creative Commons Attribution 3.0 licence. Any further distribution of this work must maintain attribution to the author(s) and the title of the work, journal citation and DOI.
Published under licence by IOP Publishing Ltd

Naturally, fatigue can reduce attention, lessen the information received, and decrease the ability to provide a response or the ability to assess a condition. This situation may lead to inaccurate decision making while driving. The worst-case scenario can happen when the driver is drowsy and thus fall asleep, losing control of the vehicle. To prevent the occurrence of this scenario, we need to monitor driving activities by detecting whether the driver is drowsy or not. Such system could be used to generate some warning alarms to alert the driver or send a signal to family relatives or friends. Indeed, detecting driver's drowsiness is challenging due to the variety of ambient illumination and condition of the driver (sleep apnea, talking, yawning, wearing attributes like glasses).

This study aims to develop a drowsiness detection method based on a video based on face feature. This requires the method to recognize the presence of a face and subsequently the eye of the driver for each frame. The detected eye is used to measure the state of the eye closed. By using a predetermined threshold, the drowsiness can be recognized by measuring the percentage of total time that eye is closed.

2. Related Works

There are several approaches have been proposed to measure the state of drowsiness driver while driving. According to [3], these approaches can be categorized into three categories as the following:

1. Vehicle-based metrics – These metrics are measured based on vehicle signals including steering wheel angle, pressure on acceleration pedal, and lane position signals.
2. Physiological metrics – The body signal of the driver can be used to detect drowsiness. These metrics include an electrocardiogram, electromyogram, electrooculogram, and electroencephalogram signals.
3. Behavioral metrics – By attaching a camera within the vehicle, the driver's behavior can be monitored. The behavior metrics include eye closure, eye blinking, head pose, and yawning.

The vehicle-based and physiological metrics are contact-based method whereas sensors need to be attached to vehicle or driver's body. This is not an easy task to be done in real-world setting and not maybe uncomfortable for the driver itself. Therefore, a noncontact-based method like behavioral metrics is more flexible and suitable for the real-world environment. Most behavioral metrics is measured based on facial feature analysis. Furthermore, they can be divided into three kinds: appearance-based, template-based, and feature-based methods. Appearance-based methods require a huge amount of data for training classifiers while template-based methods require expensive computational cost due to model matching process. In contrast, feature-based methods can monitor driver drowsiness state in a more direct way. The features can be measured based on eye state estimation, mouth state estimation, and head pose analysis. Different ambient light conditions like poor lighting and driving at night can interfere drowsiness detection process. In general, any kind of object detection schemes, such as boosting [4], neural network [5], and support vector machine [6], can be used to extract facial features.

In [7], the authors utilized infrared illuminator and infrared-sensitive camera to acquire a frame of the driver. The eye was detected based on support vector machine. Then, they used Kalman filter and mean shift to track driver's pupil. To characterize driver's fatigue based on eyelid movement, they used the percentage of eye closure during a certain time interval (PERCLOS). They verified that PERCLOS is valid in quantifying fatigue. It was clear that PERCLOS measurement was significantly lower when the driver is alert than when driver is fatigued. Besides eye state, they also used gaze, head movement, and facial expression. However, infrared-based illuminator requires additional setting in addition to camera and thus limiting the potential of the method. Less complex instrument was used by [8]. They simply used common USB camera without any illuminator. Instead of tracking pupil, they locate face using Viola-Jones method. Then, eye was located and tracked using projection technique and unscented Kalman filter respectively. They also used PERCLOS to detect driver's fatigue. Another eye detection that combines Viola-Jones method, template matching, and support vector machine was proposed by [9]. To detect eye state condition, they used features such as PCA-LDA, sparseness, and kurtosis. A More direct method to detect drowsiness detection was proposed by [10]. The authors used Viola-Jones method to detect both face and eye. From the detected eye, iris contour is extracted using OpenCV function. To detect drowsiness, the also used PERCLOS metric. In summary, PERCLOS has been used as the best potential measure of drowsiness compared to other ocular metrics. However, most of these methods used their own dataset and thus it is difficult to compare their accuracy only from their reports.

Beside eye state condition, [11] used mouth state condition. The authors used a normal video camera in daylight driving without requiring additional light sources. They extract the face from video's frame based on support vector machine method. Then, driver's mouth was detected based on circular Hough transform. Instead of using PERCLOS, the driver's fatigue is detected based on a high yawning frequency. In [12], a region with the maximum area within the segmented mouth region was used to classify whether a frame as yawn frame or otherwise. The Viola-Jones method was also used for mouth detection by [13]. They used projection theory to measure the rate and number of mouth changes to detect yawning. In contrast to ocular-based methods ([7], [8], [9], [10]) that using their own dataset, the methods of mouth state condition ([12], [13]) has been evaluated using public dataset i.e. YawDD dataset [14].

3. Methods

This section provides the overall stages in our drowsiness detection method as shown in Figure 1. It begins with the video input, then face detection, eye detection, iris region extraction, morphological operations, and finally PERCLOS estimation.

Input image – the input data is a video of drivers with various conditions. More detail will be explained in data collection subsection. Each frame in the video is processed one-by-one as the following processes.

Histogram equalization – since the input videos were taken from real setting. The illumination condition is changing due to the outdoor condition. Therefore, the frame within the video needs a preprocessing method for image enhancement. Histogram equalization method was used to adjust image intensities by redistributing the intensities distribution towards histogram boundaries. Each input pixel is transformed to new intensity value based on the cumulative histogram and scale factor.

Face and eye detection - to perform fast and efficient computation, we employed Viola-Jones method for both face detection and eye detection. This method used integral image representation to compute Haar-like features rapidly at many scales. The integral image at any point *(x,y)* contains the sum of all pixels above and to the left of *(x,y)* as follows:

$$ii(x,y) = \sum_{x'<x,y'<y} i(x',y')$$

where $ii(x,y)$ is the integral image and $i(x',y')$ is the original image. The rapid computation is achieved by constructing a cascade of classifiers which trained using Adaptive Boosting (AdaBoost) algorithm as shown in Figure 2.

Iris region extraction – from the eye image, circular Hough transform was used to isolate the iris that is darker than the other areas of the eye that has a circle shape. If Hough algorithm found iris, then edge detection method (Otsu method, Sobel, and Canny) was used to detect the boundaries of iris object. We found that Sobel is better than other edge detection methods.

Morphological filters – edges output from iris extraction contain line and curves. To measure the area of the iris, there are several morphological operations employed. First, dilation operator was used to adding more pixels to the boundaries of the iris object. This operator uses a structuring element to connect pixels of iris edges. Then, we employed flood-fill operation to fill existing holes to reconstruct a better iris shape. Since the dilation operation tends to introduce more pixel boundaries than its original shape, then we employed erode operator. This operator is the inverse of dilation operation that reduces the size of the iris area.

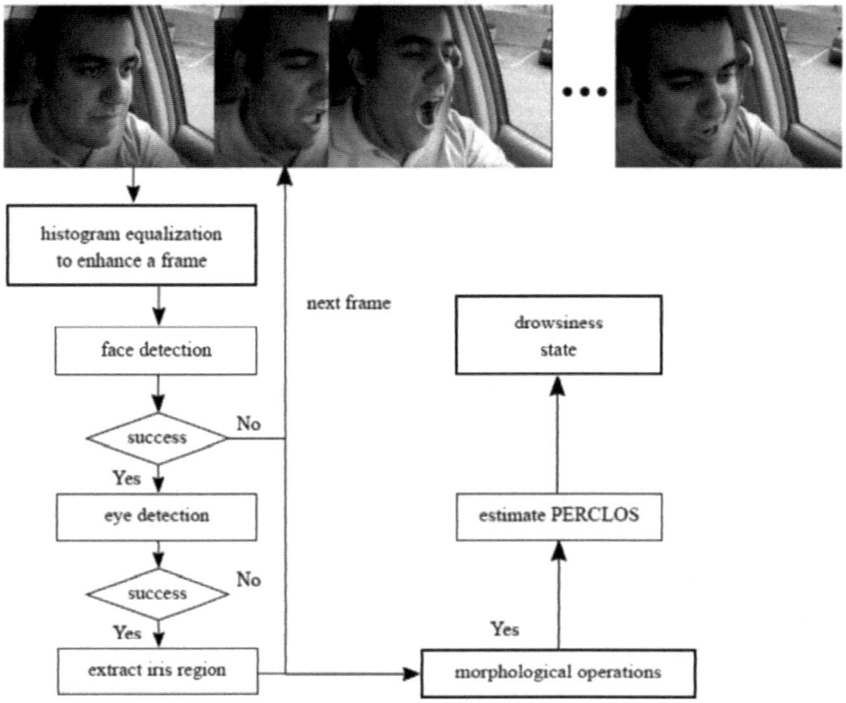

Figure 1. The proposed drowsiness detection method

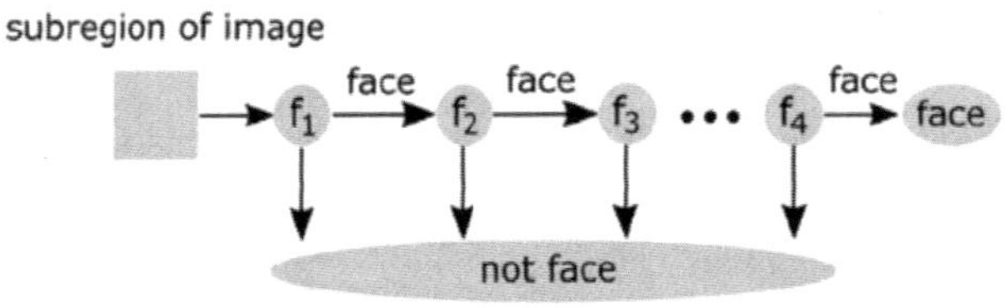

Figure 2. Face (and eye) detection using Viola-Jones method

Iris area estimation – before estimating the iris area, we ensure there is no noise or small blob existed within the iris region using Hampel identifier. Then, the area of the iris is estimated as the total number of pixels within the iris region.

PERCLOS measurement – PERCLOS is a percentage of the total frame that eye is closed during a certain time interval. The eye is defined as closed based on predetermined PERCLOS values. The PERCLOS measure is calculated based on the area estimation of iris. We used left iris area, right iris area, and the total of both. Three thresholds values (60%, 70%, and 80%) were tested in this study and they are referred as P60, P70, and P80. P60 means the percentage of total time that eye is closed at least 60%. Similar meaning also true for P70 and P80. The PERCLOS is calculated as follows:

$$PERCLOS\,(\%) = \frac{the\ sum\ of\ frames\ when\ eye\ is\ closed}{interval\ of\ frames}\ x\ 100\ \%$$

Dataset – The proposed approach was tested using Yawning Detection Dataset (YawDD) [14]. There are two datasets; the first dataset contains 322 videos and the second dataset includes 29 videos taken from real and varying illumination conditions. It consists both male and female drivers, with and without talking, and from different ethnicities. We tested our approach using selected 50 videos from the first dataset. This dataset was collected from 30 males and 20 females videos with different conditions as given in Table 1. In total, there are five video categories including male-normal, male-yawning, male-talking, female-normal and female-yawning and each category contain 10 videos.

Table 1. There are 50 videos used in this study

male	normal	10
	yawning	10
	talking	10
female	normal	10
	yawning	10

The format of input video (YawDD dataset) is 640x480 pixels and 24-bit true color (RGB) at 30 frames per second. The experiment was done on a CPU Intel Core i3-2330M, 2.20GHz with 2.0GB RAM using Matlab R2017a.

4. Experimental Results

Figure 3 shows the results of the proposed method for a frame within a video from a male driver. If the method detects a face then subsequent process continue to detect eye, iris detection using Hough transform, iris segmentation, until the measurement of the area in terms of pixel number.

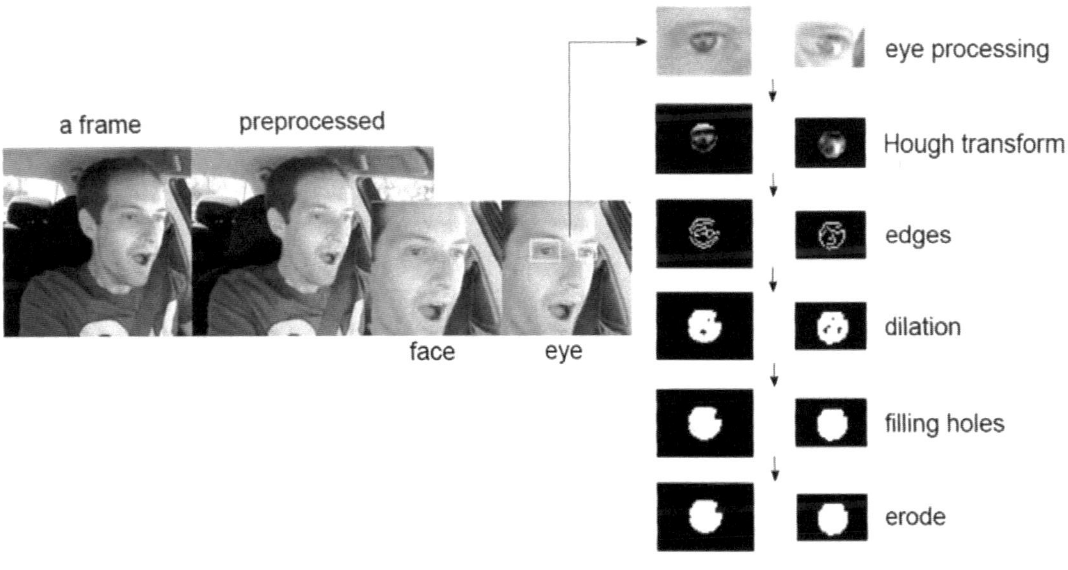

Figure 3. The results of the proposed method as explained in the method section

The illustration above happens when the face is detected. It shows each area of both iris side (from the left and right eye) was extracted. Figure 4 shows the results of the sum of both iris areas for the same subject. The left figure characterizes the condition when the driver is alert while the right figure characterizes when the driver is yawning or drowsy. It shows the difference between eye blinking and the eye being closed. Eye blinking happens within short interval while closed eye happens for a longer duration. However, iris area could appear low but not because of the drowsy condition. If the head is tilted to the side, the Viola-Jones method may not be able to detect the face as it works perfectly when the head is upright.

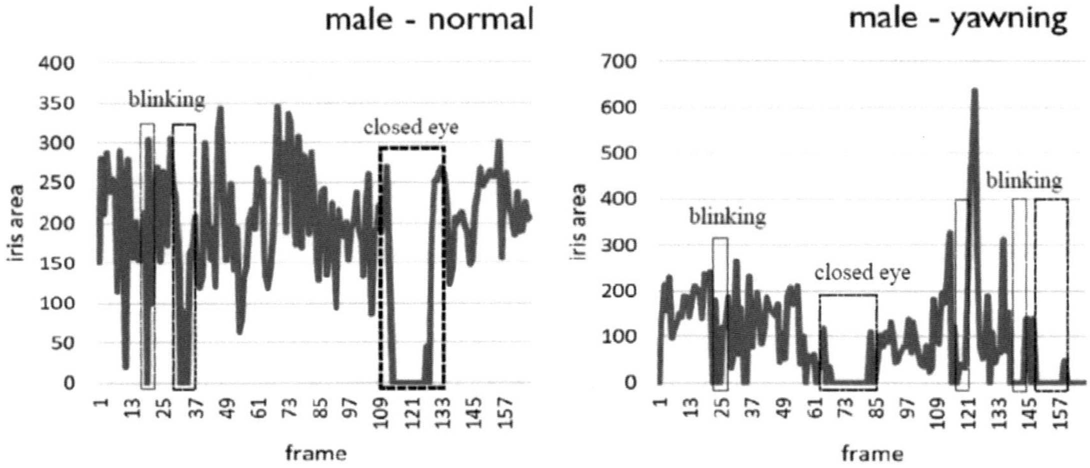

Figure 4. The iris area comparison when the driver is normal (left) and yawning (right) over time

This indicates that iris area cannot be used to characterize driver's drowsiness. However, PERCLOS was found to be the better potential measure of drowsiness. In this study, we used three values of PERCLOS threshold i.e. P60, P70, and P80. Table 2 shows the results of these PERCLOS thresholds for left iris, right iris, and a pair of both iris. It is evident that the pair works better for on all video categories because the PERCLOS values of the normal drivers are lower than the yawning drivers. However, PERCLOS values tend to go higher in talking drivers. This is because drivers tend to look sideways while talking. As a result, our method was not able to detect the face and thus no iris area can be measured (high PERCLOS value). Therefore, it is very important to know whether the driver is distracted when the face was not detected in the frame. Detecting whether driver drowsy or distracted would improve the reliability of drowsiness detection.

Table 2. Result detection of PERCLOS P60, P70, P80

driver		Left Iris			Right Iris			Pair		
		P60	P70	P80	P60	P70	P80	P60	P70	P80
male	normal	77.27	64.55	53.45	85.35	73.4	58.46	70.22	58.78	46.2
	yawning	77.41	70.29	63.15	84.8	79.89	73.83	73.18	63.44	55.02
	talking	89.18	80.12	65.03	89.02	86.07	79.14	83.66	73.74	60.12
female	normal	81.08	71.83	64.88	84.22	72.6	61.28	70.26	58.45	44.22
	yawning	82.97	73.73	61.02	82.94	75.23	61.83	74.41	61.25	46.49

5. Conclusion

This study has presented a drowsiness detection method based on face and eye detection within a video input. PERCLOS was calculated based on the area of iris region. The experimental results suggest that the proposed method potentially can detect drowsiness based on PERCLOS as it was found that PERCLOS value is lower when the driver is drowsy (compared to PERCLOS value when the driver is alert). Better results were found when both iris region (from left and right eyes) was used to measure

PERCLOS values. The limitation of the proposed method is when the face is unrecognized. This is happens when drivers are talking or distracted and thus having tendency to rotate their face sideways. In future, a distraction detection algorithm to detect tilted face is highly demanded.

References

[1] Jusuf A., Nurprasetio I P and Prihutama A 2017 Macro Data Analysis of Traffic Accidents in Indonesia *Journal of Engineering and Technological Sciences* **49** pp 132-143

[2] Williamson A., Lombardi D A, Folkard S., Stutts J., Courtney T K and Connor J L 2011 The link between fatigue and safety *Accident Analysis & Prevention* **43** pp 498-515

[3] Sahayadhas A., Sundaraj K and Murugappan M 2012 Detecting driver drowsiness based on sensors: a review *Sensors* **12** pp 16937-16953

[4] Viola P and Jones M 2001 Rapid object detection using a boosted cascade of simple features *Proc. Int. Conf. in Computer Vision and Pattern Recognition* (IEEE) pp I-I

[5] Akbar H, Suryana N and Sahib S 2011 Training neural networks using Clonal Selection Algorithm and Particle Swarm Optimization: A comparisons for 3D object recognition *Proc. Int. Conf. in Hybrid Intelligent Systems* (IEEE) pp 692-697

[6] Osuna E, Freund R and Girosit F 1997 Training support vector machines: an application to face detection *Proc. Int. Conf. in Computer vision and pattern recognition* (IEEE) pp 130-136

[7] Ji Q, Zhu Z and Lan P 2004 Real-time nonintrusive monitoring and prediction of driver fatigue. *IEEE transactions on vehicular technology* **53** pp 1052-1068

[8] Zhang Z and Zhang J 2010 A new real-time eye tracking based on nonlinear unscented Kalman filter for monitoring driver fatigue *Journal of Control Theory and Application* **8** pp 181-188

[9] Jo J, Lee S J, Jung H G, Park K R and Kim J 2011 Vision-based method for detecting driver drowsiness and distraction in driver monitoring system *Optical Engineering* **50** pp 127202-127202

[10] Nguyen T P, Chew M T and Demidenko S 2015 Eye tracking system to detect driver drowsiness *Proc. Int. Conf. in Automation, Robotics and Applications* (IEEE) pp 472-477

[11] Alioua N, Amine A and Rziza M 2014 Driver's fatigue detection based on yawning extraction *International journal of vehicular technology*.

[12] Reddy V K and Swathi K S 2017 Investigation of Effectiveness of Simple Thresholding for Accurate Yawn Detection *In Advances in Computer and Computational Sciences* (Springer) pp 81-89

[13] Omidyeganeh M, Shirmohammadi S, Abtahi S, Khurshid A, Farhan M, Scharcanski J, Hariri B, Laroche D and Martel L 2016 Yawning detection using embedded smart cameras *IEEE Transactions on Instrumentation and Measurement* **65** pp 570-582

[14] Abtahi S, Omidyeganeh M, Shirmohammadi S and Hariri B 2014, March. YawDD: A yawning detection dataset *Proc. Int. Conf. of 5th ACM Multimedia Systems* (ACM) pp 24-28

IOP Conf. Series: Journal of Physics: Conf. Series **1090** (2018) 012038 doi:10.1088/1742-6596/1090/1/012038

Dynamics, Circuit Design and Fractional-Order Form of a Modified Rucklidge Chaotic System

A Sambas[1,2], S Vaidyanathan[3], M Mamat[2], M Sanjaya WS[4], S H Yuningsih[4], K Zakaria[4] and Subiyanto[5]

[1]Department of Mechanical Engineering, Universitas Muhammadiyah Tasikmalaya, Indonesia (Corresponding author)
[2]Faculty of Informatics and Computing, Universiti Sultan Zainal Abidin, Kuala Terengganu, Malaysia
[3]Research and Development Centre, Vel Tech University, Chennai, Tamil Nadu, India
[4]Department of Physics, Universitas Islam Negeri Sunan Gunung Djati Bandung, Indonesia
[5]Department of Marine Science, Faculty of Fisheries and Marine Science, Padjadjaran University, Indonesia

Corresponding author's e-mail address: acengs@umtas.ac.id

Abstract. Rucklidge chaotic system is a nonlinear mechanical model of a double convection process. In this paper, we modify the dynamics of a Rucklidge chaotic system by adding a nonlinear term and derive a new chaotic system. The nonlinear dynamics of the proposed chaotic system is described through numerical simulations which include the stability analysis of equilibrium points, phase portraits, Lyapunov exponents, Kaplan-Yorke dimension, bifurcation diagram and a Poincarè map. For specific values of the parameters, the proposed system displays periodic and chaotic behaviour. In addition, a new circuit implementation of the modified Rucklidge chaotic system is reported and examined in MultiSIM. A good qualitative agreement is shown between the simulations and the MultiSIM results. Furthermore, the fractional-order form of the modified Rucklidge chaotic system is numerically studied. By tuning the commensurate fractional order, the new chaotic system displays chaotic and periodic attractors, respectively.

1. Introduction

Since Lorenz found the 3-D model for atmospheric convection in 1963 [1], chaos has been intensively studied and many chaotic systems have been discovered in the literature. In [2], Rössler constructed a seven-term chaotic system with just one quadratic nonlinearity. In [3-4], Rikitake discovered a new chaotic system to explain the irregular polarity switching of the earth's geomagnetic field. A model to explain the irregular variability in the luminosity of stars was presented in 1976 [5] and this system is known as the Moore-Spiegel system.

In [6], Rucklidge discovered a double convection chaotic model in a horizontal layer of Boussinesq fluid with lateral constraints. In 1994, Sprott described 19 algebraically simple chaotic systems [7]. In 2000, Malasoma proposed the simplest dissipative jerk equation that is parity invariant [8].

Content from this work may be used under the terms of the Creative Commons Attribution 3.0 licence. Any further distribution of this work must maintain attribution to the author(s) and the title of the work, journal citation and DOI.
Published under licence by IOP Publishing Ltd

In 2009, Sun and Sprott constructed a piecewise exponential jerk system [9]. In the last few years, several chaotic systems with an infinite number of equilibrium points have been proposed [10-14].

Chaos has been widely applied to many scientific disciplines such as physics [15], biology [16], ecology [17], economy [18], random bit generators [19], psychology [20], lasers [21], astronomy [22], chemical reactions [23], memristors [24], neural networks [25], robotics [26], encryption [27], secure communication systems [28-31], etc.

In this paper, we modify the dynamics of a Rucklidge chaotic system by adding a nonlinear term and derive a new chaotic system. Section 2 describes the nonlinear dynamics of the proposed chaotic system. We describe the phase portraits of the modified Rucklidge chaotic system using MATLAB and discuss the stability analysis of equilibrium points, Lyapunov exponents, Kaplan-Yorke dimension, bifurcation diagram and a Poincarè map for the new chaotic system. In Section 3, a new circuit implementation of the modified Rucklidge chaotic system is reported and examined in MultiSIM. A good qualitative agreement is shown between the simulations and the MultiSIM results. In Section 4, the fractional-order form of the modified Rucklidge chaotic system is numerically studied. By tuning the commensurate fractional order, the new chaotic system displays chaotic and periodic attractors, respectively. Section 5 contains the concluding remarks.

2. A modified Rucklidge chaotic system

Recently, Rucklidge has introduced a famous model for nonlinear double convection [6]. Rucklidge model has the following nonlinear dynamics

$$\begin{cases} x = -kx + \lambda y - yz \\ y = x \\ z = -z + y^2 \end{cases} \quad (1)$$

where x, y, z are state variables and k, λ are system parameters.

In [6], it was shown that the Rucklidge system (1) is *chaotic*, when the system parameters take the values

$$k = 2, \quad \lambda = 6.7 \quad (2)$$

In this work, we derive a new chaotic system by modifying the Rucklidge dynamics (1) as

$$\begin{cases} x = -ax - b|z| + cy - yz \\ y = x \\ z = -z + y^2 \end{cases} \quad (3)$$

where x, y, z are state variables and a, b, c are system parameters.

We shall show that the system (3) is *chaotic* when the system parameters take the values

$$a = 1.5, \quad b = 0.2, \quad c = 5 \quad (4)$$

For numerical simulations, we take the initial conditions as

$$x(0) = 0.1, \quad y(0) = 0.1, \quad z(0) = 0.1 \quad (5)$$

With the parameter values as in (4) and the initial conditions as in (5), the Lyapunov exponents of the new system (3) are calculated using Wolf algorithm [32] as

$$L_1 = 0.1559, \quad L_2 = 0, \quad L_3 = -2.6559 \quad (6)$$

This shows that the new modified Rucklidge system (3) is chaotic.

Also, the sum of the Lyapunov exponents of the system (3) is found as

$$L_1 + L_2 + L_3 = -(a+1) = -2.5 < 0 \quad (7)$$

Hence, the system (3) is a dissipative chaotic system with a strange chaotic attractor.

The phase portraits of the new modified Rucklidge system are shown in Figure 1. The Lyapunov exponents of the new modified Rucklidge system are shown in Figure 2.

The Kaplan-Yorke dimension of the modified Rucklidge system (3) is calculated as

$$D_{KY} = 2 + \frac{L_1 + L_2}{|L_3|} = 2.0587 \tag{8}$$

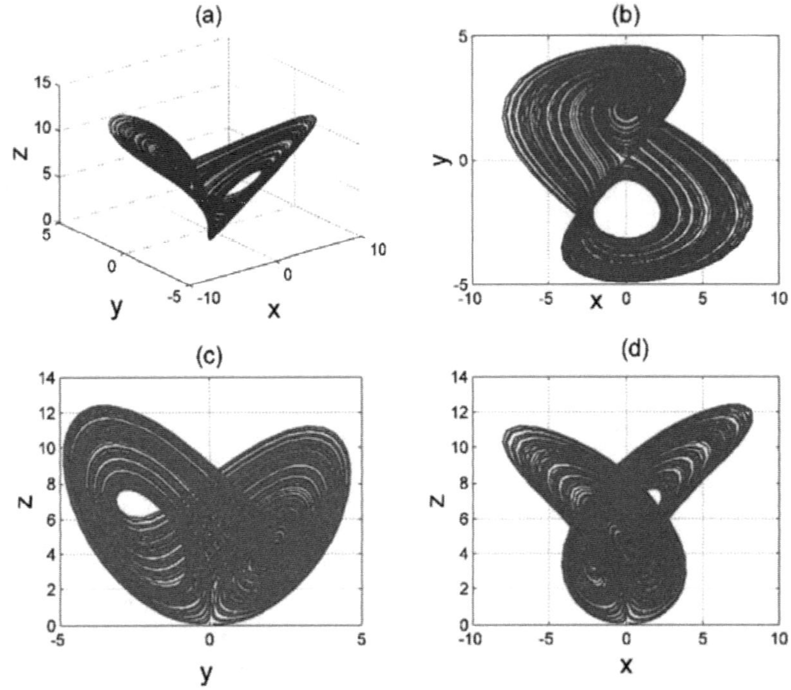

Figure 1. Phase portraits of the modified Rucklidge chaotic system

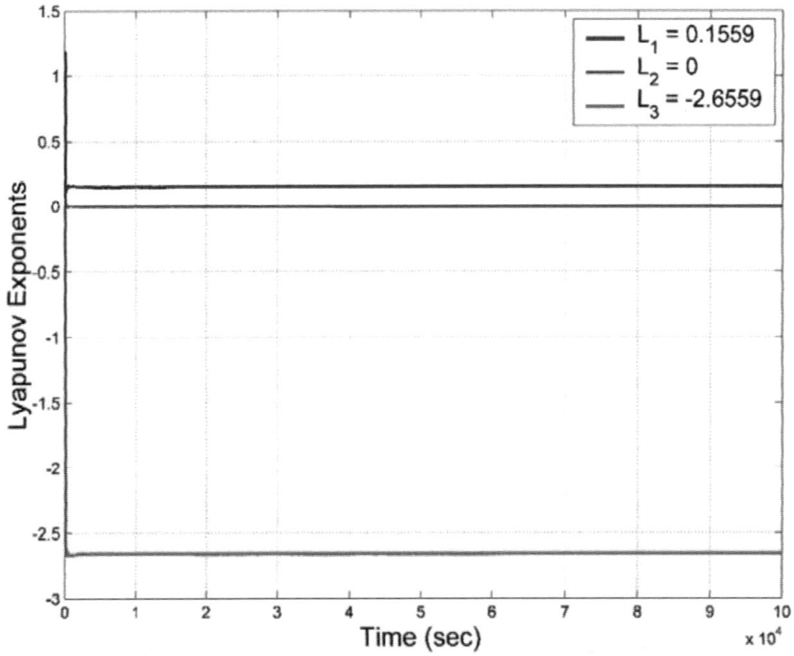

Figure 2. Lyapunov exponents of the modified Rucklidge chaotic system

The equilibrium points of the new chaotic system (3) are obtained by solving the following system

$$\begin{cases} -ax - b|z| + cy - yz = 0 \\ x = 0 \\ -z + y^2 = 0 \end{cases} \tag{9}$$

We take the parameter values as in the chaotic case (4). A simple calculation shows that the new chaotic system (3) has three equilibrium points given by

$$E_0 = \begin{bmatrix} 0 \\ 0 \\ 0 \end{bmatrix}, \quad E_1 = \begin{bmatrix} 0 \\ 2.1383 \\ 4.5723 \end{bmatrix}, \quad E_2 = \begin{bmatrix} 0 \\ -2.3383 \\ 5.4677 \end{bmatrix} \tag{10}$$

To determine the stability type of the equilibrium points of the system (3), we calculate the Jacobian matrix of the system (3) at any point $x \in R^3$:

$$J(x) = \begin{bmatrix} -1.5 & 5-z & -y-0.2\,\text{sign}(z) \\ 1 & 0 & 0 \\ 0 & 2y & -1 \end{bmatrix} \tag{11}$$

The eigenvalues of the matrix $J_0 = J(E_0)$ are determined as

$$\lambda_1 = -1, \quad \lambda_2 = -3.1085, \quad \lambda_3 = 1.6085 \tag{12}$$

This shows that E_0 is a saddle point, which is unstable.

The eigenvalues of the matrix $J_1 = J(E_1)$ are determined as

$$\lambda_1 = -3.1330, \quad \lambda_{2,3} = 0.3165 \pm 1.7191i \tag{13}$$

This shows that E_1 is a saddle focus, which is unstable.

The eigenvalues of the matrix $J_2 = J(E_2)$ are determined as

$$\lambda_1 = -3.0046, \quad \lambda_{2,3} = 0.2523 \pm 1.8494i \tag{14}$$

This shows that E_2 is a saddle focus, which is unstable.

Thus, all three equilibrium points of the system (3) are unstable.

Next, to have a detailed view of the new system (3), which is a modified Rucklidge chaotic system, the behavior of the system with respect to the bifurcation parameter b is studied. For the chosen value of $b \le 0.28$ the system (3) displays the expected chaotic behavior and for $b > 0.28$ periodic behavior is noted in the system. Figure 3 shows the bifurcation diagram for the new system (3). Figure 4 shows the Poincarè map of the new system (3). Figures 3 and 4 describe the chaotic properties of the modified Rucklidge chaotic system (3).

IOP Conf. Series: Journal of Physics: Conf. Series **1090** (2018) 012038 doi:10.1088/1742-6596/1090/1/012038

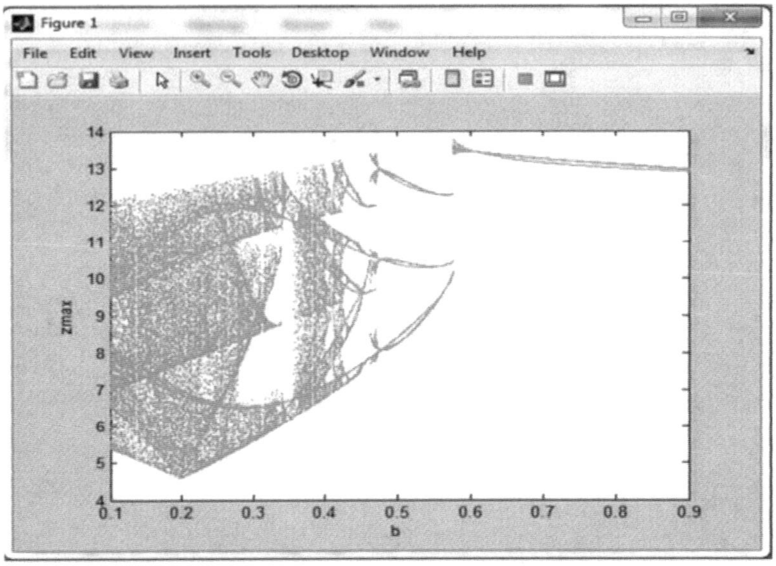

Figure 3. Bifurcation diagram of the new chaotic system (3) with b as varying parameter

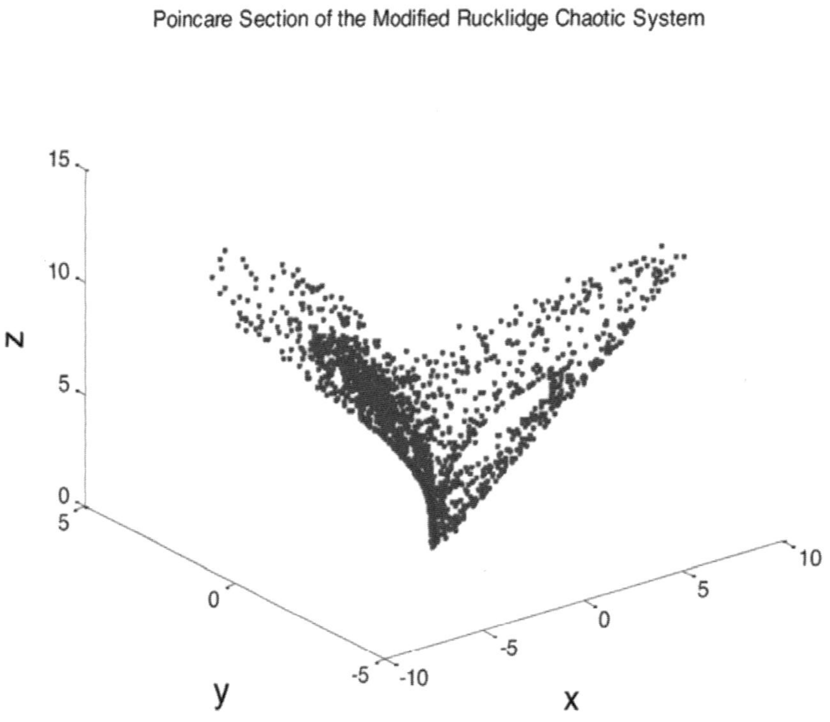

Figure 4. Poincarè map for the new chaotic system (3) when $a = 1.5, b = 0.2$ and $c = 5$

3. Circuit realization of the modified Rucklidge chaotic system

In this section, an electronic circuit which emulates the modified Rucklidge chaotic system (3) is described to show its feasibility. The circuit design in Figure 5 has been described following an approach based on operational amplifiers [33-35], where the state variables x, y, z of the new system (3) are associated with the voltages across the capacitors C_1, C_2 and C_3 respectively.

By applying Kirchhoff's circuit laws, the corresponding circuital equations of the designed circuit can be written as

$$\begin{cases} \dot{x} = -\dfrac{1}{C_1R_1}x - \dfrac{1}{C_1R_2}\,|\,z\,| + \dfrac{1}{C_1R_3}y - \dfrac{1}{10C_1R_4}yz \\ \dot{y} = \dfrac{1}{C_2R_5}x \\ \dot{z} = -\dfrac{1}{C_3R_6}z + \dfrac{1}{10C_3R_7}y^2 \end{cases} \tag{15}$$

We choose the values of the circuital elements as

$$\begin{cases} R_1 = 6.66K\Omega, R_2 = 50K\Omega, R_3 = 2K\Omega, R_4 = R_7 = 1K\Omega \\ R_5 = R_6 = R_8 = R_9 = R_{10} = R_{11} = R_{12} = R_{13} = R_{14} = R_{15} = R_{16} = 10K\Omega \\ C_1 = C_2 = C_3 = 10nF \end{cases} \tag{16}$$

The circuit has three integrators by using Op-amp TL082CD in a feedback loop and two multipliers IC AD633. In addition, two diodes (D_1, D_2) are used, which provide the signal $|\,z\,|$. The supplies of all active devices are ± 15 volt. With MultiSIM 10.0, we obtain the experimental observations of new system (3) as shown in Figures 6-8. The agreement between the experimental phase portraits with MultiSIM simulation (Figs. 6-8) and the numerical simulation with MATLAB (Fig. 1) confirms the feasibility of our new modified Rucklidge chaotic system (3).

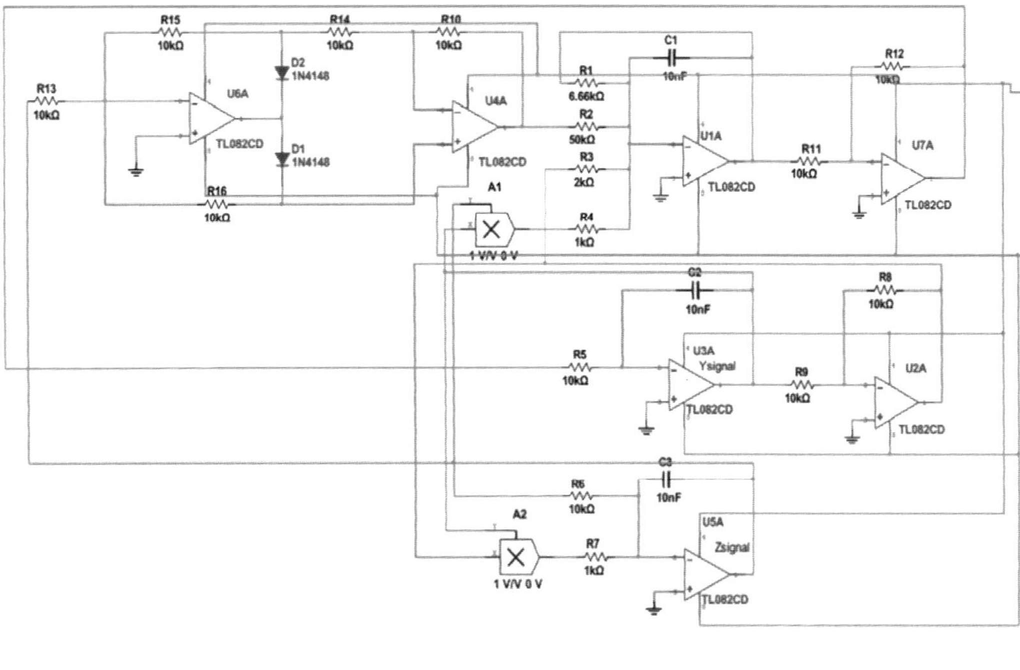

Figure 5 Schematic of the proposed new chaotic system by using MultiSIM

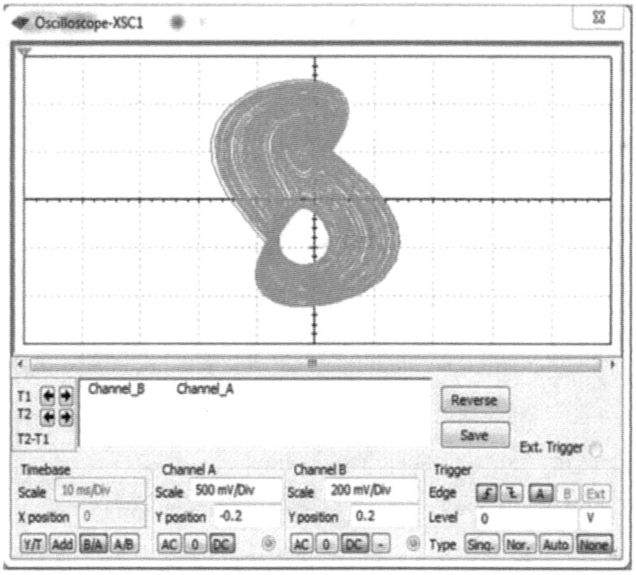

Figure 6 2-D projection of the new chaotic system on the (x, y) plane

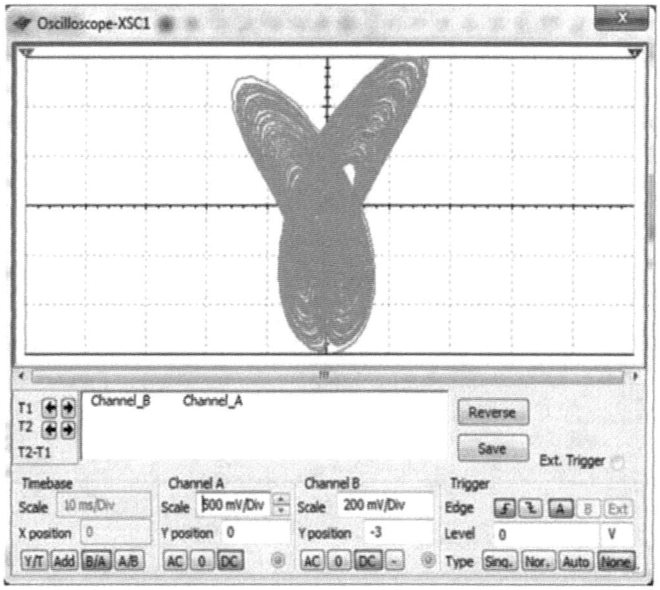

Figure 7: 2-D projection of the new chaotic system on the (x, z) plane

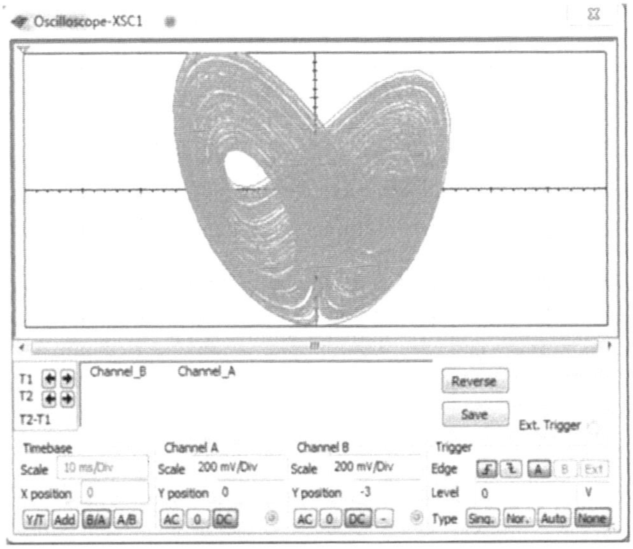

Figure 8 2-D projection of the new chaotic system on the (y, z) plane

4. A fractional order model of the modified Rucklidge chaotic system

In this section, we describe a fractional order model of the modified Rucklidge chaotic system (3). Faraji and Tavazoei [36] describe the results of the simulation of a fractional capacitor in the model of chaotic circuit. The current flowing through a capacitor and the voltage to which it is subjected are related by the following fractional-order equation

$$i = C_\alpha \frac{d^\alpha V}{dt^\alpha} \tag{17}$$

where $0 < \alpha < 1$ and α tends to 1 (note that $\alpha = 1$ for an ideal capacitor) [37-38]. Also, [36] gives various values of α and C_α commonly used in circuitry implementation of chaotic systems.

The mathematical description of the commensurate fractional order model of the modified Rucklidge chaotic system (3) can be expressed as follows:

$$\begin{cases} \dfrac{d^\alpha x}{dt^\alpha} = -ax - b\,|\,z\,| + cy - yz \\ \dfrac{d^\alpha y}{dt^\alpha} = x \\ \dfrac{d^\alpha z}{dt^\alpha} = -z + y^2 \end{cases} \tag{18}$$

IOP Conf. Series: Journal of Physics: Conf. Series **1090** (2018) 012038 doi:10.1088/1742-6596/1090/1/012038

We take the parameter values as in the chaotic case (4), i.e.

$$a = 1.5,\ \ b = 0.2,\ \ c = 5 \tag{19}$$

For numerical simulation of the fractional-order model (18) of the modified Rucklidge chaotic systems with parameter values as in (19), the Adams-Bashforth-Moulton predictor-corrector scheme [39-43] is used. This method is based on the Caputo definition of the fractional-order derivative given by [44] as follows:

$$\frac{d^{\alpha} X_i}{dt^{\alpha}} = \frac{1}{\Gamma(\alpha - n)} \int_0^t \frac{X_i^{(n)}(t')}{(t-t')^{\alpha-n+1}} dt' \tag{20}$$

where $n - 1 < \alpha < n$, $X_1 = x$, $X_2 = y$, $X_3 = z$ and $\Gamma(.)$ is the Gamma function. We describe the numerical simulations of fractional-order model of the Rucklidge chaotic system (18) with parameter values as in (19) for different fractional-order α ($0 < \alpha < 1$).

Figures 9-11 describe the phase portraits of the fractional-order model (18) of the new chaotic system in the planes (x, y), (x, z) and (y, z) of significant results obtained for specific values of commensurate fractional-order α. For α = 0.980, Figure 9 displays a chaotic attractor of fractional-order system (18) at a = 1.5, b =0.2 and c = 5. Also, we derive periodic attractors at α = 0.918 (see Figure 10) and α = 0.910 (see Figure 11). The periodic behavior found in Figure 10 is confirmed in Figure 11.

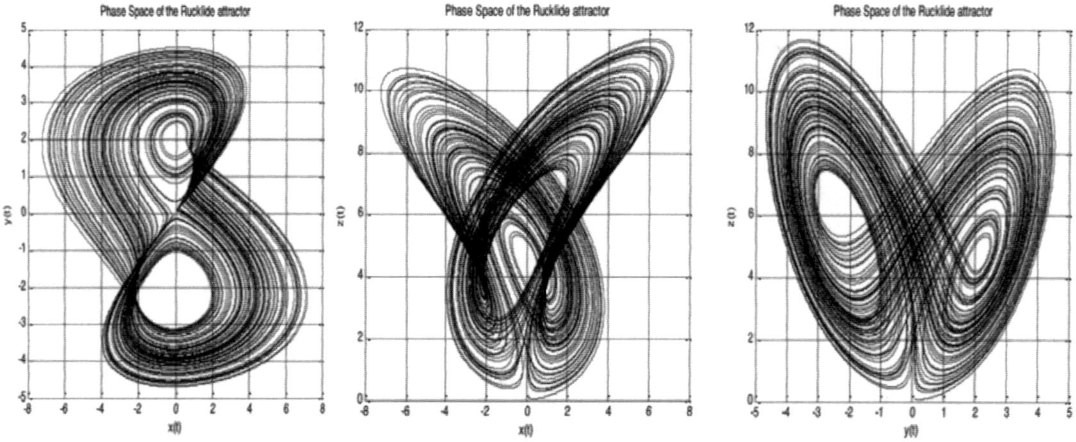

Figure 9 The phase portrait in the planes (x, y), (x, z) and (y, z) of fractional-order system (18) at a = 1.5, b = 0.2 and c = 5 for specific values of commensurate fractional order α= 0.980

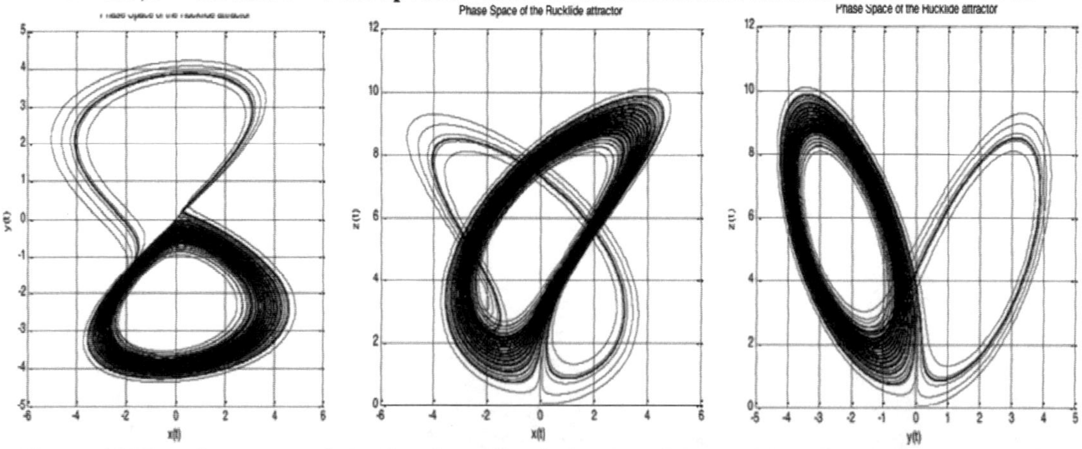

Figure 10 The phase portrait in the planes (x, y), (x, z) and (y, z) of fractional-order system (18) at a = 1.5, b = 0.2 and c = 5 for specific values of commensurate fractional order α= 0.918

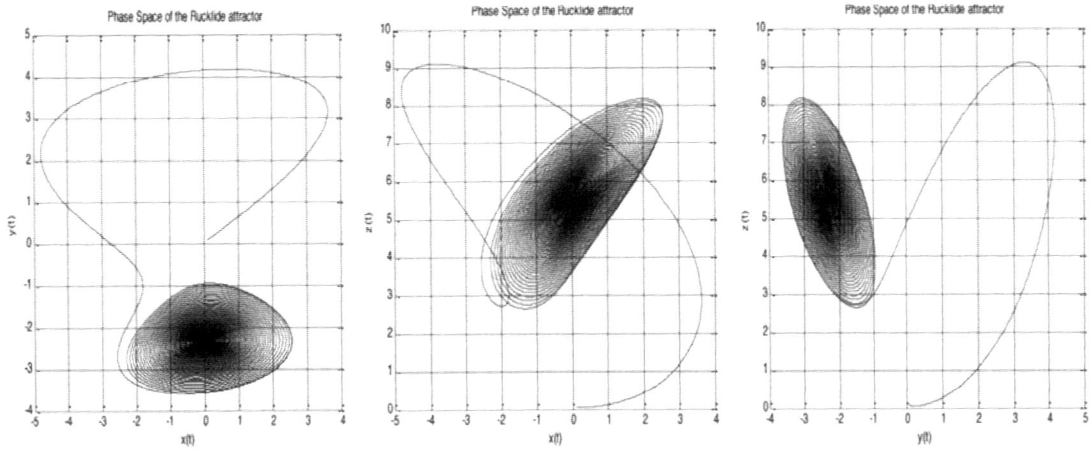

Figure 11 The phase portrait in the planes (x, y), (x, z) and (y, z) of fractional-order system (18) at $a = 1.5$, $b = 0.2$ and $c = 5$ for specific values of commensurate fractional order $\alpha = 0.910$

5. Conclusions

In this paper, we derived a new chaotic system by adding a nonlinear term to Rucklidge chaotic system model for nonlinear double convection and by changing the values of the parameters in the Rucklidge system. The dynamical behaviors of the modified Rucklidge chaotic system are analyzed, both analytically and numerically, including some basic dynamical properties, phase portraits, Lyapunov exponents, Kaplan-Yorke dimension, bifurcation diagram and Poincarè map. The designed circuit has been implemented and examined using the MultiSIM and numerical results using MATLAB. Comparison of the MultiSIM result and MATLAB simulations showed good qualitative agreement between the chaotic system and its circuitry realization. Finally, a fractional-order model of the proposed system has been investigated. As the fractional order derivative order decreases, the chaotic attractor of the modified Rucklidge chaotic system exhibits chaotic behaviour and periodic behaviour respectively.

References

[1] Lorenz E N 1963 *J. Atmospheric Sciences* **20** 130-41
[2] Rössler O 1976 *Physics Letters* A **57** 397-8
[3] Ito K 1980 *Sci. Lett.* **51** 451-6
[4] Rikitake T 1998 *Proc. Cambridge Philos. Soc.* **54** 89-105
[5] Moore D W and Spiegel E A 1986 *Astrophys. J.* **143** 871-87
[6] Rucklidge A M 1992 J. Fluid Mechanics 237 209-29
[7] Sprott J 1994 *Phys. Lett. E* **50** 647-50
[8] Malasoma J. M. 2000 *Phys. Lett. A* **264** 383-9
[9] Sun K H and Sprott J H 2009 *Internat. J. Nonlinear Sci. Num. Simulat.* **10** 1443-50
[10] Gotthans T and Petrzela J 2015 *Nonlinear Dynamics* **73** 429-36
[11] Zhou P and Yang F 2014 *Nonlinear Dynamics* **76** 473-80
[12] Pham V T, Jafari S, Volos C and Kapitaniak T 2016 *Chaos Solit. Fract.* **93** 58-63
[13] Pham V T, Volos C, Vaidyanathan S and Wang X 2016 *Advances Math Phys.* **2016** 4024836
[14] Pham V T, Jafari S, Volos C, Vaidyanathan S and Kapitaniak T 2016 *Optik* **127** 9111-7
[15] Awrejcewicz J, Supel B, Lamarque C H, Kudra G, Wasilewski G and Olejnik P 2008 *Int. J. Bifur. Chaos* **18** 2883-915
[16] Hannon B and Ruth M 2014 Modeling Dynamic Biological Systems (Berlin: Springer)
[17] Mamat M, Sanjaya W M, Salleh Z and Ahmad M F *2011 J. Sustainability Science and Management* **6** 44-50

[18] Bouali S, Buscarino A, Fortuna L, Fresca M and Gambuzza L V 2012 *Nonlinear Analysis: RWA* **13** 2459-65
[19] Akgul A, Moroz, I, Pehlivan I and Vaidyanathan S 2016 *Optik* **127** 5491-99
[20] Jafari S, Sprott J C and Golpayegani S M R H 2016 *Nonlinear Dynamics* **83** 615-22
[21] Sciamanna M and Shore, K A 2015 *Nature Photonics* **9** 151-62
[22] Daquin J, Rosengren A J, Alessi E M, Feleflie F, Valsecchi G B and Rossi A 2016 *Celestical Mechanics and Dynamical Astronomy* **124** 335-66
[23] Bringer M R, Gerdts C J, Song H, Tice J D and Ismagilov R F 2004 *Phil. Trans. Royal Society of London A* **362** 1087-104
[24] Pham V T, Vaidyanathan S, Volos C K, Jafari S, Kuznetsov N V and Hoang T M 2016 *European Physical Journal: Special Topics* **225** 127-36
[25] Lu, H 2002 *Physics Letters A* **298** 109-116
[26] Sambas A, Vaidyanathan S, Mamat M, Sanjaya W M and Rahayu D S 2016 *Studies in Computational Intelligence* **636** 283-310
[27] Volos C K, Kyprianidis I M and Stouboulos I N 2013 *J. Engineering Science and Technology Review* 6 9-14
[28] Sambas A, Sanjaya W S, Mamat M and Prastio R P 2016 *Studies in Fuzziness and Soft Computing* **337** 133-53
[29] Sambas A, Sanjaya W S and Mamat M, 2015 *J. Engineering Science and Technology Review* **8** 89-95
[30] Sambas A, Sanjaya W S and Halimatussadiyah 2012 *WSEAS Transactions on Systems* **9** 506-15
[31] Xu G, Xu J, Xiu C, Liu F and Zang Y 2017 *Neurocomputing* **227** 108-12
[32] Wolf A, Swift J B, Swinney H L and Vastano J 1985 *Physica D* **16** 285-317
[33] Li X F, Chu Y D, Zhang J G and Chang Y X 2009 *Chaos Solitons & Fractals* **41** 2360-70
[34] Li C, Sprott J C and Thio W 2014 *IEEE Trans. Circuits Sys.-II: Exp. Briefs* **61** 977-81
[35] Li C, Pehlivan I, Sprott J C and Akgul A 2015 *IEICE Electronics Express* **12** 1-12
[36] Faraji S and Tavazoei M 2013 *Open Physics* **11** 836-44
[37] Kingni S T, Pham V T, Jafari S, Kol G R and Woafo P 2016 *Circuits Systems and Signal Processing* **35** 1933-48
[38] Kingni S T, Pham V T, Jafari S and Woafo P 2017 *Chaos Solitons & Fractals* **99** 209-18
[39] Deng W 2007 *J. Comp. Applied Math.* **206** 174-88
[40] Herrmann R 2014 *Fractional Calculus: An Introduction to Physicists* (Singapore: World Scientific)
[41] Petras I, Chen Y and Vinagre, B M 2004 *Problem* **6** 208-10.
[42] Petras I 2011 *Fractional-Order Nonlinear Systems: Modelling, Analysis and Simulation* (Berlin: Springer)
[43] Caponetto R 2010 *Fractional Order Systems: Modelling and Control Applications* (Singapore: World Scientific)

IOP Conf. Series: Journal of Physics: Conf. Series **1090** (2018) 012039 doi:10.1088/1742-6596/1090/1/012039

Evaluation of downscaled near-surface wind over maritime continent based on stations measurement

Arnida L. Latifah[1], **Aryo Cokrowatianto**[1] **and Didit Adytia**[2]

[1]Research Center for Informatics, Indonesian Institute of Sciences
[2]School of Computing, Telkom University

Email: arni001@lipi.go.id

Abstract. In the maritime continent where most of the region is an ocean, wind field is the main driving force to generate ocean wave. The use of inaccurate wind fields to drive ocean model may result in significant biases. Many global reanalysis wind data are available, but the performance is subject to each region. Moreover, the resolution of global data is less representative for the maritime continent region. This paper presents the dynamically downscaled near-surface wind of NCEP/DOE re-analysis II (R2) over the maritime continent by regional climate model (RegCM4.6). We evaluate 10m wind modeled by R2 and its downscaled based on stations measurement available from Agency for Meteorology, Climatology and Geophysics (BMKG). Moreover, the results of downscaled wind are also compared with re-analysis wind from ECWMF ERA-Interim. We compare daily 10m zonal and meridional wind speed from 10 stations near the coast during 1 January 2013 until 31 December 2016. Then, we investigate the added value of the dynamical downscaling to the near-surface wind fields over the maritime continent.

1. Introduction

Ocean waves are the result of the interactions between wind and the ocean surface, in which the main energy resources to generate is so-called the wind force [1]. Moreover, the most frequently used parameters of global reanalysis products is near-surface wind (10m) [2]. The inaccuracy of wind field, which is the input of ocean wave model, may drive wrongly generated ocean wave. The use of wind data measurement is unlikely since the number of stations is limited. A number of global data provide wind field data over the globe, but the quality of the data should be evaluated. Moreover, the resolution of the global data is still relatively coarse. Study of [3] confirmed that the different resolution of the wind data gives quite differences in the generated ocean wave. Therefore, the coarse resolution of global data should be reduced to be implemented in a local region. One approach increasing the resolution of the wind data is a dynamical downscaling of the global climate data, for example the study [4] that downscale the wind over the United States.

Among the global climate data, some of the inter-comparison of climate data has been studied. Stopa and Cheung in [5] compared the wind and wave data from ECMWF Reanalysis Interim and the NCEP Climate Forecast System Reanalysis over the globe and resulted that both reanalysis data perform the spatial homogeneity with consistent levels of errors in the Northern and Southern Hemispheres. Mooney et al. in [6] compared the surface air temperature modeled by ERA-40, ERA-Interim, and NCEP/NCAR reanalysis (NNRP-1) with the observations over Ireland and the three

Content from this work may be used under the terms of the Creative Commons Attribution 3.0 licence. Any further distribution of this work must maintain attribution to the author(s) and the title of the work, journal citation and DOI.
Published under licence by IOP Publishing Ltd

reanalysis data showed good agreement with the observation and with each other. Chaudhuri et al. in [7] evaluated atmospheric fields over the Arctic from the ERA-Interim, Common Ocean-Ice Reference Experiment version 2 (CORE2), Japanese 25-yr Reanalysis Project (JRA-25), NNRP-1, NCEP Climate Forecast System Reanalysis (CFSR), and Modern-Era Retrospective Analysis for Research and Applications (MERRA) against satellite-derived and in situ observations. In their study, comparison to reference observations showed that for variables such as air temperature and humidity, all reanalysis products have similar solutions, but winds, precipitation, and radiation showed large spreads.

In this paper, we evaluate the near-surface wind of the downscaled global climate data over the Maritime Continent. We compare the downscaled wind from the global data of NCEP–DOE AMIP-II Reanalysis (R2) with the wind measurement from the meteorological stations. In order to evaluate the uncertainty of the wind field among the global reanalysis data, we also compare the global wind data from R2 and ERA-Interim over Maritime Continent. The comparison of the product of ECMWF (ERA-40) and NCEP against the observation had been done by [8]. They compared the dataset with Antarctic station observations and showed that the wind velocity fields in both datasets do not agree well with the atmospheric observations.

2. Dataset and method

2.1. Global reanalysis data

Reanalysis data is globally gridded dataset recording the changing of weather and climate over time. It incorporates observations and numerical prediction model and covers the entire globe. This paper investigates the near-surface wind, the zonal wind (u10) and meridional wind (v10) from two reanalysis data, NCEP-DOE AMIP-II Reanalysis (R2) and ERA-Interim. The R2 is an updated 6-hourly with a 2.5 x 2.5 latitude-longitude horizontal resolution global reanalysis series from 1979-present, which fixes the known processing errors in the NCEP-NCAR reanalysis (R-I) and uses an improved forecast model and data assimilation system [9]. The resolution of the data is T62 Gaussian grids, i.e 192 longitudes equally spaced and 94 latitudes unequally spaced grid points with 28 vertical sigma levels. The data assimilation system uses a 3D-variational analysis scheme and is described in more detail in [10].

The ERA-Interim is a global atmospheric reanalysis from 1979 and continuously updated in real time, produced by European Centre for Medium-Range Weather Forecasts (ECMWF) in collaboration with many institutions. The data assimilation system uses 4D-variational analysis (4D-Var) on a spectral grid with triangular truncation of 255 waves (corresponds to approximately 80 km) and a hybrid vertical coordinate system with 60 vertical levels. The ERA-Interim data used in this study is 6-hourly data on a fixed grid of 1.5° horizontal resolution.

2.2. Regional model data

The regional data is obtained from the dynamical downscaling of the global reanalysis data over the maritime continent (95°E-155°E and 11°S-8°N). This paper uses the downscaled of R2 with 0.25-degree resolution by regional climate model RegCM4.6 as proposed by [11-12]. The model is run for climatologically simulation in the period of 1980-2016, but this study only extracts 6-hourly variables u10 and v10 from the model output in the period of 1 January 2013 until 31 December 2016.

2.3. Measurement data

Measurement data from 10 meteorological stations of Agency for Meteorology, Climatology and Geophysics (BMKG) are used for the evaluation. The measurement stations in this study are chosen near the coast from various part of the maritime continent. The locations of the stations are presented in Figure 1. The measurement data are provided in a daily average and in term of the magnitude and the direction of the near-surface wind. The accuracy of the direction is quite weak since the measurement data uses the eight principal wind directions. It may result in a deviation of the wind speed up to approximately ±0.38 of the exact wind speed.

IOP Conf. Series: Journal of Physics: Conf. Series **1090** (2018) 012039 doi:10.1088/1742-6596/1090/1/012039

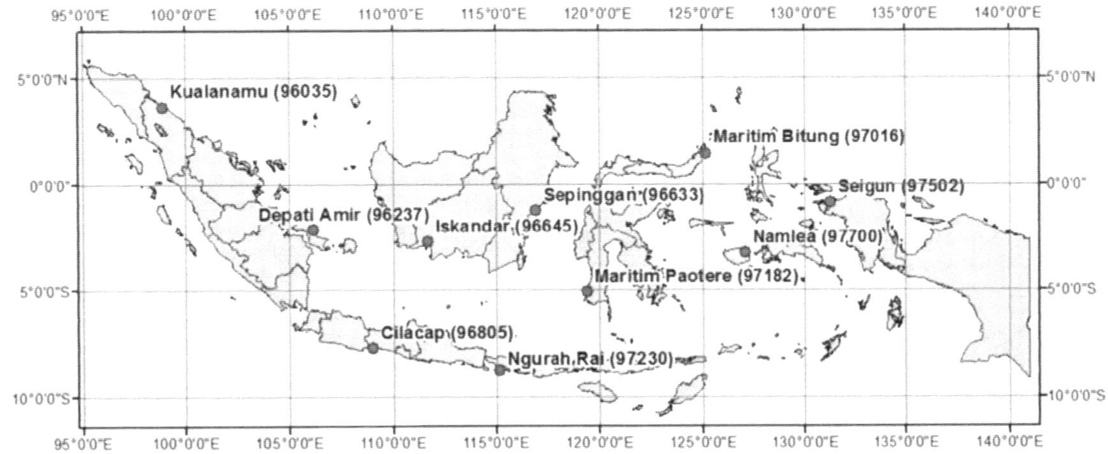

Figure 1. Map showing the location of the measurement stations used in this study.

2.4 Comparison methods

The grid size among the reanalysis datasets is different. To compare the dataset of R2 and ERA-Interim we synchronize the grid by bilinear interpolation. We compare the spatial distribution of the zonal and meridional wind in DJF and JJA season from R2 and ERA-Interim.

Further, the geographic coordinates of the measurement stations are neither equal to the grid point of the global nor regional dataset. The location of a station can be in the grid box or in the edge of the dataset. Study of [6] investigated four different approaches to compare reanalysis and observed data and resulted that there is no superior among the methods. To evaluate all the dataset we compute the correlation and the mean absolute deviation between each data with the measurement.

3. Comparison of reanalysis data

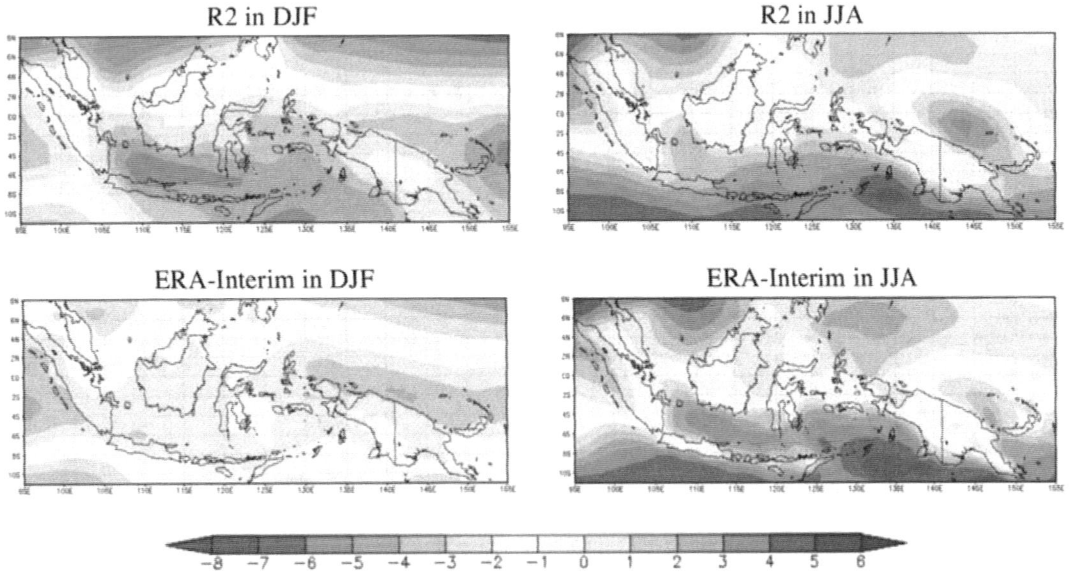

Figure 2. Zonal wind (u10) of R2 (top) and ERA-Interim (bottom) in DJF and JJA season.

Figure 2 shows that the spatial variability of the zonal wind of both R2 and ERA-Interim that have similar pattern over MC in either DJF or JJA season. However, their magnitudes of the zonal wind are slightly different. In DJF season, the R2 zonal wind are much stronger than ERA-Interim in most of the region while in JJA season, the differences are less significant. This is confirmed in Figure 4 that the trend line slope of ERA-Interim data versus R2 is mostly less than unity.

Figure 3 shows that the meridional wind of R2 is also stronger than ERA-Interim in both DJF and JJA season. The differences in the matter of the pattern and the magnitude are more significant than the zonal wind. The stronger magnitude of the meridional wind of R2 against ERA-Interim is also confirmed in Figure 4.

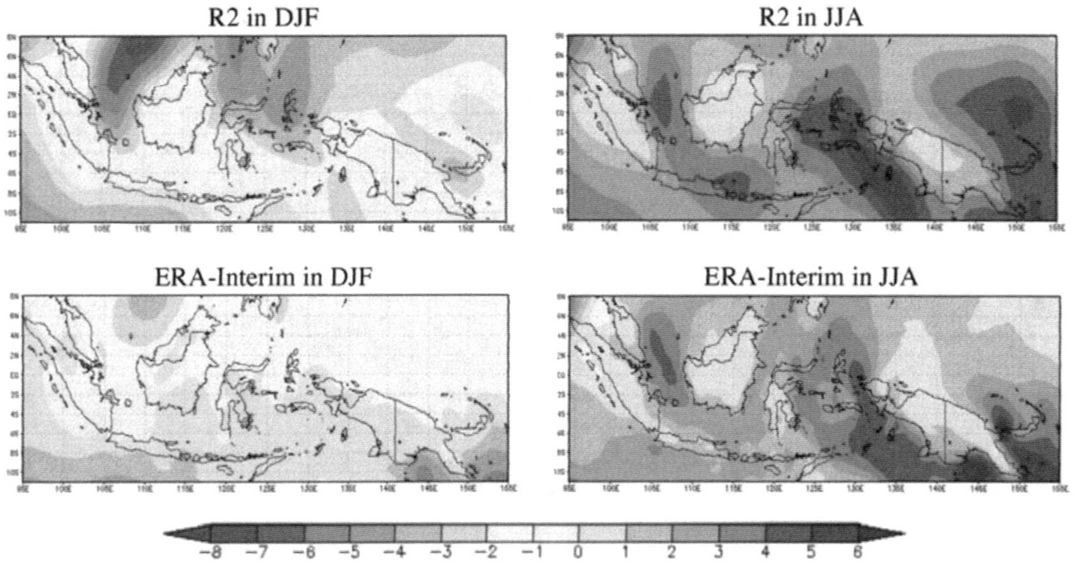

Figure 3. Meridional wind (v10) of R2 (top) and ERA-Interim (bottom) in DJF and JJA season.

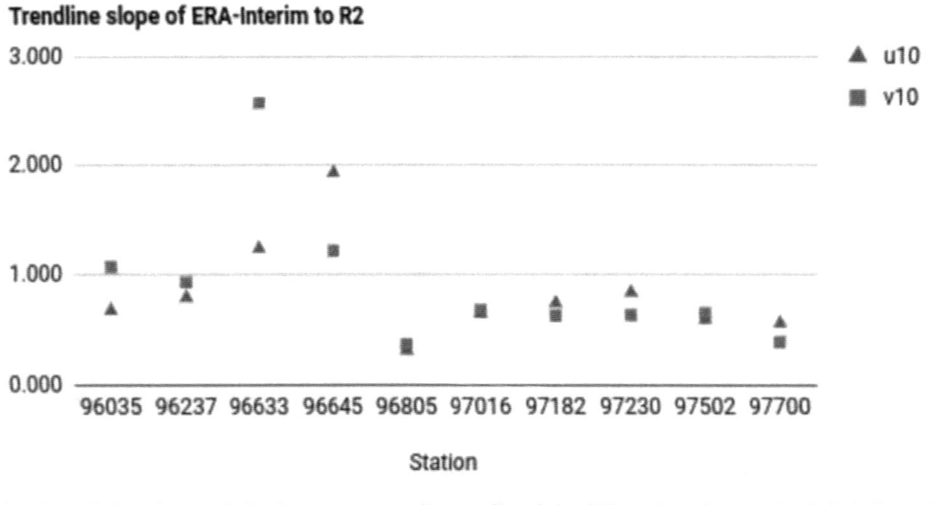

Figure 4. Plot of the slope of the least squares linear fit of the ERA-Interim to the R2 daily wind data for the time period 2013-2016.

Figure 4 shows that both zonal and meridional wind of the R2 and ERA-Interim give similar behavior at each station. Both zonal and meridional wind of the R2 are stronger than ERA-Interim, except at two stations in Borneo island (Sepinggan and Iskandar) where ERA-Interim data shows much larger than R2.

4. Comparison dataset with measurement

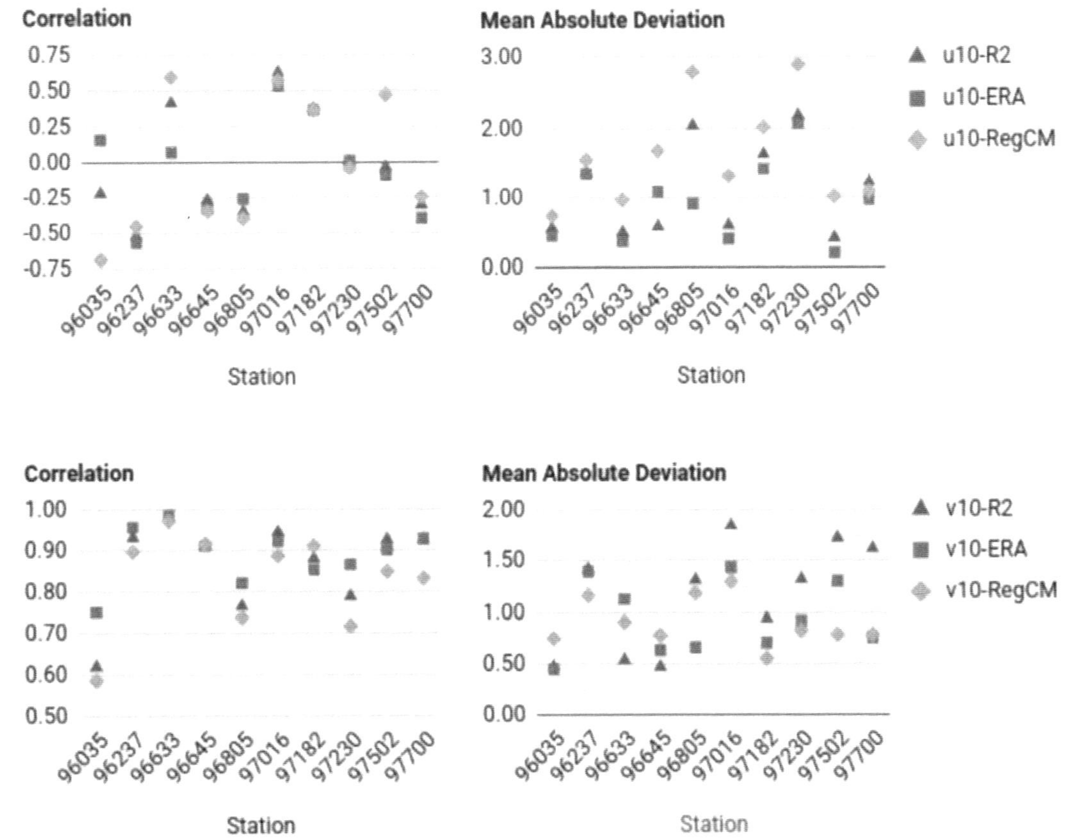

Figure 5. Correlation and mean absolute deviation to the wind measurement at 10 stations.

We analyze the performance of the annual cycle of the zonal and meridional wind from the datasets against the measurement at 10 stations. The correlation and the deviation are shown in Figure 5. Based on the correlation, the meridional wind of R2, ERA-Interim, and RegCM at most of the stations give strong correlation (> 0.5) with the measurements, while the zonal wind give poor correlation. This is also shown by the mean absolute deviation that the meridional wind of R2, ERA-Interim, and the regional data deviates relatively small from the measurements.

Meanwhile the annual cycle at three stations is shown in Figure 6. These stations are selected to show that the dynamical downscaling can improve the global data at one station, but it can also give worse performance at another station. For example, at Sepinggan station the model gives more deviation from the measurement in the zonal and meridional wind. At Seigun, the model improves only the R2 meridional wind. At Namlea, the RegCM improves both zonal and meridional of R2 data. In general, the zonal wind of the global reanalysis data or the downscaled data is still give large deviation, while the meridional wind perform the measurement well.

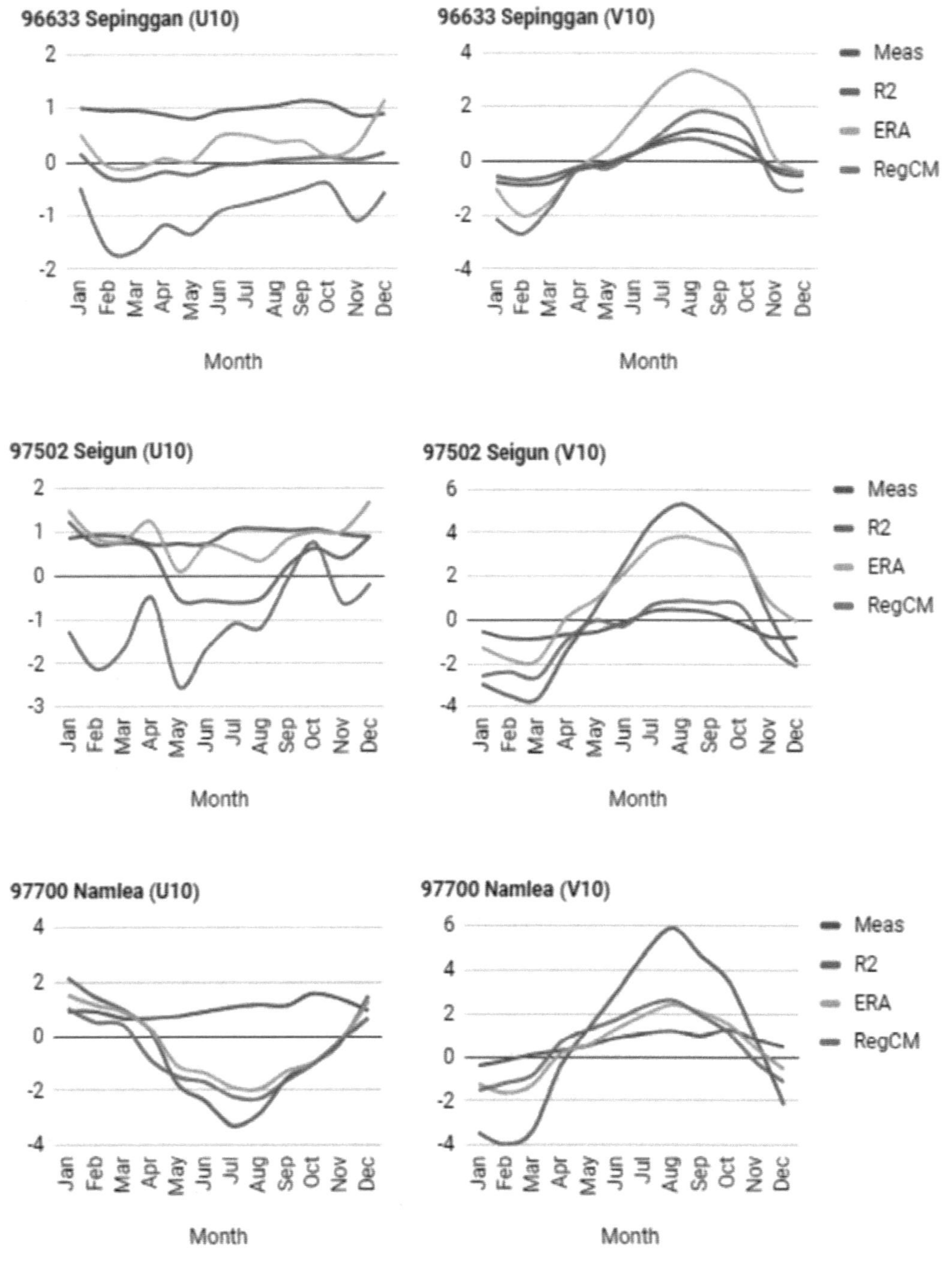

Figure 6. Annual cycle of the zonal and meridional wind during 2013-2016 at three stations.

5. Summary

A comparison of the near-surface wind between the R2 and ERA-Interim reanalysis data in Maritime Continent over the period 2013-2016 has been presented. The result shows that over most of MC the R2 gives stronger wind than ERA-Interim in both DJF and JJA season. Moreover, the ability of ERA-

Interim, R2, and the downscaled of R2 to reproduce the near-surface wind at 10 measurement stations has been investigated. There is no optimum among the three dataset. All three dataset show similar annual cycle, but at some stations it does not fit the measurement. Based on the correlation analysis, all three dataset model the meridional wind well, but not the zonal wind. Therefore, this result should be considered in the studies using the reanalysis data. Further, the performance of RegCM to downscale the global data is still need to be investigated further since the model may perform well at one region, but does not perform well at another region.

Acknowledgement

This research was supported by Research Center for Informatics, Indonesian Institute of Sciences. We would like to acknowledge to the high performance computing facilities http://grid.lipi.go.id to perform the RegCM4.6 simulations. The global reanalysis data R2 and ERA Interim provided by NOAA/ESRL are highly appreciated.

References

[1] Li Z, Wang H 2016. Ocean Wave Simulation Based on Wind Field *PLoS ONE* **11**(1) e0147123

[2] Kaspar F, Kaiser-Weiss A K, Heene V, Borsche M, and Keller J 2015 Evaluation of representativeness of near-surface winds in station measurements, global and regional reanalysis for Germany *Geophysical Research Abstracts* **17** EGU2015-11318

[3] Van Vledder G and Akpinar A 2015 Wave model predictions in the Black Sea: Sensitivity to wind fields *Applied Ocean Research* **53** 161-178

[4] Wang J and Kotamarthi V R 2013 Assessment of Dynamical Downscaling in Near-Surface Fields with Different Spectral Nudging Approaches Using the Nested Regional Climate Model (NRCM) *J. Appl. Meteor. Climatol.* **52** 1576-1591

[5] Stopa J E and Cheung K F 2014 Intercomparison of wind and wave data from the ECMWF Reanalysis Interim and the NCEP Climate Forecast System Reanalysis *Ocean Model.* **75** 65-83

[6] Mooney P A, Mulligan F J and Fealy R 2011 Comparison of ERA-40, ERA-Interim and NCEP/NCAR reanalysis data with observed surface air temperatures over Ireland *Int. J. Climatol.* **31** 545-557

[7] Chaudhuri A H, Ponte R M, and Nguyen A T 2014 A Comparison of Atmospheric Reanalysis Products for the Arctic Ocean and Implications for Uncertainties in Air–Sea Fluxes *J. Climate* **27** 5411-5421

[8] Lejiang Y, Zhanhai Z, Mingyu Z, Zhong S, Lenschow D, Hsu H, Huiding W, Bo S 2010 Validation of ECMWF and NCEP-NCAR Reanalysis Data in Antarctica *Adv. Atmos. Sci.* **27**(5) 1151-1168

[9] Kanamitsu M, Ebisuzaki W, Woollen J, Yang S K, Hnilo J J, Fiorino M, and Potter G L 2002 NCEP/DOE AMIP-II reanalysis (R-2) *Bull. Am. Meteorol. Soc.* **83** 1631-1643

[10] Kalnay E, et al 1996 The NCEP/NCAR 40-year reanalysis project *Bull. Am. Meteorol. Soc.* **77** 437-471

[11] Giorgi F, Marinucci M R, Bates G T 1993 Development of a Second-Generation Regional Climate Model (RegCM2). Part I: Boundary-Layer and Radiative Transfer Processes *Mon. Wea. Rev.* **121** 2794-2813

[12] Giorgi F, Marinucci M R, Bates G, Canio G D 1993 Development of a Second-Generation Regional Climate Model (RegCM2). Part II: Convective Processes and Assimilation of Lateral Boundary Conditions *Mon. Wea. Rev.* **121** 2814-2831

IOP Conf. Series: Journal of Physics: Conf. Series **1090** (2018) 012040 doi:10.1088/1742-6596/1090/1/012040

Optimization of time-dose fractionation radiotherapy scheme by simulated annealing with consideration of biological factor

A N Hasto, and W Rena

Department of Physics, Faculty of Mathematics and Natural Sciences, Institut Teknologi Bandung (ITB), Indonesia

hastoarief@gmail.com

Abstract. Ionizing radiation for cancer treatment also known as radiotherapy has undergone development for more than century. Radiation kills cancer by giving partially or entirely all of the energy to the tumor tissue, especially in the DNA. If tumor or a cancer is irradiated by radiotherapy, there is no way to avoid the tissues near cancer will be affected by the radiation. As a result, the healthy tissues near tumor will lose their functionality, or worse, there may appear a new tumor in that tissues. Therefore, it should be done some improvement in the quality of radiotherapy, one of the examples is optimization of time-dose fractionation radiotherapy scheme. To optimize the fractionation dose, is made a program in MATLAB using simulated annealing as the optimization method. The objective function is Biologically Effective Dose of cancer, with constraints are Biologically Effective Dose of the healthy tissues around cancer. After obtaining the optimum dose per fraction, the data will be analyzed for the total and fractionation time. Thus, it would be obtained the optimum of time scheme and dose fractionation of cancer. The result of this research will show the dose per fraction from the total time in 10 until 60 days in both of the treatment schemes, the value of Biologically Effective Dose of the cancer, and the optimum of dose per fraction from both of the treatments.

1. Introduction

For more than century, the ionizing radiation for cancer treatment or radiotherapy has undergone a massive development. Radiation is an energy emitted by the electromagnetic wave or particle stream. Radiation can break the DNA in the cell which has a function to control the growth and the division of cells. When the DNA of the cell is broken, the tumor can not grow and split, and by the time, the tumor cell will be dead. While the tumor is irradiated by radiotherapy, the healthy tissues around the tumor will be affected by the irradiation. As a result, the complication of the healthy tissues will rise. For years, the technologies have been developing on the aspect of radiation distribution that has an aim to widen the therapy window to decrease the effects in the healthy tissues while improving the effects in the target (tumor).

In general, there are three methods to improve the radiotherapy quality: (1) improve the quality of dose distribution; (2) optimize the time scheme-dose fractionation in the radiotherapy; (3) modify the response of the radiation in the tumor (radiosensitizers) and/or in the healthy tissue (radioprotection). By using the latest clinical implementation in radiotherapy, such as Intensity-Modulated Radiotherapy (IMRT), nowadays, it can be obtained the optimum dose distribution in the radiotherapy treatment. The second method, the optimization of time scheme-dose fractionation, can be done by many ways, such as using the biological aspect.[2]

Content from this work may be used under the terms of the Creative Commons Attribution 3.0 licence. Any further distribution of this work must maintain attribution to the author(s) and the title of the work, journal citation and DOI.
Published under licence by IOP Publishing Ltd

The application of the radiobiology in the field of clinical radiotherapy is still in the primitive stage. To make the optimum design of the patient treatment and the proper usage of the potential research in the radiobiology that has developed for years, it will be a good advantage to improve and develop inverse optimization technique using time-dose fractionation. This will not only be increasing the comprehension of the impact of the biological parameter in the radiotherapy treatment for the tumor, but also provide the practical tools to get the optimum result in total dose, fractionation scheme, dose per fraction and overall treatment time. One of the fundamental formulation in the radiobiology to describe the link between time-dose fractionation scheme in the tissue is linear-quadratic. Linear-quadratic is used to predict the dose isoeffect from radiotherapy when the schedule of the fractionation is changed. This formulation becomes the most commonly used formula, because linear-quadratic can describe a model that will calculate tumor control and healthy tissue complications when a dead cell occurs.

Simulated annealing or SA is an optimization method which is included in stochastic methods. A stochastic method is a mathematical method based on probability or randomness. The main advantage of SA is that the method can find the global optimum from all the evaluated area. In other words, SA can remove itself from local minima in the evaluated area. Another advantage of SA is that the method can work in a non-linear model, a model that has many noises, and has many constraints. The disadvantage of this model is that SA needs parameter calibration to get the optimum result. In addition this method is a time-consuming method. Furthermore, because SA is the one of stochastic methods, so sometimes this method do not find the true optimum, but just an approximation of the optimum value.[8],[9],[15]

This work will optimize the time scheme and dose fractionation by simulated annealing in the tumor which can be divided into two categories. The first one is called late effect tumor, this kind of tumor has very long proliferation time that can be more than a month. The second category is early effect tumor, this tumor will proliferate in short time, only a couple of days or weeks. The optimization has constraints in the form of organ at risk (OAR) around the target (tumor). Similar to the tumor, OAR also has two categories. So, this optimization will vary both of the tumor and OAR, as well as the other parameters such as kick-off time, resensitization time, and repair time to analyze the optimum time scheme and dose fractionation.

2. Materials and Methods

2.1. Linear-Quadratic Radiobiology (LQR)

Linear-quadratic is a fundamental tool in radiobiology, because this formula can calculate the tumor control probability (TCP) and normal tissue complication probability (NTCP). This formula also can predict the isoeffect dose from radiotherapy when the fractionation schedule has changed. The first formula and the famous one in the linear-quadratic is survival factor. Survival factor explains the ratio between the broken cell (dead cell) with the survived cell that happened from two unrelated processes, which is from the physical mechanism by single-hit and double hit.

$$S = \exp[-(\alpha D + \beta D^2] \quad (1)$$

The constant alpha explains about the damage caused by the ionization events that lead to lethal damage in DNA directly, whereas the beta is a constant that appears as a result of the damage in the cell caused by two ionization event which does not have any repair in the sublethal damage, so the damage becomes lethal damage. The D represents total dose which is given in the radiotherapy treatment.

In the clinical radiobiology, it is known a term called 4R radiobiology. 4R radiobiology is biological factors that affect tumor and healthy tissues responses from a fractionation radiotherapy treatment. The first one, called repair, is a biology mechanism that will fix or repair the sublethal damage caused by radiation. Secondly, the redistribution, this phenomenon is affected by the distribution in proliferation cell population that happened during the fractionation radiotherapy through cell cycle that will increase the dead cell. The next is reoxygenation from the cell which has hypoxia condition, at first the cell is resistant to radiation then as day by day the cell becomes more sensitive to the dose of radiation. The last one, called repopulation, this phenomenon happened when cell duplicates itself during the treatment of radiotherapy.

$$S = exp\begin{Bmatrix} \sum_{i=1}^{I}\left[-\alpha d_i - \beta G_i(\tau_r)d_i^2 + \frac{1}{2}\sigma^2 G_i(\tau_s)d_i^2\right] + \\ H(T_{tot}, T_k)(T_{tot} - T_k)/T_{pd} \end{Bmatrix} \tag{2}$$

$$H(x, x_0) = \begin{cases} 0, if\ 0 \le x < x_0 \\ 1, if\ x \ge x_0 \end{cases} \tag{3}$$

Equation 2 is called linear-quadratic radiobiology or LQR. This formula calculates the effects of the 4R radiobiology mentioned before. The G, usually called Lea-Cactheside time factor, will calculate the effect of the fractionation or protraction of the treatment radiotherapy. The first one, $G(\tau_r)$, is a Lea-Cactheside time factor in a function of repair and the second one, $G(\tau_s)$, is a Lea-Cactheside time factor in a function of resensitization, consisting of reoxygenation and redistribution. The $(1/2)\sigma^2$ is the variance of Gaussian Distribution of α. The last term, repopulation, is in the heaviside factor, where T_{tot} is overall treatment time, T_k is kickoff time of accelerated proliferation, and T_{pd} doubling time of tumor.

$$G_i(\tau) = 1 + \frac{2}{d_i}\sum_{j=1}^{i-1}\left[d_j \prod_{k=j}^{i-1} exp\left(-\Delta k/\tau\right)\right] \tag{4}$$

For an arbitrary fractionation with interval fractionation (Δt_1 . . ., Δt_i . . ., . . .) and radiotherapy fractional dose (d_1 . . ., d_i, . . .), which d_i, represents the ith fractional dose and Δt_i as a time interval between fraction i and i+1. For normal tissue calculation, the formula only considers the cell killing and incomplete repair of sublethal damage, or in other words only the first two in Equation 2 are included.

2.2. Objective function and constraints

In the optimization problem, the objective function is a mathematical formulation which will determine the best solution that has been obtained. The minimum possible solution (or maximum, depends on the optimization) from an objective function is called optimum solution. Whereas, a constraint is a condition on the optimization problem and the solution must satisfy some conditions.

This paper uses an objective function in the function of BED (Biologically Effective Dose), which can be written

$$F = \frac{BED_{tumor-reference}}{BED_{tumor}} \tag{5}$$

From the Equation 5, it can be said that on this research will find the minimum value of the objective function, i.e it will be found the maximum value of the BED tumor. The constraints of the problem are:

$$BED_{organ} \le BED_{organ-reference} \tag{6}$$

The reference schedule for this research is standard fractionation with a total dose 70 Gy and dose per fraction is 2 Gy/fraction. While, in this research, it will be used two kinds of schedule, standard frationation and hyperfractionation. On this paper, there are only two patterns of the interval: five fractions per week (interval=1 day) with a weekend break (standard fractionation) and ten fractions per week (interval between two fractions in the same day is 8 h) with the overnight and weekend breaks (hyperfractionation). The overall time is two weeks until eight weeks for fast proliferation, and two weeks until six weeks for slow proliferation. For each overall treatment time, the optimization program will optimize the dose per fraction using simulated annealing that will maximize the tumor BED with constant or lower healthy tissue BED.

From Equation 5 and 6, it is known that the optimization problem has constraints. The problem can be solved by linear algebra and geometry, and the result will usually be complicated and technical. In this research, it will be introduced basic technique, that will change a constraint problem into unconstraint problem. This technique is called penalty technique, it will penalize a treatment plan that has BED-organ bigger than BED-organ-reference.

$$F = \left\{ \begin{matrix} \frac{BED_{tumor-reference}}{BED_{tumor}} + \\ \sum_{n=1}^{N} k \times max\left(0, BED_{organ-n} - BED_{organ-ref-n}\right) \end{matrix} \right\} \quad (7)$$

By using the Equation 7, an optimization problem that has a constraint at first will become unconstraint. The value of k in the Equation 7 is positive, therefore if the BED organ is bigger than the BED organ reference, the objective function will be bigger and the objective function is not an optimum solution. In other words, the solution is penalized by penalize function. Otherwise, if the BED organ is less than the BED organ reference, the penalize function will be zero, and the objective function remains the same.

The organs and tumor are divided into two parts, which is early effect tissue and late effect tissue. Those two parts will have different parameter value as shown in the table below.

Table 1. Parameter of fast and slow proliferating tumor and tissues

Parameters	Fast proliferating	Slow proliferating
α tumor (Gy^{-1})	0.35	0.1
α/β tumor (Gy)	10	1.5
Tumor doubling time (day)	3	40
Tumor kick off time (day)	28	300
Tumor resensitization time (day)	1	2
Tumor repair time (hour)	0.5	1.9
Variance of gaussian distribution	0.02	1/3 β
α of late responding tissue (Gy^{-1})	0.315	0.315
α/β of late responding tissue (Gy)	3	3
Late responding tissue repair time (hour)	4	4
α of early responding tissue (Gy^{-1})	0.315	0.315
α/β of early responding tissue (Gy)	10	10
Early responding tissue repair time (hour)	0.5	0.5

3. Results and Discussion

3.1. *Fast proliferation and slow proliferation tumor*

Figure 1 shows the ability of simulated annealing to reproduce the same value of BED tumor. Simulated annealing can reprodruce with an accuracy more than 0.1% in fast proliferating tumor as illustrate in Figure 1(a). However, on the case of slow proliferating tissue in the Figure 1(b), simulated annealing fails to reproduce the value of BED tumor. As shown in Figure 1(b) the poor accuracy of reproducibility is obtained, that is less than 12%. It is happened due to the sensitivity of changing dose per fraction in slow proliferating tumor is higer than fast proliferating tumor. Therefore, if the simulated-annealing mechanism changes the dose per fraction in order to find the optimum BED tumor, a little change in dose per fraction will change large amount of BED tumor in slow proliferating tumor. Nevertheless, simulated annealing can be upgraded or improved by simple method as the result which is shown in

IOP Conf. Series: Journal of Physics: Conf. Series **1090** (2018) 012040 doi:10.1088/1742-6596/1090/1/012040

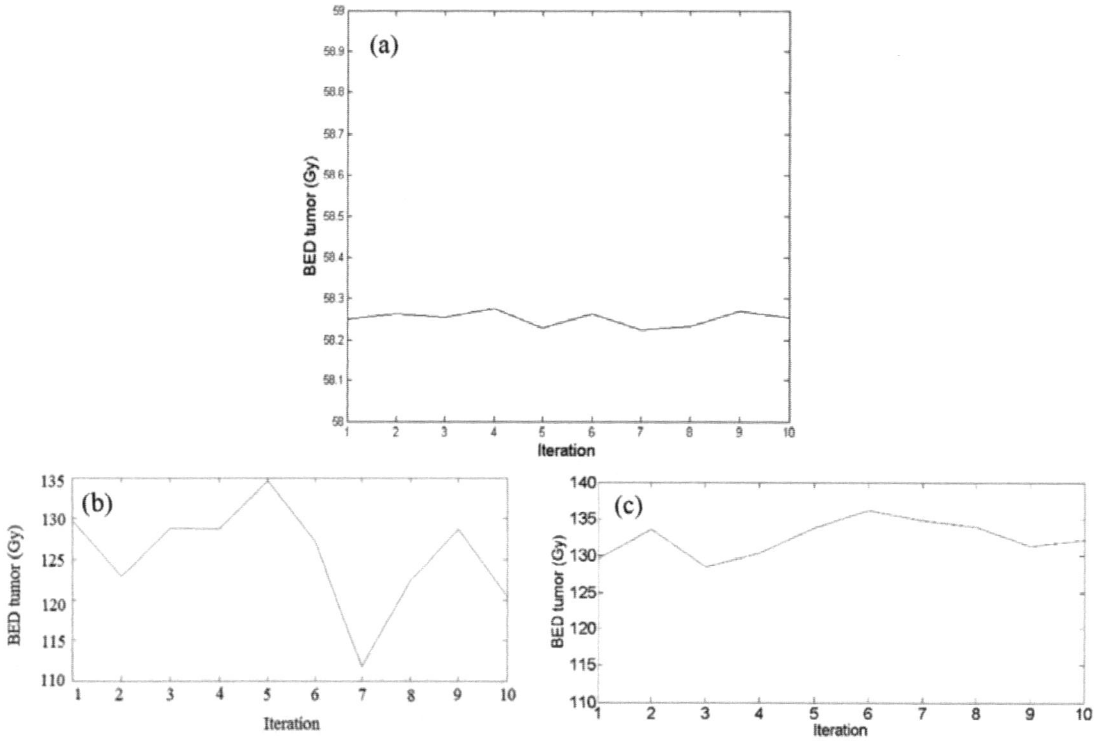

Figure 1. Reproducibility of BED tumor (a) Standard Fractionation. (b) Hyperfractionation with a normal simulated-annealing. (c) Hyperfractionation with an improve of simulated-anealing.

Figure 1(c). This improvement will re-iterate after doing the first iteration of simulated annealing with an early dose of the second iteration is the optimum dose of the first iteration, the iteration will continue until five iteration. As shown in Figure 1(c), the accuracy of reproducibility is increase, with the accuracy more than 1.1%. In spite of the accuracy is not as accurate as the fast proliferating tissue, but it is acceptable.

As illustrates in Figure 2, the BED tumor in standard fractionation will rise in the early of total treatment time and in the middle of the total treatment time, the BED tumor will decrease as the treatment time increases. It is known that the kick-off time (the time when tumor is proliferating) is the main factor of this effect. After the treatment time pass through the kick-off time, the tumor will proliferate or doubled the cell rapidly. It will decrease TCP (Tumor Control Probability) as well as the BED tumor, the TCP decreases because many cells have proliferated or doubled rapidly eventhough the cell is dead due to the treatment radiotherapy. However, the cell dead is way more few than the doubling cell mechanism, therefore the TCP will decrease as the treatment time increases. As a result of the lack of accuracy in slow proliferating tissue, the graphic in Figure 3 (slow proliferating tissue) is not as smooth as Figure 2 (fast proliferating tissue). Trend of the graphic of slow proliferating tumor is not as the same as the fast proliferating tumor. BED tumor in slow proliferating tumor will increase steeply in the early of the treatment time, but after some time the BED tumor will be more sloped. It is due to the long kick-off time of the slow proliferating tumor, hence the BED tumor will not decrease until it reaches the kick-off time of tumor which is 300 days, but as the time goes by, the increasing of the treatment time will not be more effective as in the early treatment time in order to prevent the loss of the healthy tissue functionality or the appearance of the new tumor.

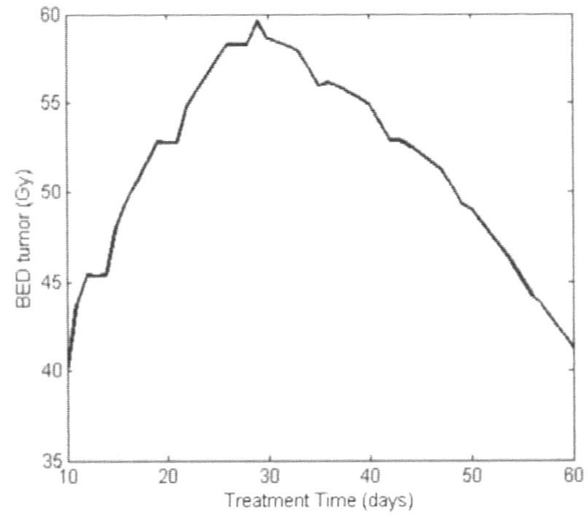

Figure 2. BED tumor vs treatment time for fast proliferating tissue

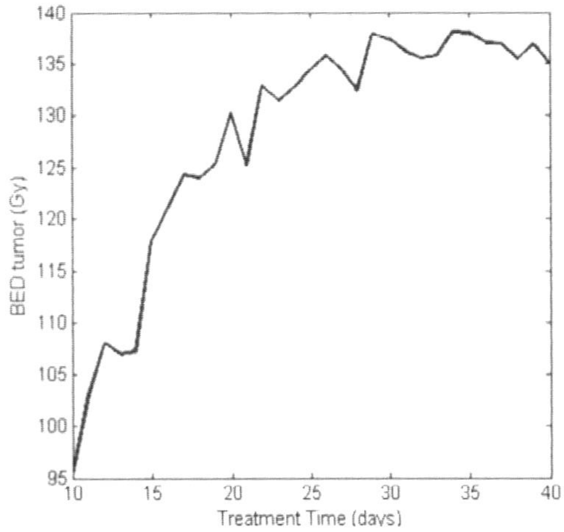

Figure 3. BED tumor vs treatment time for slow proliferating tissue

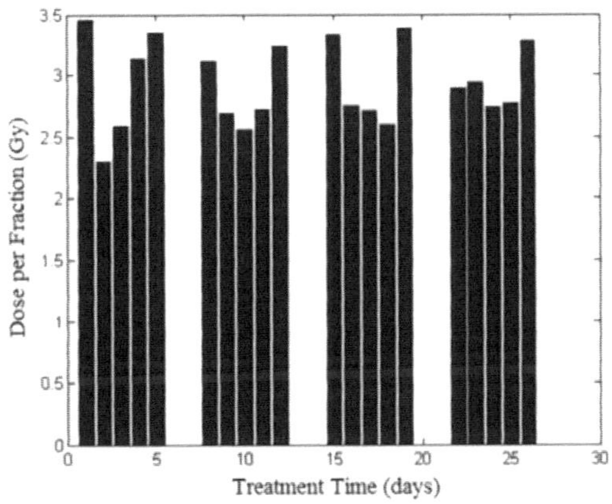

Figure 4. Dose per fraction for fast proliferatig tissue

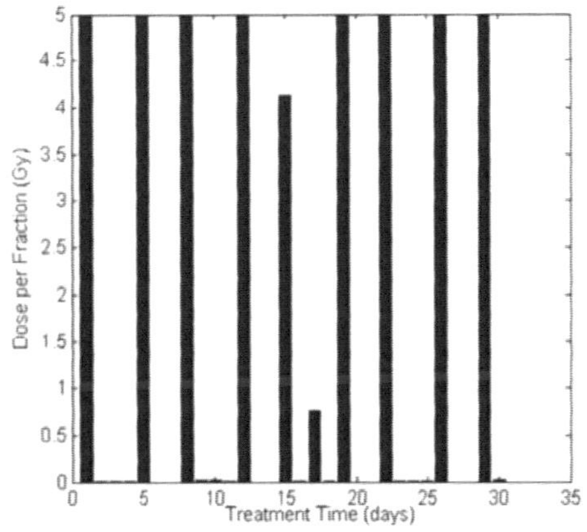

Figure 5. Dose per fraction for slow proliferating tissue

Figure 4 and 5 show the dose per fraction of the slow proliferating tissue and fast proliferating tissue. In the treatment of fast proliferating, it is shown that the dose will be delivered everyday to optimize the radiotherapy, but in the slow proliferating tumor the best dose per fraction is not delivered everyday, but 2 or 3 days per fraction, known as hyprofractionation. In the slow proliferating tumor, the given dose is a maximum dose that can be given in the linear-quadratic formulation which is 5 Gy, while in the fast proliferating tissue the dose per fraction is around 2-3 Gy per fraction which is a normal dose for standard fractionation.

3.2. Effect of parameter tumor and healthy tissue

Resensitization is a combination of repair and reoxygenation which will affect the radiosensitivity of the tumor and/or healthy tissue. This characteristic should be analyzed because it is one of the important effect of the radioterapy scheme. As can be seen in the Figure 6 and 7, the different resensitization will have the different BED tumor. As a resensitization increases, the BED tumor will decreases, both in standard fractionation and hyperfractionation. The resensitization will change the radiosensitivity of the tumor, and makes the tumor more sensitive. Hence, as the resensitization time increases, the time for the tumor to change to be more sensitive will be longer, therefore tumor will have more resistant cell if

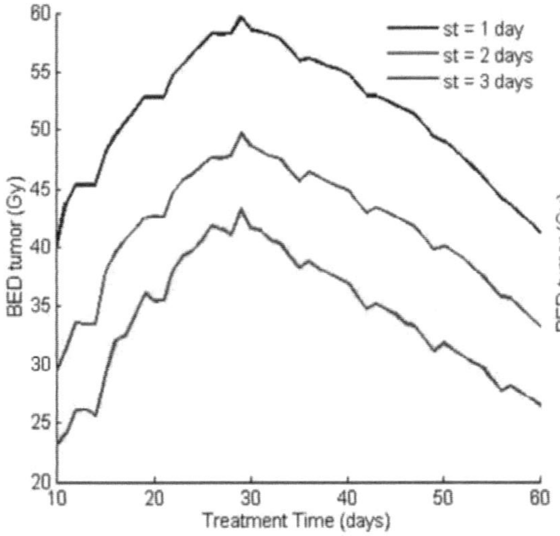

Figure 6. BED tumor vs treatment time of standard fractionation with variation of resensitization time

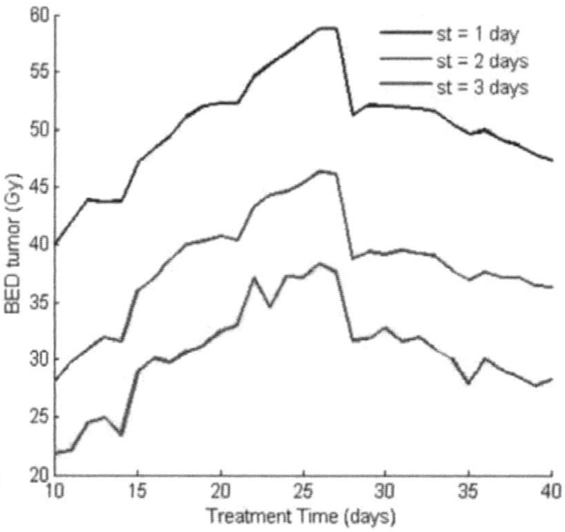

Figure 7. BED tumor vs treatment time of hyperfractionation with variation of resensitization time

the resensitization time is large. Otherwise, when the resensitization time is small, the tumor will change to be more radiosensitive rapidly, therefore many tumor cell will be dead as the treatment time increases.

Figure 8 and 9 illustrate the effect of the variance in kick-off time and doubling time of the tumor respectively. As shown in those figures, both of the parameter will have not affected the BED until the treatment time pass the kick-off time of the tumor. Kick-off time will affect the BED by decreasing the BED tumor if the treatment time exceeds the kick-off time. As kick-off time increases, the BED tumor will increase longer than if the kick-off time is small. It is obvious, because if the treatment time is shorter than the kick-off tumor, there is no doubling mechanism of the cell. However, if the treatment time is longer than the kick-off tumor, there will be doubling mechanism of the tumor cell and will decrease the TCP as well as the BED tumor. The other parameter that should be considered is doubling time of the tumor. As can be seen in the Figure 9, the doubling time affects the BED tumor. The longer the doubling time of the tumor, the larger BED tumor that will be obtained. It is due to the doubling mechanism of the tumor cell accelerates the reduction of the BED tumor. If the doubling time of the tumor is 1 day, the BED tumor will go to 0 in near treatment time of 50 days, it means that there will be no reduction of the tumor cell, instead there will be an increasement of the cell tumor.

3.3. Dose per fraction

Besides the BED tumor, another important output that should be analyzed is dose per fraction. As shown in the Figure 10, there are 4 different types of the dose per fraction scheme in the fast proliferating tumor. Figure 10(a) and 10(b) are the optimum scheme for standard fractionation, which is a schedule that gives a fractionation dose one per day. As shown in the figure, every first and last day of the week, the fractionation dose is higher than the other days. It is due to the effect of the repair time and resensitization time of the tumor and healthy tissue. In the Figure 10(a), the resensitization is zero, which means the radiosensitivity of the tumor will always increase everytime. The difference between Figure 10(a) and 10(b) is that the resensitization is not zero in Figure 10(b), but 1 day. It means that the resensitization in Figure 10(b) is slower than in Figure 10(a). As a result, the dose per fraction will be lower in the middle of the week for the Figure 10(a).

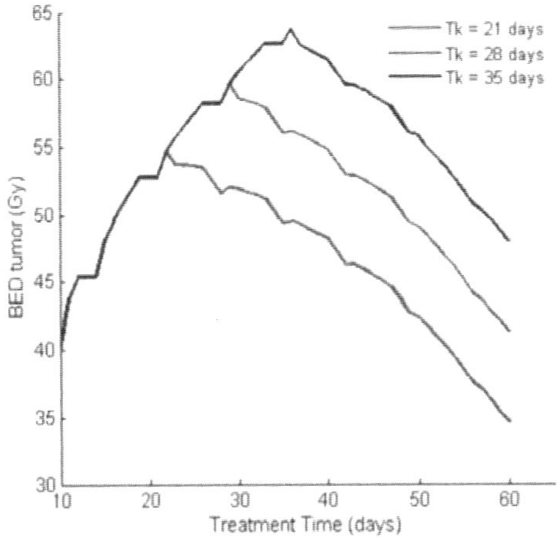

Figure 8. BED tumor vs treatment time of standard fractionation with variation of kick-off time tumor

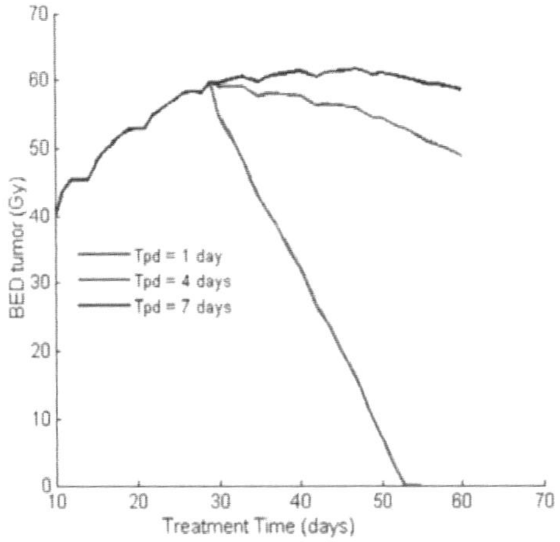

Figure 9. BED tumor vs treatment time of standard fractionation with variation of doubling time tumor

On the other hand, the Figure 10(c) and 10(d) represents hyperfractionation but it has different dose per fraction. In Figure 10(c), the dose per fraction vary randomly, with a dose per fraction in first and last week is higher than the others. In constrast, Figure 10(d) has almost constant dose per fraction every day. In the Figure 10(d), the constancy the dose per fraction is due to the zero value of all the parameters (repair time and resensitization time), which means there are no difference between the mechanism of the repair and resensitization of the tumor and healty tissue. The main factor that affects the optimization is alpha and beta of the tissue which is constant for all the treatment scheme. Whereas, Figure 10(c) is affected by the repair and resensitization time of the healthy tissue, therefore the dose per fraction vary with high dose per fraction in the first and last day of the week.

4. Conclusion

Ordinary simulated annealing is not accurate to predict the BED tumor optimum of the slow proliferating tissue. The simulated annealing is then modified to increase the accuracy of the optimization method. The increasing accuracy is about 10%. The trend of the slow proliferating tumor and fast proliferating tumor is different one and another. Fast proliferating tumor will increase at the early treatment time, and then at near the kick-off time, it will decrease because of the tumor cell will doubled rapidly. While in the slow proliferating tumor, due to the long kick-off time, there will usually be no reduction in BED tumor. However, after some time, the BED tumor will remain almost constant, in order to prevent the loss of the healthy tissue functionality or the appearance of new tumor near the healthy tissue.

The mechanism of the cell when it is irradiated by radiotherapy, such as repair, repopulation, reoxygenation, and redistribution should be considered as the parameters of optimization in radiotherapy scheme. As can be seen in the Figure 6, 7, 8, and 9, all of the mechanism affect the BED tumor. The optimum BED tumor should be analyzed and calculated with consideration of the cell mechanism in order to optimize the dose perfraction as well as the BED tumor. Besides the BED tumor, another parameter that should be anaylzed is dose per fraction. As can be seen in Figure 10, the cell mechanism affects the optimum dose per fraction and it should be optimized to get the optimum effect of radiotherapy.

(a) (b)

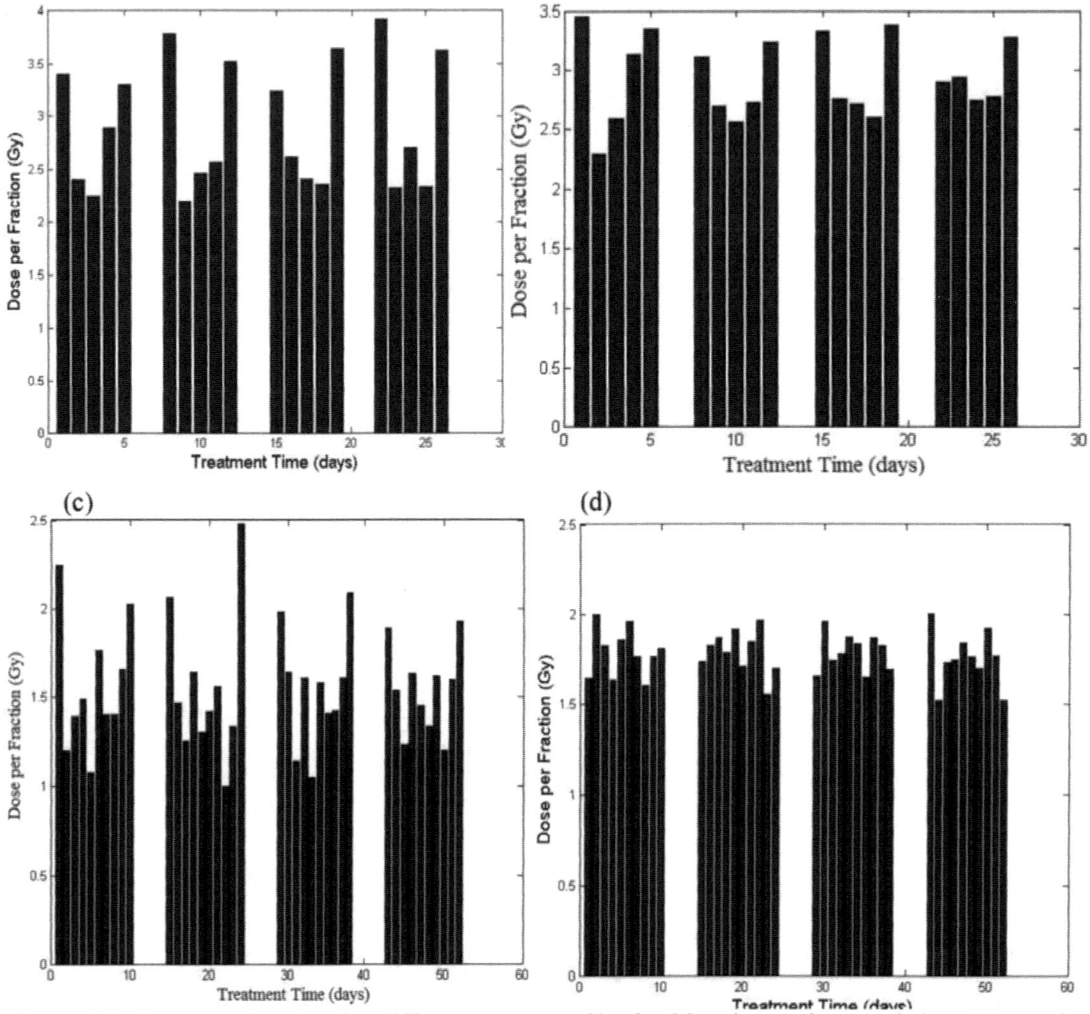

Figure 10. Dose per fraction for different tumor and/or healthy tissue characteristics. (a) standard fractionation with repair time of healthy tissue is 12 hours and resensitization time of tumor is 0 h (b) standard fractionation with repair time of healthy tissue is 12 hours (c) hyperfractionation with healthy tissue repair time is 4 hours and resensitization of tumor is 1 day (d) hyperfractionation with healthy tissue and tumor parameter (repair and resensitization time) is zero.

References

[1] Baskar, dkk. 2014. "Biological respone of cancer cells to radiation treatment". Molecular Biosciences.

[2] Yang, Y. dan Lei X. 2005. "Optimization of radiotherapy dose-time fractionation with consideration of tumor specific biology". *AAPM: The International Journal of Medical Physics Research and Practice.*

[3] Fowler, J. F. 1989. "The linear-quadratic formula and progress in fractionated radiotherapy". *The British Journal of Radiology.***62**, 679-694.

[4] Brenner, dkk. 1995. "A convenient extension of the linear-quadratic model to include redistribution and reoxygenation". *Int. J Radiation Oncology. Biol. Phys.***32 (2)** 379-390.

[5] Miasnikof, P. 2013. "Optimization of dose schedule in radiotherapy". *Toronto: University of Toronto*

[6] Mohan, dkk. 2000. "Radiobiological consideration in the design of fractionation strategies for intensity-modeulated radiation therapy of head and neck cancers". *Int. J. Radiation Oncology Biol. Phys.***46 (3)**, 619-630.

[7] Suwardi. 2015. "Kanker kepala leher terinduksi radiasi pasca radioterapi karsinoma nasofaring". *Yogyakarta: Universitas Gadjah Mada*

[8] Kirkpatrick, S., dkk. 1983. "Optimization by simulated annealing". *Science, 1983*, **220(4598)** 671-680.

[9] Bertsimas, D. dan John T. 1993. "Simulated Annealing". *Statistical Science.***8 (1)** 10-15.

[10] Bryan, K. dan Yosi S. 2005. "Penalty functions and constrained optimization". *Indiana: Dept. of Mathematics, Rose-Hulman Institute of Technology.*

[11] Majumder, dkk. 2017. "Different fractionation schedules of radiotherapy in locally advanced head and neck malignancy: A prospective randomized study to compare the results of treatment and toxicities of different protocols". *South Asian Journal of Cancer*. **2**.

[12] Podgorsak, E. B. 2005. "*Radiation oncology physics: A Handbook for Teachers and Students*". Vienna: International Atomic Energy Agency.

[13] Wang, dkk. 2011. "Organ-sparing radiation therapy for head and neck cancer". *Nature Reviews, Natural Oncology: Macmillan Publisher.*

[14] Lee, A. W. M, dkk. 1998. "Effect of time, dose, and fractionation on temporal lobe necrosis following radiotherapy for nasopharyngeal carcinoma". *Int. J. Radiation Oncology Biol Phys*. **40 (1)** 35-42.

[15] Busetti, F. 2003. Simulated Annealing Overview. *Citeseerx.*

[16] Fowler, J. F. 2001. "Biological factos influencing optimum fractionation in radiation therapy". *Acta Oncologica.* **40 (6)** 712-717.

Inertia effect of textured lubricated contact on the bearing performance using CFD approach

F Hilmy[1], M Muchammad[1,2], M Tauviqirrahman[1], J Jamari[1]

[1]Laboratory for Engineering Design and Tribology, Mechanical Engineering Department, Diponegoro University, Indonesia
[2]Laboratory for Surface Technology and Tribology, Faculty of Engineering Technology, University of Twente, Netherlands

hilmyfuad29@gmail.com

Abstract. Numerous studies reported that inertia of the lubricant has been considered as one of physical parameters which has a strong effect on the load support of textured bearing. In the present study, based on two-dimensional computational fluid dynamics (CFD) technique, the investigatation of the inertia effect on bearing performance is carried out varying the texture length. The Navier-Stokes equation coupled with the cavitation model are discretized using finite volume method and solved using the commercial software FLUENT®. The results show that the inertia increases the hydrodynamic film pressure and thus the load support. In addition, it is also found that increasing the texture length as well as Reynolds number will increase the cavitation region.

1. Introduction

As is known, the use of texturing of the lubricated contact is not a new concept. The research on surface texturing gained a new momentum since texturing technique has proven to enhance the lubrication characteristics. Considerable researches have been conducted on textured surfaces with respect to the mechanism of lift generation.

Fowell et al. [1] pointed out that any convergence between the bearing surfaces generates a significant mechanism for lift generation. Gherca et al. [2] numerically explored the effect of geometrical features such as size, density, and shape of texture on the lift generation using a mass-conserving model. Henry et al. [3] experimentally investigated the influence of surface texturing on the performance of bearings. It was found that the textured bearing reduces friction up to 30% at low loads. However, for heavy loads, the textured surface was not recommended to use because their performance is equivalent or even lower than that of an untextured bearing. Yagi et al. [4] concluded that the mechanism of lift generation was strongly affected by boundary condition. Zhang et al. [5] investigated the effect a rectangular array of circle dimples on the film thickness profile. It was highlighted that the lift generation could be improved through appropriate partial arrangement of textures. Later, Shinde and Pawar [6] proposed the optimal surface texturing parameters to improve performance of bearing.

Based on literature survey, one can find that texturing generates more positive effect compared to pure texturing with respect to the load support. However, most of the solution of the lubrication problems have been solved by of Reynolds equation which may be questionable in particular cases

Content from this work may be used under the terms of the Creative Commons Attribution 3.0 licence. Any further distribution of this work must maintain attribution to the author(s) and the title of the work, journal citation and DOI.
Published under licence by IOP Publishing Ltd

with low viscosity and high runner velocities. Therefore, the present work adopts two methods (Navier-Stokes equation and the Reynolds equation) to investigate the texturing parameter in particular case of texture depth on the hydrodynamic pressure. For obtain more accurate results, the cavitation model is also considered. In this way, the inertia effect as well as the cavitation effect can be investigated more detail.

2. Methodology

2.1. Theory

In the present study, the lubrication problem is solved by the Navier-Stokes equation and continuity equations. The Navier–Stokes (N-S) equations are solved over the domain using a finite-volume method with the commercial CFD software package FLUENT®. The Navier–Stokes and the continuity equations can be expressed, respectively,

$$\rho\frac{Du_i}{Dt}=-\frac{\partial p}{\partial x_i}+\rho G_i+\frac{\partial}{\partial x_j}\left[2\eta e_{ij}-\frac{2}{3}\eta\left(\nabla .u_i\right)\delta ij\right] \quad (1)$$

$$\nabla\bullet u=0 \quad (2)$$

Not like the previously published works in which the cavitation model is ignored, in the present study, the cavitation effect is taken into account. In FLUENT®, there are three available cavitation models: Schneer and Sauer model, Zwart-Gelber-Belamri model and Sighal et al. model [7]. In this study, the Zwart-Gelber-Belamri is employed due to their capability (less sensitive to mesh density, robust and converge quickly [7].

In addition to the Navier-Stokes equation, in this work, the lubrication problem is also solved by modified Reynolds equation (Eq. 3 as below). For more detail, the derivation of such equation used here is described in [8]. It should be noted that the isoviscous Newtonian one-dimensional Reynolds equation with slip is derived from a simple of the x-component of the Navier-Stokes equation that assumes an incompressible flow, neglecting the inertia effects in the film. However, for the analysis studied here, the slip coefficient is set to zero.

$$P_2\left[\left(h_p^3+3h_p^3K_p\right)\frac{ab+bc}{ab^2}+\left(\frac{h_0^3+3h_0^3K_0}{a}\right)\right]=P_{atm}\left[\left(h_p^3+3h_p^3K_p\right)\frac{ab+bc}{ab^2}+\left(\frac{h_0^3+3h_0^3K_0}{a}\right)\right]$$
$$-6\mu U\left[\left(h_p+h_pK_p\right)-\left(h_0+h_0K_0\right)\right] \quad (3)$$

2.2. CFD Model

Figure 1 gives the schematic illustration of a parallel textured (pocketed) sliding bearing. The assumption of the no-slip is adopted. The main characteristics of the bearing and the lubricant properties studied are presented in Table 1.

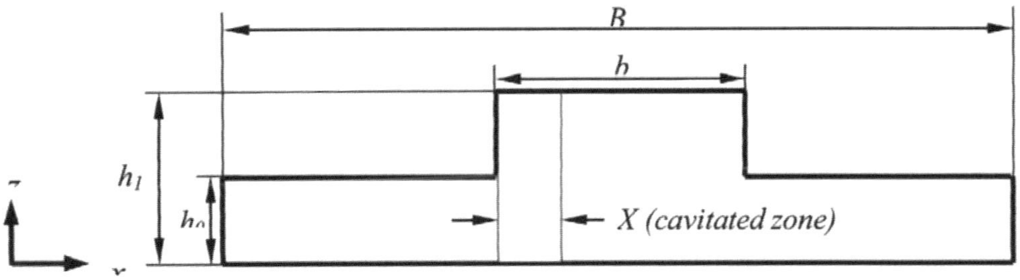

Figure 1. Geometry of slider bearing 2D model.

IOP Conf. Series: Journal of Physics: Conf. Series **1090** (2018) 012041 doi:10.1088/1742-6596/1090/1/012041

Table 1. Main bearing parameter and lubricant prperties.

Parameter	Symbol	Value	Unit
Length of bearing	B	2.00	mm
Length of texture	b	0.50	mm
Minimum film thickness	h_0	4.00	μm
Maximum film thickness	h_1	8.00	μm
Density of oil	ρ	962	kg/m^3
Density of vapor	ρ	0.013468	kg/m^3
Viscosity of oil	μ_o	0.02556	kg/ms
Viscosity of vapor	μ	$1.256\text{x}10^{-5}$	kg/ms

In the present study, the grid consists of four blocks with grid system as shown in Fig. 2. The detail grid of the mesh is shown in Table 2. The mesh quality consists of a minimum orthogonal quality of 0.994935, maximum ortho skew 0.00506503 and maximum aspect ratio of 129,357. The mesh size in the longitudinal direction (Nx) and transverse (Nz) is 1000 x 140, respectively. These grids generate 70,000 cells and 71,141 nodes based on the independent mesh study.

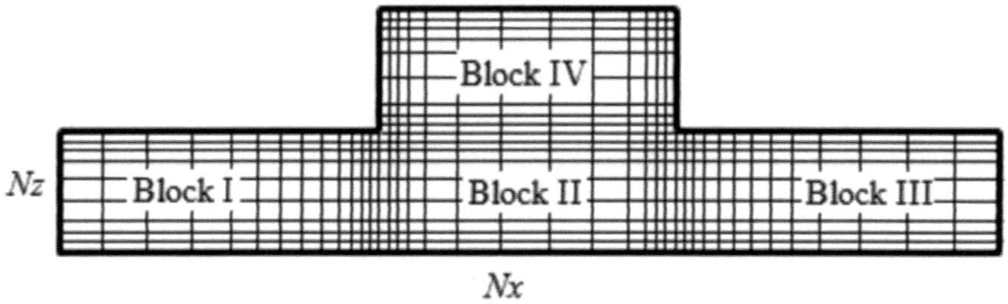

Figure 2. Grid system used in numerical analysis.

Table 2. Grid detail on each block of geometry.

Block	Direction	Mesh of edge type	Ratio	Interval count
I	Nx	First last ratio	0.1	375
	Ny	Bi-exponent	0.6	60
II	Nx	Bi-exponent	0.6	250
	Ny	Bi-exponent	0.6	60
III	Nx	Last first ratio	0.6	375
	Ny	Bi-exponent	0.6	60
IV	Nx	Bi-exponent	0.6	80
	Ny	Bi-exponent	0.6	250

The boundary condition applies pressure inlet and pressure outlet on the incoming and outgoing sides respectively. The hydrodynamic pressure is used as indicator to describe the inertia effect with respect to the Reynolds equation and Navier-Stokes equation.

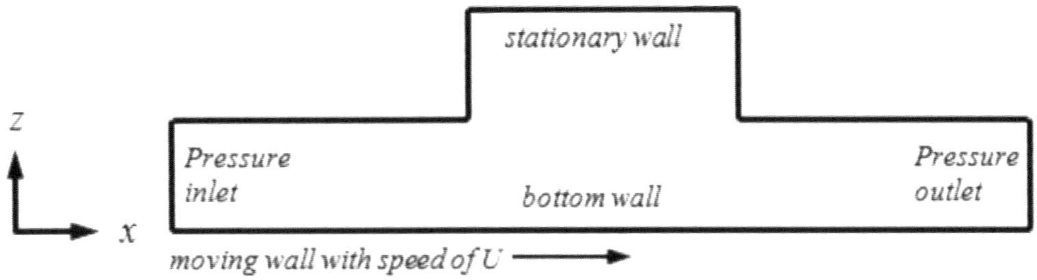

Figure 3. Model of boundary condition.

3. Result and discussion

Results are presented and analysed for inertia effect in slider with Navier-Stokes and Reynold equation. In this study inertia was investigated using two variations of conditions, low inertia and high inertia. The results of the study are represented by the pressure distribution as shown in Figure 4. Based on the results, the pressure distribution under low-inertia conditions is relatively simple and the cavitation does not not appear. The same trend is found between Navier-Stokes and Reynold equation for the case of low inertia and it indicates that both methods can be applied to this condition. This is due to low inertia conditions, slider speed is very slow. Different from low inertia, the high inertia conditions has a region cavitation and a higher value of higher pressure distribution. The trend between the prediction of Navier-Stokes and Reynolds equation is different. This difference is caused by the flow of vortex on the texture slider. Vortex flow causes the increase pressure. It is an inertia phenomenon in the texture slider bearing. While on Reynold equation, inertia is ignored. Therefore, the pressure predicted by Reynolds equation is lower when compared to Navier-Stokes. These results show that inertia has a positive effect on hydrodynamic performance.

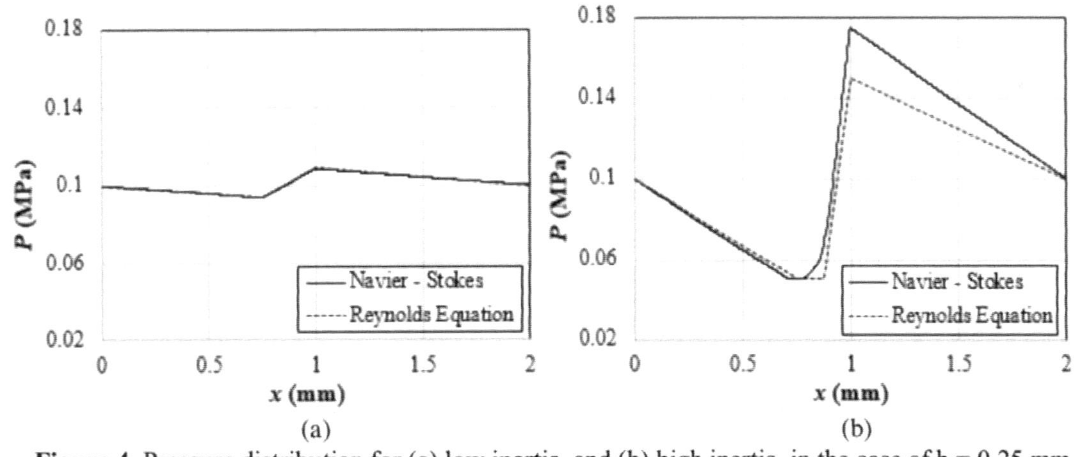

Figure 4. Pressure distribution for (a) low inertia, and (b) high inertia, in the case of b = 0.25 mm.

The vortex flow on the slider can be represented in the CFD approach. Figure 5 is a stream function on high inertia conditions with a texture length of 0.25 mm. Vortex occurs at the end position of the texture before cross section reduction. This position is where the maximum pressure occurs from the length of the slider used. While on Reynold equation, such effect is ignored. Therefore, the effect of inertia greatly affects the bearing analysis. That is the reason why inertia becomes important in slider bearing analysis.

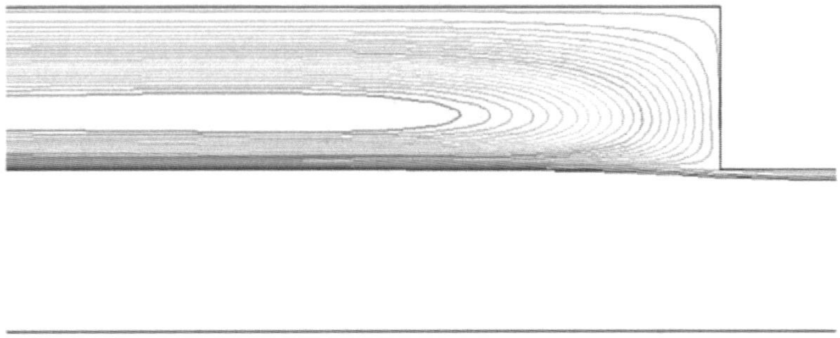

Figure 5. Streamline of fluid flow in high inertia condition.

In the present study, high inertia and low inertia conditions are also applied for the case of textured length with texture length 0.5 mm as shown in Figure 6. It can be seen that in Figure 6(a) the tendency of no cavitation is observed. While in Figure 6 (b) the occurrence of pressure is detected in the beginning of texture.On the other words, the inertia leads to the existence of the cavitation in the textured lubricated contact. With respect to the pressure profile, the prediction for the case of high inertia is greater than that for low inertia. Therefore, based on these results it can be said that inertia is able to increase hydrodynamic pressure. Furthermore, for low inertia conditions, the pressure distribution between Navier-Stokes and Reynolds have the same trend. This means that under low inertia conditions both methods can be applied. The high inertia condition can be modeled with Navier-Stokes.

A similar condition can be found when the texture length is increased to 0.75 mm as shown in Figure 7. For low inertia the small Reynolds number produces the tend in which the cavitation does not occur. While for the case of high inertia, there is a cavitation region. Other interesting result is that the inertia leads to lower hydrodynamic pressure as seen in Fig. 7(b). It indicates that the there is a critical value of texture length.

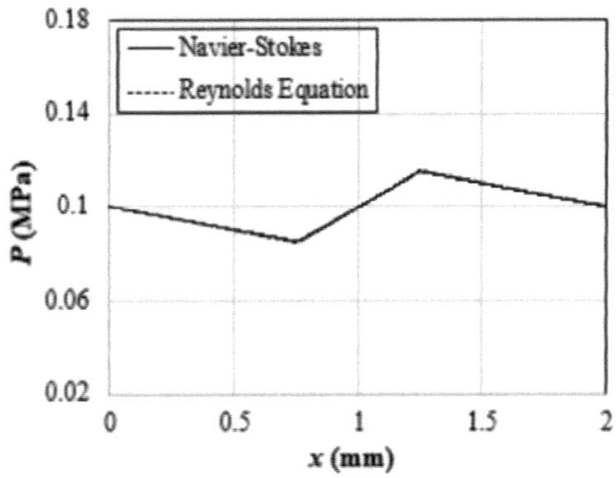

(a)

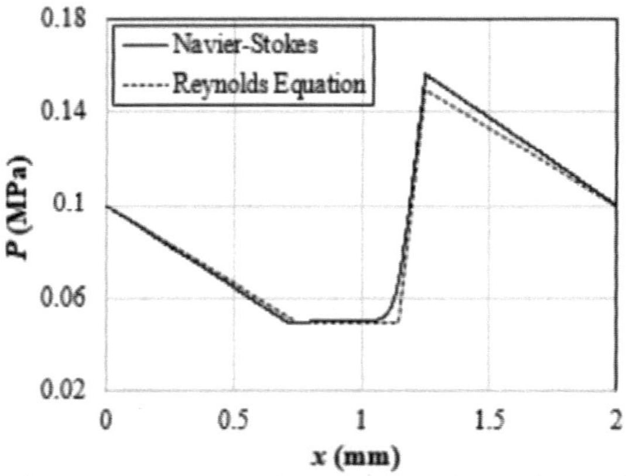

(b)

Figure 6. Pressure distribution for (a) low inertia, and (b) high inertia, in the case of b = 0.25 mm.

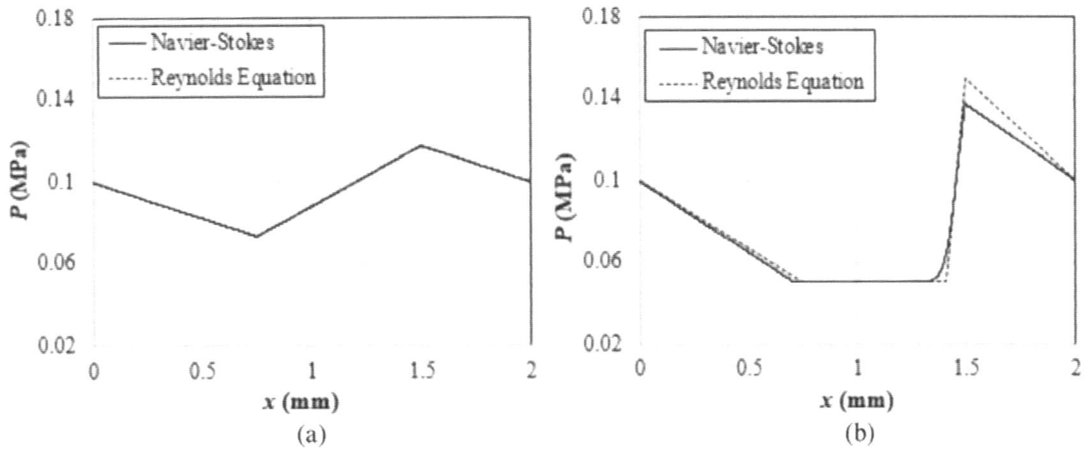

Figure 7. Pressure distribution for (a) low inertia, and (b) high inertia, in the case of b = 0.75 mm.

From the simulation results, it is also known that high inertia conditions lead to cavitation. The cavitation region increases with increasing the length of the texture. Inertia also increase the maximum pressure. But under certain conditions, by increasing the length of texture, inertia has a negative effect. It proves that inertia does not always have a positive effect on hydrodynamic pressure.

The comparison of load support is shown in Figure 8. As a note, the load support is obtained from the calculation of the integral of the area under the graph of pressure distribution. The highest load support is found when with a texture length = 0.25 mm with high inertia conditions.

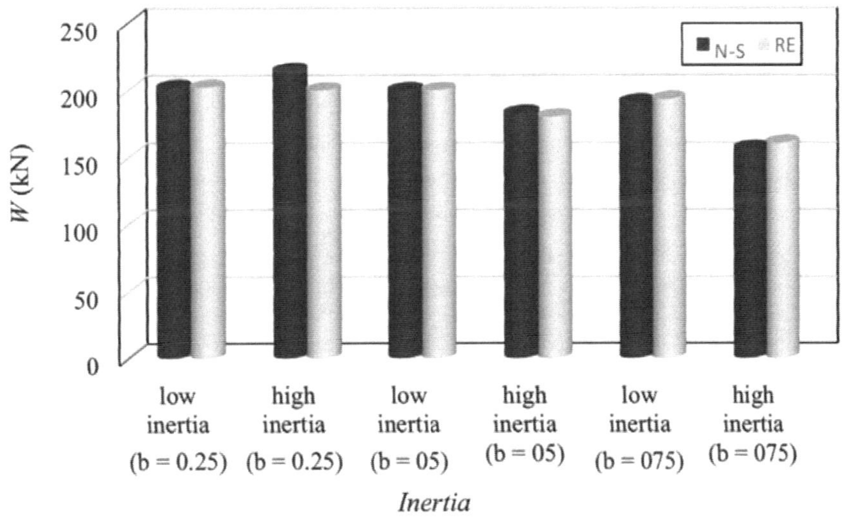

Figure 8. Comparison of the prediction of load support for various conditions.

4. Conclusions

In the present paper, two methods (i.e. Navier-Stokes and Reynolds equations) were used to to investigate the effects of inertia in a textured slider bearing. The cavitation model was adopted to model the reformation boundary. Based on the discussion above, the conclusions can be highlighted as follows:

1. Inertia brings two different effects; it can improve the pressure but on the other hand, it can also reduce the pressure depending on the texture length.
2. The increase in the texture length increases the cavitation region.
3. Cavitation region increases with increasing the Reynolds number.

References

[1] Fowell M T, Mediana S, OlverA V, Spikes H A, Pegg I G 2012 *Tribol. Int.* **52** 7
[2] Gherca A R, Maspeyrot P, Hajjam M, Fatu A 2013 *Tribol. Trans.* **56:3** 321
[3] Henry Y, Bouyer J, Fillon M 2015 *Proc. Inst. Mech. Eng. Part J: J. of Eng.* **229(4)** 362
[4] Yagi K, Sato H, Sugimura J 2015 *Tribol. Online.* **10, 3** 232
[5] Zhang H, Dong G, Hua M, Guo F, Chin K S 2015 *Ind. Lubr. Tribol.* **67: 4** 359
[6] Anil B, Shinde, Prashant M. Pawar 2017 *Tribol. Int.* **144** 349
[7] ANSYS, ANSYS Fluent, version 14.0: user manual. ANSYS, Inc., Canonsburg, USA., 2011
[8] Muchammad M, Tauviqirrahman M, Jamari J, Schipper D J 2017 *Lubr. Sci* **29** 133

IOP Conf. Series: Journal of Physics: Conf. Series **1090** (2018) 012042 doi:10.1088/1742-6596/1090/1/012042

Numerical evaluation of number of textures in slip-textured lubricated contact

E Yohana*, W K Ajie, M Tauviqirrahman

Laboratory for Engineering Design and Tribology, Mechanical Engineering Department, Engineering Faculty, Diponegoro University, Indonesia

*efnan2003@gmail.com

Abstract. Surface texturing with slip can have a significant effect on the lubrication performance in hydrodynamic bearing. In the present paper, the texture shape consisting of parabola, rectangular, and triangular for partially texture bearing is of particular interest. Number of the texture cells are also varied to obtain the optimal tribological performance in terms of load support and friction force. The slip is located at the leading edge of the contact to achieve the highest performance. The modified Reynolds equation with slip is adopted and discretized using finite volume method. The tridiagonal matrix algorithm is used to solve this equation. The numerical results show that the number of texture cells has a significant effect on the load support as well as the friction force. However, it is also found that there is no texture shape which is superior for all performances. Thus, a particular care must be chosen with respect to the number of textures as well as the texture shapes to obtain the optimal tribological performance. This finding can be used as a guideline to design the bearing with optimal performance.

1. Introduction

For more than a century, with increasing the demand for energy-saving, there has been a growing application of surface texturing due to its good behaviour in enhancing the performances of lubricated tribological pairs. A number of surface texturing researches were carried out by tribologists numerically and experimentally to solve the lubrication problem. Brajdic-Mitidieri et al. [1] used CFD analysis to investigate the friction coefficient of the pocketed pad bearing varying convergence ratios. They found that at high to medium convergence ratios, a suitably positioned pocket in the high pressure region of the bearing reduces the friction coefficient. Tala-Ighil et al. [2] analysed numerically the cylindrical texture shape effect on the performance of a hydrodynamic journal bearing. It was found that an appropriate arrangement of the textured area on the contact surface leads to a significant improvement of the performance. Later, Uddin et al. [3] based on the steady-state Reynolds equation investigated the effect of surface texture shape, geometry and orientation on hydrodynamic lubrication performance. It was found that the performance of the triangle, the chevron and the ellipse was significantly affected by their orientation with respect to the sliding direction.

In recent years, in addition to surface texturing, there is also a growing interest in creating the lubricated contact with boundary slip which is induced by (super)hydrophobic coating. In several cases, this hydrophobic coating is combined with the texturing to obtain more positive effect of the lubrication performance. It has been proven that partially slip-textured parallel slider bearing has a

Content from this work may be used under the terms of the Creative Commons Attribution 3.0 licence. Any further distribution of this work must maintain attribution to the author(s) and the title of the work, journal citation and DOI.
Published under licence by IOP Publishing Ltd

potential to generate load support and reduce coefficient of friction [4-7]. Tauviqirrahman et al. [6] confirmed that introducing the boundary slip at the leading edge of the textured contact was more efficient than extending the slip to the textured region with respect to the load support and friction force. Susilowati et al. [7] evaluated the effect of gap ratio of the lubricated contact on the performance of slip-textured bearing. Based on modified Reynolds equation, they concluded that the gap ratio appears to contribute to the generated friction force and the volume flow rate more than the boundary slip. In recent lubrication, in more detail work, Muchammad et al. [8] analytically developed new lubrication model based on first order Reynolds theory to study the texturing parameters as well as the slip parameter of lubricated contacts. Main interesting result was that adding slip over the whole surface could retard the presence of cavitation.

Extending the previous work of the authors in [6], the main objective of this paper is to investigate the effect of number of texture cells on the performance of the slip-textured contact in which the slip is located at the leading edge of the contact. Modified Reynolds equation considering cavitation is used to solve this case based on the finite volume method. The load support as well as the friction force as indicator of the lubrication performance of the slip-texture contact is of particular interest varying the texture shapes. The results may provide a more realistic guideline of the actual surface texture alternative in bearing.

2. Methodology

It is known that the rectangular is the most common shape of texture used in modeling the slider bearing. However, other texture shapes can also be used to represent the texturing technique. In the present study, the numerical analysis for the case of the multiple textured pattern (Fig. 1) varying texture shapes is carried out pointing out the effect of the number of texture cells on the lubrication performance.

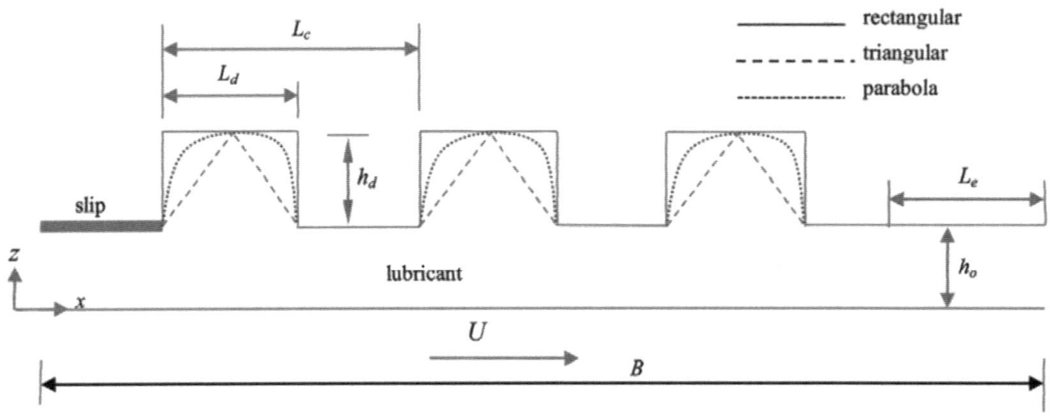

Figure 1. Multiple texture pattern with the slip at the leading edge of the contact and with three different shape textures cell (not a scale)

In the present work, the shape considered here is triangular, rectangular, and parabola as seen in Figure 1. For all following computations, the parameters used with respect to the slip-textured characteristics as well as their operating conditions are described in detail in Table 1. The bearing configuration studied here as reflected in Fig. 1 is multiple textured contact which consists of number of texture cells with slip introduced at the leading edge of the contact for improved load support [6]. In this work, the number of texture cells varying texture shapes is of particular interest. For all following computations, the texture density is set to 0.5, while for texture aspect ratio is 400. It is also assumed that the exit length is constant. In the present study, the Reynolds cavitation model is adopted.

Table 1. Multiple textured bearing configuration

Bearing length	B	0.02 m
Inlet length	a	0.0004 m (a/B = 0.02)
Pocket length	b	0.006 m (b/B = 0.3)
Sliding velocity	U	1 m/s
Viscosity	η	0.01 Pa s
Atmospheric pressure	P_{atm}	100 kPa
Minimum film thickness	h_o	1 μm
Pocket depth	h_d	2.5 μm
Texture cell length	L_c	1 mm
Dimple length	L_d	2 mm
Exit length	Lc	0.2*B
Number of texture cells	n	2-9
Slip coefficient	α	0.02 m^2s/kg

The model of lubrication presented here is based on the fact that slip at the interface between lubricant and surface will exist. It is worth noting that the slip length model is used to address the modeling of the boundary slip for the hydrodynamic analysis. The lubrication model based on modified Reynolds theory can be described as follows [6]:

$$\frac{\partial}{\partial x}\left(\frac{h^3}{\eta}\frac{\partial p}{\partial x}\left(1+\frac{3\alpha\eta}{h+\alpha\eta}\right)\right)=\frac{\partial}{\partial x}\left(6Uh\left(1+\frac{\alpha\eta}{h+\alpha\eta}\right)\right) \quad (1)$$

The physical meanings of the symbols in Equation 1 are as follows: h the film thickness (gap) at location, p the lubrication film pressure, α the slip coefficient, and η the lubricant viscosity.

For solving these problems studied here, the modified Reynolds equation (Eq. (1)) is discretized over the flow using the finite volume method, and is solved using tridiagonal matrix algorithm (TDMA). By employing the discretization scheme, the computed domain is divided into a number of control volumes using a grid with uniform mesh size. The grid independency is validated by various numbers of mesh sizes.

3. Results and Discussion

This section presents a detailed investigation of the effect of several geometrical configurations on the hydrodynamic performances (load support, and friction force) of a slip-textured parallel bearing. The most important parameters analysed are the texture cell number and texture shape. Two lubrication performances, i.e. the load support and the friction force are of particular interest. The load support is defined as the integration of pressure over surface area, whilst the friction force is obtained by integrating the shear stress along the surface area.

3.1. Effect of number of texture cells on the load support

Figure 2 presents the effect of the number of texture cells on the load support. It can be observed that there is "increase-then-decrease" characteristic of the load support with increasing the texture cell number. This trends prevails for all texture shapes studied here. It indicates that for each of texture shapes there is an optimal value of texture number. It seems that six texture cells become the optimal value which lead to the highest load support whatever the texture shape.

The numerical results based on Fig. 2 also show that the triangular shape gives the highest load support for all value of the texture numbers. On the other words, either in single texture cell or in multiple texture cells, textured bearing with triangular shape is preferable to other shapes. The interesting finding is that the rectangular shape produces the lowest load support. From the physical point of view, the most possible explanation is that the rectangular shape makes the texture cell receive more lubricant. If the lubricant supply is constant, the rupture film leading to the cavitation effect may present. This leads to the reduced hydrodynamic pressure and thus the lower load support.

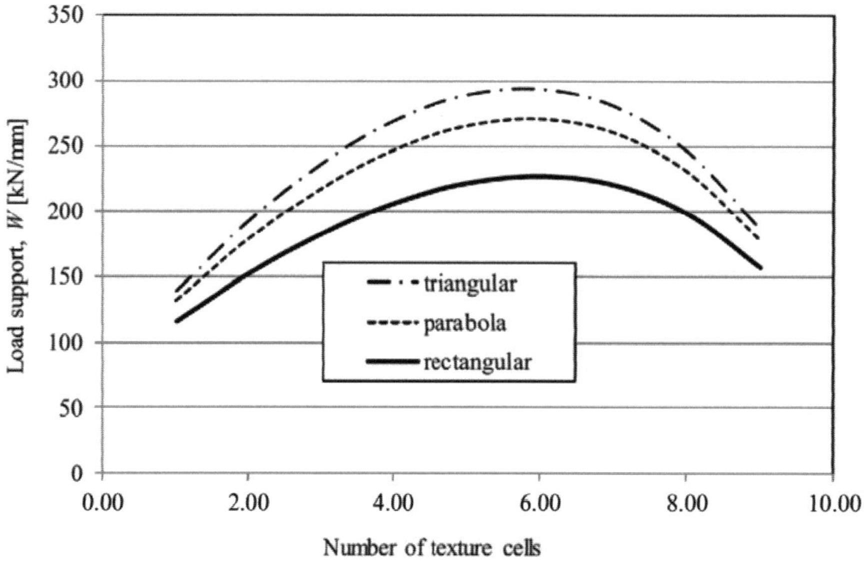

Figure 2. Effect of number of texture cell on the load support varying texture shapes

3.2. Effect of number of texture cells on the friction force

Figure 3 reflects the correlation between the friction force and the number of texture cells. It can be seen that increasing the texture cell number will reduce the friction force. This trend prevails for all texture shapes studied here. Contrary to the result in terms of load support, the texture shape of rectangular generates the positive lubrication performance in terms of friction force, that is, lowest friction force. It seems that the rectangular shape produces the lowest shear stress profile than the others. Otherwise, the triangular shape leads to highest friction force for values of texture number.

Based on the Figure 3, it is also found that the higher the number of texture cell, the lower friction force. However, it should be noted that the multiple textured pattern studied here is partial texturing, which means that the number of texture cell cannot be unlimited. There is a range value of texture cell number related to this finding, as indicated in Table 1.

Based on the numerical results mentioned earlier, it seems that from manufacturing point of view, the rectangular shape of texture may be easier to manufacture in real application compared to other shapes. By laser surface texturing, the texture cell can be created in more detail way but with high precision. Regarding with enviromental issue, the use of rectangular shape of texture can be a promosing way for saving energy. This is because the resulted friction force in bearing can be minimized so that the life time of the bearing can be longer. As a consequence, the use of resourses for manufacturing the bearing can be reduced.

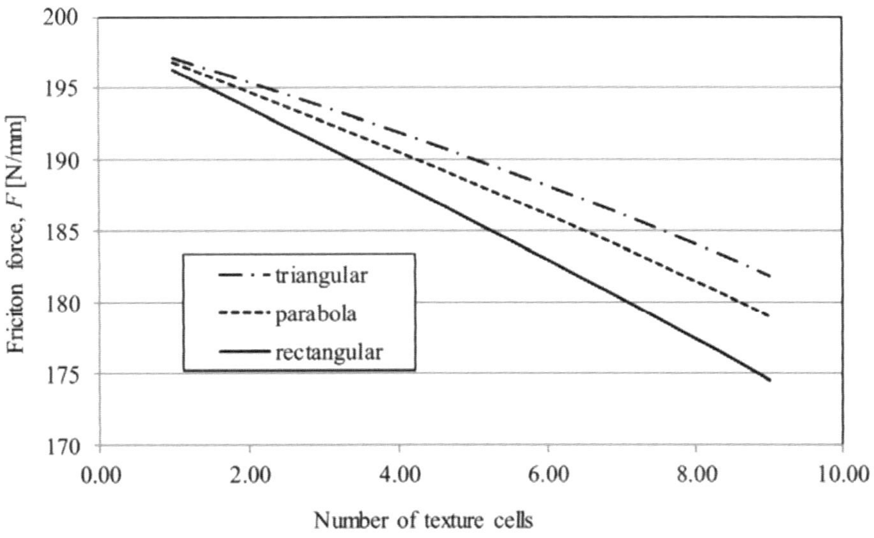

Figure 3. Effect of number of texture cell on the load support varying texture shapes

4. Conclusion

In the present work, the slip-textured contact varying the number of texture cells with three different shapes (triangular, rectangular, and parabola) was evaluated based on the modified Reynolds equation. The finite volume method combined with tridiagonal matrix algorithm was used to solve the lubrication problem. Based on the numerical analysis presented here, the conclusions can be drawn as follows:

1. The texture shape of rectangular generates the best lubrication performance in terms of friction force.
2. In terms of load support, the triangular shape is recommended to use.
3. A particular care must be chosen with respect to the number of textures as well as the texture shapes to obtain the optimal tribological performance

This findings can be used as guideline to design the textured bearing for achieving the longer life time of the bearing.

IOP Conf. Series: Journal of Physics: Conf. Series **1090** (2018) 012042 doi:10.1088/1742-6596/1090/1/012042

References

[1] Brajdic-Mitidieri P, Gosman AD, Ioannides E and Spikes HA 2005 *J. Tribol* **127** 803

[2] Uddin MS, Ibatan T and Shankar S 2016 *Lubr. Sci.* **29** 153

[3] Tala-Ighil N, Fillon M, P Maspeyrot 2011 *Tribol. Int.* **44** 211

[4] Rao TVVLN, Rani AMA, Nagarajan T and Hashim FM 2012 *Tribol. Int.* **56** 121

[5] Tauviqirrahman M, Muchammad, Jamari and Schipper DJ 2014 *Tribol. Trans.* **57** 134

[6] Tauviqirrahman M, Ajie WK, Yohana E, Muchammad M and Jamari J 2016 *Int. J. Eng. Tech.* **8**(2) 913

[7] Susilowati S, Tauviqirrahman M, Jamari J and Bayuseno AP 2016 *Tribol. - Mater. Surf. Interfaces* **10**(2), 85

[8] Muchammad M, Tauviqirrahman M, Jamari J and Schipper DJ 2017 *Lubr. Sci.* **29** 133

IOP Conf. Series: Journal of Physics: Conf. Series **1090** (2018) 012043 doi:10.1088/1742-6596/1090/1/012043

Intelligent Software Product Line For Supply Chain

Ahmad Nurul Fajar[1], Ditdit Nugeraha Utama [2], Gunawan Wang [3]

[1] Information Systems Management Department, BINUS Graduate Program-Master of Information Systems Management, Bina Nusantara University Jakarta, Indonesia 11480, E-mail : afajar@binus.edu
[2] Information System Department, UIN Syarif Hidayatullah, Jakarta, Indonesia, E-mail : dit.utama@uinjkt.ac.id
[3] Information Systems Management Department, BINUS Graduate Program-Master of Information Systems Management, Bina Nusantara University Jakarta, Indonesia 11480,E-mail:gwang@binus.edu

Abstract. Implementation of intelligent system that was utilized by research activities for numerous domains is very broad. Industry, transportation, health, and also government are several sector examples where such a system is incredibly beneficial. The condition of Indonesian government system for supply chain environment is a challenge for develop software product line which has intelligent mechanism. We proposed the Intelligent Software Product line for supply chain. We proposed conceptual framework which is a guidance for system developer to create system. We called the framework with I-SPL Framework.

1 Introduction

Software Product Line or SPL can be used to facilitate software features diversity in similar domain. The software features diversity occur while multiple developer construct the software system. The failure to develop SPL is can make wasted resources and shortened software life time. Besides that, the complexity of e-government applications is increasing significantly in the government environment. It can make an unpredictable changing in government policy .The ability to respond government policy changes is enabled by interoperability between business processes.

According to it, the variation of business process can be a challenge to reuse of parts of business processes. In order to construct intelligent system, it can supported with SOA Approach. However, People were actually getting confused about key differences between Service Oriented and Object Oriented. In fact, most of SOA development might have similarities methodology with software development. However, the quality of service that will be delivered possibly does not fully required nor support the services itself if SOA develop in software development perspective.

Service Oriented Architecture (SOA) is one of services technology provide integrity, compatibility, and support the smallest unit of business functionalities. Most of literature research, currently identified at least four methodologies for SOA development which is mostly under discussion. SOA methodologies including SOMA/RUP, SOUP Methodology, SOAF methodology, and SOADL. Furthermore, each methodology basically had been supporting business process either in bottom-up, top-down, and melt-in the-middle. It would be depends on requirement based on company needs [1]. Based on four methodologies which consider as best practices for SOA

Development, according to several literature research performs that SOADL is one of the method which is provide flexible and adaptable toward some methodologies, especially software development. SOADL methodology, when all phases implemented, will provide independent cross multi-platform services and dynamic toward any other services [2]. In other hand, when SOADL work together with a software development approach, besides providing set of service composition,

Content from this work may be used under the terms of the Creative Commons Attribution 3.0 licence. Any further distribution of this work must maintain attribution to the author(s) and the title of the work, journal citation and DOI.
Published under licence by IOP Publishing Ltd

it will deliver client application based on process-oriented [3]. Means, the service will be more flexible, even the business process must be changed both in codes or its processes. This flexibilities of business process toward services well known as orchestration—which is basically part of business logic for SOA abstraction. It's proven that SOADL methodology flexibilities toward software development in some way will deliver benefits in order to cover disadvantage while implementing SOA [2].

2 Related Work

Implementation of intelligent system that was utilized by research activities for numerous domains is very broad. Industry, transportation, health, and also government are several sector examples where such a system is incredibly beneficial. For instance in industry sector, intelligent system (specifically based on cyber-physical system) was used as a conceptual context to realize an intelligent chemical industry. It was practically used to improve the optimal control level for the whole process of the system [5]. Regarding the service oriented manufacturing system, intelligent system was also operated as an engineering framework. It was realized for becoming an intelligent automation control and execution of manufacturing system [6]. It was similarly used to optimize (as context-aware service system) in providing the information for miners according to real time situation in coal mine industry [7]. Furthermore, in transportation area, intelligent transportation system was applied every so often. It was explicitly executed for solving many kinds of problem that relate to road traffic and transportation. It was used as the way to protect the transportation infrastructure in the purpose for improving safety of transportation [8].

The intelligent system concept was adopted as well to construct the cooperative intelligent transportation system. Based on two main functions, the system can support to reduce CO2 emission and free traffic congestion by finding the best route [9]. It also was adopted to be implemented in preventing accident [10], in reducing negative impacts of road transport on the city environment [11], solving parking problems [12], or even in solving the problem of product distributing particularly in supply chain case [13]. In health sector, the intelligent system is very valuable. One of them is for diagnosing a child mental health disorder. It was applied in China for solving roughly 60 million mental health care issues [14]. The system was used also to record personal health data. It was operated in purpose to provide the health requirement analysis [15], the concept of intelligent system was adopted as well to develop the system of health care service [16]. Exclusively in government area, several researchers used intelligent system as a hot issues or methods infrequently to be studied. For example, it was operated for emergency response services [17], advisory service innovation [18], minimizing information gap in government agencies and public institution [19], and planning a national defense budget [19]. SPLE can be used to modify, customize and configure context [20].This approach can minimize cost and time to develop software systems, to improve the reusability of commonality features, to manage variability feature, time to market, and to improve product quality. SPLE consists of two processes that are domain engineering and application engineering [20].SPLE consists of domain engineering and application engineering. It can support to build a robust platform and build specific user applications [20]. SPLE requires many technical, financial, organizational, process and market considerations [20].

The benefit of SPLE compare to traditional reuse is maintenance [20]. According to [21], dynamic environment could be anticipate with the system architecture based on commonality and variability in business processes. Besides that, we can create business process model from goal model [22].Feature Model could be generated from business process model [23]. According to [24], FODA, is a method of domain analysis in order to compile the thinking process used in building software system in domain or related classes. Domain analysis supports software reuse by capturing domain expertise, including to support communication, training, tool development, software design and specification. The domain analysis offers the ability to identify and support the development of software resource reuse. The goal of FODA is to create a domain model that represents the system family to be refined into the system as desired. The feature model visually represented in a tree-like structure in which the node represents the feature and relationship between features. According to [25], in the feature model, the connection between feature and feature grouping is the variability of the software product line. According to [23], the big challenge domain during the modelling is tracing the various events of the variant and understanding its interdependence. According to ref [26], the feature diagram is widely used for variant model of software product (software product line/SPL). FODA is built by storing all available information. By analyzing the characteristics of the domain variant, all variants follow the general categorization type and all have certain behaviors.

3 Results and Discussion

This study discuss about software product line in supply chain domain. We proposed conceptual framework which is a guidance for system developer to create system. We called the framework with I-SPL Framework. This framework constructed from systematic literature review, observation and interview also. The I-SPL Framework consists of two modelling: feature model and service oriented model. The first phase is how to modelling software feature in supply chain domain. The results are (a)list of software features, (b)feature model, and (c)feature diagram. We suggest to using software product line engineering (SPLE) approach, specificity in feature modelling for execute this phase. According to it, software product line or SPL in supply chain domain area can be built by SPLE approach. The software system represent SPL implementation. This phase focused to identification, analysis, design and modelling supply chain domain environment. This phase should concern about business processes, sub business processes, tasks, activity, role, rules, and policy in supply chain domain environment. In order to complete this phase, we should using template and guidance relate to it. Feature modelling is an approach to create feature model and feature diagram. It will consists of some of software features. The software features are divided into mandatory features, optional features, and alternative features. Feature diagram could be visualization using tools and standard notation in this study. The second phase is how to modelling services using service oriented analysis and design in SOMA [27]. According to it, the results are business services, process services, and task services. This phase is derived from modelling the feature.

Feature model and feature diagram would be a baseline before modelling the services. This modelling focused on how to analysis and design in information system based on services. Business processes in supply chain domain environment are could be transform into some of business services. Process services are created from business services. Supply chain business processes having complexity in flown of information and data. In order to take granularity, we should decompose business services, process, and task into atomic and can not to be decomposed more. Besides that, mechanism in intelligent system should be embedded to system architecture. Its purpose to adding intelligent system behaviour to take user profiling before make a recommendation. While we use SOMA to modelling some of services in supply chain domain environment, we should doing service oriented analysis and design.

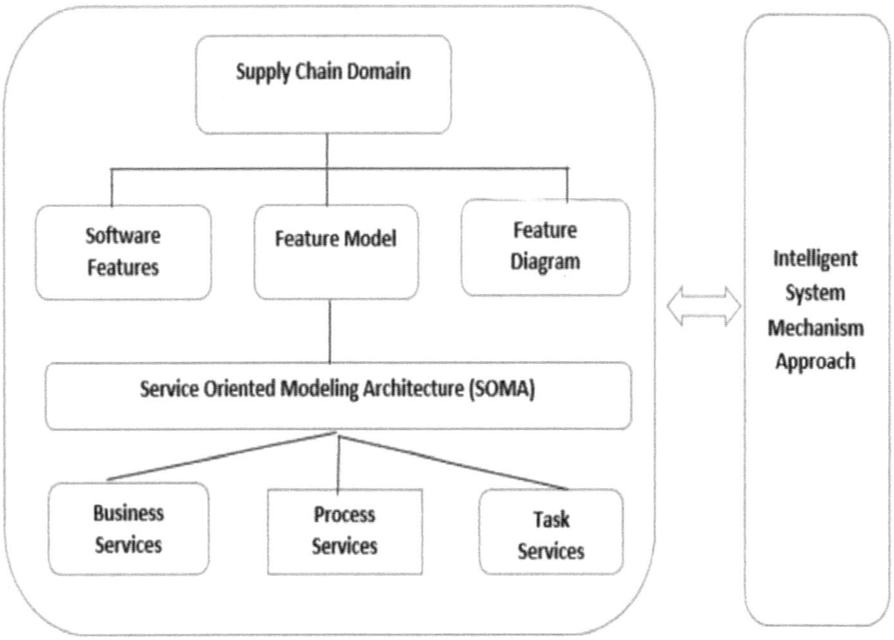

Figure 1. I-SPL Framework

Figure 1 above described our proposed framework, which is called I-SPL Framework. The second phase is focus to modelling some Decompose business processes is part of stage in modelling some of services. It will be reference to existing business processes in supply chain domain environment area. Its purpose is to represent granularity processing. After doing service oriented analysis, the next stage is service oriented design. This stage focused in task services, such as task centric design.

The guidance of to task centric design relate to [27]. The purpose of define workflow logic in task centric design is to find out which application service that will communicate with task centric. Task centric is identified based on entity centric. It would be classified as part of the business service layer. Typically, task centric is a service that could not be able to reuse. In order to implementation the services modelling, we use service oriented architecture (SOA) approach. The necessary part of SOA is interaction business process which transforms into BPEL (Business Process Execution Language). When a user would like to use web services, automatically, they will invoke the web services. The I-SPL framework also concern in intelligent system mechanism approach. There are some reason for it, such as to make a recommender system in software product line based on SOA. However, the recommender system should refer to the process of modelling the software features and modelling some services also. Its purpose to minimize the mismatch and lack of the needs of multi entities in supply chain domain environment area.

4 Conclusion

The I-SPL Framework has been propose to guidance the construction of SPL in supply chain domain environment. It has consists of 2 part in modeling, such as modeling software features and modeling some of services in software system. We use SOA Approach to support the create reusable business services, composite application and composite business processes. The I-SPL Framework also having intelligent mechanism system to support users activity in multi entities supply chain.

References

[1] Fareghzadeh, Nafise. "Service identification approach to SOA development. "*Proceedings of World Academy of Science, Engineering and Technology*. Vol. 35. 2008.

[2] T. Erl, "Service Oriented Architecture: Concept, Technology, and Design," Prentice Hall, 2005.

[3] M. Rosen, B. Lublinsky and K. T. Smith, Applied SOA, Indianapolis: Wiley Publishing, Inc, 2008.

[4] Ji, X., He, G., Xu, J., and Guo, Y. 2016. Study on the mode of intelligent chemical industry based on cyber-physical system and its implementation. *Advances in Engineering Software*, vol. 99, pp. 18-26.

[5] Giret, A., Garcia, E., and Botti, V. 2016. An engineering framework for service-oriented intelligent manufacturing systems. *Computer in Industry*, vol. 81, pp. 116-127

[6] Xue, X., Chang, J., and Liu, Z. 2014. Context-aware intelligent service system for coal mine industry. *Computers in Industry*, vol. 65, no. 2, pp. 291-305

[7] Janusova, L. and Cicmancova, S. 2016. Improving safety of transportation by using intelligent transportation systems. *Procedia Engineering*, vol. 134, pp. 14-22.

[8] Sun, L., Li, Y., and Gao, J. 2016. Architecture and application research of cooperative intelligent transportation systems. *Procedia Engineering*, vol. 137, pp. 747-753.

[9] Schmeidler, K. and Fencl, I. 2016. Intelligent transportation systems for Czech ageing generation. *Perspectives in Science*, vol. 7, pp. 304-311

[10] Malecki, K., Iwan, S., and Kijewska, K. 2014. Influence of intelligent transportation systems on reduction of the environmental negative impact of urban freight transport based on Szczecin example. *Procedia – Social and Behavioral Sciences*, vol. 151, pp. 215-229.

[11] Ashokkumar, K., Sam, B., Arshadprabhu, R., and Brito. 2015. *Procedia Computer Science*, vol. 50, pp. 58-63.

[12] Utama, D.N., Zulfiandri, Marho, F. 2012. Intelligent model for distributing product in supply chain management. *Proceeding of International Conference on Management and Artificial Intelligence*, vol. 35, pp. 55-59

[13] Chen, B.M., Fan, X.P., Zhou, Z.M., and Li, X.R. 2011. Neural network structure study in child mental health disorders intelligent diagnosis system. *Procedia Environmental Sciences*, vol. 8, pp. 669-678.

[14] Genitsaridi, I., Kondylakis, H., Koumakis, L., Marias, K., and Tsiknakis, M. 2013. Towards intelligent personal health record systems: review, criteria and extensions. *Procedia Computer Science*, vol. 21, pp. 327-334.

[15] Kaur, P.D. and Chana, I. 2014. Cloud based intelligent system for delivering health care as a service. *Computer Methods and Programs in Biomedicine*, vol. 113, no. 1, pp. 346-359

[16] Amailef, K. and Lu, J. 2013. Ontology-supported case-based reasoning approach for intelligent m-government emergency response services. *Decision Support Systems*, vol. 55, no. 1, pp. 79-97.

[17] Li, M. 2011. Online government advisory service innovation through intelligent support systems. Information and Management, vol. 48, no. 1, pp. 27-36

[18] Kim, T.H., Hong, G.H., and Park, S.C. 2008. Developing an intelligent web information system for minimizing information gap in government agencies and public institution. *Expert Systems with Applications*, vol. 34, no. 3, pp. 1618-1629

[19] Wen, W., Wang, W.K., and Wang, C.H. 2005. A knowledge-based intelligent decision support system for national defense budget planning. *Expert Systems with Applications*, vol. 28, no. 1, pp. 55-66.

[20] Pohl, K, Guinter,B, & Frank VL, Software Product Line Engineering, Foundations, Principles, and Techniques,SPRINGER, 2005

[21] Fajar, A.N., Budiardjo,E.K, Hasibuan,Z.A., "System Architecture In The Dynamic Environment Based on Commonality and Variability Business Processes, 8th ICCM, Seoul, 2012

[22] Fajar, A.N., and Shofi,I.M.,"Goal Model to Business Process Model", International Journal of Electrical and Computer Engineering (IJECE), 2016

[23] Bae, Jongsu, "A Method to Generate a Feature Model from a Business Process Model for Business Applications", *Korea Advanced Institute of Science and Technology*, Daejeon, Korea, 2004.

[24] Kyo C. Kang, Sholom G. Cohen, James A. Hess, William E. Novak, and A. Spencer Peterson, "Feature-Oriented Domain Analysis (FODA) Feasibility Study. Pennsylvania: Software Engineering Institute", 1990.

[25] Thorsten Koch, Jorg Holtmann, David Schubert, and Timo Lindemann, "Towards Feature-based

Product Line Engineering of Technical Systems", *3rd International Conference on System-integrated Intelligence: New Challenges for Product and Production Engineering, SysInt 2016*, 2016, pp. 447-454.

[26] Shamim Ripon, M. M. Piash, S. M. A. Hossain, and M. S. Uddin, "Semantic Web Based Analysis of Product Line Variant Model", *International Journal of Computer and Electrical Engineering, Vol. 6*, 2014, pp. 1-6.

[27] Erl Thomas, Service Oriented Architecture Concepts, Technology and Design. Prentice Hall 2005.

Design of Real Time Facial Tracking and Expression Recognition for Human-Robot Interaction

W.S. Mada Sanjaya[1,2], **Dyah Anggraeni**[1,2], **Atip Juwardi**[1,2], **and Madinatul Munawwaroh**[1,2]

[1]Department of Physics, Faculty of Science and Technology, Universitas Islam Negeri Sunan Gunung Djati Bandung, Indonesia
[2]Bolabot Techno Robotic Institute, CV. Sanjaya Star Group, Bandung, Indonesia

E-mail: madasws@gmail.com

Abstract. In this study, we investigate the facial tracking and expression recognition to developed Social Robot called SyPEHUL (System of Physics, Electronics, HUmanoid robot and machine Learning). This social robot contains by 12 Degree of Freedom (DoF) for actuating robotic head based on Arduino microcontroller to display four type different facial expressions and tracking of the human face. In this research, expression recognition processed by an algorithm based on Python 2.7 (with OpenCV library) using Cascade Classification and LBPH Face Recognizer. The result shows that the implementation of facial tracking and expression recognition to Social Robot show a good accuracy of recognition rate and works well for Human-Robot Interaction.

1. Introduction

Social Robot is an interactive robot which can communicate with the human. Researchers mostly developed an automation expressive Social Robot to entertaining or help human's works, for example; KISMET [1], Flobi [2], Eddie [3], Probo [4], Muecas [5], and other [6]. To build the constructions of Social Robot have many component and complex methods. One of the most important parts of Social Robot is the camera as its eyes (vision sensor) and some face detection and recognition program needed to make the robot can looking and knowing something.

Image processing need to be able to recognize an objects, whether in the form of colors, shapes, etc. Especially to detect human face, need some face detection program. Many method to get face detection, for example: Viola-Jones method [7] [8] [9], Roberts Cross method [10], and other [11] [10]. And many classifier to recognize a face detection, such as; Local Binary Pattern (LBP) [12] [13], Support Vector Machine (SVM) [14], AdaBoost [15], Hidden Markov Model (HMM) [16], Bayesian [17], and other [18] [19]. Implementation of face detection they are; social robot [1] - [6], automation attendance [20], home security [21] [22] [23], game [9], and other.

In this paper will be described a real-time face detection and recognize human expression by using Cascade Classification method (Viola-Jones method) and Local Binary Pattern Histogram (LBPH) Face Recognizer Method based on Python 2.7 and OpenCV Library. Finally, this research will be implemented to 12 Degree of Freedom (DoF) Social Robot named SyPEHUL (System of Physics, Electronics, HUmanoid robot and machine Learning) to recognize, imitating,

Content from this work may be used under the terms of the Creative Commons Attribution 3.0 licence. Any further distribution of this work must maintain attribution to the author(s) and the title of the work, journal citation and DOI.
Published under licence by IOP Publishing Ltd

IOP Conf. Series: Journal of Physics: Conf. Series **1090** (2018) 012044 doi:10.1088/1742-6596/1090/1/012044

and tracking human four type of different human expression (such as: smile, sad, angry, and surprise) based on Arduino microcontroller for Human-Robot Interaction.

The paper is organized as follows. In section 2, described the theoretical background of Face Recognition and Expression Recognition details. In section 3, describe the method and system design. In section 4, described a hardware design of Social Robot. In section 5, described the application of Face Tracking and Expression Recognition in detail. Finally, in Section 6 the concluding remarks are given.

2. Theoretical Background

2.1. Face Detection using Viola-Jones Method

The face detection in this study used the method by Paul Viola and Michael Jones published in 2001 [15] usually referred as the Viola-Jones method or only Viola-Jones. an approach done to detect objects in an Image is to combine the following four key concepts:

2.1.1. Simple rectangular feature often referred to as Haar features Images of human faces grouped on the bright side and the dark side. The existence of Haar features is determined by subtracting the average of the dark-region pixels by the average of the pixel light-region, show at Figure 1. If the difference is above the threshold, then the Haar features are said to be "exist".

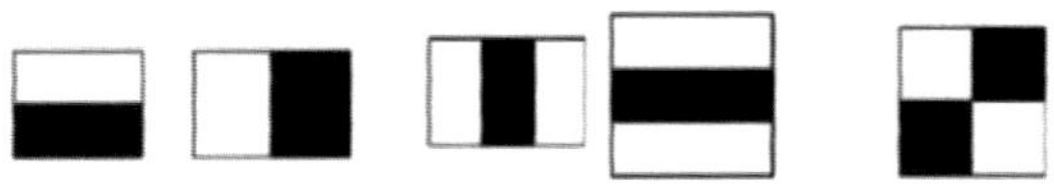

(a) (b) (c)

Figure 1. Haar Rectangular Feature [15]: (a)Edge, (b)Line, (c)Four Rectangle

2.1.2. An integrated Image for quick feature detection The integrated image is the addition of pixel values into the original image. In general, "integration" means to add all the smallest units. In this case, the smallest unit is the value. The already integrated value represents the sum of all pixels above the threshold and is to the left of the image. By starting the above procedure from the top left of the image to the bottom right of the image, the entire image can be integrated with only a few operations per pixel, show at Figure 2.

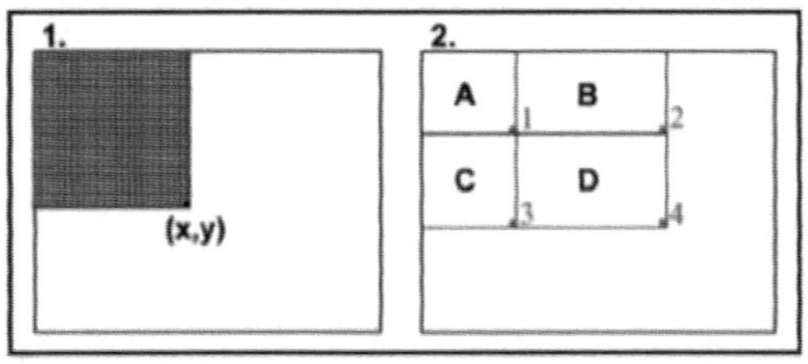

Figure 2. (1) Integration Process (2) Rectangle Divided into Multiple Segments [15]

2.1.3. Methods AdaBoost machine-learning Viola and Jones use a machine-learning method called AdaBoost method. AdaBoost classifier incorporates many "weak" to make it into a classifier that "strong". AdaBoost selects a set of "weak" classifier to combine and assign weights for each.

2.1.4. A Cascade Classifier that combines many features efficiently Viola and Jones combine a series of AdaBoost classifier as a filter chain shown in Figure 3(a). The order of filters in this cascade is based on the weight of results from the AdaBoost. The larger the filter weights in the first chain, then to eliminate the non-face image area will also be done quickly. Figure 3(b) shows the first two features of the original Viola-Jones cascade placed on a face. The first key is on the cheek area which is brighter than the eye area and the second is that the connection between the forehead and the nose is brighter than the eye.

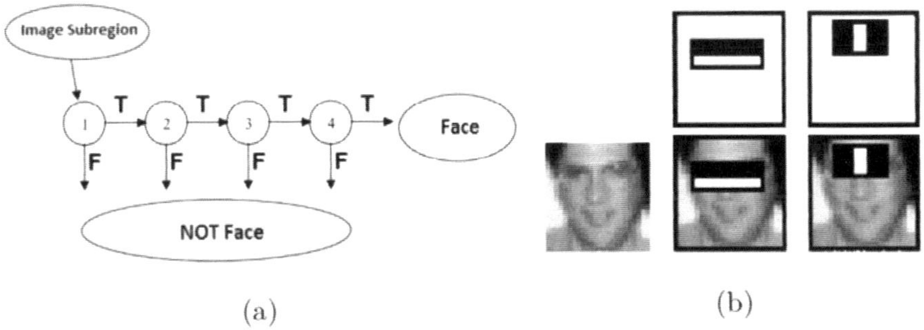

Figure 3. Cascade Classifier [15]; (a)Method, (b)Result

2.2. Face Recognizer using Local Binary Pattern (LBP) Method
Local Binary Pattern (LBP) is a texture descriptor that can be used for faces. LBP is defined as the comparison of the pixel binary value at the center of the image with 9-pixel values around it [19]. For example, in a 3x3 image, the binary value in the image center is compared with the surrounding value. By subtracting the pixel value at the center of the image by its surrounding pixel value, if the result is more or equal to 0 then it is given a value of 1 and if the result is less than 0 then it is given a value. After that, arrange 8 binary values clockwise or otherwise and convert 8 binary bits into decimal values to replace the pixel value in the center of the image, as illustrated in Figure 4.

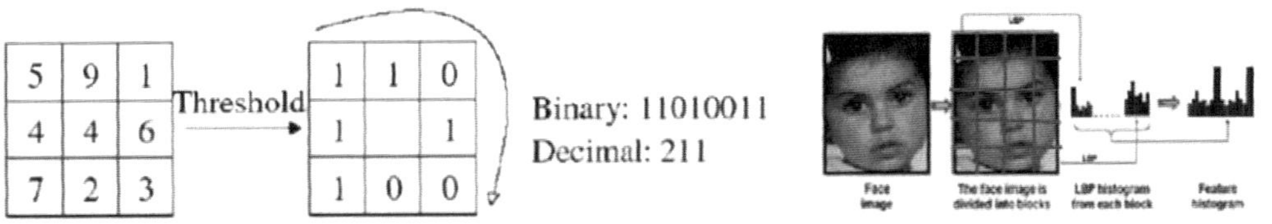

Figure 4. Illustration of Local Binary Pattern (LBP) [19]

The decimal value of 8 bits of (LBP code) can be expressed in the following equation 1 [24]:

$$LBP_{P,R}(x_0, y_0) = \Sigma_{p=0}^{p-1} s(g_p - g_0)2^p \tag{1}$$

From the function $s(x)$, define:

$$s(x) = \begin{cases} 1, x \geq 0 \\ 0, x < 0 \end{cases} \tag{2}$$

Equation 2 aims to eliminate the variability caused by the contrast illumination so that the facial image with various variations of the illumination will get almost similar output. By

IOP Conf. Series: Journal of Physics: Conf. Series **1090** (2018) 012044 doi:10.1088/1742-6596/1090/1/012044

extracting pixels using LBP, we get a new matrix value which will be changed to a histogram for facial vector features.

The LBP method can be implemented on face images to extract features which can be used to get a measure of the similarity between images. LBP code is calculated for every pixel of an image. The histogram of these patterns from the image called labels. Lable forms a feature vector and is represented the image texture. These histograms calculating the distance between the histograms can then be used to measure the similarity between the images. [24]

3. Method and System Design

In general, the main tools and component used in this research are: Personal Computer, Arduino microcontroller, Social Robot (SyPEHUL), camera, connections, and others. Program written in Arduino IDLE and Python 2.7 (with OpenCV Library). And Figure 5 shown generally the process of Facial Tracking and Expression Recognition of Social Robot SyPEHUL, below:

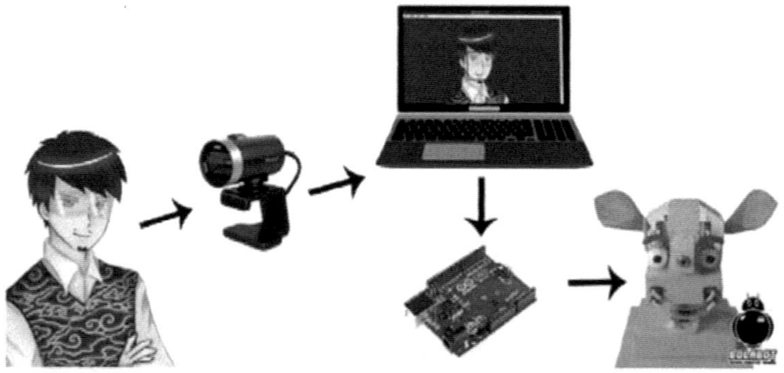

Figure 5. General Research Scheme

From the Figure 6 can be explained that the camera as a Social Robot vision sensor. The human face detection process by algorithm on the computer using Cascade Classifier Method. Then the process divided into 2 processes: the first process is making training data, consist of features extraction process using LBPH Feature and using LBPH Face Recognizer Method for classifying face expression. Second, is testing process, such as LBPH features extraction, then matching with Trained Data. The matching data be processed to obtain expression classification and face coordinates. For coordinate data is used to move the robot's neck so that the robot seems to follow our face. While the classification process Social Robot will make interaction expression (happy, sad, angry, and surprise) like the facial human expression. All processes work in real-time based on Python 2.7 (with OpenCV) and Arduino. microcontroller.

4. Hardware Design

To design a system of social robots it takes a major component especially the motor servo that serves as a robot facial driver, such as; Eyebrows, eyes, mouth, ears and neck. And in general, the materials needed the camera as a visual sensor and the Arduino microcontroller to serves as the motor servo to control the facial expression of robot. Figure 7 are the design and realization of our Social Robot named SyPEHUL which used in this research:

SyPEHUL consists of twelve motor servo (12 Dof motor servo) component which connects with Arduino microcontroller. SyPEHUL mechatronic consists of; eyebrow using 2 servos, eye using 2 servos, mouth using 4 servos, ear using 2 servos, and neck using 2 servos. From the Figure 8, each servo supplied used by the 5-volt battery which has 100 mA of current, and each ground of servo connected to the ground of Arduino Board in order get a better working. And

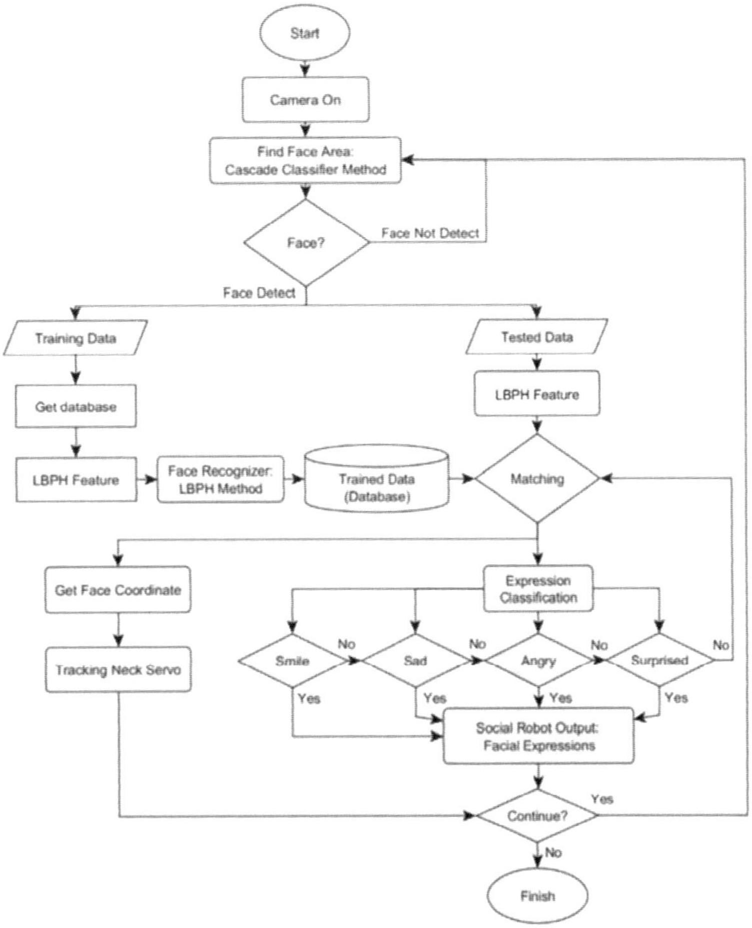

Figure 6. General System Scheme of Social Robot Algorithm

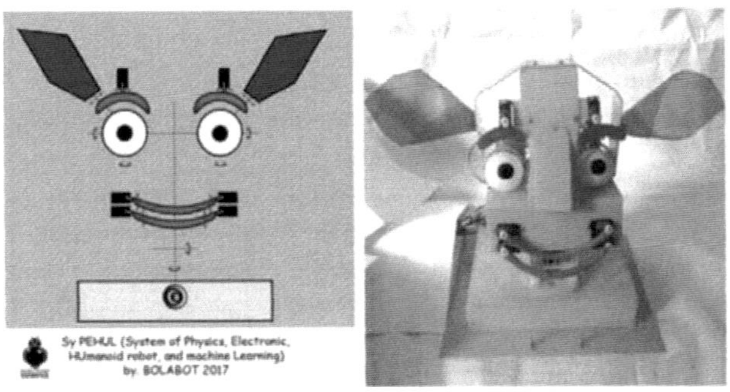

(a) (b)

Figure 7. SyPEHUL Design; (a)Design, (b)Realization

from the mechatronics, SyPEHUL Expressions divided by 4 interaction, such as; Smile, Sad, Angry, and Surprise, shown as the Figure 9.

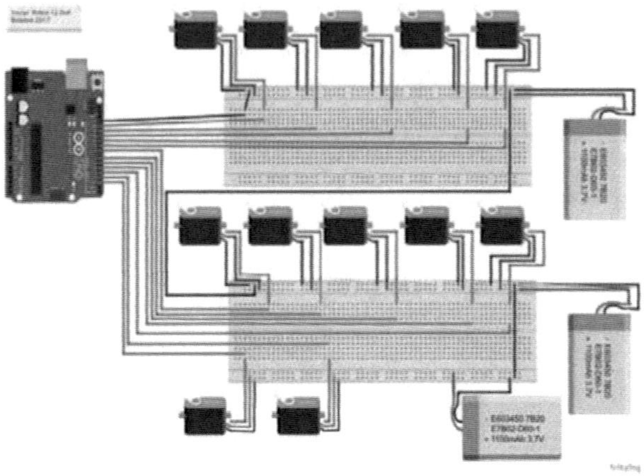

Figure 8. SyPEHUL Schematic

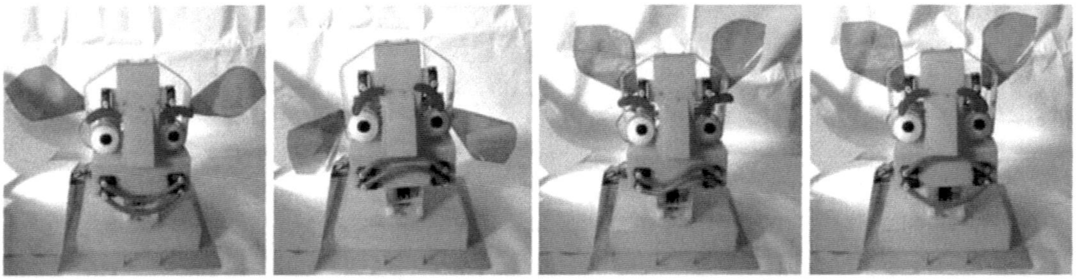

(a) (b) (c) (d)

Figure 9. SyPEHUL Expression; (a)Smile, (b)Sad, (c)Angry, (d)Surprise

5. Result and Discussion

5.1. Expression Recognition Database and System Test

The beginning of the research process is formulated ideas, concepts, objectives, problem formulation, until the study of making Social Robot. To get a Social Robot can recognize human faces, in this research used Cascade Classification method for face detection and Local Binary Pattern Histogram (LBPH) Face Recognizer to recognize the expression of human faces based on Python 2.7 and OpenCV library. Figure 10 is an example of collecting databases of some people for making training data.

Database made from 5 expression image, they are; Smile, Sad, Angry, Surprise, and Neutral. From the obtained group of data, needed for classification process using LBPH Feature Extraction and LBPH Face Recognizer, then will be called the trained data. First, the trained data is tested by trained respondent for data clarification. The test results are shown on Table 1 is the accuracy average rate of expression recognition by trained respondents (in the database). And Figure 11 show the examination of trained data by trained respondent.

Table 1 explain that the database clarification by trained respondent for human facial expression recognition shows a good accuracy. Second, the trained data is tested to the not trained respondents (outside database). The expression recognition accuracy average rate of not trained respondents shown on Table 2 and the examination by not trained respondent shown as Figure 12. And from Table 2 show that the accuracy of expression recognition by not trained data is well for some expression.

Figure 10. Database Example

Table 1. The expression recognition accuracy rate of trained respondents in Percent

Test	Smile	Sad	Surprise	Angry	Neutral
Smile	**100**	0	0	0	0
Sad	0	**90**	0	0	10
Surprise	0	0	**100**	0	0
Angry	0	0	0	**100**	0
Neutral	0	12.5	5	0	**82.5**

(a)Smile (b)Sad (c)Angry (d)Surprise (e)Neutral

Figure 11. Tested the expression recognition of trained respondent

5.2. Facial Tracking

From the programming system of face detection can get the face coordinate. In this research, the face coordinate needed to control neck motor servo of Social Robot SyPEHUL in order to track human face. The first step is set the resolution of the video image, we define as 480 x 640 pixel. Then, Equation 3 and Equation 4 shows an algorithm to find a center coordinate of the human face from camera capture.

$$xx = x + (w/2) \tag{3}$$

$$yy = y + (h/2) \tag{4}$$

Table 2. The expression recognition accuracy rate of respondents not trained data in Percent

Test	Smile	Sad	Surprise	Angry	Neutral
Smile	**80**	0	0	0	20
Sad	0	**50**	22.5	0	27.5
Surprise	0	0	**100**	0	0
Angry	0	0	0	**77.5**	12.5
Neutral	5	0	35	0	**60**

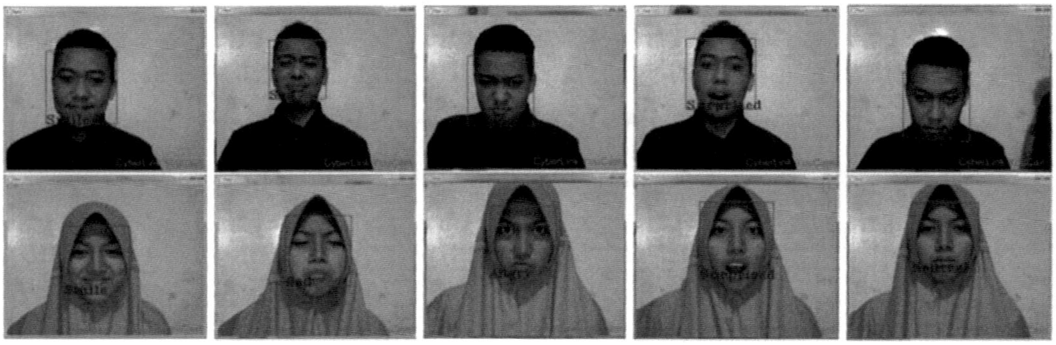

(a)Smile (b)Sad (c)Angry (d)Surprise (e)Neutral

Figure 12. Tested of not-trained respondents

x is initial horizontal face coordinate, w is width coordinate of face, xx is a center value of horizontal face coordinate, y is initial vertical face coordinate, h is height coordinate of face, yy is a center value of vertical face coordinate, and units in pixel. Figure 13 show the section of camera coordinate divided into 4 section, are; Up, Down, Right and Left.

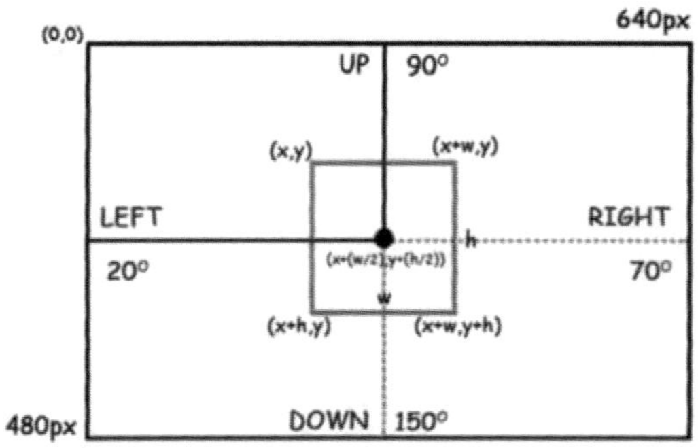

Figure 13. Camera Coordinate and angle of motor servo

Before converting the coordinate value become an angle of the motor servo, we must know the minimum and maximum angle of Social Robot neck motor servo to look up/down (vertically) or look right/left (horizontally). In this research, motor servo arrange are: (1) arrange of horizontal angle is 90° - 150°, (2) arrange of vertical angle is 20° - 70°. Figure 13 illustrated the distribution of camera pixel, direction, and angle of Social Robot neck motor servo. Thus, after the center coordinate value sent from Python to Arduino IDLE, then process to control the Social Robot neck motor servos.

5.3. Implement The System to Social Robot

The identifying and categorizing the trained data of expression recognition are successfully, then applied the trained data to be a control expression system of Social Robot SyPEHUL. Figure 14 shown a training of the Expression Recognizer:

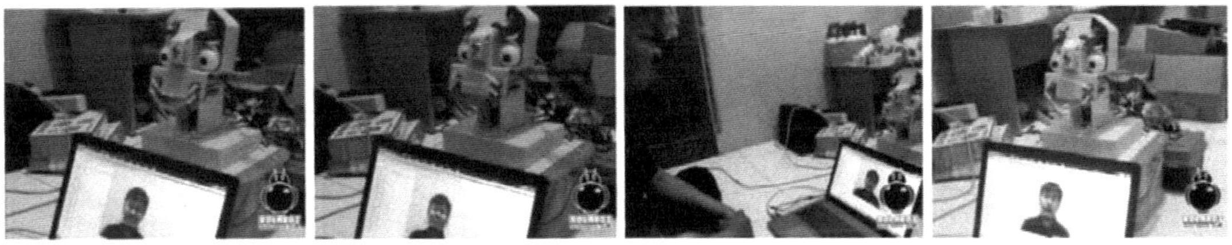

(a)Smile (b)Sad (c)Angry (d)Surprise

Figure 14. Examination of human expression recognition to Social Robot expression interaction

In addition to the classification of human facial expressions, faces detected by the camera obtained coordinates of the face position. The coordinates can define as x coordinate (as the horizontal neck) and y coordinate (as the vertical neck). Then, the coordinates can be converted as the angle to control motor servo of Social Robot by using Python and Arduino serial communication was described before. Figure 15 show the result of tracking and expression recognition for Human-Robot Interaction.

(a) (b) (c) (d)

Figure 15. Tracking and following human face; (a)Smile, (b)Sad, (c)Angry, (d)Surprise

6. Conclusion

In this research have been presented and developed a Social Robot SyPEHUL which can recognize, imitating, and tracking human face. The result of expression recognition processed by the algorithm based on Python 2.7 (with OpenCV library) using Cascade Classification and LBPH Face Recognizer method has a good accuracy rate. Also, the implementation of facial tracking and expression recognition to control 12 DoF of Social Robot SyPEHUL based on Arduino microcontroller works effectively to make four type imitating facial expression. Future works will focus on the combination of speech recognition and facial expression to enhance emotional expression of SyPEHUL for Human-Robot Interaction.

Acknowledgement

The authors would like gratefully acknowledge the financial support from DIPA UIN Sunan Gunung Djati Bandung.

References

[1] Breazeal C 2003 *Robotics and Autonomous System* **42** 167–175

[2] Hegel F, Eyssel F and Wrede B 2010 *IEEE International Symposium on Robot and Human Interactive Communication* **19** 107–112

[3] Buss M, Sosnowski S, Bittermann A and Kolja K 2006 *International Conference on Intelligent Robots and Systems* 3113–3118

[4] Doroftei I, Adascalitei F, Lefeber D, Vanderborght B and Doroftei I A 2016 *7th International Conference on Advanced Concepts in Mechanical Engineering* **147**

[5] Cid F, Moreno J, Bustos P and Nunez P 2014 *Sensors* **14** 7711–7737

[6] Shayganfar M, Rich C and Sidner C L 2012 *Int. Conf. of Intelligent Robots and Systems* **12**

[7] Tikoo S and Malik N 2016 *International Journal of Computer Science and Mobile Computing* **5** 288–295

[8] Boda R and Priyadarsini M J P 2016 *ARPN Journal of Engineering and Applied Sciences* **11** 13472–13476

[9] Zhan C, Li W, Ogunbona P and Safaei F 2008 *International Journal of Computer Games Technology* **2008** 1–7 ISSN 1687-7047

[10] Das S 2016 *International Journal of Signal Processing, Image Processing and Pattern Recognition* **9** 143–158

[11] Lakshmi H C V and PatilKulakarni S 2010 *International Journal of Computer Theory and Engineering,* **2** 552–558

[12] Shan C, Gong S and Mcowan P W 2009 *Image and Vision Computing* **27** 803–816 ISSN 0262-8856 URL `http://dx.doi.org/10.1016/j.imavis.2008.08.005`

[13] Sanjaya W S M, Anggraeni D, Zakaria K, Juwardi A and Munawwaroh M 2017 The Design of Face Recognition and Tracking for Human-Robot Interaction *ICITISEE* (Yogyakarta, Indonesia: IEEE) pp 1–6

[14] Sadek N O, Hikal N A and Zaki F W 2017 *International Journal of Advanced Computer Science and Applications* **8** 303–311

[15] Viola P and Jones M J 2001 Robust Real-time Object Detection Tech. Rep. February Cambridge Research Laboratory The Cambridge, Massachusetts

[16] Punitha A and Geetha M K 2013 *International Journal of Emerging Technology and Advanced Engineering* **3** 180–185

[17] Hedaoo S V, Katkar M D and Khandait S P 2014 *International Journal of Engineering Trends and Technology* **8** 517–521

[18] Khan M I and Bhuiyan M A A 2009 *International Journal of Computer Science and Network Security* **9** 300–306

[19] G M R and Rajesham J 2013 *International Journal of Scientific & Engineering Research* **4** 1540–1546

[20] Thakare N, Shrivastava M and Kumari N 2016 *International Journal of Computer Science and Mobile Computing* **5** 74–78

[21] Sharma N and Thanaya I 2016 *International Journal of Innovative Research in Science, Engineering and Technology* **5** 10357–10362

[22] Hajari P V and Andurkar A G 2015 *International Journal of Advanced Research in Computer and Communication Engineering* **4** 232–234

[23] Manjunatha R and Nagaraja R 2017 *International Research Journal of Engineering and Technology* **4** 437–442

[24] Rahim M A, Hossain M N, Wahid T and Azam M S A 2013 *Global Journal of Computer Science and Technology Graphics & Vision* **13**

The effect of heterogeneity in ion-exchange membrane structure on Donnan Exclusion

K. Khoiruddin[1], Danu Ariono[1], S. Subagjo[1], and I Gede Wenten[1*]

[1]Department of Chemical Engineering, Institut Teknologi Bandung, Jl. Ganesha 10, Bandung 40132, Indonesia

*Corresponding author e-mail: igw@che.itb.ac.id

Abstract. Ion-exchange membrane (IEM) can be classified into homogeneous and heterogeneous membrane according to the heterogeneity of membrane structure. Results of various studies indicated that the degree of heterogeneity significantly affects the properties and performance of IEMs, such as conductivity, permselectivity, mechanical and chemical stability, energy consumption, and limiting current density. In the case of transport properties, it may be associated with efficacy of Donnan exclusion influenced by the presence of interstitial solution and inert phase in the membrane matrix. Therefore, understanding the effect of heterogeneity on Donnan exclusion is important to provide preliminary information about IEM transport properties. In this paper, the effect of IEC/Wu on Donnan exclusion is discussed. The effect of heterogeneity on Donnan exclusion is expressed as a correlation of IEC/Wu ratio with Donnan equilibrium constant, KD, and transport number of counter-ion, t_i.

1. Introduction

Recently, membrane technology has been applied in the various field which is primarily due to its features of relatively low energy consumption, low operating cost, easy to scale-up, lower footprint, and intensive process [1–9]. Ion-exchange membrane (IEM) which has charge groups fixed in the membrane matrix is basically used for ionic separation. IEM-based processes have found many applications in fields such as water treatment, wastewater treatment, chemical synthesis, and energy conversion [10–16].

To which extent the separation efficiency in IEM-based processes depends on membrane properties i.e. electrochemical properties, chemical and thermal stabilities, and mechanical strength [11,17]. Those properties are primarily determined by water uptake and ion-exchange capacity of IEM [18]. Furthermore, the structure of IEM is also an important factor that affects the transport properties [19–22]. According to the microheterogeneous model, there is a heterogeneity in the membrane structure [23–26]. Generally, the heterogeneity is defined as the number of phases in the IEM structure. According to the microheterogeneous model, the heterogeneity of IEM is based on the fraction of gel and inter-gel phase in the membrane. The heterogeneity increases with increasing the fraction of inter-gel phase. The degree of heterogeneity can affect the ionic transport across in IEM such as conductivity and permselectivity. According to Xu et al. [18], parameters of ion-exchange capacity (IEC) and water uptake (Wu) can be used to determine the other IEC properties, qualitatively. In this paper, the effect of heterogeneity on Donnan exclusion phenomenon is discussed. The effect of

Content from this work may be used under the terms of the Creative Commons Attribution 3.0 licence. Any further distribution of this work must maintain attribution to the author(s) and the title of the work, journal citation and DOI.
Published under licence by IOP Publishing Ltd

heterogeneity on Donnan exclusion is expressed as a correlation of IEC/Wu ratio with Donnan equilibrium constant, KD, and transport number of counter-ion, t_i.

2. Donnan Equilibrium Theory and Transport-Structural Parameter

Generally, IEMs can be classified into cation- and anion-exchange membrane (CEM and AEM) according to fixed charge in the membrane structure (Figure 1). CEM has the negative charge which allows permeating cation (counter-ion) while excluding anion (co-ion). Meanwhile, AEM allows the permeation of anion (counter-ion) while excluding cation (co-ion) due to the positive charge in the membrane structure. This phenomenon is well-known as 'Donnan Exclusion'. This phenomenon has been explained by Donnan equilibrium theory [27]. In this theory, the concentration of ion in the solution will be in an equilibrium state with the concentration of ion in the IEM. This correlation is expressed as follows. The Donnan equilibrium constant, K, is defined by the equation [18]:

$$K = \left(\frac{C_{i,m}}{C_{i,s}}\right)^{1/z_i} \tag{1}$$

where $C_{i,m}$ and $C_{i,s}$ are the concentration of ions in membrane and solution, respectively, i is for cation and anion, and z_i is valence of the ion.

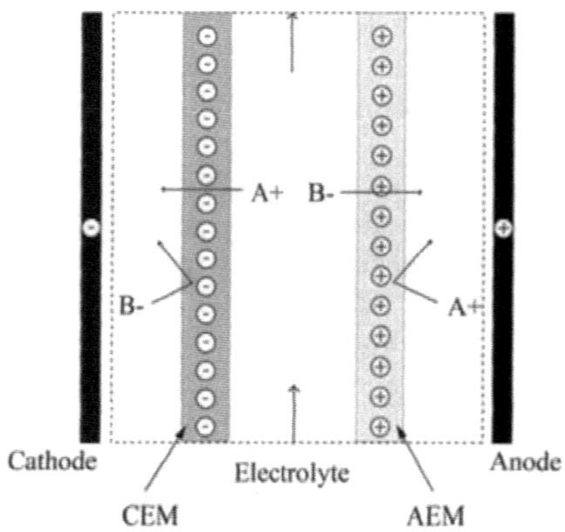

Figure 1. Schematic illustration of ion separation in an electrodialysis process.

From an electroneutrality condition both in the membrane and in solution, the K of cation (K_K) can be calculated from the concentration of solution, C_S, and ion-exchange capacity of the membrane, C_{ex} [18]:

$$K_K = \frac{-\omega C_{ex}}{2C_s} + \sqrt{1 + \left(\frac{\omega C_{ex}}{2C_s}\right)^2} \tag{2}$$

while for anion, K_A:

$$K_A = \frac{\omega C_{ex}}{2C_s} + \sqrt{1 + \left(\frac{\omega C_{ex}}{2C_s}\right)^2} \tag{3}$$

Then from equation (1) to (3), the concentration of cation and anion in the membrane ($C_{K,m}$ and $C_{A,m}$, respectively) can be calculated. The concentration of cation in IEM, $C_{K,m}$ can be determined by [18]:

$$C_{K,m} = \frac{-\omega C_{ex}}{2} + \sqrt{(C_s)^2 + \left(\frac{\omega C_{ex}}{2C_s}\right)^2} \tag{4}$$

and anion, $C_{A,m}$:

$$C_{A,m} = \frac{\omega C_{ex}}{2} + \sqrt{(C_s)^2 + \left(\frac{\omega C_{ex}}{2C_s}\right)^2} \tag{5}$$

In equation (2) to (5), the value of ω is +1 for AEM and -1 for CEM.

There are several models which have been proposed to explain the relationship between membrane structure (especially the heterogeneity) and membrane separation properties, for example, microheterogeneous, three-wire, and extended three-wire models [23,25,26,28]. In the microheterogeneous model proposed by Zabolotsky and Nikonenko [23], effective conductivity of the membrane, $L_{i,m}$, is defined as the conductivity fraction of the gel phase and the inter-gel phase (Figure 2). The relationship between the fraction of those phases is explained as follows. The effective conductivity generalized conductivity of the membrane is described as [23]:

$$L_{i,m} = \left(f_1 L_{i,g}{}^{\alpha} + f_2 L_{i,s}{}^{\alpha}\right)^{1/\alpha} \tag{6}$$

where $L_{i,g}$ and $L_{i,s}$ are generalized conductivities of gel and inter-gel phases, f_1 and f_2 are volume fractions of gel-phase and inter-gel phases, respectively, and α is a geometrical parameter which illustrates the arrangement of gel and inter-gel phases in the membrane. Furthermore, the generalized conductivities in the gel and inter-gel phases are correlated with diffusion coefficient, D [23]:

$$L_{i,g} = \frac{D_{i,g} C_{i,g}}{RT} \tag{7}$$

$$L_{i,s} = \frac{D_{i,s} C_{i,s}}{RT} \tag{8}$$

where R is gas constant and T is absolute temperature. The value of $D_{i,g}$ can be further defined from $(RT/F^2)(\sigma_{iso}/C_{ex})$ [23,29]. Here, F is Faraday constant and σ_{iso} is an isoconductance point.

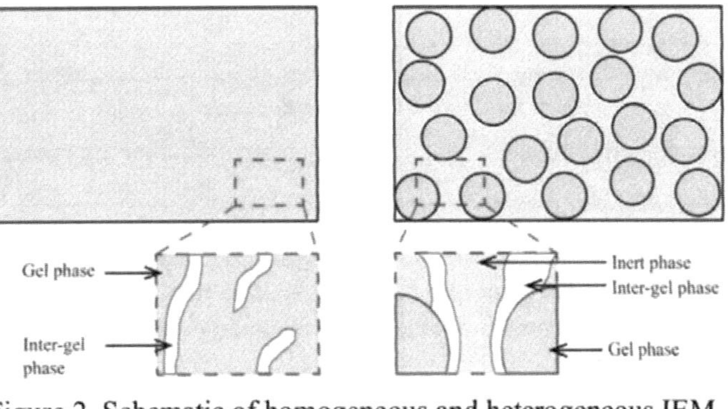

Figure 2. Schematic of homogeneous and heterogeneous IEM.

Selectivity of IEM is generally characterized by determining the transport number of counter ion, $t_{i,m}$. Transport number counter-ion is defined by [23]:

$$t_{i,m} = \frac{z_i^2 L_{i,m}}{z_K^2 L_{K,m} + z_A^2 L_{A,m}} \tag{9}$$

According to equations (4) to (9), transport number of counter-ion in the membrane is determined by solution concentration (ion concentration), ion-exchange capacity of the membrane, and the structural parameters i.e. fraction and geometrical arrangement of gel and inter-gel phases in the membrane matrix.

The conductivity of the membrane, σ_m, can also be correlated with the fraction of phases element in the membrane. For alternating current conductivity (AC), $\sigma_{m,a}$, it is defined as [23]:

$$\sigma_{m,a} = \left(f_1\sigma_g{}^{\alpha} + f_2\sigma_s{}^{\alpha}\right)^{1/\alpha} \tag{10}$$

Meanwhile, the direct current conductivity of IEM is expressed and can be simplified to [23]:

$$\sigma_{m,d} = \left(\sigma_g t_{i,g}\right)^{f_1}\left(\sigma_s t_{i,s}\right)^{f_2} \tag{11}$$

Here, the transport number of counter-ion in the gel phase can be supposed to ~1 considering the ideal selectivity of the gel phase.

3. Analysis of Donnan Equilibrium Constant

Properties of several IEMs from two classes of the membrane, i.e. homogeneous and heterogeneous are summarized in Table 1. It can be seen from the table that heterogeneous IEMs have higher inter-gel phase. It will result in the lower membrane permselectivity and counter-ion transport number (t).

It has been explained by Xu et al. [18] that ion-exchange capacity (IEC) and water uptake (Wu) can be used to qualitatively determine the others IEM properties. Wu is the amount of water adsorbed in the membrane matrix at the swelling state. Wu may also be used to indicate the void fraction in the membrane matrix filled by a solution or electrolyte. Therefore, Wu may be used as the indicator of the membrane heterogeneity. On the other hand, IEC represents the number of functional sites in the membrane matrix. Functional site concentration in the membrane is important for providing a higher selectivity. Furthermore, the ratio of functional sites to the solution in the membrane matrix may represent this fixed charge concentration. Therefore, ratio of IEC to Wu (IEC/Wu) will determine the efficacy of Donnan Exclusion as well as the membrane selectivity. This IEC/Wu value is then used in equation (2) and (3) as C_{ex} by assuming the equilibrium condition occurs in the membrane matrix.

Table 1. Properties of several IEMs[a]

Membrane	Type	Properties				Ref.
		C_{ex} (meq/cm^3)	f_2	σ_{iso} (mS/cm)	t_i	
AMV	AEM; Hm	2.09	0.06	3.79	0.986 (0.01 M)	[30]
AMX	AEM; Hm	1.43	0.10	3.04	0.999 (0.01 M)	[30]
AR-204-SZRA-412	AEM; Hm	0.34	0.10	8.9	0.93 (0.5 M)	[31–33]
CMS	CEM; Hm	2.55	0.13	2.72	0.97	[34]
CR-67-HMR	CEM; Htr	0.36	0.10	5.2	0.945 (0.01 M)	[24,31,33]
HGA	AEM; Htr	0.41	0.17	0.80	0.84 (0.1 M)	[35,36]
HGC	CEM; Htr	0.673	0.27	1.05	0.859 (0.01 M)	[35]
HJA	AEM; Htr	1.52 [b]	0.35	0.25	0.89 (0.5 M)	[32]
IPA	AEM; Hm	0.784	0.32	1.13	0.92 (0.1 M)	[35,36]
IPC	CEM; Hm	1.401	0.37	1.32	0.94 (0.1 M)	[35,36]
LNA	AEM; Htr	1.63 [b]	0.28	2.65	0.91 (0.5 M)	[32]
MA-40	AEM; Htr	3.2	0.28	-	-	[26,37,38]
MA-41	AEM; Htr	1.3	0.18	5.8	-	[31,38]
MC-40C	CEM; Htr	1.8	0.2	6.4	-	[23]
MK-40	CEM; Htr	1.7	0.16	3.5	0.87 (1 M)	[29,31,37]

[a] In NaCl solution; [b] meq/g; Hm – homogeneous IEM; Htr – heterogeneous IEM.

It can be observed from Figure 3 (a) that the value of KD is linearly increased with increasing IEC/Wu as indicated by equation (2). The value of KD is also increased by reducing solution concentration. Similar to KD vs IEC/Wu, the value of t also increases with KD and IEC/Wu (Figure 3 (b) and (c)). It is obvious since the increasing IEC/Wu results in improving the efficiency of Donnan exclusion leading to the increased membrane selectivity (shown by transport number).

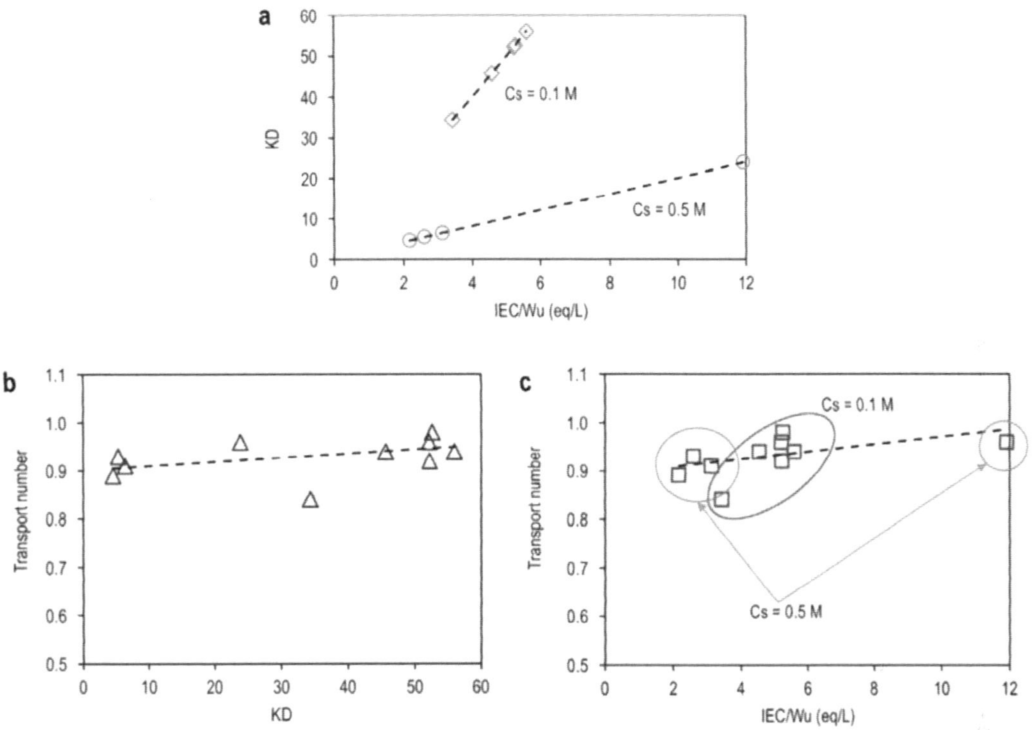

Figure 3. (a) Donnan equilibrium constant (KD) at various IEC/Wu (calculated by equation (2) and (3)), (b) transport number vs KD, and (c) transport number vs IEC/Wu. Membranes: AMX, 204SZRA, HJA, LNA, HGA, IPA, IPC, CR67-HMR-402, AR204-SZRA-412, ASM, ASM-F; their properties (IEC, Wu, t_i) are from ref: [32,33,35,36,39].

4. Conclusion

Results of various studies indicated that the degree of heterogeneity significantly affects the separation properties of IEMs such as conductivity and permselectivity. It can be associated with efficacy of Donnan exclusion which is influenced by the presence of inter-gel phase filled by an electrolyte and inert phase in the membrane matrix. Therefore, understanding the effect of heterogeneity on Donnan exclusion is important to provide preliminary information about the transport properties. In this paper, the effect of IEC/Wu on Donnan exclusion is discussed. The effect of heterogeneity on Donnan exclusion is expressed as a correlation of IEC/Wu ratio with Donnan equilibrium constant and transport number of counter-ion, t_i. Results of this study showed that membrane transport number is increased with increasing IEC/Wu ratio. The IEC/Wu may be used for preliminary determination of IEM transport number.

Acknowledgement

Financial assistance for this work has been provided by Lembaga Pengelola Dana Pendidikan (LPDP) Indonesia through 'Beasiswa Tesis dan Disertasi' 2016.

References

[1] Drioli E, Ali A and Macedonio F 2015 Membrane distillation: Recent developments and perspectives *Desalination* **356** 56–84
[2] Wenten I G, Dharmawijaya P T, Aryanti P T P, Mukti R R and Khoiruddin 2017 LTA zeolite membranes: current progress and challenges in pervaporation *RSC Adv.* **7** 29520–39
[3] Wardani A K, Hakim A N, Khoiruddin and Wenten I G 2017 Combined ultrafiltration-electrodeionization technique for production of high purity water *Water Sci. Technol.* **75** 2891–9
[4] Purwasasmita M, Kurnia D, Mandias F C, Khoiruddin and Wenten I G 2015 Beer dealcoholization using non-porous membrane distillation *Food Bioprod. Process.* **94** 180–6
[5] Purwasasmita M, Nabu E B P, Khoiruddin and Wenten I G 2015 Non dispersive chemical deacidification of crude palm oil in hollow fiber membrane contactor *J. Eng. Technol. Sci.* **47** 426–46
[6] Ariono D, Purwasasmita M and Wenten I G 2016 Brine effluents: characteristics, environmental impacts, and their handling *J. Eng. Technol. Sci.* **48** 367–87
[7] Sianipar M, Kim S H, Khoiruddin, Iskandar F and Wenten I G 2017 Functionalized carbon nanotube (CNT) membrane: progress and challenges *RSC Adv.* **7** 51175–98
[8] Wenten I G, Khoiruddin K, Hakim A N and Himma N F 2017 Chapter 11 - The Bubble Gas Transport Method - Membrane Characterization (Elsevier) pp 199–218
[9] Mangindaan D, Khoiruddin K and Wenten I G 2018 Beverage dealcoholization processes: Past, present, and future *Trends Food Sci. Technol.* **71** 36–45
[10] Strathmann H, Grabowski A and Eigenberger G 2013 Ion-exchange membranes in the chemical process industry *Ind. Eng. Chem. Res.* **52** 10364–79
[11] Strathmann H 2010 Electrodialysis, a mature technology with a multitude of new applications *Desalination* **264** 268–288
[12] Khoiruddin K, Hakim A N and Wenten I G 2014 Advances in electrodeionization technology for ionic separation - A review *Membr. Water Treat.* **5** 87–108
[13] Khoiruddin, Widiasa I N and Wenten I G 2014 Removal of inorganic contaminants in sugar refining process using electrodeionization *J. Food Eng.* **133** 40–45
[14] Khoiruddin, Aryanti P T P, Hakim A N and Wenten I G 2017 The role of ion-exchange membrane in energy conversion *AIP Conf. Proc.* **1840** 90006
[15] Wardani A K, Hakim A N, Khoiruddin, Destifen W, Goenawan A and Wenten I G 2017 Study on the influence of applied voltage and feed concentration on the performance of electrodeionization in nickel recovery from electroplating wastewater *AIP Conf. Proc.* **1805** 30004
[16] Khoiruddin and Wenten I G 2016 Investigation of electrochemical and morphological properties of mixed matrix polysulfone-silica anion exchange membrane *J. Eng. Technol. Sci.* **48** 1–11
[17] Xu T 2005 Ion exchange membranes: State of their development and perspective *J. Memb. Sci.* **263** 1–29
[18] Xu T W, Li Y, Wu L and Yang W H 2008 A simple evaluation of microstructure and transport parameters of ion-exchange membranes from conductivity measurements *Sep. Purif. Technol.* **60** 73–80
[19] Berezina N P, Kononenko N A, Dyomina O A and Gnusin N P 2008 Characterization of ion-exchange membrane materials: Properties vs structure *Adv. Colloid Interface Sci.* **139** 3 – 28
[20] Wenten I G and Khoiruddin 2016 Recent developments in heterogeneous ion-exchange membrane: Preparation, modification, characterization and performance evaluation *J. Eng. Sci. Technol.* **11** 916–34
[21] Ariono D, Khoiruddin, Subagjo and Wenten I G 2017 Heterogeneous structure and its effect on properties and electrochemical behavior of ion-exchange membrane *Mater. Res. Express* **4** 24006
[22] Khoiruddin, Ariono D, Subagjo and Wenten I G 2017 Surface modification of ion-exchange

membranes: Methods, characteristics, and performance *J. Appl. Polym. Sci.*

[23] Zabolotsky V I and Nikonenko V V 1993 Effect of structural membrane inhomogeneity on transport properties *J. Memb. Sci.* **79** 181–98

[24] Volodina E, Pismenskaya N, Nikonenko V, Larchet C and Pourcelly G 2005 Ion transfer across ion-exchange membranes with homogeneous and heterogeneous surfaces *J. Colloid Interface Sci.* **285** 247–58

[25] Demina O A, Berezina N P, Sata T and Demin A V 2002 Transport–structural parameters of domestic and foreign anion-exchange membranes *Russ. J. Electrochem.* **38** 896–902

[26] Demina O A, Kononenko N A and Falina I V 2014 New approach to the characterization of ion-exchange membranes using a set of model parameters *Pet. Chem.* **54** 515–25

[27] Donnan F G 1995 Theory of membrane equilibria and membrane potentials in the presence of non-dialysing electrolytes. A contribution to physical-chemical physiology *J. Memb. Sci.* **100** 45–55

[28] Gnusin N P, Berezina N P, Kononenko N A, Demina O A and Annikova L A 2009 The three-wire model and Lichtenecker's equation for calculations of the conductivity of ion-exchange columns *Russ. J. Phys. Chem. A* **83** 107–10

[29] Gnusin N P, Berezina N P, Demina O A and Kononenko N A 1996 Physicochemical principles of testing ion-exchange membranes *Russ. J. Electrochem.* **32** 154–63

[30] Le X T 2008 Permselectivity and microstructure of anion exchange membranes *J. Colloid Interface Sci.* **325** 215–22

[31] Gnusin N P, Demina O A, Berezina N P and Kononenko N A 2004 Modeling of mass electrotransfer in terms of the transport and structural properties of ion-exchange membranes *Theor. Found. Chem. Eng.* **38** 394–398

[32] Lee H-J, Hong M-K, Han S-D and Moon S-H 2008 Influence of the heterogeneous structure on the electrochemical properties of anion exchange membranes *J. Memb. Sci.* **320** 549–55

[33] de Lara R, Rodríguez-Postigo J, García-Herruzo F, Rodríguez-Maroto J M and Benavente J 2005 Decontamination of soils by membrane processes: characterization of membranes under working conditions *Ind. Eng. Chem. Res.* **44** 400–7

[34] Tuan L X and Buess-Herman C 2007 Study of water content and microheterogeneity of CMS cation exchange membrane *Chem. Phys. Lett.* **434** 49–55

[35] Gohil G S, Shahi V K and Rangarajan R 2004 Comparative studies on electrochemical characterization of homogeneous and heterogeneous type of ion-exchange membranes *J. Memb. Sci.* **240** 211–9

[36] Shah B G, Shahi V K, Thampy S K, Rangarajan R and Ghosh P K 2005 Comparative studies on performance of interpolymer and heterogeneous ion-exchange membranes for water desalination by electrodialysis *Desalination* **172** 257–65

[37] Pismenskaya N, Melnik N, Nevakshenova E, Nebavskaya K and Nikonenko V 2012 Enhancing ion transfer in overlimiting electrodialysis of dilute solutions by modifying the surface of heterogeneous ion-exchange membranes *Int. J. Chem. Eng.* **2012** 1–11

[38] Zabolotskii V I, Bugakov V V, Sharafan M V and Chermit R K 2012 Transfer of electrolyte ions and water dissociation in anion-exchange membranes under intense current conditions *Russ. J. Electrochem.* **48** 650–659

[39] Tanaka N, Nagase M and Higa M 2012 Organic fouling behavior of commercially available hydrocarbon-based anion-exchange membranes by various organic-fouling substances *Desalination* **296** 81–6

Speech Recognition using Linear Predictive Coding (LPC) and Adaptive Neuro-Fuzzy (ANFIS) to Control 5 DoF Arm Robot

W.S. Mada Sanjaya[1,2], Dyah Anggraeni[1,2] and Ikhsan Purnama Santika[1,2]
[1]Department of Physics, Faculty of Science and Technology, Universitas Islam Negeri Sunan Gunung Djati Bandung, Indonesia
[2]Bolabot Techno Robotic Institute, CV. Sanjaya Star Group, Bandung, Indonesia

E-mail: madasws@gmail.com, tsugumikaoru@gmail.com

ABSTRACT. This research shows the implementation of speech recognition to control arm robot. The method to identify the speech recognition using Linear Predictive Coding (LPC) and Adaptive Neuro-Fuzzy Inference System (ANFIS). LPC method used to feature extraction the signal of speech and ANFIS method used to learn the speech recognition. The data learning which used to ANFIS processed are 6 features. The examination system of speech identification using trained and not trained data. The result of the research shows the successful grade for trained speech data is 88.75% and not trained data is 78.78%. Identification of speech recognition system was applied to controlled arm robot based on Arduino microcontroller.

1. Introduction

The automation system, especially in robotics is a system which can help human works [1]. Robotics system can be controlled by human or programmed on robot processor. Speech recognition (or speech control) can be an alternative to help disability. In control system based on speech recognition needed a method to features extraction and learning method as a database (trained data) to identify the features.

They are some method to extract the features of speech, such as Linear Predictive Coding (LPC) method [2] [3] and Mel-Frequency Cepstrum Coefficient (MFCC) [4] [5]. This method can be applied in some field, such as; control smart home [6] [5], control mobile robot [7], control wheelchair [8], control robot [9], [10], biomatrix [11], speaker identification [2] , control arm robot [12] [13]. To learning and classify of speech method, have been investigating using Artificial Neural Networks (ANN) [2] [3] [14] [15], Neuro-Fuzzy [6], and other soft computing [16] [17].

In this research will be described an extraction features by using Linear Predictive Coding and learning method to classify the speech by using Adaptive Neuro-Fuzzy Inference System (ANFIS). Finally, this research will be implemented the speech recognition to control arm robot for placing an object.

The paper is organized as follows. In section 2, describe the theoretical background of LPC and ANFIS details. In section 3, describe the method and the system design of the research. In section 4, describe a hardware design of arm robot. In section 5, describe the application of speech recognition in detail. Finally, in Section 6 the concluding remarks are given.

Content from this work may be used under the terms of the Creative Commons Attribution 3.0 licence. Any further distribution of this work must maintain attribution to the author(s) and the title of the work, journal citation and DOI.
Published under licence by IOP Publishing Ltd

2. Theoretical Background

2.1 Linear Predictive Coding (LPC) Method

A signal processing is an activity to extract a signal information. Linear Predictive Coding (LPC) is a powerful speech analysis technique and facilitating a features extraction which has a good quality and efficient result for computing. In 1978, LPC uses to make a speech synthesis. LPC doing an analysis with predicting a formant decided a formant from signal called inverse filtering, then estimated an intensity and frequency from residue speech signal. Because speech signal has many variations depending on a time, the estimation will do to cut a signal called frame. Procedure to get an LPC coefficient shown in Figure 1.

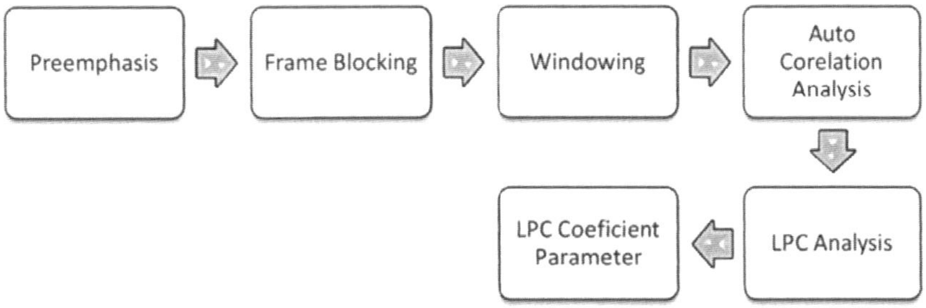

Figure 1. LPC Method Diagram

2.1.1 Preemphasis. On processing of speech signal, preemphasis filter needed after sampling process. The filtering purpose is to get a smooth spectral shape of the speech signal. A spectral which have a high value for the low-frequency field and decrease for field frequency higher than 2000 Hz. Preemphasis filter based on the relation of input/output on time domain which is shown by the equation (1),

$$y(n) = x(n) - ax(n-1), \tag{1}$$

a is a constant of preemphasis filter, ordinary have $0.9 < a < 1.0$.

2.1.2 Frame Blocking. Frame Blocking. On this process, segmented of speech signal become some frame which overlaps. So that no signal is lost (deletion).

2.1.3 Windowing. Analog signal which converts become digital signal read frame by frame and each frame is windowing with the certain window function. This windowing process purpose to minimize discontinue signal from initial to end of each frame. If window as $w(n)$, $0 \le n \le N-1$, when N is total of sample of each frame, thus result of windowing is a signal:

$$y_1(n) = x_1(n)w(n), 0 \le n \le N-1\,. \tag{2}$$

Where $w(n)$ usually use window Hamming with the form:

$$w(n) = 0.54 - 0.46.cos\left(\frac{2\pi n}{N-1}\right), 0 \le n \le N-1\,. \tag{3}$$

2.1.4 Autocorellation Analysis. The next step is autocorrelation analysis toward each frame result by windowing $y_1\,(n)$ with equation (4),

$$r_1(m) = \sum_{n=0}^{n-1-m} y_1\,(n)y_1(n+m),\ m = 0,1,2..,p\,, \tag{4}$$

Where p is ordered from LPC. LPC order which usually used is between 8 until 16.

2.1.5 LPC Analysis. This step will convert each frame from p+1 autocorrelation become compilation of "LPC parameter" $a_m = a_m^p$ for m = 1, 2... p. This compilation becomes LPC coefficient or become other LPC transformation. The formal method to change autocorrelation coefficient become parameter LPC compilation called Durbin method, the form as:

$$E^0 = r(0) \tag{5}$$

$$k_m = \frac{\left\{r(m) - \sum_{j=1}^{m-1} \alpha_j^{(m-1)} r(|m-j|)\right\}}{E^{(m-1)}}, \quad 1 \le m \le p, \tag{6}$$

$$\alpha_m^m = k_m, \tag{7}$$

$$\alpha_j^m = \alpha_j^{(m-1)} - k_m \alpha_{m-j}^{(m-1)}, \quad 1 \le j \le m-1, \tag{8}$$

$$E^m = (1 - k_m^2) E^{(m-1)}. \tag{9}$$

With $r(0)$ is a result of autocorrelation, $E(m)$ is an error, k_m is rebound of the coefficient, α_j^m is prediction coefficient for $1 \le j \le m$.

2.1.6 LPC Parameter Conversion to LPC Coefficient. LPC coefficient parameter a_m conversion to cepstral coefficient c_m to get the best performance and endure to noise, as the questions:

$$c_m = a_m + \sum_{k=1}^{m-1} (k/m) c_k a_{m-k}, \quad 1 \le m \le p \tag{10}$$

$$c_m = \sum_{k=1}^{m-1} (k/m) c_k a_{m-k}, \quad m > p \tag{11}$$

This cepstral coefficient is representation coefficient of logarithm spectrum.

2.2 Adaptive Neuro-Fuzzy Inference System (ANFIS)

Neuro-fuzzy is relation from fuzzy logic and artificial neural networks. Neuro-fuzzy is the relation between fuzzy logic and artificial neural networks. Neuro-fuzzy system based on inference of fuzzy system which tests using learning algorithm which derivate from artificial neural networks system. Thus, neuro-fuzzy has an advantage which has by the interference fuzzy system and artificial neural networks system. From the learning capability, the neuro-fuzzy system can be called as ANFIS (Adaptive Neuro-Fuzzy Inference System).

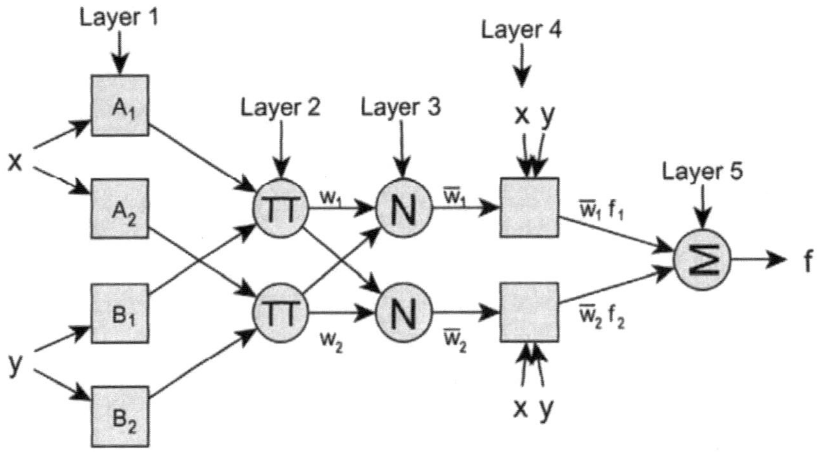

Figure 2. ANFIS Roger Jang [18]

As shown as Figure 2, the fuzzy structure which applied is the fuzzy model by Takagi-Sugeno-Kang. In ANFIS system consist 5 layers, described below;

2.2.1 Layer 1: Fuzzification. The output from node i on layer 1 can be noticed as $O_{1,i}$. Each node on layer 1 is an adaptive node, with the output:

$$O_{1,i} = \mu_{Ai}(x), \quad i = 1,2 \tag{12}$$

$$O_{1,i} = \mu_{Bi}(y), \quad i = 1,2 \tag{13}$$

Where x and y is an input value for the node, A_i and B_i are fuzzy membership. So, the function of each node on layer 1 is to generate the degree of a membership.

2.2.2 Layer 2: Product. Noticed as π. It is also known as rule layer. The function of each node on this layer is to calculate activation strange (firing strength) on each rule as the product from all input or as a t-norm (triangular norm) operator:

$$O_{2,i} = w_i = \mu_{Ai}(x)\Delta\, \mu_{Bi}(y), \quad i = 1,2\,, \tag{14}$$

So,

$$w_1 = \mu_{A1}(x)\ AND\ \mu_{B1}(y)\,, \tag{15}$$

$$w_2 = \mu_{A2}(x)\ AND\ \mu_{B2}(y)\,. \tag{16}$$

The output of this layer works as weight function.

2.2.3 Layer 3: Normalization. Noticed as N. This layer has a non-adaptive characteristic which has a function to calculate a ratio between firing strength on rule to i toward firing strength to sum from all the rule:

$$O_{3,i} = \overline{w}_i = \frac{w_i}{w_1+w_2}\,, \quad i = 1,2\,. \tag{17}$$

2.2.4 Layer 4: Defuzzification. Each node in this layer have adaptive characteristic, with the function:

$$O_{4,i} = \overline{w}_i f_i = \overline{w}_i(p_i x + q_i y + r_i) \tag{18}$$

Where $\overline{w}_i$ is a normalized firing strength from *layer* 3 and $\{x + q_i y + r_i\}$ are the parameter of membership on fuzzy Takagi-Sugeno first order model.

2.2.5 Layer 5: Total Output. Represented by a single summation node, Symbolized as Σ. This layer has a function to aggressing all output on layer 4 (summation from all input signal);

$$O_{5,i} = \sum_i \overline{w}_i f_i = \frac{\sum_i w_i f_i}{\sum_i w_i} \tag{19}$$

Totally, the five-layer will build as an adaptive-networks which functionally equivalent to fuzzy Takagi-Sugeno first order model.

3. Method and System Design

The tools which used in this research are microphone 54db+2 db (which have specification: impedance 32 ohm at 1 kHz sensitivity102db/mW), 5 DoF arm robot based Arduino microcontroller. The personal computer as speech processing using MATLAB® software. The steps in this research generally show in Figure 3:

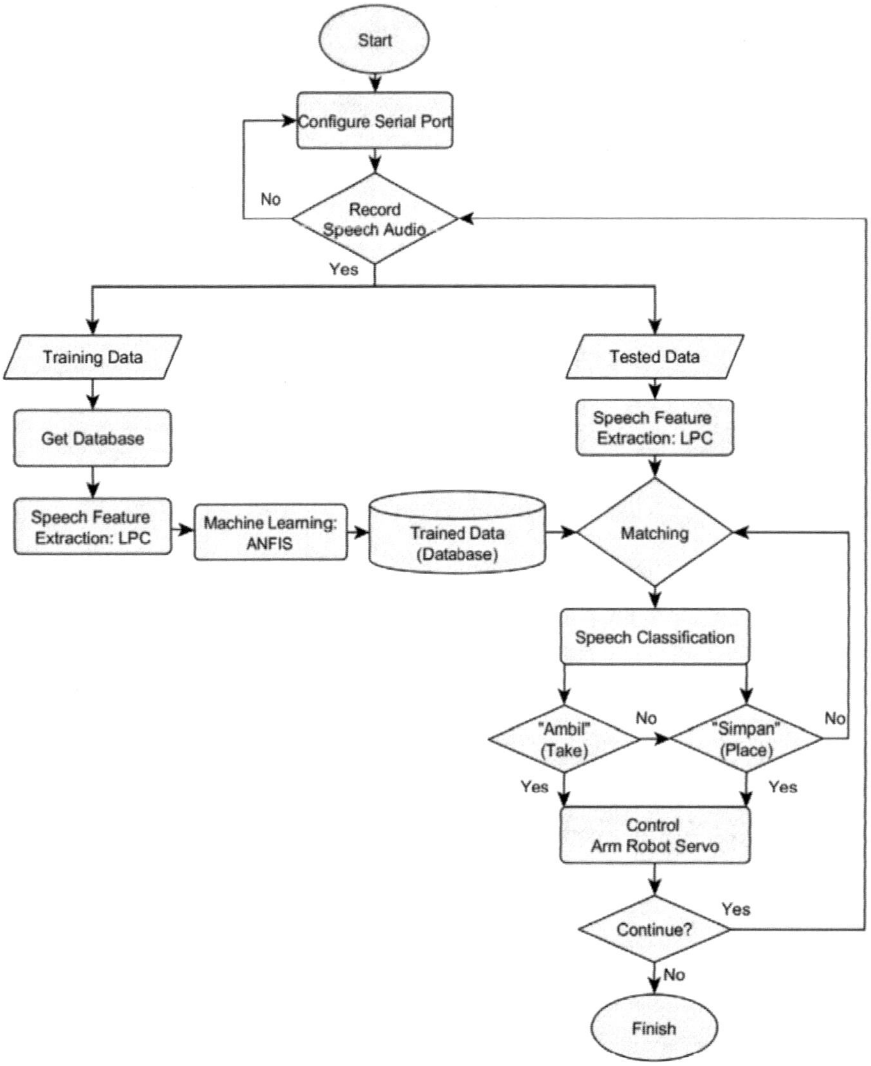

Figure 3. General scheme system

The first process is making training data, consist of features extraction process using LPC and learning process using ANFIS that will build a database as fuzzy logic formulation. The second process is the testing system, such as LPC features extraction, classifying ANFIS data process, and implementation to control arm robot for placing an object using speech command.

4. Hardware Design

Figure 4 is the schematic and realization of arm robot which used in this research. This arm robot consists of five motor servo component which connects with Arduino microcontroller.

IOP Conf. Series: Journal of Physics: Conf. Series **1090** (2018) 012046 doi:10.1088/1742-6596/1090/1/012046

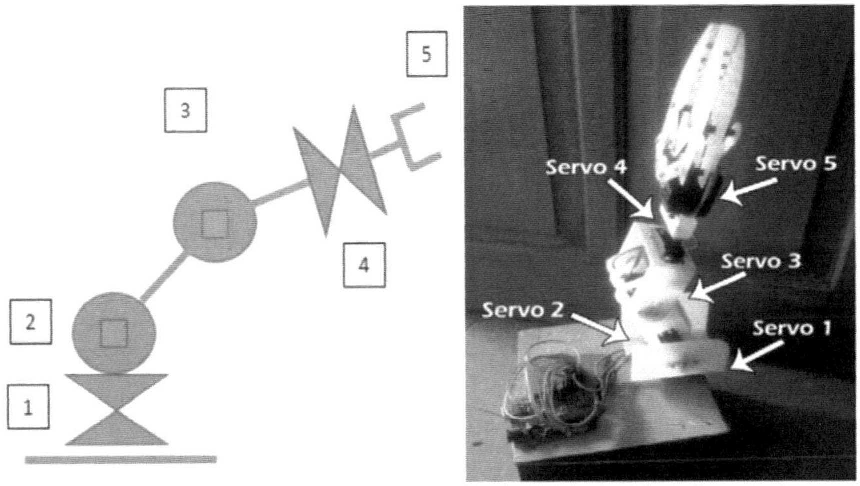

(a) (b)

Figure 4. Arm robot, (a) Schematic, (b) Realization

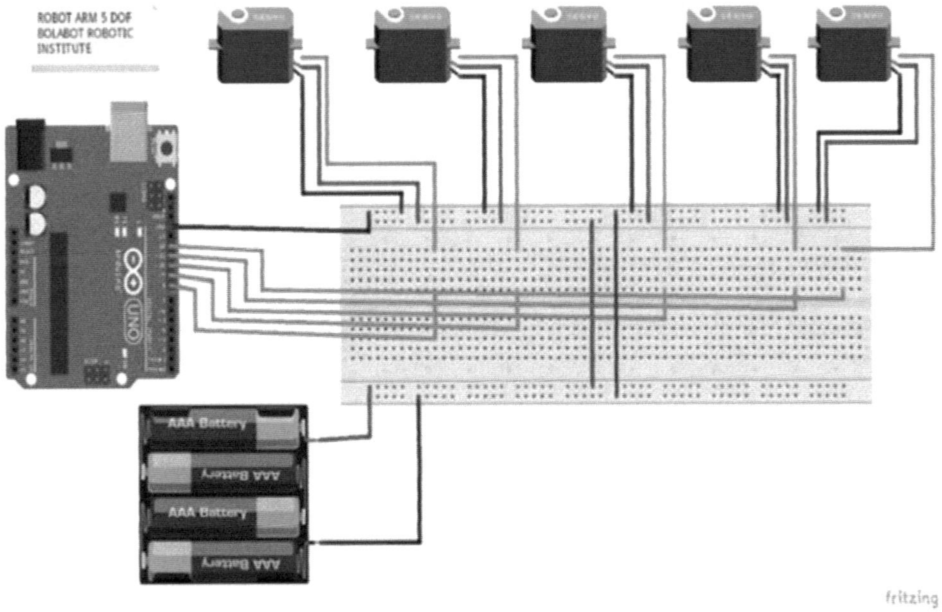

Figure 5. Arm Robot schematic circuit

Arm robot circuit is shown in Figure 5, each servo must have a supply using the 5-volt battery which has 100 mA of current in order get a better result. Each servo ground connected to the ground on Arduino microcontroller. Arm robot consists of 5 motor servo called 5 DoF (Degree of Freedom); Servo1 connect to pin 8 (base, rotate horizontally), Servo2 connect to pin 9 (work as shoulder, rotate vertically), Servo3 connect to pin 10 (work as elbow, rotate vertically), Servo4 connect to pin 11 (work as wrist, rotate horizontally), and Servo5 connect to pin 12 (work as gripper to place an object).

5. Result and Discussion

5.1 IDE Arduino Program Test

To test an IDE Arduino program, have been investigating by simulation using Proteus. The result shows a suitable between program and simulation output shown in Figure 6.

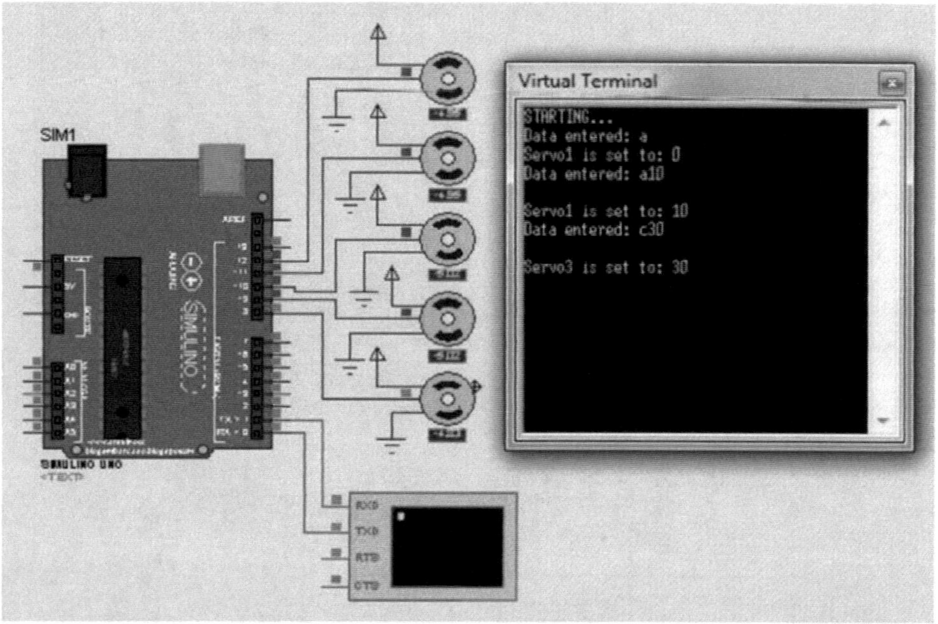

Figure 6. Serial test by IDE Arduino program to control 5 DoF arm robot.

5.2 Features Extraction Database using LPC

We use 6 features of speech signal using LPC method toward respondents with speaking a word (in Indonesian) “ambil” (pick) and “simpan” (place) with each of 10 sample, can be seen in the following table:

Table 1. LPC Features Extraction Result and Target

No	Feature 1	Feature 2	Feature 3	Feature 4	Feature 5	Feature 6	Target
1	-1.76619	-0.38658	1.854785	1.101728	-2.29431	-2.42487	1
2	-1.75599	-0.5623	2.043627	1.520301	-2.65515	-3.35814	1
3	-2.13159	0.095945	2.878341	-0.25327	-4.7726	0.496327	1
4	-2.60417	1.248259	2.836023	-2.65658	-3.70216	5.616098	1
5	-2.26193	0.756091	2.494955	-1.85409	-3.3322	4.04501	1
6	-2.52144	1.082503	2.82768	-2.35296	-4.16399	5.577152	1
7	-2.08075	0.570233	1.791693	-0.97266	-1.79269	1.671262	1
8	-2.24344	0.76278	1.953595	-0.82145	-2.60754	1.466322	1
9	-2.39999	0.617584	3.218707	-1.45934	-5.45294	3.365111	1
10	-1.93844	-0.02257	2.184292	0.417742	-3.35865	-0.9442	1
11	-1.51965	-0.31074	1.035648	0.670428	-0.92408	-0.82099	0

12	-2.52814	1.101294	3.142449	-2.87479	-4.68526	6.828066	0
13	-2.50959	1.360502	2.598425	-3.1903	-2.84935	6.621596	0
14	-2.53472	1.395844	2.705712	-3.49334	-2.81948	7.304646	0
15	-2.35065	1.401495	1.510068	-2.29127	-0.6991	3.374636	0
16	-2.60938	1.574224	2.636275	-3.70426	-2.6895	7.842864	0
17	-2.36393	0.963927	2.697491	-2.38921	-3.52772	5.093925	0
18	-1.39629	-0.56564	1.295431	1.07866	-1.40928	-1.99593	0
19	-2.3476	0.555616	3.055627	-1.05143	-5.36299	2.303807	0
20	-2.57447	1.566313	2.520161	-3.77133	-2.2918	7.974006	0

On Table 1 show the differences between word "ambil" and "simpan". Speech "ambil" as target "1", then speech "simpan" as target "0". This features will become an input for ANFIS method as database for Fuzzy logic.

5.3 Speech Recognition System Test

The test system results shown on Table 2 that the average speech recognition accuracy rate respondents trained data (in database) are 88.75% and the respondents not trained (outside database) data produces an accuracy rate of 78.78%. Errors on speech recognition can be influenced by the accent / intonation speech input and noise on the environment.

Table 2. The test result speech recognition with respondent.

Sample	Word	Target	Recognition Trained Data		Recognition Not-Trained Data	
			Output	Error	Output	Error
1	"Ambil"	1	0.97226	0.02774	1.40514	0.40514
	"Simpan"	0	-0.0273662	0.0273662	0.0781117	0.0781117
2	"Ambil"	1	1.31593	0.31593	1.00856	0.00856
	"Simpan"	0	0.109141	0.109141	-0.21035	0.21035
3	"Ambil"	1	0.87573	0.12427	0.573719	0.426281
	"Simpan"	0	0.0438407	0.0438407	-0.112783	0.112783
4	"Ambil"	1	1.06282	0.06282	1.0853	0.0853
	"Simpan"	0	0.0520409	0.0520409	0.158865	0.158865
5	"Ambil"	1	0.766645	0.233355	0.80136	0.19864
	"Simpan"	0	0.128364	0.128364	0.437931	0.437931

Figure 7 shown a GUI (Graphic User Interface) when tested the word "ambil" and "simpan". When the word "ambil" recorded, the system will recognize with output value 0.953285 approach to "1", so that the computer will transmit serial data as characters "1" to the microcontroller to drive the robot arm in a position to pick up objects. Meanwhile, when the word "simpan" recorded, then the system will recognize the output amounted to 0.232289 which is close to the value "0", so that the computer will transmit serial data as character "0" to the microcontroller to drive the robot arm in a position to place objects. Generally, the tool works fine 100% because of rounding value.

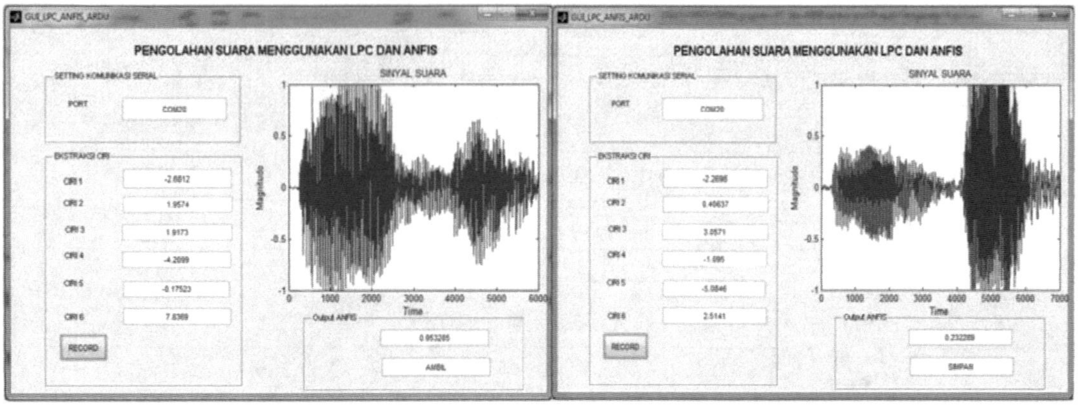

(a) (b)

Figure 7. GUI MATLAB interface to control arm robot using speech recognition.

6. Conclusion

In this research can be concluded that the system works successfully suitable the speech command. The results obtained speech recognition system implementation of LPC and ANFIS method have a high average accuracy rate of speech recognition, which is 88.75% of the respondents trained data and 78.78% of the respondents not trained data. In the experiment result to control the 5 DoF of arm robot based on Arduino microcontroller to pick and place an object works well.

Acknowledgement

The authors would like gratefully acknowledge the financial support from DIPA UIN Sunan Gunung Djati Bandung.

Reference

[1] al Jazari I a R 1974 The Book of Knowledge of Ingenious Mechanical Devices Trans and annotated by Donald R. Hill *Dordrecht: Reidel*

[2] Daqrouq K, Morfeq A, Ajour M and Alkhateeb A 2013 Wavelet LPC With Neural Network for Speaker Identification System," *WSEAS Transaction on Signal Processing* **9** **4** 216-226

[3] Thiang and Wijoyo S 2011 Speech Recognition Using Linear Predictive Coding and Artificial Neural Network for Controlling Movement of Mobile Robot *International Conference on Information and Electronics Engineering* **6** 179-183

[4] Abinayaa B, Arun D, Darshini B and Nataraj C 2015 Voice Command Based Computer Application *International Journal of Innovative Research in Science, Engineering and Technology* **4** **4** 57-63

[5] A. M. Soe, M. M. Latt, H. M. Tun and Z. Mi, "Electronic Control System Of Home Appliances Using Speech Command Words," *International Journal of Scientific & Technology Research,* vol. 4, no. 6, pp. 323-329, 2015.

[6] Sanjaya W S M and Salleh Z 2014 Implementasi Pengenalan Pola Suara Menggunakan Mel_Frequency Cepstrum Coefficients (MFCC) dan Adaptive Neuro-Fuzzy Inferense System (ANFIS) sebagai Kontrol Lampu Otomatis *Al-Hazen Journal of Physics* **1** **1** 44-54

[7] Kulji B, János S and Tibor S 2007 Mobile robot controlled by voice *International Symposium on Intelligent Systems and Informatics* **5** 189-192

[8] Srishti, Jain P, Shalu and Singh S 2015 Design and Development of Smart Wheelchair using Voice Recognition and Head Gesture Control System *International Journal of Advanced Research in Electrical, Electronics and Instrumentation Engineering* **4** **5** 4790-4798

[9] Koo Y m, Kim G B, Jang S C, Lee W S, Kim H-G and Han S H 2015 A Study on Travelling Control of Mobile Robot by Voice Commend *International Conference on Control, Automation and System* **15** 13-16

[10] Punchihewa A and Arshad Z M 2011 Voice Command Interpretation for Robot Control *International Conference on Automation, Robotics and Applications* **5** 90-95

[11] Rudrapal D, Smita Das S D and Kar N D N 2012 Voice Recognition and Authentication as a Proficient Biometric Tool and its Application in Online Exam for P.H People *International Journal of Computer Applications* **39** **12** 6-12

[12] Varalakshmi M and Raju N N 2014 Design of Speech Controlled Pick and Place Robot with Wireless Zigbee Technology *International Journal of Scientific Engineering and Technology Research* **3** **20** 4062-4066

[13] Jadhav K S and Gaikwad S M 2015 Writing Robotic Arm by Speech Recognition *International Journal of Advanced Research in Electrical, Electronics and Instrumentation Engineering,* **4** **6** 4983-4990

[14] Rai N and Rai B 2014 An ANN Based Mobile Robot Control Through Voice Command Recognition Using Nepali Language *International Journal of Applied Control, Electrical and Electronics Engineering (IJACEEE)* **2** **4** 13-22

[15] Srivastava N 2014 Speech Recognition using Artificial Neural Network *International Journal of Engineering Science and Innovative Technology* **3** **3** 406-412

[16] Thiang and Wijaya D 2009 Limited Speech Recognition for Controlling Movement of Mobile Robot Implemented on ATmega162 Microcontroller in *International Conference on Computer and Automation Engineering* 347-350

[17] El-emary I M M, Fezari M and Hamz 2011 Hidden Markov model/Gaussian mixture models(HMM/GMM) based voice command system: A way to improve the control of remotely operated robot arm TR45 *Scientific Research and Essays* **6** **2** 341-350

[18] Jang R, Sun C T and Mizutani E 1997 Neuro-Fuzzy and Soft Computing, *Prentice-Hall, Inc.*

A finite volume method for solving the gravity wave-model equations

Cecilia Heru Purwitaningsih[1] and Sudi Mungkasi[2]

[1]Department of Mathematics Education, Faculty of Teacher Training and Education, Sanata Dharma University, Yogyakarta, Indonesia
[2]Department of Mathematics, Faculty of Science and Technology, Sanata Dharma University, Yogyakarta, Indonesia

E-mail: cecilia.purwita@gmail.com, sudi@usd.ac.id

Abstract. One of models for free surface flows is the gravity wave-model equations. In this paper, we propose a finite volume numerical method for solving the gravity wave-model equations. The numerical scheme is explicit, so it is easy to implement. The scheme is consistent with the gravity wave-model equations. Numerical results show that the proposed method is stable. As cell widths and time steps are taken smaller, the errors of numerical solutions get smaller too. Therefore, we infer that the numerical solutions are convergent to the exact solution.

1. Introduction

Real problems can be modelled mathematically. The Saint-Venant equations form a mathematical model for free-surface flows. The Saint-Venant equations are often simplified for practicality reasons and faster calculation. However, the simplification must still be able to capture the physical phenomena.

The gravity wave-model equations are some simplifications of the Saint-Venant equations by neglecting the convective term in the equation of momentum conservation [1-2]. This is valid as long as the gravity effect is much more dominant than the convective effect. This model can be used for simulation of natural disasters, such as tsunami, floods [3-6] and landslides [7]. The analytical exact solution to the gravity wave-model equations is available only for some specific type of problems, such as the dam break problem studied by Martins et al. [1]. The general analytical exact solution to this model is not yet available. Therefore, in order to solve this model for the general case, a numerical method should be developed. A fast and accurate but simple numerical method is still a research gap.

In this paper, we aim to fill the mentioned research gap by proposing a finite volume numerical method for solving the gravity wave-model equations. The numerical fluxes in our numerical method are calculated using the Lax-Friedrichs formulation. We provide the numerical scheme and some numerical results. We use the dam break problem for a numerical test case. However, note that the dam break problem in this paper is assumed to be governed by the gravity wave-model equations, instead of the usual Saint-Venant equations.

The rest of the paper is written as follows. We recall the gravity wave-model equations in Section 2. We provide the finite volume method that we propose in Section 3. Numerical results are explained in Section 4. Finally, we conclude the paper in Section 5.

Content from this work may be used under the terms of the Creative Commons Attribution 3.0 licence. Any further distribution of this work must maintain attribution to the author(s) and the title of the work, journal citation and DOI.
Published under licence by IOP Publishing Ltd

2. Gravity wave-model equations

In this section we recall the mathematical model (the gravity wave-model equations) that we want to solve.

The gravity wave-model equations are [1, 8]

$$\frac{\partial}{\partial t}h + \frac{\partial}{\partial x}q = 0\,, \tag{1}$$

and

$$\frac{\partial}{\partial t}q + \frac{g}{2}\frac{\partial}{\partial x}h^2 = 0\,, \tag{2}$$

where $h = h(x,t)$ is the fluid depth, $q = q(x,t)$ denotes the unit-discharge, g is the gravitational acceleration, t is the time variable and x is the space variable. All quantities are assumed to be in SI units. Here, the conserved quantities are the water depth $h(x,t)$ for equation (1) and the unit-discharge $q(x,t)$ for equation (2). We assume the relation $q(x,t) = u(x,t)h(x,t)$, where $u(x,t)$ is water velocity.

The gravity wave-model equations are some simplifications of the shallow water equations (the Saint-Venant model). That is, the convective term in the momentum equation of the shallow water equations is neglected. This is valid as long as the convective term is assumed to be very much insignificant in comparison with the gravity term. More discussion on the gravity wave-model equations can be found in the work of Martins et al. [1-2].

Our focus in this paper is to solve the gravity wave-model equations numerically. The general exact solution to the gravity wave-model equations is not yet available. However, their exact solutions for some specific cases of the gravity wave-model equations have been identified, for example, see Martins et al. [1] and Mungkasi et al. [9]. We can use them as benchmarks for our numerical tests. In particular, we use the exact solution derived by Martins et al. [1] as the benchmark for the present paper.

3. Finite volume method

In this section, we present the Lax-Friedrichs finite volume method for solving the gravity wave-model equations.

Each of the gravity wave-model equations (1)-(2) is a conservation law in the form of

$$\frac{\partial}{\partial t}Q(x,t) + \frac{\partial}{\partial x}F(Q(x,t)) = 0\,, \tag{3}$$

where $Q(x,t)$ denotes the conserved quantity and $F(x,t)$ represents the corresponding flux function. Using the standard finite volume method for conservation law (3), we have the first order finite volume scheme in the explicit form, as follows:

$$Q_i^{n+1} = Q_i^n - \frac{\Delta t}{\Delta x}\left(F_{i+\frac{1}{2}}^n - F_{i-\frac{1}{2}}^n\right), \tag{4}$$

where $Q_i^n \approx Q(x_i, t^n)$, $F_{i+\frac{1}{2}}^n \approx F\big((x_{i+1/2}, t^n)\big)$, Δt is the time step, and Δx is the space step. The Lax-Friedrichs fluxes for equation (1) are

$$F_{i+\frac{1}{2}}^n = \frac{1}{2}(q_{i+1}^n + q_i^n) - \frac{\Delta x}{2\Delta t}(h_{i+1}^n - h_i^n), \tag{5}$$

and

IOP Conf. Series: Journal of Physics: Conf. Series **1090** (2018) 012047 doi:10.1088/1742-6596/1090/1/012047

$$F^n_{i-\frac{1}{2}} = \frac{1}{2}(q^n_i + q^n_{i-1}) - \frac{\Delta x}{2\Delta t}(h^n_i - h^n_{i-1}). \tag{6}$$

The Lax-Friedrichs fluxes for equation (2) are

$$F^n_{i+\frac{1}{2}} = \frac{1}{2}\left(\left(\frac{g}{2}h^2\right)^n_{i+1} + \left(\frac{g}{2}h^2\right)^n_i\right) - \frac{\Delta x}{2\Delta t}(q^n_{i+1} - q^n_i), \tag{7}$$

and

$$F^n_{i-\frac{1}{2}} = \frac{1}{2}\left(\left(\frac{g}{2}h^2\right)^n_i + \left(\frac{g}{2}h^2\right)^n_{i-1}\right) - \frac{\Delta x}{2\Delta t}(q^n_i - q^n_{i-1}). \tag{8}$$

The finite volume scheme (4) with numerical fluxes (5)-(8) can be iterated with respect to the space index i and the time index n, in order to obtain the numerical solution to the gravity wave-model equations. The finite volume method (4)-(8) is consistent with the conservation law (3), as explained by LeVeque [10-11].

4. Numerical results

We present our numerical results. Numerical solutions shall be judged based on its accuracy with respect to the curve matching with the analytical exact solution. All quantities are assumed to have SI units with the MKS system.

In our simulations, we use $g = 1$, $\Delta x = 0.05$, and $\Delta t = 0.25\Delta x$. This time step Δt is set to this value, in order that the numerical method is stable. Our simulations are stopped at time $t = 1$. We assume that the initial fluid depth is $h_1 = 10$ and $h_0 = 5$, as shown in Figure 1. The initial condition for fluid velocity is $u(x,0) = 0$, for all x. The boundary condition at $x = -L$ is

$$h(-L,t) = 10, \quad u(-L,t) = 0, \tag{9}$$

and at $x = L$ is

$$h(L,t) = 5, \quad u(L,t) = 0. \tag{10}$$

This is called the dam break problem. In this problem, we want to investigate how water flows for time $t > 0$ if the dam wall is removed at time instant $t = 0$. In our simulation, we take $L = 10$.

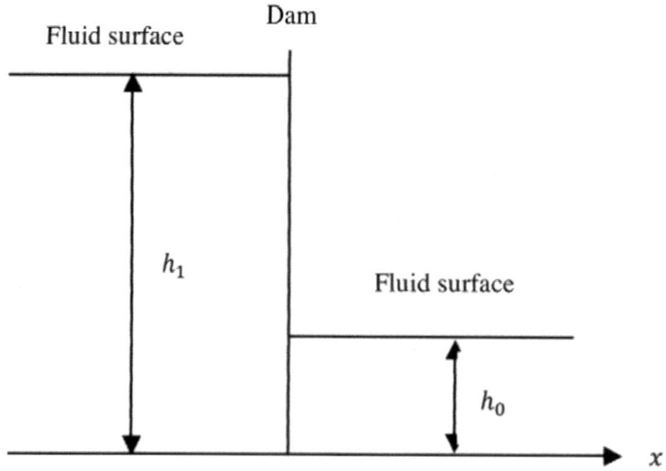

Figure 1. Initial condition of the dam break problem.

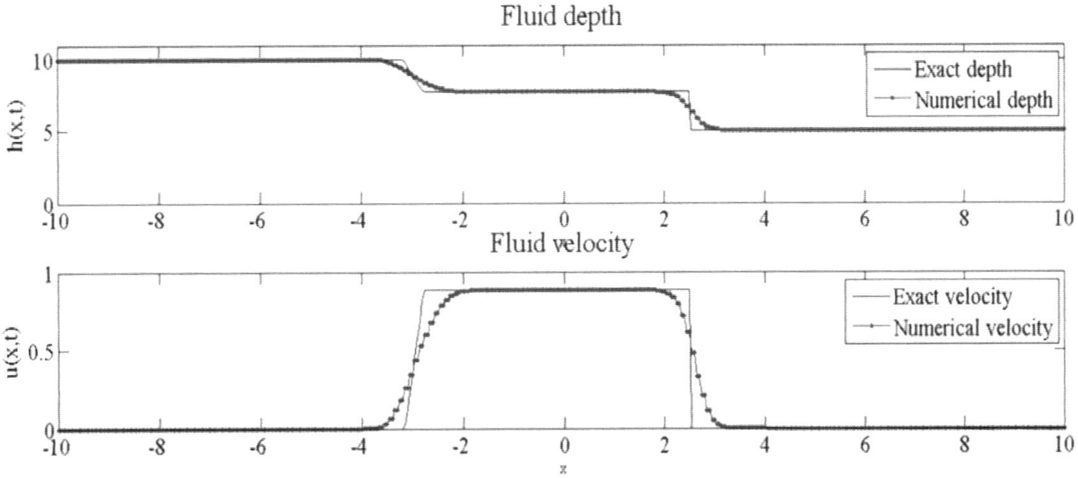

Figure 2. Graph of fluid depth and velocity. Here we compare our numerical solutions with the exact solution. The number of cells is $N = 400$, the uniform cell width is $\Delta x = 0.05$, and the uniform time step is $\Delta t = 0.25\,\Delta x$.

The exact solution to this dam break problem governed by the gravity wave-model equations was derived by Martins et al. [1]. The exact solution for the fluid depth is

$$h(x,t) = \begin{cases} h_1 & \text{if} \quad x \le -c_1 t \\ h_3 = x^2/gt^2 & \text{if} \quad -c_1 t \le x \le -c_2 t \\ h_2 = c_2^2/g & \text{if} \quad -c_2 t \le x \le \xi t \\ h_0 & \text{if} \quad x \le \xi t \end{cases} \tag{11}$$

The exact solution for the fluid velocity is

$$u(x,t) = \begin{cases} 0 & \text{if} \quad x \le -c_1 t \\ u_3 = 2(c_1^3 t^3 + x^3)/3x^2 t & \text{if} \quad -c_1 t \le x \le -c_2 t \\ u_2 = 2/3\,[(c_1^3/c_2^2) - c_2] & \text{if} \quad -c_2 t \le x \le \xi t \\ 0 & \text{if} \quad x \le \xi t \end{cases} \tag{12}$$

where c is the wave propagation speed. In addition,

$$c_1 = \sqrt{gh_1}\,, \qquad c_3 = -x/t\,, \qquad c_2^3 = c_1^3 + 3/2\,(c_0^2 - c_2^2)\sqrt{(c_0^2 + c_2^2)/2}\,, \qquad c_0 = \sqrt{gh_0}\,. \tag{13}$$

Furthermore, ξ is the shock speed given by

$$\xi = \sqrt{(c_0^2 + c_2^2)/2}\,. \tag{14}$$

A representative of numerical results is given in Figure 2. This figure shows a comparison between the numerical solution with the exact solution. The numerical solution is produced using the number of cells $N = 400$, the uniform cell width $\Delta x = 0.05$, and the uniform time step $\Delta t = 0.25\,\Delta x$. We notice that numerical solutions are very close to the exact solution. We also notice that the significant source of errors are at non-smooth regions.

Table 1. Errors and convergence rates of numerical solutions

N	Error of h	Error of u	Convergence rate for h	Convergence rate for u
100	0.1150	0.0422	-	-
200	0.7777	0.0280	0.68	0.66
400	0.0510	0.0185	0.66	0.66
800	0.0330	0.0121	0.65	0.65
1600	0.0205	0.0075	0.62	0.62

Results of those if we vary the number of cells are quantitatively written in Table 1. As the number of cells gets larger, the errors becomes smaller. Note that a large number of cells means a small cell width. We observe that numerical solutions are convergent to the exact solution, as the number of cells gets larger. We have used a first order finite volume method. This means that the convergence rate should be one if the solution is smooth. However, our solution to the dam break problem contains non-smooth regions. This makes the convergence rate is less than one, as shown in Table 1.

These findings could have implications in the development of flood flow simulations when the gravity wave-model equations are used. The proposed finite volume method based on the Lax-Friedrichs formulations are simple and fast to compute yet it is accurate with relatively small errors. The proposed method could be implemented as an alternative solver in case studies, such as that in the work of Martins et al. [2].

5. Conclusion

We have presented the Lax-Friedrichs finite volume method used to solve the gravity wave-model equations. The method is explicit, so it is simple to implement. The first order method leads to less than one for its convergence rate when we solve the dam break problem. As long as the time step is chosen sufficiently small such that the method is stable, the numerical solutions are convergent to the exact solution. Our research is limited to frictionless model of the gravity wave-model equations. Future research direction could be investigating the numerical solver for the gravity wave-model equations involving friction.

Acknowledgements

This work was financially supported by Sanata Dharma University. Both authors are very thankful to Sanata Dharma University for the financial support.

References

[1] Martins R, Leandro J and Djordjevic S 2016 Analytical solution of the classical dam-break problem for the gravity wave-model equations *Journal of Hydraulic Engineering* **142** 06016003

[2] Martins R, Leandro J and Djordjevic S 2018 Influence of sewer network models on urban flood damage assessment based on coupled 1D/2D models *Journal of Flood Risk Management* **11** S717

[3] Aronica G, Tucciarelli T and Nasello C 1998 2D multilevel model for flood wave propagation in flood-affected areas *Journal of Water Resources Planning and Management* **124** 210

[4] Bates P D, Horritt M S, Fewtrell T J 2010 A simple inertial formulation of the shallow water equations for efficient two-dimensional flood inundation modelling *Journal of Hydrology* **387** 33

[5] De Almeida G A M, Bates P, Freer J E and Souvignet M 2012 Improving the stability of a simple formulation of the shallow water equations for 2-D flood modeling *Water Resources Research* **48** W05528

[6] Seyoum S D, Vojinovic Z, Price R K and Weesakul S 2012 Coupled 1D and noninertia 2D

flood inundation model for simulation of urban flooding *Journal of Hydraulic Engineering* **138** 23

[7] Mangeney A, Heinrich P and Roche R 2000 Analytical solution for testing debris avalanche numerical models *Pure and Applied Geophysics* **157** 1081

[8] Mungkasi S and Dheno M F S 2017 Adomian decomposition method used to solve the gravity wave equations *AIP Conference Proceedings* **1788** 030103

[9] Mungkasi S, Supriyadi B and Wiryanto L H 2016 Steady flow over an arbitrary obstruction based on the gravity wave equations *Journal of Physics: Conference Series* **776** 012080

[10] LeVeque R J 1992 *Numerical Methods for Conservation Laws* (Basel: Springer)

[11] LeVeque R J 2002 *Finite Volume Methods for Hyperbolic Problems* (Cambridge: Cambridge University Press)

IOP Conf. Series: Journal of Physics: Conf. Series **1090** (2018) 012048 doi:10.1088/1742-6596/1090/1/012048

Inverse Modeling of Gravity Data with Two Layers Density in Sedimentary Basin Structure

Eko Januari Wahyudi[1] **and Wawan Gunawan A. Kadir**[1]
[1]Institut Teknologi Bandung

Abstract. Study of sedimentary basin structure and basement variation in the subsurface using gravity methods are very common in early stage of hydrocarbon exploration. Most of inverse modeling are designed with several series of juxtaposed vertical prism in the subsurface. In this study, we utilize two layers of vertical prism below each gravity stations to represent homogeneous density of sedimentary layer and basement in the subsurface. In order to reach best miss-fit data we used iterative calculation utilizing differences between observed and calculated data. Performance of iterative calculation will attempt to find solution of interface density in the subsurface. In this paper, we use two synthetic data of symmetrical sedimentary basin model for inversion performance test. The performance test will evaluate the effect of initial model, exploration well data, topographic data, and noisy data effect. Based on synthetic data test, the performance of iterative method provide small order of miss-fit data with residual solution model of noisy data in the range -44 to 35 meters. Real data application identified three shallow sedimentary layer and two deeper sedimentary layer in the study area.

1. Introduction

Study of sedimentary basin structure is very common conducted using gravity data. Density contrast between interface layer (sedimentary and basement) provide the source of gravity anomalies that we can observe in the surface. Inverse modeling of gravity data observation can be targeted to provide subsurface structure as sedimentary thickness variation. There are several publication of sedimentary inverse modeling study since 70's that we can use as references ([1]-[14]). In this paper, we will use iterative method to find the solution of subsurface interface using two layers density (sedimentary layer in the top and basement in the bottom layer). Design of the iterative calculation will be described in the next section and we will evaluate the performace using two synthetic model of symmetrical sedimentary basin. Real data application also conducted using S-N section of gravity data that observed in small area in East Java Basin [15].

2. Methodology

We design the iterative calculation start from subsurface model and simplified as ilustrated in the **figure 1**. The subsurface model consists of M-series juxtaposed vertical prism. Vertical series of subsurface model placed below each gravity stations (N-series data). Two layers of vertical prism below each gravity stations represent two layers in the subsurface (homogeneous density of sedimentary layer (ρ_s) and basement (ρ_b)). Each vertical prisms (of sedimentary layer and basement) has length in 3D cartesian coordinate system (dx, dy, and dz). The lower bound of basement vertical prisms in the subsurface model are limited with certain depth, while the upper bound of sedimentary vertical prisms in the subsurface model are limited with tophographic. Interface for each vertical series (between basement and sedimentary layer) are representation of our subsurface structure that will be updated as our solution.

Content from this work may be used under the terms of the Creative Commons Attribution 3.0 licence. Any further distribution of this work must maintain attribution to the author(s) and the title of the work, journal citation and DOI.
Published under licence by IOP Publishing Ltd

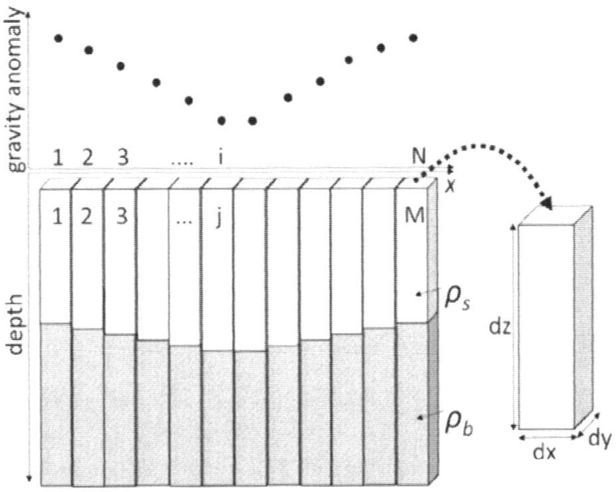

Figure 1. Ilustration of subsurface model with flat topographic to explain simplified sedimentary basin structure.

In order to minimize data miss-fit we used iterative calculation of differences between observed (g_{obs}) and calculated (g_{cal}) data. Iterative calculation attempt to modify model solution ($z(j)$) and updated using slight modification from Chakravarthi [10], as follow:

$$z^{(k+1)}(j) = z^{(k)}(j) + \frac{Rand \times (g_{obs}(i) - g_{cal}(i))}{G\rho_s}. \quad (1)$$

Where *i, j* and *k* respectively denotes index number of data observation, vertical series in subsurface model, and iteration. Random number (*Rand*), universal gravitational constant (*G*), and sedimentary density (ρ_s) used as nominator and denominator in the equation (1) in order to provide rough estimation of updated solution. Flow-chart of the program will be described in several steps. Start from the input, initialization, initial model, iterative process (update model, well constraint, miss-fit evaluation), stoping criteria, and output. The procedure of iterative calculation described with program flow-chart as seen in **figure 2**.

Input of the program consist of three categories:

1. Gravity data observation and station coordinates. In this category, program will accomodate information related with station's interval, target anomalies (residual Bouguer anomaly), N-obs data, and instrument height from the surface.
2. Subsurface model parameters. In this category, we can accomodate following information: well data, well location/coordinate, depth estimation from spectral analysis, M-vertical series of prism, length of body strike in axis-Y direction, coordinat prisms, maximum depth for subsurface model, sedimentary layer density, and basement density.
3. Iterative parameters for maximum iteration number and realization number for statistical purpose.

Initialization of the program used to introduce:

1. Default constants (universal gravitational constant, unit conversion, prism's index).
2. Empty matrix as storage space of objective function (data miss-fit), calculated (g_{cal}), and model solution.

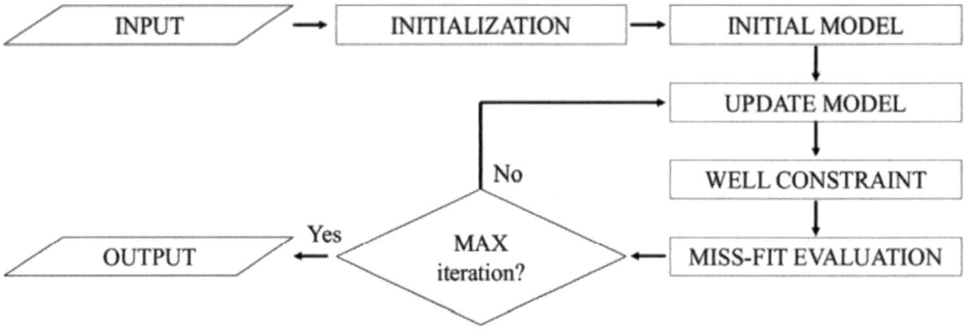

Figure 2. Flow-chart of the program to seek model solution of sedimentary basin structure.

After initialization, next step from the program is generating initial model and several calculation (Kernel matrix, g_{cal}, and miss-fit). Initial model can be generated as follow:

1. Random interface between sedimentary and basement based on random number.
2. Flat interface between sedimentary and basement based on depth estimation from spectral analysis.
3. Bowl shape interface between sedimentary and basement based on mimicking the input of gravity anomalies.
4. Subjective initial model based on user interpretation.

Iterative process to update model solution for each vertical series is following mathematic formula of equation (1). Well data also accomodate in iterative process (as well-constraint) in order to avoid updated solution in this well location. After model solution in well location is tied-up, program calculate Kernel matrix, g_{cal}, and miss-fit. Miss-fit evaluation will compare current and previous model. Miss-fit evaluation only accept down-hill changes of the objective function (miss-fit). Miss-fit (ϕ) in this study calculated with this equation:

$$\phi = \sum_{i=1}^{N} (g_{obs}(i) - g_{cal}(i))^2 \tag{2}$$

Stoping criteria the iterative process is sets-up using maximum iteration number. If maximum iteration number is met, we get model solution as our output. The program working with random number, so it is possible we will get different solution as realization. We can evaluate deviation of solution based on statistical point of view from several realization.

3. Synthetic Data

Forward calculation for synthetic data uses equation of gravitation caused by 3-D polygonal prism body. Vertical component of gravitation (g_z) in the coordinate O (0, 0, 0) caused the body, mathematically expressed with this following equation:

$$g_z = G\rho \int_{\gamma=z_1}^{\gamma=z_2} \int_{\beta=y_1}^{\beta=y_2} \int_{\alpha=x_1}^{\alpha=x_2} \frac{z}{r^3} d\alpha \, d\beta \, d\gamma. \tag{3}$$

$$g_z = G\rho \left[\begin{matrix} \alpha\, ln(\beta + r) + \\ \beta\, ln(\alpha + r) - \gamma\, tan^{-1}\left(\frac{\alpha\beta}{\gamma r}\right) \end{matrix} \right] \Bigg|_{x_1}^{x_2} \Bigg|_{y_1}^{y_2} \Bigg|_{z_1}^{z_2} . \quad (4)$$

The formula in equation (3) accomodate universal gravitational constant (G), density of prism (ρ), resultant distance between 3-D body and gravity station (r), vertical distance between 3-D body and gravity station (z) and prism edges coordinate (x_1, x_2, y_1, y_2, z_1, and z_2). Next, by integral the equation (3) we get equation (4). Equation (4) is gravitation effect from one prism body, while for the overall model that consists of several prism bodies, the calculation conducted cumulatively as described by Pluoff [16].

Synthetic model A with flat topographic shown in **figure 3a**. The model A is calculated over 11 stations with 500 m of station's interval in flat topographic. Synthetic model consists of 11 vertical series (of sedimentary and basemenet) with 2700 m in maximum depth. Length of each vertical prisms in x-direction (*dx*) and y-direction (*dy*) respectively are 500 m and 10000 m, while the length of prisms in z-direction (*dz*) are varied symmetrically as shown in numbers below the surface in **figure 3a** (900, 1020, 1140, 1260, 1380, 1500, 1380, 1260, 1140, 1020, 900). Density contrast for numerical example used in the sedimentary and basement model respectively are -0.10 g/cc and +0.10 g/cc.

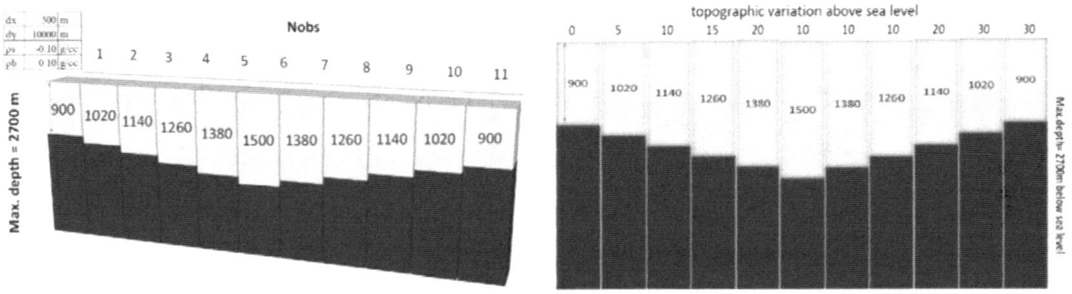

(a) (b)

Figure 3. Model ilustration: (a) model A with flat topography and (b) model B with topographic variation.

Synthetic model B with topographic variation for algorithm performance test shown in **figure 3b**. Model B is similar as the previous one, but topographic variation above sea level shown in numbers above the surface (**figure 3b**). The sedimentary thickness are varied symmetrically from mean sea level (same with model A). In this study, algorithm performance test for model B will be conducted in two types synthetic data: (1) free-noise and (2) with noise. We use random noise within the range of ± 10 μGal.

4. Performace Test

Performance test using synthetic data A will be evaluate the effect of initial model and well data. Synthetic data A is tested with three types of set-up, as follow: random initial model, flat initial model, and flat initial model with well constraint. We conducted inversion with 1000 iteration for 100 realization in order to provide statistical sample.

Random initial model introduce to the iterative calculation as simulation of no subsurface data consideration. This set-up try to simulate data driven in the iterative proces to seek model solution. Inversion performance shown as the average changes of data miss-fit over iteration number. Miss-fit is average value from 100 realization (**figure 4**). As we can see from the **figure 4**, miss-fit (of random initial model) curve changes significantly at the first 50 iteration and then the data miss-fit changes looks like as linear downward (up to 10^{-8} order).

We also can use depth estimation from spectral analysis results as flat initial model to the iterative calculation. In this example (synthetic data A), we get rough result as depth estimation from spectral analysis is 857.84 m. Average miss-fit is calculate from 100 realization (**figure 4**). As we can see from the **figure 5**, data miss-fit (of flat initial model) changes significantly at the begining of 50 iteration and then the data miss-fit continously curve down to 10^{-9} order.

Well constraint introduce to the iterative calculation can be very helpfull to reduce ambiguity of inverse model solution in the subsurface. Average miss-fit is calculate from 100 realization (**figure 4**). As we can see from the **figure 4**, data miss-fit (of flat initial model with well constraint) changes significantly at the begining of 50 iteration and then the data miss-fit continously curve down to 10^{-11} order.

Average model (solution) from inverse modeling is shown in **figure 5**. Maximum differences between synthetic and inverse model in the range of -6 to 8 meters. Using flat initial model provide smaller standard deviation of solution model compare to random initial model. Big difference between solution model and synthetic model shows in the centre-part of the section. **Figure 5a** shows the comparison of 11 solution model within depth range of 800-1500 m, while **figure 5b** shows the zoom in of comparison 3 solution model in the centre-part of the section (within depth range of 1350-1550 m). **Table 1** shows the summary of comparison from statistical point of view. This performance test of inversion with free-noise data and simple synthetic model is completely recover the true solution, because almost to zero difference between synthetic and inverse model (flat initial model with well constraint).

Performance test using synthetic data B will be evaluate the effect of topographic and noisy data. Flat initial model and well data are introduce into the calculation of 1000 iteration for 100 realization. Average and median of inverse model are shown in **table 2** to provide sedimentary thickness comparison from synthetic and inverse model. With small topographic variation in the syntethic model B (0 to 30 m above the sea level), the iterative calculation of inverse model possible to recover the true solution. The difference solutions for each vertical series between synthetic and inverse model as can be seen in **table 2** in the range of -4 to 2 meters. More realistic input data are simulated with noisy data in the synthetic data model B. Maximum difference between synthetic and average inverse model in the range of -44 to 35 meters. Comparison of solution model using synthetic data B is shown in **figure 6**.

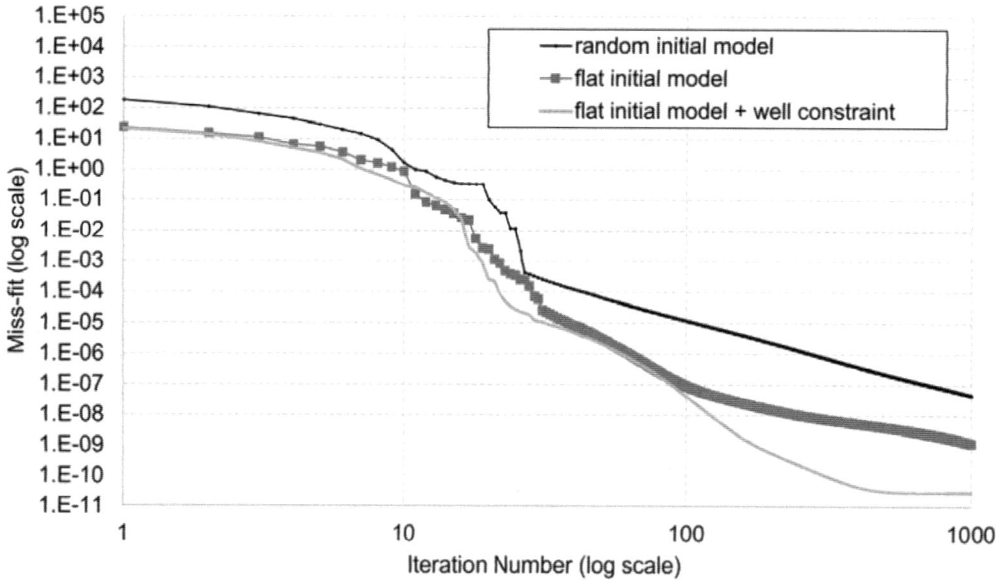

Figure 4. Comparison of inversion performance using synthetic data A describe as down-hill curves of average miss-fit over iteration number.

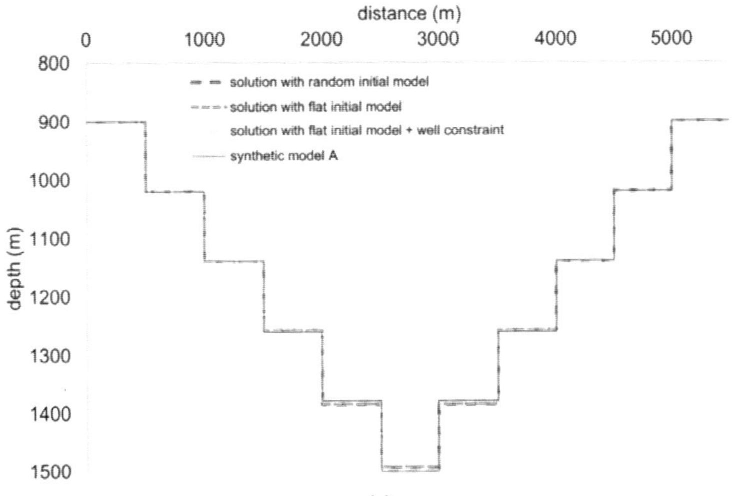

(a)

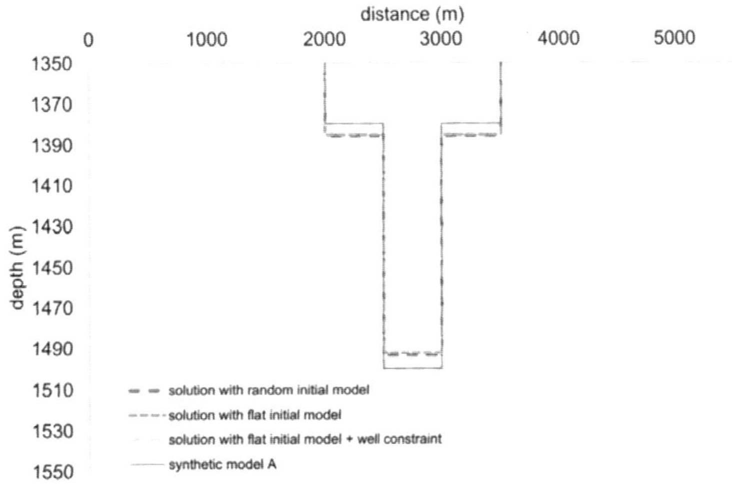

(b)

Figure 5. Comparison of solution model using synthetic data A: (a) depth interval 800-1500 m, (b) depth interval 1350-1550 m.

Table 1. Comparison of sedimentary thickness (m) between synthetic model A and inverse models.

no	synt.	random initial model				flat initial model				flat initial model w/ well const.			
		avg.	med.	min.	max.	avg.	med.	min.	max.	avg.	med.	min.	max.
1	900.00	899.94	900.06	898.37	900.42	900.08	900.08	900.07	900.08	900.00	900.00	899.95	900.00
2	1020.00	1019.74	1019.80	1016.28	1022.95	1019.75	1019.75	1019.74	1019.77	1020.00	1020.00	1020.00	1020.23
3	1140.00	1140.61	1140.70	1128.16	1151.20	1140.87	1140.87	1140.82	1140.91	1139.99	1140.00	1139.49	1140.00
4	1260.00	1257.48	1258.17	1224.78	1285.98	1257.76	1257.76	1257.66	1257.89	1260.01	1260.00	1260.00	1260.68
5	1380.00	1385.67	1385.11	1316.15	1462.74	1385.26	1385.27	1384.96	1385.51	1380.00	1380.00	1379.65	1380.00
6	1500.00	1493.66	1492.38	1390.18	1602.27	1492.42	1492.41	1492.07	1492.85	1500.00	1500.00	1500.00	1500.00
7	1380.00	1385.67	1385.58	1315.10	1465.45	1385.26	1385.27	1384.96	1385.51	1380.00	1380.00	1379.65	1380.00
8	1260.00	1257.48	1257.57	1222.63	1287.27	1257.76	1257.76	1257.66	1257.89	1260.01	1260.00	1260.00	1260.68
9	1140.00	1140.61	1140.91	1127.42	1152.36	1140.87	1140.87	1140.82	1140.91	1139.99	1140.00	1139.49	1140.00
10	1020.00	1019.74	1019.73	1015.79	1023.23	1019.75	1019.75	1019.74	1019.77	1020.00	1020.00	1020.00	1020.23
11	900.00	899.94	900.08	898.29	900.42	900.08	900.08	900.07	900.08	900.00	900.00	899.95	900.00

Table 2. Comparison of sedimentary thickness (m) from msl between synthetic model B and inverse models.

no	synt.	free-noise data		noise data added	
		avg.	Δ solution	avg.	Δ solution
1	900.00	900.49	-0.49	901.00	-1.00
2	1020.00	1019.99	0.01	1016.84	3.16
3	1140.00	1139.55	0.45	1127.22	12.78
4	1260.00	1258.99	1.01	1303.10	-43.10
5	1380.00	1383.11	-3.11	1345.48	34.52
6	1500.00	1500.00	0.00	1500.00	0.00
7	1380.00	1378.20	1.80	1381.30	-1.30
8	1260.00	1261.22	-1.22	1278.71	-18.71
9	1140.00	1140.18	-0.18	1129.43	10.57
10	1020.00	1020.27	-0.27	1017.90	2.10
11	900.00	900.26	-0.26	902.38	-2.38

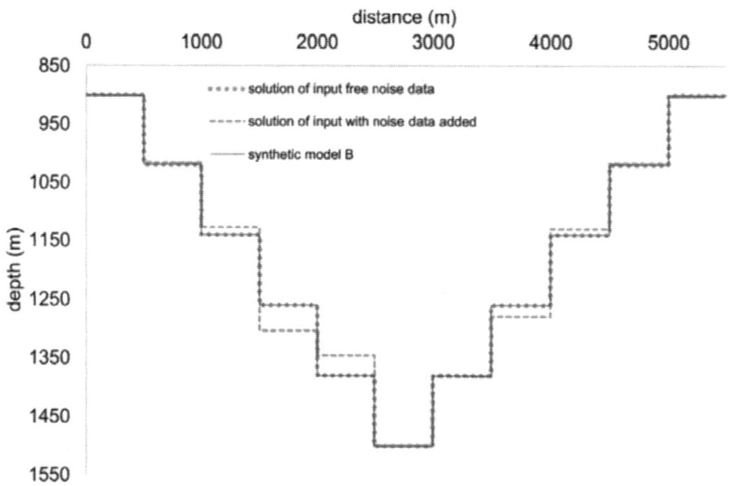

Figure 6. Comparison of solution model using synthetic data B (depth interval 850-1550 m).

5. Real Data Application

Grid survey of gravity data observation conducted in study area that located close to Blora in Central Java, Indonesia. About 250 m interval grid station over 5x5 km square are acquired in 2014 using two Scintrex CG-5 (blue and red team). Summary of the survey is shown in **table 3**. Relative gravity data from 402 stations are processed to get Complete Bouguer Anomaly, then regional-residual anomaly are separated using moving average filter. Residual anomaly and topographic map are sliced diagonaly to provide SW-NE section as the input data of inverse modeling.

The geological formations imaged on seismic profile indicate that the center part of study area is an anticline structure whose axis is continued for E-W direction [17]. Simplified surface geology map [18] (**figure 7**) also confirm anticline structure in the center part of study area. High residual anomaly in the center-part of study area expected as an anticline structure in the inverse modeling. Density contrast as an input of first (upper) layer in this study area are test in the range of 0.00 to -0.10 g/cc, while density contrast as an input of second (lower) layer in the range of 0.00 to +0.10 g/cc. Combination of first and second layer provide 121 pairs as an input to be calculated in the scheme of 100 iteration number. Strike length (in the y axis of subsurface model) as an input of first and second layer are 1 km and 1.5 km respectively, while length interval in the x axis of subsurface model is 50 m. Maximum depth (in the z axis of subsurface model) are set-up in 5 km. Total 137 vertical series of subsurface model calculated below 137 data input. Data input are tophographic and residual anomaly with 50 m interval data in the direction of SW-NE azimuth.

Table 3. Survey summary.

Survey	2014	
Team	Blue	Red
Date	19-29 Sep	18-29 Sep
Gravimeter	CG-5 822	CG-5 1092
Tot. Time Duration	37560 min.	43020 min.
Good Leveling	98.02%	98.26%
Auto Rej. > 5%	19.59%	17.91%
Average daily drift	42 μGal	46 μGal
#Day(s)	10	11
#Reading(s)	1113	1379
#Observation(s)	243	239
#Station(s)	206	202
#Repeat Obsv.(s)	37	37
#VG Station(s)	3	2

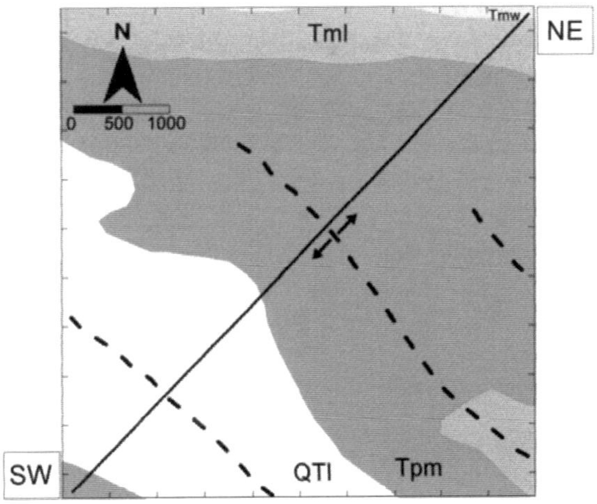

Figure 7. Simplified surface geology map [18] in study area consist of Lidah Formation (QTl), Mundu Formation (Tpm), Ledok Formation (Tml), and Wonocolo Formation (Tmw).

As the input of iterative calculation described in the previous paragraph, we will get 121 model as possible solution for our study. Miss-fit data for each model solution had been minimized over 100 iteration. The example of miss-fit data changes over the iteration number shown in **figure 8**. There are 27 model solution that have small missfit data (less then 1 mGal). We consider the 27 possible model solution to be the best solution in average value as shown in **figure 9**. The density of first and second layer based on 27 best model respectively in the range of -0.10 to -0.07 g/cc and 0.04 to 0.10 g/cc. Based on the results in **figure 9**, we get average density for the first and second layer are -0.09 and 0.07 g/cc. Simplified subsurface structure and the depth variation of sedimentary layer (first layer) described in **figure 10**. Shallow sedimentary layer (less then 1 km depth) shown in **figure 10** as 1A, 1B, and 1E, while deeper sedimentary layer (1C and 1D) located beside the center anticline structure (2B).

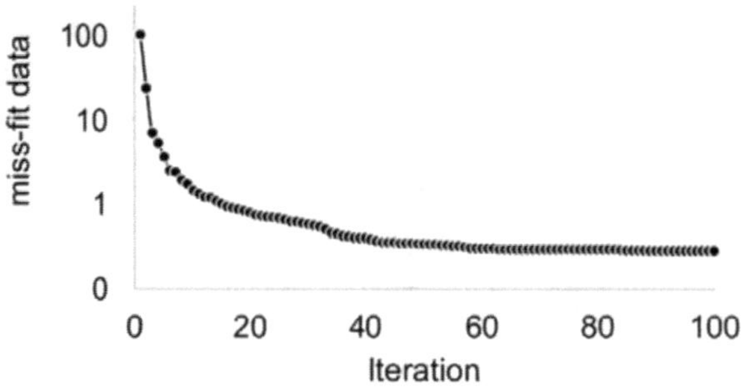

Figure 8. Example of miss-fit data changes (in the log scale) during iterative process.

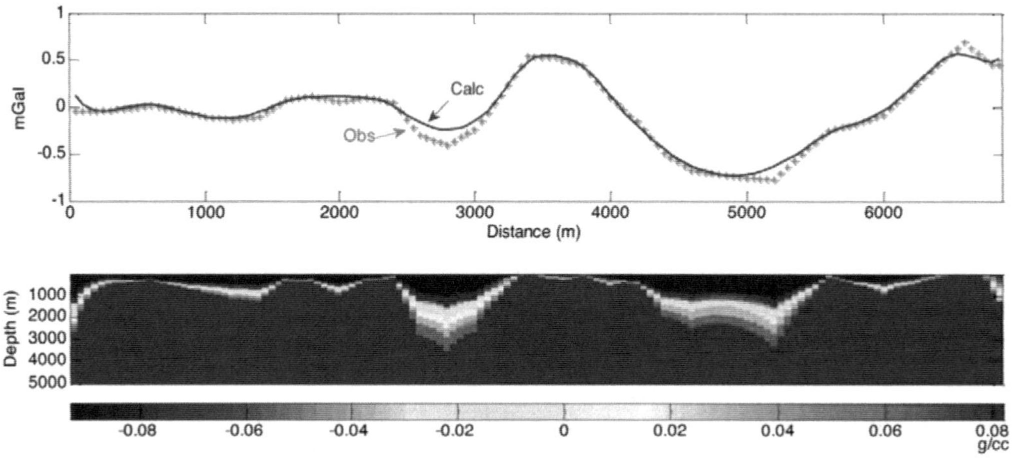

Figure 9. Average model from 27 best miss-fit data (depth interval 0-5 km).

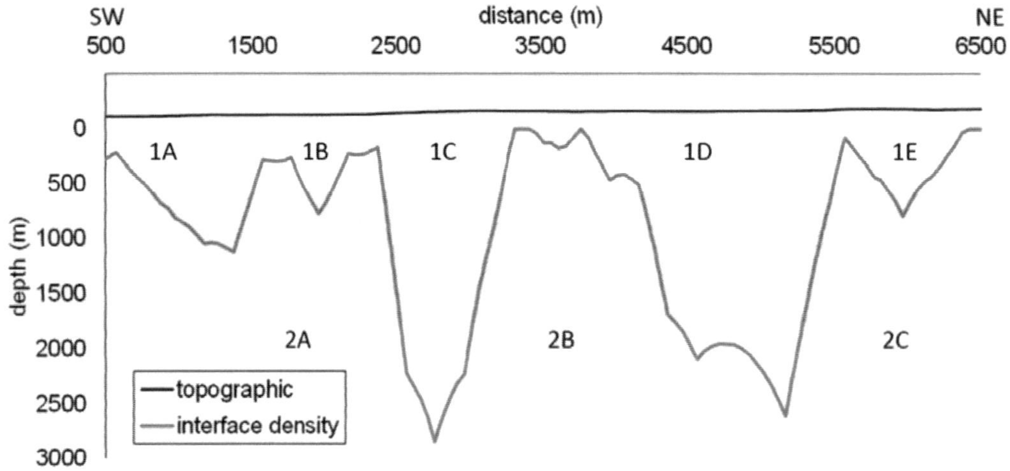

Figure 10. Simplified subsurface structure.

6. Conclusion

Iterative calculation using gravity data can be used to determine the interface density contrast in the subsurface between sedimentary layer and basement layer. Based on synthetic data test, the performance of iterative method using synthetic data of model A provide small order of miss-fit data (10^{-8} to 10^{-11}).

The performance test using synthetic data of model B (topographic and noisy data added) provide residual solution of model in the range -44 to 35 meters. Real data application identified three shallow sedimentary layer and two deeper sedimentary layer in the study area.

7. Acknowledgments

Authors acknowledge support of the P3MI-ITB, SATREPS project by JICA-JST, and other Gundih CCS pilot project team members that help data acquisition.

8. References

[1] Fournier K P and Krupicka S F 1975 A new approximate method for directly interpreting gravity anomaly profiles caused by surface geologic structures *Geophysical Prospecting* **23** 80
[2] Overmeeren R A V 1975 A combination of gravity and seismic refraction measurements, applied to groundwater near Taltal, Province of Antofagasta, Chile *Geophysical Prospecting* **23** 248
[3] Verma R K, Majumdar R, Ghosh D, Ghosh A and Gupta N C 1976 Results of gravity survey over Raniganj coalfield, India *Geophysical Prospecting* **24** 19
[4] Pedersen L B 1977 Interpretation of potential field data a generalized inverse approach *Geophysical Prospecting* **25** 199
[5] Baldi P and Unguendoli M 1978 Inversion of gravity profiles by polynomial method *Geophysical Prospecting* **26** 247
[6] Rao D B, Prakash M J and Babu N R 1990 3D and 2½D modelling of gravity anomalies with variable density contrast *Geophysical Prospecting* **38** 411
[7] Lee T C and Biehler S 1991 Inversion modeling of gravity with prismatic mass bodies *Geophysics* **56** 1365
[8] Mickus K L and Peeples W J 1992 Inversion of gravity and magnetic data for the lower surface of a 2.5 dimensional sedimentary basin *Geophysical Prospecting* **40** 171
[9] Rao V C, Pramanik A G, Kumar G V R K and Raju M L 1994 Gravity interpretation of sedimentary basins with hyperbolic density contrast *Geophysical Prospecting* **42** 825
[10] Chakravarthi V 1995 Modelling of density interface with binomial density variation *Journal of Applied Geophysics* **34** 69
[11] Abdeslem J G 2000 2-D inversion of gravity data using sources laterally bounded by continuous surfaces and depth-dependent density *Geophysics* **65** 1128
[12] Chappell A and Kusznir N 2008 An algorithm to calculate the gravity anomaly of sedimentary basins with exponential density-depth relationships *Geophysical Prospecting* **56** 249
[13] Silva J B C, Oliviera A S and Barbosa V C F 2010 Gravity inversion of 2D basement relief using entropic regularization *Geophysics* **75** I29
[14] Zhou X 2010 Analytic solution of the gravity anomaly of irregular 2D masses with density contrast varying as a 2D polynomial function *Geophysics* **75** I11
[15] Wahyudi E J, Marthen R, Fukuda Y, Nurali Y 2017 Time-lapse microgravity data acquisition in baseline stage of CO_2 injection Gundih pilot project *IOP Conf. Series: Earth and Environmental Science* **62** 012047
[16] Pluoff D 1976 Gravity and magnetic fields of polygonal prisms and application to magnetic terrain corrections *Geophysics* **41** 727
[17] Tsuji T, Matsuoka T, Kadir W G A, Hato M, Takahashi T, Sule M R, Kitamura K, Yamada Y, Onishi K, Widarto D S, Sebayang R I, Prasetyo A, Priyono A, Ariadji T, Sapiie B, Widianto E, Asikin A R and Gundih CCS project team 2014 Reservoir characterization for site selection in the Gundih CCS project, Indonesia *Energy Procedia* **63** 6335
[18] Pringgoprawiro H and Sukido 1992 Geologic map of the bojonegoro quadrangle, East Jawa *Geological Research and Development Centre*

IOP Conf. Series: Journal of Physics: Conf. Series **1090** (2018) 012049 doi:10.1088/1742-6596/1090/1/012049

Comparison of absorbed dose distribution 10 MV photon beam on water phantom using Monte Carlo method and Analytical Anisotropic Algorithm

Ridwan Ramdani[1] , Freddy Haryanto[2], Yudha Satya Perkasa[1]

[1]Department of Physics, Faculty of Science and Technology, UIN Sunan Gunung Djati, Bandung, Indonesia
[2]Department of Physics, Faculty of Mathematics and Natural Sciences, Institut Teknologi Bandung, Indonesia

E-mail: ridwan@fst.uinsgd.ac.id

Abstract. Treatment planning is a process of radiation therapy performed by medical physicists in the hospital. This process can be understood as a simulation to determine parameters such as photon energy, field size, and the absorbable dose to be received by the patient. Simulation treatment planning is usually done using TPS ECLIPSE with AAA or Acuros XB method that has been integrated into it. In this research, the comparison of the distribution of absorbent dose of 10 MV photon at water phantom using Monte Carlo-EGSnrc and AAA-ECLIPSE method was used. Monte Carlo simulations were performed with 6x6 cm^2, 10x10 cm^2, and 20x20 cm^2 field size variations and initial electron energy variations 10.1 MeV, 10.2 MeV, 10.3 MeV and 10.4 MeV. The design and simulation of Linac heads are done using BEAMnrc and the calculation of absorbent doses in this water phantom using DOSXYZnrc. The BEAMnrc simulated particle information was analyzed using BEAMDP. To obtain the depth dose distribution or PDD and PROFILE the dose is done using statdose code. Based on the data processing, the result shows that the mean deviation of PDD and D_{max} from the three field sizes is 7.09% for 10.1 MeV, 2.71% for 10.2 MeV, 2.86% for 10.3 MeV and 7.39% for 10.4 MeV. These results indicate that the initial electron energy has the least deviation of 10.2 MeV. As for the dose profile in the field size 10x10 cm^2 obtained deviation 6.65% to 10.2 MeV

1. Introduction

Treatment planning system is a process of radiation therapy planning performed by medical physicists in the hospital. This process can be understood as a simulation to determine physical parameters such as photon energy, field size and absorption dose to be accepted by patients using Analytical Anisotropic Algorithm (AAA) and Acuros XB algorithms integrated with the Eclipse Varian Medical System TPS system. AAA is a dose calculation algorithm divided into two processes namely configuration and dosage calculation, the configuration is a process to determine the physical

Content from this work may be used under the terms of the Creative Commons Attribution 3.0 licence. Any further distribution of this work must maintain attribution to the author(s) and the title of the work, journal citation and DOI.
Published under licence by IOP Publishing Ltd

parameters that can be used to obtain fluence characteristics and photon energy spectrum, this physical parameter has been previously calculated using Monte Carlo then modified in order According to clinical data, this continues throughout the configuration process and all parameters of the configuration results will be used for dose calculations, while dosing of AAA algorithm based on primary photons, extra-focal photons, electron contamination and photon scattering are all incorporated in a beamlet) [1]. Nowadays, AAA algorithm development has been developed, AAA and Pencil beam Convolution (PBC) evaluation in the case of water-lung phantom [2], comparison of dose distribution on small size phantom using AAA and Monte Carlo [3], comparison of dose distribution Using AAA and experiments [4].

Monte Carlo is a method that uses the principle of random numbers, this method is very well used for the simulation of particle interactions that occur in stochastic, very accurate in the calculation of the dose distribution to make Monte Carlo as "gold standard" in the calculation of dose distribution. Previous studies have shown that the deviation of Monte Carlo PDD results and experiments is 1.07% [5], not only to calculate dosage distributions inhomogeneous tissues but also in Monte Carlo's inhomogeneous tissues capable of calculating dose distribution very accurately [6]. In this research, the calculation of dose distribution in this water phantom using Monte Carlo BEAMnrc and DOSXYZnrc simulation which is the development of EGSnrc. BEAMnrc is used to design and simulate the particle interaction on the Linac head Varian, the result is a phase space file containing the interaction particle information and it can be analyzed using BEAMDP. DOSXYZnrc is used to design and simulate the dose distribution in water phantom the result is data of .3ddose which is then analyzed using statdose code. The results of this dose distribution will compare with the dose distribution of AAA simulation results from one of the hospitals in Bandung.

2. Method

2.1. Design and simulation of head Linear Accelerator (Linac)

Material and geometry data specifications for the design of Linac heads were obtained from the Monte Carlo High Energy Varian Clinac iX. The data information is used as input in Monte Carlo simulation on BEAMnrc.

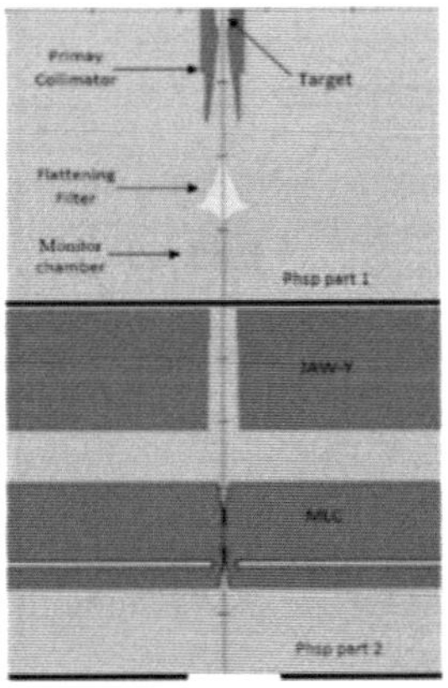

a

No.	Component Module	CM
1	Primary Collimator+target	FLATFILT
2	Vacuum Window	FLATFILT
3	Flattening Filter	FLATFILT
4	Monitor Chamber	CHAMBER
5	Jaws	JAWS
6	MLC	DYNVMLC
7	Air	SLAB

b

Figure 1. (a) Design head Linac, (b) Component module used in BEAMnrc [8]

Figure 1 shows the result of head Linac design using BEAMnrc. Linac modeled is Linac Varian with 10 MV photon beam, variation of field size 6x6 cm^2, 10x10 cm^2 and 20x20 cm^2 and variation of initial electron energy 10.1 MeV, 10.2 MeV, 10.3 MeV and 10.4 MeV. In this process the Linac design is made of two components: Patient-dependent component, using ISOURCE 19 (Elliptical beam with

Gaussian distribution in X and Y) resulting phsp part 1 and Patient-independent components, using ISOURCE 21 (full phsp beam data, incident on any CM) that produces phsp part 2, this is done to reduce the simulation time required to use different field size variations. The second phsp is used as input into DOSXYZnrc.

2.2. Design and simulation of dose distribution

Size of water phantom used is 40x40x40 cm^3, made into two design that is for the calculation of the depth dose and dose profile. The simulation is done in DOSXYZnrc with ISOURCE 2 (full phsp source file).

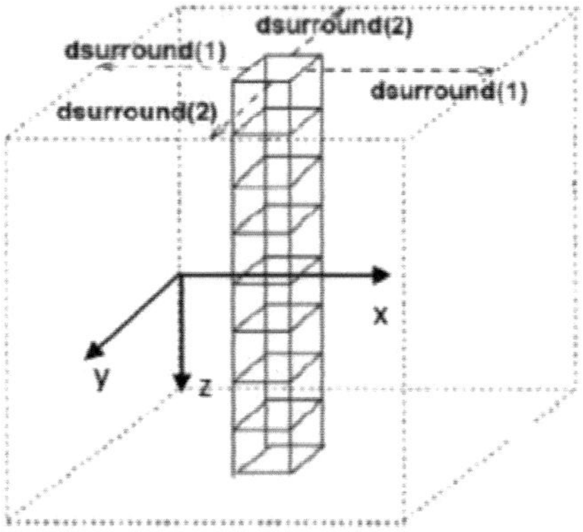

Coord	Num of Voxels	Voxel Size (cm)	
X	3	1	18.5
		2	3
		3	18.5
Y	3	1	18.5
		2	3
		3	18.5
Z	36	Group 1 (2 voxel)	0.4
		Group 2 (16 voxel)	0.2
		Group 3 (18 voxel)	2

Figure 2. Design water phantom for depth dose calculation [7]

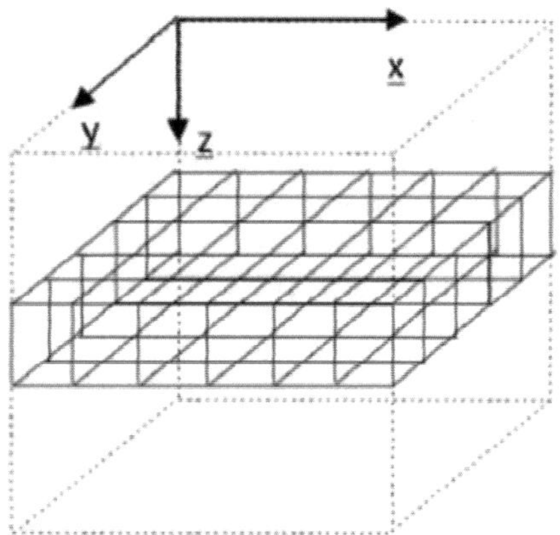

Coord	Num of Voxels	Voxel Size (cm)	
X	3	1	19
		2	2
		3	19
Y	80	Group 1 (7 voxel)	2
		Group 2 (10 voxel)	0.2
		Group 3 (8 voxel)	1
		Group 4 (10 voxel)	0.2
		Group 5 (7 voxel)	2
Z	3	1	9.5
		2	1
		3	29.5

Figure 3. Design water phantom for dose profile calculation [7]

It can be seen Fig 4. Design of water phantom for depth dose calculation is made the smaller size and number of voxel on the Z axis, while for dose profile in Y-axis direction at 10 cm depth.

3. Result

Based on data processing, the result of depth dose and dose profile for each variation of initial electron energy and field size, then compared with AAA simulation result. The graphic as follows.

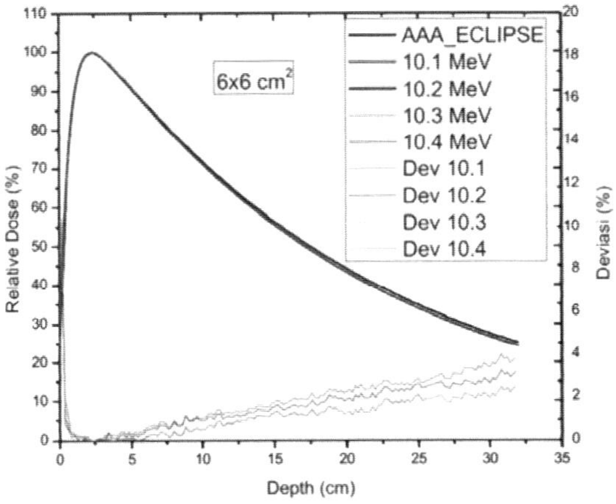

(a)

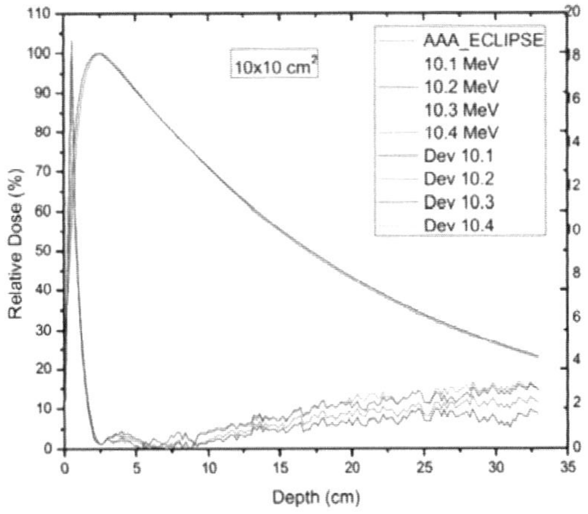

(b)

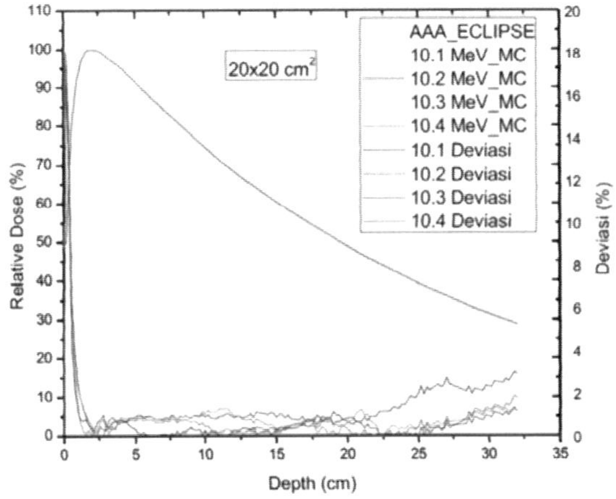

(c)

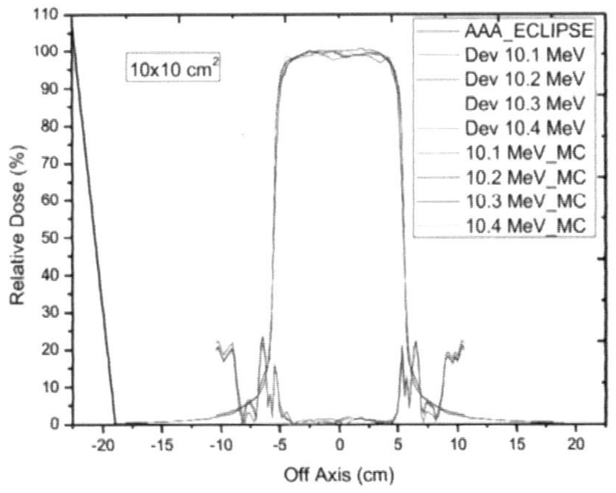

(d)

Figure 4. (a) (b) (c) Depth dose graph for field size and initial electron energy variation (d) dose profile graph with initial electron energy variation in size field 10x10 cm^2 and depth 10 cm.

Based on Figure 4 it can be seen that the depth distribution pattern has increased significant doses on the surface of the water phantom until it reaches the maximum depth of dose then undergoes decline as the depth of water phantom increases. This is because the photon beam generated from the Linac are not only composed of photon beams but also composed of an electron and positron particles that have a shorter range than photons so that will interact earlier, electron and positron particle distribution can be seen when performing the analysis Phase space file with BEAMDP. The dose decrease after D_{max} because the photon interacted by giving energy to the water phantom, more in water phantom the photon energy given will be smaller.

IOP Conf. Series: Journal of Physics: Conf. Series **1090** (2018) 012049 doi:10.1088/1742-6596/1090/1/012049

4. Discussion

Based on Figure 4, the initial electron energy of 10.4 MeV produces the smallest depth dose deviation of 1.85% in the 6x6 cm^2, 10.1 MeV with deviation 1.57 % on 10x10 cm^2 and 10.2 MeV with deviation 0.73 % on 20x20 cm^2. While the initial electron energy with a small D_{max} deviation of 10.1 MeV gives a deviation 3.91 % on 6x6 cm^2, 10.2 MeV with deviation 3.60 % on 10x10 cm^2 and 10.1 MeV with deviation 5.71 % on 20x20 cm^2. To determine the initial electron energy used can be done by looking for the least average deviation of depth dose, D_{max} and dose profile.

5. Conclusions

The mean deviation of depth dose and D_{max} from the three field sizes is 7.09% for 10.1 MeV, 2.71% for 10.2 MeV, 2.86% for 10.3 MeV and 7.39% for 10.4 MeV. As for the dose profile in the field size 10x10 cm^2 obtained deviation 6.65% with 10.2 MeV. Based on these results it can be concluded that the initial electron energy having the least deviation is 10.2 MeV with the mean deviation 4.02 %.

6. References

[1] Sievinen, Janne. at. al. AAA Photon dose calculation model in Eclipse™. Varian Medical System.
[2] Gagne, M. Isabelle. at. al. 2007. Evaluation of the analytical anisotropic algorithm in an extreme water-lung interface phantom using Monte Carlo dose calculations. Journal of Applied Clinical Medical Physics (JACMP), Vol 8, Num 1.
[3] Chow, C.L. James. at. al. 2014. Dosimetry of small bone joint calculated by the AAA: a Monte Carlo evaluation using the EGSnrc. Journal of Applied Clinical Medical Physics (JACMP), Vol 15, Num 1.
[4] Breitman, Karen. at. al. 2007. Experimental validation of the Eclipse AAA algorithm. Journal of Applied Clinical Medical Physics (JACMP), Vol 8, Num 2.
[5] Ramdani, Ridwan. at. al. 2015. Commissioning Linear Accelerator Varian Clinac iX photon beam 10 MV menggunakan simulasi Monte Carlo EGSnrc Code system. Prosiding SNIPS 2015 ITB.
[6] Chetty I, at. al. 2007. Report of the AAPM task group No.105: Issues Associated with clinical implementation of Monte Carlo Photon and electron beam treatment planning, Vol 34, Med Phys.
[7] D.W.O. Rogers, Kawrakow. I, Walters. B. 2011. DOSXYZnrc users manual. NRCC
[8] Kawrakow. I, Walters. B, D.W.O. Rogers. 2011. BEAMnrc users Manual. Ottawa: National Research Council of Canada

Numerical solutions to a parabolic model of two-layer fluids

Sudi Mungkasi and Friska Dwi Mesra Mangadil

Department of Mathematics, Faculty of Science and Technology,
Sanata Dharma University, Yogyakarta, Indonesia

E-mail: sudi@usd.ac.id

Abstract. In this paper, we consider the motion of two layers of fluids having different viscosity values. The motion is driven by a moving surface, but the bottom is fixed. An exact analytical solution for unsteady state cases is not available. Therefore, a numerical method should be used for the solution to unsteady state cases of the problem. In this work, we propose a finite volume numerical method to find the numerical velocity of the problem. We use the Lax-Friedrichs formulation for the flux calculation. Our numerical results show that fluids move following the motion of the surface. In addition, the fluid at the top layer moves faster than the bottom layer fluid. These behaviour is correct with respect to the physical problem under consideration.

1. Introduction

Numerical simulation has been widely used by mathematical modellers to investigate physical phenomena. The phenomena is first modelled into a system of equations. The system is then solved numerically, as the exact analytical solution is generally difficult to find.

This paper considers the problem of two layers of fluids moving in one direction. We assume that oil layer is on the top of water layer. The two layers are in between a moving surface and a fixed bottom. The moving surface has a constant velocity. This problem was introduced by Caldwell and Ng [1], who used a finite difference method to obtain its numerical solution.

In this work, we provide an alternative numerical solver by proposing a finite volume method to find the numerical solution to the problem. The finite volume method works well as long as the mathematical model can be written in a conservative form. We implement the Lax-Friedrichs formulation [2-3] to compute numerical fluxes of the conservative form of the model. Our numerical scheme is explicit. As time tends to infinity, the motion of fluids is steady. This is confirmed by the exact as well as the numerical solution. The numerical solution agrees quite well with the exact steady state solution.

The rest of this paper is organised as follows. We present the problem formulation in Section 2. Numerical method is proposed in Section 3. We provide numerical results in Section 4. The paper is concluded in Section 5.

2. Problem formulation

This section introduces the equations for the motion of the two-layer fluids.

Consider two horizontal plates with a distance of 10 cm and between those two plates there are oil and water layers, as shown in Figure 1. The top plate moves to the right with a constant velocity. The bottom plate is fixed. We assume that starting at time $t = 0$, the top plate moves with constant velocity 7 cm/s.

Content from this work may be used under the terms of the Creative Commons Attribution 3.0 licence. Any further distribution of this work must maintain attribution to the author(s) and the title of the work, journal citation and DOI.
Published under licence by IOP Publishing Ltd

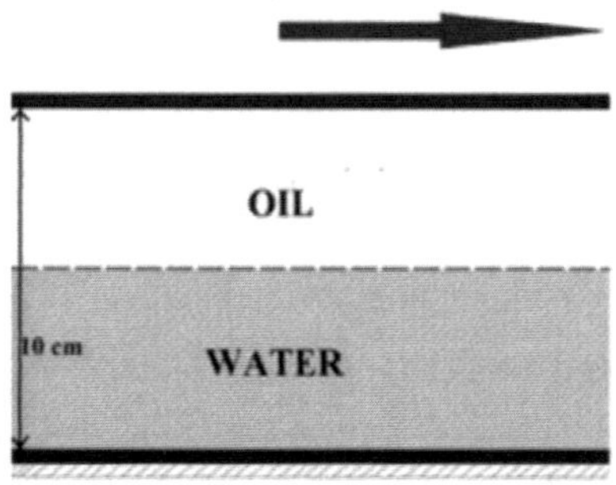

Figure 1. Two plates spaced 10 cm apart.

This problem has been modelled mathematically by Caldwell and Ng [1] as:

$$\frac{\partial v_{\text{water}}}{\partial t} = \mu_{\text{water}} \frac{\partial^2 v_{\text{water}}}{\partial x^2}, \tag{1}$$

and

$$\frac{\partial v_{\text{oil}}}{\partial t} = \mu_{\text{oil}} \frac{\partial^2 v_{\text{oil}}}{\partial x^2}. \tag{2}$$

Here $v_{\text{water}}(x,t)$ is the velocity of water, $v_{\text{oil}}(x,t)$ is the velocity of oil, μ_{water} is the viscosity of water, and μ_{oil} is the viscosity of oil. The free variables are time t and space x. The space x is the fluid height measured from the bottom. At the oil-water interface, we have the following relations:

$$v_{\text{oil}} = v_{\text{water}}, \tag{3}$$

and

$$\mu_{\text{oil}} \frac{\partial v_{\text{oil}}}{\partial x} = \mu_{\text{water}} \frac{\partial v_{\text{water}}}{\partial x}. \tag{4}$$

The simulation is conducted to determine the velocity of fluids at any time $t > 0$, especially for a large time t, when the steady state is achieved.

3. Numerical method

In this section we present a complete mathematical model and the finite volume method that we propose to solve the problem.

The mathematical model is given in the following two sets of initial-boundary value problems:

$$\begin{cases} \dfrac{\partial v_{\text{water}}}{\partial t} = \mu_{\text{water}} \dfrac{\partial^2 v_{\text{water}}}{\partial x^2}, 0 \le x \le 6 \\ v_{\text{water}}(x,0) = 0 \\ v_{\text{water}}(0,t) = 0 \\ v_{\text{water}}(6,t) = v_{\text{oil}}(6,t) \\ \mu_{\text{oil}} \left.\dfrac{\partial v_{\text{oil}}}{\partial x}\right|_{x=6} = \mu_{\text{water}} \left.\dfrac{\partial v_{\text{water}}}{\partial x}\right|_{x=6} \end{cases} \tag{5}$$

$$\begin{cases} \dfrac{\partial v_{\text{oil}}}{\partial t} = \mu_{\text{oil}} \dfrac{\partial^2 v_{\text{oil}}}{\partial x^2}, 6 \le x \le 10 \\ v_{\text{oil}}(x,0) = 0 \\ v_{\text{oil}}(10,t) = 7 \\ v_{\text{oil}}(6,t) = v_{\text{water}}(6,t) \\ \mu_{\text{oil}} \left.\dfrac{\partial v_{\text{oil}}}{\partial x}\right|_{x=6} = \mu_{\text{water}} \left.\dfrac{\partial v_{\text{water}}}{\partial x}\right|_{x=6} \end{cases} \quad (6)$$

The problem (5) and (6) can be solved separately, because there are some conditions that relate to each other.

In order to use the finite volume method, we rewrite equations (5) and (6) as:

$$\frac{\partial v_{\text{water}}}{\partial t} - \mu_{\text{water}} \frac{\partial^2 v_{\text{water}}}{\partial x^2} = 0\,, \quad (7)$$

$$\frac{\partial v_{\text{oil}}}{\partial t} - \mu_{\text{oil}} \frac{\partial^2 v_{\text{oil}}}{\partial x^2} = 0\,. \quad (8)$$

Therefore, we have

$$\frac{\partial v_{\text{water}}}{\partial t} - \frac{\partial}{\partial x}\left(\mu_{\text{water}} \frac{\partial v_{\text{water}}}{\partial x}\right) = 0\,, \quad (9)$$

$$\frac{\partial v_{\text{oil}}}{\partial t} - \frac{\partial}{\partial x}\left(\mu_{\text{oil}} \frac{\partial v_{\text{oil}}}{\partial x}\right) = 0\,. \quad (10)$$

In compact forms, we write equations (9) and (10) as

$$v_{\text{water}_t} + \left(-\mu_{\text{water}} v_{\text{water}_x}\right)_x = 0\,, \quad (11)$$

$$v_{\text{oil}_t} + \left(-\mu_{\text{oil}} v_{\text{oil}_x}\right)_x = 0\,. \quad (12)$$

Equations (11) and (12) are in the form of conservation laws, that is, in their conservative forms:

$$v_{\text{water}_t} + f(v_{\text{water}})_x = 0\,, \quad (13)$$

$$v_{\text{oil}_t} + f(v_{\text{oil}})_x = 0\,. \quad (14)$$

where $f(v_{\text{water}}) = -\mu_{\text{water}} v_{\text{water}_x}$ and $f(v_{\text{oil}}) = -\mu_{\text{oil}} v_{\text{oil}_x}$ respectively.

Equation (13) is solved using the finite volume method with an explicit numerical scheme:

$$V_{\text{water}_i}^{\ n+1} = V_{\text{water}_i}^{\ n} - \frac{\Delta t}{\Delta x}\left[F_{\text{water}_{i+\frac{1}{2}}}^{\ n} - F_{\text{water}_{i-\frac{1}{2}}}^{\ n}\right]. \quad (15)$$

Here,

$$V_{\text{water}_i^n} \approx v_{\text{water}}(x_i, t^n) \quad \text{and} \quad F_{\text{water}_{i+1/2}^n} \approx f(v_{\text{water}}(x_{i+1/2}, t^n))$$

are the conserved quantity and the flux function, respectively. Here Δt is time step and Δx is space step, i is notation for the space index and n is notation for the time index. To compute fluxes in equation (15), we use the Lax-Friedrichs formulation:

$$F_{\text{water}_{1-\frac{1}{2}}^n} = \frac{F_{\text{water}_i^n} - F_{\text{water}_{i-1}^n}}{2} - \frac{\Delta x}{2\Delta t}\left(V_{\text{water}_i^n} - V_{\text{water}_{i-1}^n}\right), \tag{16}$$

and

$$F_{\text{water}_{1+\frac{1}{2}}^n} = \frac{F_{\text{water}_{i+1}^n} - F_{\text{water}_i^n}}{2} - \frac{\Delta x}{2\Delta t}\left(V_{\text{water}_{i+1}^n} - V_{\text{water}_i^n}\right). \tag{17}$$

Analogously, equation (14) is solved using the finite volume method with an explicit scheme:

$$V_{\text{oil}_i^{n+1}} = V_{\text{oil}_i^n} - \frac{\Delta t}{\Delta x}\left[F_{\text{oil}_{i+\frac{1}{2}}^n} - F_{\text{oil}_{i-\frac{1}{2}}^n}\right]. \tag{18}$$

To compute fluxes in equation (18), we use the Lax-Friedrichs formulation:

$$F_{\text{oil}_{1-\frac{1}{2}}^n} = \frac{F_{\text{oil}_i^n} - F_{\text{oil}_{i-1}^n}}{2} - \frac{\Delta x}{2\Delta t}\left(V_{\text{oil}_i^n} - V_{\text{oil}_{i-1}^n}\right), \tag{19}$$

and

$$F_{\text{oil}_{1+\frac{1}{2}}^n} = \frac{F_{\text{oil}_{i+1}^n} - F_{\text{oil}_i^n}}{2} - \frac{\Delta x}{2\Delta t}\left(V_{\text{oil}_{i+1}^n} - V_{\text{oil}_i^n}\right). \tag{20}$$

The finite volume schemes (15)-(17) and (18)-(20) are iterated with consideration of the initial and boundary conditions given in equations (5)-(6). The iterations result in numerical solutions for the fluid velocities. Note that another approach of finite volume method is available, for example, the relaxation system [4-6], but it doubles the number of equations to be solved. Our finite volume method in this paper is simpler than those relaxation approach.

4. Numerical results

In this section we present some representatives of our numerical results. The numerical results are compared with a known analytical exact solution.

As the benchmark, we recall the analytical exact solution obtained by Caldwell and Ng [1] as follows. First considering

$$\frac{d^2 v_{\text{water}}}{dx^2} = 0\,,\ 0 \le x \le 6\,, \tag{21}$$

with

$$v_{\text{water}}(0) = 0\,, \tag{22}$$

we get

$$v_{\text{water}}(x) = Ax + B\,. \tag{23}$$

Because $v_{\text{water}}(0) = 0$, then $B = 0$. Equation (23) becomes:

$$v_{\text{water}}(x) = Ax\,,\ 0 \le x \le 6\,. \tag{24}$$

The steady state case for oil requires

$$\frac{d^2 v_{\text{oil}}}{dx^2} = 0\,,\ 6 \le x \le 10\,, \tag{25}$$

with

$$v_{\text{oil}}(10) = 7\,, \tag{26}$$

$$v_{\text{oil}}(6) = v_{\text{water}}(6)\,, \tag{27}$$

$$\mu_{\text{oil}} \left.\frac{\partial v_{\text{oil}}}{\partial x}\right|_{x=6} = \mu_{\text{water}} \left.\frac{\partial v_{\text{water}}}{\partial x}\right|_{x=6}\,. \tag{28}$$

Then we get:

$$v_{\text{oil}}(x) = Cx + D\,,\ 6 \le x \le 10\,. \tag{29}$$

The conditions (26) and (27) lead to

$$10C + D = 7\,, \tag{30}$$

$$6C + D = 6A\,. \tag{31}$$

Because $\frac{\partial v_{\text{oil}}}{\partial x} = C$ and $\frac{\partial v_{\text{water}}}{\partial x} = A$, we write

$$\mu_{\text{oil}}(C) = \mu_{\text{water}}(A). \tag{32}$$

Equation (30)-(32) can be solved to obtain:

$$A = \frac{7\mu_{\text{oil}}}{6\mu_{\text{oil}} + 4\mu_{\text{water}}} \tag{33}$$

$$C = \frac{7\mu_{\text{water}}}{6\mu_{\text{oil}} + 4\mu_{\text{water}}} \tag{34}$$

$$D = \frac{21\,(\mu_{\text{oil}} - \mu_{\text{water}})}{3\mu_{\text{oil}} + 2\mu_{\text{water}}} \tag{35}$$

Finally, the exact steady state solutions are:

$$v_{\text{water}}(x) = \frac{7\mu_{\text{oil}}}{6\mu_{\text{oil}}+4\mu_{\text{water}}} x\,,\ \ 0 \le x \le 6\,, \tag{36}$$

$$v_{\text{oil}}(x) = \frac{7\mu_{\text{water}}}{6\mu_{\text{oil}}+4\mu_{\text{water}}} x + \frac{21\,(\mu_{\text{oil}}-\mu_{\text{water}})}{3\mu_{\text{oil}}+2\mu_{\text{water}}}\,,\quad 6 \le x \le 10\,. \tag{37}$$

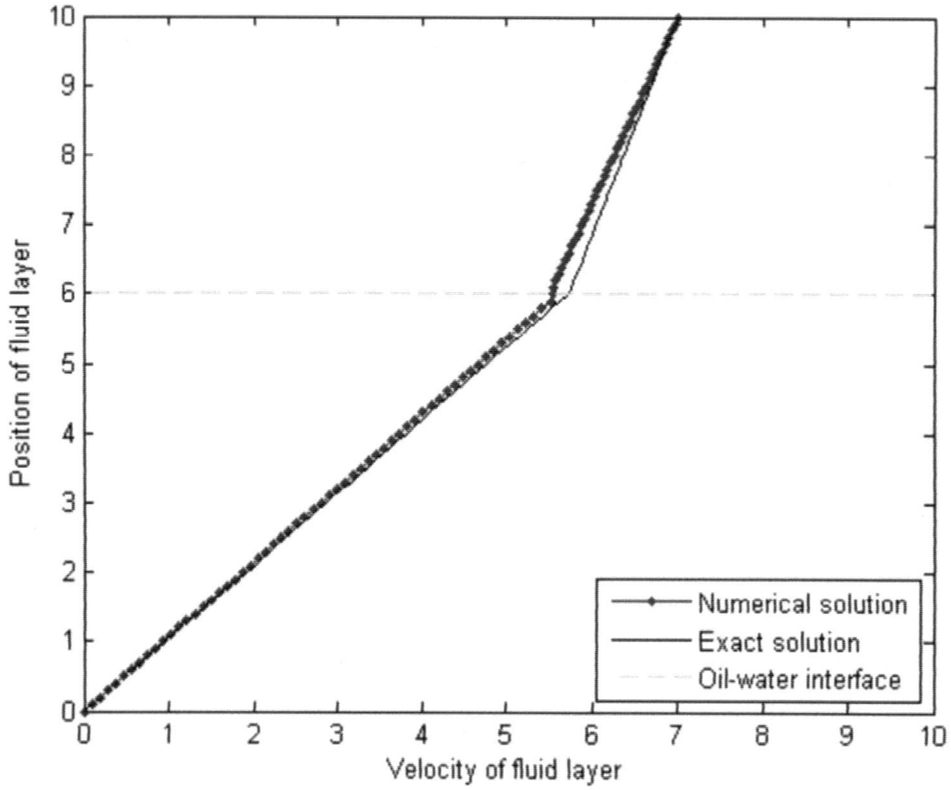

Figure 2. Large time behaviour of velocity of two fluid layers, at time $t = 100$ s.

Numerical setting in this simulation is as follows. We assume that $\mu_{\text{water}} = 1$ cp and $\mu_{\text{oil}} = 3$ cp. The cell width or also known as the space step is $\Delta x = 0.1$. The time step is $\Delta t = 0.001 \cdot \Delta x$. The simulation is stop at the final time $t = 100$ s. Here we want to see that change of velocity of two-fluid layers from $x = 0$ to $x = 10$ when the steady state is achieved. Note that the x-axis is vertical, instead of horizontal.

Our numerical results show that the numerical method solves the problem successfully. The main source of numerical error is at the non-smooth trasition between the oil and water layers, that is, the oil-water interface. At this interface, the fluid viscosity is discontinuous. However, we can have smaller error if we take smaller cell width and smaller time step. A representative of our numerical results is shown in Figure 2. The numerical results is correct physically. The velocity values get larger from the bottom to the top surface. This phenomena is identified in both the analytical and numerical solutions.

5. Conclusion

We have proposed a finite volume numerical method to solve the problem of two layers of fluids driven by a horizontally moving surface with fixed bottom. The numerical method can be used to solve both steady and unsteady state problems. Accurate solution can be obtained using a fine numerical mesh, that is, fine cell width and fine time step. Our research results are limited to one-dimensional problems. Future research direction could be an extension to higher dimensional problems.

Acknowledgements
The work of Sudi Mungkasi was financially supported by DRPM of *Ditjen Penguatan Riset dan Pengembangan* of *Kementerian Riset, Teknologi, dan Pendidikan Tinggi* of The Republik of Indonesia in year 2017 with Research Contract Number 051/HB-LIT/IV/2017 dated on 14 April 2017 (under DIPA-042.06-0.1.401516/2017 funds).

References

[1] Caldwell J and Ng D K S 2004 *Mathematical Modelling: Case Studies and Projects* (New York: Kluwer Academic Publishers)
[2] LeVeque R J 1992 Numerical Methods for Conservation Laws (Basel: Springer)
[3] LeVeque R J 2002 Finite Volume Methods for Hyperbolic Problems (Cambridge: Cambridge University Press)
[4] Banda M K and Herty M 2012 Adjoint IMEX-based schemes for control problems governed by hyperbolic conservation laws *Computational Optimization and Applications* **51** 909
[5] Jin S and Xin Z 1995 The relaxation schemes for systems of conservation laws in arbitrary space dimensions *Communications on Pure and Applied Mathematics* **48** 235
[6] Yohana E M and Banda M K 2016 High-order relaxation approaches for adjoint-based optimal control problems governed by nonlinear hyperbolic systems of conservation laws *Journal of Numerical Mathematics* **24** 45

IOP Conf. Series: Journal of Physics: Conf. Series **1090** (2018) 012051 doi:10.1088/1742-6596/1090/1/012051

Vortex Formation on Horizontal Axis Wind Turbine with Splitted winglets

Rizal Ibnu Syifa*, Setyo Nugroho

Power Plant Engineering, Electronic Engineering Polytechnic Institute of Surabaya, Surabaya, Indonesia, 60111
*rizalibnu95@gmail.com

Abstract
Horizontal axis wind turbine has developed to many combination. One of the combination is by adding variation in their blades such as variation of device. Many kind of devices that added to the blade such as addition splitted winglet to the blade. This additional device is the application device that added to the wings of recent aircraft that gives better performance than wings without this device. The purpose of using this device is to minimize the backflow air in their tip of blade. This research expectation is turbine will get better performance. However, the addition of the device do not increase the turbine performance. Comparing the turbine with the blade without splitted winglet, the turbine without splitted winglet gives better performance than the turbine that has blade with splitted winglet. Vortex formation in the splitted winglet device is one of phenomenon that cause the decrease performance of wind turbine. Vortex is air that flowing back to the device and makes the other flow velocity slower. In turbine without splitted winglet, the performance of Cl is greater than the other. When the air that hit the turbine is 4.5 m/s, the torque of the turbine with splitted winglet is 2.66 E-3 Nm. The turbine without splitted winglet gives better torque about 2.8 E-3 Nm. From variation that compared, additional of device not always give the better performance.

1. Introduction

Wind turbine is a device that is conversing energy from wind energy to kinetics and then it conversed to make an electricity. The turbine performance affected by external and internal factor. The external factor come from the ambiance condition including weather. The internal factor that affected in turbine performance come from turbine component such as blades, tail, generator, and its transmissions. Blades make the rotation of turbine to make an electricity. This device can make the turbine's performance become better by adding some variation. One of the variation that can added into the blades is splitted winglet.

Splitted winglet is the innovation from blended winglet that added by another winglet in the against direction. This additional variation is adapted from the recent aircraft. The purpose of adding this variation is to enhance the performance so that there are minimalize the back flow on the blade. From the research [1] the wings or aircraft with splitted winglet has better performance than wings of aircraft using blended winglet or single winglets.

Computational fluid dynamics simulation is part of process when designing product by some industry. Almost industries using fluid dynamics in their process to make a product [2]. Industries want to their product looks and performance before they process it into the factory. So that, Computational Fluid Dynamics simulation makes the construction of tools become efficient.

Content from this work may be used under the terms of the Creative Commons Attribution 3.0 licence. Any further distribution of this work must maintain attribution to the author(s) and the title of the work, journal citation and DOI.
Published under licence by IOP Publishing Ltd

In this computation is limited by velocity of the air, sum of the rotor of turbine, pitch angle and angle of attack. This scope of problem is described in each sub in this paper.

2. Geometry

Geometry that used in this paper is rotor of wind turbine with three blades. Wind turbine rotor with three blades give better performance[3]. This three blades rotor show better performance with cooefficient of performance Cp about 5.2 when the TSR in 4 until 5. TSR is tip speed ratio which show the ratio of speed on tip of blade to the velocity.

There are two kinds of blade in this research they are blade without winglet and blade with splitted winglets. In the previous research, The blade with single winglets give better performance [4]. Single winglets are used in the recent aircraft. The winglets advantage is to minimize the backflow on the tip of blade. So that, the lift increase and the drag decrease. This single winglets has dimension is the span angle is 70° and the length of the winglets is 2 cm.

The blade with splitted winglets dimension adapted form blade with single winglets a with additional other winglets in the against direction. The splitted winglets ratio is h2 is 0.4 times h1 [5]. Where h1 is the main winglets and h2 is the side winglets. The position of h1 or main winglets is as same as winglets in blade with single winglets with span angle 70° . The h2 or side winglets position is against the main winglets with span angle 52°.

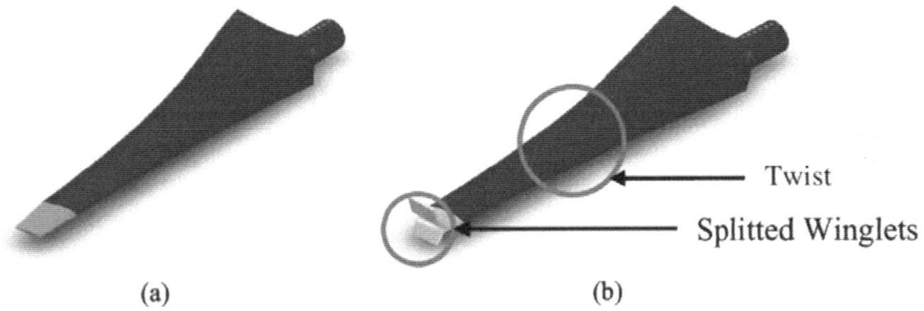

Figure 1. Blade with different additional winglets (a) blade without winglets (b) blade with splitted winglets

Other additional method is using twist. Twist is the additional method in the segment of airfoil that construct blade with different angle of twist, dimension or chord length, and distribution of the segment. Twist give variable angle of attack through the length of the blade. Twist angle is an angle that given to the each segment of airfoil. Twist angle distribution is based on aerodynamic behavior of each individual airfoil at a certain angle of attack [6].

$$\beta(r) = \frac{2}{3} arctan\frac{R}{r\lambda} - \alpha_D \quad (1)$$

The airfoil variation is defined by the chord length. The chord length is the size of chprd in every segment of chord. The airfoil with different chord and twist angle is distributed based on the length of the blade [7]. The size of each blade is 12 cm and the hub radians is 3 cm.

$$c(r) = \frac{16\pi r}{BC_l} sin^2\left(\frac{1}{3}tan^{-1}\left(\frac{R}{\lambda r}\right)\right) \quad (2)$$

So, the rotor diameter is 30 cm. From the equation above [7], The segment distribution is counted on excel and described in table below. In the table 1 from research [4] we know that longer segment position from the hub, smaller chord we get. The variation of twist angle is different in each segment.

Table 1. Detail Of Geometry Data Input

Segment	r/ (cm)	c/ (cm)	Twist β (°)
1	3,20	3,80	28,13
2	4,14	3,59	22,06
3	5,09	3,32	17,53
4	6,04	3,04	14,11
5	7,00	2,79	11,46
6	7,95	2,56	9,35
7	8,90	2,36	7,63
8	9,85	2,18	6,21
9	10,81	2,03	5,02
10	11,76	1,89	4,00
11	12,71	1,77	3,12
12	13,66	1,67	2,36
13	14,62	1,57	1,70
14	15,62	1,48	1,08
15	15,95	1,45	0,88

3. Simulation

The simulation is using ANSYS Computational Fluid Dynamics package. ANSYS Computational Fluid Dynamics package is using divided into three step they are geometry making, mesh distribution and fluent for the calculation. Geometry making is use to make the domain of the simulating and the fluids flow. The mesh is used to distribute the calculation point. The Fluent is used to calculate the fluids flow in every point that made by Mesh. This simulation is using MRF method.

3.1. Geometry Preparation

The domain is using rotor of wind turbine that covered by the tube cylinder. The tube cylinder defined as the ambient environment for the air flow. The air flow trough the input front cylinder to the output in back cylinder. The cover of cylinder tube is defined as wall. The rotor of turbine defined as the object that hit by the air. The rotor is subtracted by boolean method .The use of boolean method is to trim the blades area and makes the cylinder volume less because of the rotor and makes a hole rotor shaped in the middle of tube cylinder.

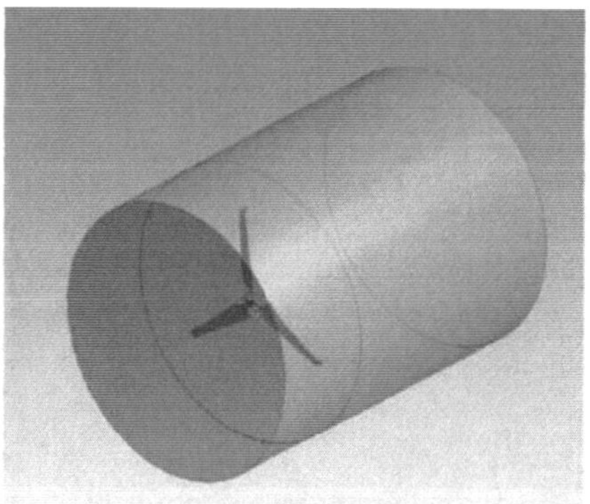

Figure 2. Geometry for simulation

3.2. Mesh Distribution

The mesh function is to makes some nodes point to calculate. The calculation is used to solve the problem in each nodes. In this paper, the mesh method is using global mesh with tetrahedron. In the figure 3 the mesh is more close when the nodes approach the blades. The different colour is caused by rotation duplicate. The generate process gives average nodes 7.0 E+5 for every body geometry of simulation.

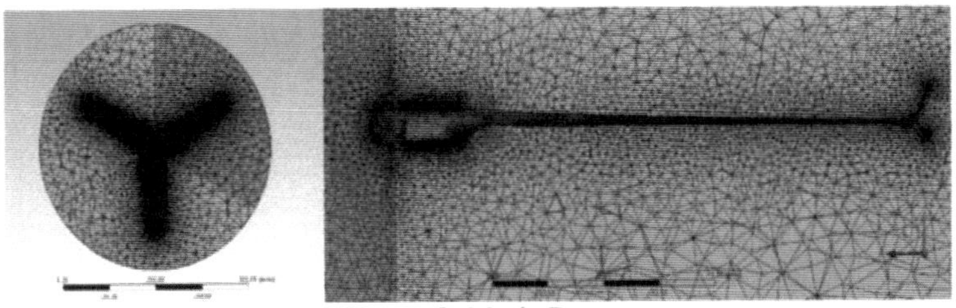

Figure 3. Mesh Geometry

3.3. Fluent Calculation

The calculation process is solving by ANSYS fluent. The fluent using Computational Fluid Dynamics fluid dynamics analysis. This simulation process using Moving Reference Frame method or MRF method. This method is The fluids flow trough the cylinder is defined by air with standard density. The air flow trough the cylinder in rotational flow. The fluids flow with turbulence flow in k-omega. In the research [2] k-omega result is closer to the test result.

4. Result and Discussion

From the simulation of wind turbine using ANSYS, data that we can obtain are torque, lift force and drag force. Torque is the force in length of force that happen on rotor. Lift force is the force that makes the rotor rotate. Drag force is the force that it is the disturbance of rotor rotation. From the data that obtained, we must calculate it into lift coefficient and drag coefficient.

$$C_L = \frac{F_L}{\frac{1}{2}\rho V_0{}^2 c} \quad (3)$$

$$C_D = \frac{F_D}{\frac{1}{2}\rho V_0{}^2 c} \quad (4)$$

Lift coefficient is is coefficient that show the capability of airfoil to give lift force in the rotor. Drag coefficient is is coefficient that show capability of airfoil to give drag force in the rotor. The result of the simulation is given by data in the table below.

Table 2. Data Result of Simulation

a. Turbine without winglets

V (m/s)	Torque (Nm)	Cd	Cl
4.5	0.002806	8.97033E-05	0.346412
5	0.003663	0.000313371	0.363378
5.5	0.004454	0.000326399	0.363325
6	0.005661	0.000576013	0.383118

b. Turbine with Splitted Winglets

V (m/s)	Torque (Nm)	Cd	Cl
4.5	0.00266	0.000270187	0.324787
5	0.003551	0.000618428	0.348624
5.5	0.004365	0.000556515	0.351798
6	0.005361	0.000692312	0.360047

From the table 2, the best torque is given by turbine rotor without splitted winglets. The coefficient lift is increasing when the velocity raise. The drag coefficient also increase when the velocity raise. This condition is proofed by figure 4 that show air flow vector on each blades. The figure 4 show blade a for blade without winglets and blade b for blade with splitted winglets. This two blades are simulated in the same condition with velocity of air 5 m/s. That two kinds of blade gives different vector that happened around the tip blades.

In the blade a the vectors show minimum blue vectors than the vectors that happened in the blade b. This blue vectors show backflow air in the tip of blades. The backflow air is the air that move back and hit another air flow. The air flow that hit will decrease the velocity of its flow. So, in the tip of blades makes the lift force decrease and increase the drag force. This condition makes the torque lower. In the wind turbine smaller drag that happen small engine that required [8].

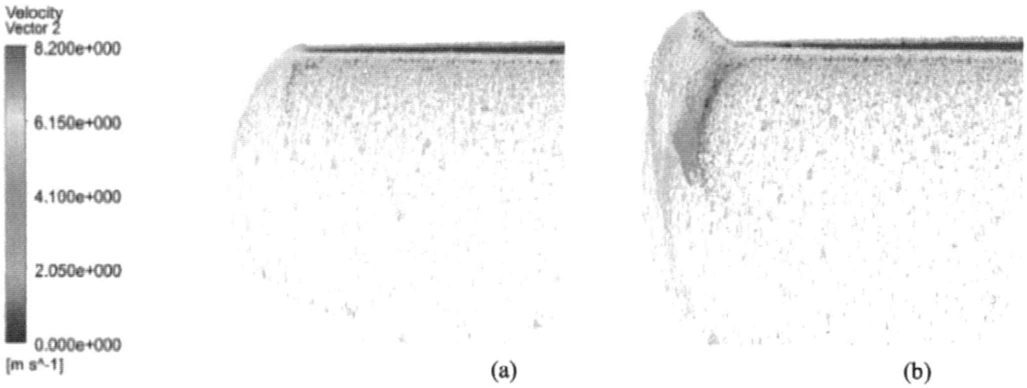

Figure 4. Vector of flow in blade tip in different additional winglets in velocity of 5 m/s (a) blade without winglets (b) blade with splitted winglets

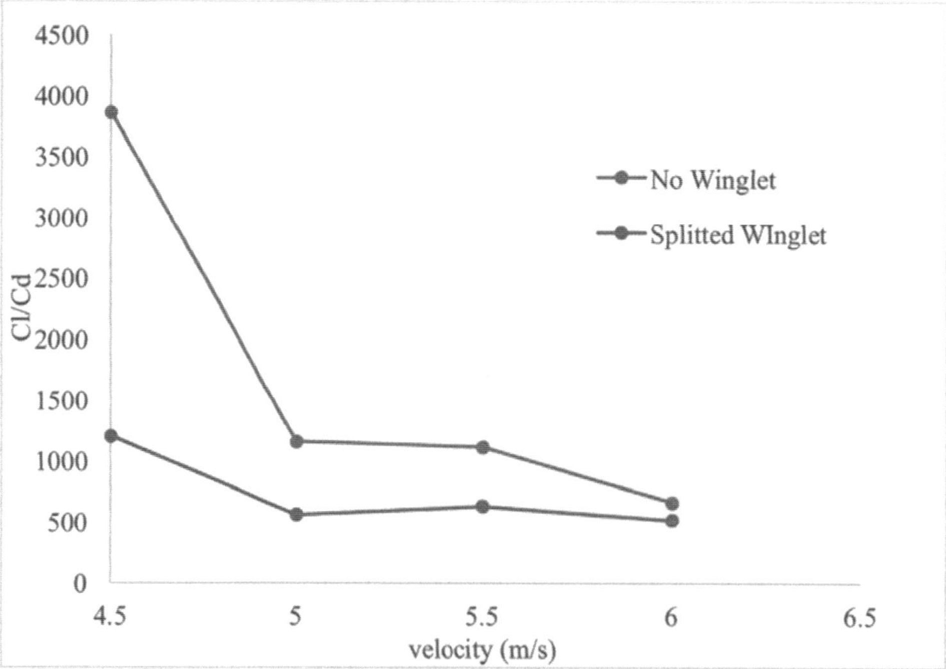

Figure 5. Variation coefficient of lift to drag as function of velocity

The graph in figure 5 that show lift to drag coefficient as function of velocity gives that the lift to drag coefficient is decrease as the increase of velocity in each typical of blade. The blue line is for blade without winglets named by blade a. The red line is for blade with splitted winglets named by blade b. The highest value of coefficient lift to drag is given by point in velocity of air 4.5 with 3.86 E+3 by blade a. The coefficient of lift to drag value is almost the same for blade a and b in the velocity of 6 m/s. The blade without splitted winglets give better performance than blade with splitted winglets in the lower air velocity.

5. Conclusion

This research is successfully by doing simulation about horizontal axis wind turbine using ANSYS Computational Fluid Dynamics package by using MRF method with result:

1. The additional of splitted winglets do not give better performance than blade without winglets.
2. The torque of the turbine with splitted winglet is 2.66 E-3 Nm when the velocity of air hit the turbine is 4.5 m/s.
3. The turbine without splitted winglet gives better torque about 2.8 E-3 Nm.
4. The backflow on the tip of blade the torque decrease.

Acknowledgement

Thank you to fluid laboratory-EEPIS for helping this research by providing data that needed in this research. Authors wish the data can be used for the next research.

References

[1] Sohail R, Helmut S, Abas A,George S, "Winglets-Multiobjective Optimization of Aerodynamic Shapes," *Department of Mechanical and Materials Engineering*, MAIDROC Laboratory, Florida International University, Miami, FL 33174, USA
[2] Mohammad Moshfeghi, Ya Jun S, Yong H, "Effects of near-wall grid spacing on SST-K-ω model using NREL Phase VI horizontal axis wind turbine," *J. Wind Eng. Ind. Aerodyn.*, vol. 94, pp. 107–108, 2012.
[3] Tony B, David S, Nick J, Ervin B, "*Wind Energy Handbook,*" John Willey & Sons Ltd., UK, 2001.
[4] Akhmad K, "Experimental Study of Effects of blade pitch angle Against Horizontal Axis Wind Turbine Performance," *EEPIS Release*, 2016.
[5] Louis B, "Split Blended Winglet" *Patent Application Publication*, Seattle, WA (US), 2012.
[6] Rainaldo G, Diego M, Juancarlo R ,"Winglets Design and Analysis for Wind Turbine," *Florida International University Press*. 2011.
[7] Soren Gundtoft, " *Wind Turbines,*" University College of Aarhus, 2009.
[8] Martin O. L. Hansen, "*Aerodynamics of Wind Turbines,*" Earthscan, UK, 2008.

Analysis of Fouling Mechanism in Polysulfone based Ultrafiltration Membrane during Peat Water Filtration

Danu Ariono* and Anita Kusuma Wardani
Department of Chemical Engineering, ITB, Jl. Ganesha 10, Bandung 40132, Indonesia

* Email: danu@che.itb.ac.id

Abstract. Polysulfone (PSf) has been widely used as ultrafiltration (UF) membrane material due to its thermal and chemical stability. However, membrane fouling becomes a severe problem limiting the potential of PSf membrane. This fouling is mainly formed by deposition of organic onto membrane surface which resulting the decrease of permeate flux. There are four fouling mechanisms proposed by Hermia (1982), namely cake formation, standard blocking, intermediate blocking, and complete blocking. In this work, the types of fouling mechanism that occurred in PSf based UF membrane during peat water filtration were investigated. In addition, the effect of trans-membrane pressure was also studied. The determination of fouling mechanism was conducted by optimization of the fouling parameters, including Kcf (cake formation), Ksb (standard blocking), Kib (intermediate blocking), and Kcb (complete blocking). The results showed that complete blocking became the dominant fouling mechanism in this work. Meanwhile, the trans-membrane pressure did not affect fouling mechanism.

1. Introduction

The stringent regulations of drinking water lead to increased attention in developing ultrafiltration (UF) membrane for water treatment. UF membrane offers several advantages such as high product quality, a small footprint area, and low energy consumption [1-6]. Furthermore, UF is able to remove viruses, bacteria, colloids, and larger particulate matter from suspensions [7-12]. One of the polymers commonly used as UF membrane material is polysulfone (PSf) since it has good chemical and thermal stability [13, 14]. However, PSf membrane is naturally hydrophobic and susceptible to natural organic matter (NOM) fouling during surface water treatment. This fouling not only leads to productivity decline, but also requires additional energy supply to keep the membrane performance constant [15-21].

Membrane fouling is a result of interaction between the membrane material and the components in the feed water. Regarding NOM fouling mechanisms, it is generally recognized that there are four main fouling mechanisms for UF membrane, as proposed by Hermia [22]. They are complete blocking, intermediate blocking, standard blocking or pore constriction, and cake formation. In complete blocking and intermediate blocking, the particles have the same size with membrane pores, thus the particles seal off pore entrances and prevent flow. However, for intermediate blocking, there is an accumulation of some particles on top of other deposited particles. This accumulation does not occur in complete blocking. Meanwhile, standard blocking results in a decrease of the membrane porosity and increase of membrane resistance since the bulk phase particles are small enough to enter the membrane pores [23]. The particles are accumulated inside membranes on the walls of cylindrical

Content from this work may be used under the terms of the Creative Commons Attribution 3.0 licence. Any further distribution of this work must maintain attribution to the author(s) and the title of the work, journal citation and DOI.
Published under licence by IOP Publishing Ltd

pores, thus the pores become constricted and the permeability of the membrane is reduced. Furthermore, cake formation occurs when bigger particles accumulate on the surface of a membrane in a permeable cake of increasing thickness that increases membrane resistance.

In this work, PSf based UF membrane was prepared by inversion phase method. Peat water filtration was used to characterize the organic fouling on the membrane. The types of fouling mechanism that occurred in PSf based UF membrane during peat water filtration were investigated by optimization of the fouling parameters, K_{cf}(cake formation), K_{sb} (standard blocking), K_{ib} (intermediate blocking), and K_{cb} (complete blocking). In addition, the effect of trans-membrane pressure was also evaluated.

2. Materials and Method

2.1 Materials

The PSf (UDEL-P3500 MB7) from Solvay Advanced polymer was used as the main material. Meanwhile, Polyethylene glycol (PEG400) was chosen as an additives, Dimethylacetamide (DMAc) with a purity of 99.9% (Shangshai Jingsan Jingwei Chemical Co. Ltd) was used as a solvent, and demineralized water was used as a coagulant to induce the formation of membrane structure. Flux analysis was conducted using peat water from Dumai river-Riau, Indonesia.

2.2 UF membrane preparation

The UF membrane was prepared by blending 20%wt of PSf with 20%wt of PEG400 and 60%wt of DMAc. The membrane solution was stirred for about 8 hours in a closed stirred tank and then left without stirring until no bubbles appeared. Then, the membrane solution was casted on the flat glass plate and immediately immersed in demineralized water. The peat water filtration was conducted in TMP of 10 and 30 psi.

2.3 Simulation of fouling mechanism

Determination of fouling mechanism based on the experimental data was conducted by considering the flux expressions relative to the fouling mechanisms, which was established by Hermia [22]. Hermia proposed a mathematical model describing fouling mechanisms, as the following equation:

$$\frac{d^2t}{dV^2} = k\left(\frac{dt}{dV}\right)^m \tag{1}$$

Here t is the filtration time and V is the total filtered volume. m denotes a parameter of cake filtration, standard blocking, intermediate blocking, and complete blocking respectively, each of which has values of 0, 3/2, 1, and 2. Meanwhile, the value of k represents fouling parameter, which varied for each UF processes.

The Flux of UF membrane could be defined by equation (2) [1].

$$J = \frac{1}{A}\frac{dV}{dt} \tag{2}$$

where A is membrane area and J is permeate flux. By combining Hermia's mathematical model with flux expression, the flux decline could be expressed by equation (3).

$$\frac{dJ}{dt} = -kJ(AJ)^{2-m} \tag{3}$$

Furthermore, the equation characterizing of flow decline due to fouling is detailed in table 1.

Table 1 Flux equation for each fouling mechanism

Fouling mechanism	Flux equation	Equation number
Cake formation *m = 0*	$J_v = \dfrac{J_{v,0}}{(2K_{cf}J_{v,0}{}^2t + 1)^{\frac{1}{2}}}$ $K_{cf} = kA^2$	(4)
Standard blocking *m = 3/2*	$J_v = \dfrac{4J_{v,0}}{(2K_{sb}J_{v,0}{}^{1/2}t + 2)^2}$ $K_{sb} = kA^{1/2}$	(5)
Intermediate blocking *m = 1*	$J_v = \dfrac{J_{v,0}}{K_{ib}J_{v,0}t + 1}$ $K_{ib} = kA$	(6)
Complete blocking *m = 2*	$J_v = J_{v,0}exp(-K_{cb}t)$ $K_{cb} = k$	(7)

3. Results and Discussion

3.1 Flux Characteristic of Polysulfone based Ultrafiltration Membrane

Figure 1 presents the peat water flux profile as a function of peat filtration time at a constant feed velocity. It shows that the peat water flux was sharply declined in the earlier of filtration process due to rapid accumulation of organic matter on the membrane surface. The rapid accumulation of organic matter on the membrane surface was led by the addition of PEG400 that contributes to the formation of an open pore in skin layer and also in sub-structure of the membrane [24].

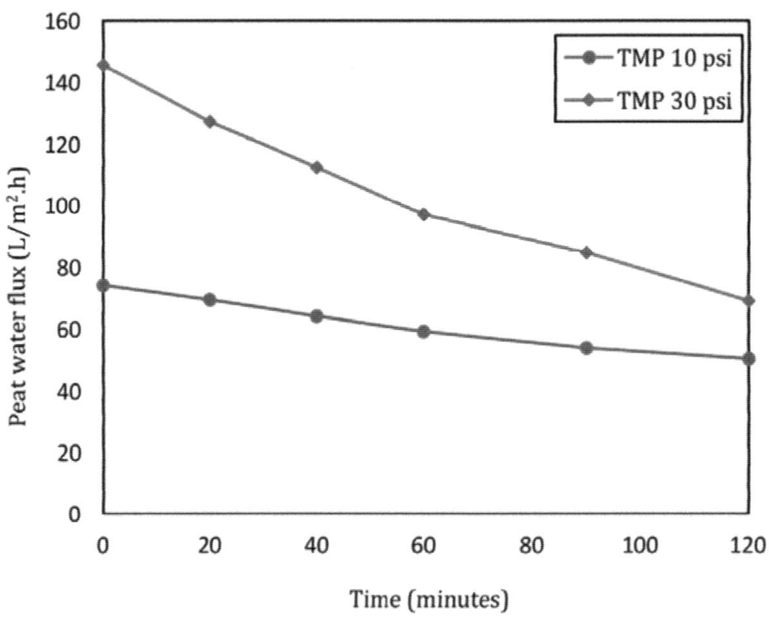

Figure 1 Effect of TMP on peat water flux

Furthermore, the flux of higher TMP (30 psi) was declined sharper than lower TMP (10 psi). At higher pressure, severe fouling of membranes occured under higher initial flux since the organic substances were transported faster to the membrane and accumulates near the membrane surface. The organic fouling layer then became more compressed that led to an increase in hydraulic resistance [25].

3.2 Determination of Fouling Mechanism

To determine the fouling mechanism of ultrafiltration process in this work, the optimization of fouling parameters (K_{cf}, K_{sb}, K_{ib}, and K_{cb}) was conducted by the least squares method programmed with Matlab R2015b. The values of fouling parameters are presented in Table 2.

Table 2 Fouling parameter for each mechanism

TMP (psi)	K_{cf}	K_{sb}	K_{ib}	K_{cb}
10	2.10E-07	4.08E-04	3.85E-05	4.30E-03
30	5.40E-07	6.10E-04	7.40E-05	4.90E-03

Figure 2 shows that the flux decline was well described by the complete blocking mechanism. In complete blocking, the particles had the same size with membrane pores, thus particles deposited on the membrane surface and block the entrances of membrane pores completely with no overlapping particles. Since the number of membrane pores was decreased by deposited particles, the filtration resistance increased and the flux decreased as a function of time [26]. This model applied to both trans-membrane pressures of 10 and 30 psi. It indicated that TMP had no effect to the fouling mechanism in PSf based UF membrane during peat water filtration.

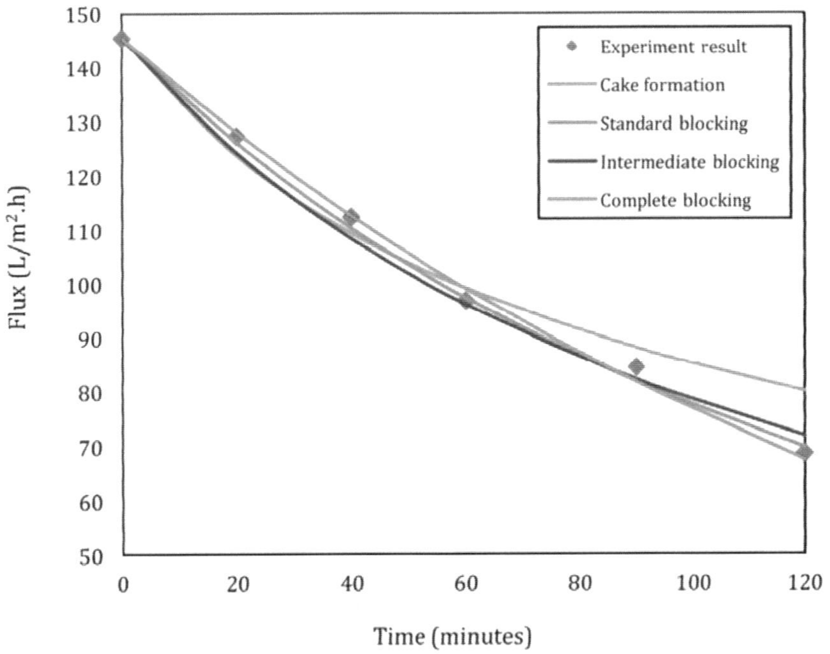

Figure 2 Simulation result of fouling mechanism for TMP 30 psi

4. Conclusion

In this work, PSf based UF membrane was prepared by phase inversion method. Peat water filtration was used to determine membrane performance. Furthermore, the types of fouling mechanism that occurred in PSf based UF membrane during peat water filtration were investigated. The results showed that complete blocking became the dominant fouling mechanism in this work. Meanwhile, fouling mechanism was not affected by the trans-membrane pressure.

Reference

[1] Cho J, Amy G and Pellegrino J 2000 *J. Membr. Sci.* **164** 89-110
[2] Aoustin E, Schäfer A I, Fane A G and Waite T D 2001 *Sep. Purif. Technol.* **22–23** 63-78
[3] Shi X, Tal G, Hankins N P and Gitis V 2014 *J. Water Process Eng.* **1** 121-38
[4] Aryanti P, Joscarita S R, Wardani A K, Subagjo S, Ariono D and Wenten I G 2016 *J. Eng. Technol. Sci.* **48** 135-49
[5] Aryanti P, Sianipar M, Zunita M and Wenten I 2017 *Membr. Water Treat.* **8** 463-81
[6] Kresnowati M T A P, Desiriani R and Wenten I G 2017 American Institute of Physics Inc.)
[7] Mulder J 2012 *Basic principles of membrane technology* (Springer Science & Business Media)
[8] Lowe J and Hossain M M 2008 *Desalination* **218** 343-54
[9] Peter-Varbanets M, Zurbrügg C, Swartz C and Pronk W 2009 *Water Res.* **43** 245-65
[10] Wardani A K, Hakim A N, Khoiruddin and Wenten I G 2017 *Water Sci. Technol.* **75** 2891-9
[11] Wenten I G, Khoiruddin K, Hakim A N and Himma N F 2017 *Membrane Characterization,* (Elsevier) pp 199-218
[12] Baker R W 2004 *Membrane technology and applications* (John Wiley & Sons, Ltd)
[13] Ariono D, Aryanti P T P, Subagjo S and Wenten I G 2017 *AIP Conf. Proceed.* **1788** 030048
[14] Ariono D, Aryanti P T P, Hakim A N, Subagjo S and Wenten I G 2017 *AIP Conf. Proceed.* **1840** 090008
[15] Zularisam A W, Ismail A F and Salim R 2006 *Desalination* **194** 211-31
[16] Du J R, Peldszus S, Huck P M and Feng X 2009 *Water Research* **43** 4559-68
[17] Choi H, Zhang K, Dionysiou D D, Oerther D B and Sorial G A 2005 *Sep. Purif. Technol.* **45** 68-78
[18] Costa A R, de Pinho M N and Elimelech M 2006 *J. Membr. Sci.* **281** 716-25
[19] Sianipar M, Kim S H, Khoiruddin, Iskandar F and Wenten I G 2017 *RSC Adv.* **7** 51175-98
[20] Aryanti P T P, Yustiana R, Purnama R E D and Wenten I G 2015 *Membrane Water Treatment* **6** 379-92
[21] Ariono D, Purwasasmita M and Wenten I G 2016 *Journal of Engineering and Technological Sciences* **48** 367-87
[22] Hermia J 1982 *Trans. Inst. Chem. Eng.* **60** 183-7
[23] Broeckmann A, Busch J, Wintgens T and Marquardt W 2006 *Desalination* **189** 97-109
[24] Aryanti P T P, Khoiruddin and Wenten I G 2013 *J. Water Sustain.* **3** 85-96
[25] Wang Z, Zhao Y, Wang J and Wang S 2005 *Desalination* **178** 171-8
[26] Briao V and Tavares C 2012 *Brazilian J. Chem. Eng.* **29** 393-407

In Silico Study of Phospholipids as An Oral Insulin Delivery System

Nanik Rahmawati Fuadah and Rukman Hertadi*
Biochemistry Research Division, Department of Chemistry, Faculty of Mathematics and Natural Science, Institut Teknologi Bandung, Jl. Ganesha 10, Bandung 40132, Indonesia.

rukman@chem.itb.ac.id

Abstract. Phospholipids have been applied as a material for an oral insulin delivery system that either in the form of solid lipid nanoparticles, liposomes, or microemulsions. However, the administration of this delivery system for diabetic treatment requires twenty times higher of dosage than the injection one. In the present study, we performed molecular dynamics simulation to evaluate how phospholipids can protect insulin at the molecular level and why its efficiency for the diabetic treatment is low. Phosphatydilcholine group, such as DPPC, OPPC, LPC and combine of DPPC-LPC, was used as the phospholipid material to model the delivery system in our simulations because *in vitro* and *in vivo* studies of these phospholipids as the delivery system have been well-conducted. In these simulations, the waters-phospholipid-insulin system was prepared by randomly mixing phospholipids molecules inside the solvent box containing an insulin molecule. NPT ensemble was used in all simulations with the setting temperature at 310 K and 1 atm for the pressure. The simulation results showed that DPPC and LPC delivery systems were able to maintain both secondary and tertiary structures of the encapsulated insulin within the time range of simulation. However, all phospholipid delivery system models failed to fully encapsulate insulin surface within 150 ns simulation. This study may explain the reason for the low success rate of using phospholipid as the delivery system of oral insulin in the experiment.

Keywords: Insulin, phosphatidylcholine, oral delivery system, molecular dynamics simulation

1. Background

Inability of pancreas to produce insulin in diabetes mellitus type I (DMI) patients make them rely on insulin therapy [1]. However, 60% DMI patients who used invasive method failed to achieve long-term glycemic control [2]. The pain arises due to needle injections and insulin treatment regimen complexity [3] as well as the lack of control of glucose in the blood [4] are the reason that contributes the incompliance of patients to do insulin therapy.

Oral insulin is a promising route for insulin therapy. Oral insulin offers many advantages: higher patient compliance, rapid hepatic insulination, and avoidance of peripheral hyperinsulinemia and other possible adverse effects, such as hypoglycemia and weight gain [3]. 80% of insulin that reached the intestine directly flow into the liver [5] thus it is more effective to perform their role as a regulator of glucose levels in the blood [6]. However, the main problem is on the journey before reaching the intestine. Insulin has a bioavailability less than 1% [7] because this protein is prone to the degradation by digestive enzymes and also due to its molecular size, it is difficult to pass through the epithelium intestine [8].

Content from this work may be used under the terms of the Creative Commons Attribution 3.0 licence. Any further distribution of this work must maintain attribution to the author(s) and the title of the work, journal citation and DOI.
Published under licence by IOP Publishing Ltd

A phospholipid is the active ingredients that widely used in the pharmaceutical industry. A phospholipid has amphipathic properties that can form micelles when positioned in the aqueous environment. Many studies have successfully carried out using phospholipid as a material for drug delivery system in the form of microemulsions [6], liposomes [9], solid lipid nanoparticles [10], self-emulsifying system [11], and disperse [12]. Phospholipid can increase bioavailability, absorption and release of the drug, protects it from degradation, reduce ill effects in the gastrointestinal tract, and reduce the bitter taste of the drug [12].
However, the success rate of using phospholipids as an oral insulin delivery system is low. It has been reported in the previous study that the administration of oral insulin requires twenty times higher of dosage than the injection one [6, 10, 13, 14]. In the present studies, we evaluate the effectiveness of using phospholipid of phosphatidylcholine group as the insulin delivery system at the molecular level by performing *in silico* analysis with molecular dynamics (MD) simulation approach. The effectiveness the delivery system was evaluated in terms of the stability of insulin structure and the condition of encapsulation. We show here that our MD simulation can provide insight into the molecular level explanation about the low success rate of phospholipid as insulin delivery system.

2. Material and Methods

2.1 Materials

MD simulations require initial coordinate of atoms of the target molecules. The initial coordinate of atoms constituted insulin structure was obtained from the crystal structure data stored in the protein data bank repository (www.rscb.org) with the accession code of 5ENA [15]. While for the atomic coordinate of the phosphatidylcholine group, DPPC, OPPC, and LPC, were obtained from gromos database and Respiratory 2.2. [16].

2.2 Procedures

All MD simulations were performed by using GROMACS software version 5.0.4 [17–21] employing GROMOS96 53a6 as a force field [22] as source of parameters for bonding and nonbonding energy calculations.

2.2.1 Preparation

The insulin-water system was constructed by randomly adding water molecules to solvate an insulin molecule that was placed in the center of 5.5 x 5.5 x 5.5 nm^3 box. While the insulin-water-phospholipid system was built by randomly added phospholipid to the insulin-water system with the ratio of phospholipid:water about 1:100. Both systems were neutralized by adding counter ions.

2.2.2 Energy Minimization

The prepared initial structures above were first energy minimized to reduce a number of bad atomic contacts and to relax the bonding interactions (bond lengths, bond angles, and dihedral). Energy minimization was divided to two stages, the first is minimization of non-protein components by restraining the protein backbone with the harmonic force with the magnitude of 1000 kJ mol^{-1} nm^{-2} for 50000 steps using steepest descent algorithm. The second stages were targeting the whole system without applying the restraining force to the protein molecule using conjugate gradient algorithm. Both of the minimization stages will stop when reaching the maximum force 1000 kJ mol^{-1} nm^{-1}.

2.2.3 Equilibration

At this stage, all of the energy components will be equilibrated at the desired conditions so it will fluctuate at a targeted value (convergent). Equilibration process is carried out in two stages, NVT and NPT. Simulated annealing was performed in this study to bring the system to desired temperature (310 K). All the conditions were analyzed using "gmx energy" to ensure the convergent of targeted value including temperature, pressure, and density.

2.2.4 Production run

The equilibrated structure prepared above was then used as the initial structure for production run. The simulation was running for 150 ns with 1 fs integration time step by maintaining temperature and pressure with a V-rescale and a Parrinello-Rahman algorithms, respectively. In the course of simulation, the coordinate of the system was stored every 100 ps resulting 1500 structures.

2.2.5 Analysis

In this study, the tertiary structure stability of insulin was analyzed in terms of radius of gyration (equation 1), while for the secondary structures analysis, we used DSSP program [23]. In order to evaluate the effectiveness of the delivery system in encapsulating insulin, we used VMD [24] to visualize the emulsion formed by the delivery system. The interaction density between the delivery system and insulin was evaluated by calculating the contact number between both entities.

$$\langle R_g \rangle_E = \left\langle \sqrt{\frac{1}{N}\sum_{i=1}^{N}\left(\left|\vec{r}_i - \vec{r}_{cm}\right|\right)^2} \right\rangle_E .$$

3. Result and Discussion

In this study, we performed MD simulation to insulin free delivery system as the control to the MD simulation of insulin encapsulated with phospholipid delivery system. Each simulation was conducted within 150 ns to see the stability of the emulsion of insulin and the delivery system. The contacts between phospholipids as a delivery system with insulin during the simulation to certain extent may give impact to the stability of the structure of insulin. Therefore, it is necessary to analyze the structure of insulin during the simulation to know its stability of tertiary and secondary structures. In addition, we also analyzed how phospholipid encapsulates insulin at the molecular level and evaluate its stability by calculating the contact number between phospholipids and insulin.

3.1 Stability of tertiary structure of insulin

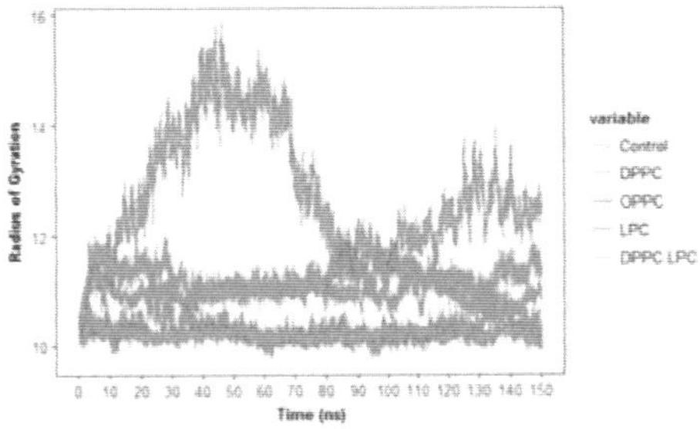

Figure 1 The profile of radius of gyration in the course of MD simulation for insulin free delivery system (control) and insulin encapsulated by phospholipid delivery system.

The tertiary structure of a protein is the 3D shape of the protein stabilized by many types of weak interactions, such as electrostatic interactions, van der Waals, hydrophobic, and hydrogen bonds. In this study, the stability of the tertiary structure of insulin during the simulation was evaluated through the change of 3D shape of the protein by calculating its radius of gyration (Figure 1). The radius of gyration is the average radius from the center mass of the protein that indicates the compactness of the protein structure.

Insulin free delivery system was unstable as indicated by the sharp increase of radius of gyration value in the first 50 ns of the simulation. Although the radius of gyration decreased after 50 ns, the shape of the protein was different from the initial one (Figure 2). Analysis at molecular level noted that the instability of insulin structure was initially caused by the hydrogen bond rupture between Gly-41 and Gly-44 that stabilized the loop connecting helix A and helix B.

Different from insulin free delivery system, insulin encapsulated by phospholipid molecules exhibited better structural stability as seen in relatively lower radius gyrations change from the initial condition. Although, the contact between insulin and phospholipids did not all provide optimum protection to the tertiary structures of insulin. Among four model of phospholipid delivery system (LPC, DPPC, OPPC, and mixed DPPC-LPC) only LPC phospholipids that seem almost optimally stabilize the insulin structure as indicated by the lowest change in its radius of gyration for the whole simulation time range. The other three models, which were DPPC, OPPC, and DPPC-LPC, showed considerable changes in insulin's radius of gyration. Based on the structure of the three phospholipids, LPC has only one hydrophobic tail, while DPPC and OPPC have two hydrophobic tails making the former is relatively has lower sterical hindrance to encapsulate an insulin molecule compare to the later.

3.2 Stability of secondary structure

Insulin is a protein's hormone that may only be functional in the form of native structure. Figure 3 shows the important residues in insulin that played an important role to bind an insulin's receptor [26]. As shown in the figure, several of those residues are the element of α-helical strands (helix A1, A2, and B). Therefore, besides three-dimensional shape, the secondary structure of this protein has to be stable as well in order to maintain the proper orientation of those key residues when interacts with the hormone's receptor.

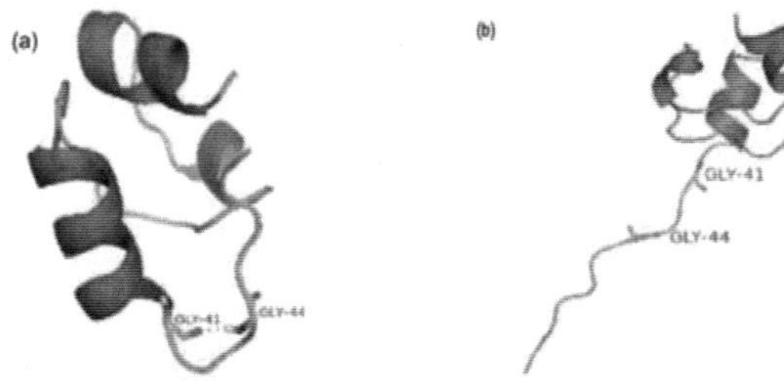

Figure 2 The structure of insulin free delivery system before and after the simulation. (A) At the initial structure of insulin, there is a hydrogen bond between the Gly-41 and Gly-44 residues making helix A and helix B close to each other. (B) After 150 ns of simulation, the hydrogen bond between Gly-41and Gly-44 residues was ruptured separating apart helix A and helix B.

The secondary structural analysis was conducted in this study by using DSSP program [23]. The result of DSSP analysis to the simulation trajectory of insulin free delivery system (Figure 4a) clearly showed the condition of the three α-helical strands in the course of simulation. All the helical strands were initially partially denatured, but after 90 ns, helix B was almost completely collapse and after 120 ns all of the helical strands were also mostly collapse untill the end of the simulation. This result explained the tertiary structure analysis above (Figure 1) that the increased of radius of gyration on insulin free delivery system was due to the collapse of secondary structure. Relatively less stability of insulin structure found in this study was also supported by the experimental evident that insulin tends to aggregate forming a quaternary hexamer molecules in order to stabilize its native structure. Therefore, a good delivery system is verified in our study based on its ability to preserve native structure of insulin.

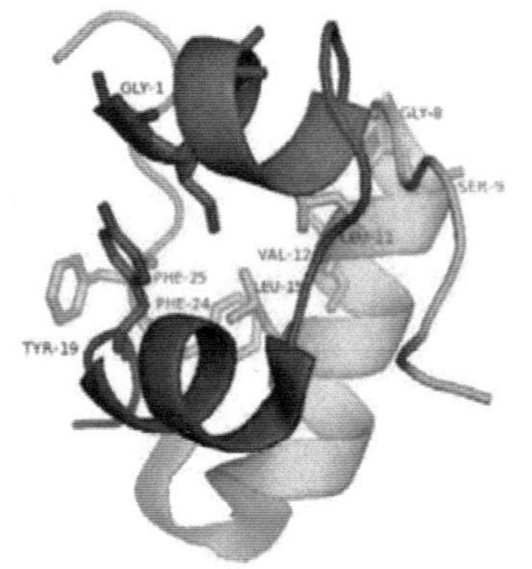

Figure 3 Important residues of insulin proteins that interact with insulin receptors [26]. The red color indicates chain A, and the green color indicates chain B.

The DSSP result to the simulation trajectory of insulin in the presence of phospholipid delivery system showed the variable effect on the secondary structure of insulin (Figure 4b-e). DPPC and LPC delivery system gave better protection to the secondary structure of insulin (Figure 4 (b) and (d)). However, DPPC showed the different effect to the secondary structure of insulin at the first 50 ns of simulation that in this period helix A1 initially collapse but then restored after 50 ns to the end of the simulation. Such behavior was not observed in LPC delivery system, the secondary structures of insulin were relatively stable from the beginning to the end of the simulation.

Figure 5, summarized the condition of insulin structure after 150 ns for insulin free and encapsulated with the delivery system. The insulin free delivery system was only experiencing partial denaturation after 150 ns, only helix B that apparently still exist while the rest is almost completely collapse (Figure 5a). At the encapsulated condition, it is clearly showed that only LPC delivery system that able to maintain the secondary and tertiary structure of insulin (Figure 5d). The worst impact to the structure of insulin was exhibited by OPPC delivery system, in which almost all secondary structures were almost collapse (Figure 5c). In contrast, DPPC, which has the same two hydrophobic tails as OPPC, gave better effect to the structure of insulin (Figure 5b). The prominent different between DPPC and OPPC is the presence of oleic acid as one of the hydrophobic tail in OPPC, while DPPC has only palmitic in both tails. Oleic acid is unsaturated fatty acid containing one cis-oriented double bond making OPPC has one non-linear hydrophobic tail. In our study such difference gave different effect to the structure of insulin.

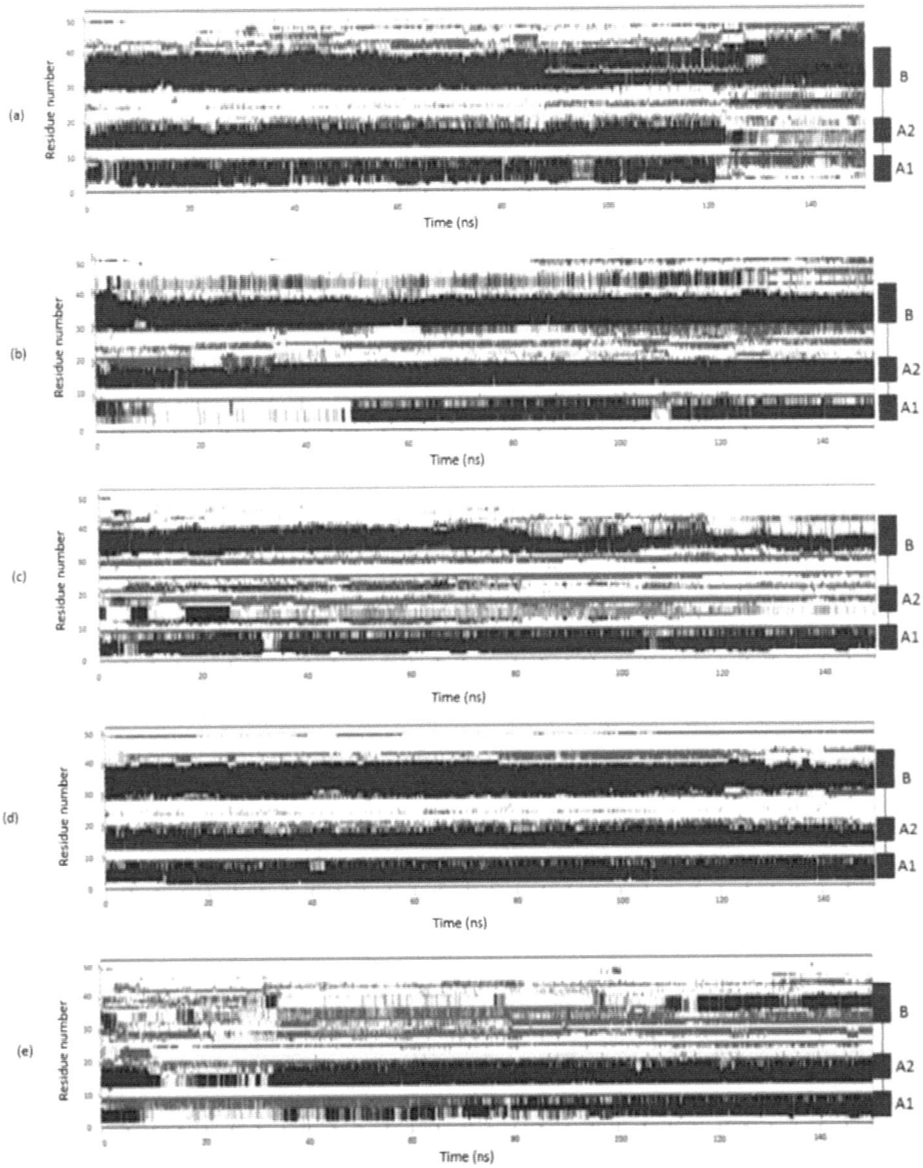

Figure 4 The result of secondary structure analysis of insulin protein (a) without delivery system and some variation of phospholipid as the delivery system: (b) DPPC, (c) OPPC, (d) LPC, (e) DPPC-LPC. The blue color in the image above shows the secondary structure of α-helix, the purple π-helical, the gray 310-helix, the red β-sheet, the yellow α-turn, the green β-turn, and the white random coil.

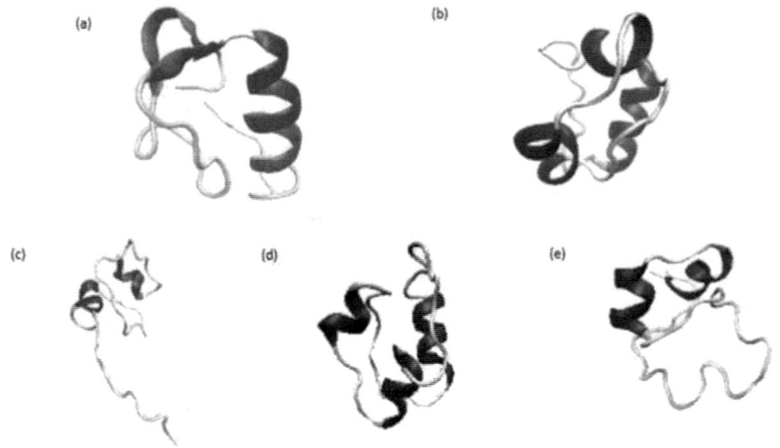

Figure 5 Secondary structure of insulin free delivery system (a) and encapsulated with DPPC (b), OPPC (c), LPC (d), and DPPC-LPC (e) after 150 ns MD simulation. Blue color represents for α-helix, purple for π-helical, gray for 3_{10}-helix, red for β-sheet, yellow for α-turn, green for β-turn, and white for random coil.

3.3 Analysis of encapsulation level of insulin by phospholipid delivery system

The previous analysis above focused only on the insulin structure point of view. In order to assess the effectiveness of phospholipid in encapsulating insulin, we evaluate the density of phospholipid covering the surface of insulin by visual inspection of the shape of emulsion and calculating the number of contact between phospholipid and insulin (Figure 6). The good delivery system should not only able to maintain the native structure of insulin but also must able to protect it from the digestive system. Therefore, the surface of insulin should be fully covered by the delivery system.

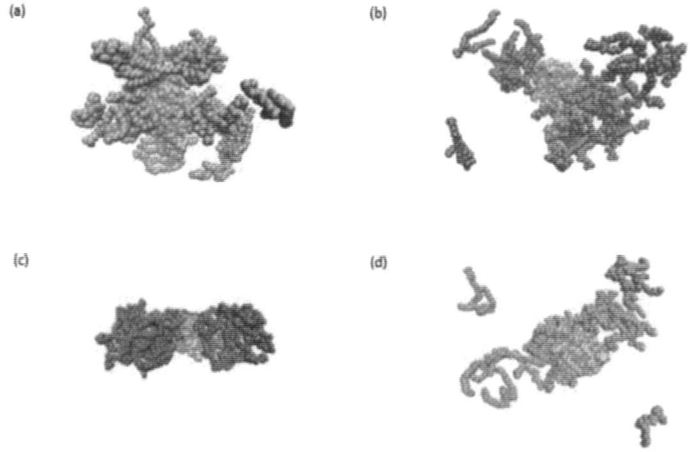

Figure 6 The shape of emulsion formed by DPPC (a), OPPC (b), LPC (c), and DPPC-LPC (d) delivery systems when encapsulated an insulin after 150 ns of MD simulation.

Our visual inspection found an interesting result, which was all phospholipids used to model the delivery system was not able to fully encapsulate the whole surface of insulin. Those phospholipids only attached to certain parts of the surface of insulin leaving uncovered part of insulin still in contact with the environment. Most parts that in contact with phospholipid molecules are the hydrophobic surfaces of insulin, the polar surfaces are mostly still free. In order to account for this behavior, we calculated the surface electrostatic potential of insulin (Figure 7). The majority of polar surfaces of insulin have a positive electrostatic potential only small fraction of the surface has the negative value. Phospholipids used in our study are negatively charged amphipathic molecule that supposed to well-attracted to the positively potential surfaces of insulin. The existence of uncovered polar surface of insulin indicating the weak interactions occurred between the polar groups of phospholipids and insulin. This is likely that the charge density of phospholipids is too low to make stable electrostatic interactions with the polar surface of insulin. The partial encapsulation of insulin by the delivery system may cause adversely affect to the stability of insulin since when it passes through the digestive system, the uncovered parts of insulin become prone to the attack of hydrolytic enzymes and extreme environments. Our study, thus, accounts for the low success rate of using phospholipid as the delivery system of oral insulin compared to the injection method [6, 10, 13, 14].

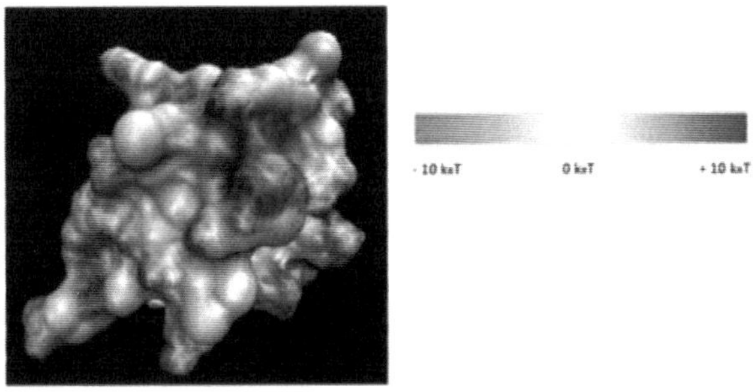

Figure 7 Surface electrostatic potential profile of insulin. Blue surface indicating positve potential value, while the negative one is depicted in red.

The result of contact number calculation for all phospholipid delivery systems after 150 ns of MD simulation exhibited a relatively small variations (Figure 8). LPC delivery system has the highest contact number with insulin, while DPPC, OPPC, and mixed DPPC-LPC showed relatively similar contact number (Figure 8a). In addition, LPC also reaches the highest contact number faster than the other phospholipids (Figure 8b). It is supported by the visual progress of insulin encapsulation by LPC delivery system depicted in Figure 9. As shown in this figure, encapsulation was already established before 30 ns of the simulation progress as indicated by similar shape of the encapsulation mode captured at 30, 60, 90, 120 and 150 ns.

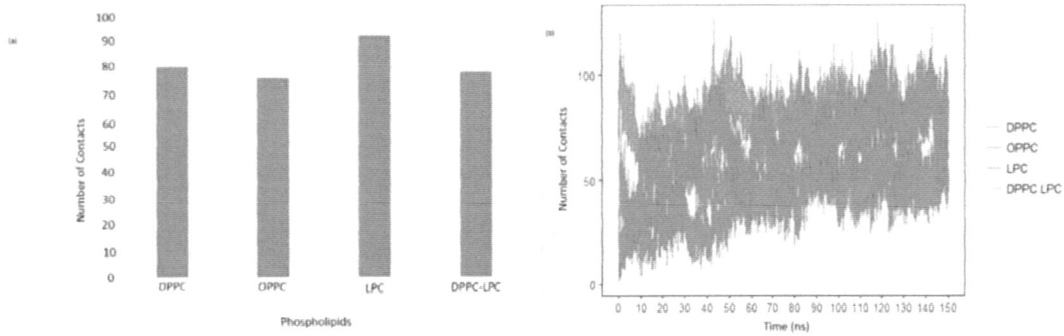

Figure 8 The number of contacts between insulin and the phospholipid delivery system. (a) The bar chart of number of contacts after 150 ns of MD simulation. (b) The graph the number of contacts during the simulation.

IOP Conf. Series: Journal of Physics: Conf. Series **1090** (2018) 012053 doi:10.1088/1742-6596/1090/1/012053

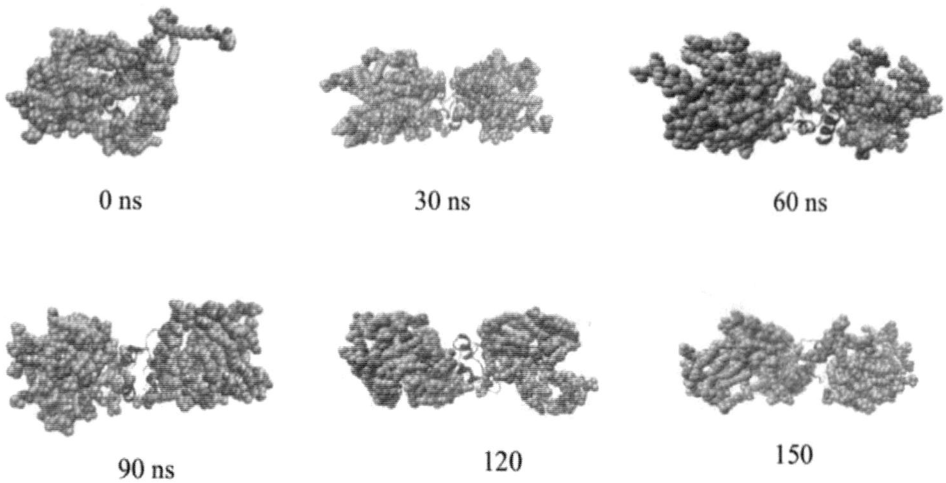

Figure 9 Encapsulation progress of insulin by LPC delivery system. The structure was created by VMD, in which LPC molecules were depicted in van der Waals balls, while insulin was displayed in cartoon.

The analysis above has shown that LPC is the best phospholipid compared to DPPC, OPPC and mixed DPPC-LPC to be developed as a delivery system for oral insulin. LPC offers better insulin protection as indicated by the highest contact number (Figure 8a) and stable encapsulation (Figure 8b). This analysis may account for the reason why LPC is providing better protection for the tertiary and secondary structures of insulin above (Figure 1 and Figure 4). In order to improve better protection, the use of phospholipid molecules only is not sufficient, the secondary amphipathic molecules that have better interaction with the uncovered surface by phospholipid molecules are required. The search of such secondary amphipathic molecules requires another study because some complexity may arise because those molecules may disturb interactions between phospholipids and insulin and also may induce structural change to the protein.

4. Conclusion

DPPC, OPPC, LPC and mixed DPPC-LPC are not fully suitable as the material for a delivery system of oral insulin because all of them fail to fully encapsulate the protein. However, among all phospholipid, LPC provide better protection to insulin structure both tertiary and secondary. Secondary amphipathic molecules require improving better encapsulation and protection of insulin to be used as the oral delivery system.

Acknowledgement

We would like to acknowledge Lembaga Pengelola Dana Pendidikan (LPDP) Indonesia for full financial support on this research.

References

[1] Hilsted J Madsbad S and Hvidberg A 1995 Intranasal insulin therapy: the clinical realities *Diabetologia* 38 680–684
[2] Lassmann Vague V and Racca, D 2006 Alternatives routes of insulin delivery *Diabetes Metabolism Research and Reviews* 32 513–22
[3] Gizurarson S and Bechgaard E 1991 Intranasal administration of insulin to humans *Diabetes Research and Clinical Practice* 12 71–84
[4] Triplitt C L Reasner C A and Isle 2008 Diabetes mellitus 1205–1241 in book: pharmacotherapy: a pathophysiologic approach 7th ed *The McGraw-Hill Companies Incorporation* New York 2597

[5] Matteucci E Giampietro O Covola V Giustarini D Fanti P and Rossi R 2015 Insulin administration: present strategies and future directions for a noninvasive (possibly more physiological) delivery *Dove Press: Drug Design, Development, and Therapy* 3109–3119

[6] Rachmawati H Haryadi B M Anggadiredja K and Suendo V 2014 Intraoral Film Containing Insulin-Phospholipid Microemulsion Formulation and In Vivo Hypoglycemic Activity Study *American Association of Pharmaceutical Scientist*

[7] Lee VH 1991 Oral route of peptide and protein drug delivery In Peptide and Protein Drug Delivery Chapter 16 *Marcel Dekker Incorporation* New York 691–738

[8] Lassmann Vague V and Racca D 2006 Alternatives routes of insulin delivery *Diabetes Metabolism Research and Reviews* 32 513–22

[9] Kisel M A Kulik L N Tsybovsky I S Vlasov A P Vorob'yov M S Kholodova E A and Zabarovskaya Z V 2001 Liposomes with phosphatidylethanol as a carrier for oral delivery of insulin: studies in the rat *International Journal of Pharmacy and Pharmaceuticals Sciences* 216 105–14

[10] Zhang Z Huixia L V and Zhou J 2009 Novel solid lipid nanoparticles as carriers for oral administration of insulin *Pharmazie* 64 574–578

[11] Cui F Shi K Zhang L Tao A and Kawashima Y 2006 Biodegradable nanoparticles loaded with insulin-phospholipid complex for oral delivery: preparation, in vitro characterization and in vivo evaluation *Journal of Controlled Released* 28 242–50

[12] Fricker G Kromp T Wendel A Blume A Zirkel J Rebmann H Setzer C Quinkert O Martin F and Müller C 2010 Phospholipids and Lipid-Based Formulations in Oral Drug Delivery *Pharmaceutical Research* 27 1469–1486

[13] Cilek A Celebi N Tirnaksiz F Tay A 2005 A lecithin-based microemulsion of rh-insulin with aprotinin for oral administration: Investigation of hypoglycemic effects in non-diabetic and STZ-induced diabetic rats *International Journal of Pharmaceutics* 298 176–185

[14] Zhang N Ping Q Guihua H Xu W Cheng Y and Han X 2006 Lectin-modified solid lipid nanoparticles as carriers for oral administration of insulin *Pharmaceutical Nanotechnology* 327 153–159

[15] Mandal K Dhayalan B Avital-Shmilovici M Tokmakoff A and Kent SB 2016 Crystallization of Enantiomerically Pure Proteins from Quasi-Racemic Mixtures: Structure Determination by X-Ray Diffraction of Isotope-Labeled Ester Insulin and Human Insulin *Chembiochem* 17 421–5

[16] Koziara K B Stroet M Malde AK and Mark A E 2014 Testing and validation of the Automated Topology Builder (ATB) version 2.0: prediction of hydration free enthalpies *Journal of Computer-Aided Molecular Design* 9713–9717

[17] Pronk S Páll S Schulz R Larsson P Bjelkmar P Apostolov R Michael R Shirts Jeremy C Spoel D Hess B and Lindahl E 2013 GROMACS 4.5: a high-throughput and highly parallel open source molecular simulation toolkit *Bioinformatics* 29 845–854

[18] Spoel D Lindahl E Hess B Groenhof H Mark E A Herman J C and Berendsen 2005 GROMACS: Fast, flexible, and free *Journal of Compitational Chemistry* 6 1701–1718

[19] Berendsen D Spoel R and Drune V 1995 GROMACS: A message-passing parallel molecular dynamics implementation *Computer Physics Communication* 91 43–56

[20] Hess B Kutzner C Spoel D and Lindahl E 2008 GROMACS 4: Algorithms for Highly Efficient, Load-Balanced, and Scalable Molecular Simulation *Journal Chemical Theory and Computation* 4 435–447

[21] Lindahl E Hess and B Spoel D 2001 GROMACS 3.0: a package for molecular simulation and trajectory analysis *Journal of Molecular Modeling* 7 306–317

[22] Oostenbrink C Villa A Mark A E and Gunsteren W F 2004 A biomolecular force field based on the free enthalpy of hydration and solvation: the GROMOS force-field parameter sets 53A5 and 53A6,*Journal of Computational Chemistry* 13 1656–76

[23] Kabsch W and Sander C 1983 Dictionary of protein secondary structure: pattern recognition of hydrogen-bonded and geometrical features *Biopolymers* 22 2577–2637

[24] Humphrey W Dalke A and Schulten K 1996 VMD - Visual Molecular Dynamics *Journal of Molecular Graphics* 14 33–38

[25] Hassiepen U Federwisch M lders T and Wollmer A 1999 The Lifetime of Insulin Hexamer *Biophysical Journal* 77 1638–1654

[26] Wirahadikusumah 2008 Biokimia: Protein, Enzim dan Asam Nukleat *ITB Bandung*

[27] Ward Colin Lawrence and Mike 2014 Insulin Binding and Activation of the Insulin Receptor [internet] *Diapedia* Available from https://doi.org/10.14496/dia.51040851468.30

IOP Conf. Series: Journal of Physics: Conf. Series **1090** (2018) 012054 doi:10.1088/1742-6596/1090/1/012054

Ab Initio Study on Electronic Structure of Some Imidazole Derivatives and Its Correlation with Corrosion Inhibition Properties

C D D Sundari[1,*], S Setiadji[2], B W Nuryadin[3], R Syafia[2], A F Huda[4] and A L Ivansyah[5]

[1]Department of Chemistry Education, UIN Sunan Gunung Djati Bandung, Jl. A.H. Nasution No.105 Bandung, West Java, 40614, Indonesia
[2]Department of Chemistry, Faculty of Science and Technology, UIN Sunan Gunung Djati Bandung, Jl. A.H. Nasution No.105 Bandung, West Java, 40614, Indonesia
[3]Department of Physics, Faculty of Science and Technology, UIN Sunan Gunung Djati Bandung, Jl. A.H. Nasution No.105 Bandung, West Java, 40614, Indonesia
[4]Department of Mathematics, Faculty of Science and Technology, UIN Sunan Gunung Djati Bandung, Jl. A.H. Nasution No.105 Bandung, West Java, 40614, Indonesia
[5]Master Program in Computational Science, Faculty of Mathematics and Natural Science, Institut Teknologi Bandung, Jl. Ganesha No. 10, Bandung, West Java, 40132, Indonesia

*citra@uinsgd.ac.id

Abstract. Imidazole and some of its derivatives have been known as a good candidate for corrosion inhibition application. In this study, ab initio calculation was carried out for several imidazole derivatives i.e. 2-phenylimidazole[1,2-α]pyridine, 6-methyl-2-phenylimidazole[1,2-α]pyridine, 6-methyl-2-(4-metoxyplenylimidazole)[1,2-α]pyridine, and 6-methyl-2-(4-chlorophenylimidazole)[1,2α]pyridine, in order to investigate their electronic properties as well as its correlation with their corrosion inhibition properties. Calculations were performed using Firefly software package with RHF method and 6-31G(d,p) basis set. Based on calculation results and analysis, it was found that 6-methyl-2-(4-metoxyphenylimidazole)[1,2α]pyridine compound has better potential as corrosion inhibitor than the rest of the compounds due to its small HOMO-LUMO energy gap, low ionization potential, low electron affinity, low electronegativity, and low global hardness i.e. 10.184 eV, 7.199 eV, -2.985 eV, 2.107 eV and 5.092 eV respectively. This compound also has the highest global softness (0.098 eV), while the greatest value of the dipole moment was owned by 6-methyl-2-(4-chlorophenylimidazole)[1,2α]pyridine with value of 6.125 Debye due to the existence of Cl atom that has larger polarity than the other substituents.

1. Introduction

Corrosion is degradation of metal surface quality due to oxidation-reduction reaction with the surrounding environment, especially corrosive media. If not promptly handled, corrosion will results in substantial material loss. The utilization of corrosion inhibitor is an effective and economic method

Content from this work may be used under the terms of the Creative Commons Attribution 3.0 licence. Any further distribution of this work must maintain attribution to the author(s) and the title of the work, journal citation and DOI.
Published under licence by IOP Publishing Ltd

to protect metal surface from corrosion by decreasing corrosion rate [1]. Organic corrosion inhibitor have more advantageous properties i.e environment-friendly and low toxicity. Various types of organic compounds have been widely used as corrosion inhibitors. One of the factors influencing the corrosion inhibition efficiency of an organic compound is the presence of heteroatom such as O, N, S and P on the molecule. The presence of an appropriate functional group will assist the formation of complexes between organic compounds and metal surfaces or assist the adsorption of inhibitor molecule on the metal surface. On the metal surface, organic inhibitor molecule forms a coating that prevents metal surfaces from contact with corrosive media [2].

Imidazole compound and some of its derivatives have been used as corrosion inhibitors [3,4,5]. One of the imidazole derivative i.e. 7-methyl-2-phenylimidazole[1,2-α]pyridine, has shown a good efficiency as a corrosion inhibitor and acts as a cathodic inhibitor [5]. The efficiency of corrosion inhibition of this compound is influenced by inhibitor concentration and temperature [5]. Other imidazole derivatives such as 4-methyl-1-phenyl imidazole, 4-methyl-1-(*p*-tolyl) imidazole, and 1-methyl-2-mercapto imidazole have also been found to be an efficient corrosion inhibitor [6-7]. Computational calculation of imidazole derivatives can help assessing (design, calculate, and analyze) potential corrosion inhibitor [8-10], which gives information complementing laboratory-scale experimental studies. There are several level of theories which is used in molecular modeling, such as DFT, ONIOM, Semiempiric Quantum, Molecular Dynamics, and Molecular Mechanics. Among them, semiempiric quantum method using PM3 and ONIOM are one of the method that has good cost-performance ratio which has been used by researcher in molecular modeling [11,12]. However, DFT theory level is the best level of theory with high accuracy for molecular modeling, especially for corrosion inhibitor application.

In this research, molecular structure and electronic properties of several imidazole derivatives i.e. 2-phenylimidazole[1,2-α]pyridine, 6-methyl-2-phenylimidazole[1,2-α]pyridine, 6-methyl-2-(4-metoxy-phenylimidazole)[1,2α]pyridine and 6-methyl-2-(4-chlorophenylimidazole)[1,2α]pyridine have been calculated computationally with ab initio method [10,13,14]. The results of the calculation with this method provide comparable results in accordance with the experimental results.

2. Computational Details

All calculations of the compounds, i.e. 2-phenylimidazole[1,2-α]pyridine, 6-methyl-2-phenylimidazole [1,2-α]pyridine, 6-methyl-2-(4-methoxyphenylimidazole)[1,2α]pyridine and 6-metil-2-(4-chlorophenylimidazole)[1,2α]pyridine were carried out using Firefly 8.2.0 [15] software package. The initial geometry of these four molecules was constructed using Avogadro 1.0 [16] software package and output of the calculations were visualized using Chemcraft 1.8 [17]. All ab initio calculations were performed at Restricted-Hartree Fock (RHF) level of theory with 6-31G(d,p) basis set for all atoms. Two dimensional structure of the compounds calculated in this research are presented in figure 1.

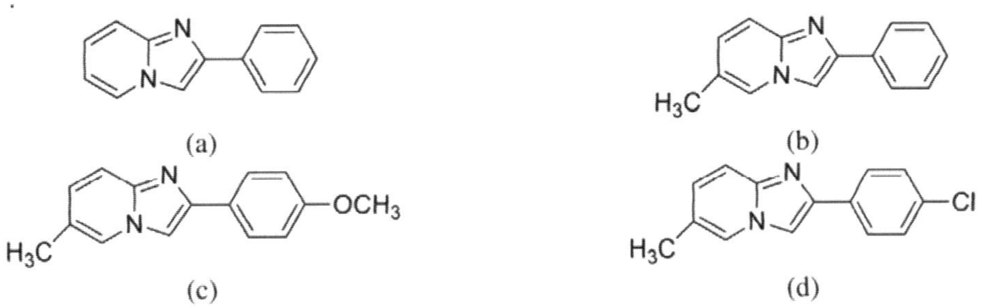

Figure 1. Structure of: (a) 2-phenylimidazole[1,2-α]pyridine, (b) 6-methyl-2-phenylimidazole[1,2-α]pyridine, (c) 6-methyl-2-(4-methoxyphenylimidazole)[1,2-α]pyridine and (d) 6-metil-2-(4-chlorophenylimidazole)[1,2-α]pyridine.

3. Results and Discussion

3.1. Optimized Structure, Atomic Charge and Electron Density Distribution in Orbitals

The optimized structure calculations were performed using RHF method with 6-31G(d,p) basis set in the gas phase. The result of structural optimization as well as bond length between atoms can be seen in figure 2. All optimized structures have planar shape with dihedral angle of 0° and 180° between atoms in phenyl and imidazole ring.

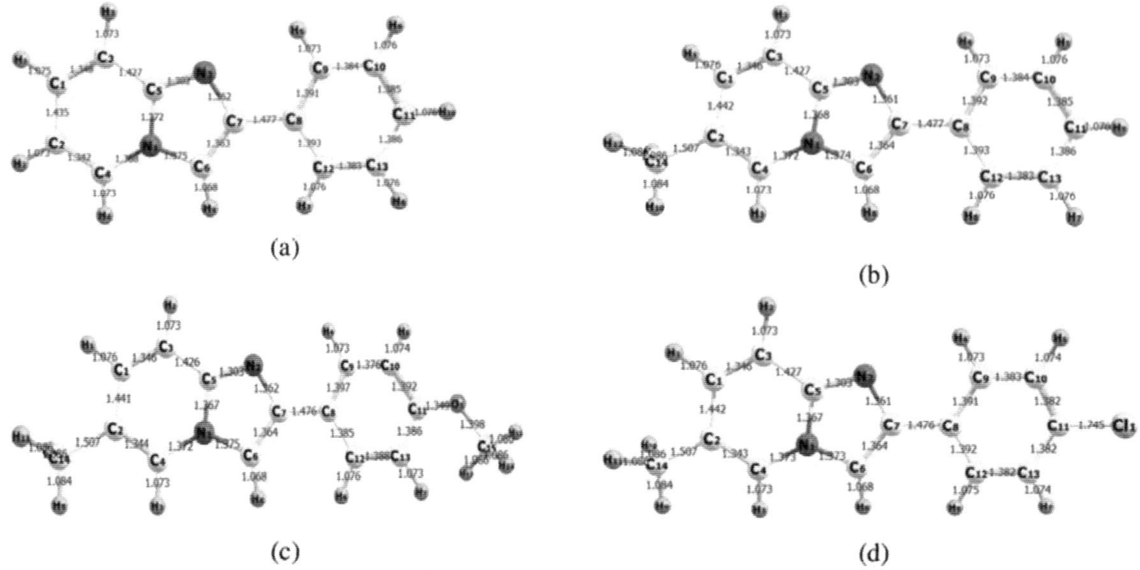

Figure 2 Optimized structure of: (a) 2-phenylimidazole[1,2-α]pyridine, (b) 6-methyl-2-phenylimidazole[1,2-α]pyridine, (c) 6-methyl-2-(4-methoxyphenylimidazole)[1,2-α]pyridine and (d) 6-metil-2-(4-chlorophenylimidazole)[1,2-α]pyridine.

Mulliken atomic charge is used to describe the electron density of each atom in the molecule. The negative sign for the atomic charge shows a high electron density, whereas a positive sign indicates that the atom lacks electrons. The negative charge of N and O atoms and the most negative carbon atom charge on the ring structure is the active part for the adsorption by the metal surface. Heteroatom is the active site that contributes to adsorption with metal atoms through electron donor-acceptor mechanism. The more negative the atomic charge will allow better adsorption to the metal surface. Table 1 lists the Mulliken atomic charge of atoms (except hydrogen atom) in the imidazole derivative molecules. Nitrogen and oxygen atom in each molecules possess the most negative charge.

Table 1. Mulliken atomic charge of atoms in: (A) 2-phenylimidazole[1,2-α]pyridine, (B) 6-methyl 2-phenylimidazole[1,2-α]pyridine, (C) 6-methyl-2-(4-methoxyphenylimidazole)[1,2-α]pyridine,and (D) 6-metil-2-(4-chlorophenylimidazole)[1,2-α]pyridine

Atomic charge	(A)	(B)	(C)	(D)
C_1	-0.119	-0.112	-0.112	-0.111
C_2	-0.244	-0.099	-0.100	-0.099
C_3	-0.178	-0.177	-0.178	-0.177
C_4	0.194	0.178	-0.178	0.178
C_5	0.626	0.622	0.623	0.624
N_1	**-0.737**	**-0.741**	**-0.740**	**-0.742**
N_2	**-0.656**	**-0.656**	**-0.661**	**-0.657**
C_6	0.066	0.068	0.062	0.071
C_7	0.181	0.176	0.187	0.177
C_8	-0.039	-0.037	-0.069	-0.044
C_9	-0.133	-0.134	-0.114	-0.125
C_{10}	-0.151	-0.150	-0.191	-0.117
C_{11}	-0.149	-0.150	0.412	-0.166
C_{12}	-0.144	-0.145	-0.131	-0.137

C_{13}	-0.155	-0.155	-0.226	-0.122
C_{14}	-	-0.340	-0.339	-0.340
O_1	-	-	**-0.663**	-
C_{15}	-	-	-0.027	-
Cl_1	-	-	-	-0.010

Electron density distribution in HOMO and LUMO orbitals of the molecules are shown in figure 3. The presence of electron donor and electron withdrawing group affects the electron density in the HOMO and LUMO orbital of the molecules. Electron in HOMO orbitals is distributed evenly in the molecules, excluding the methyl group, indicating sites with high tendency to donate electrons. Metoxy substituent in 6-methyl-2-(4-metoxyphenylimidazole)[1,2-α]pyridine effectively push electrons to nitrogen atom in imidazole as shown in electron distribution of LUMO orbital of this compound.

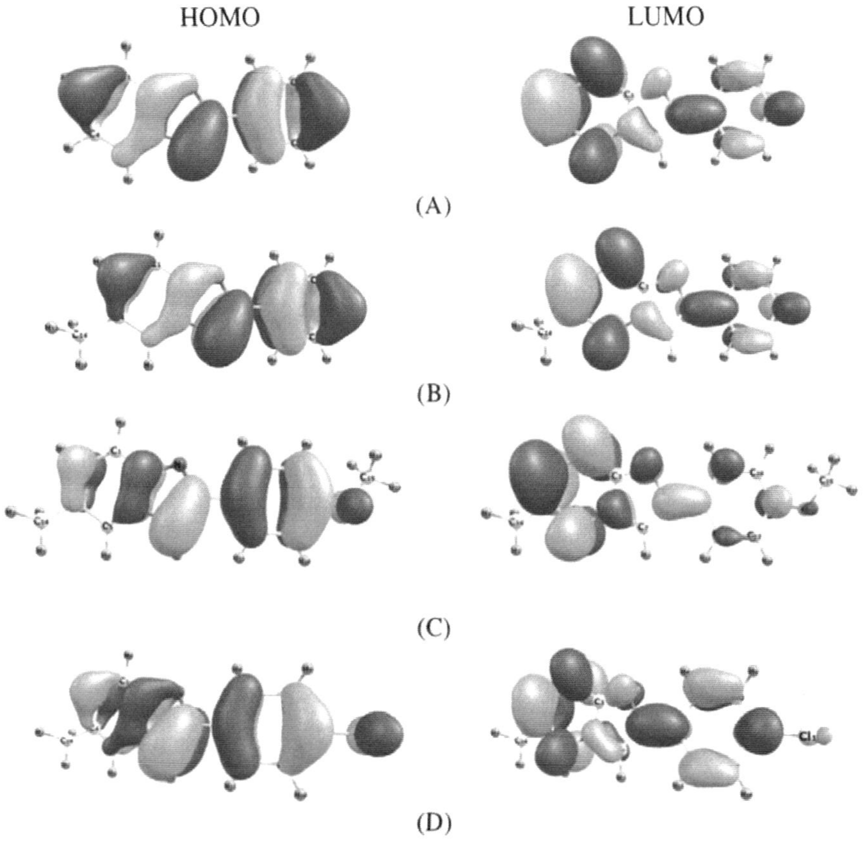

Figure 3 Electron density distribution of HOMO and LUMO orbital in: (A) 2-phenylimidazole[1,2-α]pyridine, (B) 6-methyl-2-phenylimidazole[1,2-α]pyridine, (C) 6-methyl-2-(4-methoxyphenylimidazole)[1,2-α]pyridine and (D) 6-metil-2-(4-chlorophenylimidazole)[1,2-α]pyridine

3.2. Electronic Properties

Electronic parameters give information about the ability of the inhibitor molecules to interract with metal surface. Electronic properties calculated were HOMO energy (E_{HOMO}), LUMO energy (E_{LUMO}), energy gap (E_{GAP}), ionization potential (I), electron affinity (A), electronegativity (χ), chemical potential (μ), global hardness (η), global softness (S), dipole moment (DM), and electron transfer fraction (ΔN) [1,10,14,18]. These properties are presented in table 2.

Inhibition efficiency is related with the value of HOMO energy (E_{HOMO}), LUMO energy (E_{LUMO}), and energy gap (E_{GAP}). Higher HOMO energy indicates higher ability of the molecule to donate electron. Energy gap (band gap) is the difference between LUMO and HOMO energy (E_{GAP} =

$E_{LUMO} - E_{HOMO}$). E_{GAP} value correlates to the efficiency of corrosion inhibition (%EI) through the electron transfer capability of the inhibitor molecule. Smaller E_{GAP} will results in higher inhibition efficiency. E_{GAP} value of the four molecules are similar (around 10 eV) which is consistent with previously reported calculation results for imidazole derivatives (between 8-10 eV) using RHF (Restricted Hartree Fock) method [13,14,18]. Table 2 shows that highest E_{HOMO}, smallest E_{LUMO}, and smallest E_{GAP} are possessed by 6-methyl-2-(4-metoxyphenylimidazole)[1,2-α]pyridine.

Table 2. Electronic properties of: (A) 2-phenylimidazole[1,2-α]pyridine, (B) 6-methyl-2-phenylimidazole[1,2-α]pyridine, (C) 6-methyl-2-(4-methoxyphenylimidazole)[1,2-α]pyridine and (D) 6-metil-2-(4-chlorophenylimidazole)[1,2-α]pyridine

Compound	E_{HOMO} (eV)	E_{LUMO} (eV)	E_{GAP} (eV)	I (eV)	A (eV)	χ (eV)	μ (eV)	η (eV)	S (eV)	DM (D)	ΔN (eV)
(A)	-7.574	2.774	10.348	7.574	-2.774	2.400	-2.400	5.174	0.096	3.550	0.444
(B)	-7.506	2.865	10.371	7.506	-2.865	2.321	-2.321	5.189	0.096	3.914	0.450
(C)	-7.199	2.985	10.184	7.199	-2.985	2.107	-2.107	5.092	0.098	4.416	0.481
(D)	-7.688	2.651	10.339	7.688	-2.651	2.519	-2.519	5.170	0.097	6.125	0.433

Dipole moment can also predicts inhibition efficiency related to the adsorption capability of inhibitor molecule. The electrostatic interaction between the negatively charged center of inhibitor molecule and the positively charged metal surface is a dipole interaction which illustrates physisorption. The positive sign of the dipole moment characterizes that the molecule can be adsorbed physically by metal surface. Increasing dipole moment correlates to improved inhibition efficiency. This is related to the better interaction of physisorption. Table 2 shows that the four compounds are polar, but 6-methyl-2- (4-chlorophenylimidazole)[1,2-α]pyridine has the largest dipole moment (6.125 Debye) compared to other compounds due to highly electronegative Cl substituent.

According to DFT-Koopmans theorem [1,2], the ionization potential (I) is correlated with the negative HOMO energy ($I = -E_{HOMO}$), and negative LUMO energy is correlated with the electron affinity ($A= -E_{LUMO}$). Electronegativity and chemical potential of inhibitor molecule were calculated using ionization potential (I) and electron affinity (A). Electronegativity (χ) is the ability of an atom to attract electron from other atom in a compound. Electronegativity is correlated with the average value of I and A ($\chi = (I + A)/2$) and chemical potential (μ) is correlated with negative electronegativity ($\mu = -\chi$). The electronegativity of all four inhibitor compounds are lower than iron (Fe) atom in bulk form (≈ 7 eV) [1,2] supporting the transfer of electron from inhibitor molecule to iron atom. 6-methyl-2-(4-metoxyphenylimidazole)[1,2-α]pyridine has lowest value of electronegativity (2.107 eV) and highest chemical potential (-2.107 eV), making it more reactive toward metal surface. Global hardness (η) and softness (S) also can predict chemical reactivity and structural stability of an inhibitor molecule [1,2]. Higher global softness (smaller global hardness) indicates a more reactive inhibitor molecule, which results in better inhibition efficiency. Global hardness can be expressed from HOMO and LUMO energy by equation: $\eta = (I - A)/2$. 6-methyl-2-(4-metoxyphenylimidazole)[1,2- α]pyridine has lowest value of hardness (5.092 eV), thus having greatest reactivity among the compounds studied in this paper.

The electron transfer fraction (ΔN) represents the ease of a molecule to donate electron. The value of this parameter is equal to the ratio of the electronegativity difference to the sum of hardness between inhibitor compound with metal atom in bulk phase ($\Delta N = (\chi_{Fe} - \chi_I)/2(\eta_{Fe} + \eta_I)$, $\eta_{Fe} \approx 0$) [1,2,5]. Inhibition efficiency of inhibitor molecule increases as the donation of electron to the metal surface getting easier. Large ΔN value indicates a good tendency to interact with metal atoms. 6-methyl-2-(4-methoxyphenylimidazole)[1,2-α]pyridine compound has the greatest ΔN value (0.481 eV) compared to the other compounds.

4. Conclusion

Computational calculations and studies have been performed to investigate structure and electronic properties from four imidazole derivative compounds, i.e. 2-phenylimidazole [1,2-α]pyridine, 6-methyl-2-phenylimidazole[1,2-α]pyridine, 6-methyl-2-(4-methoxyphenylimidazole)[1,2-α]pyridine and 6-methyl-2-(4-chlorophenylimidazole)[1,2-α]pyridine. The calculation results conclude that 6-

methyl-2-(4-metoxyphenylimidazole)[1,2-α]pyridine has better electronic properties than the other three compounds. This is indicated by the smallest value of energy gap (E_{GAP}), lowest electronegativity (χ), largest HOMO (E_{HOMO}) energy, lowest ionization potential (I), highest chemical potential (μ), lowest hardness (η), and highest electron transfer fraction (ΔN). In addition, the calculations predict that the 6-methyl-2-(4-chlorophenylimidazole)[1,2-α]pyridine has the greatest dipole moment, making it more polar than the other three compounds presented.

References

[1] Abelev E, Sellberg J, Ramanarayanan TA, Bernasek SL 2010 *J Mater Sci* **44** 6167
[2] Camila G. Dariva., Alexandre F. Galio. 2014, InTech, http://dx.doi.org/10.5772/57255.
[3] Nnenna Winifred Odozi, Jonathan Oyebamiji Babalola, Ekemini Bassey Ituen, Abiodun, Omokehinde Eseola 2015 *American Journal of Physical Chemistry* **4**(1-1) 1-9
[4] Ghanbari A, Attar M M and Mahdavian M 2010 *Materials Chemistry and Physics* **124** 1205–1209
[5] Anejjar A, Zarrouk A, Salghi R, Hmamou D B, Zarrok H, Al-Deyab S S, Bouachrine M, Hammouti B and Benchat N 2013 *Int. J. Electrochem. Sci.*, **8** 5961 – 5979
[6] Kabanda M M, Obot I B and Ebenso E E 2013 *Int. J. Electrochem. Sci.* **8** 10839 – 10850
[7] Rodriguez-Valdez L M, Martinez-Villafane A and Glossman-Mitnik D 2005 *Journal of Molecular Structure: Theochem* **713** 65–70
[8] Mousavi M, Mohammadalizadeh M and Khosravan A 2011 *Corrosion Science* **53** 3086–3091
[9] Bereket G et al. 2002 *Journal of Molecular Structure (Theochem)* **578** 79-88
[10] Behzadi H, Roonasi P, Momeni M J, Manzetti S, Esrafili M D, Obot I B, Yousefvand M and Mousavi-Khoshdel S M 2015 *Journal of Molecular Structure* **1086** 64–72
[11] Ivansyah A L, Martoprawiro M A and Buchari 2017 *J. Phys.: Conf. Ser.* **812** 012070
[12] Setiadji S, Sundari C D D, Ramdhani M A, Umam A B K and Ivansyah A L 2018 *IOP Conf. Ser.: Mater. Sci. Eng.* **288** 012138
[13] Momeni M J, Behzadi H, Roonasi P, Sadjadi S A S and Mousavi-Khoshdel S M. 2015 *Res Chem Intermed* **41** 6789-6802
[14] Ju H, Ding L, Sun C, and Chen J 2015 *Advances in Materials Science and Engineering* **2015** 519606
[15] a). Granovsky A A, Firefly version 8, http://classic.chem.msu.su/gran/firefly
b). Schmidt M W, Baldridge K K, Boatz J A, Elbert S T, M S Gordon, J H Jensen, S Koseki, N Matsunaga, K A Nguyen, S Su, T L Windus, M Dupuis, J A Montgomery 1993 *J.Comput.Chem.* **14** 347-1363
[16] Marcus D Hanwell, Donald E Curtis, David C Lonie, Tim Vandermeersch, Eva Zurek and Geoffrey R Hutchison 2012 *Journal of Cheminformatics* 4 17
[17] Chemcraft ver. 1.8. https://www.chemcraftprog.com
[18] M. Rajendran, A. Malkiya, P. Muthupetchi, D. Devapiriam 2016 *J Adv Sci Res* **7**(1) 32-37

IOP Conf. Series: Journal of Physics: Conf. Series **1090** (2018) 012055 doi:10.1088/1742-6596/1090/1/012055

Computational Study of Inclusion Complexes Between Omeprazole Enantiomer with Hydroxypropyl-β-Cyclodextrin

S Setiadji[1,*], C D D Sundari[2], B W Nuryadin[3], H Zayyinunnisya[1], R Cahyandari[4] and A L Ivansyah[5]

[1]Department of Chemistry, Faculty of Science and Technology, UIN Sunan Gunung Djati Bandung, Jl. A.H. Nasution No.105 Bandung, West Java, 40614, Indonesia.
[2]Department of Chemistry Education, UIN Sunan Gunung Djati Bandung, Jl. A.H. Nasution No.105 Bandung, West Java, 40614, Indonesia.
[3]Department of Physics, Faculty of Science and Technology, UIN Sunan Gunung Djati Bandung, Jl. A.H. Nasution No.105 Bandung, West Java, 40614, Indonesia.
[4]Department of Mathematics, Faculty of Science and Technology, UIN Sunan Gunung Djati Bandung, Jl. A.H. Nasution No.105 Bandung, West Java, 40614, Indonesia.
[5]Master Program in Computational Science, Faculty of Mathematics and Natural Science, Institut Teknologi Bandung, Jl. Ganesha No. 10, Bandung, West Java, 40132, Indonesia

*s.setiadji@uinsgd.ac.id

Abstract. Computational study of inclusion complexes between R/S-omeprazole as Proton Pump Inhibitor (PPI) compound with hydroxypropyl-β-cyclodextrin in three of its dimeric structure configurations (head to head, head to tail and tail to tail) has been carried out. All calculations were performed using PM3 quantum semiempirical method. Computational results showed that total binding energy (BE) of R-omeprazole complex is more exothermic than total binding energy of S-omeprazole complex. The calculated binding energy of R-omeprazole complex was -74.65 kcal/mol, while for S-omeprazole complex was -64.09 kcal/mol. This results indicate that R-omeprazole inclusion complex has higher stability compared to S-omeprazole inclusion complex. The value of ΔH, ΔS, and ΔG for R-omeprazole inclusion complex formation were -77.02 kcal/mol, -0.24 cal/mol.K and -2.62 kcal/mol respectively, while the value of ΔH, ΔS, and ΔG for S-omeprazole inclusion complex formation were -66.47 kcal/mol, -0.29 cal/mol.K and 19.29 kcal/mol respectively. These values indicate that the formation of inclusion complexes between R/S-omeprazole with hydroxypropyl-β-cyclodextrin is enthalpy driven process.

1. Introduction

PPI (Proton Pump Inhibitor) is a class of drugs used to inhibit gastric acid release. The PPI compound works by forming a covalent bond in the H^+/K^+ ATPase enzyme system present in the gastric parietal cell [1]. PPI compounds that are often used in treating gastric acid problems include lansoprazole, omeprazole, pantoprazole and rabeprazole [1]. Each of these PPI compounds has R- and S-

Content from this work may be used under the terms of the Creative Commons Attribution 3.0 licence. Any further distribution of this work must maintain attribution to the author(s) and the title of the work, journal citation and DOI.
Published under licence by IOP Publishing Ltd

enantiomer, in which has different pharmacological characteristics in treating gastric acid excessive secretion. For example, both R- and S- enantiomer of omeprazole have the same function, i.e. treating excess gastric acid, but their effectivity in treating the disease is different. S-omeprazole is more effective in treating gastric acid excessive secretion compared to R-omeprazole and its racemic mixture [2]. To obtain effective results in the treatment of the disease, separation of both enantiomers of PPI drugs was performed. The most common separation method used in this enantiomeric separation is HPLC (high performance liquid chromatography) and CE (capilary electrophoresis). To separate S-omeprazole enantiomer from the mixture, chiral stationary phase in HPLC or chiral selector in CE have important role.

One of the materials that can be used as chiral stationary phase and chiral selector is cyclodextrin. Cyclodextrin is cyclic oligosacharide which composed of glucose molecules connected with α-1,4-glicocydic bonding. The molecular geometry of cyclodextrin is similar with a truncated cone. Cyclodextrin has an ability to form inclusion complex with various guest molecule depends on its polarity and size because the inner cavity of cyclodextrin molecule is hydrophobic and the outer surface is hydrophilic. Due to this ability, cyclodextrin and its derivatives has been utilized in the field of pharmacy, cosmetic industry, agrochemical industry and analytical chemistry, especially in the separation of chiral compounds. One of cyclodextrin derivative used for this application is hydroxypropyl-β-cyclodextrin [3,4]. In the understanding of microscopic phenomena of chiral separation between PPI compound and hydroxypropyl-β-cyclodextrin, molecular modeling is one of the best method used so far. There are several level of theories which is used in molecular modeling, such as DFT, ONIOM, Semiempiric Quantum, Molecular Dynamics, and Molecular Mechanics. Even though DFT is the level of theory with high accuracy for molecular modeling [5-8], semiempiric quantum method using PM3 is one of the method that has good cost-performance ratio which has been used by some researchers to perform host-guest inclusion complex modeling [9-12]. In this research, molecular modeling calculation has been performed using semiempiric quantum PM3 to learn about microscopic phenomena in the inclusion complex between R/S-omeprazole and hydroxypropyl-β-cyclodextrin. This research aimed to study the inclusion complexes between R/S-omeprazole and hydroxypropyl-β-cyclodextrin to understand chiral separation process of R/S-omeprazole compound by hydroxypropyl-β-cyclodextrin.

2. Computational Method

The initial molecular structure of R/S-omeprazole was built using Avogadro 1.1.0 [13] software package. The molecular structure of hydroxypropyl-β-cyclodextrin used in this calculation was originated from the initial geometry of β-cyclodextrin obtained from the CSD (Cambridge Structural Database). Modification has been done by replacing the H- atom in the 6'-hydroxyl functional groups with a hydroxypropyl functional group. Furthermore, R/S-omeprazole compound and hydroxypropyl-β-cyclodextrin were optimized and the frequencies were calculated using semiempiric quantum PM3 method implemented in Gamess-US version December 5, 2014 R1 for 64 bit (x86_64 compatible) under Linux with gnu compilers [14].

The structure of hydroxypropyl-β-cyclodextrin in the dimeric forms was also made by adjusting the orientation of two hydroxypropyl-β-cyclodextrinmolecule, i.e. head to tail, head to head and tail to tail orientation. Then, molecular mechanics docking calculation was performed from the optimization result of each host and guest compound. In this process, host and guest compounds form the host-guest inclusion complex. The results of the docking process were extracted by taking the guest molecules that have the most excellent conformation. Then, the host-guest inclusion complexes from the docking results were optimized using semiempiric quantum PM3 method and the frequency calculation was also performed on them to determine the thermodynamic properties of host-guest inclusion complex.

3. Results and Discussion

3.1. Binding energy calculation

Docking can be assumed as lock and key process to the interaction of enzymes and substrates [15]. The inclusion complex itself is a guest compound surrounded by host compound. Figure 1 shows that the overall binding energy of R-omeprazole and hydroxypropyl-β-cyclodextrin inclusion complex is more negative (ΔE_R = -26.1 kcal/mol) than S-omeprazole and hydroxypropyl-β-cyclodextrininclusion complex (ΔE_S = -23.7 kcal/mol). The more negative binding energy indicates the stronger binding between guest molecule and host molecule by non-covalent interaction. It implied that S-omeprazole is going to be eluted first in HPLC, because R-omeprazole was bound stronger to hydroxypropyl-β-cyclodextrin than S-omeprazole.

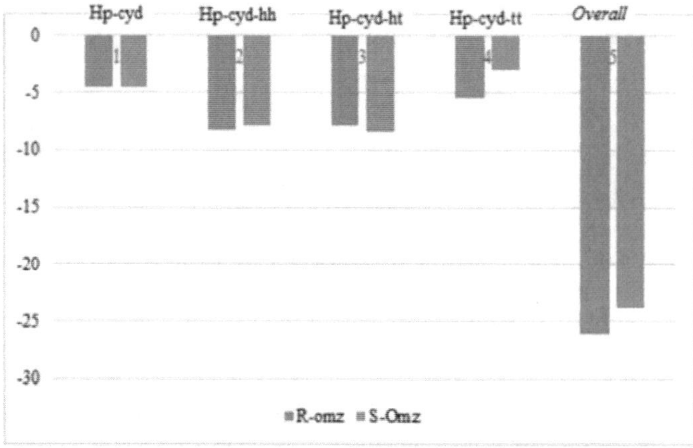

Figure 1. Binding energy comparison on chiral separation of R/S-Omeprazole using hydroxypropyl-β-cyclodextrin (Hp-cyd is hydroxypropyl-β-cyclodextrin in monomeric form; while Hp-cyd-hh, Hp-cyd-ht, and Hp-cyd-tt are dimeric form with orientation: hh = head to head, ht = head to tail and tt = tail to tail).

3.2. Molecular structure and hydrogen bonding

Optimization step forms the most stable structure of each guest, host, and host-guest compound with minimum energy. Figure 2 shows the optimized geometry of each free guest compound, figure 3 shows the optimized geometry of each free host compound, and figure 4 shows the optimized geometry of hydroxypropyl-β-cyclodextrin/omeprazole inclusion complex.

The intermolecular hydrogen bonding between the hydrogen atoms of the host hydroxyl group with the nitrogen and oxygen in omeprazole were defined as the distance of $d_{H\text{-}O}$ or $d_{H\text{-}N}$ that is less than 3.00 Å. If the $d_{H\text{-}O}$ or $d_{H\text{-}N}$ distance is more than 3.00 Å, it will be concluded that there are no intermolecular hydrogen bonding on the inclusion complex.

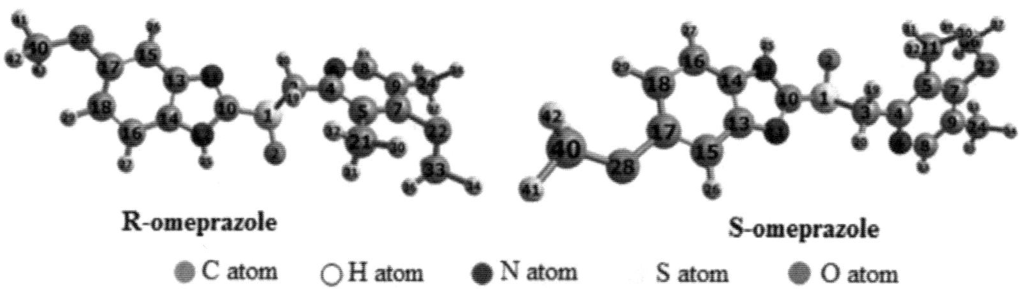

Figure 2. Optimized molecular structure of free R/S-omeprazole guest molecule

Figure 4 shows the hydrogen bonding on the interaction between R-omeprazole and hydroxypropyl-β-cyclodextrin which has $d_{H\text{-}N}$ distance of 2.970 Å between the nitrogen in the pyridine group at R-omeprazole with the hydrogen in the secondary hydroxyl group at hydroxypropyl-β-cyclodextrin; R-omeprazole with hp-cyd-hh has $d_{H\text{-}O}$ distance of 1.782 Å between oxygen in sulfonyl group with a hydrogen in secondary hydroxyl group; R-omeprazole with hp-cyd-ht has $d_{H\text{-}O}$ distance of 1.787 Å between oxygen in sulfonyl group with a hydrogen in the primary hydroxyl group; R-omeprazole with hp-cyd-tt has $d_{H\text{-}N}$ distance of 1.811 Å between nitrogen in the benzimidazole group with a hydrogen in the primary hydroxyl group. The total hydrogen bonding on the inclusion complex between R-omeprazole and hydroxypropyl-β-cyclodextrin. Thus, there are 4 hydrogen bondings.

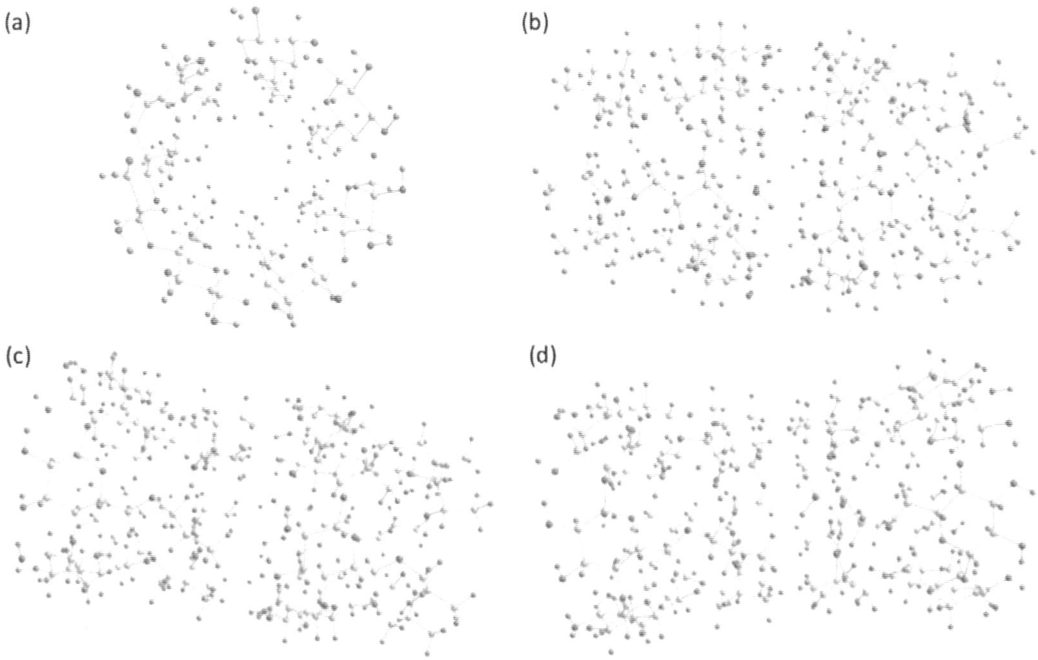

Figure 3. Optimized molecular structure of free host molecule (a) hp-cyd; (b) hp-cyd-hh; (c) hp-cyd-ht; dan (d) hp-cyd-tt.

Figure 4 also shows the intermolecular hydrogen bonding between S-omeprazole and hp-cyd-ht has a $d_{H\text{-}O}$ distance of 1.777 Å between oxygen in the sulfinyl group with hydrogen in the primary hydroxyl group. S-omeprazole with hp-cyd-tt has a $d_{H\text{-}N}$ distance of 2.257 Å between nitrogen in the benzimidazole group with a hydrogen in the primary hydroxyl group. Total hydrogen bonding on the inclusion complex between S-omeprazole and hydroxypropyl-β-cyclodextrin is 2 hydrogen bondings. Based on the total of intermolecular hydrogen bonding in each inclusion complex, it was concluded that R-omeprazole will be bound to hydroxypropyl-β-cyclodextrin stronger than S-omeprazole and S-omeprazole will be eluted faster than R-omeprazole. This result has similar tendency with the previous result that S-omeprazole will be eluted first than R-omeprazole based on binding energy value from docking result.

3.3. Thermodynamic analysis

The thermodynamic analysis was performed using frequency calculation data obtained from the output file. The analyzed thermodynamic parameters were the binding energy (taken from total energy), enthalpy, Gibbs free energy, and entropy. Statistically, this thermodynamic calculation was performed at pressure of 1 atm and temperature of 298.15 K. The results data analysis of omeprazole was used as a validation method based on experiments that have been done by Hancu et al. In the Hancu et al study, it is known that S-omeprazole is more easily separated than the R-enantiomer [10].

The complexation reaction of omeprazole with hydroxypropyl-β-cyclodextrin is exothermic, indicated by the negative value of energy and enthalpy changes. According to table 1, the thermodynamic quantities of R-omeprazole enantiomer tend to be more negative than the S-omeprazole enantiomer, which are seen from the overall quantity data of binding energy, enthalpy changes, Gibbs free energy change, and entropy changes. This proves that the R-omeprazole enantiomer is more stable than S-omeprazole enantiomer.

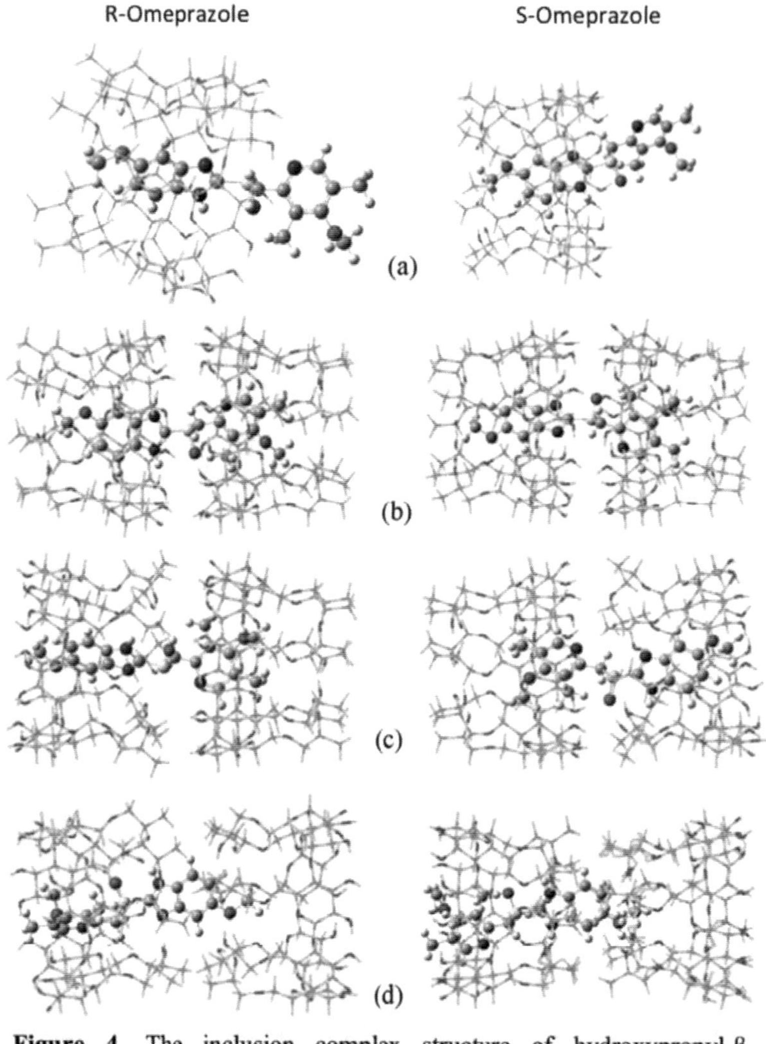

Figure 4. The inclusion complex structure of hydroxypropyl-β-cyclodextrin (hp-cyd) with R/S-omeprazole in the form of (a) monomer, (b) dimer head to head (hh), (c) head to tail (ht), and (d) tail to tail (tt)

Table 1. Thermodynamic parameters of inclusion complex of omeprazole/hydroxypropyl-β-cyclodextrin

Inclusion Complex	BE (kcal/ mol)	ΔH (kcal/mol)	ΔG (kcal/mol)	ΔS (kcal/ mol K)
R-omz : hp-cyd	-11.31	-11.90	1.64	-0.05
S-omz : hp-cyd	-2.23	-2.82	14.93	-0.06
R-omz : hp-cyd-hh	-26.72	-27.31	-3.46	-0.08
S-omz : hp-cyd-hh	-19.11	-19.70	5.16	-0.09
R-omz : hp-cyd-ht	-28.38	-28.97	-4.25	-0.09

S-omz : hp-cyd-ht	-34.69	-35.28	-12.09	-0.08
R-omz : hp-cyd-tt	-8.24	-8.84	2.07	-0.04
S-omz : hp-cyd-ht	-8.08	-8.67	11.29	-0.07
Overall				
R-omz	-74.65	-77.02	-2.62	-0.24
S-omz	-64.09	-66.47	19.29	-0.29

4. Conclusion

Molecular modeling of inclusion complex between R/S-omeprazole with hydroxypropyl-β-cyclodextrin have been performed using molecular docking and semiempiric quantum PM3 method, and the result from both method have similar tendency which implied that S-omeprazole will be eluted first, faster than R-omeprazole. Computational calculation conclude that inclusion complex formed between hydroxypropyl-β-cyclodextrin with R-omeprazole is more stable than with S-omeprazole, which is stabilized by hydrogen bonding. Moreover, the formation of the inclusion complex between R/S-omeprazole with hydroxypropyl-β-cyclodextrin is an enthalpy driven process.

References

[1] Cirilli R, Ferretti R, Gallinella B, Santis E D, Zanitti L and Torre F L 2008 *Journal of chromatography. A* **1177** 105–113
[2] Olsson J, Stegander F, Marlin N, Wan H and Blomberg L G 2006 *Journal of Chromatography A* **1129** 291–295
[3] Xie J, Yin Y X, Jin G L, He M X and Diao GW 2011 *Chinese J. Struct Chem.* **30** 706-716
[4] Bonato P S and Paias F O 2004 *J. Braz. Chem. Soc.* **15**(2) 318–323
[5] Hosni M, Kusumawati Y, Farhat S, Jouini N, Ivansyah A L, Martoprawiro M A and Pauporté T 2015 *ACS Appl. Mater. Interfaces* **7(3)** 1568-1577
[6] Sundari C D D, Martoprawiro M A and Ivansyah A L 2017 *J. Phys.: Conf. Ser.* **812** 012068
[7] Kusumawati Y, Massin J, Olivier C, Toupance T, Ivansyah A L, Martoprawiro M A, Prijamboedi B, Radiman C L and Pauporté T 2017 *Journal of Photochemistry and Photobiology A: Chemistry* **341** 69-77
[8] Sundari C D D, Setiadji S, Ramdhani M A, Ivansyah A L and Widhiasari N I T 2018 *IOP Conf. Ser.: Mater. Sci. Eng.* **288** 012033
[9] Lu Y, Guo T, Qi J, Zhang J and Wu W 2012 *AAPS Pharm Sci Tech.* **13** 4 1222–1229
[10] Hancu G, Papp L A and Rusu A 2015 *Chromathographia* **78** 279-284
[11] Ivansyah A L, Martoprawiro M A and Buchari 2017 *J. Phys.: Conf. Ser.* **812** 012070
[12] Setiadji S, Sundari C D D, Ramdhani M A, Umam A B K and Ivansyah A L 2018 *IOP Conf. Ser.: Mater. Sci. Eng.* **288** 012138
[13] Hanwell M D, Curtis D E, Lonie D C, Vandermeersch T, Zurek E and Hutchison G R 2012 *Journal of Cheminformatics* **4** 17
[14] Schmidt M W, Baldridge K K, Boatz J A, Elbert S T, Gordon M S, Jensen J H, Koseki S, Matsunaga N, Nguyen K A, Su S, Windus T L, Dupuis M and Montgomery J A 1993 *J. Comput. Chem.* **14** 1347-1363
[15] Venkatesh G, Sivasankar T, Karthick M and Rajendiran N 2013 *J Incl Phenom Macrocycl Chem* **77** 309-318

IOP Conf. Series: Journal of Physics: Conf. Series **1090** (2018) 012056 doi:10.1088/1742-6596/1090/1/012056

The reasoning for low ground magnetic anomaly reveal by model numeric and data on the field

Agus Laesanpura, Warsa, Tedy S.

Institut Teknologi Bandung

email:laesanpura1@yahoo.com

Abstract. The current practice in earth science for ground magnetic survey interpretation is analogue model based on the evidence in the other region, and make a theory. The numerical model based on principle of magnetic physical laws apply on the model running on computer and will show the clear explanation for the phenomenon. The current progress is possible to make the effect earth magnetic field, their magnetic variant and material responsible to. All the model and magnetic response as flux is overlay under the topography in 3-D and make the slicing easy. Hence we can follow directly and nearly true in the theory and the facts in East Flores island of Indonesia. On-site, low magnetic anomaly is unique. Over there the field ground magnetic, coincidence with the hot water outpouring from fracturing region. The result showing the phenomena are coherently explainable both in theory, and the fact in the field, bring the numerical software as a powerful tools.

1. Introduction

The interpretation in geosciences needs combining imagination-knowledge and numerical aids calculation. In numerical calculation the geoscientist develop the tools to predict the subsurface image based on measurement in the surface. The technique is famous as forward modeling and inversion [1][2].

Sometimes, those techniques reduces the principal of basic physical features to understand. From here this article will regain the whole, beginning with understanding ending with practical computation for subsurface imaging.

The Phenomenon, is describing besides in calculation and idea, but also introducing the fact in the field for a clue. Hence, the whole process of nature could be linked in the earth and physical parameter. The case study is an area in East Flores Island where volcanic rock exposed distinctly [3]. An example of surface measurement is capture for a model. The model will make us understand and the explanation simpler.

2. The spot of Anomaly among the high

East Flores is part of volcanic arc in Nusa Tenggara Indonesia. The Island, is part of magmatic arc of Pacific's ring of fire, about 300 km from Indonesian subduction zone. The next to volcanic island is closed to the famous, Banda in East, and Bali-Lombok in west.

The island is surrounded by volcanic product with subtle topography [4] with pyroclastic Nangapada (Tmn) to Waihekang (Tmpw) as tuff, breccias, and maybe deep intrusion Wolowaru(Tmg) Formation of Middle Miocene age. In several parts in foot of volcano and between the volcano peaks, the nearly flat topography is sometimes found.

In Theory, the cooling magmatic product will show the high magnetic anomaly, due to magnetite mineral in solidified rock of andesitic-basaltic such as found in this area. But in facts, on several parts the fumaroles found in several sites. The field observation and morphology observation is

Content from this work may be used under the terms of the Creative Commons Attribution 3.0 licence. Any further distribution of this work must maintain attribution to the author(s) and the title of the work, journal citation and DOI.
Published under licence by IOP Publishing Ltd

also predict the existence of fault zone passing by the fumaroles (hot water outpouring from earth) [4]. We note the low magnetic anomaly is nearly coincidence with the fault zone and fumaroles

On several geothermal areas, such as Kamojang geothermal field in west Java, the low magnetic anomaly is coincidence or interpreting for fumaroles activity [5] due to relationship between temperature to susceptibility or magnetic residual [8]. The current data acquire for subsurface imaging use the proton magnetometer with 500 - 1000 meter intervals.

The anomaly magnetic south to north section in equatorial southern latitude, the symptoms of negative or locally zero anomaly on the spot is based on normal behavior of positive anomaly come from susceptibility of volcanic rock. The dipole character will produce by beginning of negative curve follow to positive curve. Contrary to normal in volcanic rock, the negative anomaly is beginning by positive curve follow to negative curve. The illustration is present in figure 1.

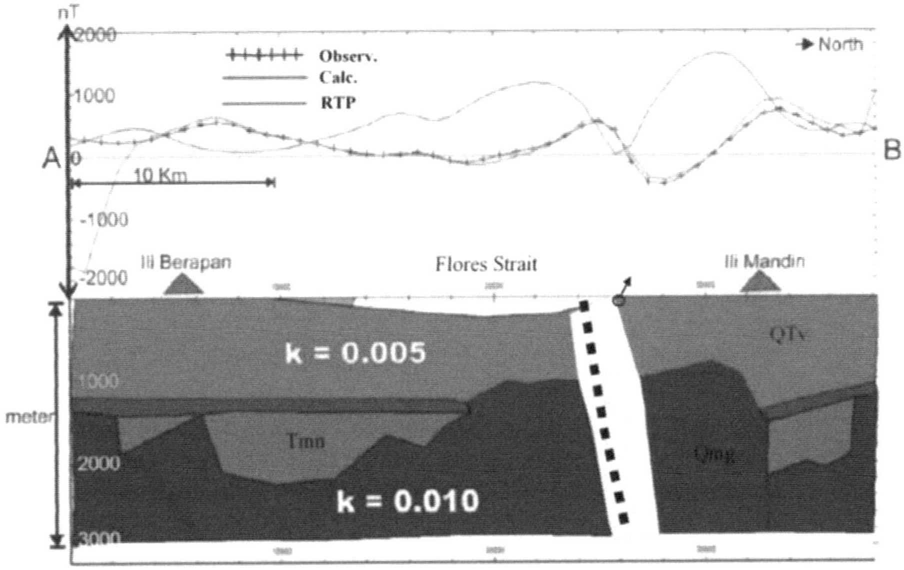

Figure 1. The interpretation section of north south, providing the perturbation local High-low anomaly magnetic due to heating liquid and providing both by fumaroles(♂) manifestation and signaling low by reduction to pole analysis, (after [6]).

3. Numerical magnetic modeling and forward modeling

The magnetic field in this case is Magneto static. It could be solved using scalar magnetic potential, Vm. The earth magnetic field, H, background is change by the presence locally intrusion or volcanic rock making higher anomaly, and the spot of fumaroles making low anomaly, dependence of permeability, μ, or susceptibility.

The general magnetic equation is beginning by magnetic field, H, and its relationship with flux density B (dependence μ of medium where H is dominance factor, and residual, Br), with scalar potential Vm as:

$$H = -V_m \tag{1}$$

$$\nabla \cdot (\mu \nabla V_m + B_r) = 0 \tag{2}$$

The Potential magnetic is separating into external factor, V_{ext}, and internal factor V_{int} (considering not influence or dissolve to V_{ext}). And flux density B, decompose to residual B_r, and background B_0, yield into the intensity, declination, and inclination value in the area. Assuming the current magnetic epoch is same for region, we adopt the value of Magnetic intensity 44476.3 nT, inclination =-33.083^0, declination 1.983^0 So the eq. 2 separating into:

$$\nabla \cdot \mu \nabla V_{red} + B_r + \mu \nabla V_{ext} = 0 \tag{3}$$

The flux density local, B_r, is proportional to B-magnetic by predicting presence permeability concentration material, say with percent (%) for magnetite or hematite equivalence.

The boundary condition for flux magnetic surrounding the box, is following Neumann's boundary condition [7], meanwhile inside the body use continuation.

The following figure show the basic magnetic influence on surface due to body in subsurface and field intensity or flux result. See how the subtle anomaly due to fumaroles hence yields the low magnetic among the high one.

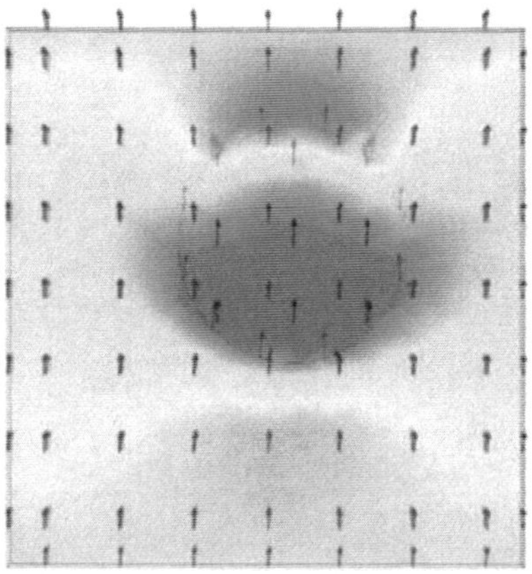

Figure 2 The reduce magnetic field in near surface igneous rock representing by cylinder body. See the dipole effect (red high, blue low) and arrow indicate its vector

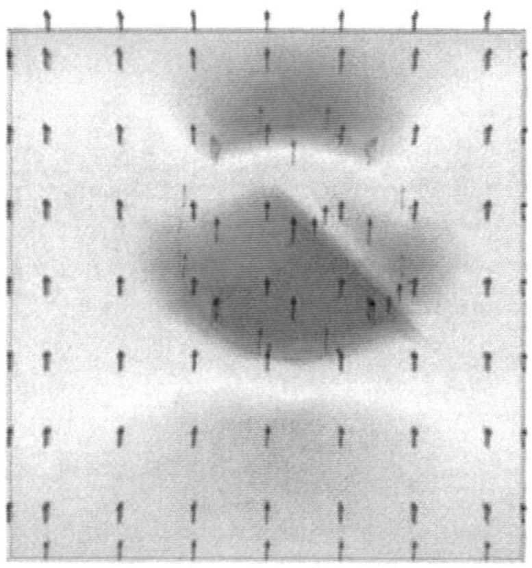

Figure 3 The reduced magnetic anomaly in near surface due to both igneous (representing by cylinder) and crack fumaroles (representing by thin ellipsoid). Compare to figure 2, on sequence three phenomena : blue (low anomaly) and cyan (slightly high) and blue toward higher one

4. Low magnetic anomaly and the quantitative Interpretation approach

Figure [2] [3], two model body is representing the causative body of rock deformed by fracturing rock where fumarole manifestation and body mounted near the surface. The rock fracture filling in fumarole represent by an ellipsoid at depth about 300 meters. While, a body mounted represent by a cylinder at base 2 km depth.

The earth magnetic field, H_O, on this area is about 44437 nT, and the contribution local Flux B, by combination of permeability, inclination (- 30 degree) and declination (1.8 degree). All of this parameter is resulting perturbation of earth magnetic field. The local magnetic effect both the positive is a cylinder of body due to high permeability value (20 percent equivalent magnetic), and an ellipsoid low permeability (2 percent equivalent magnetic). An estimation temperature we reduce the effect as low as 30% [8].

The Topography represent the area is coming from US Army Topography [9], digitize and formatting to apply in Comsol [7]. All is received and joint with numerical forming the magnetic effect of causative body and reformulating the basic parameter. The mesh generation using triangulation finite element like is producing as well by the program. The 3-D picture could maximize the observation and to see the result, figure [2] [3].

On figure [2] show the effect of magnetic field because of mounting body alone (left), compare with the combination with the crack (ellipsoid), at right. The Field magnetic show the change from positive amplitude, to wavy because of dipole effect of small ellipsoid model of crack. This effect is comparable to field observation of ground magnetic data (figure [3]). The field observation is also show the lineation zone, interpreting from Landsat, and evidence from fumaroles [4]. From here both approach show the coherence between field measurement to physical evidence, identified indirectly by cracking, fumaroles and numerical knowledge approach for the response of field magnetics curve.

5. Conclusion

The evidence of surface observation by the combination from indirect measurement of ground magnetic, surface expression, is explain by model causative body. The respond of magnetic field made by numerical-geosciences approach calculate the effect therefore comparable with the data.
The numerical based geosciences is a powerful tools to overcome mystery in subsurface and a completed approach to current routine imaging

Acknowledgments

Authors wishing to acknowledge Maison du Monde Malais of Université de la Rochelle-French, and LPPM(Empowerment Community and research) ITB.

References

[1] Henderson R and Roland G 1960 *Geophysics* **25** pp 569-585
[2] Dimri V 1992 Deconvolution and Inverse Theory *Elsevier* p.203
[3] Koesoemadinata S and Noya N 1989 Geology map of Lomblen quadrangle East Nusa Tenggara PPPG
[4] Laesanpura A 2010 Field Research of Geophysics study in the Flores Islands, *LPPM* unpublished
[5] Songkono S and Hochstein M P 1995 Application of Magnetic method to assess the excess the extent of high temperature geothermal reservoir *Pros. Twentieth workshop on geothermal reservoir engineering* p. 24-26
[6] Laesanpura et al 2017 Preliminary Study on Ground-Magnetic Data Near the Active Volcanoes in Konga Bay, East Flores Indonesia IOP Conf. Series: Earth and Environmental Science p.**62**
[7] ------------- 2013 *Comsol Multiphysics user's guide* pp. 562-626
[8] Gatacecca et al 2007 The effects of explosive-driven shocks on the natural remnants magnetization and the magnetic properties of rocks *PEPI* **162** pp 85-98
[9] -------- 1963 Topography District the Flores east (T503) *US Army*

Alternating Least Square Method for Decomposing Dance Golek Menak Tensor Data

Joko Sutopo

Department of Electrical Engineering, Universitas Teknologi Yogyakarta

Email: jksutopo@gmail.com

Abstract. Golek Menak Dance rhythms a body structure that is motion and geometric patterns that can be presented in the form of a vector, matrix and tensor. In this study the data obtained from the catcher dance Kinect sensor in the form of Bio Vision Format Hierarchy (BVH), which is a geometric data consisting of coordinate geometry (x, y, z), time as well as reinforcing the position of the body (skeleton). The motion data is compiled into a data tensor matrix decomposition process is then performed for a feature that can be obtained will be processed in motion classification. Alternating Least Square (ALS) methods are used in the data decomposition tensor.
Keywords: Golek Menak Dance; Alternating Least Square; Decomposition; Tensor data.

1. Introduction

Golek Menak Dance, which sometimes called Beksan Golek Menak is an adaption of Golek Menak puppet [1,2]. Such dance motion can be understood as motion reality which represents esthetic element and gesture movement with specific meaning [3]. Dance movement happen on the whole of body intellectually, emotionally and physically [4].

When the dance movement is happening, there is a dance element has a relation to the rhythm structure of the whole body movement pattern [3,5]. Symbol system has a closeness with art dance inside of the presentation likes motion, costume and makeup [6]. Estimating the dancer motion can be complicated which till now it is rare and not studied anywhere. Also, the dancer motion consists of array multidimensional data which needs adequate computation approach. This means that which need decomposition and normalization process in order the dance moves can be estimated. Also, since dance motion is a multidimensional array, therefore, decomposition of the dance motion data will result in high order tensor data which need to be resolved. Therefore, this study proposes an alternative method to estimate the dancer body through decomposition and normalization the dancer motion data.

2. Rudimentary

A tensor is the general form of scalar and vector [7]. Tensor has a geometric relation between vector operations, scalar in an array multidimensional process [8]. When the coordinate axis is rotated, the components change following new coordinate transformation; it is an analogue format with vector transformation. Therefore, a vector with a single subscript on the component is tensor rank 1. A tensor have rank n will have n subscript. Therefore, the tensor matrix can be represented by R-rank tensor [9].

Decomposition tensor is a method with a purpose to divide body movement horizontally and vertically [10], by combining the relation on each node. Decomposition tensor is a process of the departed frame from digital images into a smaller frame representing the components of the digital images [11]. When the digital images are divided into a smaller subsystem, it will result in different computation process, and different data consisted of complex matrices [12]. Some scholars have proposed a new approach to compute such complex matrices with a certain tensor rule which so-called

Content from this work may be used under the terms of the Creative Commons Attribution 3.0 licence. Any further distribution of this work must maintain attribution to the author(s) and the title of the work, journal citation and DOI.
Published under licence by IOP Publishing Ltd

CANDECOMP/PARAFAC (CP). Inside of CP model, matrix-based tensor data is formed from factors matrix which so-called Alternative Least Square (ALS) method [6,10].
On CP approach, a tensor X can be defined as equation 1. It contains A, B, C column, and λ which representing a component of factor matrix [1]. In geometric space, the R rank tensor has three types, e.g., Product Khatri-Rao (⊙), Kronecker Product (⊗) and Hadamard Product (*) [2]. Product Khatri-Rao (⊙) is used for dot estimation of vector dot calculation, Kronecker Product (⊗) is to measure and analyse the vector cross calculation, and Hadamard Product (*) is used for improving the quality of the matrices arrangement [13]. In our study, a combination of the three types is representing the tensor rule of CP tensor as in equation 2 to estimate the dancer motion [4,5].

$$X \approx [\![\lambda; A, B, C]\!] = \sum_{\lambda} \lambda_r a_r \circ b_r \circ c_r \quad (1)$$

Since dance motion is a multidimensional array, therefore, decomposition of the dance motion data will result in high order tensor data [7,14]. Therefore, to minimise the tensor data into simplified matrices and vectors, it needs to divide the multidimensional array into component A, B, and C. A represented the dancer body, B representing the background and C representing colour pixel [11]. The decomposition form of A, B, C is estimated with equation 2, 3 and 4.

$$A^T = (B \odot C)^{\dagger} T_{JKxI} \stackrel{\text{def}}{=} f_A(B, C) \quad (2)$$
$$B^T = (C \odot A)^{\dagger} T_{KIxJ} \stackrel{\text{def}}{=} f_B(C, A) \quad (3)$$
$$C^T = (A \odot B)^{\dagger} T_{IJxK} \stackrel{\text{def}}{=} f_C(A, B) \quad (4)$$

Where A, B and C are quadratic forms of the Alternating Least Squares (ALS) structure. The solution for finding components A, B and C are represented in Tensor mode, where A_1 is the first mode, A_2 is the second mode, and the third mode A_3 [15,16] is presented in equations 5, 6 and 7.

$$A \leftarrow A_1(C \odot B)(C^t C * B^t B)^{-1} \quad (5)$$
$$B \leftarrow A_2(C \odot A)(C^t C * A^t A)^{-1} \quad (6)$$
$$C \leftarrow A_3(B \odot A)(B^t B * A^t A)^{-1} \quad (7)$$

When the A, B and C are moving rotationally, there is a pseudo inverse implemented to each component. It is estimated with $M_A^{\dagger}, M_B^{\dagger}, M_C^{\dagger}$, as the pseudo-inverse of A^T, B^T and C^T are expressed by equations 8 and 9.

$$M_A^{\dagger} \otimes M_B^{\dagger} \otimes M_C^{\dagger} = M^{\dagger} \quad (8)$$
$$M^{\dagger} = (M^T M)^{-1} M^T \quad (9)$$

The $M^{\dagger}$ will result in a specific value which so-called the solution of Alternating Least Square (ALS) on the dance motion system with the primary function of the normalization method [17].

3. Proposed Method

Stages during the research are (1) motion capture; (2) classification and extract data; (3) data tensor X, Y, Z making; (4) decomposition and normalisation of motion dance tensor; (5) presenting data [9]. In the data extraction process, the data are extracted from the dataset using the BioVision Hierarchical (BVH) data. It is so-called data tensor X, Y, Z representing the decomposition and normalization of motion dance [10]. The flowchart of this research is given in Figure 1.

3.1. Motion Capture

On this step, the characteristic motion models are captured from the real Golek Menak dancer by sensor Kinect to get visual motion dance data. On motion capturing process [8], the appropriate set of calibration to adaptability position on motion capture process is needed. This calibration will determine the result of the captured motion data [12].

3.2. Data Classification and Extraction

The resulting capture of the whole motion dance is saved into BVH file format. Following [18] It is then conducted the classification of data sample for about ten motion dance sample which consisted 5 Jogetan motion and 5 Sabetan motion as presented in Figure 2. Based on the obtained data sample, the data extraction of the BVH file structure is conducted to estimate the X, Y, X matrix tensor which following [19].

3.3. Data Tensor X, Y, Z Making

The data tensor X, Y, Z making is the dimensioning process can be restructured from the tensor matrix of BVH file [20]. Motion dance processing is conducted by constructed multidimensional matrix or X, Y, Z, and tensor data to normalise the dance motion data based on the relation of each node to get the convergent optimum score.

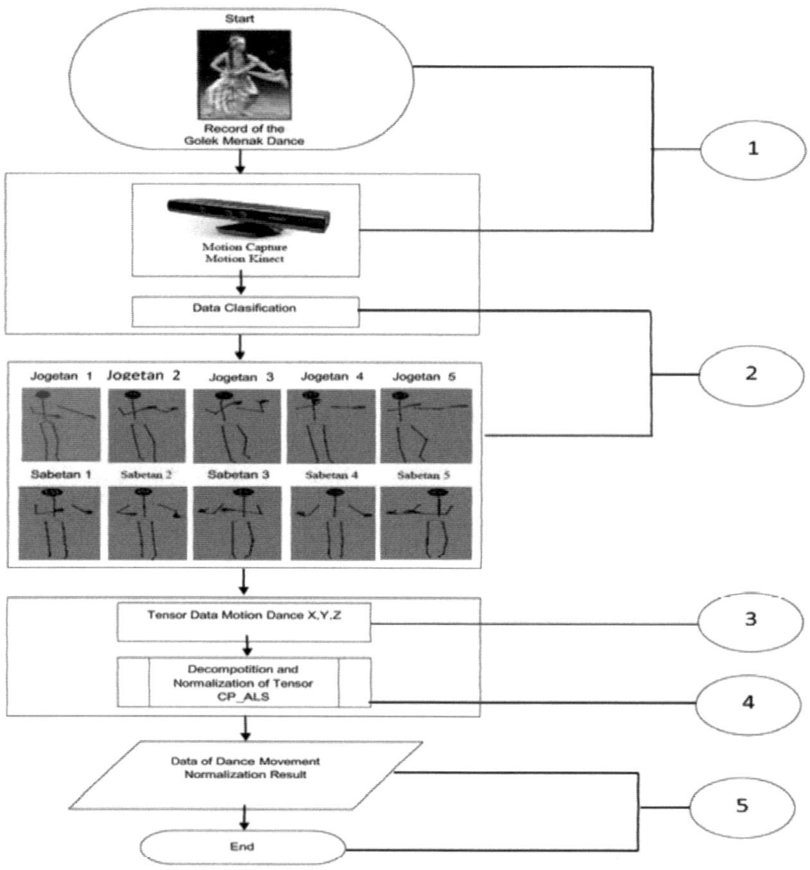

Figure 1. Research's Flow

In the flow diagram of the research in Figure 1 to generate the tensor data X, Y, Z, first performed the capture of the dance Golek terrifying motion capture motion (motion capture) Kinect sensor, which then produces data dance movement. Then from the data of dance movement is done by the classification process for Jogetan motion and Sabetan, which produce classification data. Furthermore from the data classification is done extract data to form data tensor X, Y, Z and then done the process of decomposition and normalization of tensor data by using CANDECOMP/PARAFAC_Alternating Least Square (CP_ALS) method, which then yields a result of normalization of data of dance movement.

3.4. Decomposition and Normalization of Motion Dance Tensor Data

The simplified X, Y, Z data tensor motion dance on decomposition process is conducted by dispersing the parts of tensor into some simple component [21] which divided the motion dance horizontally and vertically [19,22]. The decomposition and normalization of tensor X, Y, Z are represented by Figure 3.

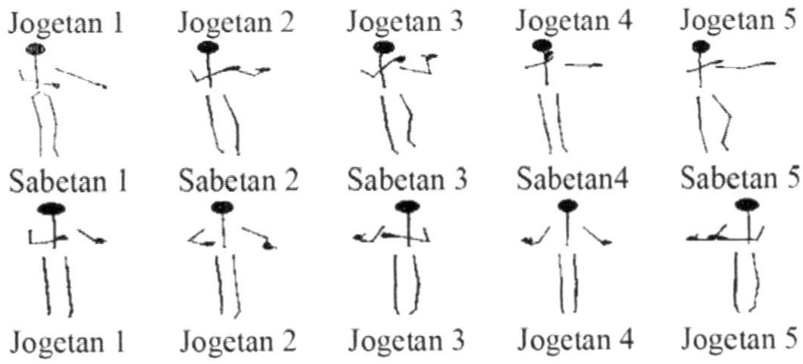

Figure 2. Data extraction process

$\mathrm{CP-ALS}\ (\mathrm{T}, 1, \mathrm{M}, \epsilon)$
$\mathrm{m} \leftarrow 0$
$\mathrm{D1} \leftarrow$ R eigen vector $\mathrm{X}_{(1)}$
$\mathrm{D2} \leftarrow$ R eigen vector $\mathrm{X}_{(2)}$
$\mathrm{D3} \leftarrow$ R eigen vector $\mathrm{X}_{(3)}$

repeat
$\mathrm{m} \leftarrow \mathrm{m}+1$
$\mathrm{D3} \leftarrow \mathrm{X}_{(3)}\ (\mathrm{D2} \odot \mathrm{D1})((\mathrm{D2}^{\mathrm{T}}\mathrm{D2}) * (\mathrm{D1}^{\mathrm{T}}\mathrm{D1}))^{\dagger}$
Column Normalization D3
$\mathrm{D2} \leftarrow \mathrm{X}_{(2)}\ (\mathrm{D3} \odot \mathrm{D1})((\mathrm{D3}^{\mathrm{T}}\mathrm{D3}) * (\mathrm{D1}^{\mathrm{T}}\mathrm{D1}))^{\dagger}$
Column Normalization D2
$\mathrm{D1} \leftarrow \mathrm{X}_{(1)}\ (\mathrm{D3} \odot \mathrm{D2})((\mathrm{D3}^{\mathrm{T}}\mathrm{D3}) * (\mathrm{D2}^{\mathrm{T}}\mathrm{D2}))^{\dagger}$
Column Normalization D1
until $\mathrm{m} > \mathrm{M}$ atau $\| \mathrm{X} - [\![\lambda; \mathrm{A}, \mathrm{B}, \mathrm{C}]\!] \| < \epsilon$
return $\lambda \in \mathbb{R}^{R}$; $\mathbf{D1} \in \mathbb{R}^{IXR}$; $\mathbf{D2} \in \mathbb{R}^{JXR}$; $\mathbf{D3} \in \mathbb{R}^{KXR}$;
Where That $\mathrm{X} \approx [\![\lambda; \mathrm{D1}, \mathrm{D2}, \mathrm{D3}]\!]$

Figure 3. Algorithm for decomposition and normalization of Tensor X, Y, Z

Where D1, D2, D3 represent dimensions of X, Y, Z. in the D3 Normalization, it is obtained from multiplication product of Khatri-Rao D2 and D1, wherein D2T and D1T is multiplication form of Hadamard's product. In D2 Normalization, the component matrix D2 is obtained from the multiplication product Khatri-Rao D3, and D1, wherein D3T and D1T is multiplication form of Hadamard's product [23]. In D1 Normalization, the component matrix D1 is obtained from the multiplication product Khatri-Rao D3, and D2, wherein D3T and D2T is multiplication form of Hadamard's product.

The last result from that solution is obtaining normalization of each component matrix D1, D2, D3 with $X \approx [\![\lambda; D1, D2, D3]\!]$ to estimate component matrix D1, D2 and D3. On the normalization process for finding the optimum solution, it needs to held iteration until reaching the excellent convergent which consisted of matrix component D1, D2, D3. The matrix T is then decomposition which create matrix T1 and being tested [24]. The normalization process used Alternating Least Square (ALS) to make diagonal vector λ for each component matrix D1, D2 and D3, that all is the component of matrix T1. The normalization process is looking for all vectors in an array of vector H as multiplier vector after normalised to find the convergent optimum score that differentiates the Jogetan and Sabetan after estimated with the algorithm in Figure 3.

3.5. Presenting Data

The last step of the decomposition process and normalization data is to process the output data into the convergent result of Sabetan and Jogetan motion.

4. Result and Discussion

4.1. BVH-Based Structured Data Representing Dance Motion

Focus object on this research is to find the method of dancer motion to be estimated into tensor data and finally the whole node of the frame representing the dancer motion in skeleton type. The decomposition and normalization of motion dance tensor have a goal to get tensor data containing X, Y, Z nodes that will decide based on offset score of the captured motion. Offset score is a pure value from the result of motion capture, and the function as the input score of X, Y, Z tensors that is obtained from data extraction of BVH structured file.

BVH structured file (Figure 2) consist of whole joints (e.g., Head, Neck, Chest, Hips, LeftHip, left knee, left ankle, right, right knee, right ankle, left shoulder, left collar, left elbow, left wrist, leftwingers, RightShoulder, RightCollar, right elbow, right wrist, RightFingers) [21]. The whole skeleton component is a skeleton frame representing a digital skeleton tensor body. The whole bone can be defined as Matrix M with the elements consisted of the whole structure joint on motion dance. The score in each motion dance's matrix element is determined by offset score of the whole motion dance.

$R_x R_y R_z$ Is homogeneous rotation matrix representing the matrix of coordinates contains X, Y, Z composite rotational vector which organised in a vector structure? Hierarchy joint value determines normalization data motion dance on extract data process on the primary hierarchy structure (parent), which is also on the translational motion in a diagonal matrix structure.

For data processing, the value of matrix element of motion dance data is determined by the offset score of the whole motion dance on BVH structure file especially for each joint [25]. The table below shows the offset score of each joint MhipsMLeftHipMRightHipMChest motion dance as presented in Table 1.

Table 1. Offset values for Joint MhipsMLeftHipMRightHipMChest

Joint	Motion	Offset		
		X	Y	Z
Hips	Jogetan 1	18.8200	24.1040	-233.820
	Jogetan 2	17.0220	23.5580	-235.0890
	Jogetan 3	14.9010	22.6860	-235.8390
	Jogetan 4	15.1860	22.4520	-235.4210

	Jogetan 5	15.6650	22.6810	-234.6420
	Sabetan 1	24.5940	27.4390	-229.3940
	Sabetan 2	24.4450	27.7390	-229.2510
	Sabetan 3	25.2620	27.4240	-229.7920
	Sabetan 4	25.6570	27.7360	-229.8660
	Sabetan 5	25.5500	27.6520	-230.2050
Left Hip	Jogetan 1	7.1180	-6.9710	1.7750
	Jogetan 2	6.8850	-7.0500	1.8220
	Jogetan 3	7.3630	-6.6990	2.5160
	Jogetan 4	7.2140	-6.7500	1.9870
	Jogetan 5	7.4560	-7.1060	1.6220
	Sabetan 1	6.6430	-6.2350	1.3300
	Sabetan 2	6.5180	-6.1110	1.2620
	Sabetan 3	6.6050	-6.1790	1.3450
	Sabetan 4	6.5690	-6.1750	1.2360
	Sabetan 5	6.6260	-6.1020	1.3900
Right Hip	Jogetan 1	-7.2790	-6.7400	1.5560
	Jogetan 2	-7.3280	-6.8460	1.4920
	Jogetan 3	-7.4200	-6.7670	2.0420
	Jogetan 4	-7.3890	-6.8800	1.9300
	Jogetan 5	-7.5990	-7.0180	1.8870
	Sabetan 1	-6.3310	-6.4970	1.5460
	Sabetan 2	-6.1820	-6.3510	1.6230
	Sabetan 3	-6.2510	-6.4420	1.5820
	Sabetan 4	-6.1340	-6.4630	1.5730
	Sabetan 5	-6.1770	-6.5280	1.4470
Chest	Jogetan 1	0.0000	5.4270	-5.4270
	Jogetan 2	0.0000	5.2800	-5.2800
	Jogetan 3	-0.0000	5.1320	-5.1320
	Jogetan 4	0.0000	5.3340	-5.3340
	Jogetan 5	0.0000	5.3960	-5.3960
	Sabetan 1	0.0000	5.9070	-5.9070
	Sabetan 2	0.0000	5.9270	-5.9270
	Sabetan 3	0.0000	5.9090	-5.9090
	Sabetan 4	-0.0000	5.8750	-5.8750
	Sabetan 5	-0.0000	5.9070	-5.9070

4.2. Decomposition and Normalization of X, Y, Z Tensors

In this study, the tensor data is represented by the T matrix, which is then decomposed to create a T1 matrix and normalized using the Alternating Least Square (ALS) method through the diagonal vector λ for each matrix component D1, D2 and D3. The normalization process is done by finding all the vectors in the 1×114 H dimension vector array in Table 2 simultaneously [26]. Whereas the vector H is the diagonal vectorization of the T1 matrix, as the multiplier vector to obtain the normalization of all components of the T1 matrix.

Table 2. Vector Normalization process

1	2	3	...	114
0.2439	0.3443	-2.4603	...	0.3382

The normalized results in Table 3 are obtained from the normalization process to find the optimum solution, by iterating to reach the optimum convergence for the matrix components D1, D2, D3 processed by the algorithm in Figure 3.

Table 3. Normalization result

λ.D1		λ.D2		λ.D3	
3x1		50x1		61x1	
1	0.0977	1	0.1439	1	0.1323
2	0.1379	2	0.1418	2	0.1313
3	-0.9856	50	0.1415	61	0.1355

Based on the test result of decomposition and normalization, the convergent optimum score is standard to score for the motion of Golek Menak dance on Jogetan and Sabetan motion from T1 matrix. Normalization score is the way to understand the effect of motion dance on 3-dimension X, Y, Z form. It is suitable for the other study [27] that decomposition and normalization can gain convergent optimum score.

Our study finding is a model which can estimate the digital pattern of dancer body movement through tensor decomposition and normalization approach. It used the tensor rule of CP-ALS method by processing the data from BioVision Hierarchical (BVH) matrix to estimate the whole structure of the dance moves. The dancer gesture movement can be estimated following new coordinate transformation to gain tensor matrix represented by R-rank tensor. The CP-ALS work on the basis that the matrix-based tensor data is formed from factors matrix of Alternative Least Square (ALS) by combining the Product Khatri-Rao (⊙), Kronecker Product (⊗) and Hadamard Product (*) to improve the quality of the matrices arrangement. The decomposition of the dancer motion has resulted in three components, e.g., A, B, and C which also used inversely with X, Y, and Z. X represented the dancer body, Y representing the background and Z representing the Color. This means we use A, B, C and X, Y, Z for the same purposes and meaning in this study.

5. Conclusion

Through five processes from motion capture to data decomposition and normalization, until it is presented, our method can process the data into decomposition and normalization process. In the data classification and extraction, we have successfully conducted the classification to get 5 Jogetan motion and 5 Sabetan motion, and then we continue to estimate the X, Y, X matrix tensor and normalize the dance motion data based on the relation of each node to get the convergent optimum score. By combining the three tensor rules, we can obtain normalized matrices with the optimum convergent result. Finally, our decomposition and normalization approach can process the output data into the converged result to differentiate two types of poses, e.g., Sabetan and Jogetan pose.

Based on the test result of decomposition and normalization, we find that our approach can differentiate both poses since the convergent optimum score is standard to score for the motion of Golek Menak dance on Jogetan and Sabetan motion from T1 matrix. We also found that the normalization score is the way to understand the effect of motion dance on 3-dimension of X, Y, Z form.

The topic of tensor decomposition will bring advantage for other researchers such as especially for those interested in studying the decomposition and normalization of dance moves. Also, our study about the use of the tensor rule of CP-ALS method also can be expanded by combining other tensor rules beside Product Khatri-Rao (⊙), Kronecker Product (⊗) and Hadamard Product (*) in other applications. We also found that the model can be improved by adding other methods such as genetic algorithm, neural network and/other artificial intelligence estimation or expert system which suitable to measure and estimate dancer body through another approach besides tensor rule to differentiate the dancer poses.

References

[1] Sutopo, J., Susanto, A., Santosa, I., & Adji, T. B. (2015). Feature Reduction of Wayang Golek Dance Data Using Principal Component Analysis (PCA). *Advances in Statistical Methods in Engineering, Science, Economy, Education and Disaster Management*, *1*, 60.

[2] Nurhayati, E., & Andayani, R. D. (2016). Javanese Traditional Menu and Bedaya Dance as the Cultural Tourism Attractions of Magersari and Yogyakarta Kingdom. *HUMANIORA*, *20* (2).

[3] Hadi, Y. S. (2016). The Continuity and Development of the Yogyakarta Style Classical Dance as the Legitimacy of the National Cultural Heritage. *Mudra*, *28*(1).

[4] Saearani, M. F. T. B., & Cahyono, A. (2016). The Inheritance of Malay Palace Dance in Malaysia and Yogyakarta Classical Dance in Indonesia. *Harmonia: Journal of Arts Research and Education*, *16*(1), 75-86.

[5] Martono, H. (2016). Environmental Choreography Enriches the Creation Methods of Dances in Indonesia. *Mudra*, *27*(2).

[6] Sumiati, L. (2016). The Transformation of Jayengrana Dance from the Perspective of Anthony Giddens's Structuration Theory. *TAWARIKH*, *5*(2).

[7] Schouten, J. A. (2013). Ricci-calculus: an introduction to tensor analysis and its geometrical applications (Vol. 10). Springer Science & Business Media.

[8] Xiao, J., Feng, Y., Ji, M., Yang, X., Zhang, J. J., & Zhuang, Y. (2015). Sparse motion bases selection for human motion denoising. *Signal Processing*, *110*, 108-122.

[9] Dai, Y., Li, H., & He, M. (2014). A simple prior-free method for non-rigid structure-from-motion factorization. *International Journal of Computer Vision*, *107*(2), 101-122.

[10] Guo, Y., Tao, D., Cheng, J., Dougherty, A., Li, Y., Yue, K., & Zhang, B. (2016). Tensor Manifold Discriminant Projections for Acceleration-Based Human Activity Recognition. *IEEE Transactions on Multimedia*, *18*(10), 1977-1987.

[11] Kopriva, I., Popovic-Hadzija, M., Hadzija, M., & Aralica, G. (2014). *U.S. Patent Application No. 14/221,017*.

[12] Adolphs, R., Nummenmaa, L., Todorov, A., & Haxby, J. V. (2016). Data-driven approaches in the investigation of social perception. *Phil. Trans. R. Soc. B*, *371*(1693), 20150367.

[13] Schatz, M. D., Low, T. M., Van De Geijn, R. A., & Kolda, T. G. (2013). Exploiting symmetry in tensors for high performance. *arXiv preprint arXiv:1301.7744*.

[14] Kolda, T. G., & Mayo, J. R. (2014). An adaptive shifted power method for computing generalized tensor eigenpairs. *SIAM Journal on Matrix Analysis and Applications*, *35*(4), 1563-1581.

[15] Bader, B. W., & Kolda, T. G. (2006). Algorithm 862: MATLAB tensor classes for fast algorithm prototyping. ACM Transactions on Mathematical Software (TOMS), 32(4), 635-653.

[16] Bhatt, V., & Kumar, S. (2014). A Cas Aided Survey Of CP Decomposition And Rank-1 Approximation Of A 3rd-Order Tensor. Palestine Journal of Mathematics, 3(2).

[17] Kotropoulos, C., & Pitas, I. (2016). Multimedia social search based on hypergraph learning. *Graph-Based Social Media Analysis*, *39*, 215.

[18] Rochani, E. (2014). Fenomena Tari Diskotik Di Era Globalisasi (Sebuah Studi Tentang Perkembangan Tari Di Yogyakarta) (Doctoral dissertation, Universitas Negeri Yogyakarta).

[19] Chen, Y., Liu, Z., & Zhang, Z. (2013). Tensor-based human body modelling. In *Proceedings of the IEEE Conference on Computer Vision and Pattern Recognition* (pp. 105-112).

[20] Vomlel, J., & Tichavský, P. (2014). On the tensor rank of conditional probability tables in Bayesian networks. *arXiv preprint arXiv:1409.6287*.

[21] Li, J., Li, C., & Cichocki, A. (2015). Canonical polyadic decomposition with assistant information for a brain-computer interface. *IEEE Journal of biomedical and health*

informatics.

[22] Sitharam, M., Tarifi, M., & Wang, M. (2016). Combinatorial rigidity of Incidence systems and Application to Dictionary learning. *arXiv preprint arXiv:1603.04109*.

[23] Yanuartuti, S. (2015). Revitalisasi Pertunjukan Wayang Topeng Jati Duwur Jombang Lakon Patah Kuda Narawangsa (Doctoral dissertation, Institut Seni Indonesia Surakarta).

[24] Tang, K.T. (2007). Mathematical Methods for Engineer and Scientists 3. Berlin: Springer.

[25] Hasan, H., & Abdul-Kareem, S. (2014). Static hand gesture recognition using neural networks. *Artificial Intelligence Review*, *41*(2), 147-181.

[26] Hidayat, Robby. 2005. Wawasan Seni Tari, Pengetahuan Praktis Bagi Guru Tari. Malang: Jurusan Seni dan Desain Fakultas Sastra UNIMA.

[27] Sari, Y. (2015). Penerapan Model Kooperatif Dalam Pembelajaran Sejarah Berbasis Nilai Budaya Lokal Untuk Meningkatkan Prestasi Belajar Dan Karakter Siswa Kelas Xi Smk N 8 Surakarta. *Candi*, *9*(1), 73-89.

IOP Conf. Series: Journal of Physics: Conf. Series **1090** (2018) 012058 doi:10.1088/1742-6596/1090/1/012058

Influence of Electrical Properties on Radiation and Emission to Pinch Radius Thermal Plasma Device

R F Syahputra[1], R Farma[1], Saktioto[1] , N D Nawi[2], N A Rashid[2], F D Ismail[2], J Ali[2]

[1]Dept. of Physics, FMIPA Universitas Riau, Pekanbaru, Indonesia
[2]Laser Centre, Ibnu Sina ISIR, Universiti Teknologi Malaysia
Johor Bahru, Malaysia

*Corresponding author's e-mail: romifadlisyahputra@yahoo.com

Abstract. Plasma focus (PF) generation has advanced the development of high-density, high-temperature plasma that emits intense radiation. The radiation output and emission spectra depend on the operating parameters of the PF device. The ability to focus plasma governs the pinch radius and electrical properties at the truncated end of the PF device. In this study, the simulation system RADPF 5.15 FIB was used to investigate the pinching properties of the neon ion beam generated using an AECS PF-1 device. The pinch radius was the dominant factor affecting the current density produced.

1. Introduction

Over the last two decades, plasma focus (PF) devices have advanced the development of energy systems by producing a strong interaction between electrons and ions in the plasma. Moreover, PF is known as a clean, safe and sustainable energy source [1]. Thus, there have been many experimental and numerical studies on fundamentals, diagnostics and simulations of PF. These studies have reported on the system set-up, dynamics of radiation, radiative collapse and industrial applications of PF [2], including coating technology [3], ion implantation [4], material technology [5,6], microelectronic lithography [7] and medical devices [8].

PF system dynamics can be optimised to produce an ion beam suitable for a wide range of applications. For medical devices, the ion beam was tailored to penetrate the body to a certain depth, quickly release high energy to the surrounding area, and then extinguish [8]. The ion beam can be adjusted so that less energy is dissipated, and intense energy is released over a localized area. Research has shown that the ion beam can deliver radiation to a cancerous focal point, achieving better treatment results than X-ray therapy [8]. In traditional therapy, side effects are related to the field of irradiation, with a wider (narrower) field resulting in more severe (milder) side effects. The PF ion beam minimises side effects by delivering radiation to a highly focussed area.

PF devices are well recognized as ion beam sources that produce characteristic radiation already used in numerous applications. These devices use the principles of electromagnetic compression and confinement, in which the Lorentz force acts on electrical current flowing in the plasma. If the magnetic pressure exceeds the pressure of the particle, a focusing effect known as pinch

Content from this work may be used under the terms of the Creative Commons Attribution 3.0 licence. Any further distribution of this work must maintain attribution to the author(s) and the title of the work, journal citation and DOI.
Published under licence by IOP Publishing Ltd

occurs [9]. High-quality focus emits intense pinch radiation, as illustrated in Fig. 1. PF devices have been shown to produce energetic ion beam emission. This study aimed to observe and report data on the pinch radius profiles and their relationship to current density. This information would be a significant contribution to the advancement of medical applications.

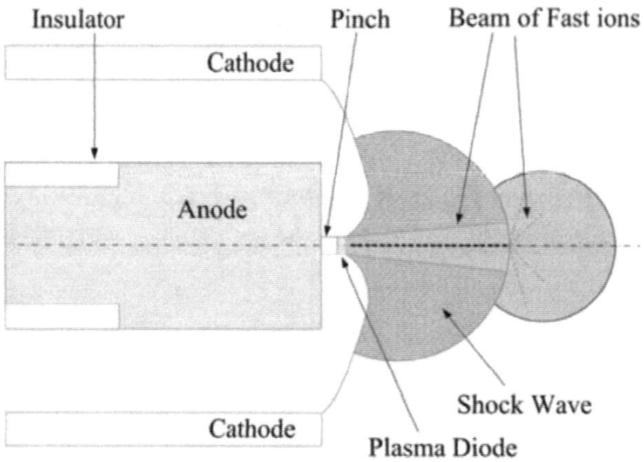

Figure 1. Schematic of radiation from the plasma focus (PF) device.

Bhuyan [10] characterized a neon ion beam using a 2.2-kJ Mather-type PF powered by a 7.1-μF high-energy storage capacitor with a charging voltage of 25 kV. In that study, neon ion beam emission was highly dependent on pressure. The optimal pressure of 0.3 Torr resulted in beam energy that ranged from 18 to 1000 keV. The maximum ion density was achieved at 5 × 10^{19} m^3, and higher ion populations occurred below 180 keV. A pressure of 0.3 Torr is sufficient for PF discharge dynamics that produce strong pinching. At this pressure, the maximum current discharge was achieved, thus transferring maximum energy to the plasma. Increasing the pressure above 0.3 Torr decreased the current sheath velocity due to a greater sheath mass; thus, the focusing was weaker, and the ion beam emission was less intense.

2. Theoretical Consideration: Electrical Discharge and Pinch Radius Profiles

It is important to understand the production of the ion beam to characterise the properties of the ions emitted from PF devices. A plasma diode is located at the instant acceleration gap (Fig. 2), where a beam of fast deuterons is generated by the diode. The beam interacts with the hot, dense plasma of the focus pinch column, producing fusion neutrons. According to Gribkov [6], the ion beam can be described by calculating the neutron yield, Y_n.

$$Y_n = Y_{n-t} = C_n n_i I_{pinch}^2 z_p^2 (\ln(b / r_p))\sigma / \sqrt{U} \tag{1}$$

The calibration constant for the neutron is C_n = 8.54 × 10^8. This constant was obtained from a graph of all available measured Y_n data at an experimental point of 0.5 MA. n_i is the pinch ion density, I_{pinch} is the current through the pinch, r_p and z_p are the final pinch radius and length, respectively, b is the cathode radius, r_p is the cross-section of the D–D fusion reaction branch, and U is the disruption-induced diode voltage [12]. Instabilities during pinching cause a strong electric field and accelerate the ion beam. Lee [13] defined the beam passing through a plasma target as having a cross-section equal to the reaction rate (i.e. beam ion number flux times number of target particles). Thus, by the definition of the beam cross-section, Equation 1 can be rewritten as

$$Y_{b-t} = (n_b v_b)[n_i(\pi r_p^2 z_p)\tau]\sigma = (J_b \tau)\sigma n_i(\pi r_p^2 z_p) \tag{2}$$

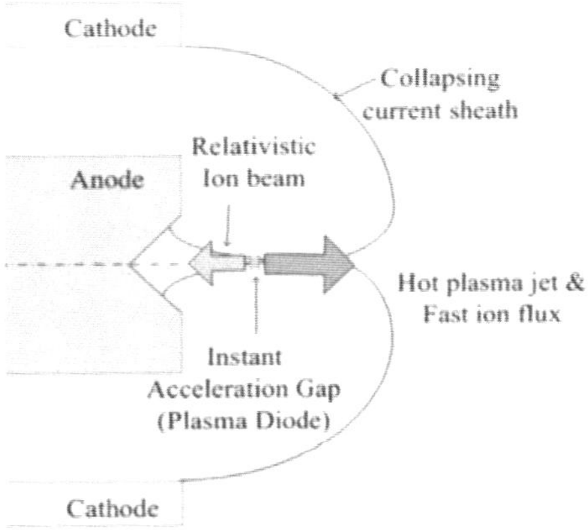

Figure 2. Particle acceleration mechanisms inside the PF device.

From Equations 1 and 2, the beam kinetic energy and pinch inductive energy can be used to determine the number of beam ions per unit plasma volume traversed, n_b, and the beam ion speed, v_b. These parameters are represented by Equations 3 and 4 [11].

$$n_b = \frac{N_b}{\pi r_p^2 z_p} = \left(\frac{\mu f_e}{2\pi^2 m_p M}\right)\left(\frac{\ln(b/r_p) I_{pinch}^2}{r_p^2 v_b^2}\right) \tag{3}$$

$$v_b = \left(\frac{2eZ_{eff}U}{m_p M}\right)^{1/2} \tag{4}$$

Equations 3 and 4 are discussed using neon gas properties. The interrelation of beam kinetic energy and pinch inductive energy are governed by f_e, the fraction of energy converted from pinch inductive energy to beam kinetic energy. These equations also include the effective charge, Z_{eff}, mass number, M, mass of a proton, m_p, charge of an electron, e, and diode voltage, U. For neon, the mass number of an ion is 20, and m_p is 1.673×10^{27} kg. Lee [11] performed extensive data fitting to equate diode voltage and maximum induced voltage of the pre-pinch radial phase, $U = V_{max}$. The number of beam ions, n_b, and the beam ion speed, v_b, are well characterized; thus, these properties can be correlated to give the definition of ion beam flux, J_b. Equations 3 and 4 are combined to give Equation 5, where J_b has units of current density for quantity of ions $m^{-2}s^{-1}$.

$$J_b = n_b v_b \tag{5}$$

The ion beam cross-section is similar to that of the pinch, with a small divergence of about 3°. The narrow beam exits along the pinch axis and is considered the post-pinch of a shock wave. This estimation of the plasma at the exit point allows the ion beam to be characterized by the ion number per unit cross-section, described by the flux, J_b, and fluence, $J_b\tau$ [11–13].

$$J_b = (2.75\times 10^{15})\left(\frac{f_e}{MZ_{eff}}\right)^{1/2}\left(\frac{\ln(b/r_p) I_{pinch}^2}{r_p^2\sqrt{U}}\right) \tag{6}$$

$$J_b\tau = (2.75\times 10^{15})\left(\frac{f_e}{MZ_{eff}}\right)^{1/2}\left(\frac{\ln(b/r_p) I_{pinch}^2}{r_p^2\sqrt{U}}\right)\tau \tag{7}$$

3. Methodology

Neon was used in the RADPF5.15 FIB with an AECS PF-1 as the working chamber. The PF device configuration is described in Table 1. The current densities and pinch radius were plotted against pressure. Four applied voltage values were used: 14, 15, 16 and 17 kV. The discharge current signal comprises the system dynamics and the information on properties of multi-radiation, so, it was identified as the points of reference. The profiles of the discharged currents start from the breakdown phase until the end of pinch phase; and along the phases, it will signify the Joule heating, radiative emissions, and the expended column current flow that abruptly transited from the narrow pinch. In instance, when the computed current profile was fitted with the measured current profile, the system ready to be run for detailed experimental calibration. The numerical experiments conducted gives significant pattern on the results in the RADPF5.15 FIB. The data obtained were then accumulated by using Microsoft visual basic.

Table 1. Plasma focus device configuration used in the RADPF 5.15 FIB simulation

PF Device Parameter	AECS PF-1
Energy, E (kJ)	2.8
Voltage, V (kV)	13–17
Inductance, L_0 (nH)	1600
Capacitor, C_0 (μF)	25
Resistance, r_0 (mΩ)	77
Cathode radius, b (cm)	3.2
Anode radius, a (cm)	0.95
Length of anode, z_0 (cm)	16

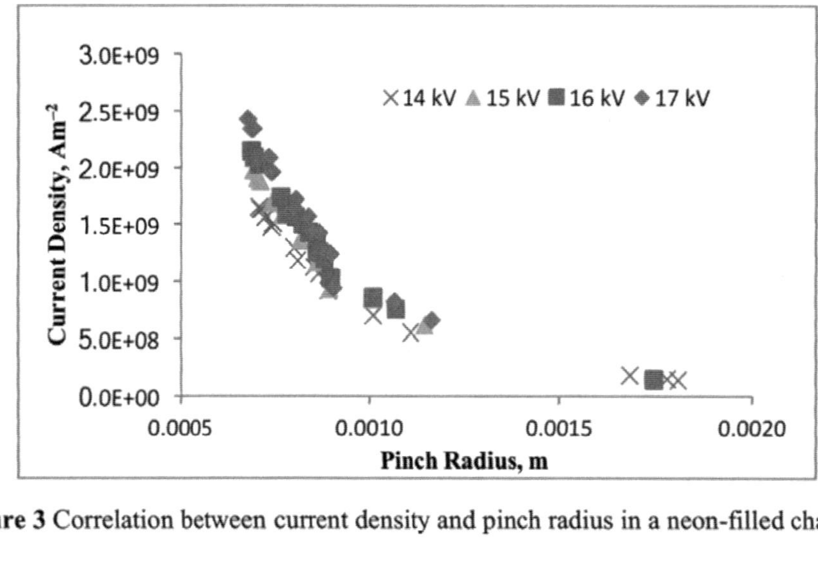

Figure 3 Correlation between current density and pinch radius in a neon-filled chamber

4. Results and Discussion

The current densities and pinch radius correlated closely. The pinch radius reached a maximum when the current density was at its minimum. This was true for all the applied voltages used in the system. The smallest pinch radius for an applied voltage of 14 kV was 0.7 mm (0.5 Torr), resulting in the highest current density of 1.7×10^9 Am^{-2}. However, at 0.85 Torr, the pinch radius increased rapidly, causing the current density to decrease. The smallest pinch radius for an applied voltage of 15 kV was 0.7 mm (0.6 Torr), and the current density was 2.0×10^9 Am^{-2}. At 16 and 17 kV, the pinch radius was also 0.7 mm, and the current density was 2.1×10^9 and 2.4×10^9 Am^{-2}, respectively. Interestingly, the smallest pinch radius for all the applied voltages was 0.7 mm. The current density varied with pressure and applied voltage. These results indicate that energy savings can be achieved by operating at lower voltages. In other words, there is no need to use a higher voltage when the same pinch radius occurs at a lower voltage.

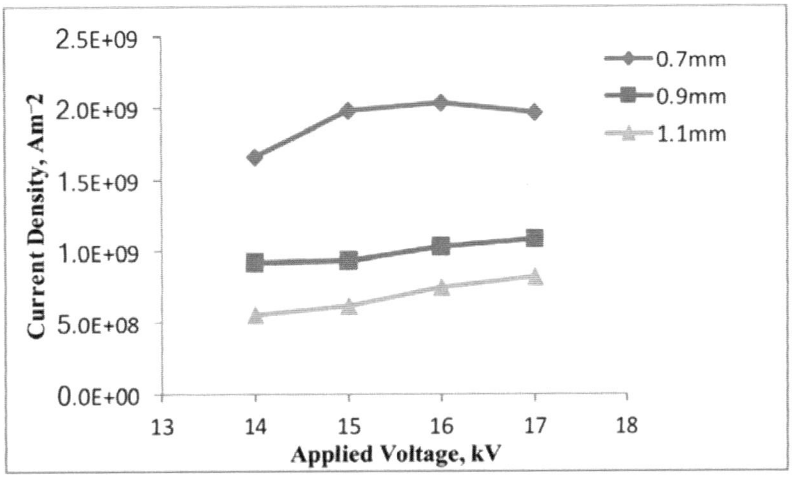

Figure 4 Correlation between current density and applied voltage at three pinch radius values for the neon-filled chamber of an AECS PF-1 device

To obtain a better understanding of the current density behaviour, the data for current density and pinch radius were correlated for applied voltages of 14–17 kV. The tabulated data show a clear trend of decreasing current density with increasing pinch radius (Fig. 3). It is notable that current density was fairly consistent at a given pinch radius, even for different applied voltages. Thus, it was expected that pinch radius would significantly affect current density, and the applied voltage can be used to optimise the pinching phenomenon. The pinch radius was observed for the AECS PF-1 operated with Neon (Fig. 4).

The results show that the variation in current density was small over the applied voltage range, for all of the selected pinch radius values. Thus, pinch radius can be used to achieve the desired current density, regardless of the applied voltage. These results also confirm that Lee's model can successfully simulate ion beam dynamics in an AECS PF-1 device.

5. Conclusion

The numerical simulation method using Lee's code is an effective tool for investigating electrical discharge properties in PF devices based on theoretical calculations. RADPF5.515 FIB showed considerable variation in ion beam behaviour as a function of pressure, as evidenced by the range of pinch radius values produced under various conditions. The numerical data obtained for the AECS PF-1 system provide insight into the correlation between pinch radius and current density.

Acknowledgment
We thank the Ministry of Research and Technology, Directorate General of Higher Education Indonesia through Lembaga Penelitian dan Pengabdian Kepada Masyarakat Universitas Riau, Pekanbaru, Indonesia, for funding, and the Laser Centre, Ibnu Sina ISIR, Univerisiti Teknologi, Malaysia, for laboratory facility support of this research in 2017.

References

[1] Bernard A *et al* 1998 Scientific status of plasma focus research *J. Moscow Phys. Soc.* **8** 93

[2] Lee S 2012 From electric birth to streaming death of the plasma focus knowledge and applications *The plasma focus - enhancing knowledge and application potentials:. Proceedings of IPFS Conference (Bangkok, Thailand, 29 March - 1 April 2012)* pp 1-14

[3] Hassan M, Qayyum A, Ahmada R, Rawat R, Lee P, Hassan S, Murtazad G, and Zakaullah M 2009 Dense plasma focus ion-based titanium nitride coating on titanium *Nucl. Instrum. Meth. B.* **267** 1911

[4] Sadiq M., Ahmad S, Shafiq M. and Zakaullah M 2006 Nitrogen ion implantation of silicon in dense plasma focus. *Nucl. Instrum. Meth. B.* **252** 219

[5] Ghareshabani E, Rawat R, Verma R, Karamat S and Sobhanian S 2010 Low energy repetitive miniature plasma focus device as high deposition rate facility for synthesis of DLC thin films. *Appl. Surf. Sci.* **256** 4977

[6] Gribkov V *et al* 2006 PF-6 an effective plasma focus as a source of ionizing radiation and plasma streams for application in material technology, biology and medicine. *Nukleonika.* 2006. **51** 55

[7] Moreno C, Vénere M, Barbuzza R, Fresno M D, Ramos R, Bruzzone H, González F P J, and Clausse A 2002 Industrial applications of plasma focus radiation *Braz. J. Phys.* **32** 20

[8] Alonso J R Medical applications of nuclear physics and heavy-ion beams. Nucleus *nucleus collisions 2000: Proceedings of 7th International Conference, NN2000* (Strasbourg, France, July 3-7, 2000) *Nucl. Phys. A* **685** pp 454-471

[9] Bing S 2000 *Comparative Study of Dynamics and X-ray Emission of Several Plasma Focus Devices* (Ph.D. Thesis. Nanyang Technological University)

[10] Bhuyan M, Neog N, Mohanty S, Rao C and Raole P 2008 Characterization of the neon ion beam emitted from plasma focus device **208** 012126

[11] Lee S and Saw S H 2013 Plasma focus ion beam fluence and flux—For various gases. Physics of Plasma, 2013. **20** 062702

[12] Saw S H 2014 Plasma focus ion beam- scaling laws *Int. J. Mod. Phys. Conf. Ser.* **32** 1460317

[13] Lee S and Saw S H 2012 Plasma focus ion beam fluence and flux—Scaling with stored energy. *Physics of Plasmas* **19** 112703

Service Desk Implementation with Information Technology Infrastructure Library Framework (Study Case Financial Company)

Dwi Handoko[1] and Abba Suganda Girsang[2]

[1] Computer Science Department, BINUS Graduate Program-Master of Computer Science, Bina Nusantara University, Jakarta, Indonesia 11480, Email: dokodwi@gmail.com

[2] Computer Science Department, BINUS Graduate Program-Master of Computer Science, Bina Nusantara University, Jakarta, Indonesia 11480, Email : agirsang@binus.edu

Abstract. This paper proposes a service desk to handle two important issues in financial company using the Information Technology Infrastructure Library (ITIL) Framework, i.e. Single Point of Contact (SPOC) and Service Level Agreement (SLA) issues. SPOC is a gateway to information needs of both users and company staff, particularly the IT staff, while SLA defines the responsibilities of the parties where such services work and provides coverage for services provided to the client to achieve client satisfaction. The service desk is built by focusing in service operation. A service desk has been successfully built in this study to make user's problem controllable. The result also shows that the user's problem can be solved faster in many cases.

Keywords: Information Technology Infrastructure Library (ITIL), Single Point of Contact , Service Level Agreement , Service Desk

1. Introduction

As IT Division and organization are shifting from technology-based to service-based management models, investing in a credible best practice framework such as ITIL (Information Technology Infrastructure Library) becomes more important for a Service Desk [1][2] . According to ITIL, Service Desk is "The single point of contact between users and IT Service Management". ITIL tasks include handling incidents and requests, as well as providing an interface for other ITIL processes while the primary functions of the Service Desk are incident control, life cycle management of all service requests, and communication with users.

There are many challenges that relate directly to ITIL implementations. Chen and Chou [3] noted in their study on information technology service management that there are ten top challenges. These are: culture shift, integration with current process, ITIL/ ITSM-related knowledge, appropriate management tool, clear measurement target, project within budget, insufficient internal professional staff, project on time, managing consultants, and incapable or inexperienced consultants. These are considered challenges due to their connection with reasons for failure. Zhu [4] pointed out that the top ten reasons for failure in ITIL implementations are ITIL is rigid, ITIL is frequently hijacked by management, ITIL takes too long, change management fails, too much IT focus/not enough business focus, ITIL is not an end itself, misunderstanding that ITIL is not mandatory, staff takes the implementation too seriously, ITIL cost too much, and ITIL drags the enterprise improvement process down. Thus, overcoming the challenges and understanding the top reasons why failure occurs will be helpful in generating strategies to minimize the risks around these challenges. Looking at the top five challenges above, culture shift, appropriate management tool, and clear measurement

Content from this work may be used under the terms of the Creative Commons Attribution 3.0 licence. Any further distribution of this work must maintain attribution to the author(s) and the title of the work, journal citation and DOI.
Published under licence by IOP Publishing Ltd

target are all directly related to Service Desk. Service desk as the single point of contact will be very useful for culture shift.

ITIL (The IT Infrastructure Library) is essentially a series of documents that are used to support the implementation of a framework for an IT Service Management. This custom is able to define how Service Management is applied within an organization. ITIL is organized into five core publications that revolve around the service lifecycle. These provide best practice guidance for an integrated approach to IT service management. The five core titles are Service Strategy, Service Design, Service Transition, Service Operation, and Continual Service Improvement.

2. Proposed Method

A Service Desk is a functional unit made up of a dedicated number of staff responsible for dealing with a variety of service activities, usually made via telephone calls, web interface or automatically reported infrastructure events.

The Service Desk is a vitally important part of an IT organization and should be the Single Point of Contact (SPOC) for IT users on a day-to-day basis. The primary aim of the Service Desk is to provide a single point of contact between the services being provided and the users.

The exact nature, type, size and location of a service desk will differ, depending upon the type of business, number of users, geography, and complexity of calls, scope of services and many other factors. There are many ways of structuring service desks and locating them. The correct solution will vary for different organizations.

The new service functions of IT Service Desk are: BMC [5-8]

a. The Service desk is the central point of contact for all clients and users of IT services. People call the service desk if an outage occurs, if they have a request for something new (such as a printer installation), or if a change is needed. The service desk resolves and fulfills incidents and requests as much as possible and coordinates and routes those requiring escalation. Figure 1 shows the new service function of IT Service Desk [9]

Figure 1 New Service Desk Function

b. Technical management knows what resources the IT organization has at its disposal. Getting the right people involved in serving the users, testing, and improving the IT services makes the service operation stage easier for the IT staff. To obtain staff with good skill in IT service desk, the value of each IT Service Desk person was identified. There some IT services values : respect, value, trust, engage, grow and team [9].

c. Operation management includes scheduling all activities and managing all resources. Operations are responsible for routine IT tasks, such as system administration and preventive and monitoring activities, categorization, and prioritization. Figure 2 shows the categorization and prioritization.

Figure 2 Categorization and Prioritization

In the operation management, the IT Service desk also maintains customer satisfaction. Figure 3 presents the example of customer satisfaction.

d. Application management maintains applications and provides technical support, as well as subject matter expertise, throughout the application lifecycle.

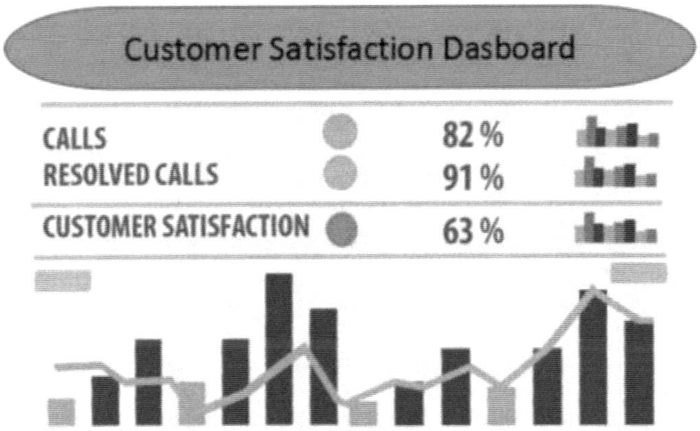

Figure 3 Customer Satisfaction

3. Analysis Result

a. Single Point of Contact

New IT service desk functional organization at ITSO Division is a part of the enhancement of the existing service to users, Figure 4 reflects the new function of service desk under the Data Processing and Recovery Center Department.

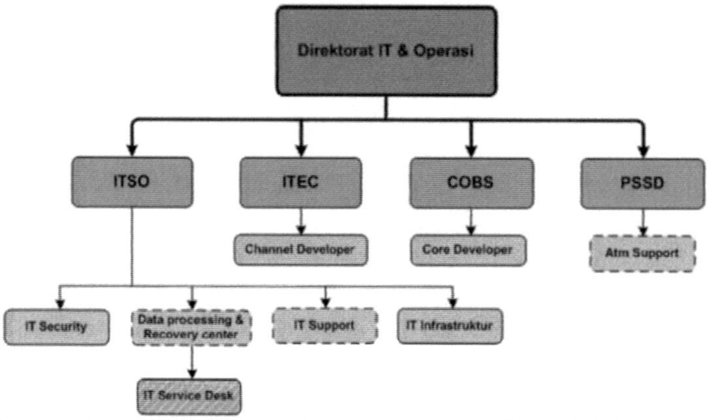

Figure 4 New Functional IT Service Desk at Financial Institution

The user of the bank contacts the IT Service Desk as the single point of contact and the other organization acts as a second level of support. With the new IT Service Desk function, the Financial institution can achieve the SLA to the user. Figure 5 presents the achievement of 100% for the availability of the 21 critical system including the core system, transactional system and support system

The availability of the critical system increasing after the Q3 are below 99.99% (Credit Card (Asscend),Treasury System(Spectrum), Internet Banking and Interface Treasury to Fast (EMX), but 3 month later the average availibility achieve the SLA Target (99.99%). This achievement makes the user more comfortable because the system always on.

No	Service	SLA/Target (% Uptime)	Achievement		
			Avg.Q1'16 (Jan-Mar)	Avg.Q2'16 (Apr-Jun)	Q3'16 (Jul)
I. AVAILABILITY CRITICAL SYSTEM					
A. CORE SYSTEM					
1	Core Banking (silver Lake)	99.95%	100.00%	100.00%	100.00%
2	Credit card (accend)	99.95%	99.94%	100.00%	100.00%
3	Treasury System (Spectrum)	99.95%	99.93%	100.00%	100.00%
4	Joint Financing (MOIF)	99.95%	100.00%	100.00%	100.00%
B. TRANSACTIONAL SYSTEM					
5	ATM Card Management	99.95%	100.00%	100.00%	100.00%
6	Proswitching Production	99.95%	100.00%	100.00%	100.00%
7	Call Center	99.95%	100.00%	100.00%	100.00%
8	Internet Banking	99.95%	99.42%	100.00%	100.00%
9	Prepaid system (OE)	99.95%	100.00%	100.00%	100.00%
10	Switching System (PCE)	99.95%	100.00%	100.00%	100.00%
11	Mobile Banking	99.95%	100.00%	99.20%	100.00%
12	SKN_GEN2	99.95%	100.00%	100.00%	100.00%
13	RTGS	99.95%	99.97%	100.00%	100.00%
14	Swift	99.95%	100.00%	100.00%	100.00%
C. SUPPORT SYSTEM					
15	Collection System (CWX)	99.95%	99.98%	100.00%	100.00%
16	Loan Origination System (CC)	99.95%	100.00%	100.00%	100.00%
17	Loan Origination System (Consumer)	99.95%	100.00%	100.00%	100.00%
18	CIS	99.95%	100.00%	100.00%	100.00%
19	Custody System (CIPS)	99.95%	100.00%	100.00%	100.00%
20	PSAK	99.95%	100.00%	100.00%	100.00%
21	Interface Treasury to fast (EMX)	99.95%	99.90%	100.00%	100.00%
Green Performance Availability System			(17/21) 81.00%	(20/21) 95.24%	(21/21) 100.00%

Note of Red Performance:
- No.11: down 1 jam 10 menit (Apr'16)
down 1 jam 48 menit (Mei'16)
down 16 jam 40 menit (Jun'16)

Performance Level per Service:
Meet SLA
Un-Meet SLA

Green Performance Level Criteria:
≥ 90%
≥ 80%
< 80%

Figure 5 SLA Achievement during Q3

b. The prioritization and categorization in Operation Management will support the second level support to analyze the root cause of the problem in an in-depth manner. The information on the table enable us to focus on the critical issue to solve. Figure 6 depicts the most frequently faced problems in a specific time. This type of table can be generated monthly.

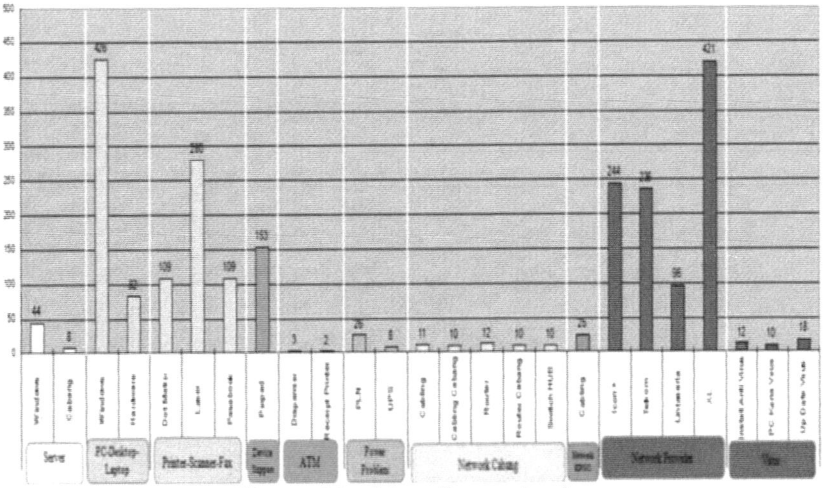

Figure 6 Incident Categorization

With the incident report generated by monitoring the IT Service Desk engine, the business users can focus their efforts towards increasing the customer transaction. The report that IT service desk produces weekly is shown in Figure 7. This report is produced by the IT service desk to help the business users speed up the fee-based income for the financial institution

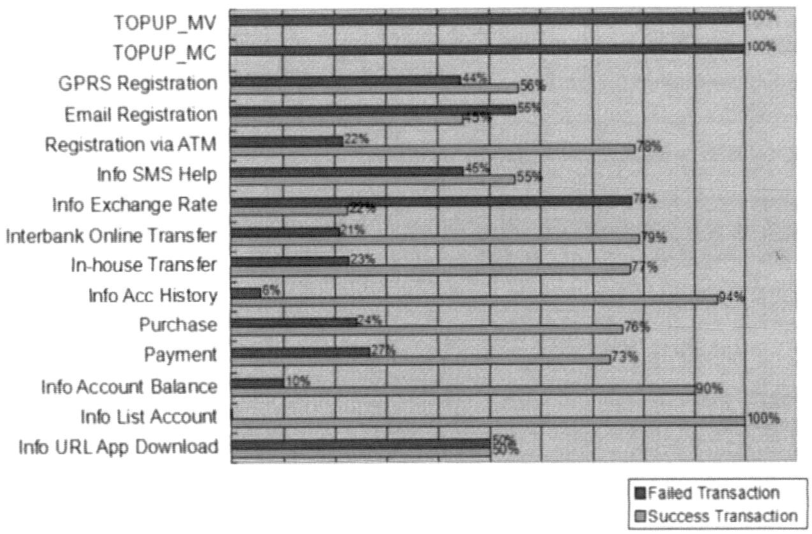

Figure 7 Internet Banking transaction monitored by IT Service Desk

Process Software Development Life Cycle for the release management is also monitored by the IT Service Desk to make sure that the application deployed in the production environment will operate smoothly without any disruption. Figure 8 presents the Service Level Agreement for the SDLC performance.

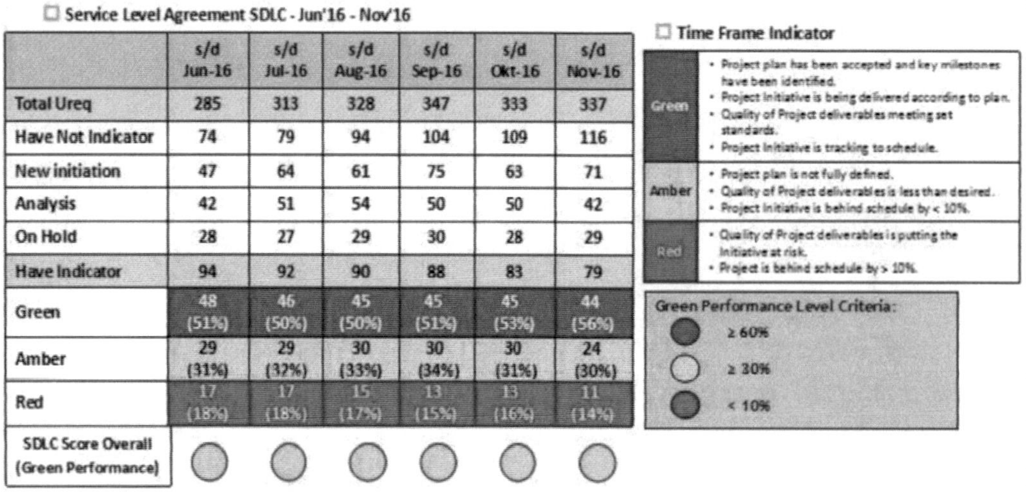

Service Level Agreement SDLC - Jun'16 - Nov'16

	s/d Jun-16	s/d Jul-16	s/d Aug-16	s/d Sep-16	s/d Okt-16	s/d Nov-16
Total Ureq	285	313	328	347	333	337
Have Not Indicator	74	79	94	104	109	116
New initiation	47	64	61	75	63	71
Analysis	42	51	54	50	50	42
On Hold	28	27	29	30	28	29
Have Indicator	94	92	90	88	83	79
Green	48 (51%)	46 (50%)	45 (50%)	45 (51%)	45 (53%)	44 (56%)
Amber	29 (31%)	29 (32%)	30 (33%)	30 (34%)	30 (31%)	24 (30%)
Red	17 (18%)	17 (18%)	15 (17%)	13 (15%)	13 (16%)	11 (14%)
SDLC Score Overall (Green Performance)						

Time Frame Indicator

Green	• Project plan has been accepted and key milestones have been identified. • Project Initiative is being delivered according to plan. • Quality of Project deliverables meeting set standards. • Project Initiative is tracking to schedule.
Amber	• Project plan is not fully defined. • Quality of Project deliverables is less than desired. • Project Initiative is behind schedule by < 10%.
Red	• Quality of Project deliverables is putting the Initiative at risk. • Project is behind schedule by > 10%.

Figure 8 SDLC Service Level Agreement

Internet Banking and mobile banking transaction are also monitored by the IT Service Desk and reported regularly in the Division meeting with the Board of Director. Figure 9 depicts the financial and non-financial internet banking transactions

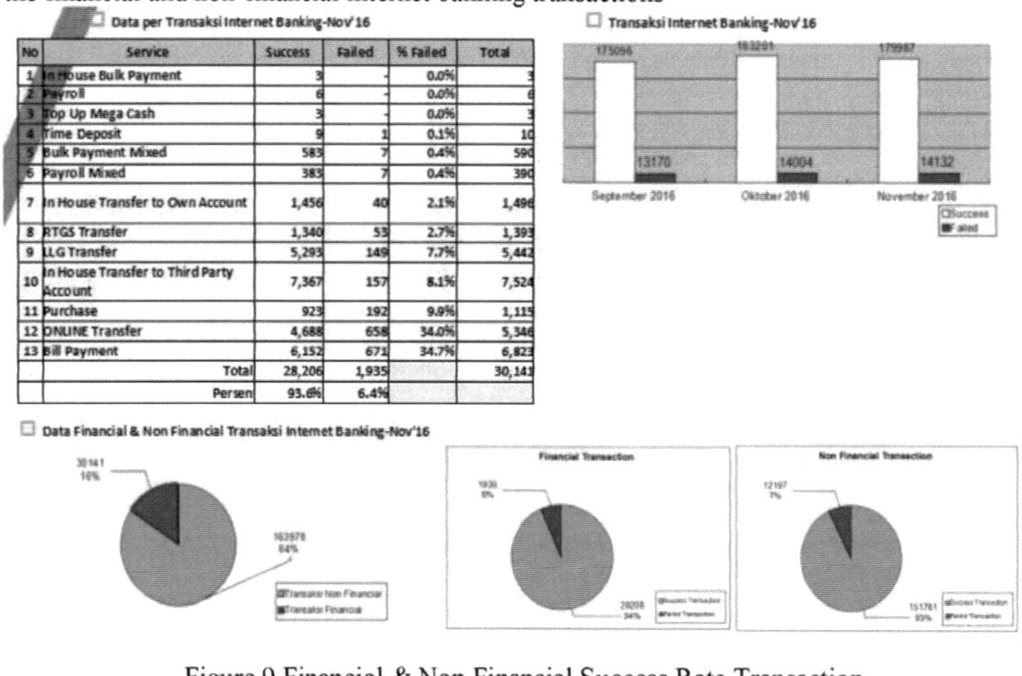

Data per Transaksi Internet Banking-Nov'16

No	Service	Success	Failed	% Failed	Total
1	In House Bulk Payment	3	-	0.0%	3
2	Payroll	6	-	0.0%	6
3	Top Up Mega Cash	3	-	0.0%	3
4	Time Deposit	9	1	0.1%	10
5	Bulk Payment Mixed	583	7	0.4%	590
6	Payroll Mixed	383	7	0.4%	390
7	In House Transfer to Own Account	1,456	40	2.1%	1,496
8	RTGS Transfer	1,340	53	2.7%	1,393
9	LLG Transfer	5,293	149	7.7%	5,442
10	In House Transfer to Third Party Account	7,367	157	8.1%	7,524
11	Purchase	923	192	9.9%	1,115
12	ONLINE Transfer	4,688	658	34.0%	5,346
13	Bill Payment	6,152	671	34.7%	6,823
	Total	28,206	1,935		30,141
	Persen	93.6%	6.4%		

Figure 9 Financial & Non Financial Success Rate Transaction

IOP Conf. Series: Journal of Physics: Conf. Series **1090** (2018) 012059 doi:10.1088/1742-6596/1090/1/012059

4. Conclusions

A good IT Service Desk allows a Financial Institution to easily connect to its business goals. Good IT Service Desk will improve user satisfaction effectively and efficiently for all stake holders of the company. The success factors that influence the successful implementation of IT Service Desk here are the commitment from the top management and also from the staff and their manager to implement ITIL Framework.

The challenge faced by this Financial Institution regarding service its users has been managed by implementing a Single Point of Contact of IT Service Desk. This service level agreement is regularly updated to better serve the users. The limitation of implementing of IT Service Desk is about the time limit and the socialization to all user of the company need more attention.

References

[1] C. Zhao, H. H. Gan and F. Gao, "A Study on the Process Model for IT Service Management," *Proceedings of Finite Element Analysis and CAD*, Peking University Press, Beijing, 1994

[2] Ekanata, A., & Girsang, A. S, "Assessment of capability level and IT governance improvement based on COBIT and ITIL framework at communication center ministry of foreign affairs, " *Proceeding of ICT For Smart Society (ICISS)"*, 2017

[3] Chen, A. & Chou, S, "Issues in implementing information technology service management. *Service Science, 1(2),* pp 13-17, 2010

[4] Zhu, P. , *ITIL framework: Value added or out-of-date.* Retrieved from: http://futureofcio.blogspot.com/2012/11/itil-framework-value-added-or-out-of.html, 2012

[5] BMC. (2011). *ITIL Best Management Practice* - 02 ITIL Service Strategy.

[6] BMC. (2011). *ITIL Best Management Practice* - 02 ITIL Service Design.

[7] BMC. (2011). *ITIL Best Management Practice* - 03 ITIL Service Transition.

[8] BMC. (2011). ITIL Best Management Practice - 04 ITIL Service Operation

[9] BMC. (2011). ITIL Best Management Practice - 05 ITIL Continual Service Improvement.

IOP Conf. Series: Journal of Physics: Conf. Series **1090** (2018) 012060 doi:10.1088/1742-6596/1090/1/012060

Evaluation of ISO 27001 implementation towards information security of cloud service customer in PT. IndoDev Niaga Internet

Ahmad Nurul Fajar[1], Hendy Christian [2], Abba Suganda Girsang[3]
[1] Information Systems Management Department, BINUS Graduate Program-Master of Information Systems Management, Bina Nusantara University Jakarta, Indonesia 11480
E-mail : afajar@binus.edu
[2] Information Systems Management Department, BINUS Graduate Program-Master of Information Systems Management, Bina Nusantara University Jakarta, Indonesia 11480
E-mail : hendy.christian@binus.org
[3] Computer Science Department, BINUS Graduate Program-Master of Computer Science, Bina Nusantara University Jakarta, Indonesia 11480,
Email : agirsang@binus.edu

Abstract. Every organization must ensure that information assets are protected and information security system has been implemented well. PT. IndoDev Niaga Internet is a provider of business solutions applications and implementation services that include application Human Resource Information System (HRIS) and Enterprise Resource Planning (ERP). In order to ensure the security of the information, then in 2015 PT. IndoDev Niaga Internet implement ISO 27001: 2013. Through the implementation of ISO 27001: 2013, it is expected that information can be properly maintained, which in turn will affect the business continuity. Companies need to know the extent to which the process has been applied and what actions can be done to improve the performance of the application of ISO 27001: 2013. Factor analysis was conducted first to determine the factors that affect to the information security. After the factors that affect to the information security known, then observation and interview conducted to gather data about PT. IndoDev Niaga Internet ISO 27001:2013 implementation according to the factors that affect. And then recommendation and corrective action developed using gap analysis method. The most influential factor to the security of customer information PT. IndoDev Niaga Internet is a factor access control and security operations. For the audit results ISO27001: 2013 from the aspect of access control, they found 11 check items that fit into the category of NC (Non-Conformance) of 33 check items in which 9 of them in the category of major and two remaining categories minor, while for operations security aspects of the 12 check items, found 5 that goes into the category of NC (Non-Conformance) and everything was included into the category of minor.

1. Introduction

Currently, businesses relied and depended on the information to operate the business daily operations to achieve the goals and objectives of the organization [1]. In this digital age, it has become common for organizations to use information systems to process information within the organization in order to get a better support to the achievement of the organization's mission [2].

Content from this work may be used under the terms of the Creative Commons Attribution 3.0 licence. Any further distribution of this work must maintain attribution to the author(s) and the title of the work, journal citation and DOI.
Published under licence by IOP Publishing Ltd

Thus, every organization must ensure that information assets are protected and information security system has been implemented well [3]. PT. IndoDev Niaga Internet is a provider of business solutions applications and implementation services that include application Human Resource Information System (HRIS) and Enterprise Resource Planning (ERP). PT. Indodev Niaga Internet has handled more than 500 customers spread across the country and abroad. In order to ensure the security of the information, then in 2015 PT. IndoDev Niaga Internet implement ISO 27001: 2013. Through the implementation of ISO 27001: 2013, it is expected that information can be properly maintained, which in turn will affect the business continuity PT. IndoDev Niaga Internet. But over the implementation of ISO 27001: 2013, companies need to know the extent to which the process has been applied and what actions can be done to improve the performance of the application of ISO 27001: 2013. Empirical data in this place shown the trigger to concern about information security management implementation. Therefore, the research undertaken will discuss the evaluation of the application of ISO 27001: 2013 in PT. IndoDev Niaga Internet.

2. Literature Study

The purposes of information security is to protect organization resources, such as, information, hardware, and software [4]. Information security means protecting information and information systems from unauthorized access, use, disclosure, disruption, modification, or destruction [5].Three main concepts in information security are confidentiality, integrity, and availability (CIA) [6]The purpose of security management information is information confidentiality, integrity, and availability [3].Information security management system is needed because of the threat to the confidentiality, availability, and integrity of information that organizations large and likely to increase [7].The process of information security management is defined as the cycle of PDCA (Plan-Do-Check-Act), in which management identifies the process that should be implemented, to investigate whether all the processes that have been implemented have been investigated and operations that take into account the feedback and results [3]. ISO 27001: 2013 provides specifications for information security management systems along with practice [7].ISO 27001 also adopted a process model PDCA (Plan-Do-Check-Act) [8]. ISO 27001: 2013 has 14 security control clauses that contain a total of 35 control objectives and 114 controls [9]. The 14 security control clauses as following: Information security policies, Organization of information security, Human resource security, Asset management, Access control, Cryptography, Physical and environmental security, Operations security, Communications security, System acquisition, development, and maintenance, Supplier relationships, Information security incident management, Information security aspects of business continuity management, Compliance. The indicators used in this study refers to the implementation guidance on the control of the ISO 27001: 2013 are set out in the code of practice for information security controls from ISO 27002: 2013. The indicators can be seen in Appendix A.

3. Research Method

Factor analysis was conducted first to determine the factors that affect to the PT. IndoDev Niaga Internet information security. The questionnaire used for gathering the data that will be used in factor analysis. Questionnaire developed using information security indicator that found from literature study. The scale used in this study is the Likert scale (1-5). Likert scale is intended to find out how to agree and disagree respondents to the statements in the questionnaire. After the factors that affect to the PT. IndoDev Niaga Internet information security known, then observation and interview conducted to gather data about PT. IndoDev Niaga Internet ISO 27001:2013 implementation according to the factors that affect.

And then recommendation and corrective action developed using gap analysis method according to the target that has been established by PT. IndoDev Niaga Internet and the current achievement of that target. Literature study also used to develop recommendation and corrective action.

4. Results and Analysis

This research was conducted at PT. IndoDev Niaga Internet by distributing questionnaires to 119 employees in IT-related departments. Reliability analysis to test the validity and reliability of the questionnaire study was conducted prior to the data before performing factor analysis. Results of reliability analysis states that the Alpha value is equal to 0.722 Thus, the research questionnaire can be accepted and can be said to be reliable or trusted as an instrument for collecting research data [10].Furthermore, a factor analysis performed on samples collected research. Confirmatory Factor Analysis (CFA) was done in this study. We create Appendix B that displays the regression results of the factor analysis conducted, these results indicate that B, IM, SU, CO, OP, P, C, AC, A, O, I have a relationship with the latent variables (information security) where the value of P < 0:05 and CR> 1.96, while for the CM, S, H showed no relationship with latent variables (information security) Since the value of P> 0.05 and CR <1.96. From these results thus obtained groups of variables that affect the security of the information that is B (Information Security Aspects of Business Continuity Management), IM (Information Security Incident Management), SU (Supplier Relationships), CO (Communications Security), OP (Operation Security), P (Physical Security), C (Cryptography), AC (Access Control), A (Asset Management), O (Organization of Information Security), I (Information Security Policies).Then to get the significance of variables - variables that can be obtained through standardized before loading factor as in Appendix C. Loading factor latent variable to the observed variables must be greater than 0.50 [11]. Based on the analysis in Appendix C, there are two variables observed as having a loading factor> 0.50 namely OP and AC, it indicates that the OP and AC included into values - core values are determined, so that the OP (Operation Security) and AC (Access Control) the variables that most influence on the latent variables. Evaluation of security operations at PT. IndoDev Niaga Internet referring to the performance targets for IT Support department for the period from July to December 2016 and the results of audits of ISO 27001: 2013 for the period of August 2016. In the performance targets for IT Support department, there are 4 KPI (Key Performance Indicator), namely: System availability, Compliance incidence, IT infrastructure enhancement plan, Ticket resolution time. From the results of the performance of system availability during the period from July to December 2016 it is known that the target of 90% can be achieved during the July-December. This indicate that the performance of the system availability is quite good. The results of performance in terms of compliance incidence during July-December, 2016 was also able to meet the targets set. Performance incidence also includes good compliance. The results of the performance targets for the IT infrastructure enhancement plan, in the period July, August, September, and October did not meet the targets set for July IT infrastructure enhancement plan that is achievable only by 87.5%, and for the month of August amounted to 80%, in September also by 80%, for the month of October by 85%, while for November and December have reached the target of 90% for the months of November and 100% for the month of December, so that the average - average achievement for the IT infrastructure enhancement plan in the period July - December 2016 was 87%, it would still be below the set target of 90%, though below target, but the average achieved is still within the limits of tolerance, so that they are in good category. Scaling back significantly in August and September occurred because the employees of the department who resigned during the period, so there is some plan IT infrastructure enhancement is not running Because of reduced resources needed, but this can be increased in the next period is in October.

The results of the performance of the ticket resolution time during July-August 2016 of the results obtained during this period nothing to meet the target, during that period the average - average obtained is 50.86%, and this result is still quite far with the targets set, where the difference amounted to 29.14% and that means if in one month there are 60 ticket that must be completed on time, only 31 ticket that can be settled in accordance with the SLA (Service Level Agreement) agreed on the amount of the targeted 48 ticket. Through the results of observations conducted on the ticket is assigned to the department IT Support, it is known that during the period from July to December, 2016 there were 511 ticket that must be completed by the department of IT Support.

The majority of the ticket that must be done is related to security operations (backup & restore, malware protection, configuration, file operations, installation, set up a new instance, SSL certification, query support). If the total ticket then there are 269, or about 53% of the total ticket that existed at that period. By dominating ticket relating to operations security, then the performance targets of the ticket resolution time should be a concern for management to be improved, so that the SLA has been agreed can be always achieved, which in turn will affect the operations of security of the company and ensure that the proper process and safely performed for the operation of information processing device in accordance with the scope and policies of the security operations of PT. IndoDev Niaga Internet. From interviews and observations, it is known that some of the causes of the ticket resolution time failed to meet the target are as follows: Lack of supervision of the direct supervisor of the ticket that must be resolved, lack of resources, division of tasks not done well, and less aware of the targets to be achieved. Evaluation of operations security also looked at the results of the audit of ISO 27001: 2013 for the period of August 2016.In audits conducted in the period August 2016, from 12 items of audit checklist relating to the operations of security, there are 5 items in the category of Non-Conformance (not appropriate), while the third item into the category of Conformance (as appropriate), 2 items into the category of observation (need further observations), and two items fit into the category of follow Up (require follow-up). Of particular concern in this evaluation are 5 items that fit into the category of Non-Conformance. Non-Conformance that occur are within the category of minor nonconformities. Of non-conformance that occur are then conducted further investigation and observation to determine the cause of the nonconformity (root cause). For the mismatches that occur in the check item A.12.4.1, A.12.4.2, A.12.4.3 and A.12.6.2 known to occur due to negligence of staff IT support, while for noncompliance with A.12.5.1 happen because documentation is not up-to-date. Non-conformance minor can be defined such that a mismatch does not have a serious impact on information security. Evaluation of access control refers to the results of the audit of ISO 27001: 2013 for the period of August 2016. The results of the audit of the period for matters relating to access control, found that of the 33 items into audit checklist contained 11 items that fall into the category of Non-Conformance (not appropriate), 9 items belonging to the category of observation (need further observations), 2 items in a category follow Up (require follow-up), and 11 items are included in the category of Conformance (as appropriate).In contrast to the results of the audit of the security operations in which discrepancies that occur are all entered into the minor category, the access control of 11 findings of nonconformity occurs, 9 of them entered into a major Non-Conformance. This means, a Non-Conformance that occurs a serious impact on the achievement of information security. Further observation and investigation conducted to determine the cause of such discrepancy (root cause). Non-Conformance that occur in A.9.1.1.1, A.9.1.1.2 and A.9.1.1.3 caused by the same thing, namely because of the procedures outlined in the working document of instruction for access control is not effective, because the approval process is still going outside the system or occur orally, so that from the IT Support difficult to know whether the request has been submitted has been approved by or not. Then to A.9.1.1.4 and A.9.2.5.1 more due to negligence of staff IT support. A.9.2.1.1, A.9.2.1.2, A.9.2.1.3 and A.9.2.2.2 due to a combination of the negligence of staff IT Support and procedures for access control that is in working instruction is not complete and therefore become ineffective, A.9.2.3.1 due to lack of inspection / review of access to shared folders. A.9.2.4.1 due to the absence of documents used to record the change password.

Table 1. Comparison between current achievement and target.

KPI	Target	Current achievement
System availability	>= 99 %	99.63%
Compliance incidence	<= 3	1.67
IT infrastructure enhancement plan	>= 90%	87%
Ticket resolution time	>=80%	50.86%
Operations security conformance	= 100%	25%
Access control conformance	= 100%	33%

From Table 1 it can be seen that for ticket resolution time, operations security conformance, and access control conformance remained well below the target set. For the ticket resolution time issue which led to the collapse reached the target set is due to aspects of IT service management is not maximized, where IT service management refers to planning, provisioning, and management of IT services in the framework of alignment with the business enterprise [12]. Referring to [13] some misunderstandings often occur in the IT service management, misunderstandings, among others, the problem can be solved if established a good relationship with the user, whereas the service antecedent deliver very closely related to the quality of the service itself, because the quality of these services should continue to be evaluated in order to achieve the expected quality. In the model of IT service management there are five major interrelated processes namely Service Level Management, Event Management, Problems Management, Help Desk Management, and Customer Satisfaction Management [13]. These five processes are interrelated and have an effect between one another, if the company is focusing only on some of the processes and ignore the impact of the processes influence each other, then the IT services provided will be disturbed, therefore, on an action plan to do as in Appendix D. When viewed from the causes of discrepancies that occur in an audit of operations security aspects can be seen that the majority of problems arise due to negligence of staff IT support, in this case means is the aspect of human error.In order to overcome the problem of human error, of course, important to first identify the type of human error that occurred. [14] classify human errors into three groups, namely skill-based errors, rule-based error, knowledge-based error. That where in this case, negligence occurred into the category of skill-based errors, because these errors occur for activities under routine and actually individuals who have the knowledge, skills and experience to do the job properly, but when attention is diverted then this error can occur [15]. This type of error can be quickly and efficiently to be restored, one of the ways that can be taken is through giving feedback [15]. Therefore, the recommendation action plan that can be done as shown in Appendix E.NC Results found in the aspect of access control most of the major discrepancies, as this would threaten the security of the information from PT. IndoDev Niaga Internet if it is not followed up. Root cause this discrepancy happens by human error as the operations of security, but the type of human error that occurs is different, in this case as the problems review the access control is never done, it is not routinely performed, this omission occurred because of a lack of information or knowledge to do it, human error is included into the knowledge-based error [15]. Furthermore, there is also a mismatch happens because the procedures contained in the working instruction ineffective as it did in A.9.1.1.1, A.9.1.1.2, A.9.1.1.3, A.9.2.1.1, A.9.2.1.2, A .9.2.1.3, A.9.2.2.2. Access control is fundamental in information security [16], the procedure of access control must have the following 3 components: Identification, so that access control can be effective, individuals involved should be identified, authentication, identification requires authentication. This is to ensure that the identity used is authentic (used by people who fit), authorization, a set of actions that are allowed for specific authorization.

Referring to the existing problems, it is mentioned that the IT Support team has not been able to determine whether the request is filed in ticketing system has been approved or not, this means that the authentication process cannot be executed properly. Besides, the other thing is access control procedures regarding authorization, which was also found that the review of user access rights cannot be done because it has no tools for reviewing it. Roled-Based Access Control (RBAC) can be used to manage roles and access rights of the user, rather than giving access only to certain users [17]. By using roled-based, then the company can undertake a review of user access rights more effectively and efficiently [17]. On the other hand, relate to [18],it is a mechanism to transform goal model to business process model. This mechanism could be used to describe business process model in PT IndoDev Niaga Internet . The system architecture has proposed by [19] to anticipate dynamic environment. Therefore, considering this, the action plan recommendations that can be done as shown in Appendix F.

5. Conclusion

Conclusion of this study are:

1.The most influential factor to the security of customer information PT. IndoDev Niaga Internet is a factor access control and security operations.

2.For the audit results ISO27001: 2013 for the period of August 2016, from the aspect of access control, they found 11 check items that fit into the category of NC (Non-Conformance) of 33 check items in which 9 of them in the category of major and two remaining categories minor, while for operations security aspects of the 12 check items, found 5 that goes into the category of NC (Non-Conformance) and everything was included into the category of minor.

3.Recommendation improvements made include improvements to IT Service Management and procedures, to detail can be seen in Appendix D, E, and F.

Appendices

Appendix A. Security indicators

Security control clauses	Indicator	Ref
Information security policies	Information security policy reduces security breaches	[18]
	Provide assurance about the enforcement of the security functions	[19]
	Information security policy is a prerequisite for effective information security	[20]
Organization of information security	Support and guidance from the executive gives a positive value to the preservation of confidentiality, integrity, and availability of corporate information	[21]
	Establishing and documenting responsibility for information security to ensure that there is no violation of responsibility that result in the information security	[22]
	Separation of duties and responsibilities to reduce the risk of errors, theft, and unauthorized changes to information	[8]
Human resource security	Effective information security awareness can improve an organization's information security posture	[23]
	Determination of the employee's responsibility in terms of information security in work rules reduce the risk of unauthorized access to information	[24]
	Provision of information security training for employees can improve awareness and understanding of the threats, risks, and information security policy	[25]
Asset management	Identify the organization and asset protection measures against these assets into the key applications of information security	[26]
	Classification information provides an indication of how to protect and handle information	[27]
	Media management actions to prevent modification, deletion and destruction of the information stored on illegally	[28]
Access control	Access control enables organizations to control and protect the availability, integrity and confidentiality of resources	[29]
	Access control ensures only authorized access of users who are entitled to the system or a particular service	[30]
	Prevent unauthorized access to a system or a particular service	[31]
Cryptography	Cryptography protects information from unauthorized	[32]
	Cryptography protects the confidentiality of information	[33]
	Cryptography ensure the integrity (accuracy) of information	[5]
Physical and environmental security	By setting a secure area and access cards can prevent unauthorized access to documents, equipment or data	[34]
	Clean desk and clear screen policy reduces the risk of information security breaches	[4]
	Line of defense (perimeter) were plated complicate the theft of information assets	[35]
Operations security	*Capacity management is better able to reduce down time and increase availability (availability)*	[36]
	Policy official ban on the use of software (software restriction policy) prevent information leakage, loss of integrity, as well as a violation of intellectual property rights	[37]
	Backup process and tested according to the procedure to cover the possibility of losses due to loss of data / corrupt	[38]
Communication security	Managed network and properly controlled can protect the information stored in the system and applications	[39]
	Segregation improve network security against network so that information security increased	[40]
	The establishment of policies and procedures to protect the information transfer from the information leakage	[41]
System acquisition, development and maintenance	By setting the policy development secure the integrity, availability and confidentiality of the information generated, manipulated and disseminated by the application system can be guaranteed	[42]
	Protection of the data testing is performed with the same controls and procedures as operational data to protect the information from unauthorized access.	[43]
	Change control procedures to change the software used in the operation of information systems ensure that information systems are not disrupted by the change.	[43]

Supplier relationships	Signing an NDA (Non-Disclosure Agreement) with third parties in order to prevent the occurrence of a security breach related to a third party	[44]
	Determination of SLA (Service Level Agreement) in agreement with a third party to make sure that the service in accordance with the level of security and service level continuity	[45]
	Monitoring and review of third party services on a regular basis to make sure the services provided maintain compliance with security requirements and audit	[46]
Information security incident management	Establishment of an information security incident management enables rapid and effective response to information security incidents	[21]
	The experience and knowledge gained from the analysis and resolution of information security incidents can be used to reduce the likelihood or impact of future incidents	[47]
	Reporting information security events quickly and responsiveness made possible a rapid response to the incident to be investigated	[48]
Information security aspects of business continuity management	Business Continuity Planning ensures availability (availability) when the incident occurred	[49]
	System redundancy sufficient and appropriate to the needs of being able to maintain the system to keep it running	[6]
	Testing business continuity plan regularly to ensure the provision of roles and responsibilities are appropriate when the incident occurred	[31]
Compliance	Legislative requirements, legal or contract for any information system should be explicitly defined	[50]
	Implementation of controls to ensure compliance with the rules relating to intellectual property to protect the intellectual property rights of information and software company	[4]
	The protection of government records in accordance with the rule of law can avoid the loss, damage of counterfeiting and unauthorized access	[51]

Appendix B. Regression weights analysis.

			Estimate	S.E.	C.R.	P	Label
CM	<---	InformationSecurity	-.161	.801	-.201	.841	
B	<---	InformationSecurity	2.752	1.344	2.048	.042	
IM	<---	InformationSecurity	3.542	1.766	2.006	.045	
SU	<---	InformationSecurity	2.580	1.194	2.161	.032	
S	<---	InformationSecurity	-.076	1.296	-.573	.567	
CO	<---	InformationSecurity	6.554	3.078	2.129	.037	
OP	<---	InformationSecurity	12.775	6.423	1.989	.016	
P	<---	InformationSecurity	2.745	1.338	2.052	.031	
C	<---	InformationSecurity	11.473	5.537	2.072	.023	
AC	<---	InformationSecurity	12.733	6.336	2.010	.017	
A	<---	InformationSecurity	1.851	.931	1.988	.048	
H	<---	InformationSecurity	-.102	.951	-.685	.493	
O	<---	InformationSecurity	3.795	1.929	1.967	.041	
I	<---	InformationSecurity	1.909	.923	2.068	.046	

Appendix C. Standardized Regression weights analysis.

			Estimate
CM	<---	InformationSecurity	-.022
B	<---	InformationSecurity	.007
IM	<---	InformationSecurity	.142
SU	<---	InformationSecurity	.080
S	<---	InformationSecurity	-.076
CO	<---	InformationSecurity	.254
OP	<---	InformationSecurity	.675
P	<---	InformationSecurity	.145
C	<---	InformationSecurity	.473
AC	<---	InformationSecurity	.733
A	<---	InformationSecurity	.051
H	<---	InformationSecurity	-.102
O	<---	InformationSecurity	.195
I	<---	InformationSecurity	.109

Appendix D. Recommendation to solved ticket resolution time issue.

Problems	Recommendation
Lack of supervision of the direct supervisor of the ticket that must be resolved	Perform regular weekly meetings to discuss and monitor about SLA realization.
Less aware of the targets to be achieved	
Resources needed less	Doing man power planning in accordance with the estimated work

	load job
There is already a division of tasks in the IT department support, but in execution the division of tasks is less walk, because the assignment of the ticket is done directly by the user and the user himself does not know about the distribution of these tasks	To disseminate returned to the user from another department, division of tasks in IT support staff and make the mapping between the type of work with IT Support staff responsible for such work.

Appendix E. Recommendation to solved operations security performance issue.

Problems	Recommendation
Log Document Review Checklist examined yet have the document number for the period August 2015 and February 2016	The examination should always be done by direct supervisor and things - things that the correction directly delivered and repaired as soon as possible.
There are some server specs are not in accordance with the standards contained in documents.	Routine inspection should be done by direct supervisor to determine whether the spec listed in the document with the original physical was appropriate.
From the samples examined laptop office inventory, they found access to the admin, there should never have admin access.	Conduct periodic audits laptop, to ensure that there is no admin access attached to laptop office inventory

Appendix F. Recommendation to solved access control performance issue.

Problems	Recommendation
Based on the working instruction that has been documented, before the user makes a request to the ticketing system is available, the user must obtain prior approval from certain parties, but at this time IT Support is not able to control whether the request has been approved or not.	1. Revise working instruction, adding owner, admin, as well as approval workflow access. 2. Updating form used. 3. Reviewing access each month and turned into a monthly routine IT Support team
Access Approval process is not appropriate, there should have prior approval from the Technology Development Manager or HR Development only after that access can be granted	
REVOKE access process should be made after reviewing the access control on the form FM-IT-RAC. But the form FM-IT-RAC itself is not valid (inactive)	
Currently, the IT Support has not been able to control whether the request is filed in already approved or not.	
Review access control is never done	Conduct a review of monthly and annual access control for all access control.
Found some errors permissions	
Deregistration process has not always been carried out, they found employees who had resigned but still active domain user whether they are still active	*Update directory user registration and de-registration process.*
Found redundant email account (one employee has more than one email account).	*Update email registration and de-registration process.*
Found users who have resigned or obsolete (not known for certain that account)	
No statement letter to preserve the confidentiality of personal or group.	Formulate a statement letter.
Not yet have the tools to do a review of user access rights and have never done a review of existing user.	Using tools RBAC (Rolled-Based Access Control)

References

[1] A. M. Padyab, "Towards More Structured Information Asset Identification Approach for Risk Assessment Methods," 2013.

[2] G. Stoneburner, A. Goguen, and A. Feringa, "Risk Management Guide for Information Technology Systems Recommendations of the National Institute of Standards and Technology," 2002.

[3] M. Kazemi, H. Khajouei, and H. Nasrabadi, "Evaluation of information security management system success factors : Case study of Municipal organization," vol. 6, no. 14, pp. 4982–4989, 2012.

[4] R. Thomas, *Information Security Fundamentals, Second Edition*. 2014.

[5] J. Vacca, *Managing Information Security*. 2014.

[6] J. Andress, *The Basic Of Information Security*. 2011.

[7] A. Calder and S. Watkins, "IT GOVERNANCE," 2008.

[8] H. Hai and K.-M. Wang, "The critical success factors assessment of ISO 27001 certification in computer organization by test-retest reliability," vol. 8, no. 17, pp. 705–716, 2014.

[9] I. 27001:2013, "INTERNATIONAL STANDARD ISO / IEC Information technology — Security techniques — Information security management systems — Requirements," vol. 2013, 2013.

[10] R. . DeVellis, *Scale development: Theory and applications*. Sage, 2012.

[11] B. M. Byrne, *Structural Equation Modeling with AMOS. Second edition*. Taylor & Francis Group, 2010.

[12] M. Marrone, "The Business Benefits of Effective ITSM," 2011.

[13] J. Wan, "Evaluation on Information Technology Service Management Process with AHP," *Technol. Invest.*, vol. 2, no. 1, pp. 38–46, 2011.

[14] J. Reason, *Human Error*. Cambridge University Press, 2010.

[15] D. Embrey and H. Lane, "Understanding Human Behaviour and Error The Skill , Rule and Knowledge Based Classification," *System*, pp. 1–10, 2010.

[16] C. Perrin, "The three elements of access control," 2007.

[17] A. Antonopoulos, "Role-based access control for effective security management," 2011.

[18] Fajar, A.N., Shofi,I.M., (2016).Goal Model to Business Process Model. International Journal of Electrical and Computer Engineering (IJECE) Vol 6 No 6.

[19]. Fajar, A.N., Budiardjo, E.K., Hasibuan , Z.A. "System Architecture In The Dynamic Environment Based on Commonality and Variability Business Processes, 8thICCM, Seoul, 2012

IOP Conf. Series: Journal of Physics: Conf. Series **1090** (2018) 012061 doi:10.1088/1742-6596/1090/1/012061

Fuel Breeding Analysis On Low Moderated Fuel Ratio Based On Actinides Closed Water-Cooled Thorium Reactor

Sidik Permana[1*], Syeilendra Pramuditya[1], Dwi Irwanto[1]

[1]*Nuclear Physics and Biophysics Research Division, Department of Physics, Institut Teknologi Bandung, Jl. Ganesha 10, Bandung, 40132, Indonesia*

**E-mail: psidik@fi.itb.ac.id*

Abstract. Utilization of spent nuclear fuel and some fuel breeding capabilities of nuclear fuels to extend the sustainability aspect of nuclear fuel become more important issues to be optimized. Thorium fuel utilization based on water-cooled reactor is one of the possible options to be used and optimized as well as uranium fuel utilization. Some schemes of accumulated spent nuclear fuels can be used as recycled fuel in water-cooled reactor based on thorium fuel. In the present analysis, fuel sustainability aspect of nuclear fuel will be evaluated, which is based on a water-cooled reactor. As a fuel basis, thorium is used with can be mixed with additional recycled spent nuclear fuels. Some minor actinides (MA) as recycled fuels are used as doping material to be loaded to the water cooled reactors with thorium fuel as fuel basis and heavy water as moderator and coolant. The evaluation has been made by adopting a computational simulation of an equilibrium burnup analysis method, which was coupled with cell calculation of computer code of SRAC with JENDL.32 as nuclear data library. Several survey parameters have been evaluated to evaluate some effect of MA doping rate, different moderation ratio and power density levels to the reactor performance including fuel-breeding capability and void reactivity coefficient. Effect of some actinide composition to fuel breeding capability as well as safety aspect, which is based on void reactivity coefficient have been investigated. Fuel breeding capability can be obtained by the present reactor systems; as well as negative void reactivity has been show for more moderator ratio and less power density. Low portion of moderation to fuel ratios (MFR) are used to have a better fuel breeding capability as well as some from contribution from recycled fuel of minor actinides (MA) and less power density. A negative void reactivity can be obtained in this system and it becomes less negative for doping MA and more power density as well as a positive void reactivity coefficient value for much less moderation ratio.

1. Introduction

Sustainability aspect of nuclear energy is basically coming from the optimum fuel utilizations of existing nuclear fuel resources, nuclear fuel management of the reactor including fuel cycle program and fuel breeding program. Fuel optimization programs can be improved and those optimized system of nuclear sustainability shown a similar trend with the renewable energies as a sustainable energy source [1]. One of the important issues, which is corresponding to nuclear fuel cycle, is nuclear proliferation resistance issues. The issue comes from the production of good quality of weaponable grade of spent nuclear fuel such as plutonium. A reduction strategy of the proliferation resistance level, it has been introduced some barriers such as matrial barrier of nuclear fuel based on isotopic composition. Some material barrier of isotopic composition can be estimated from critical mass, heat-generation rate, spontaneous neutron generation and radiation. Isotopic barrier of plutonium comes from the even mass number of plutonium isotope such as ^{238}Pu, ^{240}Pu and ^{242}Pu and in case of thorium cycle will depend on ^{232}U. For decades some thorium fuel technologies have being developed, for both conventional and advanced nuclear reactor as well as some thorium breeder reactor programs which

Content from this work may be used under the terms of the Creative Commons Attribution 3.0 licence. Any further distribution of this work must maintain attribution to the author(s) and the title of the work, journal citation and DOI.
Published under licence by IOP Publishing Ltd

was also developed for breeding program in uranium fuel technology. Thorium fuel cycle have some advantages including better fuel breeding capability in thermal and epi-thermal neutron energy region, higher fuel stability and higher level proliferation resistance [2-6] in comparison with others fuel cycles.

Some analyses on nuclear fuel sustainability and actinide production composition have been done based on heaevy water-cooled thorium reactor system and will be shown in the present study. Fuel cycle flow scheme will be based on the fuel recycled composition which all actinide compositions except fission products are recycled into the reactor as a basic fuel cycle scheme. Some key important parameters are evaluated such as doping composition of minor actinide (MA), volume ratio of moderator to fuel (MFR) and fuel cell power density which have some significant impact to the reactor performances such as criticality, fuel breeding ratio and void reactivity coefficient.

2. Parameter and Method

Analyses on nuclear fuel sustainability and void reactivity coefficient have been done based on heavy water as coolant and thorium as main fuel cycle. In the present study, well established reactor technology of water-cooled reactor based technology has been adopted as a basis for evaluation which is based on current LWR plant technology as general design parameter and using heavy water as coolant and moderator instead of using light water. **Table 1** shows a basic design parameters which consist of some key parameters such as moderator to fuel ratio (MFR), burnup, and doping material of minor actinide (MA) based on large 3000 MWt power output and heavy water as coolant and moderator. Employed MFR of 0.3 as very low moderation ratio and 1.0 as a moderate size for less moderation ratio are used to evaluate a moderator effect to the reactor performance in comparing to standard MFR of PWR (MFR =2) which show the neutron spectra are relatively harder than standard PWR. Minor actinide (MA) effect was used for comparative analysis, which is based on no MA doping and with about 1 % MA doping. A burnup of discharged fuel has been set to be 36 GWd/t as nearly a burnup level of standard LWR of PWR.

Table 1. Basic Design Parameters

Parameters	Values/Type
Thermal Power Output [MWt]	3000
Coolant	D_2O
Fuel Cycle Option	Actinides closed Cycle
Supply Fuel Composition	Th + MA (Doping)
Cladding	Zircaloy-4
Fuel pellet diameter [cm]	1.31
Fuel pin outer diameter [cm]	1.45
Moderator to Fuel Ratio (MFR) [-]	0,3 and 1.0
Fuel Power Density [W/cc]	140 and 280
Burn-up [GWd/t]	36
MA Content [%]	0 and 1

A regular blanket fuel diameter of the Shippingport of about 1.452 cm was used as fuel pin diameter based as a basic fuel parameter size [2,6-7]. A system for fuel cycle scheme has been used as a closed fuel cycle that all actinides are returned to the reactor. While only fission product (FPs) are sent into final disposal stream. The reactor will kept the heavy metal inside the core as all heavy metal closed cycle scheme concept. To evaluate fuel sustainability based on fuel conversion or fuel breeding capability of the reactors as well void reactivity coefficient, an iterative equilibrium method of burn-up calculation code (Equilibrium Cell Iterative Calculation System: ECICS) was used. The code was coupled by SRAC code and JENDL3.2 nuclear data library [2,7-18]. Fuel sustainability based on fuel breeding and fuel conversion ratio are based on the reaction rate of fissile and fertile nuclides and some small contributions of intermediate nuclides. When the conversion ratio is higher than unity than the definition becomes fuel breeding ratio and some surplus fuel from fuel breeding production is defined as fuel breeding gain. A void reactivity coefficient was used also to evaluate the reactor criticality safety in case some voided condition of coolant are occur and affect to the criticality condition.

3. Results and Discussions

3.1. MA Doping effect to fuel breeding capability

As mentioned in the previous section, the effect of MA material, which is loaded into the reactor, will be analysed. It used a comparative doping rate for non-doping MA case and for MA doping case of 1 % doping rate. The obtained results of the MA doping effect to the fuel conversion capability in the reactor as well as breeding gain are shown in **figure 1 and 2.** All obtained results of fuel conversion capabilities for both non-MA doping and MA doping cases are higher than unity. It means the reactor a fuel-breeding capability. Fuel breeding capabilities are becoming higher for doping material of MA as shown in figure 1. The result shows that MA doping can make the fuel breeding level increases during the reactor operation. It because of some additional fissile materials are produced from converted MA material during reactor operation. Obtained results to estimate the increase of some additional gain of fissile material of fuel breeding condition, it can be used a fuel breeding gain concept and the results are shown in **figure 2.** Fuel breeding of the reactor has some more than 1.16 % of breeding gain for non-doping MA case which means the Initial condition for thorium reactor without additional MA material as doping obtained higher than unity and it increases for MA doping case which is estimated of about 1,25 % of additional fissile materials were produced during reactor operation. Breeding gain can be achieved from some increasing fissile materials during reactor operation. These fissile materials are produced from conversion process from uranium to plutonium such as U-238 into Pu-239 and to other plutonium from capturing neutrons and also from conversion process of minor actinides (MA) into plutonium, which are mainly into even mass plutonium. These MA materials are estimated to be some fertile materials or intermediate materials from fertile to fissile after capturing neutrons.

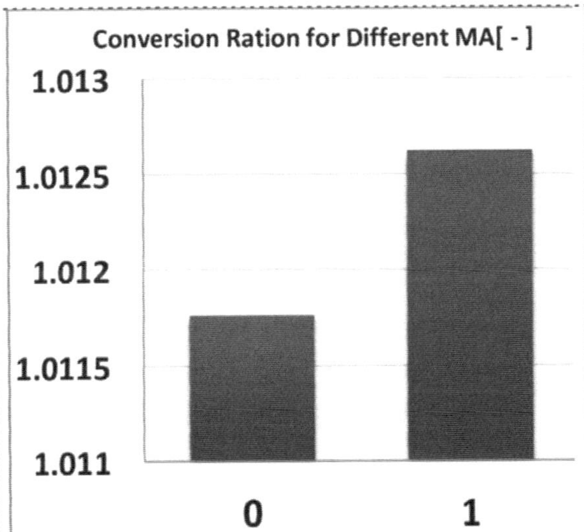

Figure 1 Fuel conversion ratio for Doping MA

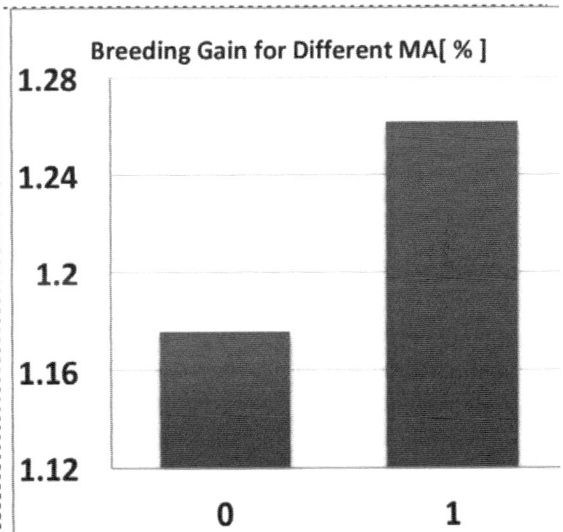

Figure 2 Fuel breeding gain for Doping MA

Some Doping MA materials are effective to increase plutonium production. The main contributions of total plutonium composition are coming from Pu-238, followed by Pu-239 or Pu-240 and Pu-241, Pu-242. MA doping is effective to increase the composition Pu-240 up to Pu-242 while it affects to decrease compositions of Pu-238 and Pu-239. Those even mass plutoniums are estimated to gives some contribution to better fel breeding condition as well as a composition of Pu-238 is suitable to protect the plutonium to be used for explosive devices based on some even number mass of plutonium isotope which can be used to some criteria of plutonium characterization such as IAEA, Pellaud and Kessler criteria (IAEA, 1972; Pellaud, 2002; and Kessler, 2004) [19-21].

3.2. Effect of very less MFR effect to fuel breeding capability

One of the reasons to have better fuel breeding capability was the hardening neutron spectrum during reactor operation. When neutron spectrum becomes harder, the potential to increase fuel conversion from capturing neutron by fertile material into fissile material become well. More neutrons are captured by fertile materials when the spectrum faster or harder. It can be estimated, if the volume ratio of moderator to fuel becomes smaller, the spectrum becomes harder because of less capability of water to moderate the neutron spectrum of the reactor. To estimate the hardening spectrum effect to the fuel conversion process, some less Moderator to fuel ratio have been used for less MFR of 1.0 and very less MFR of 0.3. MFR of 1.0 is adopted to have less moderator capability of water, which is equal to a half moderator to fuel ratio (MFR) of standard PWR. In addition, MFR of 0.3 is used to estimate the moderator ability for very low level of capability as well as in term of geometrical effect of MFR. Obtained results of fuel conversion ratio and breeding gain have been shown in **Figure 3** and **4** for different MFR values. All employed MFRs of 1.0 and 0.3 obtain more than unity which mean fuel-breeding capability is achieved for both MFRs. MFR of 1.0 give slightly more than unity and it increases with less MFR, which is shown by MFR of 0.3 that gives better fuel conversion capability, which is shown by Figure 3. In term of breeding gain or excess fuel breeding, MFR of 1.0 obtain about 2 % breeding gain and it becomes higher for MFR of 0.3 which produced almost 16% of breeding gain as shown in Figure 4. The results show that MFR of 0.1 gives a breeding capability and it will be better when the MFR is reduced. We can make the optimization for intender breeding gain, for example when we need a breeding gain of about 10 %, we can find a suitable MFR to be used for the reactor system as well as if we want to have a maximum MFR to be adopted to have a sustainable fuel breeding condition, which shows fuel breeding of about 1.0. Different MFR analysis gives a significant fuel breeding effect especially for less MFR condition. Less MFR give harder neutron spectrum and it will make the reactor more fuel breeding capability.

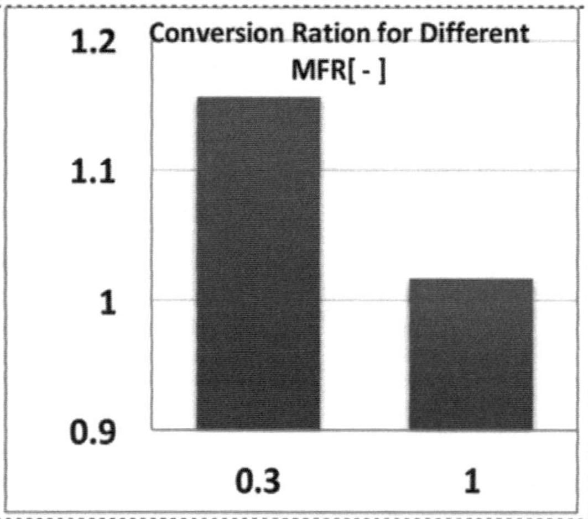

Figure 3 Fuel conversion ratio for different MFR values.

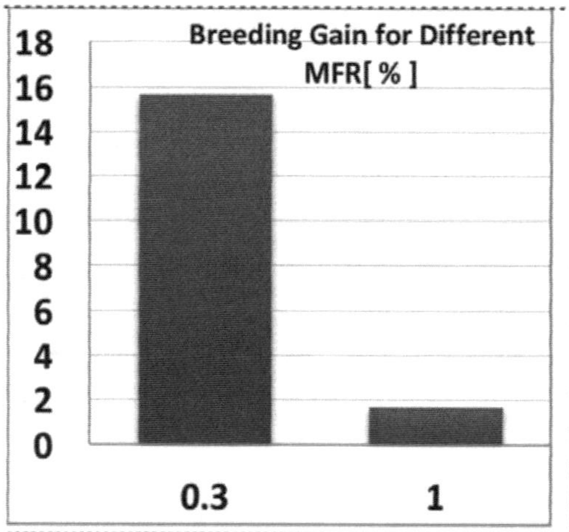

Figure 4 Fuel breeding gain for different MFR values

3.3. Fuel cell power density effect to fuel breeding capability

Reactor power output and operation are basically one of the options which is contributed to an economical point of view which also relates to capability to burn or burnup level of fuel as well as the length of reactor operation. Fuel pin cell evaluation, which regards to power density of the fuel, has been evaluated to analyse these effect to the fuel breeding capability. Power density level will shows how effective the power is produced in during operation and how many fuel are burnt during that process and what is the effect of power density level to the fuel breeding capability. Obtained results of power density effect to the fuel breeding capability and fuel breeding gain are shown in **Figure 5** and **6** for different power density of 140 and 280 W/cc. It shows less power density gives better fuel conversion capability. Power density of fuel for 140 W/cc obtains breeding condition than power density of 280 W/cc which gives less than unity for fuel breeding ratio. Breeding gain of about 3 % is given by power density of 140 W/cc and it becomes negative breeding gain for 280 W/cc which means breeding gain cannot be achieved because of fuel conversion ratio is less than unity. Less power density shows less power production is estimated to gives better fuel conversion capability while higher power density means more power production gives smaller fuel conversion ratio. It is because some fuels are converted to fissile material more for less burnt fuel to produce the power. It can be estimated to have a certain fuel power density to have some fuel breeding gain.

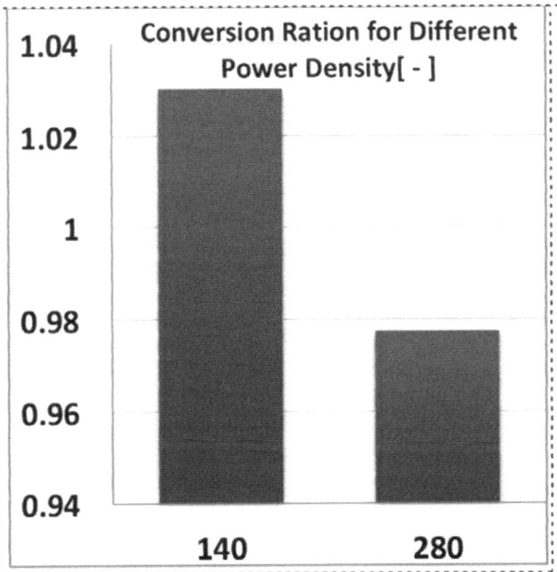

Figure 5 Fuel conversion ratio for different fuel power density

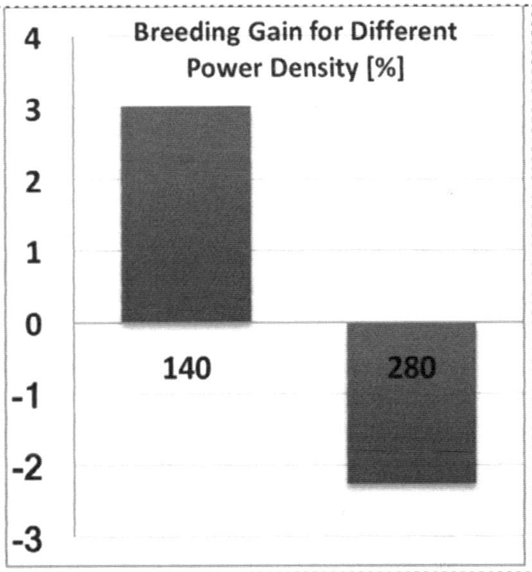

Figure 6 Fuel breeding gain for different fuel power density

3.4. MA Doping effect to void reactivity coefficient

Different fuel loadings in the reactor will affects to reactor behaviour including reactivity coefficient in term of void condition as void reactivity coefficient. A comparative analysis has been done to evaluate the MA doping effect to the void reactivity coefficient behaviour. A non-doping MA case and 1 % MA doping case are evaluated to estimate their effect. Obtained results of MA doping effect to void reactivity coefficient are shown in **Figure 7.** The figure shows all results give a negative value of void reactivity coefficient for both non-MA doping and MA doping cases. It shows MA doping case achieves less negative void reactivity coefficient than non-MA doping case. It is estimated that more MA material are inserted into the reactor will give the reactor more fission reaction when voided condition than non-MA doping case. Some converted materials into plutonium or trans-plutonium have some contributions to make less void coefficient.

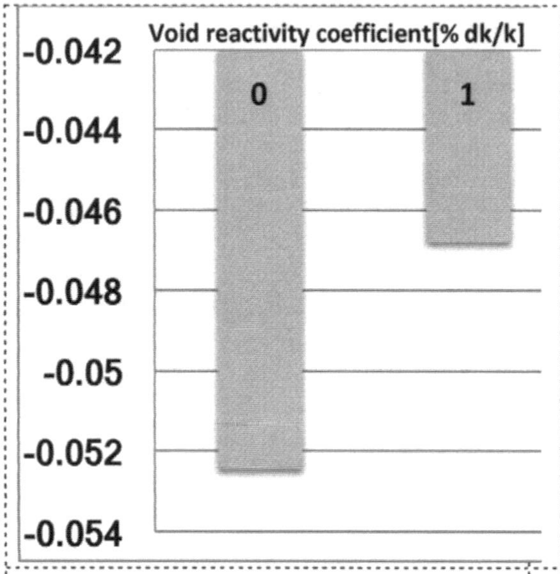

Figure 7 Void Reactivity Coefficient for different MA doping

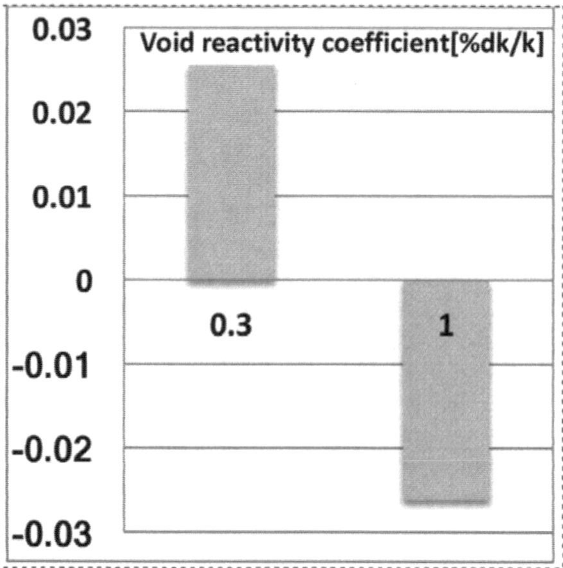

Figure 8 Void Reactivity Coefficient for different MFR

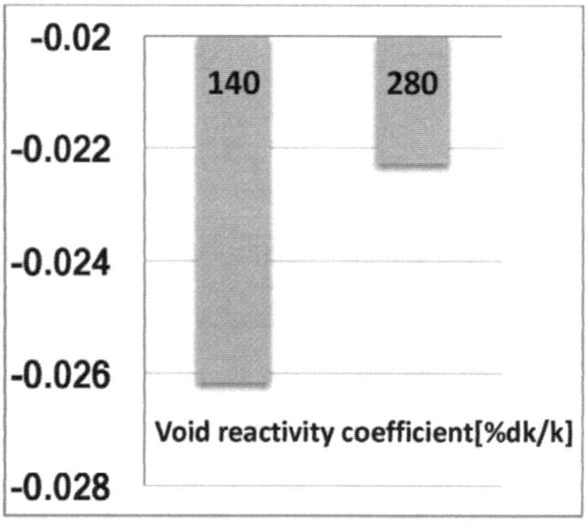

Figure 9 Void Reactivity Coefficient for different power density

3.5. MFR effects to void reactivity coefficient

As mentioned in the previous section, an effect of different moderation ratio was evaluated to the fuel conversion ratio condition. In this section these MFR different effect will be evaluated to have some significant effect to void reactivity coefficient. Obtained results of MFR effect to void reactivity coefficient are shown in Figure 8 for MFR of 1.0 and MFR of 0.3. Void coefficient of both MFRs are totally different which shows a negative void coefficient value for MFR of 1.0 and a positive value of void coefficient for MFR of 0.3. MFR 1.0 obtains negative coefficient was estimated because of the neutron spectrum is still can be thermalized and some contribution of fissile material to make less criticality or less fission reaction in voided condition. While very low MFR of 0.3 gives less thermalization factor and harder spectrum is achieved as well as some fissile materials will have some more criticality or fission reaction during voided condition. It can be estimated in certain MFR values a condition will have a threshold value for void reactivity coefficient to become positive or negative values. A negative value of void reactivity coefficients is more preferable to be adopted instead of a positive value, because of some safety concern of criticality condition.

3.6. Fuel cell power density effect to void reactivity coefficient

In this section will be shown and discussed some void reactivity coefficient from different fuel power density in the fuel cell. Obtained results of power density differences to the void reactivity coefficient are shown in **Figure 9**. It shows that all power density conditions give a negative value of void coefficient. It shows less power density gives more negative void coefficient, while higher power density obtains less void coefficient. It is estimated that more power density obtains harder neutron spectrum and some produced fissile material will make more fission reaction when voided condition in comparing with less power density condition. More power density also can be estimated to be more power is produced that gives less void reactivity for more power reactor. More power reactor also can be used for more nuclear materials especially fissile material is burnt in the reactor.

4. Conclusions

Some nuclear fuel sustainability analyses have been done based on heaevy water-cooled thorium reactor system and the fuel recycled composition which all actinide compositions except fission products are recycled into the reactor. Some key important parameters are evaluated such as doping composition of minor actinide (MA), volume ratio of moderator to fuel (MFR) and fuel cell power density which have some significant impact to the reactor performances such as criticality, fuel breeding ratio and void reactivity coefficient. As a fuel basis, thorium is used with can be mixed with additional recycled spent nuclear fuels. Some minor actinides (MA) as recycled fuels are used as doping material to be loaded to the water cooled reactors with thorium fuel as fuel basis and heavy water as moderator and coolant. Effect of some actinide composition to fuel breeding capability as well as safety aspect, which is based on void reactivity coefficient have been investigated. Fuel breeding capability can be obtained by the present reactor systems; as well as negative void reactivity has been show for more moderator ratio and less power density. Low portion of moderation to fuel ratios (MFR) are used to have a better fuel breeding capability as well as some from contribution from recycled fuel of minor actinides (MA) and less power density. A negative void reactivity can be obtained in this system and it becomes less negative for doping MA and more power density as well as a positive value of void coefficient for much less moderation ratio.

Acknowledgments

We would like acknowledge and extend our gratitude to desentralisasi research program of ministry of research, technology and higher education and research innovation program of ITB for the grant and international conference publication

References

[1] President's meeting of G8 countries, Brazil, China, India and South Africa, 2006 Moscow April 19-20, Contribution of Coal and Nuclear to Sustainable Energy Supply: Perspectives and Problems, Paper presented at *the Russian Academy of Sciences, Academies of Sciences*

[2] P. Sidik, N. Takaki, H. Sekimoto 2007 "Feasible region of design parameters for water cooled thorium breeder reactor," *J. Nucl. Sci. Technol*, **44**, 7, 946-957

[3] T. K. Kim and T. J. Downar 2002 *Nuclear Technology*, **138**,[1] 17-29

[4] L. Michael, G. Otto 1998 *Nucl. Eng. Des*. **180**, 2, 133

[5] U. Turan 2000 *Prog. Nucl. Energy*, **37**, 137

[6] L. B. Freeman et al., 1989 *Nucl. Sci. Eng*., **102**, 341

[7] P. Sidik, N. Takaki, H. Sekimoto, 2008 *J. Nucl. Sci. Technol*.,. **45**, 7, 589-600

IOP Conf. Series: Journal of Physics: Conf. Series **1090** (2018) 012061 doi:10.1088/1742-6596/1090/1/012061

[8] P. Sidik, N. Takaki, H. Sekimoto, 2006 *Ann. Nucl. Energy*, **33**, 561

[9] H. Sekimoto, N. Takaki, 1991 *J. Nucl. Sci. Technol.*, **28**, 941

[10] A. Waris, H. Sekimoto 2001 *J. Nucl. Sci. Technol.*, **38**, 7, 517

[11] A. Waris, H. Sekimoto 2001 *Ann. Nucl. Energy*, **28**, 153

[12] A. Mizutani, H. Sekimoto 1998 *Ann. Nucl. Energy*, **25**, 9, 623

[13] A. Mizutani, H. Sekimoto 1998 *Ann. Nucl. Energy*, **25**, 13, 1011

[14] A. Mizutani, H. Sekimoto 1998 *Prog. Nucl. Energy*, **32**, 713

[15] A. Mizutani, H. Sekimoto 1997 *J. Nucl. Sci. Technol.*, **34**, 6, 596 (1997).

[16] T. Seino, H. Sekimoto 1998 *Ann. Nucl. Energy*, **25**, 4–5, 223

[17] K. Okumura et al., 1996 *Japan Atomic Energy Research Institute (JAERI)*

[18] T. Nakagawa et al., 1995 *J. Nucl. Sci. Technol.*, **32**, 1259

[19] International Atomic Energy Agency, I972, Information Circular, INFCIRC/153

[20] B. Pellaud, 2002 *J. Nucl. Mater. Managements*, XXXI, No. 1

[21] G. Kessler, 2004 The First Int. Sci. Technol. Forum on Protected Plutonium Utilization for Peace and Sustainable Prosperity, March 1–3, Tokyo, Japan

IOP Conf. Series: Journal of Physics: Conf. Series **1090** (2018) 012062 doi:10.1088/1742-6596/1090/1/012062

Implementation of Rivest Shamir Adleman Cryptographic Algorithms and Techniques of Steganography First of File for Message Security

Dian Rachmawati[1] *, Yeni Rosalin Munthe[1*]

[1] Departemen Ilmu Komputer, Fakultas Ilmu Komputer dan Teknologi Informasi, Universitas Sumatera Utara, Jl. Universitas No. 9-A, Medan 20155, Indonesia.

*Email: dian.rachmawati@usu.ac.id, yenrosm@gmail.com

Abstract. Accessing data by an unauthorized person often happens these days. Addressing this case, the techniques of cryptography and steganography was necessary. The science of cryptography use randomizes/ disguise the data, whereas the science of steganography to hide data in a way. The cryptographic algorithms to be used is the Rivest Shamir Adleman (RSA) which utilizes two prime numbers to generate a public key and private key. To further enhance the security of message, the algorithm has combined with the First of File (FoF) as steganography techniques. The technique First of File (FoF) works by inserting a message at the beginning of the file image horizontally to produce a stego image that will extract to get the previous ciphertext. Then the ciphertext decrypted to be returned to the original message. By using FoF, in the cover image with large pixel size, the insertion process with FoF is entirely invisible, so the stego image and original image cannot be distinguished, whereas in small pixels there is a black line above the image. A combination of RSA and the FOF will increase data security because when a message has scrambled into a ciphertext and the ciphertext will be inserted into the image file.

1. Introduction

The primary purpose of data security is to secure data/information from abuse by irresponsible parties. With data security, actions such as sabotage and unauthorized data exchange can be prevented. But data security has not been a top priority for most people at the moment. It frequently found incidents related to data security disorders. Therefore, to overcome these circumstances can be used some science from data security majors such as cryptography and steganography.

Overall the concept of cryptography is to scramble or disguise the meaning of a message but not hide it so it can still be seen and can cause suspicion. The technique of steganography has developed to avoid suspicion. In general, steganography is an art or science that is used to hide secret messages in any way that other than the intended person, others will not be aware of the existence of the secret message. Secret hidden data can also be image, audio, text, or video. Steganography demonstrates an option method to conceal the presence of information in the cover picture. The mystery message is recovered negligible by approved individual [1]. There are a lot of steganography techniques for hiding secret

Content from this work may be used under the terms of the Creative Commons Attribution 3.0 licence. Any further distribution of this work must maintain attribution to the author(s) and the title of the work, journal citation and DOI.
Published under licence by IOP Publishing Ltd

information in a cover object. Some methods are more complicated than others, but all of them have respective strong and weak points [2].

The use of a cryptographic method or even a steganography method can no longer guarantee the security of a message. This is due to the rapid development of cryptanalysis science. In this research used the combination of cryptography and steganography method, so the use of a combination of these two methods is expected to make the message security becomes more powerful. In this research, the cryptographic algorithm used is Rivest Shamir Adleman (RSA) and steganography method used is First of File (FoF).

2. Method

2.1. *Rivest Shamir Adleman*

RSA (from Rivest-Shamir-Adleman) is a public-key cryptosystem based on exponentiation in finite fields over integers modulo N where N is a composite integer of two significant factors (i.e., a semi-prime) [3]. It has based on the mathematical operation that is very easy to compute in one direction, calculating the product of two large primes. On the other hand, the different course is tough to do, factoring the modulus equal to the production of two of large primes[4].

RSA cryptographic public key algorithm is illustrated by a sender and receiver with Alice as the sender and Bob as the recipient of the message [5].

Step 1. Generating RSA keys

(1) Bob generates two large primes p and q at random, where $p \neq q$.

(2) Calculated the value of p and q to produce $n = p \times q$ and searched for its totient value $\phi(n)$ = (p-1) (q-1). The n number of RSA is called the modulus.

(3) Then Bob specifies a random integer for the encryption key (e) where $1 < e < \phi(n)$ and the value is relatively primed with $\phi(n)$. GCD(e, $\phi(n)$ = 1). RSA is called enciphering exponent.

(4) With the Euclidean algorithm, Bob gets the decryption key (d) where $e \times d \equiv 1 (\mathrm{mod}\ \phi(n))$ with $1 < d < \phi(n)$. The encryption value d in RSA is called deciphering exponent.

Bob publishes (n, e) as the public key and stores the d, p, q, and (n) values as private keys.

Step 2. Enciphering Stage

Alice has a message or plaintext (m) where m <n, and GCD (m, n) = 1.Alice gets Bob's public key (n, e) from the database and encrypts m into ciphertext with computation c≡m ^ e (mod n). Then Alice sends ciphertext (c) to Bob.

Step 3. Deciphering Stage

When Bob gets the ciphertext (c) from Alice, then Bob will use the private key d to convert the ciphertext to plaintext. M≡c ^ d (mod n).

2.2. *First of File (FoF)*

First of File is a method or steganography technique to hide or insert messages at the beginning of the file horizontally. This technique can be used to add messages to the short length or long length. The file size which the message has inserted is larger than the original file. Images created from the pixel, pixel generated by using these three colors red, green and blue have called RGB [6]. An example of the RGB value of an image before and after the message inserted with the FoF method has shown in table 1.

Table 1. RGB value of an image before and after the message inserted

R : 90	R : 87	R : 68
G : 43	G : 89	G : 72
B : 78	B : 48	B : 34
R : 56	R : 68	R : 90
G : 93	G : 72	G : 43
B : 56	B : 34	B : 78

R : 11	R : 15	R : 16
G : 14	G : 10	G : 23
B : 17	B : 17	B : 16
R : 90	R : 87	R : 68
G : 43	G : 89	G : 72
B : 78	B : 48	B : 34
R : 56	R : 68	R : 90
G : 93	G : 72	G : 43
B : 56	B : 34	B : 78

The left of table 1 shown the RGB value of the original image and the right of table 1 shown the RGB value of stego image after hiding message which has displayed in the red amount.

3. Results and Discussions

The combined use of cryptography (RSA) and steganography (FoF) in this study are shown in Figure 1 below:

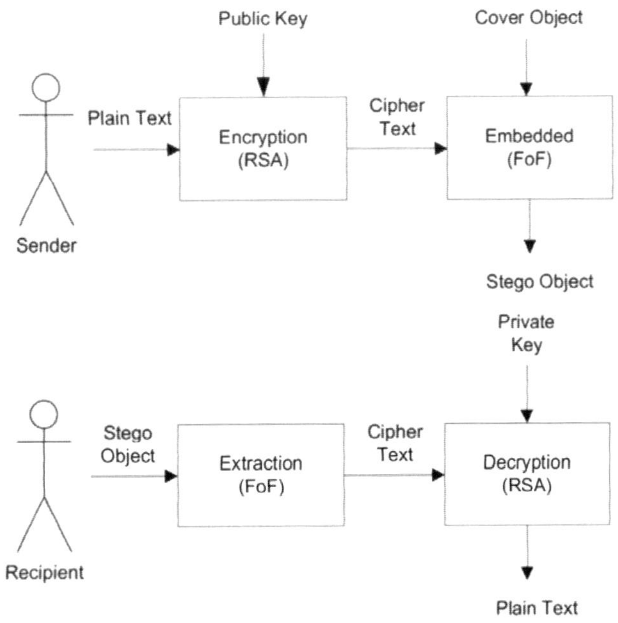

Figure 1. Combination of RSA and FoF

Figure 1 has shown the combination of RSA and FoF process which is sender do encryption by using public key and embedded operation, and the receiver does extraction and decryption by using private key.

The process of RSA shown by the calculation below : the key will be generated by taking 2 primes for example p and q are P = 23, Q = 13, N = p • q , N = 299, $\phi(n)$= (p - 1) • (q - 1) , $\phi(n)$= 264. Find e value with formula 1 <e < $\phi(n)$, gcd (e, $\phi(n)$)=1, e = prime number Let e = 137, Looking for value d with the formula $d = e^{-1} \bmod \varphi(n)$, d= 185. RSA algorithm is used to encrypt messages. The character will be changed first to the decimal form according to the ASCII table, after which the characters will be encrypted per each character. For example, there is a message = Dian Rachmawati. This message will be converted to integer number based on ASCII table. The message numbers are 068 105 097 110 032 082 097 099 104 109 097 119 097 116 105. By using the formula $c \equiv m \wedge e \pmod{n}$. The message will changed to 22 27 158 210 54 257 158 86 156 44 158 58 158 116 27. This integer numbers will be embedded to cover image by using FoF as shown by table 2.

Table 2. RGB value of an image after the message Dian Rachmawati inserted

R=22	R=210	R=158	R=44	R=158	R=0	R=0	R=0	R=0	R=0	R=0	R=0	R=0
G=27	G=54	G=86	G=158	G=116	G=0	G=0	G=0	G=0	G=0	G=0	G=0	G=0
B=158	B=257	B=156	B=58	B=27	B=0	B=0	B=0	B=0	B=0	B=0	B=0	B=0
R=52	R=42	R=48	R=61	R=80	R=97	R=52	R=52	R=52	R=80	R=52	R=61	R=80
G=89	G=79	G=82	G=93	G=111	G=120	G=89	G=89	G=89	G=111	G=89	G=93	G=111
B=118	B=106	B=107	B=116	B=132	B=134	B=118	B=118	B=118	B=132	B=118	B=116	B=132
R=58	R=49	R=46	R=50	R=54	R=75	R=58	R=58	R=75	R=58	R=50	R=50	R=54
G=86	G=77	G=78	G=86	G=98	G=111	G=86	G=86	G=111	G=86	G=86	G=86	G=98
B=107	B=98	B=101	B=110	B=125	B=133	B=107	B=107	B=133	B=107	B=110	B=110	B=125
R=69	R=60	R=62	R=56	R=52	R=73	R=69	R=73	R=62	R=56	R=46	R=56	R=52
G=65	G=87	G=91	G=87	G=89	G=102	G=65	G=102	G=91	G=87	G=78	G=87	G=89
B=32	B=106	B=109	B=107	B=108	B=116	B=32	B=116	B=109	B=107	B=101	B=107	B=108

Extracting process do by collecting the RGB values from the First of File which is contained numbers 22 27 158 210 54 257 158 86 156 44 158 58 158 116 27. By using private key d and formula $m \equiv c \wedge d \pmod{n}$. The message will be back to 68 105 97 110 32 82 97 99 104 109 97 119 97 116 105 and has converted to ASCII table as a string Dian Rachmawati. The combination of RSA and FoF can protect the message and also hide it from an intruder.

The experimental result by using message Dian Rachmawati Program Studi Ilmu Komputer Universitas Sumatera Utara. The time that it takes to do encryption is 7.54 seconds, and embedding is 0,67 seconds. The time that it takes to do extraction is 0.43 seconds and decryption is 8.2 seconds. The experimental result by using message Dian Rachmawati. The time that it takes to do encryption is 4.13 seconds, and embedding is 0,65 seconds. The time that it takes to do extraction is 0.41 seconds and decryption is 4.7 seconds.

Other experiments are using various types of pixel image as a cover image to hide string Dian Rachmawati Program Studi Ilmu Komputer Universitas Sumatera Utara. The results have shown in Table 3.

Table 3. Various Types of Pixel Image as a Cover Image

No.	*Pixel*	*Cover Image*	*Stego Image*
1	1800 x 1350	 7311 KB	 9761 KB
2	900 x 650	 1873 Kb	 2437 Kb
3	300 x 255	 214 Kb	 284 Kb
4	75 x 56	 14 Kb	 19 Kb

4. Conclusions

The conclusion of this research is the combination of RSA and FoF able to secure the message by encrypting and hiding, so it is not to arouse suspicion for those who see the image. The combination of both algorithms is also able to restore the message into its original form. In the cover image with large pixel size, the insertion process with FoF is entirely invisible, so the stego image and original image cannot be distinguished, whereas in small pixels there is a black line above the image (the place which the message hide). The time required for encryption and decryption is directly proportional to the length of the message.

Acknowledgments

The authors gratefully acknowledge that the present research was supported by Fund Dissemination IPTEKS Research Results for Lecturers / Researchers Universitas Sumatera Utara.

References

[1] M. Jain, R. C. Choudhary and A. Kumar, *Secure medical image steganography with RSA cryptography using decision tree*, 2016 2nd International Conference on Contemporary Computing and Informatics (IC3I), Noida, 2016, pp. 291-295. doi: 10.1109/IC3I.2016.7917977

[2] M. Saritha, V. M. Khadabadi and M. Sushravya, *Image and text steganography with cryptography using MATLAB,* 2016 International Conference on Signal Processing, Communication, Power and Embedded System (SCOPES), Paralakhemundi, Odisha, India, 2016, pp. 584-587. doi: 10.1109/SCOPES.2016.7955506

[3] M. Kiviharju, *On the fog of RSA key lengths: Verifying public key cryptography strength recommendations*, 2017 International Conference on Military Communications and Information Systems (ICMCIS), Oulu, Finland, 2017, pp. 1-8. doi: 10.1109/ICMCIS.2017.7956481

[4] K. Somsuk, *The improving decryption process of RSA by choosing new private key*, 2016 8th International Conference on Information Technology and Electrical Engineering (ICITEE), Yogyakarta, 2016, pp. 1-4. doi: 10.1109/ICITEED.2016.7863242

[5] Mollin, R. A. 2002. RSA and Public-Key Cryptography. 2nd Edition. Chapman & Hall/CRC: Florida.

[6] D Rachmawati, A Amalia, and J Surya, *Combination of huffman coding compression algorithm and least significant bit method for image hiding,* Journal of Physics: Conference Series 801 (2017), no. 1, 012059.

IOP Conf. Series: Journal of Physics: Conf. Series **1090** (2018) 012063 doi:10.1088/1742-6596/1090/1/012063

Isogeometric Analysis for Thin Square Bending Plate using Collocation Method

Irwan Katili, Budianto Jaya, Imam Jauhari Maknun, Mulia Orientilize

Universitas Indonesia, Civil Engineering Department, Depok 16424, Indonesia

E-mail : irwan.katili@gmail.com

Abstract. This paper discusses the evaluation of rectangular thin plate problem with uniform load using Isogeometric analysis (IGA) collocation method. The implementation details of the proposed method for problems with different boundary conditions are shown. The numerical results confirm that the proposed method represents an efficient formulation and shows good behavior for thin plate structures.

1. Introduction

Finite Element Analysis (FEA) is the most popular and powerful numerical method that is widely used in many engineering applications. The results of our research in Finite Element Method have been published in [1-13]. In general, finite element analysis (FEA) is unable to represent exact geometry of structural problems. It needs to approximate the geometry of structure to obtain representative results with a little deviation from the exact solutions. The shape functions and meshes which is used in FEA are the important factor that are able to increase the accuracy of FEA results. Shape functions of FEA can be either linear or quadratic which requires finer meshes to represent actual geometry. Meanwhile, isogeometric analysis uses a high-regularity property of its basis functions which makes IGA able to approximate the exact geometry faster and minimize the usage of finer meshes by standard FEA.

Isogeometric analysis (IGA) is proposed by Hughes and Cottrell [14] in 2005 by adapting model of geometry by using NURBS [15]. The basic IGA consists of the same basis functions used for geometry representations in CAD systems. In symmetrical problems such as rectangular plate, NURBS is the same as B-Splines. Development of IGA using MATLAB is also done by Nguyen *et.al.* [16,17], Auricchio *et.al.* [18] developed IGA collocation method. The basic idea consists of the discretization of the governing partial differential equations in strong form. Reali and Hughes [19] improved collocation approach and Reali and Gomez [20] using collocation method for Bernoulli-Euler beams and Kirchhoff plates. Beirão da Veiga *et.al* [21] have considered an IGA collocation approach for the approximation of initially straight planar Timoshenko beams. Kiendl *et.al.* [22] proposed IGA collocation for the numerical simulation of Reissner–Mindlin plate problems. In 2017, the development of Isogeometric Galerkin based on UI approach have been proposed by Katili [23]

In this paper, we focus in discussing IGA collocation for the solution of thin plate structure described by fourth-order differential equations of Kirchhoff plate models. The numerical evaluation is focused on displacement and bending moment at the center of plate. The purpose of the numerical tests is to study the convergence speed of the method to the reference solutions by increasing the polynomial degree (p = 4, 6, 8, 10) with only one element and increasing the number of elements with p = 4. In this paper three types of boundary condition will be evaluated. The implementation of IGA collocation for rectangular thin plates with several boundary conditions confirm the good behavior of the proposed formulations.

Content from this work may be used under the terms of the Creative Commons Attribution 3.0 licence. Any further distribution of this work must maintain attribution to the author(s) and the title of the work, journal citation and DOI.
Published under licence by IOP Publishing Ltd

2. Principal of Kirchhoff Plate theory

In Kirchhoff plate theory the relation between vertical displacement and rotation is

$$\{\beta\} = \begin{Bmatrix} \beta_x \\ \beta_y \end{Bmatrix} = \begin{Bmatrix} -w_{,x} \\ -w_{,y} \end{Bmatrix} \tag{1}$$

Where w is the total displacement (deflection) of the points on the middle plane ($z = 0$), β_x is the rotation in the plane z-x and β_y is the rotation in the plane z-y. The first derivative of total displacement with respect to x and y are denoted by $w_{,x}$ and $w_{,y}$ respectively. The curvatures

$$\{\chi\} = \begin{Bmatrix} \chi_x \\ \chi_y \\ \chi_{xy} \end{Bmatrix} = \begin{Bmatrix} \beta_{x,x} \\ \beta_{y,y} \\ \beta_{x,y} + \beta_{y,x} \end{Bmatrix} \tag{2}$$

Where $\beta_{x,x}$ and $\beta_{x,y}$ are the first derivative of β_x with respect to x and y, respectively, while $\beta_{y,x}$ and $\beta_{y,y}$ are the first derivative of β_y with respect to x and y, respectively. Substituting (1) into (2) gives

$$\{\chi\} = \begin{Bmatrix} \chi_x \\ \chi_y \\ \chi_{xy} \end{Bmatrix} = -\begin{Bmatrix} w_{,xx} \\ w_{,yy} \\ 2w_{,xy} \end{Bmatrix} \tag{3}$$

The bending moment $\{M\}$ are defined as the following constitutive equations

$$\{M\} = \begin{Bmatrix} M_x \\ M_y \\ M_{xy} \end{Bmatrix} = [H_b] \begin{Bmatrix} \chi_x \\ \chi_y \\ \chi_{xy} \end{Bmatrix} \tag{4}$$

Where $[H_b]$ is

$$[H_b] = D_b \begin{bmatrix} 1 & \upsilon & 0 \\ \upsilon & 1 & 0 \\ 0 & 0 & \frac{1-\upsilon}{2} \end{bmatrix} \quad ; \quad D_b = \frac{Eh^3}{12(1-\upsilon^2)} \tag{5}$$

D_b is the generalized bending constitutive matrix, E is the Young's Modulus, υ is the Poisson's ratio, and h is the plate thickness. The equilibrium equations of plate

$$M_{x,x} + M_{xy,y} - T_x = 0 \tag{6}$$

$$M_{y,y} + M_{xy,x} - T_y = 0 \tag{7}$$

$$T_{x,x} + T_{y,y} + f_z = 0 \tag{8}$$

Where f_z is distributed load. Differentiating the shear forces in (6) and (7) with respect to x and y respectively, and substituting the derivatives into (8) gives

$$M_{x,xx} + 2M_{xy,xy} + M_{y,yy} = f_z \tag{9}$$

This is the second order of differential equation in Kirchhoff plate model. In order to use collocation method in isogeometric analysis, it is necessary to obtain relationship between uniform loading and displacement. The relationship is called strong form. Substituting equation (3) into equation (4) and then substituting the result into (9) give the fourth order differential equation relating to the Kirchhoff thin plate theory for isotropic material.

$$D_b\left(w_{,xxxx} + 2w_{,xxyy} + w_{,yyyy}\right) = f_z \tag{10}$$

Equation (10) is the strong form equation. Since uniform loading is fourth derivative of displacement, the minimum polynomial degree suitable for collocation method is 4. Boundary conditions are needed to solve the problem. The boundary conditions are

$$\bar{w} = w \rightarrow \Gamma_w \tag{11}$$

$$\bar{\beta}_x = -w_{,x} \rightarrow \Gamma_{\beta_x} \tag{12}$$

$$\bar{\beta}_y = -w_{,y} \rightarrow \Gamma_{\beta_y} \tag{13}$$

$$\bar{M}_x = -D_b\left(w_{,xx} + \upsilon w_{,yy}\right) \rightarrow \Gamma_{M_x} \tag{14}$$

$$\bar{M}_y = -D_b\left(w_{,yy} + \upsilon w_{,xx}\right) \rightarrow \Gamma_{M_y} \tag{15}$$

$$\bar{T}_x = -D_b\left(w_{,xxx} + w_{,xyy}\right) \rightarrow \Gamma_{T_x} \tag{16}$$

$$\bar{T}_y = -D_b\left(w_{,yyy} + w_{,xxy}\right) \rightarrow \Gamma_{T_y} \tag{17}$$

3. Isogeometric analysis

3.1. B-Splines function

Basis function is the foundation of isogeometric analysis. B-splines as basis function are generated from knot vectors. Knot vectors value start from 0 and end with 1.

Knot vector is relying on two factors that is polynomial degree and number of elements. Polynomial degree affects knot vector in terms multiplicity. Multiplicity only applies at the first and the last knot value (knot zero and one) to amount of p+1 for each knot. Multiplicity has open knot vector behavior since B-splines is a recursive function. Number of elements affects knot span value. Increasing the number of elements will increase the number of knot span and increase number of knots in knot vector. B-spline is a recursive function meaning in order to obtain higher polynomial results, it is necessary to calculate using lower polynomial degree first. For polynomial degree of four (p = 4), we need to calculate B-spline function from polynomial degree of zero and recalculate it up to polynomial degree of four. A knot vector in direction ξ and η

$$\Xi_\xi = \left[\xi_1, \xi_2, \quad, \xi_{n+p+1}\right] \quad ; \quad \Xi_\eta = \left[\eta_1, \eta_2, \quad, \eta_{m+q+1}\right] \tag{18}$$

Where:

m and n are the number of B-Splines function in each direction.

p and q are polynomial degree of B-Splines in ξ and η directions, respectively.

In each direction, the number of knot in knot vector depends on both multiplicity and knot span. According to open knot vector properties applied in (18), there are multiplicity of the first (ξ_1 and η_1) and the last knot vector (ξ_{n+p+1} and η_{m+q+1}), as many as (p + 1) for ξ axis and (q + 1) for η axis. On the other hand, knot span is a range of value between knot ξ_i and ξ_{i+1}. The number of knot span depends on the number of element in each direction. When the number of element increases, the number of knot span increases, which also increases the number of knot in knot vector. The number of knot in knot vector is (n + p + 1) for Ξ_ξ and (m + q + 1) for Ξ_η. From the number of knot in knot vector, the number of basis functions can be known.

The zeroth degree of B-Splines [15] in ξ direction is defined by:

$$N_{i,0}(\xi) = \begin{cases} 1, & \text{if } \xi_i \le \xi < \xi_{i+1} \\ 0, & \text{otherwise} \end{cases} \tag{19}$$

The p-th degree of B-Splines in ξ direction is defined recursively using the relation

$$N_{i,p}(\xi) = \frac{\xi - \xi_i}{\xi_{i+p} - \xi_i} N_{i,p-1}(\xi) + \frac{\xi_{i+p+1} - \xi}{\xi_{i+p+1} - \xi_{i+1}} N_{i+1,p-1}(\xi) \tag{20}$$

The zeroth degree of B-Splines $\{M_j, 0\}$ in η direction is defined as

$$M_{j,0}(\eta)=\begin{cases}1, & \text{if } \eta_i \le \eta < \eta_{j+1}\\ 0, & \text{otherwise}\end{cases} \tag{21}$$

The *q-th* degree of B-Splines in η direction is defined recursively using the relation

$$M_{j,q}(\eta)=\frac{\eta-\eta_j}{\eta_{j+q}-\eta_j}M_{j,q-1}(\eta)+\frac{\eta_{j+q+1}-\eta}{\eta_{j+q+1}-\eta_{j+1}}M_{j+1,q-1}(\eta) \tag{22}$$

Thereafter, in rectangular plate, we need B-Splines in two dimension. The two-dimension B-Splines can be defined by taking tensor products of their B-Splines in one-dimensional counterparts i.e. $N_i(\xi)$ in ξ direction and $M_j(\eta)$ in η direction.

$$R_k = R_{i,j}(\xi,\eta) = N_i(\xi)M_j(\eta) \tag{23}$$

Where:

$i=1,...,n$; $j=1,...,m$; $k=i+n(j-1)$; m and n is the number of basis function in ξ and η directions

3.2. Displacement function and control variable

Control variable in isogeometric analysis is analogous to degree of freedoms in finite element analysis. However, the control variable values are not the real displacements but the values are required to obtain displacement function needed to get the real displacement value at specified point. This is described as:

$$w(\xi,\eta)=\langle R(\xi,\eta)\rangle\{\hat{w}_n\} \tag{24}$$

where $R(\xi,\eta)$ is basis function for rectangular plate and $\{\hat{w}_n\}$ are control variables. The function above shows us that basis function multiplies by control variable will give displacement function. This means that this function is the connection between isogeometric analysis and finite element.

3.3. Collocation method

The main idea of collocation method is discretizing strong form equations (10) and collocating them on a set of suitable points. For collocation points, we use Greville abscissae which are knot averages defined by knot vector and polynomial degrees. Collocation points are related to boundary of element in parametric space. The number of collocation points are the same as the number of control variables. This is described as:

$$\bar{\xi}_i=\frac{1}{p}\sum_{j=1}^{p}\xi_{i+j}\ ;\ i=1,...,n \qquad \bar{\eta}_j=\frac{1}{q}\sum_{l=1}^{q}\eta_{k+l}\ ;\ k=1,...,n \tag{25}$$

Collocation points are stated in term of coordinate $(\bar{\xi}_i,\bar{\eta}_i)$ to be used in strong form and boundary condition function. Using equation (10–25), we are able to obtain control variable which will be used to get real displacement, rotation, moment or shear values at any points.

4. Numerical Examples

This section shows the numerical results of IGA collocation in thin plate (for $L/h = 100$). The numerical results are focused on displacement and bending moment at the center of plate. The purpose of the numerical tests is to study the convergence speed of the method to the reference solutions using two ways:

1. Increasing the polynomial degree (p = 4, 6, 8, 10) with only one element (NELT = 1×1)
2. Increasing the number of elements (NELT = 1×1, 2×2, 4×4, 8×8) with $p = 4$ three types of boundary conditions are considered
   a. Fully clamped plate (CCCC)
   b. Simply supported plate (SSSS)
   c. Simply supported plate along $x = 0$ and $x = L$ (SSFF)

4.1 *Fully clamped thin plate (CCCC)*

The first case is Fully clamped thin plate (CCCC) as in Figure 1. To solve this case, we need to apply some boundary conditions that are displacement, and rotations in x and y direction.

From Figure 2 and Table 1, we are able to obtain convergence either by polynomial degree escalation or by number of element addition. Analyzing based on polynomial degree, p = 4 gives 38.14% error compared to the reference solution given by Batista [24], and by increasing polynomial degree to p = 10 the error decreases to 1.47 %. Analyzing based on number of element addition, by increasing element to 8×8, we are able to obtain 0.94% error with respect to the reference by Batista [24].

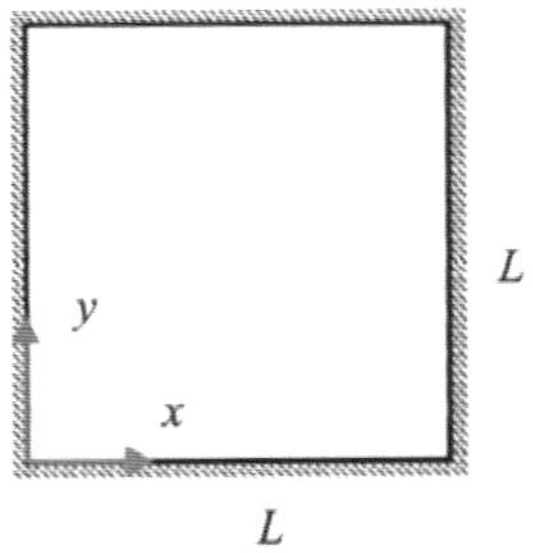

Dimension

L = 1000

h = 10

Material Properties

E = 10.92

υ = 0.3

Uniform Load

$f_z = -1$

Figure 1. CCCC plate

Table 1. Displacement at the center of CCCC thin plate

p Escalation (NELT = 1×1)			NELT Addition (*p* = 4)		
p	*w*	Error (%)	NELT	*w*	Error (%)
4	-782522.32	38.14	1×1	-782522.32	38.14
6	-1236782.82	28.30	2×2	-988552.39	21.85
8	-1268146.62	2.23	4×4	-1232275.00	2.59
10	-1267844.27	1.47	8×8	-1276935.65	0.94
Ref. -1265000 (Batista [24])					

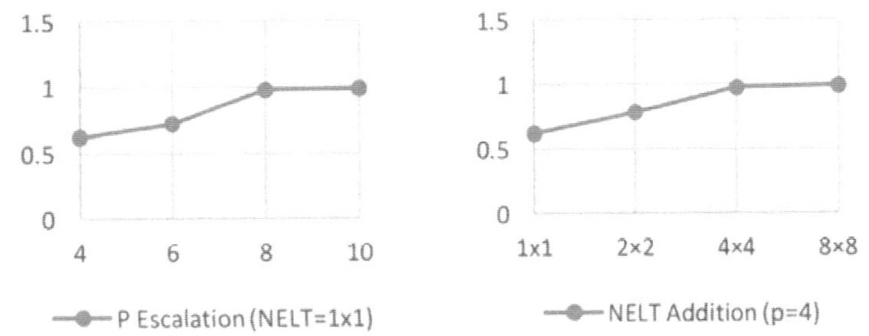

Figure 2. Convergence of normalized displacement at the center of CCCC thin plate

Table 2. Moment at the center of CCCC thin plate

p Escalation (NELT = 1×1)			NELT Addition (*p* = 4)		
p	$M_x = M_y$	Error (%)	NELT	$M_x = M_y$	Error (%)
4	-16250.00	29.65	1×1	-16250.00	29.65
6	-22571.68	2.29	2×2	-20533.07	11.11
8	-22916.49	0.79	4×4	-23448.50	1.51
10	-22904.58	0.85	8×8	-23219.41	0.52
Ref. -23100 (Batista [24])					

Figure 3 and Table 2 present the convergence of moment in the center of plate for the CCCC case. Convergence of moment can be obtained either by polynomial degree escalation or by number of element addition. Convergence by polynomial degree escalation is reached by p = 6 and higher. Satisfying results of number of element addition are reached when 4×4 elements or finer meshes are applied.

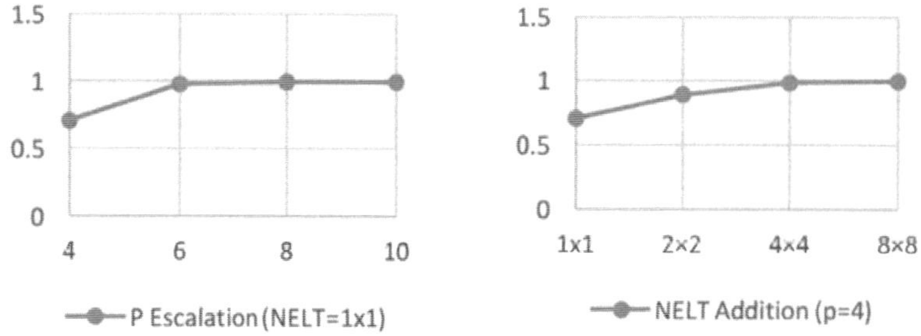

Figure 3. Convergence of normalized moment at the center of CCCC thin plate

4.2 *Simply supported thin plate (SSSS)*

To solve this case as in Figure 4, we need to apply some boundary conditions that are displacement, and moment in x and y direction. The results of displacement in the center of plate are given below in Tables 2 – 3 and Figure 5 – 6.

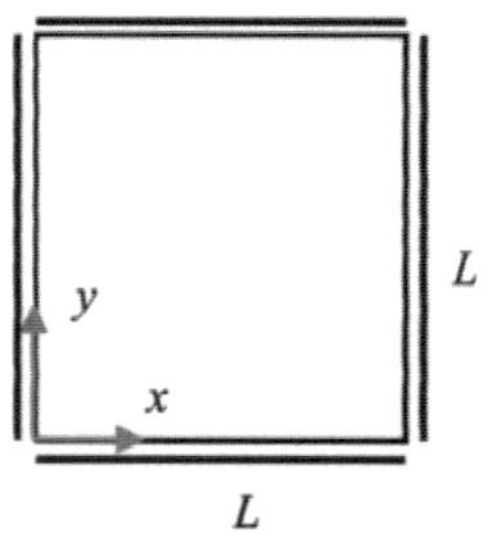

Dimension

L = 1000

h = 10

Uniform Load

$f_z = -1$

Material Properties

E = 10.92

υ = 0.3

Figure 4. SSSS plate

Table 3. Displacement at the center of SSSS thin plate

p Escalation (NELT = 1×1)			NELT Addition		
p	w	Error (%)	NELT	w	Error (%)
4	-2960904	27.11	1×1	-2960904	27.11
6	-3958021	2.56	2×2	-3289114	19.03
8	-4057473	0.11	4×4	-3866878	4.80
10	-4062993	0.02	8×8	-4042372	0.48
Ref. -4062000 (Batista [24])					

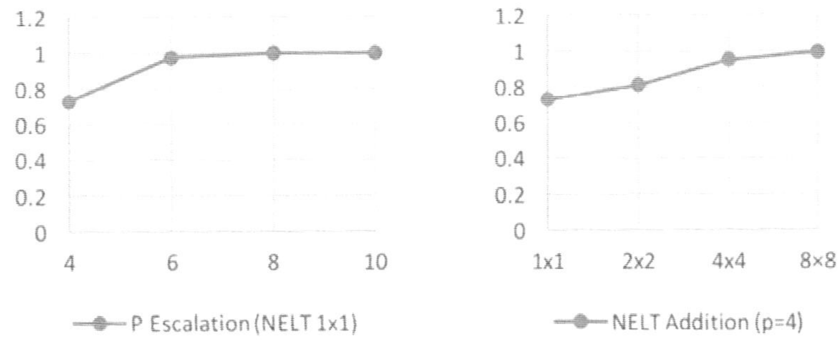

Figure 5. Convergence of normalized displacement at the center of SSSS thin plate

We can see in Table 3 and Figure 5 the convergence of displacement in the center of plate. The good results given by polynomial escalation can be obtained from $p = 6$ with error 2.56 % to $p = 10$ with error 0.02%. By number of elements addition, good results can be obtained starting from 4×4 elements with error 4.8%. Based on the results, p escalation gives better convergence than number of element addition does.

Figure 6 and Table 5 present the convergence of moment in the center of plate for SSSS case. The results show the same behavior as the results of displacement in the center of plate. Degree of polynomial escalation gives better convergence than number of element addition does.

Table 4. Moment in the center of SSSS thin plate

p Escalation (NELT = 1×1)			NELT Addition (p = 4)		
p	$M_x = M_y$	Error (%)	NELT	$M_x = M_y$	Error (%)
4	-36932	22.90	1×1	-36932	22.90
6	-46937	2.01	2×2	-41026	14.35
8	-47834	0.14	4×4	-46526	2.87
10	-47876	0.05	8×8	-47799	0.21
Ref. -47900 (Batista [24])					

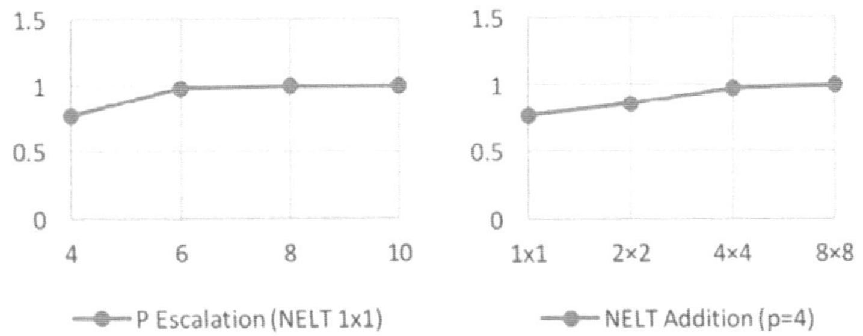

Figure 6. Convergence of normalized moment at the center of SSSS thin plate

4.3 Thin plate with simply supported sides along x = 0 and x = L (SSFF)

To solve this case as in Figure 7, we need to apply some boundary conditions that are displacement, and zero moment in x and y direction for simply supported sides.

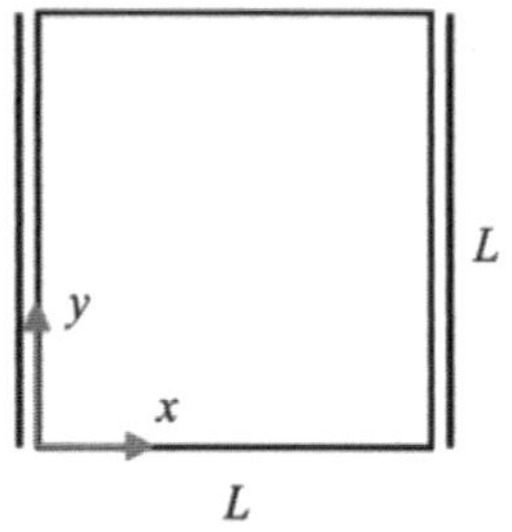

Dimension	Material Properties
L = 1000	E = 10.92
h = 10	υ= 0.3
Uniform Load	
$f_z = -1$	

Figure 7. SSFF plate

As can be seen in Figure 8 and Table 5, both results show that convergence can be obtained using even the lowest possible polynomial degree $p = 4$ with NELT = 1×1. Both p escalation and number of element addition show good results. However, increasing polynomial degree or NELT do not change the results.

Table 5. Displacement in the center of SSFF thin plate

p Escalation (NELT = 1×1)			NELT Addition (p = 4)		
p	w	Error (%)	NELT	w	Error (%)
4	-13024405	0.53	1×1	-13024405	0.53
6	-13024405	0.53	2×2	-13024405	0.53
8	-13024405	0.53	4×4	-13024405	0.53
10	-13024405	0.53	8×8	-13024405	0.53
Ref. -13093680 (Batista [24])					

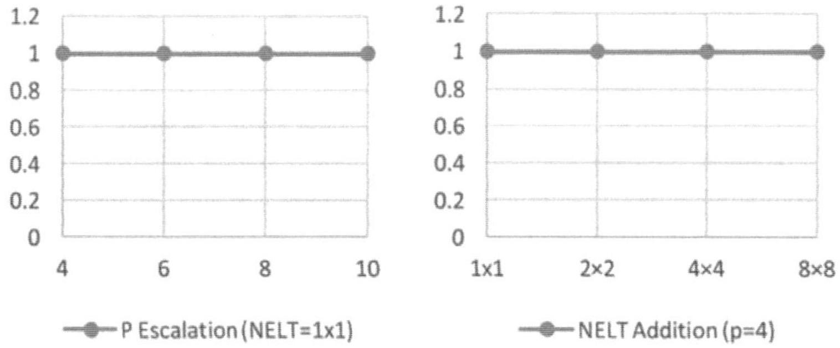

Figure 8. Convergence of normalized displacement at the center of SSFF thin plate

Figures 9 – 10 and Tables 6 – 7 present the convergence of moment in the center of plate for SSFF case. Both M_x and M_y results show the same behavior as displacement in the center of plate. Convergence can be obtained with p = 4 and NELT = 1×1. Higher polynomial or higher number of elements still give good results.

Table 6. Moment M_x in the center of SSFF thin plate

p Escalation (NELT = 1×1)			NELT Addition (p = 4)		
p	M_x	Error (%)	NELT	M_x	Error (%)
4	-125000	2.25	1×1	-125000	2.25
6	-125000	2.25	2×2	-125000	2.25
8	-125000	2.25	4×4	-125000	2.25
10	-125000	2.25	8×8	-125000	2.25
Ref. -122545 (Batista [24])					

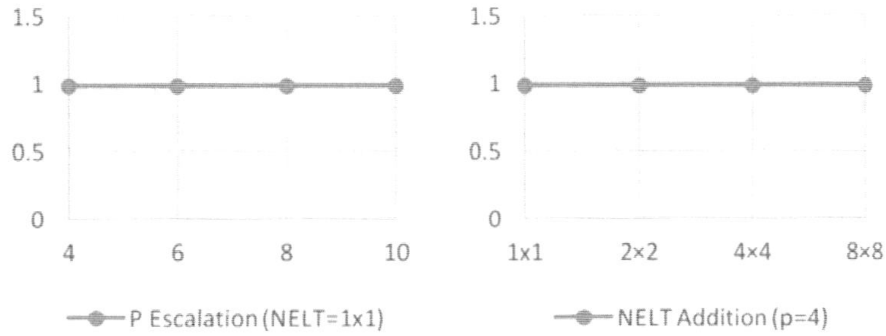

Figure 9. Convergence of normalized moment M_x at the center of SSFF thin plate

Table 7. Moment M_y in the center of SSFF thin plate

p Escalation (NELT = 1×1)			NELT Addition (*p* = 4)		
p	M_y	Error (%)	NELT	M_y	Error (%)
4	-26250	3.06	1×1	-26250	3.06
6	-26250	3.06	2×2	-26250	3.06
8	-26250	3.06	4×4	-26250	3.06
10	-26250	3.06	8×8	-26250	3.06
Ref. -27078 (Batista [24])					

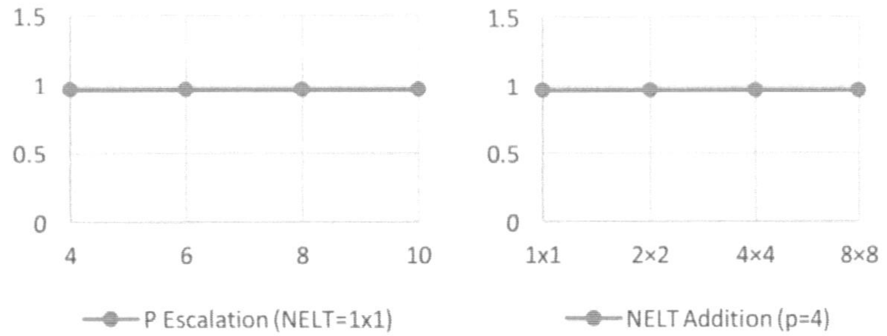

Figure 10. Convergence of normalized moment M_y at the center of SSFF thin plate

5. Conclusions

This paper presents an IGA collocation method for thin structural plate governed by Kirchhoff model. Based on all numerical tests, we can conclude that IGA collocation method using both polynomial degree escalation and number of element addition are able to obtain good approximation of the exact solution even with relatively coarse mesh. The numerical results show the accuracy, stability, and robustness of the method.

Acknowledgments

The financial support from Universitas Indonesia through the Hibah PITTA program (No.786/UN2.R3.1/HKP.05.00/2017) is gratefully acknowledged.

References

[1] Batoz J.-L., Katili I., On a simple triangular Reissner/Mindlin plate element based on incompatible modes and discrete constraints. *International Journal for Numerical Methods in Engineering* 1992; 35:1603-1632.

[2] Katili I., A new discrete Kirchhoff-Mindlin element based on Mindlin-Reissner plate theory and assumed shear strain fields - part I: An extended DKT element for thick-plate bending analysis. *International Journal for Numerical Methods in Engineering* 1993; 36:1859-1883.

[3] Katili I., A new discrete Kirchhoff-Mindlin element based on Mindlin-Reissner plate theory and assumed shear strain fields - part II: An extended DKQ element for thick plate bending analysis. *International Journal for Numerical Methods in Engineering* 1993; 36:1885-1908.

[4] Hamdouni A., Katili I., Maknun I.J., Millet O., Analytical and experimental analysis of an asymptotic thin-walled beam model. *European Journal of Environmental and Civil Engineering* 2013; 17(1): 1-18.

[5] Katili I., Batoz J.-L., Maknun I.J., Hamdouni A., Millet O., The Development of DKMQ Plate Bending Element for Thick to Thin Shell Analysis Based on Naghdi/Reissner/Mindlin Shell Theory. *Finite Elements in Analysis and Design* 2014; 100: 12-27.

[6] Maknun I.J., Katili I., Purnomo H., Development of DKMT element for error estimation in composite plate structures. *International Journal of Technology* 2015; 6(5): 780 -789.

[7] Katili I., Maknun I.J., Millet O., Hamdouni A., Application of DKMQ element for composite plate bending structures. *Composite Structures* 2015; 132: 166-174.

[8] Mahjudin M., Lardeur P., Druesne F., Katili I., Stochastic finite element analysis of plates with the certain generalized stresses method, *Structural Safety* 2016; 61; 12-21.

[9] Maknun I.J., Katili I., Millet O., Hamdouni A., Application of DKMQ element for twist of thin-walled beams: comparison with Vlasov theory, *International Journal for Computational Methods in Engineering Science and Mechanics* 2016; 17: 391-400.

[10] Wong F.T., Erwin, Richard A., Katili I., Development of the DKMQ element for buckling analysis of shear-deformable plate bending, *Procedia Engineering* 2017; 171: 805 – 812.

[11] Irpanni H., Katili I., Maknun I.J., Development DKMQ Shell Element with Five Degrees of Freedom per Nodal, *International Journal of Mechanical Engineering and Robotics Research* 2017; 6: 248-252.

[12] Katili I. Unified and integrated approach in a new Timoshenko beam element. *European Journal of Computational Mechanics* 2017; 26: 282–308.

[13] Katili, I., Composite Structures (2018), https://doi.org/10.1016/j.compstruct.2018.01.043.

[14] Hughes, T.J.R, Cottrell, J.A., Bazilevs, Y., *Isogeometric Analysis: CAD, Finite Elements, NURBS, Exact Geometry, and Mesh Refinement,* Comput. Method Appl. Mech. Engrg., 2005;194, 4135-4195.

[15] Rogers, D.F. *An Introduction to NURBS with Historical Perspective.* Academic Press, San Diego, CA, 2001.

[16] Nguyen, V.P., Bordas, S.P.A., Rabezuk, T., *An Introduction to Isogeometric Analysis with MATLAB Implementation: FEM and XFEM formulations.* Cardiff University, 2012.

[17] Nguyen, V.P., Bordas, S.P.A., Rabezuk, T. *Isogeometric Analysis: An Overview and Computer Implementation Aspects,* Cardiff University, 2013.

[18] Auricchio, F., Beirao da Veiga, L., *Isogeometric Collocation Methods.* Math. Models Methods Appl. Sci., 2010; 20, 2075-2107.

[19] Reali, A., Hughes, T.J.R., *An Introduction to Isogeometric Collocation Methods.* ICES REPORT 14-30, The Institute for Computational Engineering and Sciences, The University of Texas at Austin, September 2014.

[20] Reali, A., Gomez, H., *An Isogeometric Collocation Approach for Bernoulli-Euler Beams and Kirchhoff Plates.* Comput. Methods Appl. Mech. Engrg. 2015; 284, 623-636.

[21] Beirão da Veiga, L., Lovadina, C., Reali, A., *Avoiding Shear Locking for The Timoshenko Beam Problem via Isogeometric Collocation Methods*. Comput. Methods Appl. Mech. Engrg. 2012; 241–244, 38–51.
[22] Kiendl, J., Auricchio, F., Beirão da Veiga, L., Lovadina C., Reali, A., *An Isogeometric Collocation Methods for the Reissner-Mindlin Plate Bending Problem*. Comput, Methods Appl. Mech. Engrg. 2015; 284, 489–507.
[23] Katili I., Aristio R., Isogeometric Galerkin in rectangular plate bending problem based on UI approach, *European Journal of Mechanics - A/Solids* 2017; 67: 92-107.
[24] Batista, M., *Uniformly Loaded Rectangular Thin Plates with Symmetrical Boundary Conditions*. University of Ljubljana, 2010.

IOP Conf. Series: Journal of Physics: Conf. Series **1090** (2018) 012064 doi:10.1088/1742-6596/1090/1/012064

Numerical Simulation on Solving Three-Dimensional Global Optimization Problems in cooperation of Filled Function and Search Direction

Herlina Napitupulu[1], Ismail Bin Mohd[2], Yuyun Hidayat[3], Sukono[4], Sudradjat Supian[5]

[1)]School of Informatics and Applied Mathematics
Universiti Malaysia Terengganu
Kuala Terengganu, Malaysia
[2)]Institute for Mathematical Research (INSPEM)
Universiti Putera Malaysia
Selangor, Malaysia
[3)]Department of Statistics
FMIPA Universitas Padjadjaran
Sumedang, Indonesia
[1,4,5)]Department of Mathematics
FMIPA Universitas Padjadjaran
Sumedang, Indonesia

napitupuluherlina@gmail.com

Abstract. Global optimization problem still becomes an interest due to the challenge of locating the global optimum of nonlinear objective function with multiple local minima. Two challenges on solving global optimization problem are; firstly how to reach the better minimizer from the current minimizer, and secondly how to decide that the obtained minimizer is the desired global minimizer. One of the recent considered deterministic easy applied methods, which concerned in the mentioned problems, is the filled function method. The basic concept of filled function method is to build such an auxiliary function to locate a point with lower function value than the current minimizer. One of the keys to the successfully filled function method is how to decide the search direction to reach and locate a better local minimizer. In this paper, a three-dimensional filled function method and its search direction are introduced. The algorithm is presented and implemented to some benchmark test function. The numerical performance of the method on solving three-dimensional global optimization problems is presented.

1. Introduction

The objective of global optimization is to obtain the best solution in the possible presence of a multitude of local suboptimum. Theory and methods of optimization have wide applications in most all branches of science, engineering, business management, military, space technology, finance, and so forth. Research of global optimization has become a challenging topic, since the existence of multiple optima that differ from the global solution. This multimodal case cannot be solved by classical nonlinear programming techniques directly. General two difficulties of global optimization are; how to

Content from this work may be used under the terms of the Creative Commons Attribution 3.0 licence. Any further distribution of this work must maintain attribution to the author(s) and the title of the work, journal citation and DOI.
Published under licence by IOP Publishing Ltd

leave from present local minimizer to another minimizer with the lower value and how to decide that the present minimizer is a global solution.

The main difficulty for global optimization is to escape from the current local minimizer and to find a better one. One of the most efficient methods to deal with this issue is the filled function method [1]. The filled function method is an approach to find the global minimizer of multi-modal functions. Compared to the other methods, filled function method has an outstanding advantage that is simple while its realization is relatively easy and effective at successively finding smaller local minima [2, 3, 4].

The first filled function with two adjustable parameters was proposed for smooth optimization by Ge in 1983 and finally published in 1990, [5] which were used to solve the global minimizer of unconstrained. Some disadvantages persist in this former function due to the existence of exponential term and two adjustable parameters. Many modifications are proposed as the revision to the former filled function; however, the existence of adjustable parameter(s), exponential, and the logarithmic term still become difficulties on the computational process. However, there is no efficient criterion to determine the appropriate parameter. It is better if it is possible to build a filled function without parameter, as well as neither exponential nor logarithmic term. Moreover, many filled functions need to choose more than one starting point to obtain multiple optimum solutions. Actually, the parameter free filled function with above criteria is a desired one.

Mohd et al. [8] proposed a one dimensional parameter free filled function that is theoretically and numerically succeeded for solving multimodal general global optimization problems using one starting point. The extension of one dimensional filled function [8] into two-dimensional filled function [9] is so-called Ismail-Herlina-Ridwan filled function (in short, TIHR filled function). Napitupulu and Mohd [10] introduced types of vector direction of the second phase of filled function of their algorithm method on solving two dimensional unconstrained global optimization problems.

In this paper, we extend parameter free filled function [8] into three-dimensional which can be used to solve an unconstrained global optimization problem. Secondly, we introduce some types of search direction, to be used in the second phase of filled function algorithm method. We implement the algorithm method into benchmark test function to test the efficiency of our algorithm. Numerical and computational results show how the performance of both the algorithm and types of proposed search direction. The conclusion is given in the rest of the paper.

2. Parameter Free Filled Function

This paper considers the following global optimization problem:

$$\min_{\mathbf{x}\in D} f(\mathbf{x}) \tag{1}$$

where $D \subset R^n$ is a closed bounded domain containing all global minimizers of $f(\mathbf{x})$ in its interior. It is assumed in this paper that $f(\mathbf{x})$ has only a finite number of local minimizers. When $f: R^n \to R$ is coercive, that is, $f(\mathbf{x}) \to +\infty$ as $\|\mathbf{x}\| \to +$, then a global optimization problem (2) can be always reduced into an equivalent problem (1). Let x^* denote the isolated local minimizer of $f(\mathbf{x})$.

$$\min_{\mathbf{x}\in R^n} f(\mathbf{x}). \tag{2}$$

A non-parametric filled function of one-dimensional in [8] basically from the idea of integration, the formula is as follows

$$F_{IG}(x,x^*) = \begin{cases} -\int_{x^*}^{x} \left(f(x) - f(x^*)\right) ds & (x \ge x^*) \\ -\int_{x}^{x^*} \left(f(x) - f(x^*)\right) ds & (x < x^*) \end{cases} \tag{3}$$

Others researchers are proposed the n-dimensional parameter free filled functions as done in [6], [4], and [7]. The features of filled function [5, 6, 7] are no parameter included and no exponential or logarithmic term exists in the formula. However, these functions method has disadvantages, that is, always needs more than one starting point to obtain more than one global minimizers.

In this paper we proposed a three-dimensional parameter free filled function method, which is extension from (3), together with its search direction. The attention is to find the global minimizer(s) of multimodal function $f(\mathbf{x})$ on domain $D \subset R^3$. The properties of three dimensional parameter free filled function are explained in the following section.

3. Three-Dimensional Parameter Free Filled Function

Consider a function $f: X \subset R^3 \Rightarrow R$ where X is a box defined by

$$\begin{aligned} X &= \left\{ \mathbf{x} = (x_1, x_2, x_3)^T \mid x_{1I} \le x_1 \le x_{1S} \wedge x_{2I} \le x_2 \le x_{2S} \wedge x_{3I} \le x_3 \le x_{3S} \right\} \\ &= \left(\underline{x}_1 = [x_{1I}, x_{1S}], \underline{x}_2 = [x_{2I}, x_{2S}], \underline{x}_3 = [x_{3I}, x_{3S}] \right) \end{aligned} \tag{4}$$

and consider a function $F: D \subseteq X$ where a box D is defined by

$$\begin{aligned} D &= \left\{ \mathbf{x} = (x_1, x_2, x_3)^T \mid x_1^{Z-} \le x_1 \le x_1^{Z+} \wedge x_2^{Z-} \le x_2 \le x_2^{Z+} \wedge x_3^{Z-} \le x_3 \le x_3^{Z+} \right\} \\ &= \left(\underline{x}_1 = [x_1^{Z-}, x_1^{Z+}], \underline{x}_2 = [x_2^{Z-}, x_2^{Z+}], \underline{x}_3 = [x_3^{Z-}, x_3^{Z+}] \right) \end{aligned} \tag{5}$$

where x_{jI} and x_{jS} $(j = 1,2,3)$ are the infimum (denoted by I) and supremum (denoted by S) of the interval $\underline{x}_j = [x_{jI}, x_{jS}]$ respectively. Assume that $\mathbf{x}^* = (x_1^*, x_2^*, x_3^*)^T \in D$ is a current local isolated minimizer of f. The domain D can be divided into eight sub domains $D_1, D_2, D_3 ..., D_8$ such that $D = \bigcup_{i=1}^{8} D_i = D_1 \cup D_1 \cup ... \cup D_8$, where

$$\left. \begin{aligned} D_1 &= ([x_1^*, x_1^{Z+}], [x_2^*, x_2^{Z+}], [x_3^*, x_3^{Z+}]) \\ D_2 &= ([x_1^*, x_1^{Z+}], [x_2^*, x_2^{Z+}], [x_3^{Z-}, x_3^*]) \\ D_3 &= ([x_1^*, x_1^{Z+}], [x_2^{Z-}, x_2^*], [x_3^*, x_3^{Z+}]) \\ D_3 &= ([x_1^*, x_1^{Z+}], [x_2^{Z-}, x_2^*], [x_3^{Z-}, x_3^*]) \\ D_5 &= ([x_1^{Z-}, x_1^*], [x_2^*, x_2^{Z+}], [x_3^*, x_3^{Z+}]) \\ D_6 &= ([x_1^{Z-}, x_1^*], [x_2^*, x_2^{Z+}], [x_3^{Z-}, x_3^*]) \\ D_7 &= ([x_1^{Z-}, x_1^*], [x_2^{Z-}, x_2^*], [x_3^*, x_3^{Z+}]) \\ D_8 &= ([x_1^{Z-}, x_1^*], [x_2^{Z-}, x_2^*], [x_3^{Z-}, x_3^*]) \end{aligned} \right\} \tag{6}$$

Three-dimensional free filled function F, defined on sub domain D_i $(i = 1,2,..,8)$, is given by

$$F_i(\mathbf{x}, \mathbf{x}^*) = \begin{cases} F_{i1}(\mathbf{x}, \mathbf{x}^*) = -\left(\int_{a_1}^{b_1} \left(f(s_1, x_2, x_3) - f(x_1^*, x_2^*, x_3^*) \right) ds_1 \right) & (a_1 \le b_1) \\ F_{i2}(\mathbf{x}, \mathbf{x}^*) = -\left(\int_{a_2}^{b_2} \left(f(x_1, s_2, x_3) - f(x_1^*, x_2^*, x_3^*) \right) ds_2 \right) & (a_2 \le b_2) \\ F_{i3}(\mathbf{x}, \mathbf{x}^*) = -\left(\int_{a_3}^{b_3} \left(f(x_1, x_2, s_3) - f(x_1^*, x_2^*, x_3^*) \right) ds_3 \right) & (a_3 \le b_3) \end{cases} \tag{7}$$

where a_j and b_j $(j=1,2,3)$ are described in Table 1.

Table 1. Determination for boundary of integration a_j and b_j

	i							
j	**1**	**2**	**3**	**4**	**5**	**6**	**7**	**8**
a_1	x_1^*	x_1^*	x_1^*	x_1^*	x_1	x_1	x_1	x_1
b_1	x_1	x_1	x_1	x_1	x_1^*	x_1^*	x_1^*	x_1^*
a_2	x_2^*	x_2^*	x_2	x_2	x_2^*	x_2^*	x_2	x_2
b_2	x_2	x_2	x_2^*	x_2^*	x_2	x_2	x_2^*	x_2^*
a_3	x_3^*	x_3	x_3^*	x_3	x_3^*	x_3	x_3^*	x_3
b_3	x_3	x_3^*	x_3	x_3^*	x_3	x_3^*	x_3	x_3^*

The definition of three-dimensional parameter free filled function is given in Definition 1 and its validity is proved by Theorem 1- 3

Definition 1 $F(\mathbf{x},\mathbf{x}^*)$ is called a three-dimensional parameter free filled function of $f(\mathbf{x})$ at local isolated minimizer $\mathbf{x}^*$ of f iff

(i). $\mathbf{x}^*$ is a maximizer of $F=F(\mathbf{x},\mathbf{x}^*)<0=F(\mathbf{x}^*,\mathbf{x}^*)$

(ii). F has no stationary point in a set $H_1=\{\mathbf{x}\,|\,f(\mathbf{x})>f(\mathbf{x}^*),\mathbf{x}\in D\}$

(iii). There is a point $\mathbf{x}\in H_2=\{\mathbf{x}|f(\mathbf{x})\le f(\mathbf{x}^*),0=w\in W,\ \mathbf{x}\in D\backslash\{\mathbf{x}^*\}\}\neq\varnothing$ such that $\mathbf{x}$ is a stationary point of F, where a_j and b_j are as in Table 1 and

$$W=\left\{\frac{\partial}{\partial x_l}\left(\int_{a_j}^{b_j} f(\mathbf{x})ds_j\right)\middle|\forall l\neq j,(l=1,2,3)\right\} \tag{8}$$

Proof of Theorem 1-3 is become evidence that F is a filled function according to Definition 1.

Theorem 1

If $\mathbf{x}^*=(x_1^*,x_2^*,x_3^*)^T\in R^3$ is an isolated minimizer of the objective function $f(\mathbf{x})$ on $D=\bigcup_{i=1}^{8}D_i$, then $\mathbf{x}^*$ is a maximizer of filled function F.

Proof.

Since $f(x_1,x_2,x_3)-f(x_1^*,x_2^*,x_3^*)\ge 0$ and Table 1, then for all sub domains D_i $(i=1,2,..,8)$, we have

$$\left.\begin{aligned}&\int_{a_1}^{b_1} f(s_1,x_2,x_3)-f(x_1^*,x_2^*,x_3^*)ds_1\ge 0 \text{ for } a_1\le b_1\\&\int_{a_2}^{b_2} f(x_1,s_2,x_3)-f(x_1^*,x_2^*,x_3^*)ds_2\ge 0 \text{ for } a_2\le b_2\\&\int_{a_3}^{b_3} f(x_1,x_2,s_3)-f(x_1^*,x_2^*,x_3^*)ds_3\ge 0 \text{ for } a_3\le b_3\end{aligned}\right\} \tag{9}$$

These imply for all $\mathbf{x}\in D_i$, $F_i(\mathbf{x},\mathbf{x}^*)\le 0\,(i=1,..,8)$. Therefore $F(\mathbf{x},\mathbf{x}^*)\le 0=F(\mathbf{x}^*,\mathbf{x}^*)$. □

Theorem 2

If $(x_1^*,x_2^*,x_3^*)^T$ is a local minimizer of $f(x_1,x_2,x_3)$ on $D=\bigcup_{i=1}^{8}D_i$ and

$$H_1=\{\mathbf{x}\,|\,f(\mathbf{x})>f(\mathbf{x}^*),\ \mathbf{x}\in D\backslash\{\mathbf{x}^*\}\} \tag{10}$$

then F *has no stationary point in the set* H_1.

Proof.

Suppose that $\mathbf{x} \in H_1$ satisfies $f(\mathbf{x}) > f(\mathbf{x}^*)$. Consider $F(\mathbf{x}, \mathbf{x}^*)$ in sub domain D_1. Since f is a twice differentiable function, then

$$\partial F_{11}/\partial x_1 = -(f(x_1,x_2,x_3) - f(x_1^*,x_2^*,x_3^*)) < 0$$

$$\partial F_{11}/\partial x_2 = -\left(\int_{x_1^*}^{x_1} \frac{\partial}{\partial x_2}\big(f(s_1,x_2,x_3) - f(x_1^*,x_2^*,x_3^*)\big) ds_1 \right) \neq 0$$

$$\partial F_{11}/\partial x_3 = -\left(\int_{x_1^*}^{x_1} \frac{\partial}{\partial x_3}\big(f(s_1,x_2,x_3) - f(x_1^*,x_2^*,x_3^*)\big) ds_1 \right) \neq 0$$

Therefore, $\nabla F_{11} \neq 0$, means that F_{11} has no stationary point in H_1.

For the gradient of F_{12}, also

$$\partial F_{12}/\partial x_1 = -\left(\int_{x_2^*}^{x_2} \frac{\partial}{\partial x_1}\big(f(x_1,s_2,x_3) - f(x_1^*,x_2^*,x_3^*)\big) ds_2 \right) \neq 0$$

$$\partial F_{12}/\partial x_2 = -(f(x_1,x_2,x_3) - f(x_1^*,x_2^*,x_3^*)) < 0$$

$$\partial F_{12}/\partial x_3 = -\left(\int_{x_2^*}^{x_2} \frac{\partial}{\partial x_3}\big(f(x_1,s_2,x_3) - f(x_1^*,x_2^*,x_3^*)\big) ds_2 \right) \neq 0$$

Thus $\nabla F_{12} \neq 0$, means that F_{12} has no stationary point in H_1. Therefore, $\nabla F_{12} \neq 0$, means that F_{12} has no stationary point in H_1.

For the gradient of F_{13}, also

$$\partial F_{13}/\partial x_1 = -\left(\int_{x_3^*}^{x_3} \frac{\partial}{\partial x_1}\big(f(x_1,x_2,s_3) - f(x_1^*,x_2^*,x_3^*)\big) ds_3 \right) \neq 0$$

$$\partial F_{13}/\partial x_2 = -\left(\int_{x_3^*}^{x_3} \frac{\partial}{\partial x_2}\big(f(x_1,x_2,s_3) - f(x_1^*,x_2^*,x_3^*)\big) ds_2 \right) \neq 0$$

$$\partial F_{13}/\partial x_3 = -(f(x_1,x_2,x_3) - f(x_1^*,x_2^*,x_3^*)) < 0$$

Thus $\nabla F_{13} \neq 0$, means that F_{13} has no stationary point in H_1. Therefore, in sub-domain D_1 there is no stationary point of F in H_1. The similar proof can be applied to $F_2, .., F_8$. □

Theorem 3

If $H_2 = \left\{ \mathbf{x} | f(\mathbf{x}) \le f(\mathbf{x}^*), 0 = w \in W,\ \mathbf{x} \in D \backslash \{\mathbf{x}^*\} \right\} \neq \varnothing$ where

$$W = \left\{ \frac{\partial}{\partial x_l}\left(\int_{a_j}^{b_j} f(\mathbf{x}) ds_j \right) \middle| \forall l \neq j, (l = 1,2,3) \right\}$$

for specified j and a_j, b_j are as in Table 1, then there is a point $\mathbf{x} = (x_1, x_2, x_3)^T \in H_2$ such that $\mathbf{x}$ is a stationary point of $F(\mathbf{x}, \mathbf{x}^*)$.

Proof.

Consider F_{11} in sub domain D_1. Suppose that for $\mathbf{x} \in H_2$, $f(x_1,x_2,x_3) = f(x_1^*,x_2^*,x_3^*)$ holds and satisfies

$$\frac{\partial}{\partial x_2}\left(\int_{x_1^*}^{x_1} \big(f(s_1,x_2,x_3)\big) ds_1 \right)\Bigg|_{\mathbf{x}=\mathbf{x}} = 0 = \frac{\partial}{\partial x_3}\left(\int_{x_1^*}^{x_1} \big(f(s_1,x_2,x_3)\big) ds_1 \right)\Bigg|_{\mathbf{x}=\mathbf{x}}$$

Then

$$\nabla F_{11}\left(\mathbf{x},\mathbf{x}^*\right)\Big|_{\mathbf{x}=\mathbf{x}} = \begin{bmatrix} -\left(f(x_1,x_2)-f(x_1^*,x_2^*)\right) \\ \dfrac{\partial}{\partial x_2}\left(-\int_{x_1^*}^{x_1} f(s_1,x_2,x_3)ds_1\right) \\ \dfrac{\partial}{\partial x_3}\left(-\int_{x_1^*}^{x_1} f(s_1,x_2,x_3)ds_1\right) \end{bmatrix} = \begin{bmatrix}0 & 0 & 0\end{bmatrix}^T$$

Therefore, there exists a stationary point of F_{11} for $\mathbf{x}\in H_2$. The analog proof also applied for F_{12} and F_{13} for $\mathbf{x}\in H_2$ there is a stationary point of F. Similar proof can be applied to $F_2, F_3, .., F_8$. □

4. Algorithm Method

In this section the algorithm method is presented which is in cooperation of three dimensional parameter free filled function and the search direction. The calculation method consist of two main phases, that are the minimization of objective function and filled function construction.

Data (initialization) specify initial point x_0, domain D_i, real number d>0and set, $i=1$, $j=1$, $m=1$

step 1 specify initial step size of steepest descent $\alpha_0>0$, minimize $f(x)$ starting at $\mathbf{x}_0$ and to obtain a local minimizer $\mathbf{x}_m^*$

step 2 construct F function $F_{ij}(\mathbf{x},\mathbf{x}_m^*)$ at $\mathbf{x}_m^*$

step 3 if $i\le n$ then choose direction e_i for respected defined dimension and go to step 4 else stop

step 4 set $c:=1$

step 5 set initial point $\mathbf{x}_0^F=\mathbf{x}_m^*+c\mathrm{d}e_i$.
 if $\mathbf{x}_0^F\in D_i$ then go to step 6
 else if $j<n$ then $j:=j+1$, go to step 4
 else $j=1; i:=i+1$; and go to step 3

step 6 if $f(\mathbf{x}_0^F)<f(\mathbf{x}_m^*)$ then set $\mathbf{x}_0:=\mathbf{x}_0^F$, go to step 1
 else if one of cond$_1$ true then go to step 7
 else $c:=c+1$ and go to step 5

step 7 by Newton's method with initial $\mathbf{x}_0^F$ solve $\nabla F_{ij}(\mathbf{x},\mathbf{x}_m^*)=0$ to obtain $\mathbf{x}^F$

step 8 if cond$_2$ and cond$_3$ hold then set $\mathbf{x}_0:=\mathbf{x}^F$, $m:=m+1$, go to step 1
 else $c:=c+1$ go to step 5

Note:

cond$_1$: at least there is exists a j ($j=1,2,3$) such that the following condition holds

$$\left(\frac{\partial F_{ij}}{\partial x_j}(\mathbf{x}=\mathbf{x}_m^*+c\mathrm{d}e_i)\right)^2 < \left(\frac{\partial}{\partial x_j}F_{ij}(\mathbf{x}=\mathbf{x}_m^*+(c-1)\mathrm{d}e_i)\right)^2$$

cond$_2$: at least one of following conditions holds for a specified j

$$\text{(i). } \frac{\partial^2}{\partial x_j^2}F_{ij}(\mathbf{x}=\mathbf{x}^F)>0 \text{ and (ii). } \left|\frac{\partial^2}{\partial x_j^2}F_{ij}(\mathbf{x}=\mathbf{x}^F)\right|<10^{-4}$$

cond$_3$: $\left|f(\mathbf{x}^F)-f(\mathbf{x}_m^*)\right|<10^{-2}$

The explanation of this algorithm is as follows. The initial point is chosen for iteration steepest descent to obtain the current minimizer $\mathbf{x}^*$. After $\mathbf{x}^*$ is obtained then entering phase 2, built a filled function according to its domain. Symbol *m* represent number of current minimizer obtained. After that, set the initial point to minimize the filled function i.e., $\mathbf{x}_0^F = \mathbf{x}_m^* + cde_i$ such that satisfies $\mathbf{x}_0^F \in D_i$ and if not thus generate c and repeat to checking the initial point, if this point out of its domain, thus do $c := c+1$ or evaluated other filled function in the same domain i, if $j>2$, do $i:=i+1$ means calls the next filled function in the next domain, and do repeat to check the initial point again, until this point satisfies the conditions given, if $j>2$ then stop. If the initial point satisfy the condition, doing minimizing filled function till obtained the point which has the equal value with current minimizer. If it is not satisfied but still in its domain, generate the integer number c again and return to step 11. If the point $\mathbf{x}^F$ satisfies the conditions go to step 2, set $m:=m+1$, and do iteration steepest descent again. If the next minimizer obtained is has the lower or equal value compared to the current minimizer obtained, thus set it to be the current minimizer to entering second phase.

5. The Search Direction

How to find the right direction is one of the important things in the filled function method. Various functions may have different pattern of location of the local or global minimizer. For two-dimensional case, it can be easily seen through contour plot. In this research we use search direction to locate another better point, which is outside the current minimizer to another the desired initial point of Newton's method. It is well known that Newton's iteration will well behave if the initial point is near to the desired point. If search direction is not properly chosen, Newton's method will not work well, then may causing execution time on searching minimizer or saddle point of filled function becoming longer, or make the desired point does not obtain even if the desired point(s) exist in the evaluated sub domain.

Now let us consider the second phase of filled function's algorithm. The proposed search direction is focused on finding a new point, say $\mathbf{x}_0^F$, which exist outside the neighborhood of current minimizer, $\mathbf{x}^*$, but near to the minimizer or saddle point of filled function, say $\mathbf{x}^F$. Since our consideration is on three-dimensional function, then it is possible that will exist four triple of a sign of axis direction. In the section of algorithm method, we use following search direction for each respected sub-domain D_i with $i = (1,2,...,8)$.

$$\begin{aligned}\mathbf{e} &= \{e_1, e_2, e_3, ..e_i, .., e_8\} \\ &= \{(1,1,1)^T, (1,1,-1)^T, (1,-1,1)^T, (1,-1,-1)^T, \\ &\quad (-1,1,1)^T, (-1,1,-1)^T, (-1,-1,1)^T, (-1,-1,-1)^T\}\end{aligned} \tag{11}$$

In this paper, we introduce the search direction to be implemented to algorithm method, in the form of equation (12).

$$V^{(p)} = \left(sign_1 v_1^{(p)}, sign_2 v_2^{(p)}, sign_3 v_3^{(p)}\right)^T \tag{12}$$

The $sign_j$ $(j = 1,2,3)$ in (12) symbolizing a positive or negative sign of $v_j^{(p)}$ in search direction $V^{(p)}$ for each respected sub-domain D_i $(i = 1,2,..,8)$. Let n is the number of variable of f. Suppose that $\mathbf{D}(i)$ is decimal (denoted by $(\mathbf{D}(i))_{10}$) and b_j is a binary number with value only 1 or 0 (denoted by $(b_1 b_2 b_3)_2$). Then it can be written that $(\mathbf{D}(i))_{10} = (b_1 b_2 b_3)_2$. Let $\mathbf{D}(i) = i-1$, then we define $sign_j (j = 1,..,n)$ for the given i as

$$b_j \begin{cases} 0, sign_j = + \\ 1, sign_j = - \end{cases}$$

For example, for i=1 then $sign_j$ for all $j = 1,2,3$ are positive, and for i=2 the $sign_1 = sign_2$ is positive and $sign_3$ is negative, and so on.

In this paper, we introduce the search direction in the form of equation (12), where $v_j^{(p)} = \sin\theta_j$ for $0 \le \theta_j \le \pi$. To see the performance of the filled function algorithm method, we restrict restricted the simulation by only use the search direction with combination θ value are $\theta = \pi/6$ and $\theta = \pi$. We present the 36 combination of search direction, and the explanation on how search directions defined is given in the following explanation and algorithm.

Case 1: $(1 \le p \le 6)$
Suppose that $p(i)$ is decimal (denoted by $(p)_{10}$) and q_j is a binary number with value only 1 or 0 (denoted by $(q_1q_2q_3)_2$). Then it can be written that $(p)_{10} = (q_1q_2q_3)_2$. Thus the defined $v_j^{(p)}(j = 1,..,n)$,

$$\text{if } q_j \begin{cases} 0, \text{then } v_j^{(p)} = \sin(\pi) \\ 1, \text{then } v_j^{(p)} = \sin(\pi/6) \end{cases}$$

Case 2: $(7 \le p \le 36)$
define $p := 7$
for $(p1 := 1; p1{+}{+}, p1 \le 6)$ do
  for $(p2 := 1; p2{+}{+}, p2 \le 6)$ do
    if $(p2 == p1)$ then $(p2 := p2{+}{+})$

$$V^{(p)} \begin{cases} V^{(p1)}, \text{if } c = c_o \\ V^{(p2)}, \text{if } c = c_e \end{cases}$$

    $p := p{+}{+}$;
  end do
end do

The previous algorithm shows that the search direction with $7 \le p \le 36$ are the combination of the search direction with $1 \le p \le 6$. The c_o and c_e respectively means an even and an odd value of c (a constant value of initial point in step 5 algorithm method).

Now we have 36 type of search direction to replace former search direction e_i in Step 5 of algorithm. In further, the numerical and computational results are explained in the following section.

6. Results

To test the performance of algorithm method with several given search directions we use benchmark test function 3-dimensional sine square function, with initial point $\mathbf{x}_0 = (-2,-2,-2)^T$.

$$f(\mathbf{x}) = \frac{\pi}{3}\left\{10\sin^2(\pi x_1) + \sum_{i=1}^{n-1}(x_i - 1)^2\left(1 + 10\sin^2(\pi x_{i+1})\right) + (x_n - 1)^2\right\} \tag{13}$$
$$-10 \le x_i \le 10, i = 1,2,3$$

The program is written in visual C++ program, Processor Intel ® Core i3, 2048 MB RAM. In minimizing $f(\mathbf{x})$, steepest descent method with elimination line search step size (see [11]) is used, which its step size is made to be absolute. Stopping criteria for steepest descent method that we use is $|\nabla f| \le 10^{-3}$, while for Newton's method is $|\nabla F| < 2 \cdot 10^{-4}$.

6.1. Numerical Results

The detail numerical results of the given problem are presented in Table 2, with the explanation of symbols given as follows.
k : number of iteration

F_{ij} : the evaluated Three-Dimensional Parameter Free Filled Function

α_0 : initial step size of steepest descent method

c : integer number to be multiplied to search direction

$\mathbf{x}_0$: initial point for steepest descent method in $f(\mathbf{x})$

$\mathbf{x}_k^*$: k -th obtained minimizer

$\mathbf{x}_0^F$: initial point for Newton's method in F_{ij}

$\mathbf{x}^F$: point obtained by Newton's method or satisfied $f(\mathbf{x}^F) < f(\mathbf{x}_k^*)$

f_0, f_k^*, f_0^F, f^F : function value of $\mathbf{x}_0, \mathbf{x}_k^*, \mathbf{x}_0^F, \mathbf{x}^F$

Table 2. Numerical Results with given **e**

k	α_0, F_{ij}, c, d	**Results**
0	-	$x_0=(-2,-2,-2)^T$, $f_0=28.287$
	$\alpha_0=0.001$	$x_1^*=(-1.96892,-1.99575,-1.99582)^T$, $f_1^*=28.1544$
1	F_{11}, $c=7$,	$x_0^F=(-1.26892,-1.64575,-1.29582)^T$, $f_0^F=114.62$
	$d=0.1$	$x^F=(-1.96892,-1.10412,-0.0864502)^T$, $f^F=28.155$
	$\alpha_0=0.001$	$x_2^*=(-1.96894,-0.997303,0.00253152)^T$, $f_2^*=14.5577$
2	F_{11}, $c=5$,	$x_0^F=(-1.46894,-0.747303,0.502532)^T$, $f_0^F=84.6167$
	$d=0.1$	$x^F=(-1.96894,-0.0605696,0.919134)^T$, $f^F=14.5577$
	$\alpha_0=0.001$	$x_3^*=(-1.96895,0.00114707,0.999601)^T$, $f_3^*=10.3752$
3	F_{11}, $c=10$,	$x_0^F=(-1.46895,1.00115,1.4996)^T$, $f_0^F=17.0296$
	$d=0.1$	$x^F=(0.554693,0.999599,1)^T$, $f^F=10.3752$
	$\alpha_0=0.001$	$\mathbf{x_4^*=(0.999608,0.99999,1)^T}$, $\mathbf{f_4^*=1.71823x10^{-7}}$

6.2. Comparison

The general computational results are presented in Table 3. The following symbols are used in the table. The SD refers to types of search direction.

VD: type of search direction
iter : number of succeeded iteration
time/time1 : execution time until global minimizer obtained/ program terminated in second
nf/nf1 : number of function evaluation of f until global minimizer obtained/ program terminated
nf'/nf'1 : number of ∇f until global minimizer obtained/ program terminated
nF'/nF'1 : number of ∇F_{ij} until global minimizer obtained/ program terminated
T/T1 : total number of *nf+nf'+nF'* / total number of *nf1+nf'1+nF'1*

Table 3. General Numerical Results

VD	*iter*	*time/ time1* (second)	*nf/nf1*	*nf'/nf'1*	*nF'/nF'1*	*T/T1*
e	10	12.558/41.449	3004/3336	1502/1668	21672/38062	26178/43066
$V^{(1)}$	8	6.708/15.1	6904/7566	3452/3783	9733/27247	20089/38596

$V^{(2)}$	3	1.762/10.483	2766/3526	1383/1763	437/18552	4586/23841
$V^{(3)}$	5	8.985/50.918	4790/6150	2395/3075	19271/39028	26456/48253
$V^{(4)}$	8	7.051/16.427	2770/4080	1385/2040	8698/26520	12853/32640
$V^{(5)}$	7	11.419/20.186	5394/6522	2697/3261	17614/35073	25705/44856
$V^{(6)}$	4	5.741/15.709	1210/2066	605/1033	441/19882	2256/22981
$V^{(7)}$	6	10.342/17.596	2832/2832	1416/1416	6603/23212	10851/27460
$V^{(8)}$	9	14.71/50.653	13472/14026	6736/7013	23452/39674	43660/60713
$V^{(9)}$	8	11.856/19.671	13162/14034	6581/7017	7292/23194	27035/44245
$V^{(10)}$	7	7.285/14.398	2592/2940	1296/1470	9166/24618	13054/29028
$V^{(11)}$	6	5.116/17.69	1786/2232	893/1116	4888/20100	7567/23448
$V^{(12)}$	5	49.077/58.546	4766/5428	2383/2714	4447/21726	11596/29868
$V^{(13)}$	7	7.02/16.006	6604/7356	3302/3678	12740/29962	22646/40996
$V^{(14)}$	5	6.302/16.52	358/542	179/271	6313/23284	6850/24097
$V^{(15)}$	3	2.059/10.14	674/1068	337/534	239/15810	1250/17412
$V^{(16)}$	3	4.071/14.055	138/170	69/85	228/17807	435/18062
$V^{(17)}$	4	5.335/15.147	182/1008	91/504	5615/24022	5888/25534
$V^{(18)}$	4	246.511/256.37	2236/2412	1118/1206	6582/25737	9936/29355
$V^{(19)}$	4	2.512/11.591	1434/2288	717/1144	554/18184	2705/21616
$V^{(20)}$	6	10.576/21.356	8756/9538	4378/4769	15929/33080	29063/47387
$V^{(21)}$	3	6.286/15.225	138/736	69/368	341/18396	548/19500
$V^{(22)}$	4	3.276/12.667	2886/3416	1443/1708	3631/20916	7960/26040
$V^{(23)}$	4	1.965/9.188	784/1190	392/595	714/17622	1890/19407
$V^{(24)}$	4	3.182/12.105	1408/1996	704/998	1253/16753	3365/19747
$V^{(25)}$	8	7.363/15.256	6200/7002	3100/3501	8534/24860	17834/35363
$V^{(26)}$	6	8.97/18.174	2964/3786	1482/1893	11420/29018	15866/34697
$V^{(27)}$	8	12.511/31.527	2914/4172	1457/2086	18964/35722	23335/41980
$V^{(28)}$	5	5.647/14.898	1744/2678	872/1339	6012/22048	8628/26065
$V^{(29)}$	6	7.55/13.743	1910/2864	955/1432	13254/29460	16119/33756
$V^{(30)}$	6	11.388/20.732	1256/2908	628/1454	11962/28951	13846/33313
$V^{(31)}$	4	2.558/11.918	734/1002	367/501	294/16430	1395/17933
$V^{(32)}$	5	4.555/13.588	4184/4716	2092/2358	4096/20953	10372/28027
$V^{(33)}$	3	2.231/12.09	3826/4862	1913/2431	574/18307	6313/25600
$V^{(34)}$	8	15.849/67.501	12736/12840	6368/6420	21683/38765	40787/58025
$V^{(35)}$	4	4.258/14.071	4968/5738	2484/2869	3363/21016	10815/29623
$V^{(36)}$	6	5.319/10.452	8166/8650	4083/4325	6183/22517	18432/35492

According to Table 3, the algorithm method success to obtain the global minimizer of given problem, for all types of search directions. For this problem function, the least number of *iter* =3 is given in $V^{(2)}$, $V^{(15)}$, $V^{(16)}$, $V^{(21)}$, $V^{(33)}$. Based on execution time, the least total time until global minimizer obtained is performed by $V^{(2)}$ with 1.762 second, while the least total time until program terminated is performed by $V^{(23)}$ with 9.188 second. Based on number of T, $V^{(21)}$ performs the least number with total is 548. Based to total number of $T1$, $V^{(15)}$ performs the least number with total is 17412.

7. Conclusion

Three-dimensional parameter free filled function together with some types of search direction are introduced and explained in this paper. The objective is on solving the three-dimensional unconstrained global optimization problem. According to the numerical performance discussed previously, by applying particular type of search direction in filled function algorithm method, the method is more efficient. Besides, it is observed from the results of numerical experiment that the search direction of filled function method does works better for some case and does not for another. Based on number of T, $V^{(21)}$ performs the least number with total is 548. Based to total number of $T1$, $V^{(15)}$ performs the least number with total is 17412. However, this chapter gives insight, that, it is possible to look at the search direction as the way of speeding up the search on this algorithm method.

Acknowledgments
This work was supported by: 1) Grant research of *Riset Kompetensi Dosen Unpad* (RKDU) under project leader of Dr. Sukono MM, MSi.; 2) Unpad Academic Leadership Grant (ALG) under the project leader of Prof. Dr. Sudradjat Supian, MS.; 3) Fundamental Research Grant Scheme (FRGS Vot. No. 59255). Ministry of Higher Education Malaysia

References

[1] Lin H, Gao Y, and Wang Y 2013 A continuously differentiable filled function method for global optimization, *Numerical Algorithm* (doi: 10.1007/s11075-013-9746-3)

[2] Wang X, and Zhou G 2006 A New Filled Function for Unconstrained Global Optimization, *Applied Mathematics and Computation* 174 pp 419–429.

[3] Wei F and Wang Y, A 2013 New Filled Function Method with One Parameter for Global Optimization, Hindawi: Mathematical Problems in Engineering. (http://dx.doi.org/10.1155/2013/532325)

[4] Ma S, Yang Y, and Liu H 2010 A Parameter Free Filled Function For Unconstrained Global Optimization, *J. Applied Mathematics and Computation*, 215, pp. 3610-3619.

[5] Ge R 1990 A Filled Function Method For Finding A Global Minimizer of A Function of Several Variables", *J. Mathematical Programming*, **46** (2), pp.191-204.

[6] An L, Zhang L, Chen M 2004 A Parameter-Free Filled Function for Unconstrained Global Optimization *J. Shanghai Univ. Engl. Ed.* **8** (2) pp 117–123.

[7] Shang Y, Li P, Xie H, 2010 A New Parameter Free Filled Function for Unconstrained Global Optimization *The Third Int. Joint Conf. on Computational Sciences and Optimization* pp 65-69 (doi: 10.1109/CSO.2010.59)

[8] Mohd I B, Wen G K, and B. D. Yosza 2008 Fungsi Pengisian Sebagai Satu Kaedah Pilihan Dalam Pengoptimuman Sejagat *Jurnal KALAM (Karya Asli Lorekan Ahli Matematik)* **1**(1) pp 6-14.

[9] Napitupulu H, Mohd I B, and Pandiya R 2014 *Proc. of the 3rd int. Conf. on Global Optimization and Its Applications* Kuala Terengganu: Penerbit UMT pp 218-140

[10] Napitupulu H, and Mohd I B 2016 Vector Direction of Filled Function Method on Solving Unconstrained Global Optimization Problem, *AIP Conf. Proc.* 1707, 050011 (doi: 10.1063/1.4940843)

[11] Antoniou A, and Lu W S 2007 *Practical Optimization, Algorithm and Engineering Application*. New York: Springer

Numerical simulation of breaking regular and irregular wave propagation above a sloping bottom

Didit Adytia[1, a)], **Semeidi Husrin**[2], and **Adiwijaya**[1]

[1]School of Computing, Telkom University, Jl. Telekomunikasi 1 Bandung 40257, Indonesia

[2]Research & Development Centre for Marine and Coastal Resources, Agency for Marine & Fisheries Research & Development, Ministry of Marine Affairs & Fisheries - The Rep. of Indonesia, Jl. Pasir Putih II Ancol Timur, Jakarta 14430, Indonesia

[a)]Corresponding author: adytia@telkomuniversity.ac.id

Abstract. An accurate water wave model with breaking capability is important for accurate prediction of wave evolution especially when the wave approaching shallower water. Not only accuracy in the dispersion relation of the wave model that is important, the nonlinearity of the model also plays significant role in representing effect of wave shoaling. In this paper, we simulate propagation of breaking regular and irregular wave above a sloping bottom by using a dispersive wave model so-called variational Boussinesq model. Results of simulation are compared with experimental data from a hydrodynamic laboratory. We discuss effects of wave energy transfer from main wave frequencies to higher order frequencies when wave reaching shallower water and breaking.

1. Introduction

A good understanding of water wave behavior is essential for design activities related with water wave in coastal and offshore engineering, such as design of harbour, coastal defense, offshore platform, etc. With an increase of computer performances in the several last decades, it has been a long aim of coastal engineers to create an accurate and efficient wave model and its numerical implementation to help their work, especially to understand water wave behavior for design process. The hardest challenge for the wave model is to simulate wave propagation accurately from deep water to shallower water. Physical aspects such as dispersion, nonlinearity, shoaling, refraction, diffraction, breaking should be modelled accurately in the wave model and its numerical implementation.

The most popular type of water wave model among coastal engineering community is Boussinesq-type models (BTMs). Its popularity started when Peregrine in 1967 [1] improved the original wave equations by J.V. Boussinesq in 1987 from originally only for flat bottom to be depth dependence equations. Most noticeable BTMs introduced in 90's are the Boussinesq of Madsen & Sorensen [2] and the Boussinesq of Nwogu [3]. At that time, these BTMs were limited only for weakly dispersive and weakly nonlinear wave. The ideas of these two BTMs were soon very popular among wave modelers. Many researchers extend the capabilities of these BTMs to be applicable for wide range of waves such as for deep water (very dispersive wave) and also fully nonlinear waves. Nevertheless, the extension of these models into a strong dispersion and fully nonlinear wave model results in a long and a complex set of equations, i.e. number of terms in the equations are significantly increased and also introduce higher order derivatives (up to 6th order spatial derivatives). These complexities may

Content from this work may be used under the terms of the Creative Commons Attribution 3.0 licence. Any further distribution of this work must maintain attribution to the author(s) and the title of the work, journal citation and DOI.
Published under licence by IOP Publishing Ltd

increase difficulties for numerical implementation. Review related to the development of these BTMs can be read in [4,5].

In this paper, we use a rather different Boussinesq type of model, so-called the Variational Boussinesq (VB) model that is derived from a variational approach (see [6, 7]). Instead of introducing higher order derivative terms in the equations, the VBM only introduced 2nd order spatial derivatives and has no mixed temporal-spatial derivative. The model has been implemented numerically using pseudo-spectral method in [8], and using Finite Element Method (FEM) in [9, 10, 11]. The model is capable for simulating long wave such as tsunami [12, 13] and short wave such as wind waves [7, 11]. Nevertheless, the capability of the VB model [7, 11] was limited for simulating non-breaking waves.

Wave breaking process is a crucial physical phenomenon that has to be included in a wave model, especially for simulating wave propagation from deep to shallower water, where the wave become steeper and eventually breaks. In this paper, we extend the capability of the model especially for simulating breaking waves. To that end, we use an eddy-viscosity breaking mechanism as proposed by [14] to be incorporated in the Pseudo-Spectral (PS) implementation of the VB model.

The content of this paper is as follows. In the next section, we describe the basic derivation of the VB model. The section is followed by a description of the eddy-viscosity breaking mechanism as proposed [14]. To show the performance of the implementation of the breaking mechanism in the PS implementation of the VB model, we simulate propagation of breaking regular and irregular waves above a sloping bottom. Experimental data is available from a hydrodynamic laboratory to compare with the results of simulation. Finally, some conclusions are described in the last section.

2. Variational Boussinesq Model

The variational Boussinesq (VB) model is derived via so-called variational principle [15]. Here the water is assumed to be incompressible, inviscid and the flow is irrotational. Zakharov [16] and Broer [17] show independently that water wave phenomena can be exactly described as a Hamiltonian system, as $\partial_t \eta = \delta_\phi \mathcal{H}$, and $\partial_t \phi = -\delta_\eta \mathcal{H}$, where $\mathcal{H}$ is so called the Hamiltonian or the total energy (sum of kinetic and potential energy), η and ϕ are the canonical variables of the Hamiltonian system that denoting surface elevation and surface potential, respectively. $\delta_\varphi \mathcal{H}$ and $\delta_\eta \mathcal{H}$ are the variational derivatives of $\mathcal{H}$ with respect to η and ϕ, respectively.

In this paper we restrict the model to be 1D problem. The horizontal and vertical axes are denoted by x and z, and the time is denoted by t. As derived in [6], in the VB model, the fluid potential $\Phi(x,z,t)$ is approximated by the following expression: $\Phi(x,z,t) = \phi(x,t) + \Sigma_m F_m(z).\psi_m(x,t)$, where F_m is a priory chosen vertical function and ψ_m is an horizontally dependent function that should satisfy an elliptic equation. By substituting the approximation of $\Phi(x,z,t)$ for the VB model above, we can obtain the equations of VB model as follows

$$\partial_t \eta = -\partial_x \left((h+\eta) u \right) - \partial_x \left(\beta \partial_x \psi \right) \tag{1}$$

$$\partial_t u = -g \partial_x \eta - u \partial_x u \tag{2}$$

$$-\partial_x (\alpha \partial_x \psi) + \gamma \psi = \partial_x (\beta u) \tag{3}$$

The first and second equations are the continuity and momentum equation, whereas the third equation is an elliptic equation for searching function ψ_m that has to be solved every time step. Here α, β and γ are coefficients that depend on vertical profile F_m. Note that the vertical profile F_m denote completely the dispersion quality of the VB model. The optimization of vertical profile F_m for broad band wave can be read in [6, 7]. For a parabolic vertical profile as proposed by [7], the expression for α, β and γ are as follows $\alpha = 2H^3/15$, $\beta = -H^2/3, \gamma = H/3$, where $H = h + \eta$ is the total depth.

In this paper, the systems (1-2) are implemented numerically by using the Pseudo-Spectral (PS) method, while the (linear) elliptic system (3) is solved by using Finite Element Method or FEM [6]. The combination of PS method and FEM is investigated in [8]. The main idea of the PS method is to calculate spatial derivation in the Fourier space by exploiting Fourier and Invers Fourier transform, while the multiplication operation in the nonlinear terms are performed in the real space [8].

3. Breaking Mechanism

In order to incorporate wave breaking mechanism in the wave model, we use the eddy-viscocity breaking mechanism of Kennedy *et al* 2000 [14]. To that aim, an eddy viscosity term R_b is added in the momentum equation of the VB model, such that the equation (5) become $\partial_t u = -g\partial_x \eta - u\partial_x u + R_b$. where R_b is defined as $R_b = \frac{1}{H}\partial_x F$, with $H = \eta + h$ is the total depth. Based on [14] the F is defined as $F = \nu.\partial_x(Hu)$, where ν is the so-called eddy-viscocity coeffcient, i.e. $\nu = b.H\partial_x(Hu)$. The eddy viscocity model of Kennedy is actually derived based on the Shallow Water Equations (SWE) where the term $\partial_x(Hu)$ is the normal velocity N, i.e. from the continuity equation of the SWE; $\partial_t \eta = -\partial_x(Hu) = -N$. The coefficient b in the expression of ν contains a mixing length coefficient δ_B and a smooth function B that determine the cessation of the wave breaking [14] for the details). In this paper, we use $\delta_B = 1.5$ for two simulations in the next section.

4. Test Cases

For testing the PS-implementation and the Kennedy's eddy viscosity breaking mechanism in the VB model, we perform two test cases, i.e. breaking regular and irregular waves propagating above a sloping bottom. These two cases are based on a physical experiment that was performed in the twin flumes of Leichtweiss - Institute (LWI), TU Braunschweig, Germany. The flume consists of $1m$ and $2m$ wide parallel flumes, with approximately $90m$ long and $1.2m$ high. The original aim of the experiment is to investigate energy wave dissipation by an artificial parameterized mangrove forest that is placed in the $2m$-wide flume. For this paper, we only use the experiment results in the $1m$-wide flume, where there is no obstacle in the basin. In the middle of the flume, it is installed a sloping bottom with gradient 1:20 that is started from depth h_0=0.415m up to h_1=0.2m. The layout of the experiment is shown in Figure 1. There are 6 wave gauges (WG) installed in the basin as shown in Figure 1. To perform numerical simulations, instead of using the signal at the wave flap at x=0m, we use the signal at *WG1a*=21.415m as the influx signal for the wave model. The signal is influxed into the domain of computation by using a method so-called embedded influxing as proposed by [18].

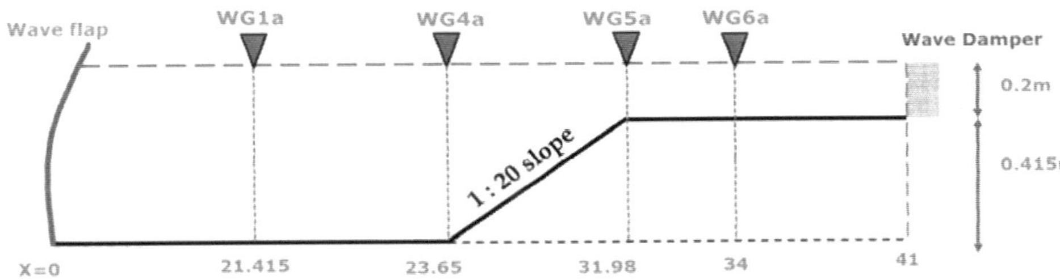

Figure 1. Layout of the experiment setup.

4.1. Breaking Regular Wave

The first case that is simulated is a breaking regular wave (with case number 2009051403). In the experiment, monochromatic (regular) waves are influxed by the wave flap with wave height of 0.2m (above 0.615m depth) and with wave period of 2.5s (or frequency of 0.4Hz). As stated previously, to perform the simulation, the signal at WG1a is used as an influx signal for the simulation. For the numerical simulation we use a uniform grid with grid size dx=0.08m. The simulation is performed until time t=200s.

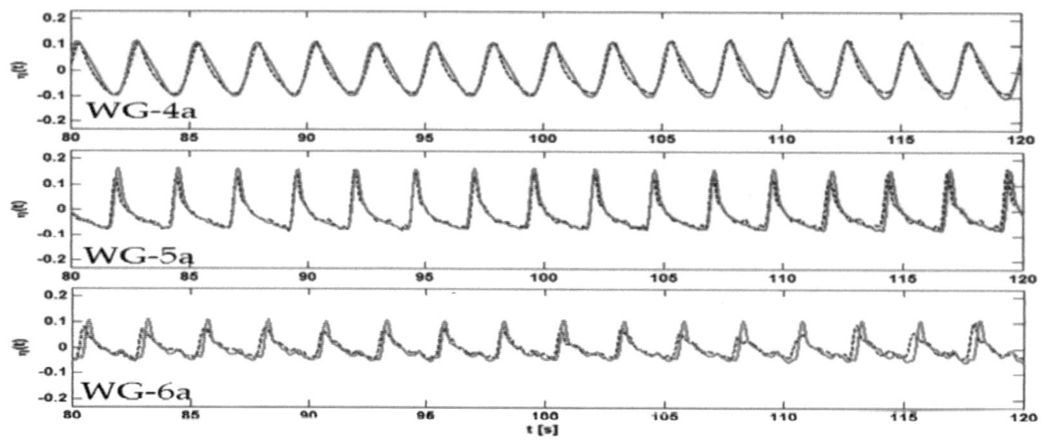

Figure 2. Signal comparison between the simulation (solid red lines) and the measurement (dashed blue lines) at positions WG-4a, WG-5a, and WG-6a for the regular wave case 2009051403.

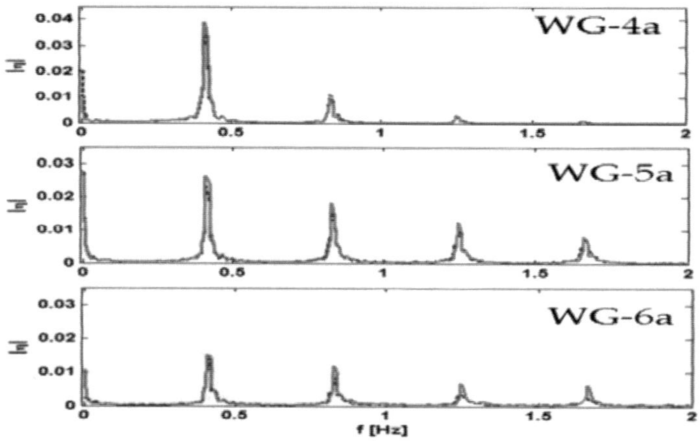

Figure 3. Comparison of amplitude spectrum between simulation (solid red lines) and measurement (dashed blue lines) at positions WG-4a, WG-5a, and WG-6a for the regular wave case 2009051403.

In Figure 2, results of numerical simulation are compared with signals from the experimental data at WG4a (at the foot of the slope), WG5a (at the top of the slope) and WG6a (at the shallower area). The comparison in signal shows a good agreement with the experimental data. The signal plots in WG4a and WG5a show that the wave become steeper as the wave propagate from deep water into shallower water. The signal in WG6a shows that a significant amount of energy is released as the wave breaks between WG5a and WG6a that is shown by an amplitude reduction, i.e. from $0.15m$ in WG5a to $0.09m$ in WG6a.

Since the wave propagates from deep area (WG4a) into shallower area (WG5a), there are effects of wave shoaling and wave nonlinearity. These effects can be shown from plots of amplitude spectrum of signal in WG4a to WG6a in Figure 3. In the WG4a, the signal already contains relatively small bound waves (low and high frequencies waves that are formed by the free wave, i.e. wave with frequency $f_0 = 0.4Hz$). As the wave propagates into shallower area, in WG5a, the energy of bound waves become larger. In other words, there are transfer of energy from the free wave ($f_0 = 0.4Hz$) into the bound waves, i.e. wave with frequencies of $f = 2f_0, 3f_0, 4f_0$, etc (or higher order waves). In the

WG6a, the amplitude spectrum plot shows that there are a significant amount of energy is released for the free wave as well as the bound waves.

4.2. Breaking Irregular Wave

The second test is the breaking irregular wave that is propagate in the same domain setting as in the previous case. The case number of the experiment is 2009051509. In this experiment, an irregular wave is influxed in the wave flap with significant wave height Hs of $0.2m$ and with peak wave period Tp of $2.5s$ (or corresponds with $f_p = 0.4Hz$). Similar to the regular case, we use a uniform grid with grid size of dx=0.08m. The simulation is performed until time t=400s.

Just as in the previous case, results of the numerical simulation are compared with the experimental data at WG4a, WG5a and WG6a, which are shown in Figure 4 for the signal and in Figure 5 for the amplitude spectrum. Different than the previous case, in this irregular wave case, the wave propagate in a more complex behaviour, where dispersion plays significant role besides the nonlinearity, i.e. lower wave frequency (long wave) propagates faster than higher wave frequency (short wave). Nevertheless, the comparison in Figure 4 and 5 show a good agreement between the simulation and the experimental data.

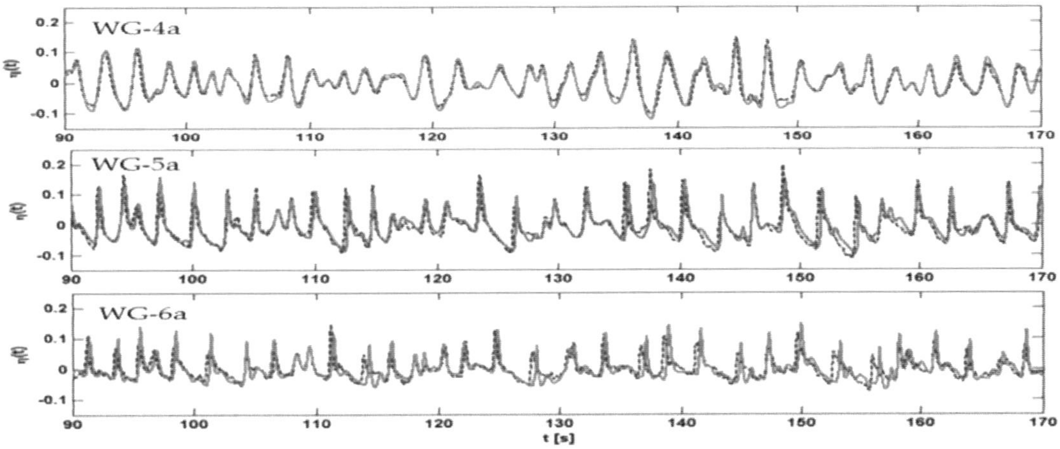

Figure 4. Signal comparison between the simulation (solid red lines) and the measurement (dashed blue lines) at positions WG-4a, WG-5a, and WG-6a for the irregular wave case 2009051509.

Similar to the regular wave case, the irregular wave breaking position is between WG5a and WG6a. This can be seen by the waveheight reduction in the WG6a in comparison to the waveheight in WG5a (Figure 4). An energy transfer from the main (free) wave frequencies into higher and lower frequencies can also be observed in the amplitude spectrum plots of signal in WG4a to WG5a in Figure 5. The amplitude spectrum plot in the WG6a shows a significant wave energy reduction that is caused by wave breaking.

5. Conclusion and discussion

An extension of a Pseudo-Spectral implementation of the Variational Boussinesq (VB) model with Kennedy's eddy-viscocity breaking mechanism is presented. The accuracy of the implementation is tested by comparing results of simulations with laboratory experiment of breaking regular and irregular wave propagation above a sloping bottom. The comparison in signal and amplitude spectrum of the simulation show a good agreement with the experimental data. The simulation can reconstruct the experiment accurately, for effects of dispersion, nonlinearity and breaking process. Effects of wave breaking, i.e. wave energy reduction, is clearly seen in the plot of amplitude spectrum. Besides that, in the regular wave simulation, a tranfer of energy from main frequency to higher frequencies can be seen clearly from the plot of amplitude spectrum.

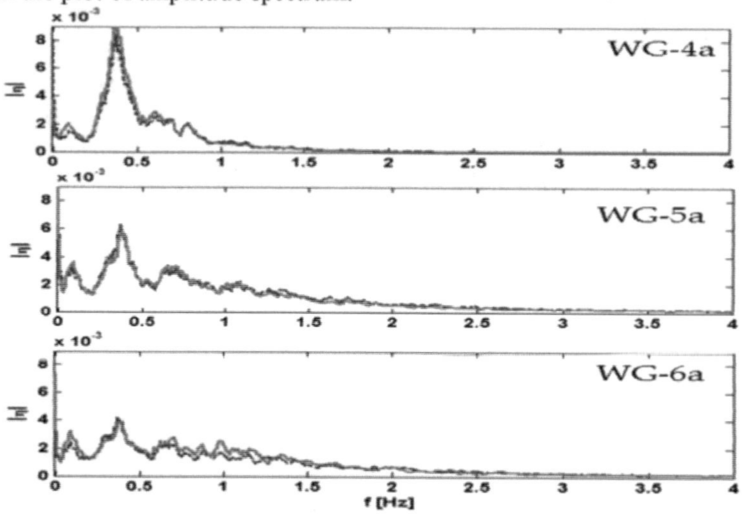

Figure 5. Comparison of amplitude spectrum between simulation (solid red lines) and measurement (dashed blue lines) at positions WG-4a, WG-5a, and WG-6a for the irregular wave case 2009051509.

Acknowledgments

The authors would like to acknowledge the Leichtweiss Institute for Hydraulic Engineering (LWI) of TU Braunschweig - Germany for the access to the experimental data of TAPFOR project.

References

[1] Peregrine, DH, 1967. *J. Fluid Mech.* **27** pp 815–827.
[2] Madsen, PA, and Sorensen, OR, 1992. *Coast. Eng.* **18** pp 183–204.
[3] Nwogu, O, 1993. *J. Waterw. Port Coast. Ocean Eng.* **119** (6) pp 618– 638.
[4] Brocchini, M, 2013. *Proc R Soc A* **469** 20130496.
[5] Kirby, JT, 2016. *J. Waterway, Port, Coastal, Ocean Eng.*, 03116005.
[6] Adytia, D and van Groesen, E 2012. *Coast. Eng.* **64** pp 139–150.
[7] Lawrence, C, Adytia, D, and van Groesen, E, 2018, *Wave Motion* **76** pp 78-102.
[8] Adytia, D, 2017. *Indonesian Journal on Computing (Indo-JC)* **2** pp 101-108.
[9] Adytia, D, and van Groesen, E, 2011. *Proc. International Conference on Developments in Marine CFD*, in Chennai, India. RINA, pp 25–29.
[10] Adytia, D 2014. *Proc. of The Twenty-fourth International Ocean and Polar Engineering Conference*. International Society of Offshore and Polar Engineers, pp 912–918.
[11] Adytia, D, and Lawrence, 2016. *Proc. of the ASME 2016 35th International Conference on Ocean, Offshore and Arctic Engineering*, American Society of Mechanical Engineers, pp V007T06A106-V007T06A106.
[12] Adytia, D, and van Groesen, E, 2010. Asia and Pasific Coasts 2009, pp 122-128.
[13] van Groesen, E, Adytia, D, and Andonowati, 2008, *Natural Hazards and Earth System Science*, **8** (2), pp 175-185.
[14] Kennedy, AB, Chen, Q, Kirby, JT, Dalrymple, RA, 2000. *J. Waterw. Port Coast. Ocean Eng.* **126** pp 39-47.
[15] Luke, J, 1967. *J. Fluid. Mech.* **27** pp 395–397.
[16] Zakharov, V, 1968. *J. Eppl. Mech.Tech. Phys.* **9** (2) pp 190–194.
[17] Broer, L, 1974. *Appl. Sci. Res.*, **29** (1), pp 430–446.
[18] Liam, LS, Adytia, D, and van Groesen, E, 2014. *Ocean engineering*, **80**, pp 73-83.

IOP Conf. Series: Journal of Physics: Conf. Series **1090** (2018) 012066 doi:10.1088/1742-6596/1090/1/012066

Thermal analysis of composite disc-brake based on finite element method

S Riyadi*, Sulistyo, M Tauviqirrahman

Laboratory for Engineering Design and Tribology, Mechanical Engineering Department, Engineering Faculty, Diponegoro University, Semarang, Indonesia

*riyadislamet11@gmail.com

Abstract. Disc-brake is designed to decelerate the vehicle by sliding friction between the contact surface of the pad and the disc. This braking process will convert the mechanical energy of the vehicle to the heat energy. In the present paper, based on finite element method, the thermal behaviour characteristic induced by heat energy due to the contact between the pad and the brake surface is investigated varying the rotational velocities. The results show that the highest value of thermal distribution on the second surface of the disc and pad occurs during a high rotational speed. The rated thermal distribution on the surface of the disc is higher than the surface with the pad.

1. Introduction

The brake system as a part of the motorcycle safety system has an important role in protecting the passengers and driver. However, one big issue of the use of disc brakes is temperature rise due to long repetitive braking. The increase in thetemperature during braking leads to the decrease in the performance of the brake system. Related to thermal characteristics during braking, some workers have conducted investigations to this issue.

Adamowicz and Grzes[1] based on fully three-dimensional finite element model evaluated an impact of convective mode of heat transfer on the thermal behavior of a disc brake system during repetitive braking process with the constant velocity. It was found that during single braking the convective cooling has insignificant effect on the temperature distributions of a disc brake, consequently is not able to prevent overheat problem. Yevtushenko and Grzes[2] used a finite element analysis to investigate transient temperature fields using the three dimensional contact model of a disc brake focusing on temperature coefficient of friction. In their case, the temperature fields, braking distances, and braking times were calculated and confronted with the cases at the constant coefficient of friction.Grzes et al. [3] conducted the analysis of a temperature field at single braking in a pad-disc braking system of a railway vehicle experimentally as well as numerically. their main result was that at single braking the disc is not heated uniformly within the depth until the moment of standstill. With respect to the thermal characteristic, Arifin et al. [4], investigated the effect of the presence of ventilation as well as the hole number of the temperature distribution during brake. It was found that the larger the number of the ventilation holes, the lower the maximum temperature variation. In recent lubrication, Nong et al. [5] conducted thermal and thermal stress investigations considering air cooling during emergency braking phase using finite element (FE) and computational fluid dynamics (CFD) methods. It was shown that low temperature was obtained, as well as distributed uniformly on the

Content from this work may be used under the terms of the Creative Commons Attribution 3.0 licence. Any further distribution of this work must maintain attribution to the author(s) and the title of the work, journal citation and DOI.
Published under licence by IOP Publishing Ltd

brake disc, which attributes to high thermal conductivity of the disc material together with excellent cooling ability of the brake disc.

Based on the literature survey above, it is highlighted that most of the previously published works the pad geometry is not modeled in computational domain. While considering this scientific context, the main objective of the present paper is to investigate the temperature distribution of discbrake induced by the contacting surface using a finite element analysis by modeling the pad geometry. In this way, the heating due to friction between contact between the pad and the disc can be explored, and thus more realistic results can be obtained.

2. Methodology

The procedures for the disc brake simulation were carried out in two stages; the first stage focused on the disc pad operating in steady state and the second stage included operating in moving condition varying the rotational speeds of the disc. The measurement of the temperature of the pad against the disc and the rotation speed of the disc in ABAQUS can be determined through a load module.

For solving the problems dealing with the previousley mentioned stages above, several steps were performed in this study. The first step was to prepare a structure FEM model of the brake disc with pads including types of material used. This step was carried out using finite element software ABAQUS. Then, the boundary condition as well as the loading is applied on the components. The meshing conducted here used automatic feature in ABAQUS to obtain the best configuration of mesh. For final step, the initialization as the assignment of an initial value for a data object or variable was conducted. The manner in which initialization is performed depends on kind of steps as mentioned earlier.

2.1. FEM model

Figure 1 shows the geometric model of the disc brakes of motorcycle. The disc brakes consist of two main components; namely the disc and the brake pad. In the following computation, the contact between the disc and the brake pad occurs due to the applied load. This will generate the friction leading to the occurence of thermal profile.

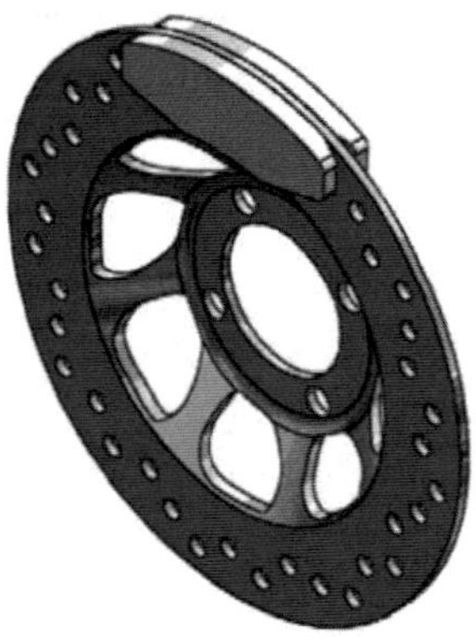

Figure 1. Geometry modelling of disc brake with pad of motorcycle

2.2. Material

The material used in the disc brake models consisted of two types, i.e. composite for pad component and stainless steel for disc component. The properties of the two materials in detail are shown in Table 1.

Table 1. Thermal Properties of Composite Materials and Steel

	Pad	Disc
Materials	Composite	Steel
Density [kg/m^3]	1450	7,800
Modulus Elasticity [GPa]	2.2	209
Poisson's ratio	0.24	0.29
Thermal Expansion [1/C]	1.1 x 10^{-5}	1.26 x 10^{-5}
Thermal Conductivity [W/mK]	1.1	48
Specific Heat [J/kgK]	1200	452

2.3. Boundary condition and loading

Figure 2 shows the position of the loading and boundary conditions as applied to a series of disc brakes. Pressure is applied to the outer surface of the second pad, while the boundary conditions were applied to the four pin holes. The pressure applied to the outer surface of the second pad is 1 MPa. The boundary conditions were applied by assuming that the four pin holes cannot move in the direction of x, y and z, but can rotate on the rotary axis z. Rotating speedsare varied in three conditions, namely low speed of 120 rad/s, medium speed of 150 rad/s and highest of 200 rad/s.

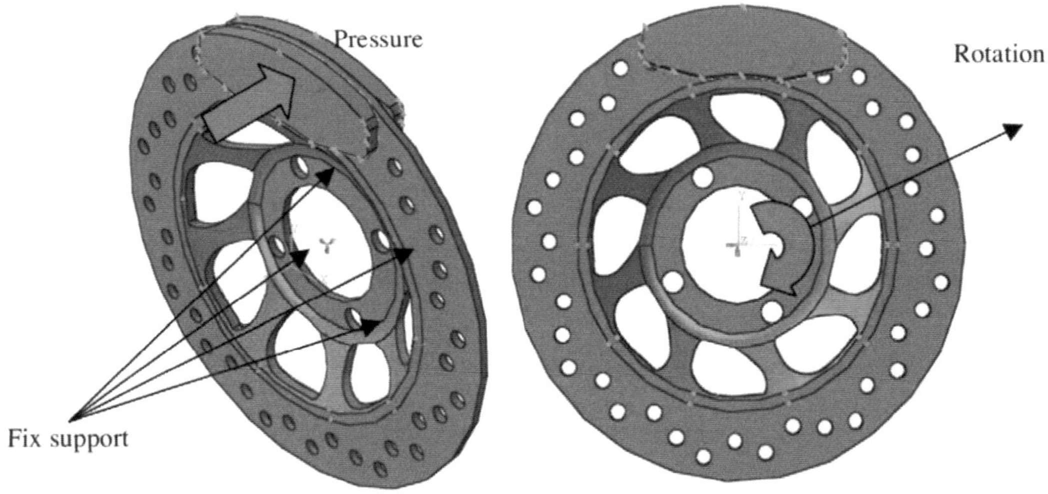

Figure 2. Boundary conditions and loading

2.4. Meshing

Figure 3 shows the mesh as applied to this model. The kind of element used was the C3D8T type in which the element type consisted of 8-nodes capable of thermal analysis due to friction and displacement in three directions. The total nodal applied to this model was about 22,195, while the number of elements was around 30,178.

IOP Conf. Series: Journal of Physics: Conf. Series **1090** (2018) 012066 doi:10.1088/1742-6596/1090/1/012066

Figure 3. Mesh of model with the C3D8T Element

3. Results and Discussion

3.1. On the stationary condition

Figure 4 shows the thermal distribution on the contact surface between the pad and the disc when the disc is not moving. There are two thermal distributions on both sides of the surface, while the pad is only on one side of the contact surface. The maximum thermal value on the disc surface is larger than that on the side with the pad. It indicates that a rise in surface temperature due to the interaction of contact and friction is highlighted.

It is also found form Fig. 4 that the thermal profile is relatively low during stationary condition even there is contact between the pad and the disc. This is as expected because there is no high friction between them. As a note, the brightness of the contour colour in Fig. 4 can be seen on the surface of the disc and pad, where the red colour indicates a high value while the blue colour indicates a low value. The relationship value and colour can be seen in the legend next to the picture of the disc and pad.

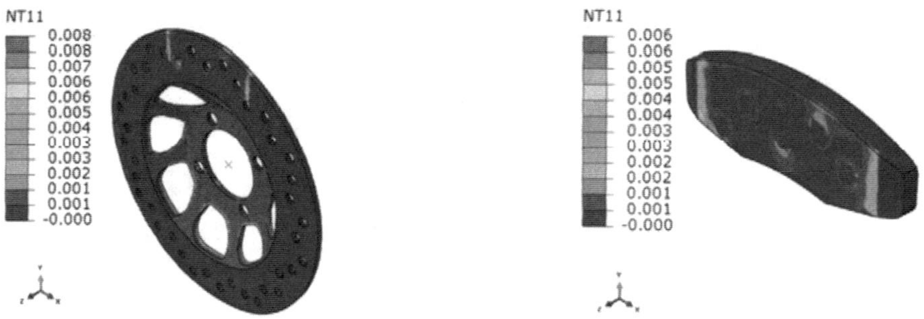

(a) (b)

Figure 4. Thermal distributionon the stationary condition for (a) disc, (b) pad

3.2. On the moving condition

Figure 5 shows a comparison of the values of the thermal distribution on the contact surface of the disc as a result of contact interaction between the pad and the disc while spinning. This condition is based on spinning the disc at varying speeds. Of the three variations of speed, it can be seen that the value of the highest thermal distribution occurs during high speed. The thermal distribution value here means a rise in surface temperature due to the interaction of contact and friction.

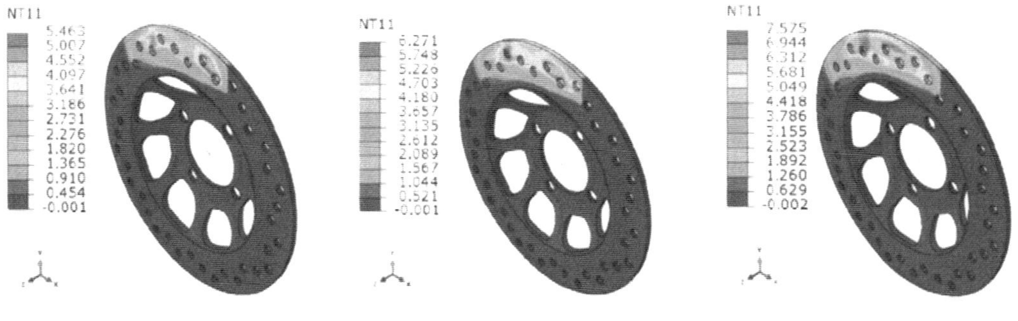

(a) (b) (c)

Figure 5. Thermal distribution on the disc during rotational movement, (a) low speed, (b) normal speed, (d) high speed

Figure 6 shows a comparison of the value of the thermal distribution on the contact surface due to the interaction of contact between the pad and the disc during the condition of movement. Figure 10 (a) shows the thermal contact distribution on the pad surface when the disc rotates at a low speed, Figure 10 (b) at the time of medium speed, and Figure 10 (c) at high speed. Of the three variations of speed, it can be seen that the value with the highest thermal distribution occurs at high speed. This is similar to thermal distribution value that occurs on the disc component.

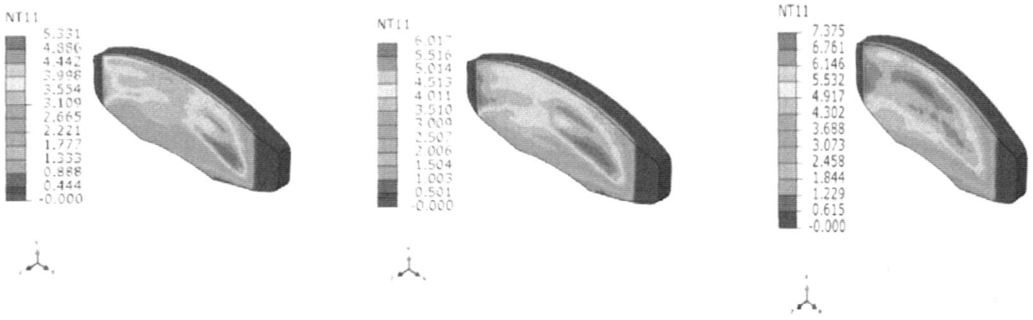

(a) (b) (c)

Figure 6. Thermal distribution on the pad when the system is moving, (a) low speed, (b) moderate speed, (c) high speed

Figure 7 shows the comparison of the maximum temperature values on the surfaces of the disc and pad while the disc is spinning. In this case, the disc rotational velocity ranges from slow, medium, and high. Based on Fig. 2, it can be seen that the highest value of temperature distribution on the second surface of the disc and pad occurs during a high rotational speed. It is also found that the rated temperature distribution on the surface of the disc is higher than the surface with the pad. On the other words, it can also be concluded that the higher the speed, the higher the temperature distribution.

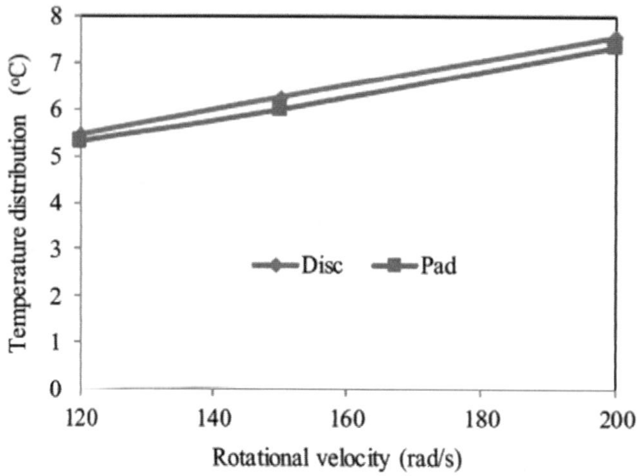

Figure 7. Comparison of the temperature distribution between the disc and pad at varying speeds.

4. Conclusion

The main objective of the investigation was to investigate the thermal characteristic of the disc brake induced by the contact with composite pad due to braking. A numerical simulation in steady state was performed using finite element method based software ABAQUS. Based on the explanation above, the conclusions can be drawn as follows:

1. Rotating velocity has a significant effect on the thermal behaviour of disc brake resulted during the braking. It was found that the higher the velocity, the higher the temperature distribution.
2. The highest value of thermal distribution on the second surface of the disc and pad occurs during a high rotational speed.
3. The temperature distribution on the surface of the disc is higher than the surface with the pad.

These findings are useful for the study of the performance of the composite pad-disc brake with respect to the temperature rise during the braking.

References

[1] Adamowicz A, Grzes P 2011 *Appl. Therm. Eng.* **31** 2177

[2] Yevtushenko AA and Grzes P 2016 *Int. J. Therm. Sci.***102** 254

[3] Grzes P, Oliferuk W, Adamowicz A, Kochanowski K, Wasilewski P and Yevtushenko AA 2016 *Int. J. Heat Mass Transf.***75** 1

[4] Arifin, Tauviqirrahman M, Muchammad, Jamari J and Bayuseno AP 2016 *Int. J. Eng. Tech***8**(6) 3073

[5] Nong XD, Jiang YL, Fang M, Yua L and Liu CY 2017 *Int. J. Heat Mass Transf.***108** 1374

IOP Conf. Series: Journal of Physics: Conf. Series **1090** (2018) 012067 doi:10.1088/1742-6596/1090/1/012067

CFD analysis of texture depth effect on the performance of hydrodynamic lubricated bearing

M Muchammad[1,2*], F Hilmy[1], M Tauviqirrahman[1], J Jamari[1], DJ Schipper[2]

[1]Laboratory for Engineering Design and Tribology, Mechanical Engineering Department, Diponegoro University, Indonesia.
[2]Laboratory for Surface Technology and Tribology, Faculty of Engineering Technology, University of Twente, The Netherlands.

m_mad5373@yahoo.com

Abstract. Surface modification of the lubricated bearing such texturing has proven to improve the hydrodynamic performance. The present paper examined the effect of texture depth as well as the texture length on the tribological performance using computational fluid dynamic (CFD) approach. The cavitation model was also considered to obtain more realistic characteristic of bearing. It was shown that by increasing texture depth of the surface of bearing, the enhanced hydrodynamic pressure was achieved. Moreover, other interesting result is that the length of cavitation region decreases with the increase in texture depth and the decrease in texture length.

1. Introduction

For more decades, the research of texturing of the lubricated contact has obtained an explosion interest by researchers around the world. It is well known that texturing technique has proven to improve the lubrication characteristics. Numerous researchers based on hydrodynamic lubrication theory have introduced several mechanisms of lift generation of textured surfaces.

Fowell et al. [1] using a mass-conserving numerical analysis concluded that any convergence between the bearing surfaces provides a significant mechanism for lift generation. Gherca et al. [2] evaluated the effects of the texture geometry using a mass-conserving model. It was found that the load support was strongly dependend on the geometrical features such as size, density, and shape. Henry et al. [3] experimentally studied the effect of surface texturing on the steady-state behavior of bearings.They concluded that the textured bearings reduces friction up to 30% at low loads while for heavy loads, their performance is equivalent or even lower than that of an untextured bearing. Yagi et al. [4] studied the value of the load support enabled by textured pattern in hydrodynamic lubrication regime. They concluded that the mechanism of lift generation was strongly affected by boundary condition. Zhang et al. [5] investigated the effect a rectangular array of circle dimples on the film thickness profile. It was concluded that the lift generation could be improved through appropriate partial arrangement of textures. Later, Shinde and Pawar [6] proposed the optimal surface texturing parameters to improve performance of bearing.

Based on published works, it can be highlighted that texturing produces more positive effect compared to pure texturing with respect to the load support. However, from literature survey, most of the solution of the lubrication problems have been solved by of Reynolds equation which may be questionable in particular cases with low viscosity and high runner velocities.

Content from this work may be used under the terms of the Creative Commons Attribution 3.0 licence. Any further distribution of this work must maintain attribution to the author(s) and the title of the work, journal citation and DOI.
Published under licence by IOP Publishing Ltd

Following this main frame, in the present paper a CFD method as well as the Reynolds equation have been applied to investigate the texturing parameter in particular case of texture depth on the hydrodynamic pressure. For obtain more accurate results, the cavitation model is also considered.

2. Methodology

2.1. Theory

In the present study, the lubrication problem is solved by the Navier-Stokes equation and continuity equations. The Navier-Stokes (N-S) equations are solved over the domain using a finite-volume method with the commercial CFD software package FLUENT®. The Navier–Stokes and the continuity equations can be expressed, respectively.

$$\rho\frac{Du_i}{Dt} = -\frac{\partial p}{\partial x_i} + \rho G_i + \frac{\partial}{\partial x_j}\left[2\eta e_{ij} - \frac{2}{3}\eta\left(\nabla u_i\right)\delta ij\right] \tag{1}$$

$$\nabla \bullet u = 0 \tag{2}$$

Not like the previously published works in which the cavitation model is ignored, in the present study, the cavitation effect is taken into account. In FLUENT®, there are three available cavitation models: Schneer and Sauer model, Zwart-Gelber-Belamri model and Sighal et al. model [7]. In this study, the Zwart-Gelber-Belamri is employed due to their capability (less sensitive to mesh density, robust and converge quickly [7].

In addition to the Navier-Stokes equation, in this work, the lubrication problem is also solved by modified Reynolds equation (Eq. 3 as below). For more detail, the derivation of such equation used here is described in [8, 9]. It should be noted that the isoviscous Newtonian one-dimensional Reynolds equation with slip is derived from a simple of the x-component of the Navier-Stokes equation that assumes an incompressible flow, neglecting the inertia effects in the film. However, for the analysis studied here, the slip coefficient is set to zero.

$$P_2\left[\left(h_p^3 + 3h_p^3K_p\right)\frac{ab+bc}{ab^2} + \left(\frac{h_0^3 + 3h_0^3K_0}{a}\right)\right] = P_{atm}\left[\left(h_p^3 + 3h_p^3K_p\right)\frac{ab+bc}{ab^2} + \left(\frac{h_0^3 + 3h_0^3K_0}{a}\right)\right]$$
$$-6\mu U\left[\left(h_p + h_pK_p\right) - \left(h_0 + h_0K_0\right)\right] \tag{3}$$

2.2. CFD Model

Figure 1 gives the schematic illustration of a parallel textured (pocketed) sliding bearing. The assumption of the no-slip is adopted. The main characteristics of the bearing and the lubricant properties studied are presented in Table 1 and 2 respectively.

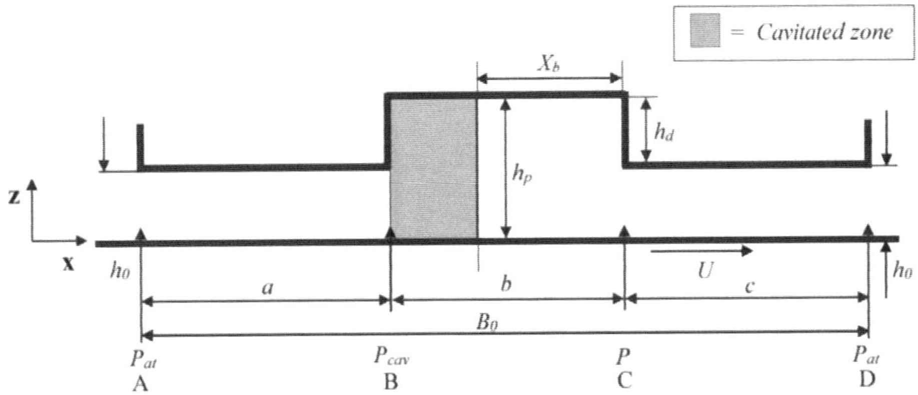

Figure 1. Geometry of slider bearing 2D model.

Table 1. Characteristics of main parameter bearing analyzed.

Parameter	Symbol	Value	Unit
Length of bearing	B_o	2.00	mm
Length of inlet	a	0.75	mm
Length of texture	b	0.50	mm
Length of outlet	c	0.75	mm
Minimum film thickness	h_o	4.00	μm
Depth of texture	h_d	2.00 ; 8.00	μm
Inlet pressure	Pa	100	kPa
Cavitation pressure	Pc	50	kPa
Sliding velocity	U	1	m/s

Table-2. Characteristics of lubricant properties.

Parameter	Symbol	Oil	Oil Vapor	Unit
Density	ρ	962	0.02556	kg/m^3
Viscosity	μ	0.013468	$1.256\text{x}10^{-5}$	kg/ms

In the present study, the grid consists of three blocks with grid system as shown in Fig. 2. The number of grids in the longitudinal direction (Nx) and transverse (Nz) is 1000 x 100. These grids results in 90,000 cells, faces 18,112 and 91,121 nodes based on the independent mesh study.

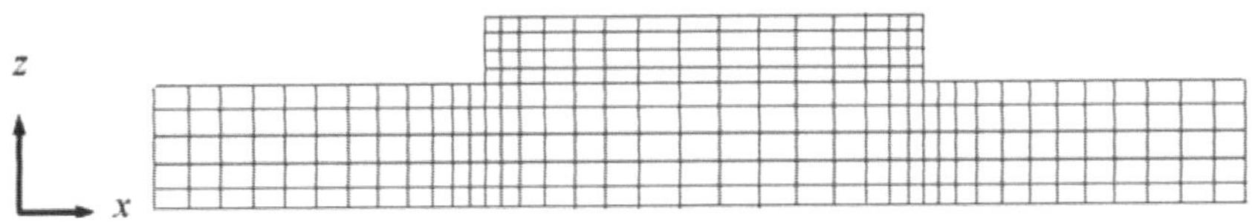

Figure 2. Grid system analyzed in numerical analysis.

In the present work, the texture depth is of particular interest. The pocket depth is categorized to two kinds, that is, low texture depth and high texture depth, as shown in Fig. 3. The hydrodynamic pressure is used as indicator to describe the texture depth effect with respect to the Reynolds equation and Navier-Stokes equation..

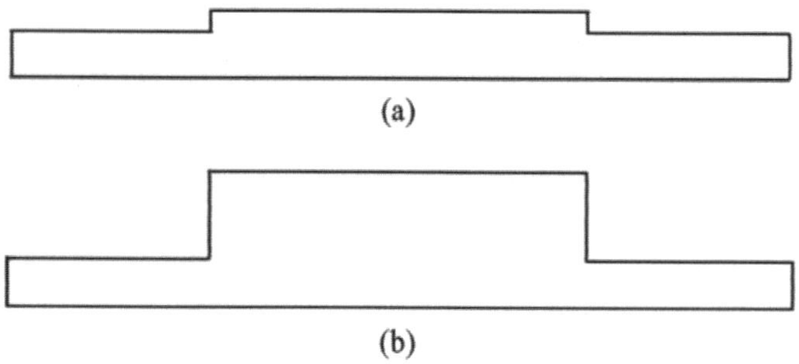

Figure 3. Pocket depth analyzed (a) $h_d = \frac{1}{2}\, h_o$ (low texture depth), (b) $h_d = 2\, h_o$ (high texture depth).

3. Result and discussion

In this section, the results of the application of different texture depths with respect to pressure distribution graph are shown. In Fig. 4, the pressure distribution with b = 0.25 at low and high texture depth is presented. It can be observed from Fig. 4a that the N-S solution gives a larger prediction of pressure compared to Reynolds approach for low texture depth. It indicates that the inertia effect exists in the case of low texture depth. As a note, this inertia term cannot be predicted by Reynolds equation. The interesting finding is that when texture depth is increased, the inertia effect has gone. As indicated in Fig. 14b, the prediction of the hydrodynamic pressure by the N-S equation and the Reynolds approach is exactly the same.

Related to the cavitation behavior, textured contact with low texture depth leads to the generation of the cavitation area. However, the N-S equation tends to predict the cavitation area smaller than the Reynolds approach. On the other hands, the Reynolds theory underestimates the hydrodynamic pressure when the inertia effect has a dominant role in the texture contact.

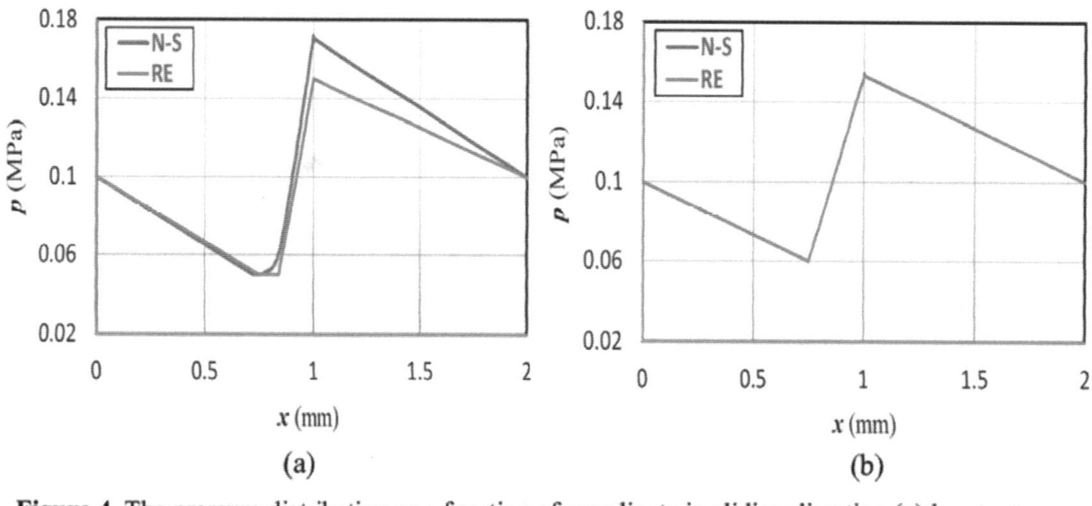

Figure 4. The pressure distribution as a function of coordinate in sliding direction (a) low texture depth, and (b) high texture depth, of interest with b = 0.25 mm.

Figure 5 shows the hydrodynamic pressure predicted by the N-S approach and the Reynolds equation for the case of length of contact $b = 0.5$ mm. It can be seen that for the case of high texture depth, the pressure distribution predicted by the N-S approach is larger than the Reynolds equation. It means that the inertia effect exists in the texture contact with larger length of contact b. For low texture depth, the prediction of the pressure by two approaches is the same, in particular to the cavitation region. On the other words, the inertia and the cavitation term is independent. The profile of the hydrodynamic pressure is strongly affected by the texture depth as well as the texture length. This finding is strengthened by the simulation result as indicated in Figure 6. For high b, in this case b = 0.75, the prediction by two approaches is similar both for low and high texture depth.

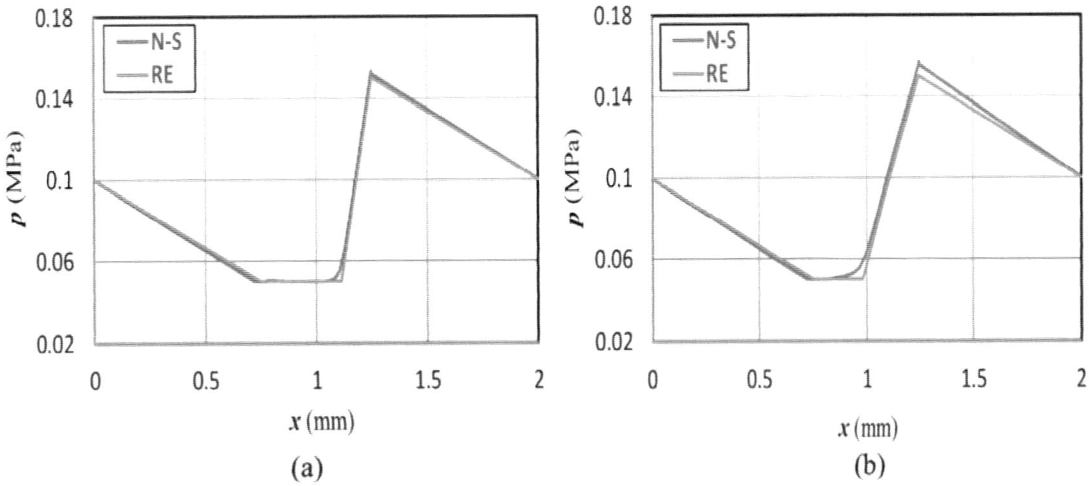

Figure 5. The pressure distribution as a function of coordinate in sliding direction (a) low texture depth, (b) high texture depth, of interest with b = 0.5 mm.

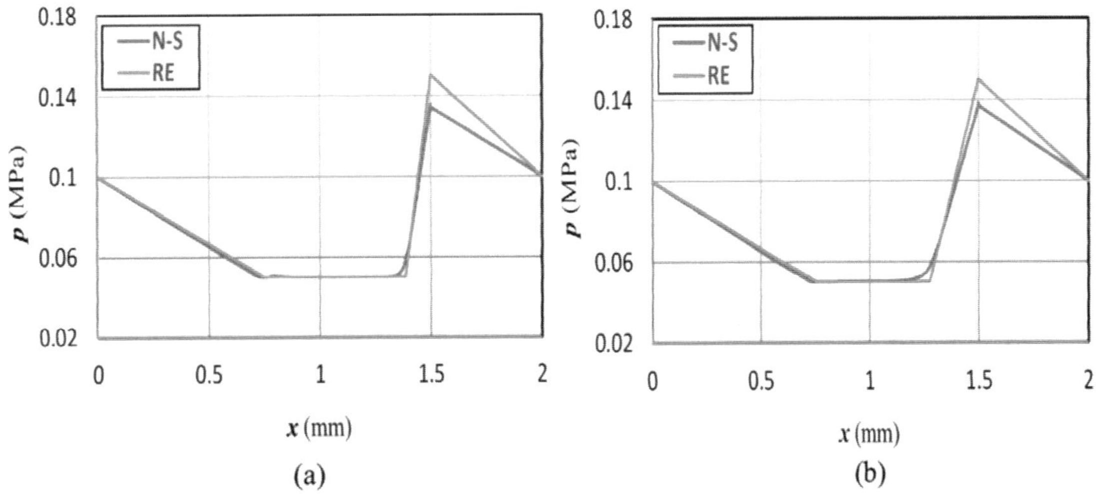

Figure 6. The pressure distribution as a function of coordinate in sliding direction (a) low texture depth, (b) high texture depth, of interest with b = 0.75 mm.

Figure 7 shows the load support for each texture depth and bearing texture length configuration, either obtained by the CFD method with the Navier-Stokes equation or by the Reynolds equation method. In general, load support decreases with increasing texture length and increases with increasing texture depth except for the case of high texture depth with texture length of 0.25 mm. Under these conditions the support load decreases. A part from that phenomenon, generaly the high texture depth gives a positive effect on bearing performance. This is influenced by the higher cavitation region at large texture length. Conversely, a high texture depth actually makes the cavitation region smaller. This indicates that the occurrence of cavitation causes a negative effect on bearing performance. Thus, this study indicates that the maximum bearing performance corresponds to the smallest texture length configuration of 0.25 mm with low texture depth.

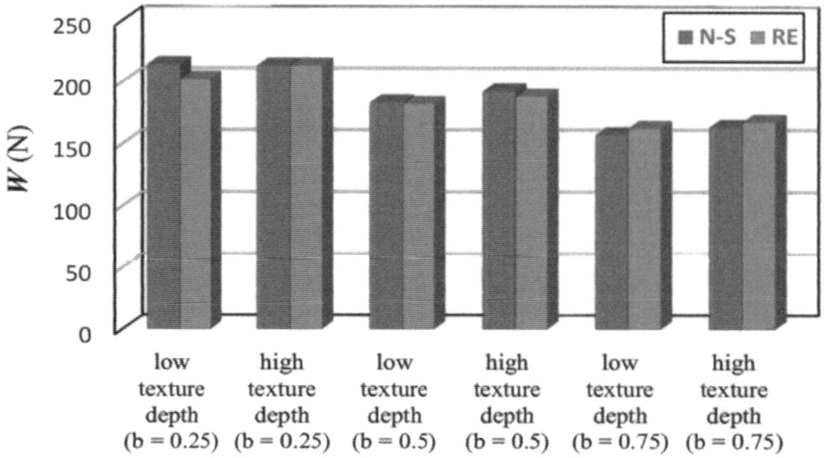

Figure 7. The load support as a function of texture depth.

4. Conclusions

In the present study, the lubrication analysis of the single textured contact was solved by two approaches; the Navier-Stokes Equation and the Reynolds equation. Based on the explanation discussed earlier, the conclusion can be drawn as follows:

1. The increasing texture depth leads to the enhanced of the hydrodynamic pressure.
2. The increasing texture depth decreases the cavitation area.
3. The decreasing texture length decreases the cavitation area.

References

[1] Fowell M T, Medina S, Olver A V, Spikes H A, Pegg I G 2012 *Tribol. Int.* **52** 7
[2] Gherca A R, Maspeyrot P, Hajjam M, Fatu A 2013 *Tribol. Trans.* **56(3)** 321
[3] Henry Y, Bouyer J, Fillon M 2015 *Proc. Instn. Mech. Eng. Part J: J. of Eng. Tribology* **229(4)** 362
[4] Yagi K, Sato H, Sugimura J 2015 *Tribol. Online* **10(3)** 232
[5] Zhang H, Dong G, Hua M, Guo F, Chin K S 2015 *Ind. Lubr. Tribol.* **67(4)** 359
[6] Shinde A B, Pawar P M 2017 *Tribol. Int.* **144** 349
[7] ANSYS, ANSYS Fluent, version 14.0: user manual. ANSYS, Inc., Canonsburg, USA., 2011
[8] Muchammad M, Tauviqirrahman M, Jamari J, Schipper D J 2017 *Lubr. Sci.* **29(3)** 133
[9] Muchammad M, Tauviqirrahman M, Pratomo A W, Jamari J, Schipper D J 2017 *IJSurfSE* **11(2)** 100

IOP Conf. Series: Journal of Physics: Conf. Series **1090** (2018) 012068 doi:10.1088/1742-6596/1090/1/012068

Numerical optimization of load support of the single textured lubricated contact with boundary slip

M Tauviqirrahman*, M F Noorkarim, Arijanto, Muchammad

Laboratory for Engineering Design and Tribology, Mechanical Engineering Department, Engineering Faculty, Diponegoro University, Semarang, Central Java, 50275, Indonesia

*mohammad.tauviqirrahman@ft.undip.ac.id

Abstract. Textured surfaces containing macro texture in one of the surfaces are the subject of a great deal of research to enhance tribological characteristic (higher low support but lower friction) of the bearing. In this paper, a general parametric model of the groove texture profile of single textured contact with boundary slip is developed and the exact optimization method is adopted to obtain the global-optimum texture profile of the lubricated sliding contact. The optimization objective is the maximization of the load support of the lubricant. The results show that there is a shift of the texture location to the centre of the contact with lower texture length. It is also found that the optimal single textured bearing can help to improve the load support up to 75%.

1. Introduction

Recent technological advances have presented some new methods such as grafting, coating, laser surface texturing, and micro machining to improve the tribological characteristic (higher low support but lower friction) of the bearing. Within the broad area of tribology, the researches relating to the lubrication have paid much attention to surface texturing, as is reflected in many recent papers. Buscaglia et al. [1] studied the effect of one-dimensional and two-dimensional textures on the load support of a bearing with a "homogenized" effect and a homogenization error. Scaraggi et al. [2] investigated the friction properties of the lubricated laser textured surface by tuning microhole depth. With the aim of minimizing the friction, different surfaces characterized by different micro hole depths, but same void ratio have been explored.

Related to the optimization of the texture shape, many researchers have explored new method to improve the lubrication performance. Uddin and Liu [3] introduced the design and optimization of what-so-called star-like texture shape for improving tribological performance. The main result was that the triangle effect is the most dominant in reducing the friction. The other result proposed by Zhang et al. [4] stated that either bullet or fish shape was the preferable texture shape than those of circular shape based on genetic algorithm. In recent publication, Wang et al. [5] used the GA-SQP hybrid method to obtain the global-optimum profile of the groove texture bottom. The maximization of the load support of the oil film was of particular interest. The authors also highlighted the superiority of the proposed GA-SQP hybrid method.

It is also known that in addition to surface texturing, another technique to enhance the bearing performance is by introducing the boundary slip on the lubricated surface via coating technique. However, the available literature survey indicates that the studies related to optimization of the

Content from this work may be used under the terms of the Creative Commons Attribution 3.0 licence. Any further distribution of this work must maintain attribution to the author(s) and the title of the work, journal citation and DOI.
Published under licence by IOP Publishing Ltd

combined effect of slip and texture of lubricated contact are rather very limited. In addition, most of the previous studies on the slip-texture shape optimization were based on pre-determined shapes and distributions, for example [6,7]. In these references [6,7], the global-optimum shapes were uncertain.

Therefore, more work is required in the area of slip-textured lubricated contact to provide the needful information for the design process. The contribution of this paper is to numerically study the texturing geometry (i.e. pocket texture profile) based on a general parametric model. In particular, single texture cell is investigated with maximizing the load support. The modified Reynolds equation with slip is proposed to analyze the lubrication characteristic.

2. Methodology

2.1. Lubrication model

A schematic representation of the single textured lubricated sliding contact is shown in Fig. 1. The non-textured lower surface slides with a uniform velocity. The upper surface with groove texture remains stationary. The bearing is fully lubricated by a Newtonian, incompressible lubricant, with a constant viscosity. The modified Reynolds boundary condition is used to solve the lubrication problem. It reads:

$$\frac{\partial}{\partial x}\left(\frac{h^3}{12\eta}\frac{\partial p}{\partial x}\frac{h^2+4h\eta\alpha}{h(h+\eta\alpha)}\right)=\frac{U}{2}\frac{\partial}{\partial x}\left(\frac{h^2+2h\alpha\eta}{h+\eta\alpha}\right)-U\frac{\alpha\eta}{h+\alpha\eta}\frac{\partial h}{\partial x}+\frac{h}{2\eta}\frac{h\alpha\eta}{h+\alpha\eta}\left(\frac{\partial p}{\partial x}\frac{\partial h}{\partial x}\right) \tag{1}$$

The physical meanings of the symbols in Eq. (1) are as follows: h the film thickness (gap), p the lubrication film hydrodynamic pressure, α the slip coefficient on the stationary surface, U = Sliding velocity and η the lubricant viscosity.

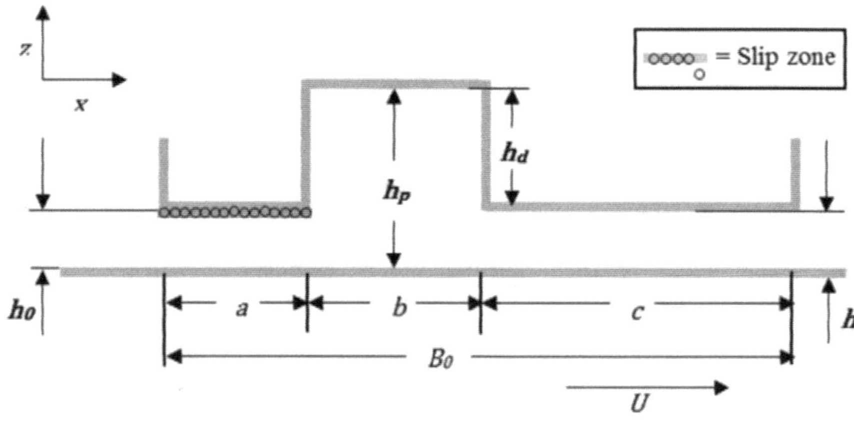

Figure 1. Schematic representation of slip-textured bearing (Note: a = inlet length; b = texture length; c = exit length; h_d = texture depth; h_p = total texture depth; h_o = land film thickness; B_o = total bearing length; U = sliding velocity)

2.2. Discretization

In this paper the modified Reynolds equation is solved numerically using a finite difference equations obtained by means of the micro control volume approach. The entire computed domain is assumed as a full fluid lubrication. The lubricant fluid is in a laminar flow since the Reynolds number is really small. By employing the discretization scheme, the computed domain is divided into a number of control volumes.

The mesh number obtained from grid independent study is approximately 5,000 nodes. In the present study, as seen in Fig. 1, the boundary slip is located at the leading edge of the contact to obtain the maximum lubrication performance [8]. In this work, the inlet length a, the texture length b and the exit length c are varied, while the total texture depth h_p as well as the texture depth h_d is assumed to constant.

2.3. Optimization

In order to conduct the optimization performance of the configuration of slip-textured lubricated contact, the same basic parameters (i.e., working parameters and boundary conditions) are used in the calculations. These parameters are listed in Table 1. Recently, in mathematical programming several methods have been introduced to solve nonlinear constrained optimization problems for the most generic subjects regarding the texture profile in the case of parallel sliding contact. In this section, the exact optimization algorithm is developed to optimize the texture profile of the lubricated sliding contact. The detail parameters for the optimization process is shown in Table 2.

Table 1. Parameters for computations

Total bearing length	B_o	0.02 m
Sliding velocity	U	1 m/s
Viscosity	η	0.01 Pa s
Atmospheric pressure	P_{atm}	100 kPa
Land film thickness	h_o	1 μm
Pocket depth	h_d	2.5 μm
Slip coefficient	α	0.02 m^2s/kg

Table 2. Parameter settings for optimization procedure

Objective Function	Maximize the load support
Constrain	Total bearing length
Design Variable	*Inlet Pocket* Lower bond: 0.001 m Upper bond: 0.01 m *Pocket* Lower bond: *inlet* + 0.001 m Upper bond: *inlet* + 0.01 m *Outlet* Lower bond: *inlet* + *pocket* + 0.001 m Upper bond: *inlet* + *pocket* + 0.01 m

In the present study, the object of the optimization is to maximize the load support. The load support satisfies two main functional purposes: (1) carry the external load, and (2) reduce the contact between the solids and thus reducing the wear. The optimization analysis attempts to satisfy both functional requirements with various design parameters. i.e. slip inlet length (a) texture length (b) and exit length (c), as seen in Fig. 1. The computer code contains the finite volume method used to solve the hydrodynamic pressure and thus the load support, in combination with a numerical optimization library based on the exact optimization method.

IOP Conf. Series: Journal of Physics: Conf. Series **1090** (2018) 012068 doi:10.1088/1742-6596/1090/1/012068

3. Results and Discussion

In the present study, for optimization process, the influence of texture dimension as well as texture position should be exempted. Optimization predictions for this study are initiated by presetting the parameters of: (i) inlet length a as 0.004 m (Fig. 1), (ii) texture length b as 0.006 m, (iii) outlet length c as 0.010 m. In order to ensure that the lubricated interface was under hydrodynamic lubrication regime, the film thickness parameter (h_o = 1 μm) is kept constant throughout the computation processes.

Figure 2 depicts the comparison between the hydrodynamic pressure of the initial configuration and that of the optimized pattern. It can be observed that based on the optimization process, the hydrodynamic pressure profile increases significantly along the contact area. In the textured area, it can also be highlighted that the pressure peak in the case of optimized pattern increases up two times compared to the initial pattern.

Based on Figure 2, it can also be reflected that there is a shift of the starting point of the texture profile. On the other words, the inlet length containing slip becomes larger, that is, up to 150% compared to initial configuration of the bearing. It means that the slip condition has more role to alter the flow characteristic. From the physical point of view, the slip boundary decreases the pressure gradient along the slip area (i.e. inlet length), and this generates the increased hydrodynamic pressure. Other interesting result is that the texture length of the optimized pattern is half the texture length of the initial pattern. Again, it strengthens the previous statement that the slip has more dominant effect in enhancing the tribological performance than the texturing effect (i.e. texture length in this case). This result is in a good agreement with recent publication [9].

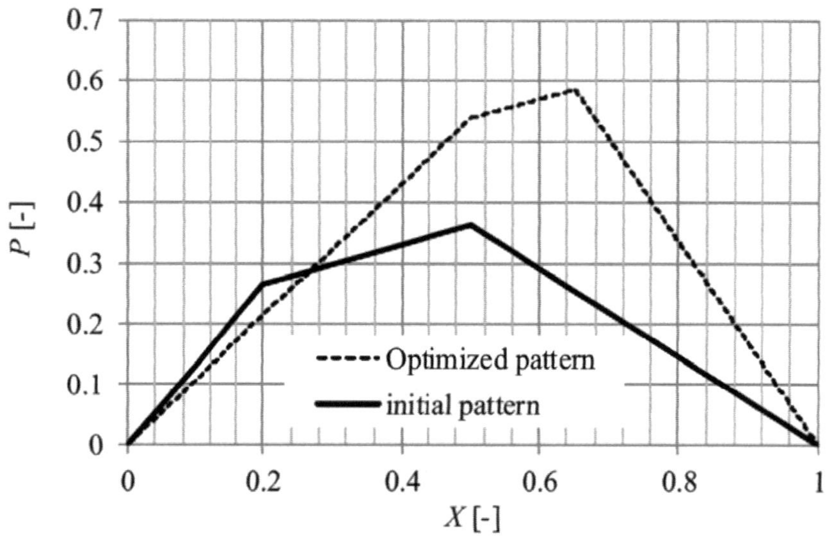

Figure 2. Hydrodynamic pressure profile before and after optimization

The comparison of the textured bearing configuration between "before" and "after" optimization is presented in Fig. 3. It can be observed that at the optimal textured lubricated contact, the texture length becomes shorter than the initial pattern of textured contact ($b' < b$ as seen in Fig. 3). In addition, the value of slip inlet length of the optimal contact becomes larger than the initial condition ($a' > a$ as shown in Fig. 3). It can be said that the combined effect of longer slip area and shorter texture length makes the hydrodynamic pressure increase significantly.

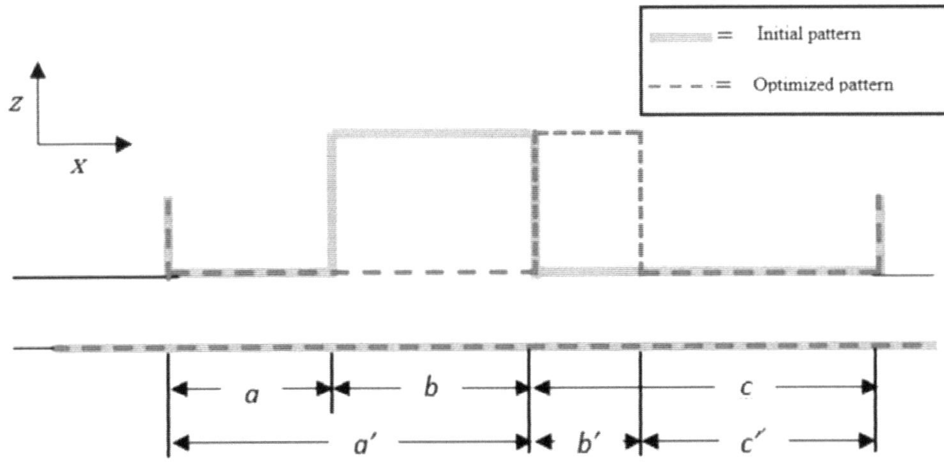

Figure 3. Comparison of the textured pattern profile for the case of "before" and "after" optimization

Generally, by design the improvement in pressure generation at textured lubricated contact (i.e. higher load support) is wanted condition of lubrication. The load support is obtained by integrating the hydrodynamic pressure over the surface area. In this section, the simulation result will be presented in the dimensionless form, i.e. the dimensionless load support W, in which $W = wh_o^2/U\eta B_o^2$ (where w is the load support, h_o is the land film thickness, U sliding velocity, and B_o is total length contact). Figure 4 shows the global solution of dimensionless load support W for each textured pattern during optimization process. It can be seen that the load support increases with increasing the number of combination data (i.e. slip and texture pattern) and finally reaches the steady condition.

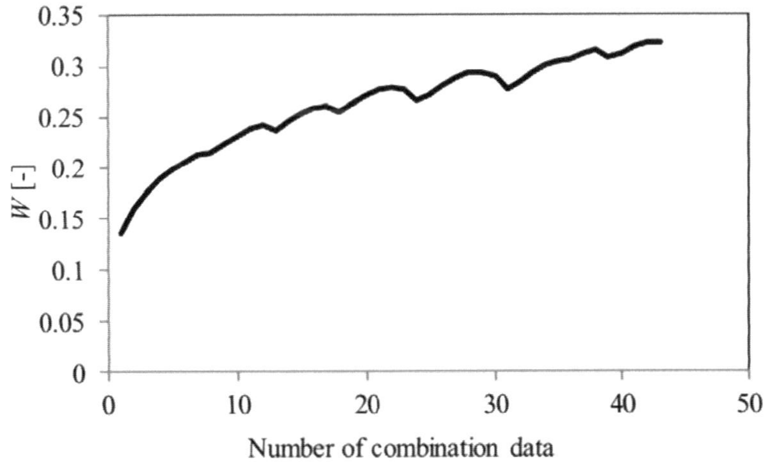

Figure 4. Prediction of dimensionless load support W during optimization process

Finally, it is necessary to show the characteristic of the lubricated textured contact for the case of initial pattern and the case optimized pattern in more detail with respect to the load support as shown in Table 3. It can be observed that the load support of the optimized pattern can be increased to up 75% compared to initial condition.

Table 3. Comparison of contact characteristics between "before" and "after" optimization

	Before optimization	After Optimization
Inlet length [m]	0.004	0.010
Pocket length [m]	0.006	0.003
Outlet length [m]	0.010	0.007
Load support [N]	8.60×10^5	1.20×10^6

4. Conclusion

The numerical optimization of a single textured lubricated contact with boundary slip was considered. Textured pattern was optimized pointing out the importance of the position of the texture cell as well as the texture length to increase the load support. The flow analysis based on Modified Reynolds equation has been linked with an optimization strategy with the exact optimization method. The main conclusions involved are as follows:

1. There exists a shift of the texture location to the center of the contact with lower texture length, which leads to improved load support.
2. The optimization results show that that the load support can be enhanced with as much as 75% compared to the initial pattern.

References

[1] Buscaglia GC, Ciuperca I and Jai M 2007 *J. Math. Anal. Appl.* **335** 1309
[2] Scaraggi M, Mezzapesa FP, Carbone G, Ancona A, Sorgente D and Lugarà PM 2014 *Tribol. Int.* **75** 123
[3] Uddin MS and Liu YW 2016 *Biosurf. Biotribol.* **2** 59
[4] Zhang H, Hua M, Dong GZ, Zhang DY, Chen WJ and Dong GN 2017 *Tribol. Int.* **115** 222
[5] Wang W, He Y, Zhao J, Li Y and Luo J 2017 *Tribol. Int.* **109** 69
[6] Rao TVVLN, Rani AMA, Nagarajan T and Hashim FM 2012 *Tribol. Int.* **56** 121
[7] Tauviqirrahman M, Muchammad, Jamari and Schipper DJ 2014 *Tribol. Trans.* **57** 134
[8] Tauviqirrahman M, Ajie WK, Yohana E, Muchammad M and Jamari J 2016 *Int. J. Eng. Tech.* **8** 913
[9] Muchammad M, Tauviqirrahman M, Jamari J and Schipper DJ 2017 *Lubr. Sci.* **29** 133

Optimization of Enemy's Behavior in Super Mario Bros Game Using Fuzzy Sugeno Model

Nanang Ismail[1*], Husnul R M[1], M Nurul Subkhi[2], Miftahudin[1], Eki A Z Hamidi[1]
[1]Electrical Engineering Departement, UIN Sunan Gunung Djati Bandung, Indonesia
[2]Physics Departement, UIN Sunan Gunung Djati, Bandung, Indonesia

* E-mail: nanang.is@uinsgd.ac.id

Abstract. One of the key factors to make the action game more interesting is the smart behavior of an enemy. This paper describes the enhancement of smart enemy's behavior in the game that similar to Super Mario Bros using Sugeno Fuzzy Model. Sugeno Fuzzy Model is applied to regulate the enemy's behavior and determine the steps to be taken. The enhancement of the enemy's behaviors performed in the process of chasing down moving player and enemy's behaviors when meeting the player. Tests on the chasing player process, performed with 4 scenarios and shows achievement of 75% success, while in the game before adding Fuzzy Sugeno Model, the success is only 20%. Meanwhile, testing for enemy's behavior in this game is 100% successful, while in previous games only 33.3%. An enemy behavior logic test is performed to obtain defuzzification results as a Cartesian axis graph. Testing is done using Matlab®. Testing yielded 3 Cartesian axis graphs. Each graph has a behavioral decision. Therefore, there are three enemy's behavior decisions in this game, while Super Mario Bros game only produces 1 graph of Cartesian axis that does not result in enemy's behavior decisions.

1. Introduction

Video games take a lot of developments from simple concepts to complex technology products. One of the most attractive fields to be a research until now is artificial intelligence in games. This technology has been successfully implemented in various types of games, one of them is a strategy game [1].

Action games or the strategy games apply artificial intelligence systems in determining attitudes and actions of "enemies" in the game. A smart enemy due to implementing an artificial intelligence system (AI), will give the challenges to players [2]. Generally, the users of the game want the challenge and experience of playing which is better. Therefore, it is needed to continue the development of research on the creation of non-player characters with unpredictable behaviors [3].

Research on the development of games by applying various methods of artificial intelligence had been done by Cole, et al by implementing a genetic algorithm for regulating and developing Bot's behavior pattern, so it becomes more intelligent [4]. Its artificial intelligence is applied to "the non-player enemy" in the game against player strategy in Counter-strike games. Increased intelligence of "enemy" characters in a game can also help to complete the journey on the map [5]. References [5] use the genetic algorithm in the Lemmings game, and the result shows that the use of genetic algorithms can solve complex paths on the map. Fuzzy logic has also been used in improving the "enemy" intelligence in a Pac-Man game, so that the "enemy" in this game acts as a smart agent [6].

This paper describes the optimization of the enemy's behavior in an action game. The action game is a kind of game that emphasizes physical challenges and it needs fast reaction with coordination

Content from this work may be used under the terms of the Creative Commons Attribution 3.0 licence. Any further distribution of this work must maintain attribution to the author(s) and the title of the work, journal citation and DOI.
Published under licence by IOP Publishing Ltd

between eyes and hands. The chosen game concept as a comparator is Super Mario Bros Game. This game features a plumber named Mario who has to save the princess of the Mushroom Kingdom who was abducted by a King of Koopas named Bowser. The gameplay is displayed with platforming levels theme, where the player must avoid the different enemies and obstacles at each level. Based on the same concept and story with Super Mario Bros Game, then a new game is developed which will optimize the enemy's behavior as well as establish the steps will be taken using the fuzzy logic of Sugeno's method. All of these approaches will be implemented using Phaser framework as a media of game creation based on the web [7]. The Phaser framework is an open source platform and directly integrated with the web. In addition, with Phaser, games can be developed into a mobile game. Phaser also allows the user to create an online game without having to download or install additional applications [8]. Furthermore, this game is named Chase and Conquer Game (ChaCon Game).

2. Method

The steps of research development activities refer to the following diagram.

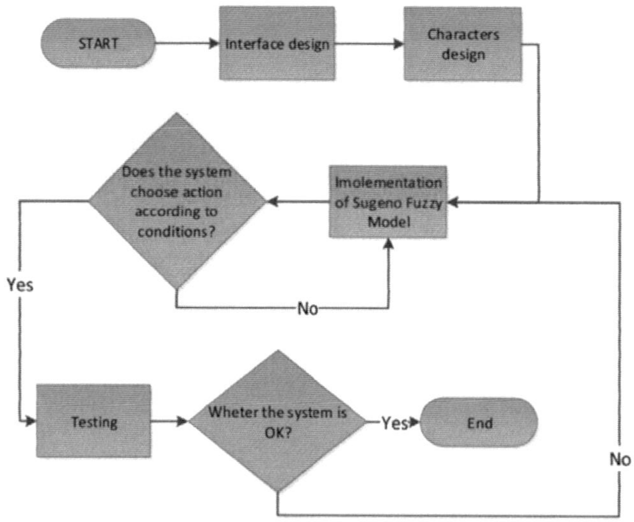

Figure 1. The steps of development

ChaCon Game is designed with 1 player and 3 enemies. The character of player is shown in Figure 2 and the character of enemies are shown in Figure 3.

Figure 2. Player.

Figure 3. Enemy.

2.1. Implementation of Sugeno Fuzzy Model

The fuzzy logic implementation of ChaCon Game is divided into 2 categories, they are fuzzy logic for enemy chasing/following the player, and for "enemy's behavior".

"Enemy chasing/following player" is a setting of the enemy to find the location of the player, then to follow it anywhere. This command is important for artificial intelligence because an enemy can move like a player which is controlled by a human. The "enemy chasing/following player" command is created using the fuzzy Sugeno rules presented in Table 1.

Table 1. Fuzzy rules for "enemy chasing/following player".

		Distance		Velocity		Decision-Making
1.	IF	Enemy X < Player X	AND	25 px	THEN	Move to right
2.	IF	Enemy X > Player X	AND	-25 px	THEN	Move to left
3.	IF	Enemy Y > Player Y	AND	250 px	THEN	Jump

In Fuzzy for Enemy's Behavior Implementation, "enemy's behavior" will be regulated with fuzzy logic Sugeno model, either when the health condition is fit or injured. The pattern of "enemy's behavior" that has a health point in a fit condition will be different from "enemy's behavior" that has a health point in injured. The pattern of "enemy's behavior" is presented in Table 2.

Table 2. Fuzzy rules for enemy's behavior.

		Distance		Health		Decision-Making
1.	IF	Near	AND	Fit	THEN	Attack
2.	IF	Near	AND	Injured	THEN	Avoid
3.	IF	Far	AND	Fit	THEN	Attack
4.	IF	Far	AND	Injured	THEN	Avoid

Meanwhile, Fuzzy for Enemy Chasing at Super Mario Bros Game uses the rule as in Table 3.

Table 3. Fuzzy Rules for enemy chasing in Super Mario Bros game.

		Distance		Velocity		Decision Making
1	IF	Enemy X > Player X	AND	-20 px	THEN	Move to left

Afterward, the fuzzy logic for "enemy's behavior" in Super Mario Bros Game has a function as an attacking command from an enemy to the player. The "enemy's behavior" command in Super Mario Bros game is shown in table 4.

Table 4. Fuzzy Rules for "enemy's behavior" in Super Mario Bros game

		Distance		Decision Making
1	IF	Near	THEN	Attack
2	IF	Far	THEN	Attack

Based on Table 4, the enemy's behavior is not distinguished by health condition, since there is only the condition which is represented by values 1 and 0. The value of 1 means alive, the value of 0 is dead.

3. Result and Analysis

ChaCon Game involves 1 player and 3 enemies. The number of enemies are not created as much as Super Mario Bros game, for this reason, the analysis is focused on "enemy's behavior" that is not associated with the number of enemies.

3.1. Game Interface

The game display visually is shown in Figure 4.

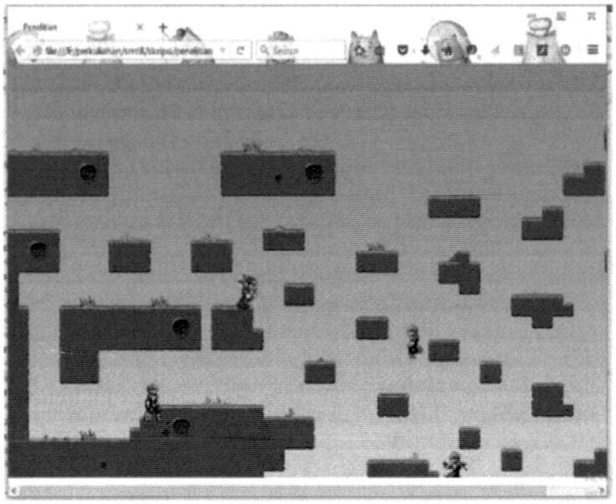

Figure 4. The Interface of Game.

3.2. Testing

There are two parameters will be tested, they are "enemy chasing" and "enemy's behavior". The parameter "enemy chasing" is tested by the black-box method. Meanwhile, the testing of "enemy's behavior" uses Matlab software to ensure defuzzification values.

3.2.1. Black-Box Testing. The fuzzy testing results for "enemy chasing/following player" are found in Table 5.

Table 5. The fuzzy testing for "enemy chasing" parameter.

	The Test Scenario	Expected results	Conclusion on ChaCon Game	Conclusion on Super Mario Bros game
1.	When the player is on the right side	Enemy moves along with the player	Successful	Unsuccessful
2.	When the player is on the left side	Enemy moves along with the player	Successful	Successful
3.	When the player is on top of the left side/right	Enemy jumps	Successful	Unsuccessful
4.	When the player is on top/ bottom	Enemy moves along with the player	Unsuccessful	Unsuccessful

Meanwhile, the fuzzy testing results for the parameter "enemy's behavior" are listed in Table 6.

Table 6. The fuzzy testing for "enemy's behavior" parameter.

	The Test Scenario	Expected results	Conclusion on ChaCon Game	Conclusion on Super Mario Bros game
1.	When the player is on the right and close to the enemy	Enemy attacks to right	Successful	Unsuccessful
2.	When the player is on the left and close to the enemy	Enemy attacks to left	Successful	Successful
3.	When enemy's health 50%	Enemy jumps to avoid player's attack	Successful	Unsuccessful

3.2.2. Testing with Matlab. The next step is to test a fuzzy algorithm using the Matlab on the ChaCon game and it is compared with the Super Mario Bros game. There are several steps carried out for testing using this Matlab, they are fuzzification step, entering the fuzzy input variable and creating the defuzzification output.

Fuzzification is the process of mapping a numerical value into a fuzzy set and determining the degree of membership. Fuzzification on Chacon Game has 3 inputs, namely distance, health, and avoid.

The distance variable of ChaCon Game is divided into 2 sets, namely "near" and "far". The range of value for distance variable between 0-1000 in accordance with map length, with the criteria presented in equations (1) and (2),

$$\mu[Near] = \begin{cases} 0; x \geq 700 \\ 1; x \leq 300 \\ \frac{700-x}{700-300}; 300 < x \leq 700 \end{cases} \tag{1}$$

$$\mu[Far] = \begin{cases} 0; x \leq 300 \\ 1; x \geq 700 \\ \frac{x-300}{700-300}; 300 < x \leq 700 \end{cases} \tag{2}$$

The result of calculation, then it is presented in a graph form as shown in Figure 5.

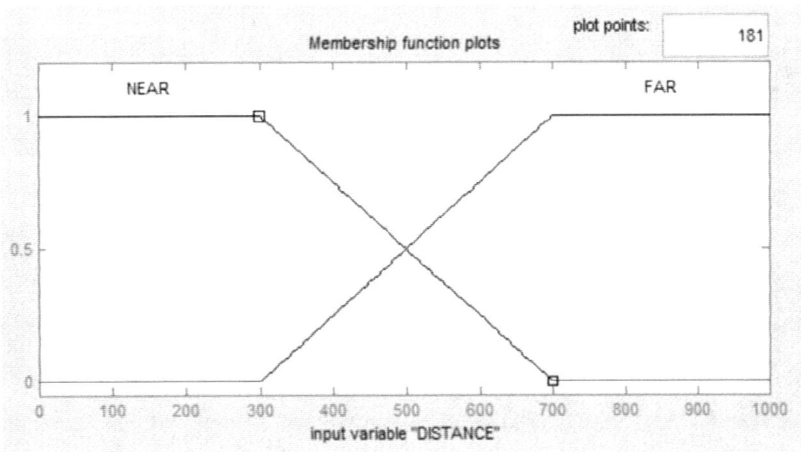

Figure 5. Input value of distance variable on ChaCon Game.

The health variable on ChaCon Game is divided into 2 sets, namely, fit and injured. The range of value for health variable between 0-200 in accordance with health input on the program, with the criteria presented in equations (3) and (4),

$$\mu[Injured] = \begin{cases} 0; x \geq 140 \\ 1; x \leq 60 \\ \frac{140-x}{140-60}; 60 < x \leq 140 \end{cases} \tag{3}$$

$$\mu[Fit] = \begin{cases} 0; x \leq 60 \\ 1; x \geq 140 \\ \frac{x-60}{140-60}; 60 < x \leq 140 \end{cases} \tag{4}$$

Then, the result of calculation is presented in graph form as shown in Figure 6.

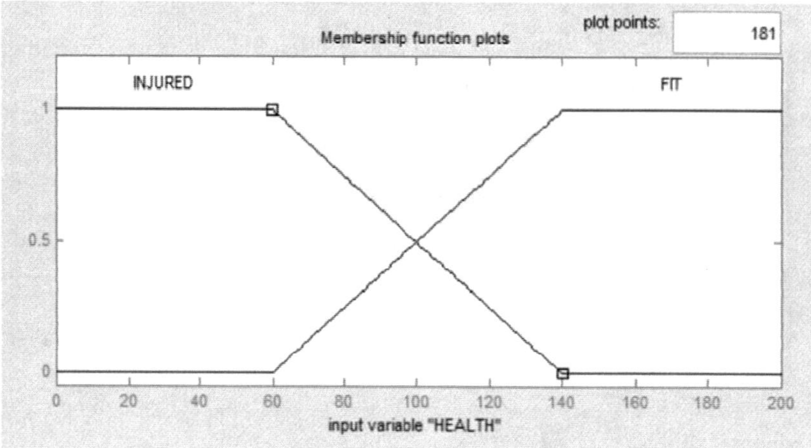

Figure 6. An Input value of health variable on ChaCon Game.

The avoid variable on ChaCon Game has a range of value 60 – 140, with the criteria presented in equation (5),

$$\mu[Avoid] = \begin{cases} 0; x \leq 60 \text{ } or \text{ } x \geq 140 \\ \frac{x-60}{100-60}; 60 < x \leq 100 \\ \frac{140-x}{140-100}; 100 \leq x < 140 \end{cases} \tag{5}$$

The result of the calculation is presented in graph form as shown in Figure 7.

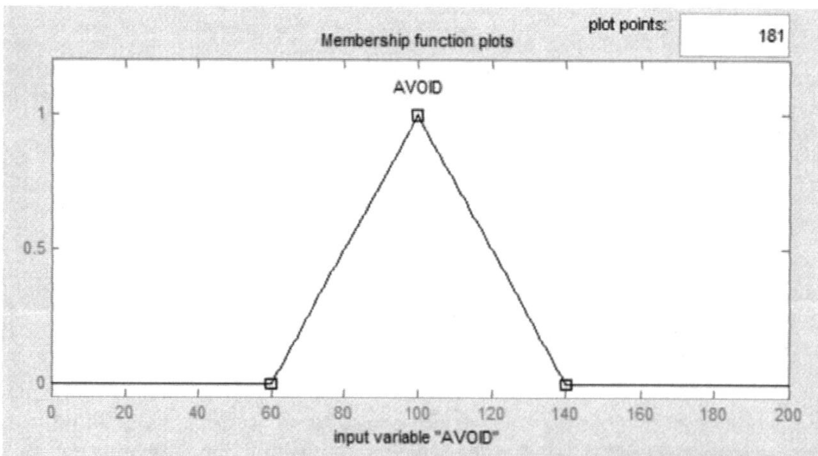

Figure 7. Input value of avoid variable on ChaCon Game.

The input variables of the Super Mario Bros game are distance and health. The distance variable of an enemy on Super Mario Bros is divided into 2 sets, they are 'near' and 'far'. The range of value for distance variable between 0 - 1000 in accordance with map length will be explained as follows:

1. Near = 0 - 700
2. Far = 300 - 1000

Manually calculated fuzzification values for near and far functions are performed using equations (2) and (3). The result of the calculation is presented in graph form as shown in Figure 8.

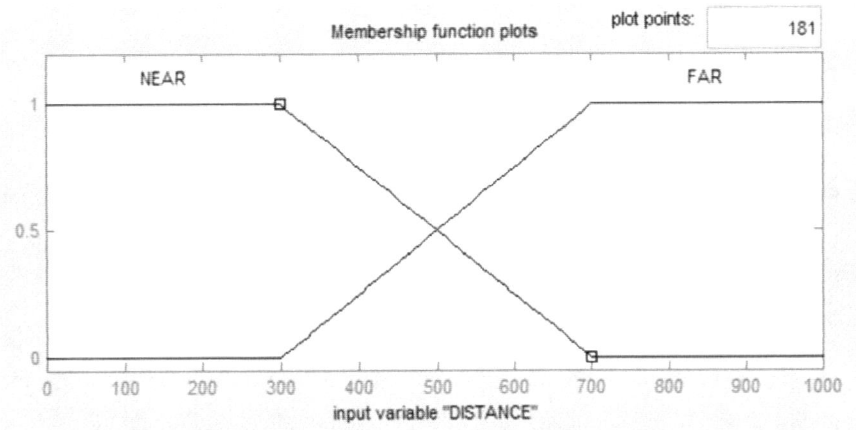

Figure 8. Input value of distance variable on Super Mario Bros game.

The Health variable on Super Mario Bros game is dead. The range of value for health variable is between 0 – 1. The dead curve is formulated by the following equation (6),

$$\mu[Dead] = \begin{cases} 0; x = 0 \\ 1; x = 1 \\ \frac{x-0}{1-0}; 0 < x \leq 1 \end{cases} \quad (6)$$

The result of the calculation is presented in graph form as shown in Figure 9.

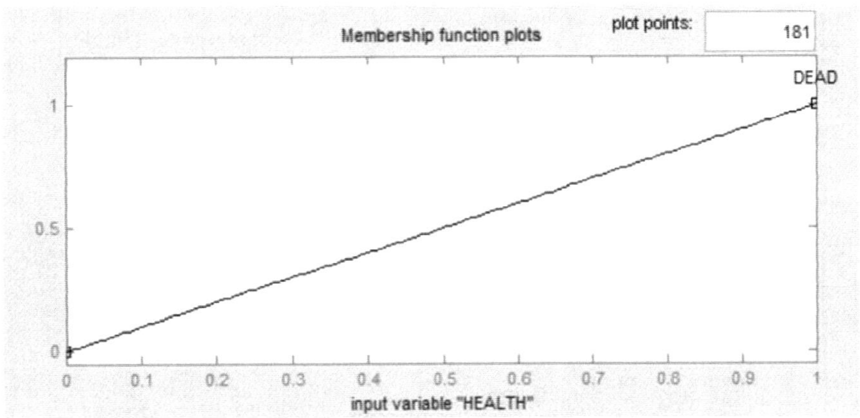

Figure 9. Input variable of health on Super Mario Bros game.

After all input variables are entered, then the defuzzification process is done to get the output result of ChaCon game as well as Super Mario Bros game. In ChaCon Game, the results obtained are as follows:

- IF distance = 0-500 AND health = 100-200 THEN decision = 0

- IF distance = 0-500 AND health = 0-100 THEN decision = avoid
- IF distance = 500-1000 AND health = 100-200 THEN decision = 0
- IF distance = 500-1000 AND health = 0-100 THEN decision = avoid

In Super Mario Bros game, the results obtained from Matlab output are as follows:

- IF distance 0-500 AND health 0-1 THEN decision = 0
- IF distance 500-1000 AND health 0-1 THEN decision = 0

3.3. Analysis

Based on testing of the parameter "enemy chasing", it can be concluded that there is only one failed test scenario in ChaCon game, it is when the player is on the top/bottom of the player. When the player position is on the top/bottom of the enemy, then the enemy cannot follow the player, for this reason, in the design there is no input of the command for the condition. While on testing for "enemy chasing" in Super Mario Bros game, there is only one successful test scenario that is when the player is on the left. When the player is on the left, the enemy will follow the player to the left. The testing for other positions, the enemy failed to follow the player's movement. This failed due to fuzzy input for parameter "enemy chasing" in Super Mario Bros game is only available when the player is only on the left.

Meanwhile, the testing for parameter "enemy's behavior" in ChaCon game shows that all test scenarios are successful, for this reason, all the fuzzy inputs entered in the game do not have an error. While the testing for parameter "enemy's behavior" in Super Mario Bros game, it is only one successful test scenario when the player is on the left. When the player is on the left, the enemy will attack the player because the fuzzy input for "enemy's behavior" in Super Mario Bros game does not provide input to avoid.

The testing applies Matlab for parameter "enemy's behavior" will be reviewed using a cartesian axis graph to determine the relation between input-output of the Fuzzy Sugeno method. Based on the defuzzification result, this ChaCon game produces 3 input-output relations, they are the relation between inputs of health and distance, the relation between inputs of avoid and distance, and the relation between inputs of avoid and health.

The "enemy's behavior" decision by entering the value of health = 0-200 (Z-axis), and the distance value = 0 - 1000 (X-axis), produce the decision value of the enemy = 1 (Y-axis). The input-output relationship for health and distance variables in ChaCon game can be seen in Figure 10.

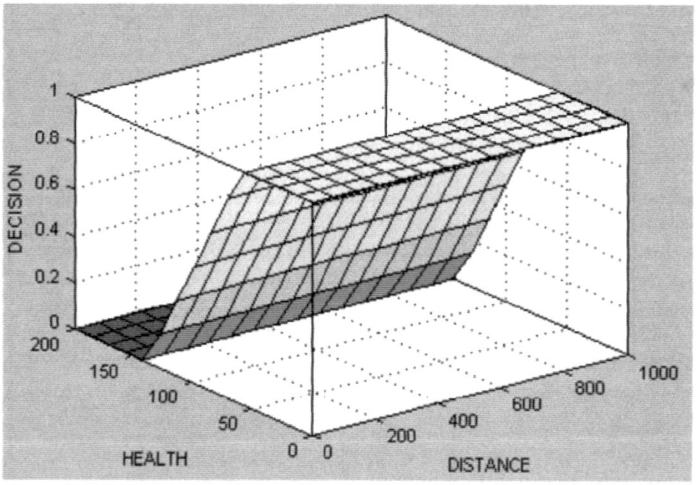

Figure 10. The relation of input-output for health and distance variables on ChaCon Game.

The decision of "enemy's behavior" by entering the value of avoid = 60 – 140 (Z-axis), and distance value = 0 – 1000 (X-axis), produce the decision value of the enemy = 0.5 (Y-axis). The input-output relationship for avoid and distance variables in ChaCon Game can be seen in Figure 11.

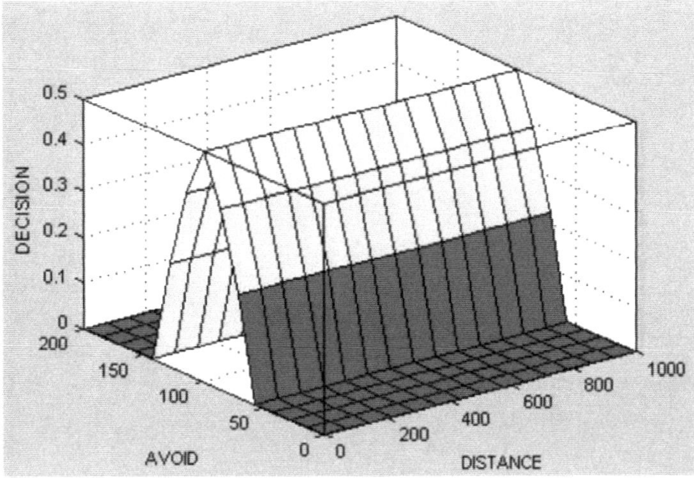

Figure 11. The relation of input-output for avoid and distance variables on ChaCon Game.

The decision of "enemy's behavior" by entering the value of avoid = 60 – 140 (Z-axis), and health value = 0 – 200 (X-axis), produces the decision value of the enemy = 1 (Y-axis). The relation of input-output for avoid and health variables on ChaCon Game can be seen in Figure 12.

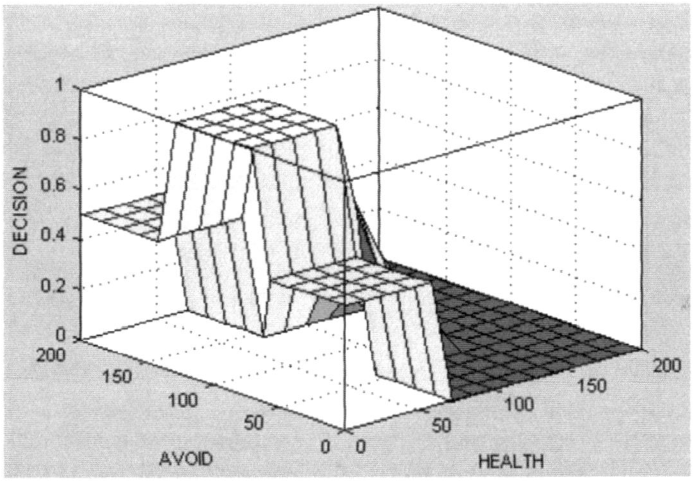

Figure 12. The relation of input-output for avoid and health variables in ChaCon Game.

The defuzzification results in Super Mario Bros game produce 1 graph of the Cartesian axis, that is the relationship between input of health and distance. The "enemy's behavior" decision by entering the value of health = 0 - 1 (Z-axis), and the distance value = 0 - 1000 (X-axis), produce the decision value of the enemy = 0.5 (Y-axis). The relation of input-output for health and distance variables in Super Mario Bros game can be seen in Figure 13.

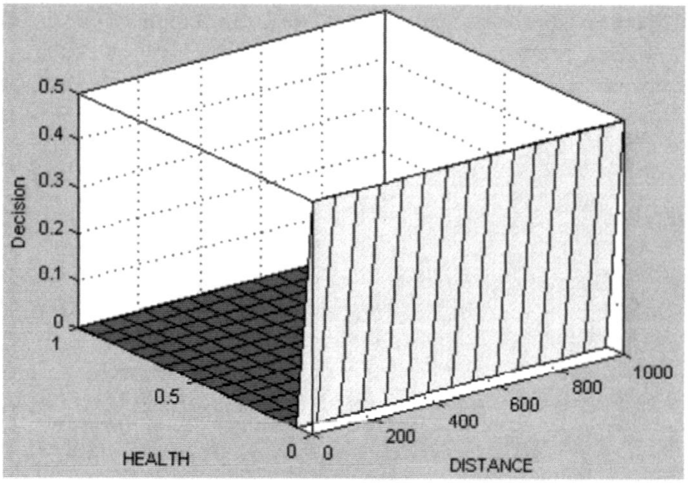

Figure 13. The relation of input-output for health and distance variables in Super Mario Bros game.

Based on the input-output relation graph in both games, it can be concluded that ChaCon Game produces the decision value = 2.5 and Super Mario Bros game produces the decision value = 0.5. These results prove that artificial intelligence on ChaCon Game is better than Super Mario Bros game, for this reason, the ChaCon Game produces a new value that is an input of avoid.

4. Conclusion

Based on the research which is conducted, testing of the "enemy chasing" of the player on ChaCon Game, which is carried out with 4 scenarios, it shows the success of 75%. Meanwhile, in Super Mario Bros game, the success only reaches 20%. Then, the testing for "enemy's behavior" on ChaCon Game shows the success of 100%, while in Super Mario Bros game only reaches 33.3%. The testing for "enemy's behavior" using Matlab on ChaCon Game produces 3 cartesian axis graphs with the decision value = 2.5, while in Super Mario Bros game only produces 1 graph of the Cartesian axis with the decision value = 0.5. The results show that artificial intelligence on ChaCon Game is better than Super Mario Bros, and it has an impact on enemy's behavior in facing the player.

References

[1] M. A. Akbar, A. Muttaqin and E. M. A. J, "Designing Word Slash Games Using Kinect as A Game's Control Sensor With Hand Movement (In Bahasa Indonesia)," *PTIIK DORO,* 2013.

[2] A. Pamungkas, E. P. Widiyanto and R. Angreni, "Application of Algorithm A * In The Android-Based Maze Island Educational Game (In Bahasa Indonesia)," STMIK MDP, 2014.

[3] A. M. Mora, J. J. Merelo, P. García-Sánchez, P. A. Castillo, M. S. Rodríguez-Domingo and R. M. Hidalgo-Bermúdez, "Creating autonomous agents for playing Super Mario Bros game by means of evolutionary finite state machines," *Evolutionary Intelligence,* vol. 6, no. 4, pp. 205-218, 2014.

[4] S. J. L. a. C. M. N. Cole, "Using a genetic algorithm to tune first-person shooter bots," in *in Evolutionary Computation, 2004. CEC2004*, Portland, 2004.

[5] G. Kendall and K. Spoerer, "Scripting the game of Lemmings with a genetic algorithm," in *Evolutionary Computation, 2004. CEC2004. Congress on*, Portland, 2004.

[6] A. Shaout, B. W. King and L. A. Reisner, "Real-Time Game Design of Pac-Man Using Fuzzy Logic," *Int. Arab J. Inf. Technol,* vol. 3, no. 4, pp. 315-325., 2006.

[7] J. Pranata, K. R. Purba and Liliana, "Strategy Game Application Using Genetic Algorithm Methods for Making Troops (in Bahasa Indonesia)," *Jurnal Infra,* vol. 4, no. 1, pp. 199-204, 2016.

[8] A. Katmada, "Towards Building An Affective Educational Game Framework: A Proof of Concept," Aristotle University of Thessaloniki , Thessaloniki, 2015.

IOP Conf. Series: Journal of Physics: Conf. Series **1090** (2018) 012070 doi:10.1088/1742-6596/1090/1/012070

Colored Object Sorting using 5 DoF Robot Arm based Artificial Neural Network (ANN) Method

W. S. Mada Sanjaya[1,2*], Dyah Anggraeni[1,2], Madinatul Munawwaroh[1,2], M. Yusuf S. Nurasyidiek[1,2], Darmawan Setia Rahayu[1,2], Ahmad Samsudin[1,2], Ikhsan Purnama Santika[1,2], and Endah Kinarya Palupi[1,2]

[1]Department of Physics, Faculty of Science and Technology, Universitas Islam Negeri Sunan Gunung Djati Bandung, Indonesia
[2]Bolabot Techno Robotic Institute, CV. Sanjaya Star Group, Bandung, Indonesia

E-mail: madasws@gmail.com

Abstract. Automation process is very important to help any industry activity especially to sorting an object. In this research, we build 5 DoF of Robot Arm based on Arduino microcontroller to pick and place colored object intended sorting task in real-time. The algorithm of interfaces and image processing to detect the colored object based on Python 2.7 with OpenCV library. From image processing system, the detected colored object based on Red-Green-Blue (RGB) identifier has a coordinate. The coordinate data is collected to build the inverse kinematic model and trained by using Artificial Neural Network (ANN) method to control Robot Arm. Finally, the algorithm implemented to Robot Arm to do pick and place a colored object show a good accuracy rate. The advantages of this Colored Object Sorting system are Open Source and Low Cost.

1. Introduction

On this decade, the high technology needed to get an efficiency of Industrial works especially an automation of sorting objects. Industrial robots are programmed based on sensory information with accurate and precise work [1]. The sensor is required because of the object have any characteristic to sorting by a machine, such as color, shape, size, etc. And algorithm must be presented to classifying the feature of an object.

Beside geometry method [2] from the image processing, the inverse kinematic model (from the object coordinate to the angle of the motor servo) and various method of Artificial Intelligence is needed to control Robot Arm. The various method of Artificial Intelligence needs, such as Fuzzy Logic [3] [4], Artificial Neural Networks (ANN) [2] [5] [6], Adaptive Neuro-Fuzzy Inference System (ANFIS) [7] [8], and other [9]. The application of Colored Object Sorting in any field, for example: Engineering [10] [11], Harvesting [12] [13] [14], Industrial [15] [16], and other [17]. To make an Industrial Colored Object Sorting certainly very expensive to build. But, at this time the Low-Cost system of Colored Object Sorting can build by everyone using Arduino microcontroller [18] [19] or Raspberry Pi microprocessor [11] [20] [21].

In this research described an image processing of color object detection, inverse kinematic model and Artificial Neural Networks method as machine learning based on Python 2.7 with

Content from this work may be used under the terms of the Creative Commons Attribution 3.0 licence. Any further distribution of this work must maintain attribution to the author(s) and the title of the work, journal citation and DOI.
Published under licence by IOP Publishing Ltd

OpenCV library. Finally, this research will be implemented to 5 DoF Robot Arm based on Arduino microcontroller to pick and place the colored object in real-time.

The paper is organized as follows: In section 2, described the general System Overview of Colored Object Sorting. Section 3, describe color detection of the object. Schematic and hardware of Robot Arm describe in section 4. Building database from object color detection and machine learning which used in this paper describe in section 5. The succeed training, classifying and evaluating between trained and tested data implemented to Robot Arm doing the task to pick and place the colored object described in section 6. Finally, the concluding remarks are given in Section 7.

2. System Overview

In this study, the concept design and the realization of Colored Object Sorting system shown as Figure 1. In general, the design explains the main materials are; personal computer, webcam, Arduino board and 5 DoF Robot Arm.

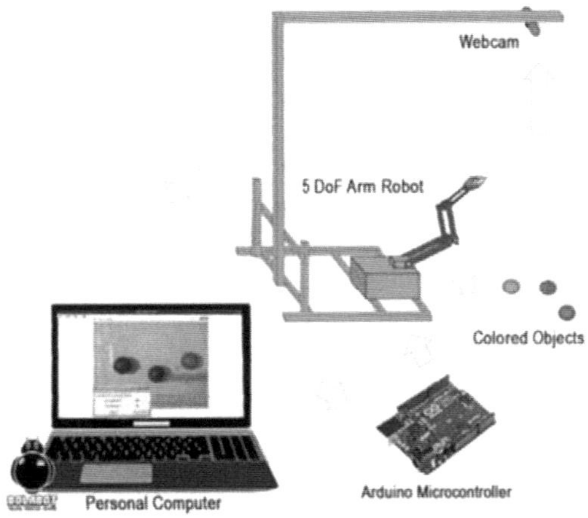

(a)Design

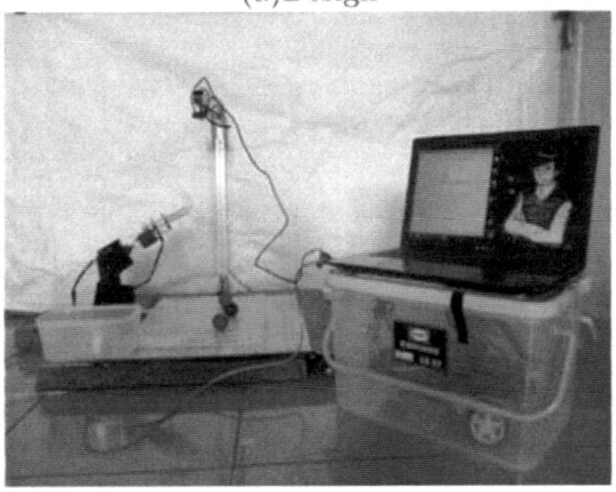

(b)Realization

Figure 1. 5 DoF Robot Arm Design

Figure 1 and Figure 2 describe the general system of Colored Object Sorting. Especially from Figure 2 explained that after selecting the color the webcam will starting to detect the colored object. Then, divided into 2 processes: the First process is making training data, consists by getting the centroid coordinate of the colored object and collecting data of angle for the servo to doing pick and place the object as Trained Data. The second process is testing the system after getting the centroid coordinate of the colored object, then the testing data of colored object matching with Trained Data. The matching data be processed to obtain the angle of servo based on Artificial Neural Network method, which used to drive the motor servo Robot Arm. All processes work in real-time based on Python (with OpenCV library) and Arduino microcontroller.

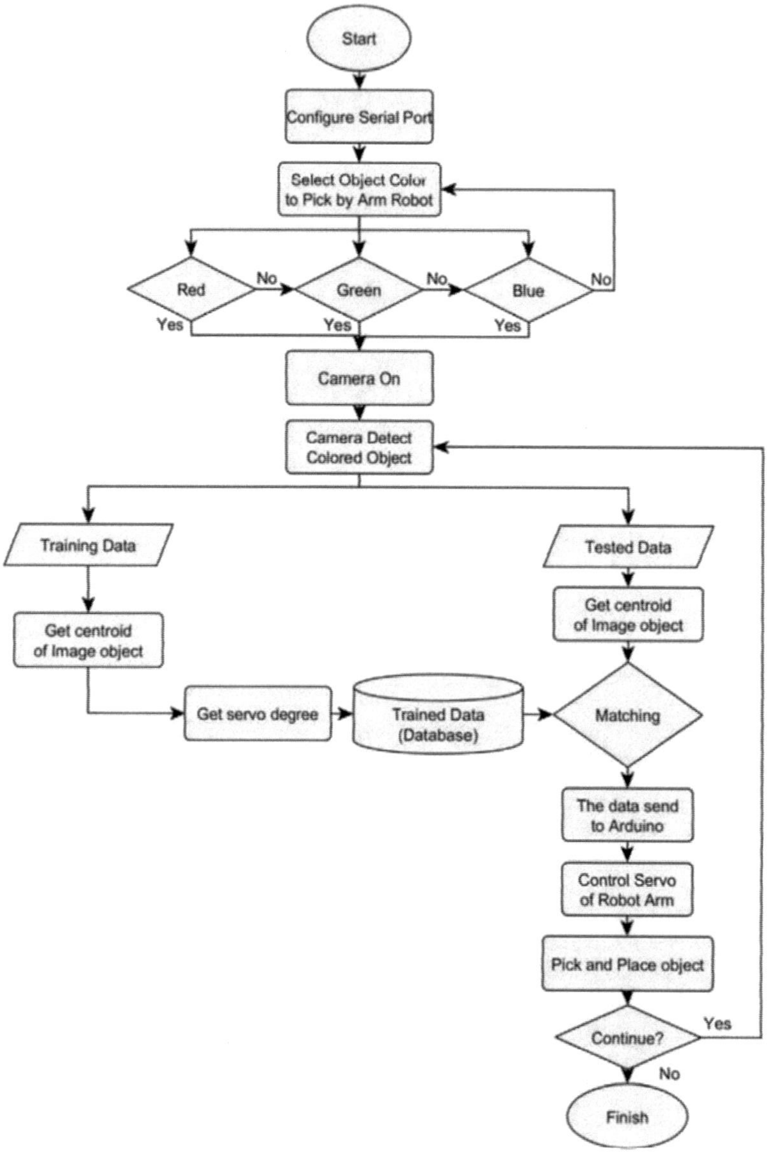

Figure 2. General System Scheme of Colored Object Sorting

3. Color Detection

The first step in designing the concept of Colored Object Sorting is designing algorithms for detecting colored objects. The algorithm and interfaces build based on Python 2.7 and OpenCV library. To determine the color to be selected in this system in a way to configure the value of RGB color (Red, Green, and Blue).

The colored objects can be detected by image processing by capturing images and classifying color types by value of RGB. HSV (Hue, Saturation, and Value) used to reduce the influence light of the colored object. The edge of the image is optimized with "threshold", the pixels which are stronger than the high threshold value are preserved, but the weaker than the lower threshold value is suppressed. Then the image converted to binary to get the edge in a given image gets the morphological filter. Then the image filtered binary get the contour detection. Finally, only the selected color object will be marked by rectangle illustrated as Figure 3 (d). [17] In addition the detected object have a centroid coordinate (x, y) position of colored object. Figure 3 show the evaluation of color detection with coordinate value of RGB object.

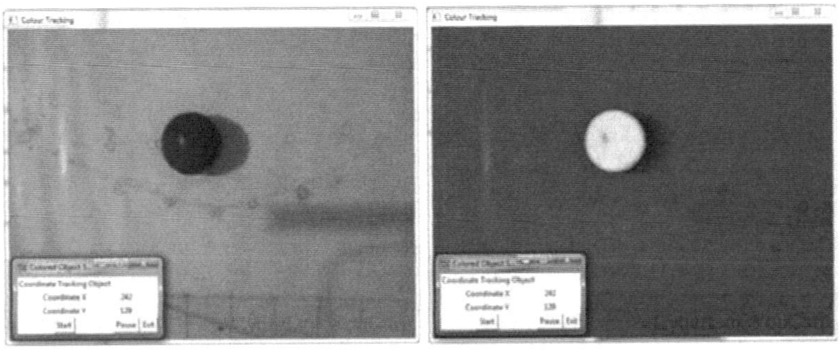

(a)Real Capture (b)HSV image

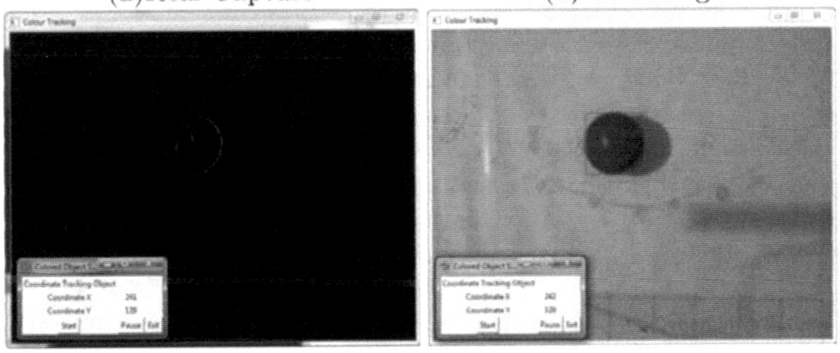

(c)Binary image (d)Marked colored object

Figure 3. Image Processing Process

4. Hardware of Robot Arm

To build the Robot Arm system, the main component is Arduino board, motor servo, battery, cables and Robot Arm hardware construction (Figure 5 (b)). The schematic of Robot Arm shown as Figure 5 (a) described that each servo supplied by the 5-volt battery with 100 mA of current, and each servo's ground connected to the ground of Arduino Board. 5 DoF Robot Arm consists of 5 motor servo connect to Arduino PWM pin, are; Servo1 connect to pin 5 (base, rotate horizontally), Servo2 connect to pin 6 (work as shoulder, rotate vertically), Servo3 connect to pin 9 (work as elbow, rotate vertically), Servo4 connect to pin 10 (work as wrist, rotate horizontally), and Servo5 connect to pin 11 (work as gripper to place an object).

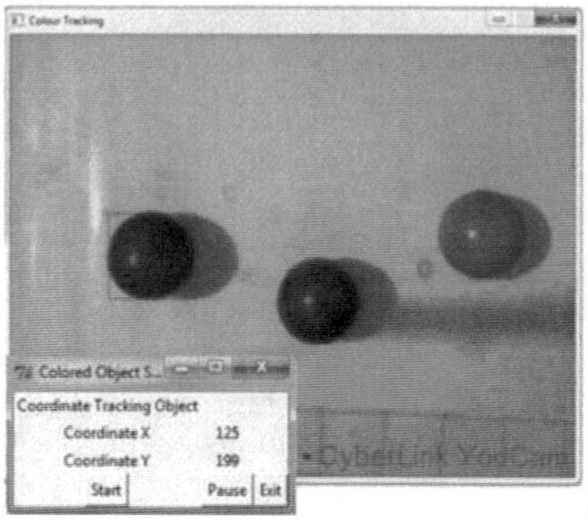

(a)Red

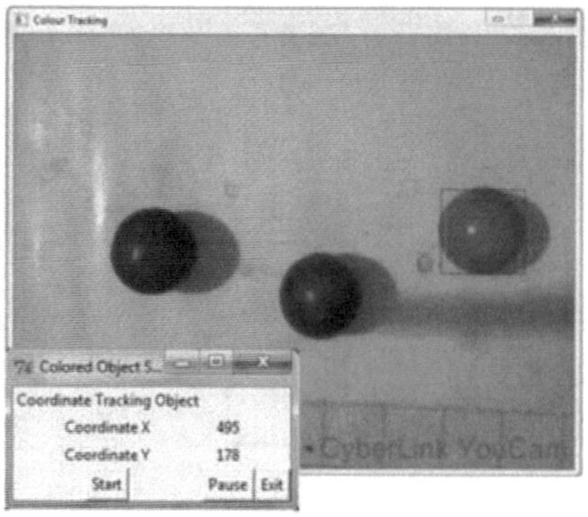

(b)Green

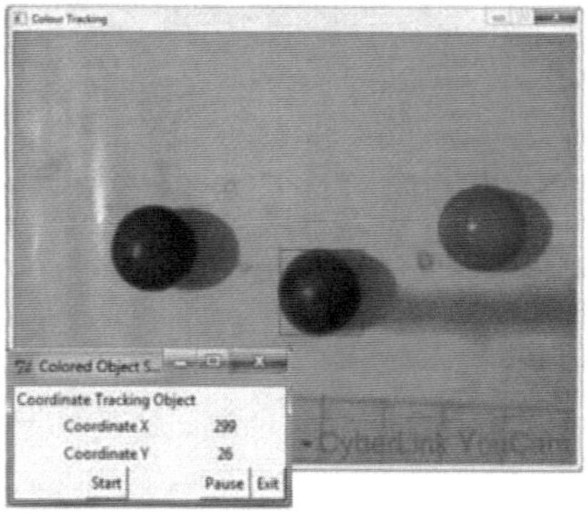

(c)Blue

Figure 4. Interface of Color Detection with Coordinate

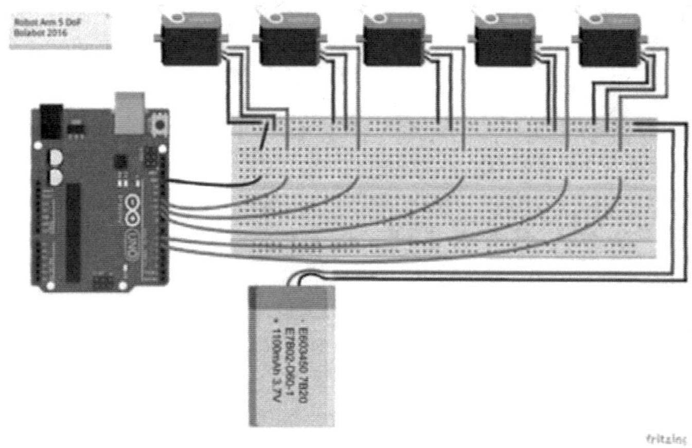

(a)Schematic of Robot Arm

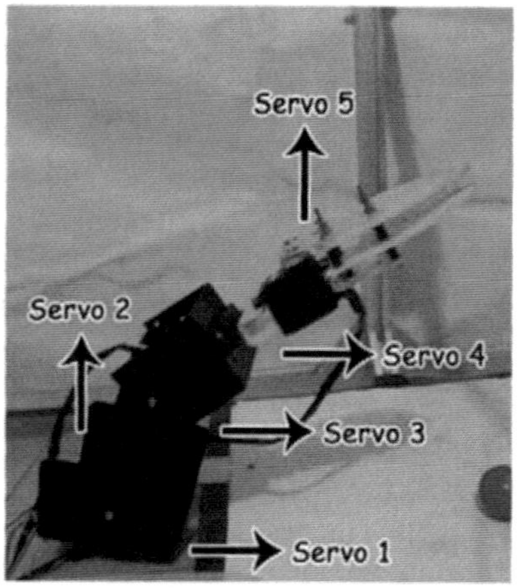

(b)Robot Arm Hardware

Figure 5. Schematic of Robot Arm

5. Inverse Kinematic Model and Artificial Neural Network

Kinematics in this study mean the conversion from Cartesian coordinates (x, y, z) to the moving angle of the joint $(\theta_1, \theta_2, \theta_3, ...)$ of the mechanical Robot Arm. Kinematic divided into two section are Forward Kinematic (from joint angle to coordinate) and Inverse Kinematic (from coordinate to joint angle). [3]

In this study, Inverse Kinematic Model used to build the database of 5 DoF Robot Arm to pick and place a colored object on certain coordinate. The data consist of input data as x and y coordinate, and the output data of servo's angle as Servo1-Servo5 shown at Table 1. The coordinate data received from the evaluation of colored object detection which captures by webcam. The video capture configure as 480 x 640 pixel, but the usable workspace coordinate is 467 x 624 pixel shown as Figure 6 and the Robot Arm position in the middle of the workspace. For the angle of the servo obtain when the object is in a certain coordinate, and the servo angle must be able to move to reach the object. The collected data will be used become Training data at Artificial Neural Network.

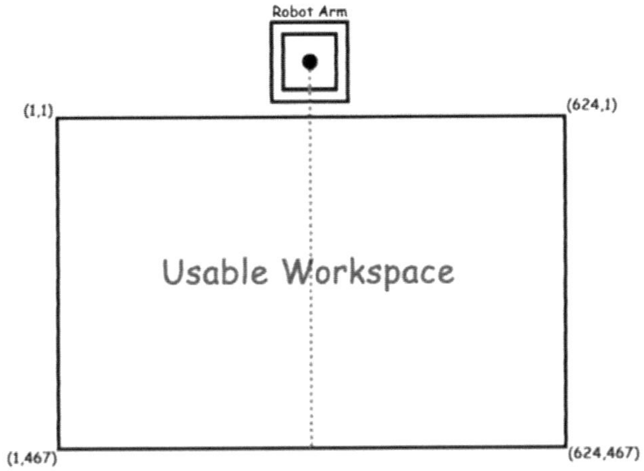

Figure 6. Usable Workspace Cartesian Coordinates (in pixel)

Artificial Neural Network (ANN) method is a system that works similar to the human brain. MLP (Multi-Layer Perceptron) is one of the ANN method using supervised learning method. MLP is also known as Multilayer Feedforward Networks. As the name implies, MLP can have one or more hidden layers. In this research 20 hidden layers used for ANN, illustrated as Figure 7.

All the neuron of input, hidden and output layer connected each other. The first layer (input layer) receives an input from the external environment. The between layer of input and output layer (hidden layers) invisible to the external environment. And the last layer (output layer) transmits the result from the network. [5]

Table 1 is the Inverse Kinematic model data consist by the coordinate data (x and y) of the colored object, and the angle of 5 servos with trained data (show as Cal=Calibration) and tested data (show as ANN=Artificial Neural Network data training) as the evaluation. The Inverse Kinematic model data trained by using ANN method. After enter to ANN training, the database becomes "the trained data". The trained data is a reference database to matching with the new external environment (coordinate) data until getting the new target data (angle of the motor servo).

From the Table 1, the bold data is an error value between data of Calibration and ANN result is 1.3%. The accuracy result gets from average successful data show the good accuracy rate to show the coordinate and angle of servos, and can implement to control Robot Arm to pick and place the colored object.

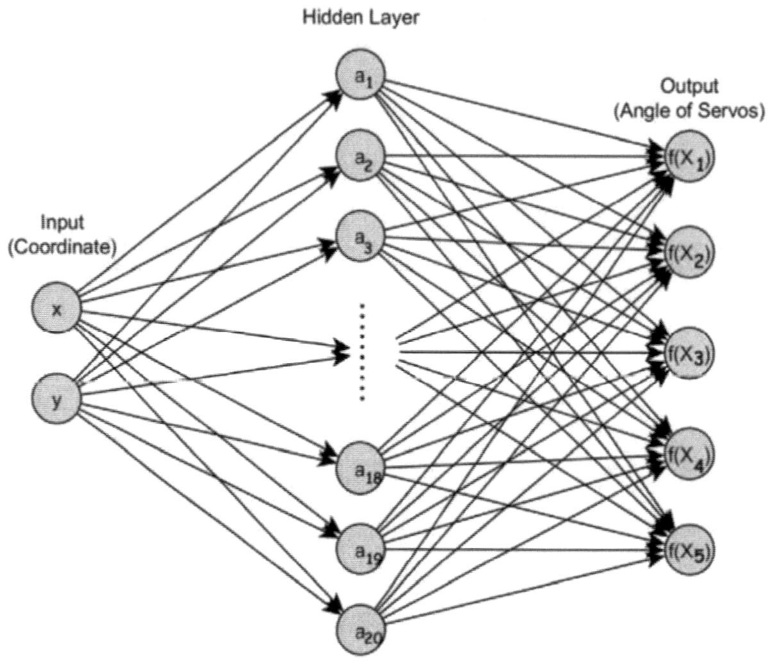

Figure 7. ANN with Hidden Layer illustration

Table 1. The database of Inverse Kinematic Model to Control Robot Arm

Coordinate		**Servo1**		**Servo2**		**Servo3**		**Servo4**		**Servo5**	
x	y	Cal	ANN	Cal	ANN	Cal	ANN	Cal	ANN	Cal	ANN
1	116	70	70	100	100	116	116	175	175	0	0
21	147	84	84	100	100	116	116	175	175	0	0
27	67	81	81	107	107	133	133	175	175	0	0
123	131	89	89	**120**	**100**	**140**	**116**	175	175	0	0
143	167	95	95	100	100	116	116	175	175	0	0
210	132	**99**	**110**	**120**	**100**	**141**	**116**	175	175	0	0
223	231	100	100	100	100	116	116	175	175	0	0
260	157	**102**	**110**	118	118	146	146	175	175	0	0
283	245	104	104	100	100	116	116	175	175	0	0
339	179	110	110	120	120	144	144	175	175	0	0
367	228	110	110	100	100	116	116	175	175	0	0
429	135	**117**	**110**	125	125	148	148	175	175	0	0
447	214	**108**	**110**	100	100	116	116	175	175	0	0
494	126	110	110	118	118	144	144	175	175	0	0
508	197	110	110	100	100	116	116	175	175	0	0

6. Implementation of Color Detection to Sorting Object

After the database successfully trained by ANN method, then applied the system as control Robot Arm to pick and place the object. Figure 8 show the interface of color detection with coordinate data and angle of servo prediction. And Figure 9 show the Robot Arm motion to pick and place the colored object. From the figures show that the colored object was successfully detected, thus the Robot Arm effective can pick and place the object.

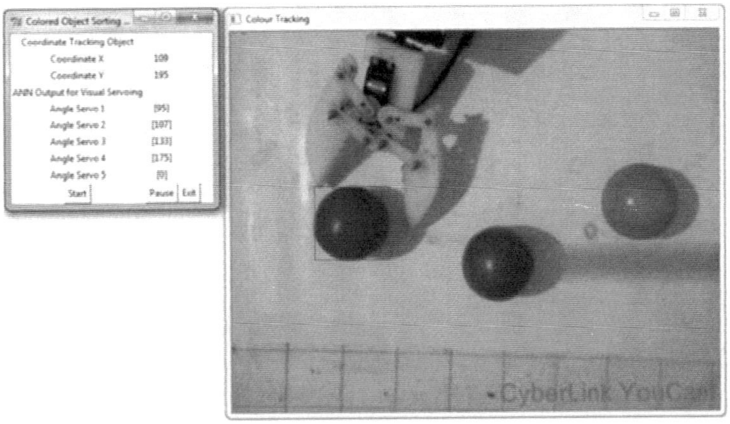

(a)Red

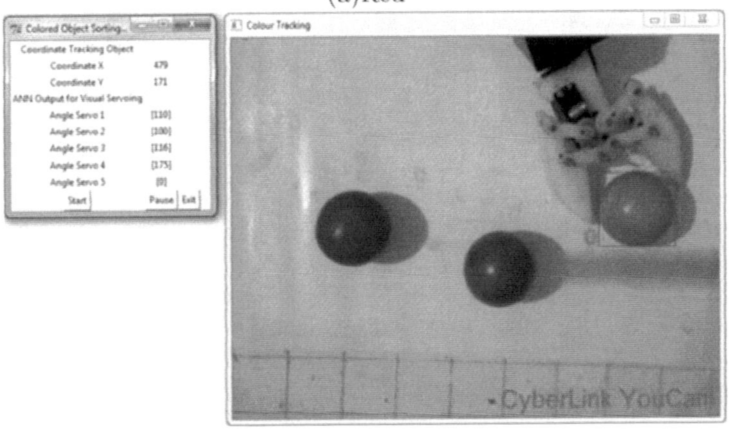

(b)Green

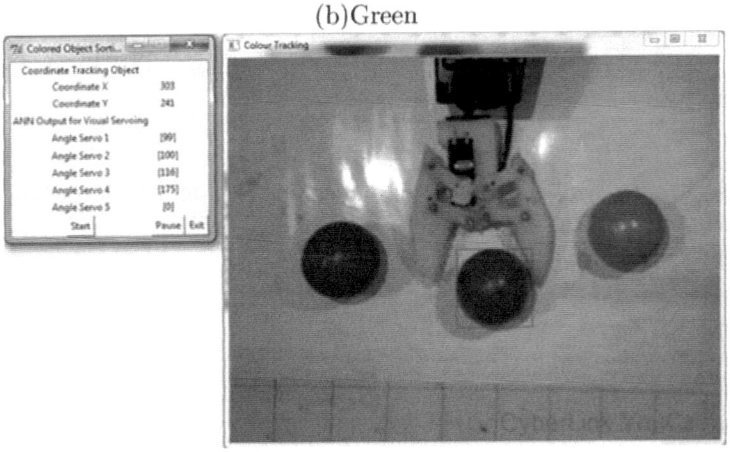

(c)Blue

Figure 8. Experiment Color Detection to Control Robot Arm

(a)Find colored object (b)Pick colored object (c)Place colored object

Figure 9. Evaluating the system

7. Conclusion

The study has been presented about the development of Colored Object Sorting which can pick and place colored object. The image processing has been processed by the algorithm based on Python 2.7 (with OpenCV library) to detect the colored object. The inverse kinematic database is trained by Artificial Neural Network to matching with internal database show the good accuracy. Finally, the implementation of color detection and inverse kinematic model to control 5 DoF of Robot Arm based on Arduino microcontroller works effectively to pick and place the colored object. Future works will be focus on Low Cost and Open Source system of Colored Object Sorting based on Raspberry Pi microprocessor.

Acknowledgement

The authors would like gratefully acknowledge the financial support from DIPA UIN Sunan Gunung Djati Bandung.

References

[1] Devalla V, Singh R, Mondal A K and Kaundal V 2012 *International Journal of Science and Advanced Technology* **2** 30–35
[2] Daya B, Khawandi S and Akoum M 2010 *J. Software Engineering & Applications* **3** 230–239
[3] Juang J G, Tsai Y J and Fan Y W 2015 *Appl. Sci.* **5** 851–880
[4] Huang S J 2011 Intelligent Robotic System with Fuzzy Learning Controller and 3D Stereo Vision *International Conference on Systems* vol 15 (WSEAS) pp 123–130 ISBN 9781618040237
[5] Wu H, Tizzano W, Andersen T T, Andersen N A and Ravn O 2013 Hand-Eye Calibration and Inverse Kinematics of Robot Arm using Neural Network Hand-Eye Calibration and Inverse Kinematics of Robot Arm using Neural Network *Proceedings of RITA* (Springer) pp 581–591 ISBN 9783319055824
[6] Duka A V 2014 *Procedia Technology* **12** 20–27 ISSN 2212-0173 URL `http://dx.doi.org/10.1016/j.protcy.2013.12.451`
[7] Hendarto H A 2014 ANFIS Application for Calculating Inverse Kinematics of Programmable Universal Machine for Assembly (PUMA) Robot *Infonnation Technology, Computer and Electrical Engineering (ICITACEE) ANFIS* vol 1 (IEEE) pp 35–40 ISBN 9781479964321
[8] Aghajarian M and Kiani K 2011 Inverse Kinematics Solution of PUMA 560 Robot Arm Using ANFIS *Ubiquitous Robots and Ambient Intelligence (URAI)* vol 8 pp 574–578 ISBN 9781457707230
[9] Tajdari F and Rad N F 2017 *Artificial Intelligence and Robotics (IRANOPEN)* 97–101
[10] Szabó R and Lie I 2012 *Electronics and Telecommunications (ISETC)* **10** 1–4
[11] Pereira V, Fernandes V A and Sequeira J 2014 *GHTC-SAS* 1–6
[12] Han K S, Kim S C, Lee Y B, Kim S C, Im D H, Choi H K and Hwang H 2012 *Journal of Biosystems Engineering* **37** 65–74
[13] Hemming J, Ruizendaal J, Hofstee J W and Henten E J V 2014 *Sensors* **14** 6032–6044
[14] De-An Z, Jidong L, Wei J, Ying Z and Yu C 2011 *Biosystems Engineering* **110** 112–122 ISSN 1537-5110 URL `http://dx.doi.org/10.1016/j.biosystemseng.2011.07.005`
[15] Nageswari K L, Sujana M and Polaiah B 2016 *IJARMATE* **2** 2–5

[16] Ata A A, Rezeka S F, El-shenawy A and Diab M 2013 *International Journal of Mechanical, Aerospace, Industrial, Mechatronic and Manufacturing Engineering* **7** 2457–2464
[17] Lee M M and Slater A P 2016 Robotic Candy Sorter URL `https://courses.cit.cornell.edu/ece5990/`
[18] Hrishikeshan G S, Kusuma B V, Lahari C, Sowmya V R and S C K N 2015 *International Journal of Combined Research and Development (IJCRD)* **4** 622–626
[19] Sachdeva A, Gupta M, Pandey M and Khandelwal P 2017 *The International Journal of Engineering and Science* 56–59
[20] Szabó R and Gontean A 2016 Industrial Robotic Automation with Raspberry PI using Image Processing *International Conference on Applied Electronics (AE)* (Pilsen: IEEE) pp 265–268 ISBN 9788026106029
[21] Siagian P, Sisingamangaraja J, Laguboti S, Samosir T, Sisingamangaraja J, Laguboti S, Samosir T and Pi A R 2015 *ICSITech* 192–196

Neutronic Comparison Study Between Pb(208)-Bi and Pb(208) as a Coolant In The Fast Reactor With Modified CANDLE Burn up Scheme.

Nina Widiawati[1], Zaki Suud[1], Dwi Irwanto[1], Hiroshi Sekimoto[2]

[1]Nuclear Physics and Biophysics Research Group, Faculty of Mathematics and Natural Science, Bandung Institute of Technology, Indonesia.
[2]Nuclear Engineering, Tokyo Institute of Technology, Japan

ninawidiawati@gmail.com

Abstract. Neutronic study of Pb(208)-Bismuth as a coolant in the Lead Fast Reactor (LFR) with Modified CANDLE burn up scheme has been conducted. Lead cooled fast reactor (LFR) is one of the fourth-generation reactor designs. The reactor is designed with 500 MW thermal power output. Modified CANDLE burn-up scheme allows the reactor to have long life operation by supplying only natural uranium as fuel cycle input. This scheme introducing discrete region, the fuel is initially put in region 1, after one cycle of 10 years of burn up it is shifted to region 2 and region 1 is filled with fresh natural uranium fuel. The reactor is designed for 100 years with 10 regions arranged axially. The neutronic calculations were performed by SRAC code using nuclear data library based on JENDL 4.0. Level burn up of Pb(208)-Bi cooled fast reactor is 530.688 GWD/MTU at BOC and 433.051 GWD/MTU at EOC whereas 190.790 GWD/MTU at BOC and 433.051 GWD/MTU at EOC for Pb(208) . The effective multiplication factor of Pb(208)-Bi Cooled Fast Reactor is 1.0554 at BOC and 1.05958 at EOC whereas 1.06703 at BOC and 1.06816 at EOC for Pb208.

1. Introduction

The accident of Chernobyl, Three Miles Island (TMI) and Fukushima Daichi has led to an emphasis on passive safety in the design of advanced reactors. Generation IV International Forum (GIF) has chosen six reactor systems for further development. The six reactor system chosen were: Gas Cooled Fast Reactor (GFR), Lead Cooled Fast Reactor (LFR), Molten Salt Reactor (MSR), Sodium Cooled Fast Reactor (SFR), Supercritical Water Cooled Reactor (SCWR), and Very High-Temperature Reactor (VHTR). Generation IV reactor systems has several specific goals, they were: inherent safety, sustainability, non-proliferation, and economical.

1.1. Lead Cooled Fast Reactor (LFR)

LFR is one of the six selected reactor systems. Some advantages of the LFR system [12]:

1. The potential for high outlet liquid metal temperature (up to 850^oC) would enable the LFR to meet the hydrogen and electricity production missions.
2. The fast neutron spectrum would facilitate achieving the actinide fissioning mission

Content from this work may be used under the terms of the Creative Commons Attribution 3.0 licence. Any further distribution of this work must maintain attribution to the author(s) and the title of the work, journal citation and DOI.
Published under licence by IOP Publishing Ltd

3. By full recycling of actinides in a fast spectrum, the LFR would minimize the discharge of long-lived waste to HLW repositories and could maximize the utilization of the energy content of uranium (including depleted uranium) by transmutation of U-238 into fissionable transuranics.

1.2. Coolant

The choice of coolant for Fast Breeder Reactor (FBR) is very important because the coolant strongly influences the neutronic behavior of the core. The most recognizable effect of coolant selection is its influence on major components such as pumps and steam generators [12].

The coolant for FBR it must:

1. Minimize neutron moderation
2. Remove heat adequately from a high-power-density system
3. Minimize parasitic neutron absorption
4. Has low melting point and high boiling point
5. Has Low neutron absorption cross-section

1.3. Pb208 and Bismuth

Pb208 is one of the stable isotopes of Pb-nat. It content in Pb-nat above 52%, is high enough.

Some advantages of Pb-208 as the coolant of FBR are:

1. Has small neutron absorption in Fast Reactor (FR). Due to excess of neutrons, it can save more the fuel loading.
2. Increasing the fuel breeding in Fast Reactors
3. Hardening of FR neutron spectra. Meanwhile, hard spectrum of neutrons in FR core is preferable for incineration of minor actinides (MA).
4. Pb-Bi eutectic as a coolant gives better performance for extreme accidents than Pb because of the lower operation temperature (Zaki Suud, 1996). Pb-Bi eutectic has the low melting point but has the high boiling point.
5. The use of Pb208-Bi eutectic as a coolant expected to contribute significantly to the safety of the reactor

2. Design Concept

CANDLE (Constant Axial shape of Neutron flux, nuclide densities, and power shape During Life of Energy production) reactor system is originally developed by Prof. Hiroshi Sekimoto from Tokyo Institute of Technology, Japan. The several advantages of CANDLE reactor system, are:

1. Not required enriched fuels after the second cycle:
2. The burn-up of the spent fuel is about 40% (400 MWd/tHM): This value is competitive to the value of the presently expected FR system with reprocessing plant.
3. Can be designed also as a long-life reactor, since the burning region velocity is very low.
4. Requires more neutrons than conventional fast reactors since it does not employ reprocessing. Therefore, it is important to use a neutronically superior coolant.

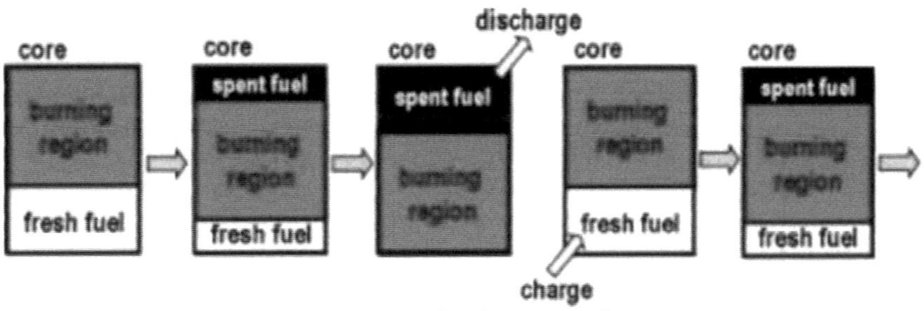

Figure 1. CANDLE burn up scheme

IOP Conf. Series: Journal of Physics: Conf. Series **1090** (2018) 012071 doi:10.1088/1742-6596/1090/1/012071

Modified CANDLE Reactor System

Modified CANDLE burn-up scheme allows the reactor to have long life operation by supplying only natural uranium as fuel cycle input. This scheme introducing discrete region, the fuel is initially put in region 1, after one cycle of 10 years of burn up it is shifted to region 2 and region 1 is filled with fresh natural uranium fuel. This concept is basically applied to all regions. The reactor is designed for 100 years with 10 regions arranged axially.

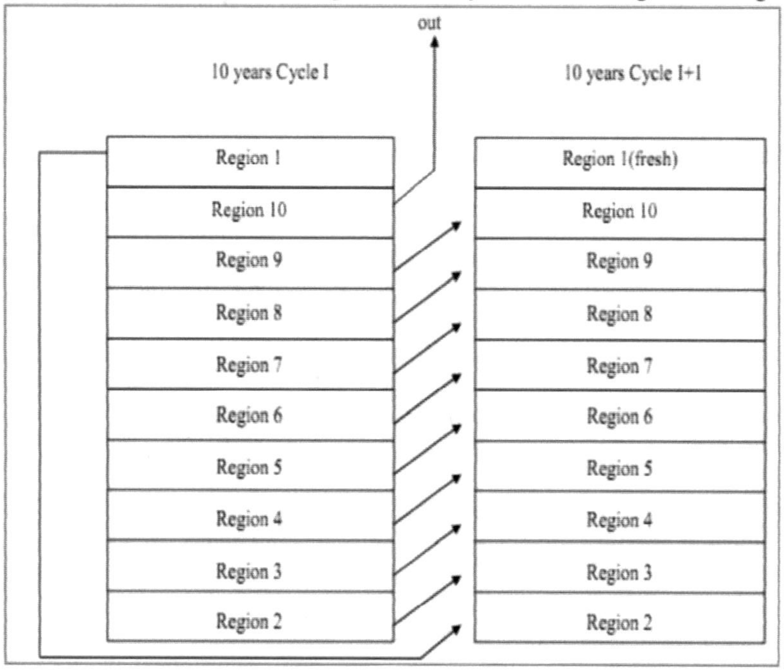

Figure 2. CANDLE burn up scheme

Table 1. Reactor Design Parameter

Parameter	Description
Thermal power	500 MWt
Number of equal volume region in core	10
Sub cycle length (years)	10
Fuel type	Uranium Nitride (UN)
Fuel volume fraction	60%
Cladding volume fraction	10%
Coolant volume fraction	30%
Fuel diameter	1.2 cm
Coolant type	Pb-208
Axial width of each region	15 cm
Active core radial width	110 cm
Reflector radial width	50 cm
fuel cell geometry	cylindrical
Core geometry	cylindrical

3. Calculational Method

The neutronic calculations were performed by SRAC (Standard thermal Reactor Analysis Code system) code using nuclear data library based on JENDL 4.0. The SRAC (Standard Reactor Analysis Code) system is designed and developed at Japan Atomic Energy Research Institute (JAERI). The system covers production of effective microscopic and macroscopic group cross-sections, cell and core calculations including burn-up and fuel management.

For neutronic calculation, we use PIJ module for calculating burnup of the fuel cell and CITATION module for multi groups diffusion calculation. First, we assume the power density level in each region. Then we perform the burnup calculation using the assumed data. The burn-up calculation is performed using cell burn-up in SRAC code which then gives eight energy group macroscopic cross section data to be used in two-dimensional R-Z geometry multi groups diffusion calculation. The average power density in each region resulted from the diffusion calculation is then brought back to SRAC code for cell burn-up calculation. This iteration is repeated until the convergence is reached.

4. Results and Discussion

4.1. Burn up Level

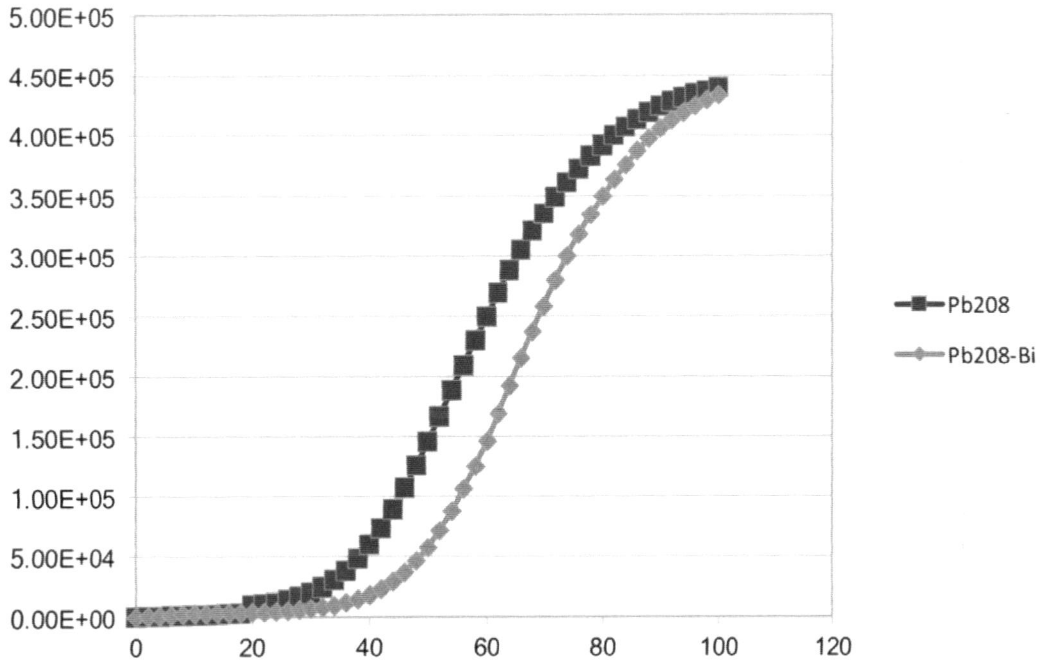

Figure 3. Burn up Level comparison between different coolant

Fig. 3 shows the burn up level of the reactor with different coolant. Reactor with Pb208 as the coolant has burn up level higher than Pb208-Bi as the coolant. It is mean that reactor with Pb208 produces fissile fuel faster and gain in core fuel loading due to excess of neutrons.

4.2. The infinite neutron multiplication factor

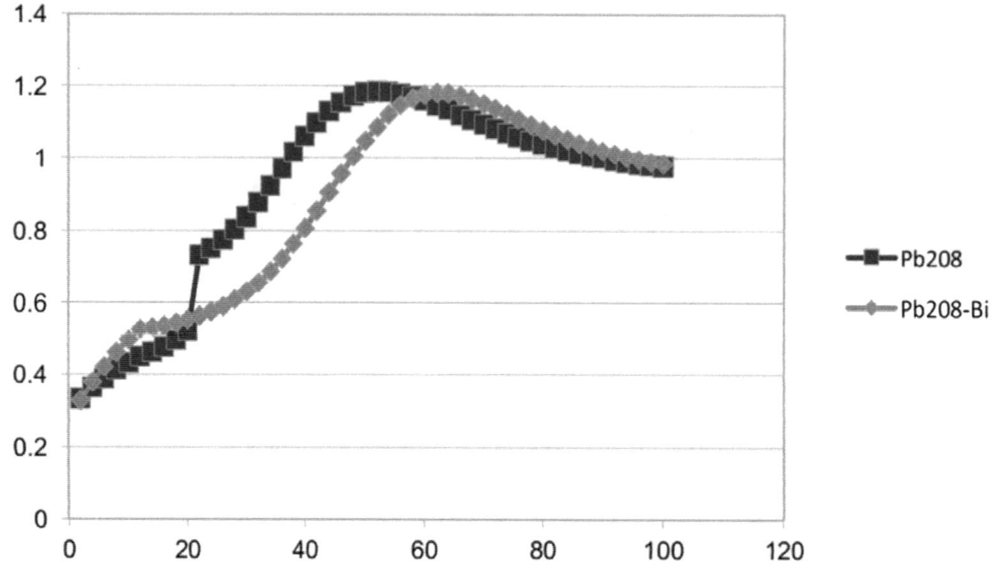

Figure 4. infinite neutron multiplication factor comparison between different coolant

Fig. 4 shows the infinite neutron multiplication factor of reactor with different coolant. Reactor with Pb208 as coolant has k-inf lower than Pb208-Bi as the coolant in the beginning. It is because the neutron is used to breeding. After 20 years, K-inf for Pb208 as coolant higher than Pb208-Bi as the coolant because of its Low neutron absorption cross-section. Then decrease because of fission product accumulation.

4.3. The change of atomic density during operation

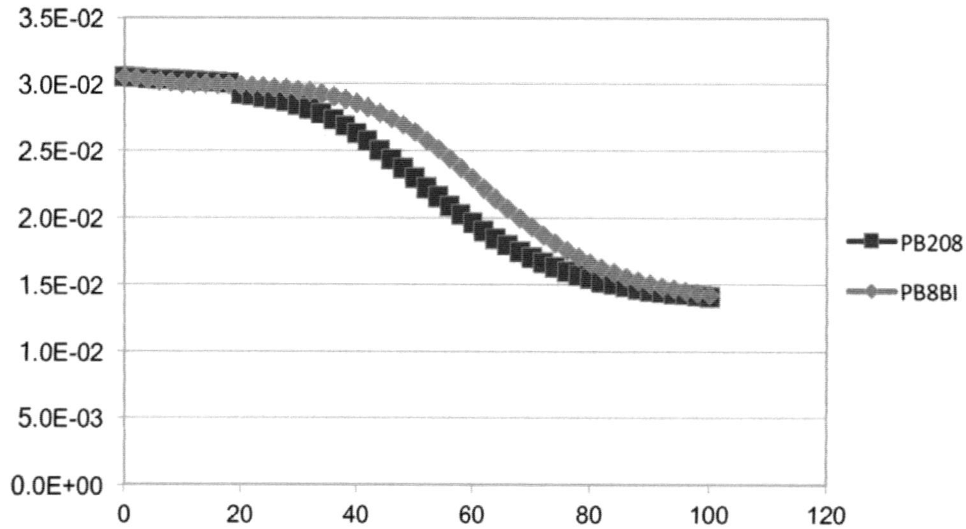

Figure 5. the change of atomic density during operation comparison between different coolant

Fig. 5 shows the change of atomic density during operation. Reactor with Pb208 as coolant has an atomic density of U-238 lower than Pb208-Bi as the coolant during operation. It is because the excess neutron is used to breeding U-238.

4.4. The change atomic density of Pu-239 during operation

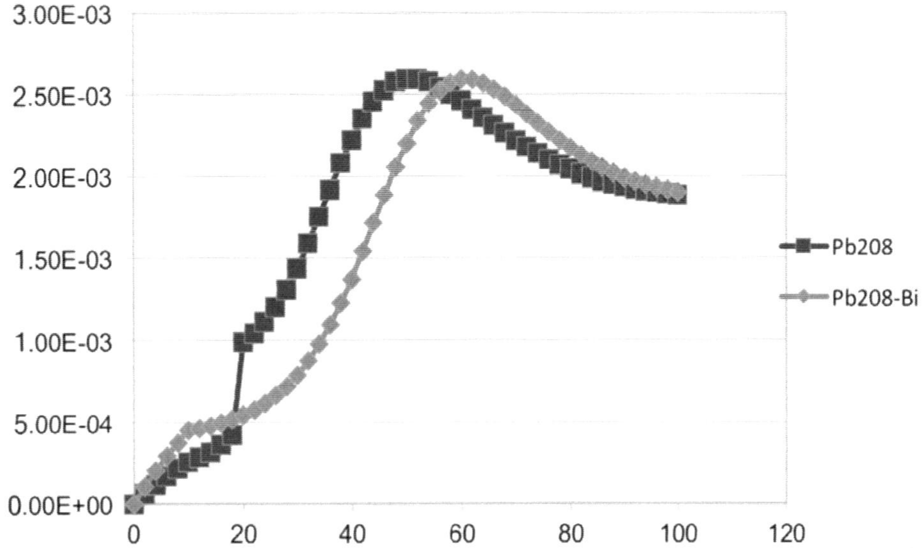

Figure 6. the change of atomic density during operation comparison between different coolant

Fig. 6 shows the change atomic density of Pu-239 during operation. Average Reactor with Pb208 as coolant has the atomic density of Pu-239 higher than Pb208-Bi as the coolant in the beginning. It is because Pb208 makes the production of Pu-239 faster.

4.5. The change integrated conversion ratio during operation

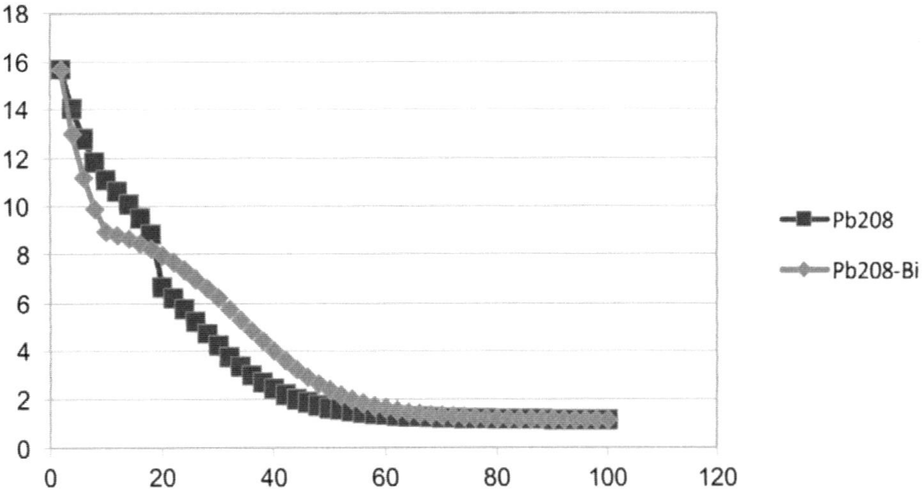

Figure 7. the change of atomic density during operation comparison between different coolant

Fig. 7 shows the change integrated conversion ratio (Inte-CR) during operation. Pb-208 has higher Inte-CR than Pb208-Bi at BOL because of its capability to produces than consume. After almost 20 years Pb208 burn the fuel lower than Pb208-Bi

4.6. *The effective neutron multiplication factor*

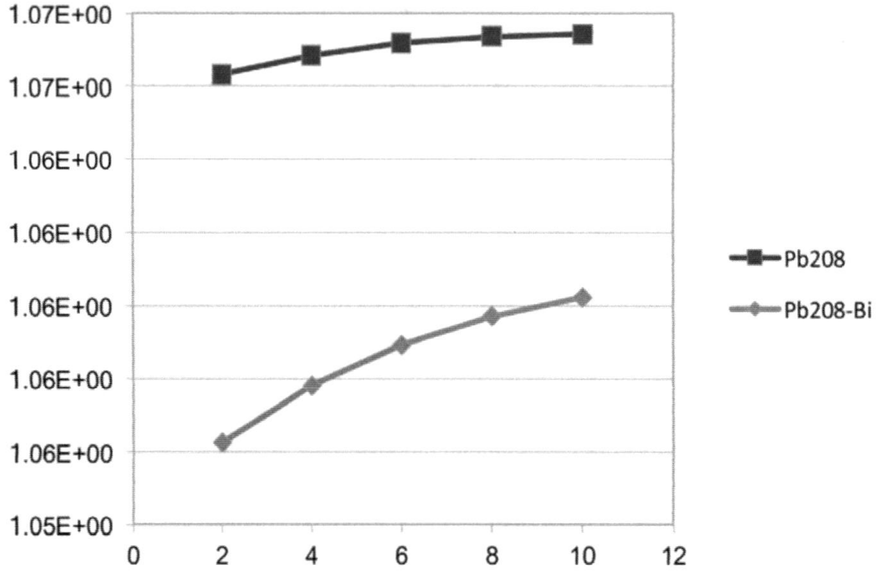

Figure 8. the change of atomic density during operation comparison between different coolant

Fig. 8 shows the effective neutron multiplication factor of the reactor with different coolant. Reactor with Pb208 as coolant has k-eff higher than Pb208-Bi as coolant at BOL. It is because Pb208-Bi as the coolant has Low neutron absorption cross-section.

5. Conclusion

- The neutronic aspect between Pb-208 and Pb208-Bismuth eutectic is not too much difference
- The k-eff of Pb208 as a coolant higher than Pb208-Bismuth eutectic
- Pb208-Bismuth eutectic probably gives better performance for extreme accidents than Pb208 because of the lower operation temperature. It is because Pb208-Bi eutectic has low melting point but high boiling point (almost same as Pb-208)

Acknowledgment

The authors express their gratitude for the support from The Ministry of Research Technology and Higher Education of The Republic of Indonesia through the scheme of Penelitian Unggulan Perguruan Tinggi (PUPT) RISTEKDIKTI.

References

[1] BPPT. (2016): Indonesia Energy Outlook 2016. Agency for The Assessment and Application of Technology, Jakarta.
[2] Duderstadt, James J., Hamilton, Louis J. (1976): Nuclear Reactor Analysis, first edition, John Willey and Sons, Ltd, 105-117, 124-140
[3] GIF. (2014): Technology roadmap update for generation IV nuclear energy systems. Nuclear Energy Agency (NEA)
[4] Khorasanov, G.L. Ivanov, A.P. Korobeinikov, V.V(2009): Lead Coolant For Fast Reactor-Burner With Hard Neutron Spectrum. State Scientific Center A.I. Leypunsky Institute for Physics and Power Engineering Bondarenko Sq., Obninsk Kaluga Region, 249020 Russia.
[5] Khorasanov, Georgy L., Blokhin, Anatoly I. (2009): Benefits in Using Lead-208 Coolant for Fast Reactors and Accelerator Driven Systems. State Scientific Centre of the Russian Federation – Institute for Physics and Power Engineering named after A.I. Leypunsky (IPPE), Russia.
[6] Khorasanov, G.L. Korobeynikov, V.V. Ivanov, A.P. Blokhin, A.I. (2009): Minimization of an initial fast reactor uranium–plutonium load by using enriched lead-208 as a coolant. Nuclear Engineering and Design 239 (2009) 1703–1707
[7] Okumura, K. et.al. (2007): SRAC2006: A Comprehensive Neutronics Calculation Code System, JAERI report.
[8] Sekimoto, H., Ryu, K., Yoshimura, Y.(2001): CANDLE: The New Burnup Strategy, Nuclear Science and Engineering Vol. 139 Number 3 pp 306-317
[9] Sekimoto, H, Okawa, T. Nakayama, S. -, Pb-208 Coolant for Small Long-life CANDLE Reactors, Tokyo Institute of Technology.
[10] Stacey, Weston M.(2007): Nuclear Reactor Physics 2nd edition, Wiley-VCH Verlag GmbH & Co. KgaA
[11] Su'ud, Zaki., Sekimoto, H., (2007), "Modified CANDLE Burnup Scheme and Its Application for Long Life Pb-Bi Cooled Fast Reactor with Natural Uranium as Fuel Cycle Input", Proceeding of International Conference on Advantages in Nuclear Science and Engineering, November 13-14, Bandung-Indonesia.
[12] Waltar, Alan E. Reynolds, Albert B. (1981): Fast Breeder Reactor. Springer.

On The Bernoulli Mixture Model for Bidikmisi Scholarship Classification with Bayesian MCMC

W Suryaningtyas[1,2], N Iriawan[3], K Fithriasari[3], BSS Ulama[3], I Susanto[1], AA Pravitasari[1]

[1] Doctoral Student at Department of Statistics, Institut Teknologi Sepuluh Nopember, Surabaya, Indonesia.
[2] Mathematic Education Program Study, Faculty of Teacher Training and Education, Muhammadiyah University of Surabaya, Surabaya, Indonesia.
[3] Department of Statistics, Institut Teknologi Sepuluh Nopember, Surabaya, Indonesia.

E-mail: wahyuni.pendmat@fkip.um-surabaya.ac.id; nur_i@statistika.its.ac.id

Abstract. This research has a purpose to develop Bernoulli Mixture model for Bidikmisi data modelling using Bayesian approach. Model development is done by considering the specificity in the data acceptance of Bidikmisi scholarship prototype in East Java Province. Bidikmisi acceptance status having a binary type (0 and 1) coupled with the main criteria factor of parent income and the number of dependents family produces a structure of Bernoulli mixture distribution with two components. The characteristics of each component can be identified through the Bernoulli Mixture modelling by involving the covariates of Bidikmisi scholarship recipients. The estimating parameter was performed using Bayesian Markov Chain Monte Carlo (MCMC) couple with the Gibbs Sampling algorithm. This model is applied to data registrants Bidikmisi districts/cities in the province of East Java as many as 44,489 students. This model shows the smallest value of Deviance Information Criteria (DIC) compared with Bayesian binary logistic regression.

1. Introduction

Bernoulli Mixture Model (BMM) is a model used to analyze the Bernoulli Mixture distributed data. Most of the references on binary data, the BMM research are mostly applied in the area of text mining [1]. BMM was first performed by Duda and Hart [2]. In its development, some research related to BMM were performed by Grim et al. [3], González, et al. [4], and Vidal [5,6], Patrikainen and Manilla [7], Zhu et al. [8], Sun et al. [9], Tikka et al. [10], Hollmen and Tikka [11], Myllykangas et al. [12], Bouguila [13] and Saeed et al. [14].

In this study, the development of BMM using Bernoulli Mixture regression analysis was applied in the social field by using local data of Bidikmisi scholarship. Bidikmisi tuition assistance program is an education assistance program by the Indonesian government through the Directorate General of Higher Education, which was launched in 2010. The government through Bidikmisi program aims to achieve equitable access and learning opportunities at university level and produce the independent, productive and having social care graduates who can play a role to solve the poverty chains to fill the needs of the Indonesian human resources and are ready for competing in the ASEAN Economic Community (MEA) [15].

This research employed the Bidikmisi prototype data of East Java Province. In 2015, the registrants of Bidikmisi in East Java were 44,489 students. The results of the data exploration showed that only 24.07% were awarded the scholarships, while those who were not successful were 75.93%, ie 33,780 student applicants. Bidikmisi grantee status with Binary type (0 and 1) was coupled with the main criteria factor, namely parent income and total of dependent. This combination produced the Bernoulli mixture data distribution with two components. Characteristics of each component of Bernoulli mixture

Content from this work may be used under the terms of the Creative Commons Attribution 3.0 licence. Any further distribution of this work must maintain attribution to the author(s) and the title of the work, journal citation and DOI.
Published under licence by IOP Publishing Ltd

can be identified through the Bernoulli Mixture modelling by involving the founding covariates of Bidikmisi scholarship grantee.

This study aims to perform the classification analysis of acceptance of Bidikmisi based on acceptance conditions with indicators of family ability. The classification is done by competing the Bayesian Bernoulli Mixture regression and Bayesian binary logistic regression. Both methods are analyzed using openBUGS software. Furthermore, the results of the analysis are used to compile the accuracy of Bidikmisi acceptance classification.

2. Literature Review

2.1. Bernoulli Mixture Model

If random sample Y is independently distributed Bernoulli Mixture derived from i-units, then there will be a vector $\mathbf{y}=\begin{bmatrix} y_1 & y_2 & \cdots & y_n \end{bmatrix}^T$ with $i=1,2,\ldots,n$, which can contain L groups with the proportion $\boldsymbol{\pi}=\ \pi_1,\pi_2,\ldots,\pi_L$. The finite mixture model of Y with L number components could have the density functions as follows [16]:

$$p(y_i)=\sum_{l=1}^{L}\pi_l p_l(y_i), \tag{1}$$

in which L is the number of mixture components and for each l, $p_l(y_i)$ is the mixture density component and π_l is the non-negative quantity which amounts to one, that is:

$$\sum_{l=1}^{L}\pi_l=1. \tag{2}$$

BMM based on model (1), if $\mathbf{Y}=(Y_1,Y_2,\ldots,Y_n)$ is a random sample. The goal is to partition $\mathbf{Y}$ into L (might be unknown, but limited) groups. The finite mixture density with L components can be written as [17]:

$$p(\mathbf{Y}\,|\,L,\boldsymbol{\pi},\boldsymbol{\Theta})=\sum_{l=1}^{L}\pi_l p_l(\mathbf{Y}\,|\,\theta_l), \tag{3}$$

where $p_l(.)$ is called with the mixture density l^{th} component, $\boldsymbol{\Theta}=(\theta_1,\theta_2,\ldots,\theta_L)$ is a mixture density component parameter, and $\boldsymbol{\pi}=(\pi_1,\ldots,\pi_L)$ is the mixture proportion.

2.2. Bayesian MCMC

Bayesian Markov Chain Monte Carlo (Bayesian MCMC) is an approach, which applies Gibbs sampling process conducted through sampling by means of a series of Gibbs random variables based on the basic properties of Markov Chain [18]. Inference using the Bayesian approach to the parameter estimation process is done by integrating the posterior distribution. Numerically can be done integration through a simulation procedure commonly known as Markov Chain Monte Carlo (MCMC) method. The following are given the work steps of Markov Chain Monte Carlo method in general [19], namely:

1. Choosing a starting value $\boldsymbol{\theta}^{(0)}$.
2. Generating value $\boldsymbol{\theta}^{(m)}$, $m=1,\ldots,M$ until they reached convergence in distribution.
3. Diagnosing the convergence of the chain of $\boldsymbol{\theta}$ used for monitoring. If the convergence diagnosis fails for length of M, then do some additional observation by increasing the larger M.
4. Cutting the first B observations as burn-in observations and dispose them.
5. Sampling some data $\left\{\boldsymbol{\theta}^{(B+1)},\boldsymbol{\theta}^{(B+2)},\ldots,\boldsymbol{\theta}^{(M)}\right\}$ as the estimated posterior parameters distribution.
6. Plotting each posterior parameter distribution (as the univariate marginal distribution each).
7. Finally, obtain summaries of the characteristics of the posterior distributions of $\boldsymbol{\theta}$.

3. Methodology

3.1 Source of Data

The data used in this research was sourced from Database of Ministry of Research, Technology and Higher Education through Bidikmisi channel, that was Bidikmisi data of all districts in East Java Province in 2015.

3.2 Variable Research

Research variables used in this study consisted of the response variable (Y) and the predictor variable (X).

Y = The acceptance Status of Bidikmisi Scholarship (1 = accepted, 0 = not accepted)

X_1 = Father's job with four dummies – b_{11} as an agricultural sectors, b_{12} as the government employee, b_{13} as an entrepreneur, and b_{14} as a private employee;

X_2 = Mother's Job with four dummies b_{21}, b_{22}, b_{23}, and b_{24} defined as in Father's job;

X_3= Father's Education with three dummies – b_{31} as non educated, b_{32} as elementary to senior high school education, and b_{33} as higher education;

X_4 = Mother's Education with three dummies – b_{41}, b_{42}, and b_{43} defined as in Father's Education.

Each dummy variable has the value of 0 or 1.

3.3 Research Design

Classification analysis procedures using Bayesian Bernoulli mixture regression model and Bayesian binary logistic regression model are given the following research flows in Figure 1:

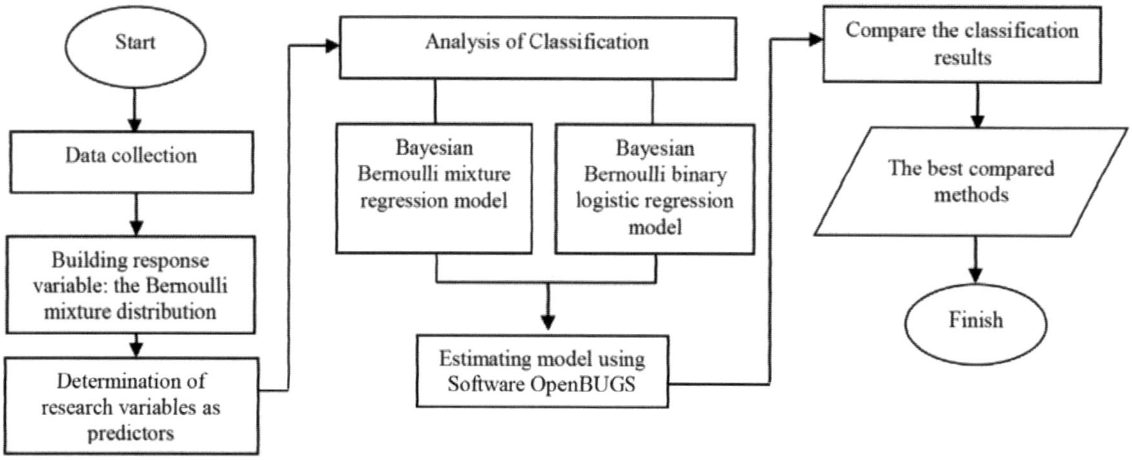

Figure 1. Flowchart Classification Bidikmisi using Bayesian Bernoulli Mixture Regression Model and Bayesian Binary Logistic Regression Model

4. Research Result

4.1. Pre-processing

The explanations of the techniques used in the pre-processing stages of identification for building the Bernoulli mixture distribution are as follows:

Step 1. take response variable (Y)

Step 2. select covariate "father's income", "mother's income" and "family dependent"

Step 3. Create a new covariate by counting the amount of "dad's income" and "maternal income" divided by "the number of family dependents", then name it with "Code Category (CC)".

Step 4. coding the covariate "CC" with the following criteria:

0 = if CC> Rp. 750,000 per head in the family included in the category of wealthy family

1 = if CC <Rp. 750,000 per head in the family fall into the category of poor families.

Step 5. match the response variable (Y) to the CC in Step 4 to the AC (Acceptance Condition) with the

Bidikmisi acceptance classification table of "wrong" and "right" which are given as follows:

Table 1. Identification Components Mixture of Bidikmisi Scholarship 2015

Y	CC	AC	Condition	Interpretation
1	0	0	Wrong	Acceptance Condition is wrong (AC = 0) if the grantee (Y = 1) is followed with the category of wealthy family (CC = 0)
0	1	0	Wrong	Acceptance Condition is wrong (AC = 0) if the grantee (Y = 0) is followed with the category of poor family (KK = 1)
1	1	1	Right	Acceptance Condition is right (AC = 1) if the grantee (Y = 1) is followed with the category of poor family (CC = 1)
0	0	1	Right	Acceptance Condition is wrong (AC = 0) if the grantee (Y = 0) is followed with the category of wealthy family (CC = 0)

The pre-processing result by involving founder covariate of the Bidikmisi scholarship shows response data of the Bernoulli mixture distribution with two components, namely component of wrong acceptance condition and component of right acceptance condition. Based on Table 1, it is obtained that *Bernoulli*-1 with θ_1=0.62 (AC is wrong) and *Bernoulli*-2 with θ_2=0.38 (AC is right).

4.2. Bayesian Binary Logistic Regression

The doodle of Bayesian binary logistic regression to model the acceptance of the Bidikmisi scholarship is presented in Figure 2.

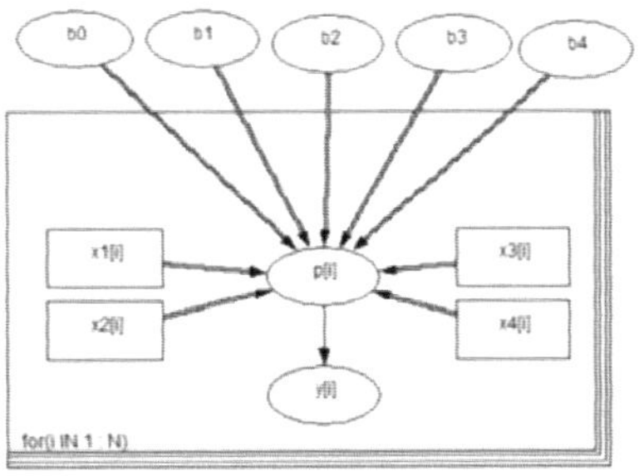

Figure 2. *Doodle* of Bayesian Binary logistic regression

After compiling the doodle and its syntax, the next step is running program to get the estimated regression model. The serial historical sample values of each posterior parameter show a stable random pattern in a fixed domain, indicating the fulfillment of irreducible, aperiodic and recurrent properties. The historical sample series of the first six posterior estimation values (b_0, b_{11}, b_{12}, b_{13}, b_{14}, and b_{21}) are shown through the serial plot is presented in Figure 3. While their significant parameter estimates of the binary logistic regression using a link function logit is provided in Table 2.

Based on the significant parameter estimation by using Bayesian binary logistic regression in Table 2, the model can be expressed as follows:

$$\hat{\pi}(x)=\frac{\exp(-1,6540+0,2612\,b_{11}-0,4487\,b_{12}+0,2486\,b_{13}+0,1235\,b_{14}+0,3779\,b_{21}+0,3335\,b_{23}-0,1147\,b_{32}+0,0623\,b_{33}-0,1661\,b_{41}-0,1823\,b_{42}+0,0469\,b_{43})}{1+\exp(-1,6540+0,2612\,b_{11}-0,4487\,b_{12}+0,2486\,b_{13}+0,1235\,b_{14}+0,3779\,b_{21}+0,3335\,b_{23}-0,1147\,b_{32}+0,0623\,b_{33}-0,1661\,b_{41}-0,1823\,b_{42}+0,0469\,b_{43})}$$

4.3. Bernoulli Mixture Bayesian Regression

The doodle of regression of Bernoulli Mixture Bayesian to model the Bidikmisi scholarship acceptance is presented in Figure 4.

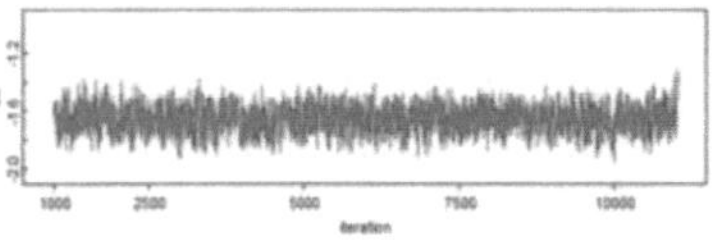

(a) b_0

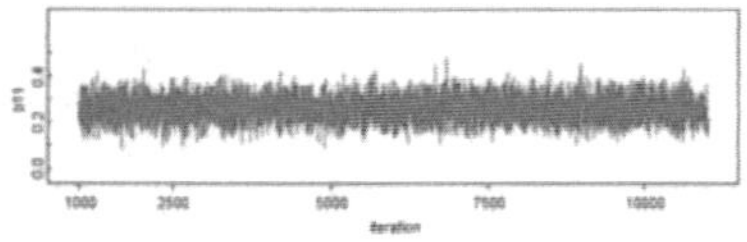

(b) b_{11}

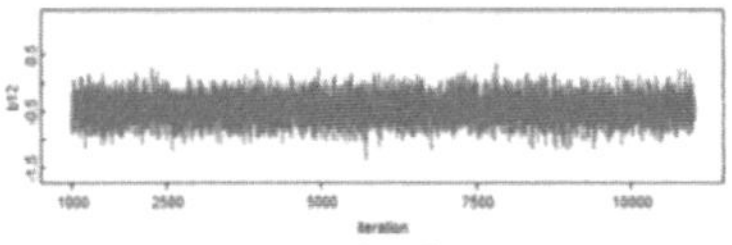

(c) b_{12}

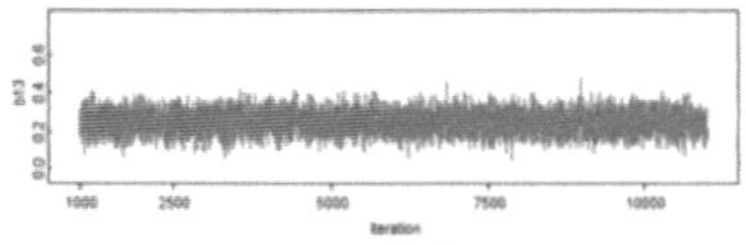

(d) b_{13}

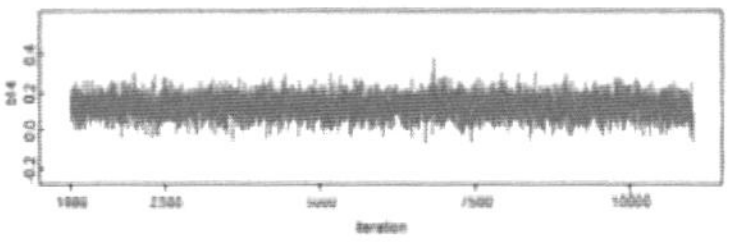

(e) b_{14}

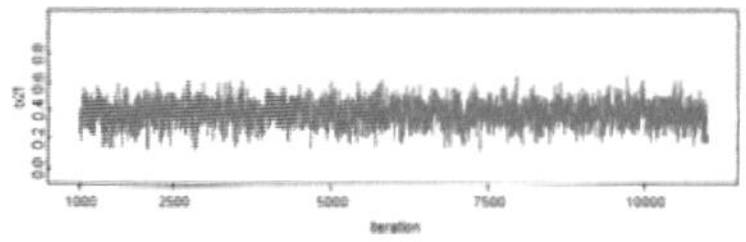

(f) b_{21}

Figure 3. Serial Plot of the first six posterior parameters b_0, b_{11}, b_{12}, b_{13}, b_{14}, b_{21} with 100.000 Iterations (10.000 thin 10) for Bayesian Binary Logistic Regression

Table 2. Parameters Estimation of Bayesian Binary Logistic Regression

Parameter	**Dummy Parameter**	**Mean**	**Standar Deviasi**	**2,5%**	**97,5%**	**Significant**
Constant		-1,6540	0,0851	-1,8210	-1,4880	√
X_1	b_{11}	0,2612	0,0492	0,1641	0,3562	√
	b_{12}	-0,4487	0,2133	-0,8687	-0,0315	√
	b_{13}	0,2486	0,0535	0,1448	0,3528	√
	b_{14}	0,1235	0,0526	0,0219	0,2259	√
X_2	b_{21}	0,3779	0,0776	0,2277	0,5306	√
	b_{22}	-	-	-	-	-
	b_{23}	0,3335	0,0827	0,1725	0,4967	√
	b_{24}	0,1656	0,0862	-0,0027	0,3375	No
X_3	b_{31}	-0,0587	0,0483	-0,1533	0,0356	No
	b_{32}	-0,1147	0,02042	-0,1549	-0,0751	√
	b_{33}	0,0623	0,0201	0,0232	0,1012	√
X_4	b_{41}	-0,1661	0,0455	-0,2556	-0,0758	√
	b_{42}	-0,1823	0,0198	-0,2208	-0,1437	√
	b_{43}	0,0468	0,0209	0,0060	0,0877	√

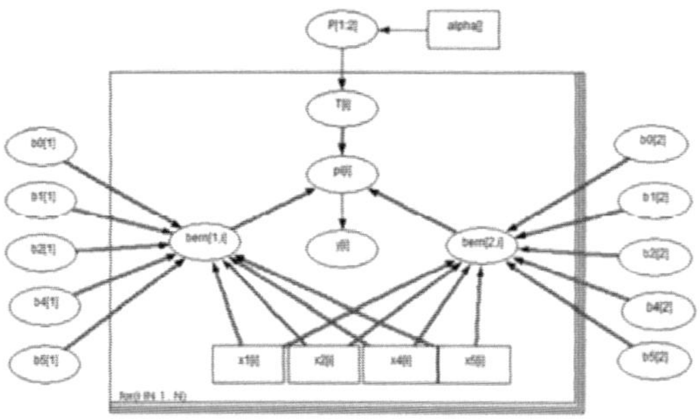

Figure 4. *Doodle* of Bernoulli Mixture Bayesian Regression

Figure 4 tells that the predicted parameters bern[1,i] and bern[2,i] have the same link function but are connected to different parameters. Node T[i] functions to separate data based on the acceptance condition of Bidikmisi that contains value 1 if the acceptance condition is false, and 2 if the acceptance condition is correct. Therefore, data in group 1 was analyzed on the bern[1,i] and data in group 2 was analyzed on the bern[2, i]. Prior parameters were normal distributions because the modeling method used a link function. Node P[1,2] was used to determine the proportions of group 1 and group 2 in the model. The total value of P[1] and P[2] is one. The parameter estimation is shown in Table 3.

Table 3. Parameters Estimation of Bernoulli Mixture Bayesian Regression

Parameter	Dummy Parameter	Mean	Standar Deviasi	2,5%	97,5%	Significant
P[1]		0,6148	0,0023	0,6103	0,6194	√
P[2]		0,3852	0,0023	0,3806	0,3897	√
b0[1]		1,0490	0,1256	0,8015	1,2930	√
b0[2]		-1,7490	0,0875	-1,9130	-1,5750	√
X_1	$b_{11}[1]$	-1,3060	0,0711	-1,4450	-1,1640	√
	$b_{11}[2]$	0,8516	0,0525	0,7482	0,9546	√
	$b_{12}[1]$	-1,1700	0,3535	-1,8760	-0,4949	√
	$b_{12}[2]$	-0,6488	0,2256	-1,0970	-0,2113	√
	$b_{13}[1]$	-1,1340	0,0792	-1,2890	-0,9791	√
	$b_{13}[2]$	0,5607	0,0591	0,4451	0,6775	√
	$b_{14}[1]$	-0,7730	0,0757	-0,9218	-0,6228	√
	$b_{14}[2]$	-0,0154	0,0578	-0,1298	0,0968	No
X_2	$b_{21}[1]$	-2,0240	0,1143	-2,2470	-1,7960	√
	$b_{21}[2]$	1,1030	0,0801	0,9451	1,2580	√
	$b_{23}[1]$	-1,5540	0,1213	-1,7920	-1,3140	√
	$b_{23}[2]$	0,5299	0,0879	0,3573	0,7015	√
	$b_{24}[1]$	-1,3330	0,1272	-1,5820	-1,0820	√
	$b_{24}[2]$	0,0041	0,0951	-0,1828	0,1900	No
X_3	$b_{31}[1]$	-0,6911	0,1137	-0,9169	-0,4709	√
	$b_{31}[2]$	0,0688	0,0560	-0,04112	0,1781	No
	$b_{32}[1]$	-0,2939	0,0378	-0,3692	-0,2201	√
	$b_{32}[2]$	-0,0917	0,0239	-0,1387	-0,0445	√
	$b_{33}[1]$	0,2806	0,0343	0,2139	0,3484	√
	$b_{33}[2]$	0,0415	0,0233	-0,0038	0,0870	No
X_4	$b_{41}[1]$	-0,2338	0,0963	-0,4225	-0,0467	√
	$b_{41}[2]$	-0,0854	0,0536	-0,1901	0,0197	No
	$b_{42}[1]$	-0,1130	0,0355	-0,1833	-0,0432	√
	$b_{42}[2]$	-0,1479	0,0228	-0,1921	-0,1028	√
	$b_{43}[1]$	0,0900	0,0354	0,0208	0,1598	√
	$b_{43}[2]$	0,0046	0,0241	-0,0429	0,0514	No

Historical serial sample values for parameters estimation of Bernoulli Mixture Bayesian Regression show a stable random pattern in a fixed domain. The historical sample series of the first 8 posterior estimation values (p_1, p_2, $b_{0[1]}$, $b_{0[2]}$, $b_{11[1]}$, $b_{11[2]}$, $b_{12[1]}$, $b_{12[2]}$) indicating the fulfillment of irreducible, aperiodic and recurrent properties as their convergence has been reached.

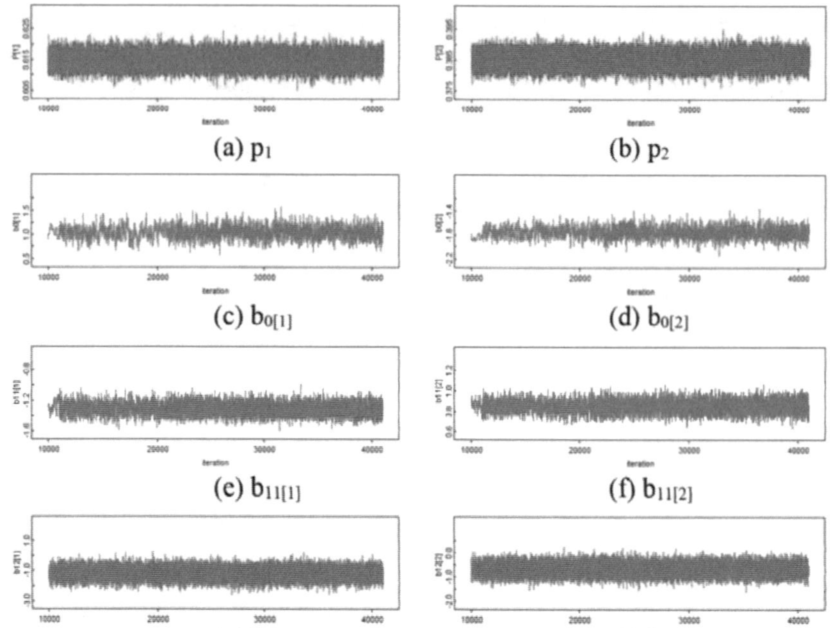

(a) p_1 (b) p_2

(c) $b_{0[1]}$ (d) $b_{0[2]}$

(e) $b_{11[1]}$ (f) $b_{11[2]}$

(g) $b_{12[1]}$ (h) $b_{12[2]}$

Figure 5. Serial Plot of the first 14 posterior estimation values (p_1, p_2, $b_{0[1]}$, $b_{0[2]}$, $b_{11[1]}$, $b_{11[2]}$, $b_{12[1]}$, $b_{12[2]}$) with 10.000 Iterations thin 30 for Bernoulli Mixture Bayesian Regression

4.4. Classification Comparison

The best combination of significant parameters for Bayesian Bernoulli mixture regression model and for Bayesian binary logistic regression model are compared to see which model can be more representative to explain the acceptance of the Bidikmisi data. The comparison is done using their classification values. The bigger classification percentage, the better the model explain the acceptance. Their classification results are tabled in Table 4. It can be seen that Bayesian Bernoulli mixture regression model with 71.70 percents shows more accurate than Bayesian binary logistic regression, which is just only able to accurately classify as big as 56.50 percents.

Tabel 4. Accepted Qualification Percentage

Model	**% Classification**
Bayesian binary logistic regression	56.50
Bayesian Bernoulli mixture regression	71.70

5. Conclusion

The new development approach, a Bayesian Bernoulli mixture regression model coupled with MCMC approach, has successfully worked to classify the Bidikmisi acceptance. This new development approach and the Bayesian binary logistic regression model can reach their posterior parameters estimation convergence perfectly. This proposed Bayesian Bernoulli mixture regression model coupled with MCMC approach can give a higher percentage of acceptance classification accuracy than the Bayesian binary logistic regression model. This model is more representative for explaining the classification of Bidikmisi acceptance.

Acknowledgements
The Authors are grateful to Directorate for Research and Community Service (DRPM) Ministry of Research, Technology and Higher Education Indonesia which support this research under PUPT research grant no. 608/PKS/ITS/2017.

References

[1] Wang X, Kabán A. Finding Uninformative Features in Binary Data. Intelligent Data Engineering and Automated Learning - IDEAL 2005. 2005; 3578: p. 40–47.

[2] Duda RO, Hart PE. Pattern Classification and Scene Analysis Wiley; 1973.

[3] Grim J, Pudil P, Somol P. Multivariate Structural Bernoulli Mixtures for Recognition. In ; 2000; Proceedings. 15th International Conference on, vol. 2, pp. 585–589.

[4] González J, Juan A, Dupont P, Vidal E, Casacuberta F. A Bernoulli Mixture Model for Word Categorization. In ; 2001; Benicassim, Spain: Proceedings of the IX Spanish Symposium on Pattern Recognition and Image Analysis.

[5] Juan A, Vidal E. On The Use of Bernoulli Mixture Models for Text Classification. Pattern Recognition. 2002; 35(12): p. 2705–2710.

[6] Juan A, Vidal E. Bernoulli Mixture Models for Binary Images. In ; 2004: Proceedings of the 17th International Conference on Pattern Recognition (ICPR'04).

[7] Patrikainen A, Mannila H. Sub Space Clustering of High-Dimensional Binary Data-A Probabilistic Approach. In ; 2004; Workshop on Clustering High-Dimensional Data and Its Applications: SIAM International Conference on Data Mining.

[8] Zhu S, Takigawa I, Zhang S, Mamitsuka H. A Probabilistic Model for Clustering Text Documents with Multiple Fields. In ; 2007; Berlin, Heidelberg: Advances in Information Retrieval, 29th European Conference on IR Research (ECIR2007).

[9] Sun Z, Rosen O, Sampson A. Multivariate Bernoulli Mixture Models with application to Postmortem Tissue Studies in Schizophrenia. Biometrics. 2007; 63: p. 901-909.

[10] Tikka J, Hollmen J, Myllykangas S. Mixture Modelling of DNA Copy Number Amplification Patterns in Cancer. In ; 2007; Berlin, Heidelberg: Proceedings of the 9th International Work-Conference on Artificial Neural Networks (IWANN' 2007), Springer-Verlag.

[11] Hollmen J, Tikka J. Compact and Understandable Descriptions of Mixture of Bernoulli Distributions. In ; 2007; Springer-Verlag, Berlin, Heidelberg: Proceedings of the 7th International Symposium on Intelligent Data Analysis (IDA2007).

[12] Myllykangas S, Tikka J, Böhling T, Knuutila S, Hollmén J. Classification of Human Cancers Based on DNA Copy Number Amplification Modelling. BMC Med. Genomics. 2008; 1: p. 1-13.

[13] Bouguila N. On Multivariate Binary Data Clustering and Feature Weighting. Comput. Stat.Data Anal. 2010; 54: p. 120-134.

[14] Saeed M, Javed K, Babri HA. Machine Learning Using Bernoulli Mixture Models: Clustering,Rule Extraction and Dimensionality Reduction. Neurocomputing. 2013; 119: p. 366–374.

[15] Direktorat Jenderal Pembelajaran dan Kemahasiswaan KRTdPT. Pedoman Penyelenggaraan Bantuan Biaya Pendidikan Bidikmisi Tahun 2016 Jakarta: Belmawa, Kemeristek Dikti; 2016.

[16] McLachlan G, Peel D. Finite Mixture Models New York: John Wiley and Sons, Inc.; 2000.

[17] Nadif M, Govaert G. Clustering for Binary Data and Mixture Models-Choice of The Model. Appl. Stochastic Models Data Anal. 1998; 13: p. 269-278.

[18] Casella G, George EI. Explaining Gibss Sampler. The America Statistical Association. 1992; 46(3): p. 167-174.

[19] Ntzoufras I. Bayesian Modeling Using WinBUGS New Jersey, USA: Wiley; 2009.

IOP Conf. Series: Journal of Physics: Conf. Series **1090** (2018) 012073 doi:10.1088/1742-6596/1090/1/012073

Preliminary Study of Long-life GFR 100 and 150 MWth

S.C. Pattipawaej[1,a*] , Z. Su'ud[1,b]

[1]*Nuclear Physics and Biophysics Research Division, Physics Department, Faculty of Mathematics and Natural Science, Bandung Institute of Technology Jl. Ganesha 10, Bandung 40132, Indonesia*

Abstract. The breeding capability of fast reactor allow the reactor to operate in a long life time, but with the risk of a large reactivity swing and high power peaking in the center of the core. Both of these factors can be reduced by adjusting the fuel with a right configuration. This is down in this study, where optimization is performed on 100 and 150 MWth reactors, with the aim to getting the reactor to operate for more than 20 years with Power Peaking factor (PPF) is less than 1.3 and reactivity swing less than 1%. Based on the optimization result, a maximum configuration of fuel in inner, mid, and outer core of reactor I and II is given by 6.6%, 11.6%, 12.4% and 6.6%, 9.8%, 14.0%. Neutronic analysis results of this design indicate that the reactor II can only operate for 29 years, while reactor I can operate for more than 30 years because it has a larger conversion ratio (CR). The PPF value of reactor II is greater than reactor I caused by fission rate in the center of the core of this reactor is larger as a result of the short buildup process. Reactivity swing of reactor I and II are 0.25% and 0.16%, where the maximum reactivity of reactor I is 1.02% and occur in BOL, and for reactor II, the maximum reactivity is 1.03% and occurred in the 18th year.

1. Introduction

Seventy percent of commercial nuclear reactors are used until this time is generation two reactor like Pressurized Water Reactor (PWR) and Boiling Water Reactor (BWR). This reactor had several drawbacks, such as low uranium utilization and low efficiency of reactors. Efficiency of reactors is less than 35%, caused by the maximum temperature output is less than 330 ^{0}C [1]. Ninety percent of fuel waste of reactors still have the same form with original fuel, composed of 35% fission products, 1% waste of ^{235}U and 1% plutonium as transmutation yield [2].

This weakness becomes a problem in the nuclear industry, thus needed new nuclear technology is more promising. One of the more modern nuclear technology were able to overcome this problem is Gas Cooled Fast Breeder Reactor (GFR) is a fourth generation reactor. This reactor takes the advantage of breeding process to produce plutonium from ^{238}U that the reactor can operate within a

Content from this work may be used under the terms of the Creative Commons Attribution 3.0 licence. Any further distribution of this work must maintain attribution to the author(s) and the title of the work, journal citation and DOI.
Published under licence by IOP Publishing Ltd

longer time. This means uranium utilization and radioactive isotopes destroyed of this reactor is more than efficient [3]. This reactor also uses helium gas as a coolant, so possible to produce high-temperature gas output until 850^{0}C and net thermal efficiency is 42% [4]. As the pay, neutronic analysis of this reactor become more complicated due to high conversion ratio (high breeding) of the core.

Breeding process in the cores depends on the amount of fertile material and hard neutron spectrum. To keep the fast neutrons and the fissile materials in sufficient quantity, fuel in the active core must be arranged in a particular configuration where fuel with the lowest percentage of plutonium is placed at the center of core whereas a material with the highest percentage of plutonium is located at outer of the core. Fuel with a low percentage of plutonium serves as a provider of the fertile material needed for breeding in a long time. Fuel in outer core is used as driver fuel.

When the reactor operates, most fission reactions occur in the outer region that has the higher amount of fissile material. Neutrons produced by this fission reaction are then used to subsequent fission reaction and also for breeding. The highest probability of breeding occurs in the inner core due to the greater number of neutron and fissile material. High breeding in the inner core has an impact on increasing of power distribution in this region as a result of an increase of the amount of fissile material. The increase of power distribution in a long time at this region can result in the increasing value of Power Peaking Factor (PPF) of the core until 1.5, this is dangerous for the reactor. In addition to PPF, the other neutronic problems were appeared as a consequence of the high breeding velocity of GFR, especially for small power GFR is the values of maximum reactivity and reactivity swing more than 1%.

To keep the reactor operable for a long time with neutronic parameters such as PPF, maximum reactivity, and reactivity swing remains within safe limits, the fuel in the core must be arranged in the optimum configuration.

In this paper, optimization is done to get small power GFR which capable of operating more than 20 years with a low PPF and small values of reactivity swing and maximum reactivity. As a comparison, do the neutronic analysis of the reactor with a power 100 and 150 MWth. These reactors are suitable to use in the remote area like Indonesia islands which has a relatively low electricity requirements. In addition, due to the high outlet temperature, these reactors can be used to produce hydrogen gas and useful in chemical industry.

2. Design

2.1 Assembly Design.

The reactor core is composed of some several types of assemblies such as fuel, control, shutdown, reflector, and shielding assembly.

Fuel assemblies were designed base on allegro 75 MWth assembly design [5]. This assembly is filled by 91 cylinder fuel pin with arranged in six line hexagonal concentric and surrounded by hexagonal duct assembly with thickness 3.0 mm and distance of duct flat-to-flat are 10.4 cm. Every fuel pin is filled by fuel pellet with diameter 8.56 mm, and wrapped cladding with thickness 0.91 mm. 50% of this assembly is fill up by fuel, 23.6% is cladding and the remaining part is coolant.

The material of fuel pellet is (U,Pu)C, where natural uranium mixed with plutonium waste from PWR with a certain percentage plutonium. Plutonium used as driver fuel and ^{239}Pu and ^{241}Pu which result of breeding from ^{238}U and ^{240}Pu would be used as next fissile fuel. The cladding and assembly

IOP Conf. Series: Journal of Physics: Conf. Series **1090** (2018) 012073 doi:10.1088/1742-6596/1090/1/012073

duct are formed of the same material is SiC. The design of fuel pin and fuel assembly is given by figure 1.

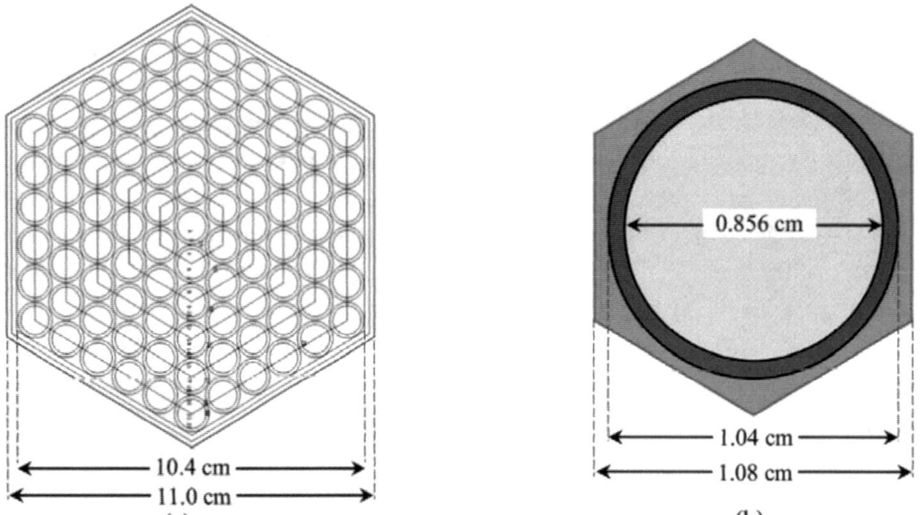

(a) (b)

Figure 1. a) Fuel assembly design. b) Fuel pins design

The reflector and shielding assembly have the same form but different compound of material. Material of reflector is Zirconium Carbide (ZrC) while shielding is Boron Carbide. Any assemblies are filled by 19 pin rods with radius size is 2.0 cm. The design of shielding/reflector assembly and shielding/reflector pins is given by figure 2.

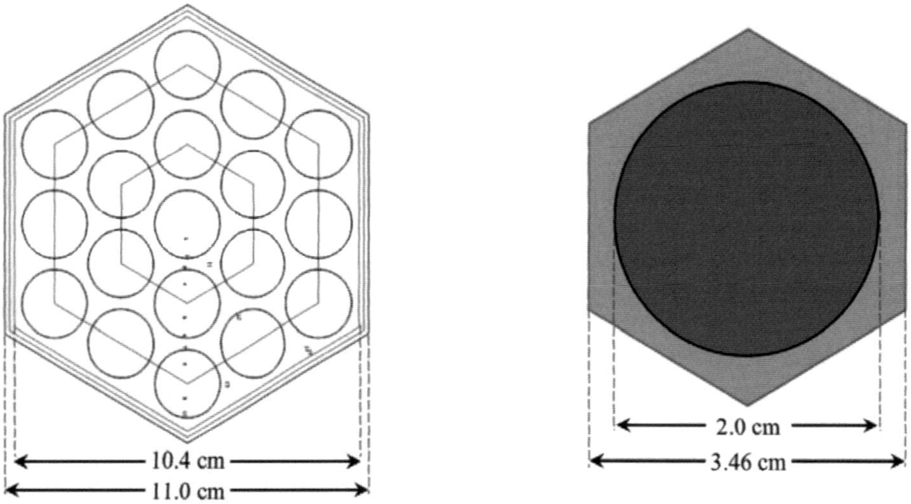

(a) (b)

Figure 2. a) Reflector/Shielding assembly design. b) Reflector/Shielding pins design

2.2 Core Design

The reactor core is designed in three-dimensional hexagonal, where the core is consists of three main part: active core, reflector, and shielding. Active core is located in the center of the core and covered by reflector and shielding in the outer of the core.

Active core is filled by 127 assemblies where it is arranged in 7 lines hexagonal concentric. Active core is divided in III region based on percentage of plutonium of fuel assembly. Line 1-3 of active core is region I (inner core) consisting of one inner shut down assembly (S1) was placed in the center of region, six inner control assemblies (C1) was placed in vertex in line three, and twelve inner fuel assemblies (F1) was filled in another place in line 2 and 3. Region II (mid core) is filled by six outer shut down assemblies (S2), and thirty-six mid fuel assembly (F2). Outer shut down assemblies were placed in the middle of every side of line 5, and another area of line 4 and line 5 is filled by mid fuel assembly. In region III (outer core) outer fuel (F3) was placed in line six and seven except in every vertex in line seven was filled by outer control assemblies (C2). Reflector (R) of core in the radial direction is filled by one hundred and forty-four Assemblies which arranged in three line hexagonal concentric. In the axial direction, there is a reflector at the top and bottom of active core with the same thickness of 30 cm. the shielding (SH) was placed as wrappers of the reflector. In axial direction it has a thickness 50 cm, and in radial direction is arranged in four line assemblies. Number of shielding assembly is two hundred and seventy six assemblies. Core views in radial and axial directions is given by **figure 3**.

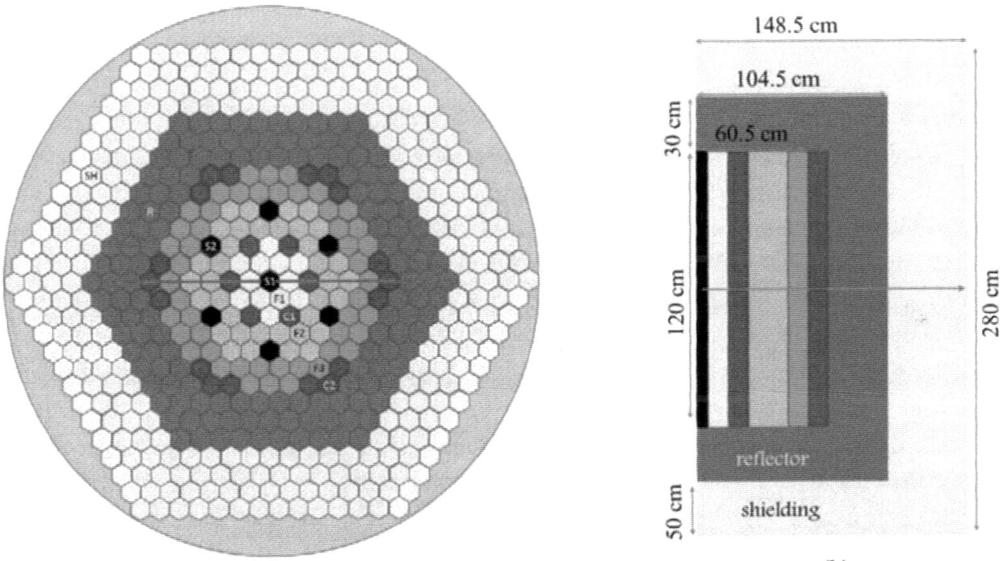

(a) (b)

Figure 3. Core design. a) Radial view. b) Axial view

This reactor was designed to operate more than 20 years with reactivity and swing reactivity is less than 1.0%, and Power peaking factor is less than 1.3 and applies to two different power 100 MWth and 150 MWth. Some design limitation of the reactor is must be operated with average power density 50 - 100 MW/m^3 [4], with inlet coolant temperature is less than 500^0C and outlet 850^0C or more in pressure 7 MPa. General specification of reactor is described in Table.1.

Table 1. Some result of neutronic analysis

Parameter	Specification
Thermal power	100 – 150 MWth
Refueling period	20 – 30 years
Core geometry	Hexagonal 3D
Coolant	Helium gas
Fuel type	(U,Pu)C
Number of fuel assembly	96
Duct flat-to-flat of fuel assembly	104 mm
Duct thickness of fuel assembly	3 mm
Fuel pin pitch	10.8 mm
Fuel pin OD	8.56 mm
Cladding thickness	1.84 mm
Number of control assembly	24
Number of shut down assembly	7

3. Analysis Method

In this paper, neutronic analysis of reactor design calculation was analyzed by used SRAC (Standard Reactor Analysis Code). JENDL-40 from JAEA (Japan Atomic Energy Agency) is used as a library of this code. The primal data of this library is diffusion coefficient, and the cross section data of 406 nuclides with incident neutron energy range is from 10^{-5} eV ~ 20 MeV in the discretization to 107 groups energy (fine groups)[6].

In SRAC, the calculation is divided into two steps those are the lattice calculation and core calculation [7]. The purpose of lattice calculation is to prepare a few-group homogenized macroscopic cross section of two-dimensional (2D) infinite arrangement of assemblies. This cross section would be used to solve diffusion equation in core calculation. The dependence of energy has been calculated in here so that calculation in the core will be easier because only focus on the region such as power distribution in the core.

In the lattice calculation, the calculation started from the preparation of effective microscopic cross section of any group energy in every segment area of the lattice from the self-shielding factor and infinite dilution cross section where the self-shielding factor are calculated from background cross section. All of the parameters are calculated based on reactor design like temperature or other characteristic of the material and calculated by used Narrow Resonance (NR) approximation. Base on the value of atomic density number and effective microscopic cross section can be calculated the value of the effective macroscopic cross section. This value will be used to solve neutron transport equation to find neutron flux in any group energy and any segment of the lattice. This equation solved iteratively use Collision Probability Method (CPM) with convergence criterion of neutron flux is 0.0001. The iteration known as inner iteration.

To reduce the calculation cost in times and memory due to a huge amount of neutron flux and microscopic cross section, do homogenization to minimize the space-dependent information and group-collapsing to reduce the energy-dependent information. Homogenization of neutron flux is performed at the first step, and then this result can be used in homogenization calculation of microscopic cross section and macroscopic cross section. In the same way, group-collapsing calculation of result of neutron flux, microscopic cross section, and macroscopic cross section homogenization was started from neutron flux, where 107 groups energy would grouping to 10 group energy. This result will be used to calculate the few-group homogenization microscopic cross section and macroscopic cross section.

The next step in lattice calculation is burnup calculation. In this step, the cycle calculation of few-group homogenized cross section will be repeated for the different atomic number density of fuel material in every time burnup. The atomic number density in every time burnup was calculated used burnup equation.

To solve the multi-group diffusion equation in the core calculation is required macroscopic cross section that has been obtained in lattice calculation and also needed the diffusion coefficient was contained in the JENDL-40 and transmutation few-group macroscopic cross section.

This diffusion equation solved iteratively to get the value of neutron flux and effective multiplication factor that convergence. This iteration process is known as outer iteration. In this paper, the convergence criterion of neutron flux and effective multiplication factor are 0.00001 and 0,001. After the neutron flux of all segment in reactor core obtained and also the value of multiplication factor, the calculation continued to get both value in another burnup steps along 30 years. Furthermore, based on the result of multiplication factor effective at BOL and EOL can be calculated the reactivity swing of the reactor and then power in every segment in the reactor core can be calculated based on neutron flux in each segment as a function of fission reaction rate. Diagram block of neutronic calculation with SRAC is given by figure 4.

The power density in every mesh of core can be averaged in axial direction so that the radial power density distribution in the core is easier to be analyzed. Based on this value, the power peaking factor can be calculated as a ratio of peak power to average power. The CR value of reactor is also can calculated from CR value and power in regions.

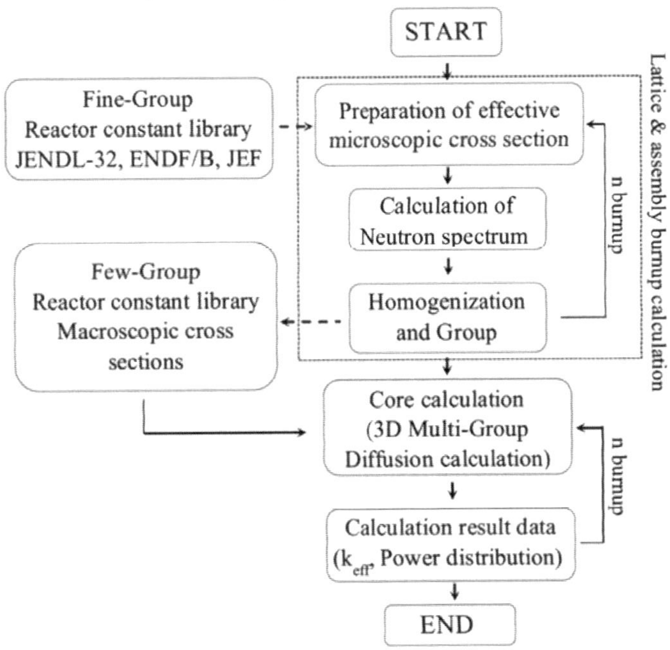

Figure 4. Diagram block of neutronic calculation with SRAC

IOP Conf. Series: Journal of Physics: Conf. Series **1090** (2018) 012073 doi:10.1088/1742-6596/1090/1/012073

4. Result

The optimization result of reactor 100 MWth and 150 MWth provide, the maximum fuel configuration of reactor 100 MWth is 6.6%, 11.6% and 12.4% percentage of plutonium in the inner core, mid core, and outer core. Maximum fuel configuration of reactor 150 MWth in the inner, mid, and outer region is 6.6%, 9.8% and 14.0%.

Some result of neutronic calculation of both reactors is given by table 2.

Table 2. Some result of neutronic analysis

Reactor parameter	Reactor I	Reactor II
Power (MWt)	100	150
Refueling time (years)	30	28
Average average power density (MW/m^3)	62.61	93.92
Reactivity swing (%)	0.25	0.16
Power peaking factor at BOL, MOL, EOL	1.09, 1.19, 1.26	1.16, 1.18, 1.30

Besides this result are also given details of the result such as atomic density change every burnup, neutron flux, distribution of average power density, conversion ratio and a multiplication factor (k_{eff}) of any reactor.

The atomic density change and neutron flux in BOL, MOL, and EOL, and CR of reactor I is given by figure 5, 6 and 7.

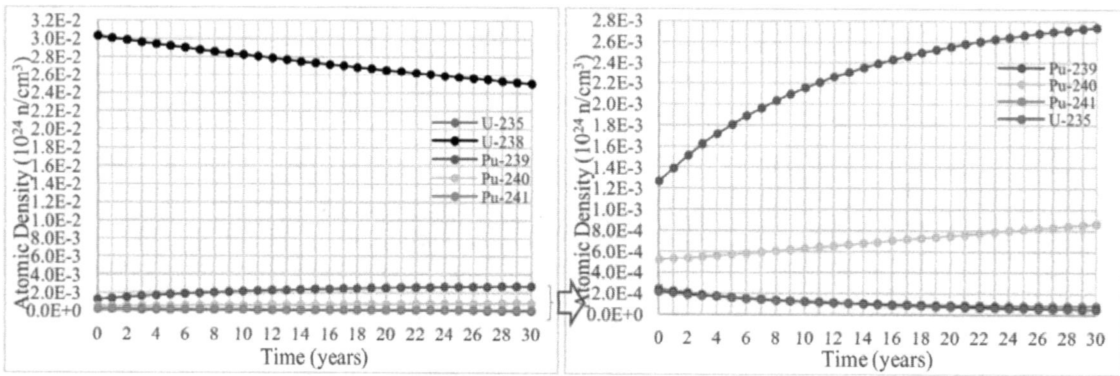

(a)

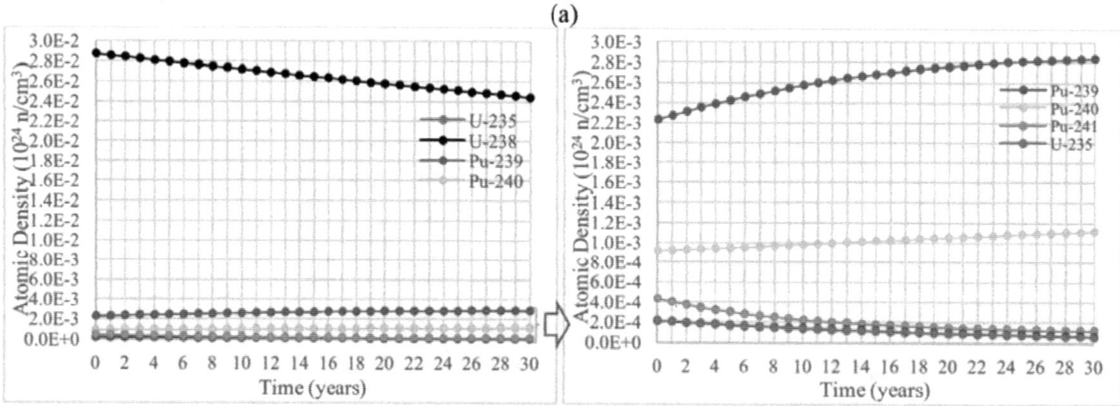

(b)

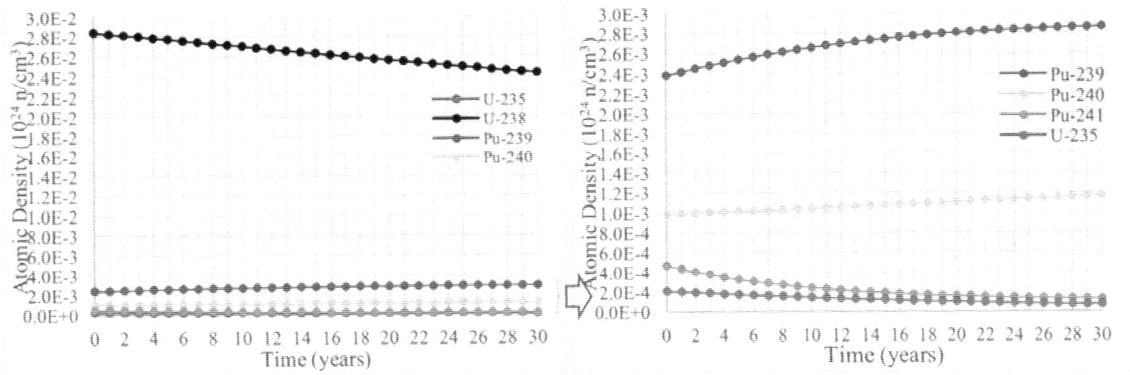

(c)

Figure 5. Atomic density change of reactor I. a) Inner core. b) Mid core. c) Outer core

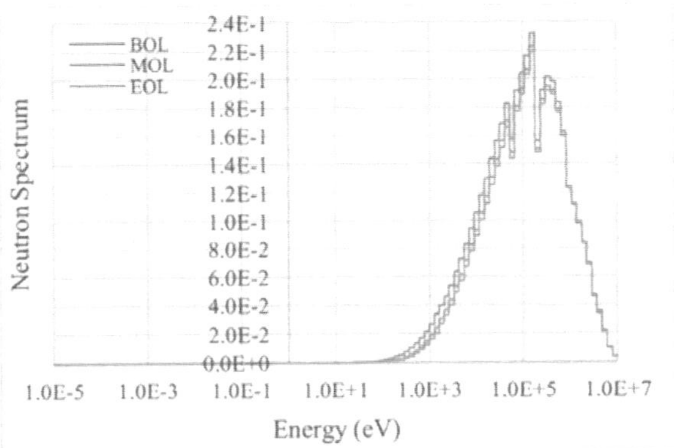

(a)

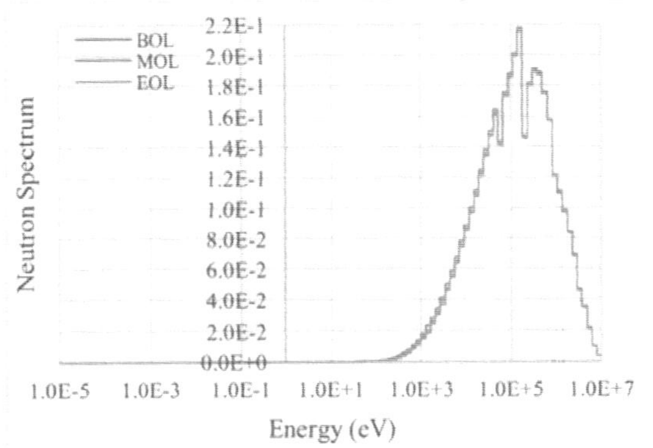

(b)

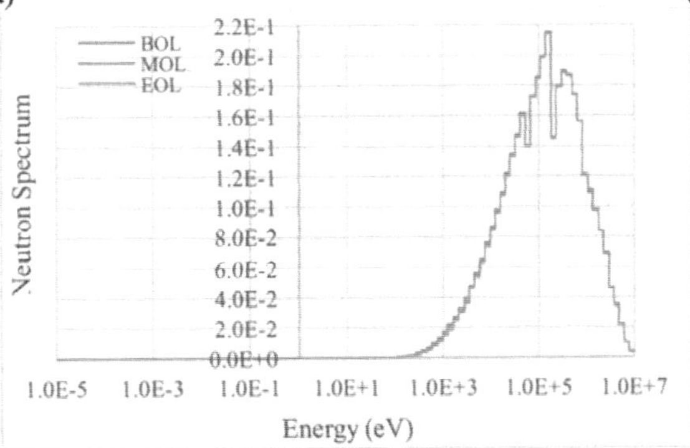

(c)

Figure 6. Neutron flux of reactor I. a) Inner core. b) Mid core. c) Outer core

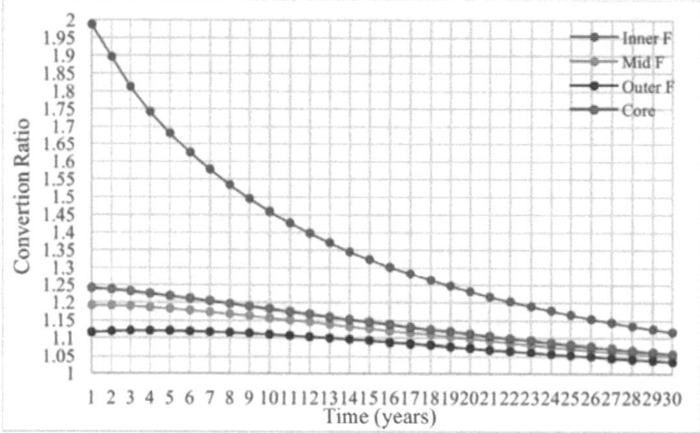

(a)

Figure 7. Conversion ratio of reactor I. a) All assembly type. b) Core

As shown in figure 5, the atomic density number in BOL is proportional to the percentage of plutonium who used in each region. Based on this figure, it is known that the atomic density number of uranium is highest in the inner core, and the atomic density number of plutonium is highest in the outer core where the atomic density change of every nuclide is dependent on breeding, and fission reaction rate in the core, and vice versa, the core fission and breeding reaction is dependent on atomic density number of every nuclide and neutron flux each time burnup.

Base on figure 5.a and 6.a, due to the high number of atomic density of ^{238}U and the high value of neutron flux especially from epithermal to the fast region, allow to happen high breeding at the start of the operation. The high breeding resulted in decreased atomic density number of ^{238}U at BOL to MOL is also impact on the increase of ^{239}Pu. From BOL to MOL the number density of ^{239}Pu not only increase due breeding of ^{238}U but also decrease due to fission reaction and neutron capture to become ^{240}Pu, but because the breeding rate is more dominant so that the amount of ^{239}Pu is significant to grow. The significant grow of ^{239}Pu is a direct impact on increasing the number density of ^{240}Pu due to capture and it is more dominant than the breeding of ^{240}Pu to ^{241}Pu. Different with ^{239}Pu and ^{241}Pu, atomic density number of ^{241}Pu is decrease due to fission rate is higher than breeding rate. The same thing also happens on ^{235}U who not have breeding.

From MOL to EOL, neutron flux and atomic number density do not change drastically, the significant changes occur only in a view years after MOL and after that, it became saturation condition. In this state, the atomic density number of ^{239}Pu has slightly increased due to an increase of fission reaction of ^{239}Pu as a consequence of the declining amount of fission reaction of ^{235}U and ^{241}Pu.

The change of atomic density of this region also seen in CR (figure 7), where have high value in the BOL but drastically change towards MOL and in EOL the value to 1.116. The change of CR is accordance with the rate of reduction atomic density of ^{238}U or the rate of increase atomic density ^{239}Pu which is the dominant nuclides of inner core.

Different with the inner core, in the BOL the atomic density of uranium in the mid-core is slightly little but the atomic density of plutonium is slightly larger (figure 5.b). The neutron flux of this region (figure 6.b) was much smaller than the previous region resulted in CR of this region is becomes smaller as shown in figure 7. This was resulted in the rate of atomic density change to be slower but still with the same principle as in the inner core, where the atomic density has significant change in BOL to MOL and change be slower from MOL to EOL. It is also seen in neutron flux, where the neutron flux from BOL to MOL have the little decrease and from MOL to EOL the reduction to be very slightly. The significant change of CR only happen in BOL and view years after MOL and after that CR be more stable until in EOL value to 1.049. Another thing that can be seen from this region is the nuclides number of ^{235}U is so small that the main source of fission is become from ^{239}Pu and ^{241}Pu, where the burner of ^{241}Pu in the operation started is dominant but drastically decrease after 11 years operating.

Different from the previous two regions, atomic density of uranium is far fewer lead to the breeding capability of the outer core to be much lower. On the other contrary, the fission reaction rate of this region in BOL is higher. The lower breeding capability of this region seen in slowly rate change of atomic density and neutron flux value which almost unchanged from BOL to EOL.

Overall it appears that each region requires a long time to buildup and only in five years before EOL all regions in equilibrium states. In addition, it is also seen that CR value of every region is more than one until the end of the operation so that allows the reactor to operate more than 30 years. It also looks at the result of the CR calculation of reactor I, where at EOL reactivity is still positive and CR value still greater than 1.05 as shown in figure 8 and 7.

From figure 8, the buildup process takes a long time resulting value of the reactivity swing reaches 0.25%, where maximum reactivity is 1.02% in the beginning of the operation and minimum reactivity occurs in year 7. This reactor needs 25 years to buildup cause high breeding process and low power of reactor.

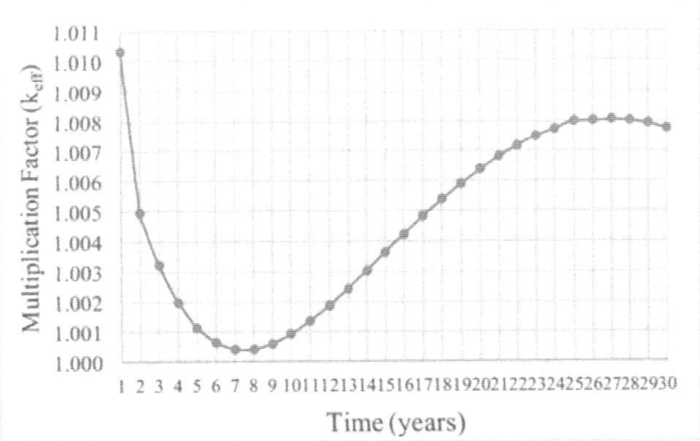

Figure 8. Multiplication factor of reactor I.

Another impact of the high breeding of reactor is a high transition of average power density distribution of reactor as shown in figure 9. In figure 9, average power density distribution is given in radial direction as shown at red line in figure 3. The minimum of average power density in the radial direction is at the center of the core and at a distance of 22 and 66 cm from the center which is the shutdown and control assemblies. The power density of this reactor is 62.61 MW/m^3.

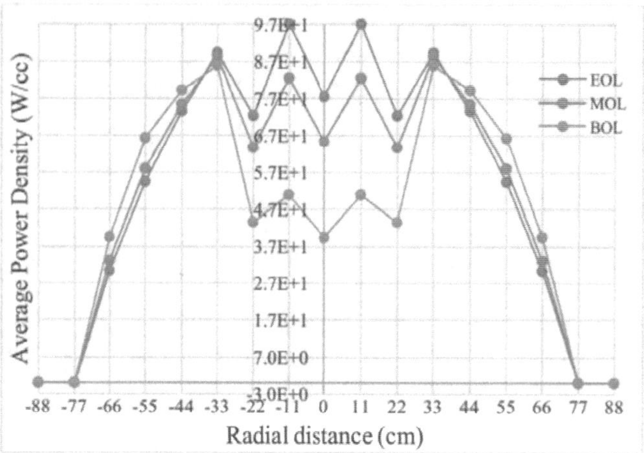

Figure 9. Average power density distribution of reactor I.

In this figure, in BOL the greatest power contributed by mid core. This is slightly different when compared with the analysis result of atomic density change in every core regions in the figure 5, but this is acceptable if the effect of neutron leakage at outer core is calculated.

From BOL to MOL the transition of average power density are significant in the inner core and outer core. In inner core, the number of fission reactions increased due to the growing of number fissile nuclides as a consequence of the high breeding of this region thus resulting in accumulation of power in the center of the core. In contrast to this region, the power in the outer core is decreasing because the number of fission decreased due to the reduce amount of fissile material as the slow process of breeding. In mid of core, there was a difference between the line 4 and line 5 assembly arrangement, where has a decrease of power at the outermost regions with the highest possibility of neutron leakage and otherwise has additional power on the inner side of mid-core were has a number of more neutrons. The transition of power distribution from MOL to EOL has the same principle like from BOL to MOL but with no significant changed. It can be explained by figures 5, 6 and 7, where the breeding rate in this time range is not as high as in BOL to MOL. Furthermore, can be calculated PPF of this reactor in BOL, MOL, and EOL are 1.09, 1.19, and 1.26. This result shows that even though the reactor can operate more than 30 years, but the PPF in the inner core will continue to increase do to the accumulation of excess power in this region so it is not allowed in the thermal hydraulic analysis.

Atomic density change, neutron flux and the conversion ratio of reactor II have the same principle like in reactor I. These parameters are given by figure 10, 11 and 12.

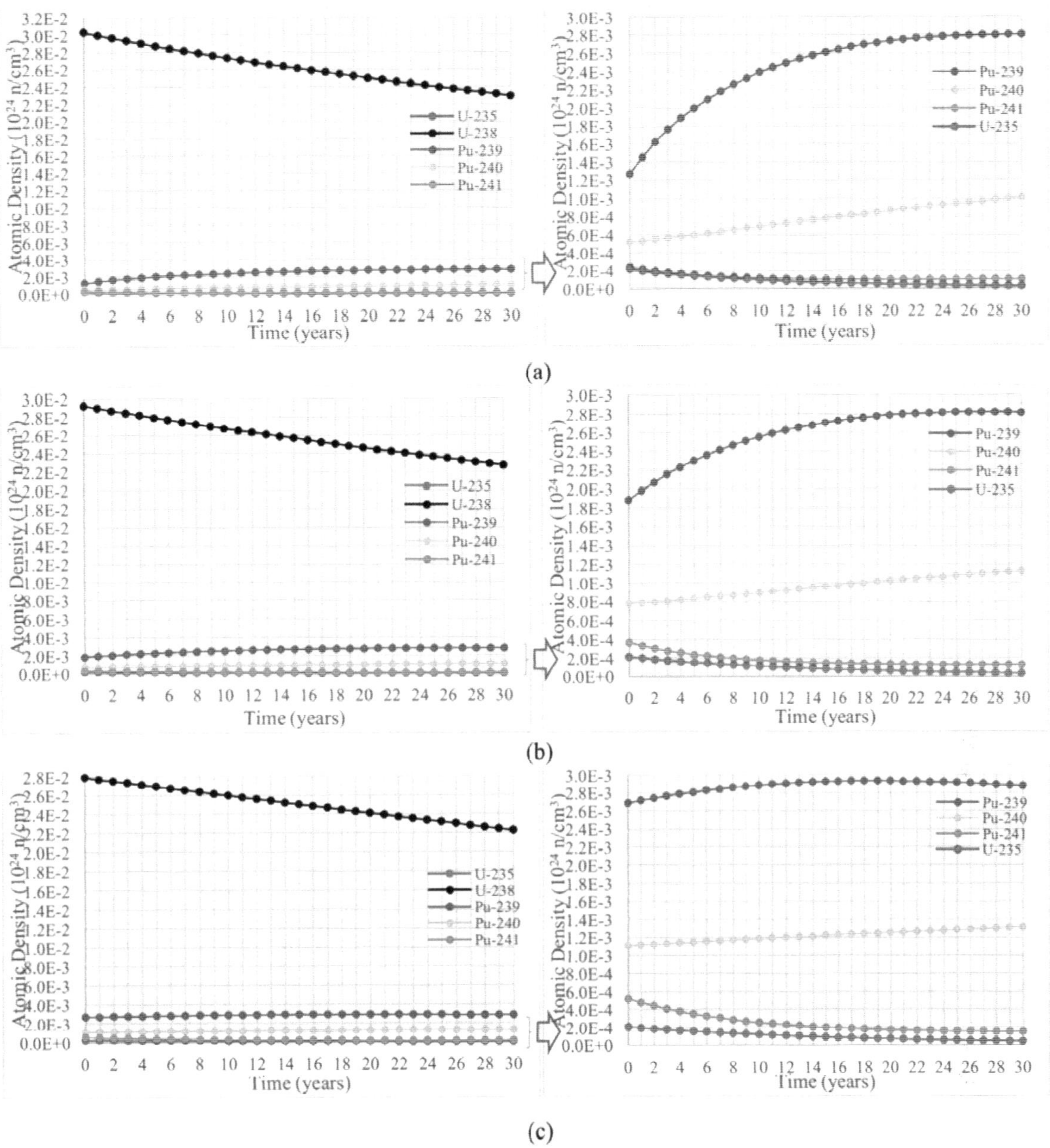

Figure 10. Atomic density change of reactor II. a) Inner core. b) Mid core. c) Outer core

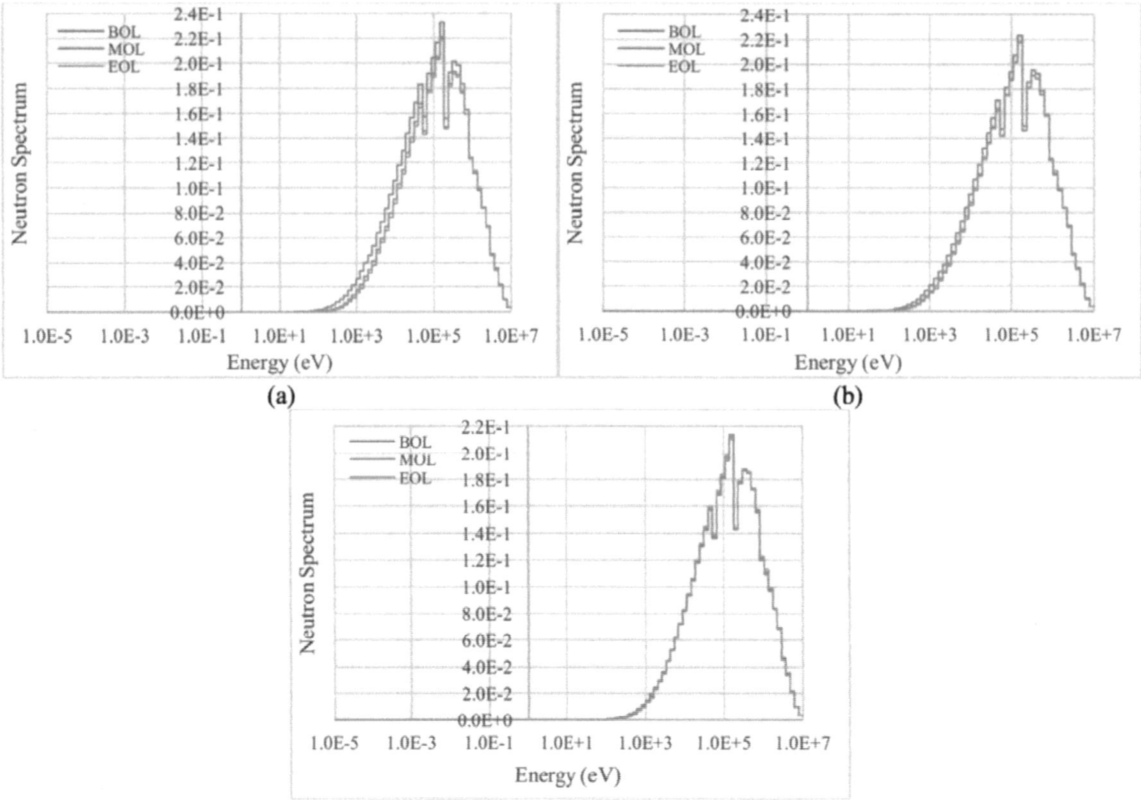

Figure 11. Neutron flux of reactor II. a) Inner core. b) Mid core. c) Outer core

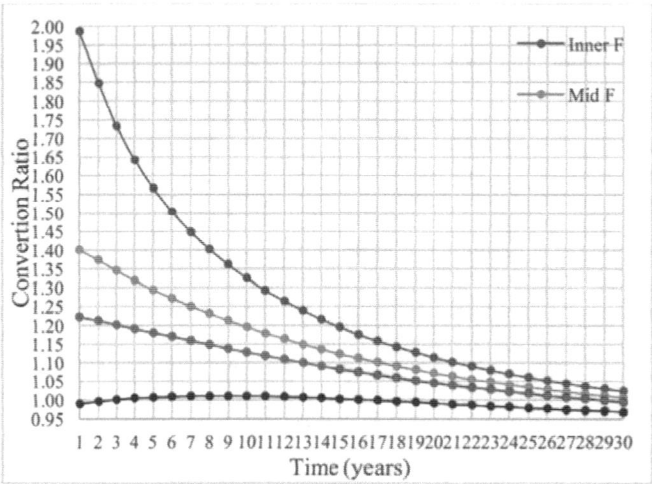

(a)

Figure 12. Conversion ratio of reactor II. a) All assembly type. b) Core

The inner core of reactor II has the same fuel percentage with the same region of reactor I so that the atomic density, neutron flux, and CR of both reactors are same, as a shown in figures 10.a, 11.a, and

12. When the reactor operates, the reduction rate of atomic density ^{238}U and the increase rate of atomic density ^{239}Pu of reactor II is higher than reactor I, also seen the neutron flux and CR value of this reactor are drastically decrease in the period from BOL to MOL. This is due to the fission reaction needs of reactor II is higher than reactor I, so that the number of neutrons generated per reactor period is also higher, but most of this neutrons are used for capture ^{238}U and for fission reactions ^{235}U, ^{239}Pu, and ^{241}Pu consequently the neutron flux of this region is decreased sharply in each period. The rapidly decreasing of the neutron flux and the atomic density ^{238}U result in a rapidly decreasing of CR value of the region.

From MOL to EOL, the breeding capability of this region is still decrease like shows at atomic density change of ^{239}Pu which significant only along 5 years after MOL, and thereafter the addition rate of this nuclide being very slow and finally reaches saturation state. The same thing happened to the neutron flux which not changes from MOL to EOL when comparable to the first core.

The situation of mid core is as same as like in inner core of this reactor, where atomic density and neutron flux is higher than the same region in core I, but due to greater power, the rate of change of neutron flux, atomic density, and also CR value of this region is higher than reactor I. The rate of change of three parameters is highest from BOL to MOL. From BOL to MOL, the shifting value of neutron flux in mid core of reactor II is greater than reactor I. From MOL to EOL, the neutron flux values do not to seem to shift. However, overall neutron flux values of this region from BOL to EOL is higher than in the same region of reactor I.

Atomic density change of ^{238}U from BOL to MOL is significant, in MOL an amount of Atomic density ^{238}U is less than in mid core of reactor I. In EOL this number is much less than in reactor I. Instead, the atomic density growth of ^{239}Pu of this region is very vast, even the saturated of this region is 5 years faster than saturated of the inner core of the same reactor.

The impact of the change of neutron flux and atomic density ^{238}U to CR value of this region is shown in figure 12, where the most drastic change of CR occurs in interval BOL to MOL. In MOL the value of CR has been lower than the CR of the same region in reactor I in the same time, and in EOL this value becomes much lower.

Much different from two pervious regions, the outer region of reactor II has higher percentage of plutonium than another region so that the breeding rate of this region is very small and only occurred in some years in the beginning of operation. The low breeding rate plus the higher fission rate of this region resulting ^{239}Pu growth is slow. Only in some years, this region has reached saturation, and even the amount of atomic density number of ^{239}Pu has decreased at the end of the operation. This corresponds to a neutron flux which is almost no change over time burnup, and the value of CR is less than 1 after MOL.

The combination of these three regions in reactor II resulted in the buildup process is shorter compared to the first reactor where the buildup process in this reactor occurred for 17 years as shown in figure 13. From this figure also seen that the reactivity in BOL is lower than reactor I, where minimum reactivity occurs in the third year and maximum reactivity occurs in the 18th year at 1.03%. This reactor only able to operate for twenty-nine years with the reactivity swing of this reactor is 0.16%. The negative reactivity value at year 30 in figure 13 is proportional to the CR value at the same time as shown in figure 12.

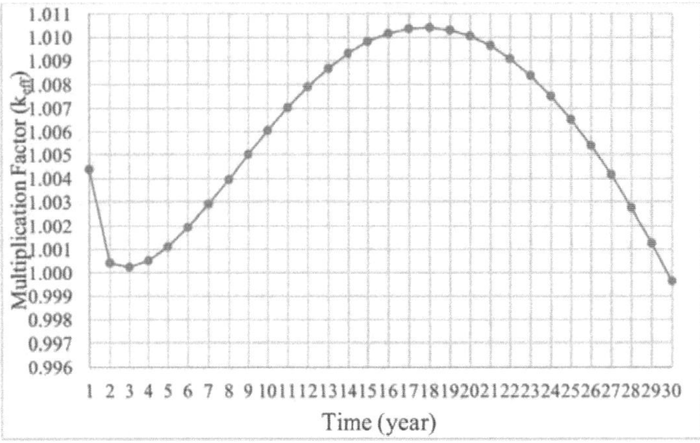

Figure 13. Multiplication factor of reactor II.

The high of the breeding rate in the beginning of the operation and high fission reaction along the process direct impact on the power distribution of reactor. The power density of this reactor is 93.92 MW/m^3, it is bigger than the previous reactor. The power distribution in BOL, MOL, and EOL is given by figure 14.

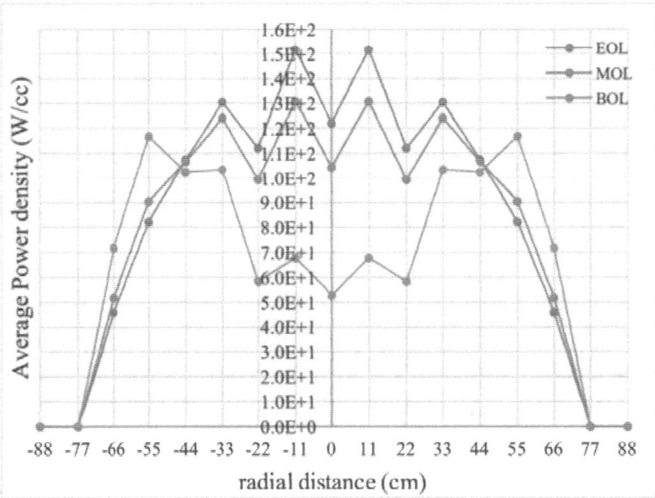

Figure 14. Average power density distribution of reactor II.

As shown in figure 14, the peak power of this reactor at BOL occurred in the outer core. This is caused by the percentage of plutonium in inner and mid-core is very low if compared with at outer core, so that although the neutron leakage is maximum in the outer core region, this region still become the main fission reaction source when the reactor in operating.

Furthermore, the higher breeding rate in the beginning of the operation, especially in the inner and mid-core resulted in an accumulation of fissile material like ^{239}Pu and ^{241}Pu in these regions. Otherwise, in the outer core, the amount of fissile material continues to decrease due high fission process and low breeding rate in this region. Consequently, when MOL power in the inner and mid-core rose sharply, while the outer core is significantly reduced power. In this time, the peak power occurred in the second line of the inner region.

From MOL to EOL, the impact of low breeding rate in each region of reactor II looks at the power distribution in the EOL which had a slight increase of distribution power in MOL. The PPF of the reactor in BOL, MOL, and EOL is given by 1.16, 1.18, and 1.30.

5. Summary and Conclusions

The high breeding velocity of fast reactor impacted on the high value of reactivity swing and significant change of power density distribution especially for small power reactor who can operate more than 20 years. However, by adjusting fuel configuration in a good composition this effect can be reduced as was done in this study. As a comparison, optimization is performed on the reactor power 100 MWth (reactor I) and 150 MWth (reactor II), where maximum fuel configuration of any reactor in inner, mid and outer core are given by 6.6%, 11.6%, and 12.4% plutonium for reactor I, and 6.6%, 9.8%, 14.0% for reactor II. Based on the optimization result, it is known that the both of reactor have high breeding rates in the BOL, but due to the greater power, the buildup process in the second reactor takes place shorter. A shorter of buildup process on reactor II resulted in the decrease rate of CR value of this reactor being sharper than reactor I, where in the thirty years the CR value of this reactor had been less than 1, while the CR value of reactor I remained higher than 1. A CR value of more than 1 allows reactor I to operate more than thirty years, while the second reactor can only operate for twenty-nine years.

The shorter buildup process in reactor II also impacts the power distribution shift of reactor, where the shift of power distribution of reactor II from BOL to MOL is greater than reactor I, whereas the high rate of breeding in the MOL to EOL period in reactor I results in a power distribution shift over large compared to reactor II. Over all the PPF value of reactor II is higher than reactor I, but due to the reactor I can continue to operate, its PPF value can increase even bigger than reactor II so it is dangerous for the reactor.

Maximum reactivity of both reactors is still more than 1%. Maximum reactivity of reactor I occur on BOL of 1.02% so that reactivity swing of this reactor is greater than reactor II. The reactivity value of reactor I is 0.25%. The maximum reactivity of reactor II occurred in the year 18 is equal to 1.03% with reactivity swing of 0.16%.

References

[1] Rajan KS. Comparison of characteristics of fast and thermal reactors, Role of fast reactors in Indian Nuclear [Internet]. [cited 2017 Feb 28]. Available from:
http://www.nptel.in/course/103106101/Module - 12/Lecture - 2.pdf

[2] Processing of Used Nuclear Fuel – World Nuclear Association [Internet]. [2016 Nov; 2017 Feb 2]. Available from:
http://www.world-nuclear.org/information-library/nuclear-fuel-cycle/fuel-recycling/processing-of-used-nuclear-fuel.aspx

[3] Won JH, Cho NZ, Park HM, Jeong YH. Sodium-cooled fast reactor (SFR) fuel assembly design with graphite-moderating rods to reduce the sodium void reactivity coefficient. Nuclear Engineering and Design. 2014; 280:223-232.

[4] Rouault J, Wei TYC. The GEN IV Gas Cooled Fast Reactor: Status of Studies [Internet]. 2016 Feb 16 [cited 2016 Aug 8]. Available from :

http://www.oecd-nea.org/science/meetings/ARWIF2004/2.01.pdf

[5] Zajac R. Serpent Code Using in ALLEGRO Project [Internet]. 2014 Sep [cited 2016 Mar 23]. Available from :

http://www.montecarlo.vtt.fi/mtg/2014_Cambridge/Radoslav_Zajac.pdf

[6] Japan Atomic Energy Agency. JENDL [Internet]. [2017 Aug 1]. Available from:

http://wwwndc.jaea.go.jp/jendl/ /j40/j40..html

[7] Oka Y, Madarame H, Uesaka M, editors. An Advanced Course in Nuclear Engineering. Tokyo: Springer; 2014.

Flory-Huggins Based Model to Determine Thermodynamic Property of Polymeric Membrane Solution

P T P Aryanti[1], D Ariono[2], A N Hakim[2], and I G Wenten[2]

[1]Chemical Eng. Dept., Universitas Jenderal Achmad Yani, Jl. Terusan Jend. Sudirman, Cimahi, Indonesia
[2]Chemical Eng. Dept., Institut Teknologi Bandung, Jl. Ganesha 10, Bandung 40132, Indonesia

*Corresponding author e-mail: igw@che.itb.ac.id

Abstract. Flory-Huggins (F-H) model has been widely used to analyze the thermodynamic behavior of polymeric membrane solution during the formation of membrane structure in a phase inversion process. The F-H model involves concentration and binary interaction parameter of components in membrane solution. Those parameters are used to calculate the composition of membrane solution at equilibrium which is then plotted on ternary diagram known as a binodal curve. The binodal curve is used to determine the type of demixing of the polymeric membrane solution and to predict the structure of resulted membrane. Several modifications of F-H model have been conducted in relation to composition (ternary or quaternary) and temperature of the membrane solution. In this paper, the development of F-H model used in designing polymeric membrane by phase inversion method, particularly immersion precipitation method, will be reviewed. The structure and performance of membranes that are governed by the thermodynamic property of the membrane solution based on F-H model will be discussed.

Keywords: Flory-Huggins model, immersion precipitation, polymeric membrane, thermodynamic

1. Introduction

Phase inversion method is the most used technique in industrial membrane manufacturing to produce a polymeric asymmetric membrane, both flat-sheet and hollow fiber membrane type [1-6]. This method involves transformation from a homogenous polymeric membrane solution to the final membrane structure (solid) through a phase separation of polymer solution. There are two basic techniques in membrane preparation by phase inversion, i.e. dry and wet techniques [7-11]. In the dry phase inversion technique, solvent is evaporated from the membrane solution to the atmosphere and/or absorbing a vapor phase (non-solvent) from the air moisture. Meanwhile, in the wet technique, membrane structure is formed by immersing the membrane solution into a non-solvent-containing bath. This method is known as immersion precipitation method, which has been widely used in the manufacture of commercial membranes.

The phase separation behavior, also known as liquid-liquid (L-L) demixing, is strongly influenced by thermodynamic and kinetic property of the membrane system. A ternary diagram is commonly used as a tool to analyze the thermodynamic and kinetic behavior of the membrane system (Figure 1). The phase separation of the membrane solution is mentioned as binodal curve (or also called as cloud point curve) in the ternary diagram. When the binodal composition is reached, phase separation (L-L demixing) of the membrane solution begins to occur. Every composition at the boundary conditions

Content from this work may be used under the terms of the Creative Commons Attribution 3.0 licence. Any further distribution of this work must maintain attribution to the author(s) and the title of the work, journal citation and DOI.
Published under licence by IOP Publishing Ltd

(binodal curve) begins to separate into two liquid phases, i.e. rich polymer (point A) and lean polymer (point B), which differ in compositions but remains in thermodynamic equilibrium with each other [12]. A pair of equilibrium compositions in the phase diagram is connected by a tie line while area within the binodal curve is called as miscibility gap. The thermodynamic properties of membrane solution can be evaluated in terms of degree of shift in binodal curve and miscibility gap area in ternary diagram, which depends on several parameters such as composition of the membrane solution and interaction parameters of each component. Different composition in membrane solution results in different membrane structure and performances. The influence of those parameters on the membrane structure have been explained in some literature [13-15]. The solidification process of the membrane solution is presented by precipitation path (line from point C to D), which describes the composition changes in polymer solution with respect to the solvent–non-solvent exchange rate (or mass transfer) in the membrane solution. During the evaporation of solvent from the polymer solution, the polymer concentration increases and then reaches a very high packing density until all solvent is evaporated (point D). Overall, the resulted membrane morphology is strongly related to the initial composition of membrane solution, binodal curve position, and the precipitation path [16-19]. In this paper, the mathematical model of thermodynamics to predict binodal curve or phase separation behavior of the membrane solution will be discussed.

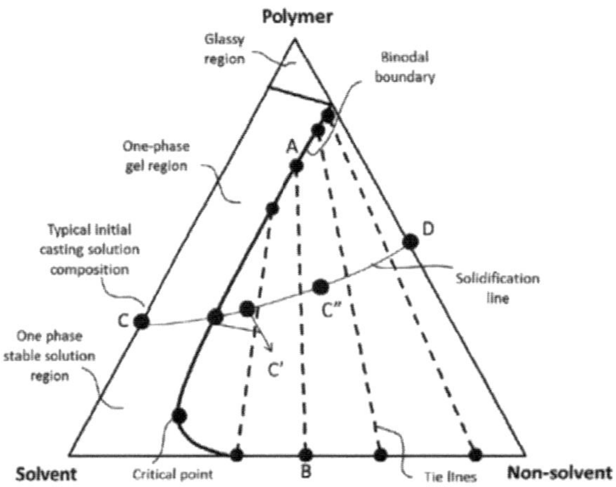

Figure 1. Ternary diagram of three component membrane system [12]

Several thermodynamic theories have been developed and used to analyze the membrane formation by phase inversion method (Table 1). Among of these theories, Flory-Huggins (F-H) theory has been widely used due to its relative simplicity, low number of input parameters, and provides a good prediction of the phase behavior of polymeric solutions [20]. Although the other models showed better thermodynamic analysis of the membrane solution compared to F-H model, the higher mathematical complexity and less predictive capability become a barrier to the completion of the model [21]. The F-H model is based on the Gibbs free energy (ΔG_m) of the membrane solution, which involves binary interaction parameters (χ_{ij}) and volume fraction (ϕ) of each component in membrane solution.

In general, ΔG_m in a ternary membrane system is formulated by the following equation based on theory developed by Tompa [22]:

$$\frac{\Delta G}{RT} = \left(\sum \frac{\phi_i}{m_i} \ln \phi_i + \sum \chi_{ij} \phi_i \phi_j \right) \sum m_i n_i \tag{1}$$

where m_i are molar volume ratio reflecting size different of components ($m = V_i/V_1$), n_i is mole fraction, ϕ_i is volume fraction, and χ_{ij} is binary interaction parameters of each component in membrane solution. R and T are the gas constant and temperature. In F-H theory, the binary interaction parameter (χ_{ij}) is an excess free energy parameter involving all non-ideality in the membrane system, which contains an enthalpic as well as an entropic contribution [14]. In the original F-H theory, the χ_{ij} is assumed to be concentration independent, which means that the value does not depend on the number of components in the system. However, some experimental studies have shown that the said assumptions cannot be used in certain cases.

Table 1. Thermodynamic models to analyze the membrane formation by phase inversion method

Year	Researchers	Model	Computation Method	Ref.
1982	Altena F.W. dan Smolders C. A.	Flory-Huggins	Least square procedure	[23]
1987	Shuguang L., et al.	Flory-Huggins	Least square procedure	[24]
1995	Termonia Y.	Difusi Monte Carlo	-	[25]
1997	J. Young K., et al.	Flory-Huggins	Newton Raphson	[26]
2005	Akthakul A., et al.	Kisi Boltzmann (BK)	-	[27]
2006	Yong M. W., et al.	Flory-Huggins	-	[28]
2007	Barzin J., et al.	Flory-Huggins	Newton Raphson	[29]

When two components are mixed, the free enthalpy of mixing is determined by chemical potential (μ) of both components. The chemical potential of a component is defined by the following equation:

$$\Delta\mu_{m,i} = (\partial\Delta G_m / \partial n_i)_{P,T,nj,nk...} \tag{2}$$

where μ_i is chemical potential of component containing n_i moles, while the pressure (P), temperature (T), and the number of moles of other components (n_j, n_k, and so on) are held constant. The conditions in liquid-liquid equilibrium (or composition on the binodal curve) are:

$$\Delta\mu_{i,A} = \Delta\mu_{i,B} \qquad i = 1, 2, 3 \tag{3}$$

Subscripts *A* and *B* refer to the rich-polymer and lean-polymer. The following sections will discuss the development of F-H models for ternary and quaternary systems, along with the determination of binary interactions in membrane systems.

2. Flory Huggins Theory for Ternary Membrane System

Thermodynamic analysis of ternary membrane systems, consisting of polymer (P)/solvent (S)/non-solvent (NS), have been studied with some restrictive assumption to reduce the complexity of phase diagram calculations [22]. In ternary membrane system, there are three interaction parameters, i.e χ_{13} (non-solvent/polymer), χ_{23} (solvent/polymer), and χ_{12} (solvent/non-solvent), which in the early application the F-H model are considered constant (concentration independent). Suh et al. [30] assumed that the polymer concentration in lean phase polymer is near zero or negligible. This assumption is used only if the equilibrium exists between a mixed solvent and a polymer or if the polymer confined to one phase. In other literature, interaction parameter of non-solvent and polymer is assumed as concentration independent (constant).

Further development of F-H model for ternary membrane system is proposed by Altena and Smolder [23]. Two membrane forming systems are used, i.e cellulose acetate/solvent/water and polysulfone/solvent/water. They suggested that the change of solvent and non-solvent concentration in membrane solution gave a major effect on the formation of membrane structure, thus the interaction parameter between the two components is concentration dependent (denoted as g_{12}). Meanwhile, the other two interactions are considered constant over the concentration changes. They found that the use of the dependent concentration (g_{12}) showed a better agreement between experimental cloud point curves and theoretical binodals. Based on Eq. (1), the Gibbs energy of ternary membrane system is given by the following relation:

$$\frac{\Delta G_m}{RT} = n_1 \ln\phi_1 + n_2 \ln\phi_2 + n_3 \ln\phi_3 + g_{12}(u_2)n_1\phi_2 + \chi_{13}n_1\phi_3 + \chi_{23}n_2\phi_3 \tag{4}$$

The chemical potential equations derived from the above equation are:

$$\Delta\mu_1 = RT\left[\ln\phi_1 + 1 - \phi_1 - s\phi_2 - r\phi_3 + (g_{12}\phi_2 + \chi_{13}\phi_3)(\phi_2 + \phi_3) - s\chi_{23}\phi_2\phi_3 - \phi_2 u_1 u_2 \frac{dg_{12}}{du_2}\right] \tag{5}$$

$$\Delta\mu_2 = \frac{RT}{s}\left[s\ln\phi_2 + s - \phi_1 - s\phi_2 - r\phi_3 + (g_{12}\phi_1 + s\chi_{23}\phi_3)(\phi_1 + \phi_3) - \chi_{13}\phi_1\phi_3 - \phi_1 u_1 u_2 \frac{dg_{12}}{du_2}\right] \tag{6}$$

$$\Delta\mu_3 = \frac{RT}{r}\left[r\ln\phi_3 + r - \phi_1 - s\phi_2 - r\phi_3 + (\chi_{13}\phi_1 + s\chi_{23}\phi_2)(\phi_1 + \phi_2) - \chi_{12}\phi_1\phi_2\right] \tag{7}$$

where r and s are molar volume ratio V_1/V_3 and V_1/V_2, while $u_1 = \phi_1/(\phi_1+\phi_2)$ and $u_2 = \phi_2/(\phi_1+\phi_2)$. The subscripts (1,2, and 3) refer to non-solvent (1), solvent (2), and polymer (3).

To determine the composition of each component in the equilibrium condition, 6 (six) parameters ($\phi_{1,A}$, $\phi_{2,A}$, $\phi_{3,A}$, $\phi_{1,B}$, $\phi_{2,B}$, and $\phi_{3,B}$) should be known. Therefore, some assumptions are proposed to simplify the completion of the F-H equation. Two mass balance equations ($\sum\phi_i = 1$) is used for the rich and lean phase of membrane solution. By choosing one (1) variable, the mass balance equations can be substituted into equations (5) to (7). The procedure can also be simplified by assuming that the polymer concentration in lean-phase ($\phi_{3,B}$) is near to zero (10^{-30} up to 10^{-90}) or negligible [23, 24, 31]. The binodal curve is calculated numerically by minimizing the following functions:

$$F = \sum_{i=1}^{4}\left[\frac{\Delta\mu_{i,A}}{RT} - \frac{\Delta\mu_{i,B}}{RT}\right]^2 \tag{8}$$

The calculated binodal curves using the above calculation method compared to the cloud point measurement are shown in Figure 2 [23].

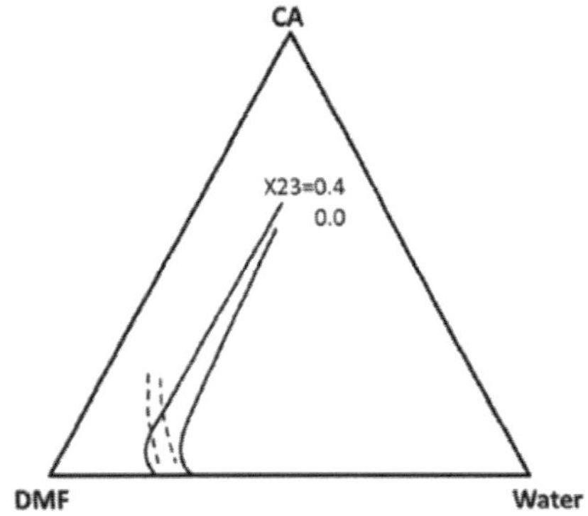

(A)

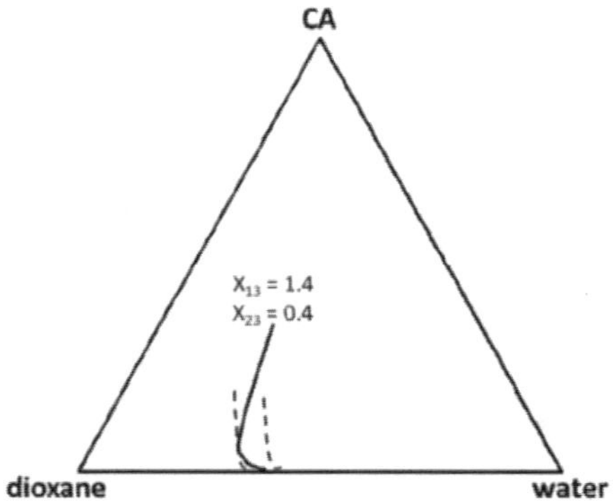

(B)

Figure 2. Calculated binodal (full curve) compared to cloud point measurements (dashed line) [23]

Yilmaz et al. [31] studied the effect of variants of interaction parameters on phase diagram characteristics as well as their relevance to membrane formation. They found that F-H theory can be empirically extended using concentration-dependent interaction parameters (g_{ij}), particularly for g_{12} and g_{23}. The ΔG_m of mixing becomes:

$$\frac{\Delta G_m}{RT} = n_1 \ln\phi_1 + n_2 \ln\phi_2 + n_3 \ln\phi_3 + g_{12}(u_2)n_1\phi_2 + g_{13}(\phi_3)n_1\phi_3 + g_{23}(\phi_3)n_2\phi_3 \quad (9)$$

The following penalty function is used to compute the binodal composition, which is similar to the Eq. (8).

$$F_i = \frac{(\Delta\mu_{i,A} - \Delta\mu_{i,B})/RT}{(\phi_{i,A} - \phi_{i,B})^r} \quad (10)$$

with r = 1 or 2. The $\phi_{3,B}$ is assumed as 10^{-90} and selected as the independent variable.

3. Binary Interaction Parameter in Flory-Huggins Model

Binary interaction (χ_{ij}) parameters play an important role in the F-H theory, which indicates the solubility of each component in the membrane solution and affects the binodal curve position in the ternary diagram (Figure 2) [32]. It shows that an increase in χ_{13}, which indicates a low interaction between non-solvent and polymer, will shift the binodal curve to the left corner of the ternary diagram. It means that the less non-solvent concentration required for the phase separation of the membrane system. The same tendency is obtained when the parameter interaction value of χ_{23} is enhanced. Membrane structures that are formed tend to be finger-like. Conversely, the increase of χ_{12} value will move the binodal curve to the right side of the ternary diagram, which tends to form sponge-like membrane structure. Therefore, the calculation and evaluation of the correct χ_{ij} of components are necessary to study the thermodynamic property of membrane solution to predict the resulted membrane structure. In F-H model, the binary interaction parameter can be calculated from activity data or vapor-liquid equilibrium data in some literature. The challenges in determining χ_{ij} are the limited data of the components used in membrane solution.

Besides the type of components in the membrane system, the value of interaction parameters depends on temperature, concentration, and molecular weight of the components, especially the molecular weight of polymer. Altena et al. [23] found that the demixing behavior of the water/dimethyl formamide (DMF)/polysulfone (PSf)/ system is highly affected by interaction parameter value of non-solvent/polymer (χ_{13}) and non-solvent/solvent (χ_{12}). Meanwhile, interaction parameter of solvent/polymer (χ_{23}) gave a small effect to the demixing behavior. The X_{ij} values of different membrane systems have been reported in the literature [20, 33-36]

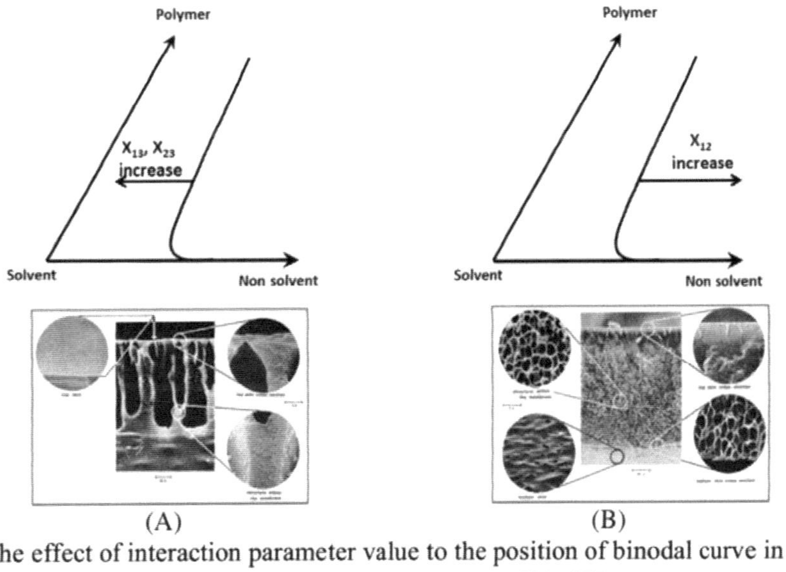

Figure 3. The effect of interaction parameter value to the position of binodal curve in a ternary diagram: (A) χ_{13} / χ_{23} and (B) χ_{12} [31, 37]

3.1. Interaction parameter non-solvent-polymer (χ_{13})

Interaction parameter χ_{13} can be measured using equilibrium swelling experiment at different temperature and high concentration of polymer [38]. When a non-solvent is contacted with a polymer, the non-solvent will diffuse into the polymer matrix until the equilibrium is reached. Simultaneously, the polymer matrix will be swelled. When the interaction between polymer and penetrant is relatively low, the χ_{13} can be calculated by Eq.(11).

$$\chi = -\frac{\ln(1-\phi_3)+\phi_3}{\phi_3^2} \tag{11}$$

where ϕ_3 is volume fraction of polymer.

Li et al [39] calculated the χ_{13} of three membrane forming systems based on g_{12} and intrinsic viscosity of polymer in membrane solution, as shown in Table 1 (No. 2). They found that the intrinsic viscosity measurement can be applied in various solvent systems without limitation of the refractive index of the solvents. The calculated χ_{13} was nearly the same as that obtained from the light scattering measurements. The proposed models to calculate χ_{13} are shown in Table 2.

Table 2. Some equations to calculate interaction parameter of non-solvent/polymer (χ_{13})

No	Equations	Remarks	Reference
1	$\chi = -\frac{\ln(1-\phi_3)+\phi_3}{\phi_3^2}$	ϕ_3 is volume fraction of polymer	[14]
2	$Y = -\chi_{13} + (1-2a_\chi)g_{12}\phi_2$	χ_{13} is determined as intercept of Y against $g_{12}\phi_2$. $(1-2a_\chi)$is a constant related to intrinsic viscosity of polymer and interaction parameter g_{12}.	[39]
3	$\ln a_{1,g} = \ln(\phi_1) + \left(1-\frac{v_1}{v_3}\right)\phi_3 + \chi_{13}\phi_3^2$	Using F-H equation in Eq. (4), by specifying n_2, $\phi_2 = 0$. $\ln a_{1,g}$ is vapor phase activity of the non-solvent.	[34]
4	$\chi = \frac{\Delta\mu_1}{RTv^2} - \frac{\left[\ln(1-v)+v\left(1-V_1^o/V^o\right)\right]}{v^2}$	T is temperature, V_1^o/V^o is the ration molar volumes for liquid and polymer, while v is volume fraction.	[40]
5	$g_{13}(\phi_3) = a + b\phi_3$	System H_2O/PES 49w (T=32°C): a = 2.00 and b = 0.10, System H_2O/PSf 39w (T=32°C): a = 2.30 and b = 0, ϕ_3 is volume fraction of polymer	[41]

Note: PES = polyethersulfone; PSf = polysulfone

3.2. Interaction parameter non-solvent-sovent (χ_{12})

The concentration dependent interaction paremeter between non-solvent and solvent, which noted as g_{12}, is generally determined from excess Gibbs free energy data, as shown in the following equation [14]:

$$g_{12} = \frac{1}{x_1\phi_2}\left[x_1 \ln\frac{x_1}{\phi_1} + x_2 \ln\frac{x_2}{\phi_2} + \frac{G^E}{RT}\right] \tag{12}$$

G^E can be determined from activity coefficient data or experimental data, which are available in the literature. Meanwhile, several researchers suggested the following polynomial equation to calculate g_{12} based on the equilibrium data of non-solvent/solvent:

$$g_{12} = a_o + a_1\varphi_2 + a_2\varphi_2^2 + ...\tag{13}$$

Different g_{12} values have been reported with different values of a in the literature [23, 24, 29]. The differences resulted from different sources of equilibrium data. A change in solvent and non-solvent type will change the g_{12} value. Meanwhile, [41] calculated g_{12} by the following equation, based on experimental data obtained by means of Headspace-Gas Chromatography (HSGC):

$$g_{12}(\phi_2) = \alpha + \frac{\beta}{1-\gamma\phi_2}\tag{14}$$

For binary system of H_2O/DMF at 32°C, the value of α is 0.3223, while β and γ are 0.0751 and 0.9604, respectively.

3.3. *Interaction parameter sovent-polymer (χ_{23})*

Experimentally, the χ_{23} value can be measured by light scattering, vapor pressure depression, and membrane osmometry [14, 35]. In membrane osmometry technique, the dilute solvent in polymer solution moves to the concentrated phase (pure solvent) through a tight membrane. Osmotic pressure (π) of the pure solvent will increase, and then used to calculate χ_{23} by the following equation:

$$\pi = \frac{RT}{V_2}\phi_2 + \frac{RT}{V_1}(0.5-\chi)\phi_2^2\tag{15}$$

Other method is proposed by Karimi et al. [42]. They measured χ_{23} of various polymer solutions based on solvent activity that measured by vapor pressure osmometry (VPO) data.

$$\frac{\ln(a_2/\phi_2)-\phi_3}{\phi_3} = \chi_{23}\phi_3\tag{16}$$

a_2 is measured solvent activity from VPO data. The interaction parameter χ_{23} is obtained as slope of a fitted straight line if (ln (a_2/ϕ_2) - ϕ_3) data are plotted versus ϕ_3. Besides of VPO method, the activity values of the solvents can be determined by other methods that have been reported in the literature [43, 44].

The χ_{ij} value can also be determined based on solubility data of each component in membrane solution. Shuguang et al. [24] combined Hildebrand-Scatchard theory with F-H theory to calculate the χ_{23} of water/ dimethylacetamide (DMAc)/polysulfone (PSf) system, as the following equation:

$$\chi_{23} = (\delta_2 - \delta_2)^2 V_1 / RT\tag{17}$$

where V_1 is molar volume of the solvent, while δ_1 and δ_2 are the solubility parameters of solvent and polymer ($cal^{1/2}.cm^{-3/2}$). The calculated χ_{23} by the above equation had a different value with those obtained from gas-liquid chromatography measurement. Therefore, they suggested that the χ_{13} and χ_{23} should be concentration dependent. Table 3 shows several models to calculate the χ_{13}.

4. Development of Flory-Huggins Theory for Quaternary Membrane Solution

The F-H model for a quaternary system has been developed by Boom et al [45]. They modified F-H model, as shown in Eq. (1), for a membrane system consisting of non-solvent (1), solvent (2), membrane forming polymer (3), and second polymer or additive (4). The modified F-H model gave the Gibbs energy of mixing as a function of concentration-dependent interaction parameters and then used to calculate the binodal composition, as follows:

$$\begin{aligned}\Delta G/RT &= n_1\ln\phi_1 + n_2\ln\phi_2 + n_3\ln\phi_3 + n_4\ln\phi_4 + n_1\phi_2 g_{12}(u_2) + n_1\phi_3\chi_{13} + n_1\phi_4 g_{14}(u_4)\\ &\quad + n_2\phi_3 g_{23}(v_2) + n_2\phi_4 g_{24}(w_2) + n_3\phi_4 g_{34}(v_3)\end{aligned}\tag{18}$$

where n_i and ϕ_i are mol and volume fraction, respectively. χ_{13} is considered as independent-concentration interaction parameter due to low affinity between the weak affinity between the two binary components. Meanwhile, the other interaction parameters are dependent on the following variables:

- g_{12} is dependent on u_2, where $u_2 = \phi_2/(\phi_1+\phi_2)$.
- g_{23} is dependent on v_2, where $v_2 = \phi_2/(\phi_2+\phi_3)$.
- g_{14} is dependent on u_4, where $u_4 = \phi_4/(\phi_1+\phi_4)$.
- g_{24} is dependent on w_2, where $w_2 = \phi_2/(\phi_2+\phi_4)$.
- g_{34} is dependent on v_3, where $v_3 = \phi_3/(\phi_3+\phi_4)$.

Table 3. Equations to calculate interaction parameter of non-solvent/polymer (χ_{13})

No	Equations	Remarks	References
1	$\frac{\ln(a_2/\phi_2)-\phi_3}{\phi_3} = \chi_{23}\phi_3$	a_2 = solvent activity, ϕ_2 = volume fraction of solvent, ϕ_3 = volume fraction of polymer	[29]
2	$\chi_{23} = (\delta_1 - \delta_2)^2 V_1 / RT$	V_1 is molar volume of the solvent, while δ_1 and δ_2 are the solubility parameters of solvent and polymer	[24]
3	$A_2 = \frac{\left(1/2 - \chi_{23}\right)}{\rho_2^2 V_1}$	ρ_2 = density of polymer (g/cm^3), V_1 = molar volume of solvent (cm^3/mol). A_2 is a second virial coefficient, which is a function of intrinsic viscosity and molecular weight of polymer.	[40]
3	$\chi_{23} = \chi_1 + \chi_2 v + \chi_3 v^2 + \ldots$	Coefficient $\chi_1, \chi_2, \chi_3, \ldots$, are function of molecular characteristics and temperature. v is polymer volume fraction at equilibrium (maximum) degree of swelling	[46, 47]
4	$g_{23}(\phi_3) = a - 3b - 5c - b(2\phi_3 - 1) + c(2\phi_3 - 1)^2$	Membran system (T=32°C): - *H_2O/DMF/PES 49w* *a = 0.3719; b = 0.0793; c = -0.0236* *H_2O/DMF/PSf 39w* *a = 0.4685; b = 0.1472; c = -0.0728* ϕ_3 is volume fraction of polymer	[41]

Note: DMF = dimethyl formamide

The chemical potentials of the quaternary system ($\Delta\mu_i$) are obtained by deriving the Gibbs energy in Eq. (18) with respect to the number of moles of each component:

$$\frac{\Delta\mu_1}{RT} = \ln\phi_1 - s\phi_2 - r\phi_3 - t\phi_4 + (1 + g_{12}\phi_2 + \chi_{13}\phi_3 + g_{14}\phi_4)(1-\phi_1) - sg_{23}\phi_2\phi_3 - rg_{34}\phi_3\phi_4 - sg_{24}\phi_2\phi_4 \quad (19)$$

$$s\frac{\Delta\mu_2}{RT} = s\ln\phi_2 - \phi_1 - r\phi_3 - t\phi_4 + (s + g_{12}\phi_1 + sg_{23}\phi_3 + sg_{24}\phi_4)(1-\phi_2) - \chi_{13}\phi_1\phi_3 - g_{14}\phi_1\phi_4 - rg_{34}\phi_3\phi_4 - \phi_1 u_2(1-u_2)\left(\frac{dg_{12}}{u_2}\right) - s\phi_3 v_2\left(\frac{dg_{23}}{dv_2}\right) - s\phi_4 w_2(1-w_2)\left(\frac{dg_{24}}{dw_2}\right) \quad (20)$$

$$r\frac{\Delta\mu_3}{RT} = r\ln\phi_3 - \phi_1 - s\phi_2 - t\phi_4 + (r + \chi_{13}\phi_1 + s\chi_{23}\phi_2 + rg_{34}\phi_4)(1-\phi_3) - g_{12}\phi_1\phi_2 - g_{14}\phi_1\phi_4 - sg_{24}\phi_2\phi_4 - s\phi_2 v_2(1-v_2)\left(\frac{dg_{23}}{dv_2}\right) - r\phi_4\phi_3(1-v_3)\left(\frac{dg_{34}}{v_3}\right) \quad (21)$$

$$t\frac{\Delta\mu_4}{\mathrm{RT}} = t\ln\phi_4 - \phi_1 - s\phi_2 - r\phi_3 + (t + g_{14}\phi_1 + sg_{24}\phi_2 + rg_{34}\phi_3)(1-\phi_4) - g_{12}\phi_1\phi_2 - \chi_{13}\phi_1\phi_3 - sg_{23}\phi_2\phi_3 + \phi_1 u_4(1-u_4)\left(\frac{dg_{14}}{u_4}\right) - s\phi_2 w_2(1-w_2)\left(\frac{dg_{24}}{dw_2}\right) - r\phi_3 v_3(1-v_3)\left(\frac{dg_{34}}{dv_3}\right) \quad (22)$$

The Eq. (19) to (22) is solved using an objective function as described in Eq. (8) through Newton-type algorithm.

Thermodynamic properties of a quaternary system, consisting of water/N-Methylpyrrolidone (NMP)/ polyethersulfone (PES)/Poly(vinylpyrrolidinone) (PVP), has been studied by Boom et al. [48]. They studied the role of polymeric additive on the binodal curve position in pseudo-ternary diagram, as shown in Figure 4. They found that the presence of additive decreased the size of demixing gap, which means that the solution more compatible with the non-solvent and led to delay demixing. At high concentration of additive with no movement between the two polymers, the non-solvent concentration can reach 50% by volume to disturb the equilibrium of membrane solution. In this condition, smaller membrane pore size is formed.

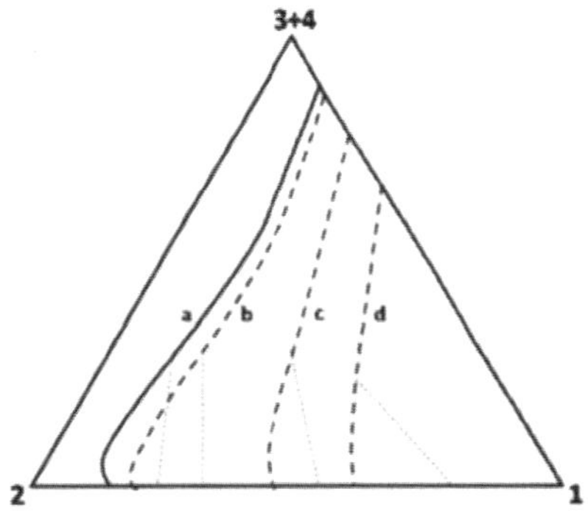

Figure 4. The influence of additive concentration on binodal curve position in pseudo-ternary diagram: (a) $\phi_3/\phi_4 = 0$, (b) $\phi_3/\phi_4 = 0{,}1$, (c) $\phi_3/\phi_4 = 0{,}5$, (d) $\phi_3/\phi_4 = 1{,}0$.Composition: (1) Water, (2) NMP, (3) PES, and (4) PVP [49]

It has been reported that the binodal curve is not affected by the molecular weight of polymer as long as the polymer is a macromolecule in nature. The polymer-solvent interactions, g_{23} and g_{24}, are not important as long as the miscibility of both polymers occurs [45]. When using two (2) polymers in the membrane systems, polymer-polymer interaction (χ_{34}) can be calculated using Eq. (17). Emerson et al. [33] determined χ_{34} for poly(3-hexylthiophene) (P_3HT) and polystyrene (PS). A correction factor based on solvent vapor swelling data is used, thus the Eq. (17) becomes:

$$\chi_{34} = \frac{V_3}{RT}(\delta_3 - \delta_4)^2 + 0.34$$

Ariono et al. [12] and Aryanti et al. [50] found a different result. They added PEG400 instead of PVP as an additive into membrane solution (water/DMAc/PSf/PEG400) and then studied the thermodynamic behavior of the membrane system. The χ_{ij} of the components is calculated based on the cloud point test data. They found that the increase of PEG400 concentration shifted binodal composition to the left corner of the ternary diagram. Instantaneously demixing is occurred, which form a longer finger-like pore in the final membrane structure (Figure 5). It means that the binodal curve is not only influenced by the concentration of the additive but also affected by the type of additive. It suggested that PVP has higher molecular weight than PEG400, which inhibited the membrane pore growth [51, 52]. Therefore, smaller pore size is produced in the final structure. During the thermodynamic analysis, an interaction between additive-solvent and additive-non-solvent should also be considered.

Up to this time, the development of F-H equation and model or method to calculate interaction parameters are continued to minimize poor agreement between theoretical and experimental result. Further development is needed, in which the current membrane modification already involves more than four compositions. Several factors should be considered before analyzing the thermodynamic property of membrane solutions, such as molecular weight and type of each component in membrane solution, temperature, concentration, and binary interaction between the components. Those parameters are used to evaluate the theoretical result, which may produce different characteristics of the membrane with the experimental results.

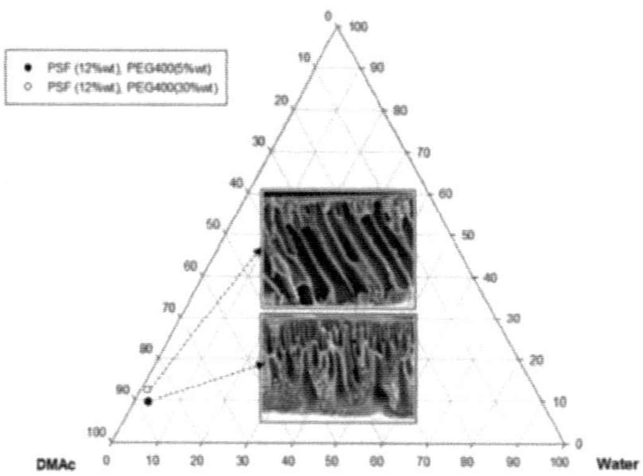

Figure 5. The influence of PEG400 concentration in polysulfone membrane solution to binodal composition

5. Conclusions

In this paper, the development of F-H model used in designing polymeric membrane by phase inversion method, particularly immersion precipitation method, has been reviewed. The F-H model is based on the Gibbs free energy of the membrane solution to determine the thermodynamic equilibrium of the membrane solution. This model involves binary interaction parameters (χ_{ij}) and volume fraction of each component in membrane solution. Those parameters are used to calculate the composition of membrane solution at equilibrium which is then plotted on ternary diagram known as a binodal curve. The binodal curve is used to determine the type of demixing of the polymeric membrane solution and to predict the structure of resulted membrane, whether finger-like or sponge-like.

Up to this time, several modifications of F-H model have been conducted in relation to composition (ternary or quaternary) and temperature of the membrane solution to minimize the poor agreement between theoretical and experimental data. Several parameters should be considered before analyzing the thermodynamic property of membrane solutions, such as molecular weight and type of each component in membrane solution, intrinsic viscosity of the solution, temperature, concentration, and binary interaction between the components. It is suggested that dependent-concentration interaction parameters result in better agreement with the experimental data.

References

[1] Khulbe KC, Matsuura T. Recent progress in polymeric hollow-fibre membrane preparation and applications. Membrane Technology. 2016;2016(7):7-13.

[2] Himma NF, Wardani AK, Wenten IG. Preparation of Superhydrophobic Polypropylene Membrane Using Dip-Coating Method: The effects of solution and process parameters. polymer-plastics technology and engineering. 2017;56(2):184-94.

[3] Khoiruddin, Aryanti PTP, Hakim AN, Wenten IG. The role of ion-exchange membrane in energy conversion. AIP Conference Proceeding. 2017;1840: 90006.

[4] Aryanti PTP, Sianipar M, Zunita M, Wenten IG. Modified membrane with antibacterial properties. Membrane Water Treatment. 2017;8(5):463-81.

[5] Himma NF, Prasetya N, Anisah S, Wenten IG. Superhydrophobic membrane: progress in preparation and its separation properties. Reviews in Chemical Engineering, 2018; DOI: https://doi.org/10.1515/revce-2017-0030

[6] Makertihartha IGBN, Khoiruddin, Hakim AN, Aryanti PTP. Wenten IG. Hydrogenation of Maltose in Catalytic Membrane Reactor for Maltitol Production. MATEC Web of Conferences. 2018;156:08008.

[7] Himma NF, Wardani AK, Wenten IG. The effects of non-solvent on surface morphology and hydrophobicity of dip-coated polypropylene membrane. Materials Research Express. 2017;4(5):054001.

[8] Ariono D, Khoiruddin, Subagjo S, Wenten IG. Heterogeneous structure and its effect on properties and electrochemical behavior of ion-exchange membrane. Materials Research Express. 2017;4(2):024006.

[9] Himma NF, Wenten I G, Surface engineering of polymer membrane for air separation. AIP Conference Proceedings. 2017;1840: 090005.

[10] Ariono D, Khoiruddin, Subagjo S, Wenten IG. The Influence of Polymerization Degree on Morphology and Electrochemical Properties of PVC-Based Heterogeneous Ion-Exchange Membrane. Advanced Science Letters. 2017;23(6):5762-4.

[11] Ariono D, Khoiruddin, Prabandari D, Wulandari R, Wenten IG. Preparation and characterization of polysulfone/PEG heterogeneous ion exchange membrane for reverse electrodialysis (RED). Journal of Physics: Conference Series. 2017;877:012075

[12] Ariono D, Aryanti PTP, Hakim AN, Subagjo S, Wenten I G. Determination of thermodynamic properties of polysulfone/PEG membrane solutions based on Flory-Huggins model. AIP Conference Proceedings. 2017;1840(1):090008.

[13] Baker RW. Membrane technology: Wiley Online Library; 2000.

[14] Mulder M. Basic principles of membrane technology: Springer Science & Business Media; 1996.

[15] Wenten I G, Dharmawijaya P, Aryanti PTP, Mukti R, Khoiruddin. LTA zeolite membranes: current progress and challenges in pervaporation. RSC Advances. 2017;7(47):29520-39.

[16] Sadrzadeh M, Bhattacharjee S. Rational design of phase inversion membranes by tailoring thermodynamics and kinetics of casting solution using polymer additives. J Membr Sci. 2013;441:31-44.

[17] Wenten IG, Khoiruddin. Recent developments in heterogeneous ion-exchange membrane: preparation, modification, characterization and performance evaluation. Journal of Engineering Science and Technology. 2016;11(7):916-34.

[18] Himma NF, Anisah S, Prasetya N, Wenten IG. Advances in preparation, modification, and application of polypropylene membrane. Journal of Polymer Engineering. 2016;36(4):329-62.

[19] Khoiruddin K, Wenten IG. Investigation of Electrochemical and Morphological Properties of Mixed Matrix Polysulfone-Silica Anion Exchange Membrane. Journal of Engineering and Technological Sciences. 2016;48(1):1-11.

[20] Kools WFC. Membrane formation by phase inversion in multicomponent polymer systems. Mechanisms and morphologies: Universiteit Twente; 1998.

[21] Maghsoud Z, Famili MHN, Madaeni SS. Phase diagram calculations of water/tetrahydrofuran/poly (vinyl chloride) ternary system based on a compressible regular solution model. Iran Polym J. 2010;19:581-8.
[22] Tompa H. Polymer solutions: Butterworths Scientific Publications; 1956.
[23] Altena FW, Smolders CA. Calculation of liquid-liquid phase separation in a ternary system of a polymer in a mixture of a solvent and a nonsolvent. Macromolecules. 1982;15(6):1491-7.
[24] Shuguang L, Chengzhang J, Yuanqi Z. The investigation of solution thermodynamics for the polysufone—DMAC—water system. Desalination. 1987;62:79-88.
[25] Termonia Y. Fundamentals of polymer coagulation. Journal of Polymer Science Part B: Polymer Physics. 1995;33(2):279-88.
[26] Kim JY, Lee HK, Baik KJ, Kim SC. Liquid-liquid phase separation in polysulfone/solvent/water systems. Journal of applied polymer science. 1997;65(13):2643-53.
[27] Akthakul A, Scott CE, Mayes AM, Wagner AJ. Lattice Boltzmann simulation of asymmetric membrane formation by immersion precipitation. Journal of membrane science. 2005;249(1):213-26.
[28] Wei Y-M, Xu Z-L, Yang X-T, Liu H-L. Mathematical calculation of binodal curves of a polymer/solvent/nonsolvent system in the phase inversion process. Desalination. 2006;192(1):91-104.
[29] Barzin J, Sadatnia B. Theoretical phase diagram calculation and membrane morphology evaluation for water/solvent/polyethersulfone systems. Polymer. 2007;48(6):1620-31.
[30] Suh KW, Liou DW. Phase equilibria in polymer–liquid–liquid systems. Journal of Polymer Science Part A-2: Polymer Physics. 1968;6(5):813-23.
[31] Yilmaz L, McHugh A. Analysis of nonsolvent–solvent–polymer phase diagrams and their relevance to membrane formation modeling. Journal of applied polymer science. 1986;31(4):997-1018.
[32] Mulder M, Hendrikman JO, Wijmans J, Smolders C. A rationale for the preparation of asymmetric pervaporation membranes. Journal of applied polymer science. 1985;30(7):2805-20.
[33] Emerson JA, Toolan DT, Howse JR, Furst EM, Epps III TH. Determination of solvent–polymer and polymer–polymer Flory–Huggins interaction parameters for poly (3-hexylthiophene) via solvent vapor swelling. Macromolecules. 2013;46(16):6533-40.
[34] Karimi M, Albrecht W, Heuchel M, Kish M, Frahn J, Weigel T, et al. Determination of water/polymer interaction parameter for membrane-forming systems by sorption measurement and a fitting technique. Journal of membrane science. 2005;265(1):1-12.
[35] Xu L, Qiu F. Simultaneous determination of three Flory–Huggins interaction parameters in polymer/solvent/nonsolvent systems by viscosity and cloud point measurements. Polymer. 2014;55(26):6795-802.
[36] Etxabarren C, Iriarte M, Uriarte C, Etxeberrıa A, Iruin J. Polymer–solvent interaction parameters in polymer solutions at high polymer concentrations. Journal of Chromatography A. 2002;969(1):245-54.
[37] Strathmann H, Kock K, Amar P, Baker R. The formation mechanism of asymmetric membranes. Desalination. 1975;16(2):179-203.
[38] Zeman L, Tkacik G. Thermodynamic analysis of a membrane-forming system water/N-methyl-2-pyrrolidone/polyethersulfone. Journal of Membrane Science. 1988;36:119-40.
[39] Li Z, Jiang C. Determination of the nonsolvent–polymer interaction parameter χ 13 in the casting solutions. Journal of Membrane Science. 2000;174(1):87-96.
[40] Kok CM, Rudin A. Prediction of Flory–Huggins interaction parameters from intrinsic viscosities. Journal of Applied Polymer Science. 1982;27(2):353-62.
[41] Barth C, Wolf BA. Quick and reliable routes to phase diagrams for polyethersulfone and polysulfone membrane formation. Macromolecular Chemistry and Physics. 2000;201(3):365-74.

[42] Karimi M, Albrecht W, Heuchel M, Weigel T, Lendlein A. Determination of solvent/polymer interaction parameters of moderately concentrated polymer solutions by vapor pressure osmometry. Polymer. 2008;49(10):2587-94.
[43] Bonner DC. Vapor-liquid equilibria in concentrated polymer solutions. Journal of Macromolecular Science—Reviews in Macromolecular Chemistry. 1975;13(2):263-319.
[44] Durant YG, Sundberg DC, Guillot J. A comparison of methods for the estimation of polymer/monomer interaction parameters: The polystyrene/n- butyl methacrylate system methacrylate system. Journal of applied polymer science. 1994;52(12):1823-32.
[45] Boom R, Van den Boomgaard T, Smolders C. Equilibrium thermodynamics of a quaternary membrane-forming system with two polymers. 1. Calculations. Macromolecules. 1994;27(8):2034-40.
[46] Bahar I, Erbil HY, Baysal BM, Erman B. Determination of polymer-solvent interaction parameter from swelling of networks: the system poly (2-hydroxyethyl methacrylate)-diethylene glycol. Macromolecules. 1987;20(6):1353-6.
[47] Marzocca A. Evaluation of the polymer–solvent interaction parameter χ for the system cured styrene butadiene rubber and toluene. European polymer journal. 2007;43(6):2682-9.
[48] Boom RM. Membrane formation by immersion precipitation: the role of a polymeric additive: University of Twente; 1992.
[49] Boom RM, Wienk IM, van den Boomgaard T, Smolders CA. Microstructures in phase inversion membranes. Part 2. The role of a polymeric additive. Journal of Membrane Science. 1992;73(2):277-92.
[50] Aryanti PTP, Joscarita SR, Wardani AK, Subagjo S, Ariono D, Wenten IG. The Influence of PEG400 and Acetone on Polysulfone Membrane Morphology and Fouling Behaviour. Journal of Engineering and Technological Sciences. 2016;48(2):135-49.
[51] Chakrabarty B, Ghoshal A, Purkait M. Preparation, characterization and performance studies of polysulfone membranes using PVP as an additive. Journal of Membrane Science. 2008;315(1):36-47.
[52] Aryanti PTP, Yustiana R, Purnama R. Wenten I G. Performance and characterization of PEG400 modified PVC ultrafiltration membrane. Membrane Water Treatment. 2015;6(5):379-392.

Tidal Current Circulation in Westhern Bali Sea Using a 2-D Hydrodynamic Model

D A Widiawan[1] and S Nurdjaman[1,2,3]

[1]Department of Oceanography, Faculty of Earth Sciences and Technology, Bandung Institute of Technology, Indonesia
[2]Department of Civil, Faculty of Civil and Environmental Engineering, Bandung Institute of Technology, Indonesia
[3]Kyushu University, Japan

E-mail : destika1808@gmail.com and susanna@fitb.itb.ac.id

Abstract. The circulation of tidal currents in the northern part of the Bali Sea was created using two main steps, namely a large model simulation and a small model simulation. The first step, a large model simulation detailing the circulation of tidal currents in the Bali Strait, uses a 2D barotropic hydrodynamic model that utilizes tidal-generating forces and Gebco bathymetry data with a 30-second resolution. In this large model simulation, tidal heights are procured which are then used as data for the small model. The second step, a small model circulation which focuses on the northern part of the Bali Sea, uses a 2D barotropic hydrodynamic model and utilizes tidal-generating forces from the large model and BIG (Geospatial Information Agency) bathymetry data with a 50-meter resolution. This smaller resolution is used to observe the effects of the cape in the northern Bali Sea. The tidal simulation is done for 29 days (May 2016) with inputted data in the form of bathymetry and tidal data at four points with open boundary conditions. The results of the tidal simulation indicate that tidal currents follow the phase of the tides. During high tides, the tidal currents move North and during low tides, the currents move South. Meanwhile, for the area around the Gondol Cape in Northern Bali, an eastward-westward current is observed.

Keywords: Tidal Currents, 2-D Hydrodynamic Model, Western Bali Sea.

1. Introduction

Ocean currents are an important hydrodynamic parameter to observe when defining the characteristics of a body of water. Ocean currents are the movement of seawater mass, also called mass movements, in either the vertical or horizontal direction [1]. This movement of seawater is caused by various generative forces. Ocean currents are the movement of seawater mass from one place to another, which means a change in position. Ocean current movements happen throughout the world and the movements influence each other, either vertically or horizontally [2].

Everything connected to maritime processes and events is not exempt from the role of ocean currents, including marine resource exploration in the industries of fishing, minerals, hydrography and ocean mapping, meteorology and climate, and the connection between the ocean and coastal building construction [3].

Content from this work may be used under the terms of the Creative Commons Attribution 3.0 licence. Any further distribution of this work must maintain attribution to the author(s) and the title of the work, journal citation and DOI.
Published under licence by IOP Publishing Ltd

The island of Bali is known as a good tourist location in Indonesia. Besides that, the western area of the Bali Sea is also used for aquacultures such as the cultivation of pearls and tuna. There is also dolphin conservation areas. The western part of the Bali Sea is a mixture of waters from the Java Sea, Flores Sea, and Makassar Strait. The western part of the Bali Sea is very productive but its characteristics are still unknown, including the characteristics of its tidal currents. Tidal currents have a very close relationship with the tides in these waters [4]. The tides in the Badung Strait in Bali have the dominant harmonic components of M2 (Lunar Semidiurnal) and S2 (Principal Solar Semidiurnal) [5]. The pattern of the tidal currents is vital to figuring out the characteristics of the western part of the Bali Sea. Because of that, the writer of this paper observed the patterns of the tidal currents in the western part of the Bali Sea using a 2D hydrodynamic model.

2. Methodology

The tidal current simulation was done for May 2016. The generative force used is tidal forcing obtained from Tide Model Driver (TMD). The two-dimensional hydrodynamic model used is the Princeton Ocean Modelling (POM). This simulation uses a barotropic model, that does not account for the difference in pressure for the vertical components. The observation area does not only encompass the Bali Strait, but also the western part of the Bali Sea which has never been specifically simulated for its tidal currents. Previous studies have been done, but only in the Bali Strait and not including the western part of the Bali Sea.

Two different bathymetry data are used. The bathymetry for the large-scale model is obtained from Gebco with a resolution of 30 seconds or around 930 meters. This bathymetry was taken from 114.3295 ° E – 115.1942 ° E and 7.7561 ° S – 8.6936° S. It was then converted into a grid with dimensions of 930 meters x 930 meters. The number of grids for this bathymetry is 104 x 113 grids. The maximum depth is 1108 meters. The bathymetry in the western part of the Bali Sea indicates that the area near the shore and near the Bali Strait has a depth of around 100 – 200 meters. These waters are rather shallow. Meanwhile, the eastern side of this large-scale model has a reasonably deep depth. The bathymetry for the small model was obtained from the Geospatial Information Agency which was then resampled. This bathymetry was taken from 114.35 ° E – 114.55 ° E and 8.10 ° S – 8.40° S. It was then converted into a grid with dimensions of 50 meters x 50 meters. The number of grids for this bathymetry is 26 x 31 grids. The maximum depth of this bathymetry is 55 meters at a basin near the Gondol Cape.

3. Results and Discussion

The large-scale model is a simulation with a large observed area that covers the Bali Strait. In this model, open boundaries are used at the north, south, and west sides. The results of the large model show that the maximum elevation is 1.2 meters and the lowest is 0.9 meters. These values are taken from the spring high tide and spring low tide.

The tidal elevation for the large-scale model was taken from TMD for May 1st 2016 to May 31st 2016. In May of 2016, field data for oceanographic parameters were taken at Gondol, Bali. The field data used to verify the simulation in this study are current data and tidal data. The tidal data were obtained from field observations for 15 days, from May 21st 2016 to June 4th 2016 at the Grand Center for Aquaculture Research and Development (BBPPBL), Gondol, West Bali Sea. Stationary current data was obtained from an ADCP at coordinates 8° 9' 10.05'' S and 114° 43' 7.11'' E. Current data were taken at depths of 1.6 meters, 2.6 meters, and 3.6 meters from the sea surface. Aside from stationary current data, scattered current data was also taken at coordinates 114.71° E – 114.73° E dan 8.14° S – 8.16° S.

The tidal current circulation can be measured from the phase of the spring and neap tides. During the spring high tide, the elevation in the western part of the Bali Sea ranges from 0.8 to 1.2 meters, while the tidal current reaches its maximum during the spring high tide with a speed of 2.6 meters/second. The current circulation shows that the dominant direction for the current is southward in the Bali Strait and westward in the western part of the Bali Sea. During the transition from the spring high tide to the spring low tide, the elevation in the western part of the Bali Sea ranges from -0.8 to 0.1 meters while the current has a maximum speed of 1.3 meters/second. The current circulation shows that the dominant direction is westward in the Bali Strait and westward in the western part of the Bali Sea as well. During the spring low tide, the elevation in the western part of the Bali Sea ranges from -0.9 to -0.7 meters, while the maximum current speed is 2.1 meters/second.

The current circulation shows that the dominant direction in the Bali Strait is northward. During the transition from the spring low tide to the spring high tide, the elevation in the western part of the Bali Strait ranges from -0.1 to 0.3 meters, while the tidal current has a maximum speed of 2.0 meters/second. The current circulation shows that the dominant direction in the Bali Strait is northward. During the transition from the neap high tide to the neap low tide, the elevation in the western part of the Bali Sea ranges from 0.1 to 0.3 meters while the current has a maximum speed of 2.2 meters/second. The current circulation shows that the dominant direction is westward in the Bali Strait and westward in the western part of the Bali Sea as well. During the neap low tide, the elevation in the western part of the Bali Sea ranges from -0.4 to -0.1 meters, while the maximum current speed is 1.0 meters/second. The current circulation shows that the dominant direction in the Bali Strait is northward. During the transition from the low tide to the neap high tide, the elevation in the western part of the Bali Strait ranges from -0.7 to -0.1 meters, while the tidal current has a maximum speed of 1.0 meters/second.

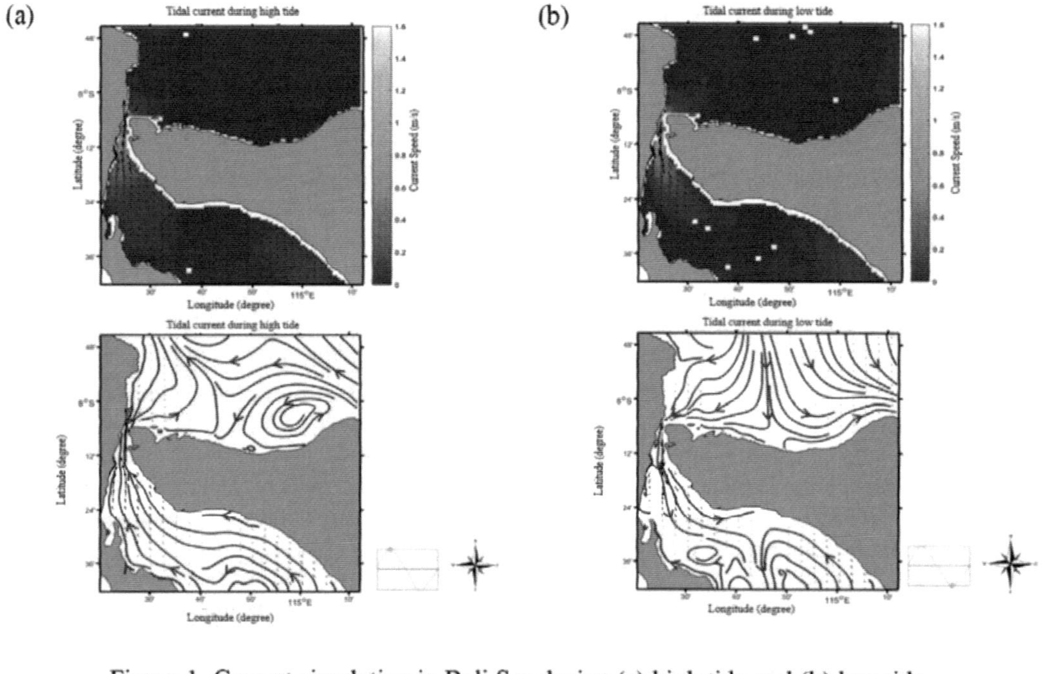

Figure 1. Current circulation in Bali Sea during (a) high tide and (b) low tide.

The current circulation shows that the dominant direction in the Bali Strait is northward. The small-scale model is a simulation with a specific observed area to get a more detailed area of the western part of the Bali Sea near Gondol. The tidal elevation data is obtained from the large-scale model. The model uses open boundaries at the north, east, and west sides.The results of the small model show that the maximum elevation is 0.302 meters and the lowest elevation is -0.855 meters. These values are taken from the spring high tide and spring low tide. The tidal current circulation can be measured from the phase of the spring and neap tides.

During the spring high tide, the elevation in the western part of the Bali Sea is 0.166 meters, while the tidal current reaches its maximum during the spring high tide with a speed of 0.03 meters/second. The current circulation shows that the dominant direction for the current is eastward in the western part of the Bali Sea. During the transition from the spring high tide to the spring low tide, the elevation in the western part of the Bali Sea is 0.044 meters while the current has a maximum speed of 0.02 meters/second. The current circulation shows that the dominant direction is westward in the western part of the Bali Sea as well.

(a) (b)

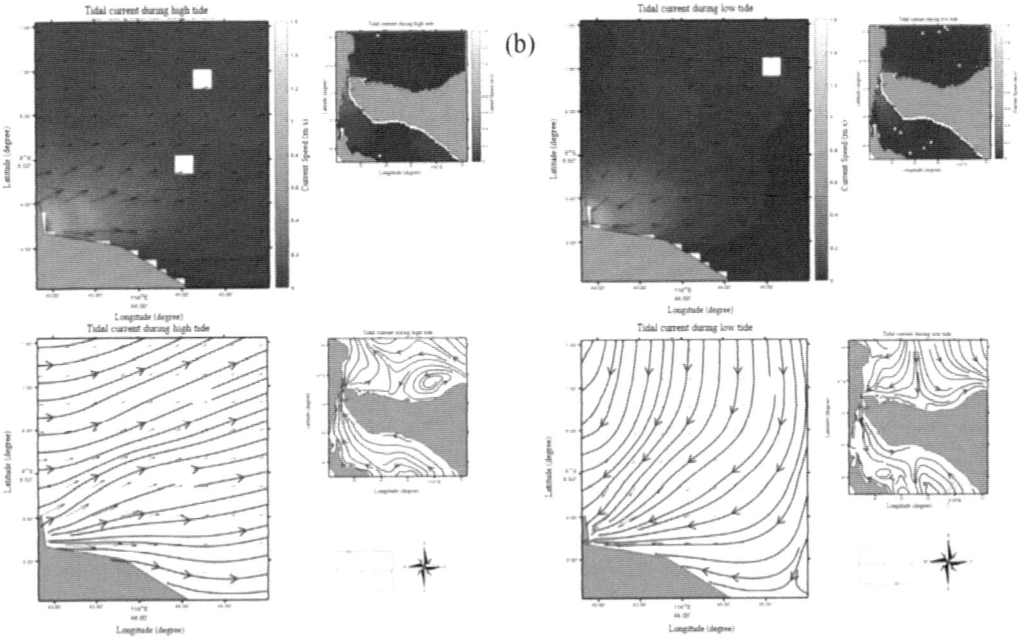

Figure 2. Current circulation in the western part of Bali during (a) high tide and (b) low tide refer to current circulation in Bali Sea until Bali Strait.

During the spring low tide, the elevation in the western part of the Bali Sea is -0.855 meters, while the maximum current speed is 0.02 meters/second. The current circulation shows that the dominant direction in the western part of the Bali Sea is northward. During the transition from the spring low tide to the spring high tide, the elevation in the western part of the Bali Strait is -0.038 meters, while the tidal current has a maximum speed of 0.02 meters/second. The current circulation shows that the dominant direction in the western part of the Bali Strait is westward on the right of the cape and eastward on the left of the cape. During the neap high tide, the elevation in the western part of the Bali Sea is -0.301 meters, while the maximum current speed is 0.02 meters. During the transition from the neap high tide to the neap low tide, the elevation in the western part of the Bali Sea is around 0.124 meters while the current has a maximum speed of 0.02 meters/second. During the neap low tide, the elevation in the western part of the Bali Sea is around -0.303 meters, while the maximum current speed is 0.02 meters/second. During the transition from the low tide to the neap high tide, the elevation in the western part of the Bali Strait is around -0.088 meters, while the tidal current has a maximum speed of 0.02 meters/second.

IOP Conf. Series: Journal of Physics: Conf. Series **1090** (2018) 012075 doi:10.1088/1742-6596/1090/1/012075

4. Conclusion

To conclude, the type of tide in the western part of the Bali Sea is a mixed predominantly semidiurnal type with a Formzahl number of 1.108 from the K1, O1, M2, and S2 components. Verification of the elevation shows a root mean square error for the large model and small model of 0.16 and 0.10, respectively (in meters). Verification of the tidal currents for the small model in the east-west direction and in the north-south direction shows a root mean square error of 0.026 and 0.022, respectively (in meters/second). The maximum current speed occurs during the spring high tide with a speed of 0.03 meters/second and the minimum current speed occurs during the neap low tide with a speed of 0.02 meters/second. The current in the western part of the Bali Sea is predominantly eastward during the high tide and westward during the low tide.

References

[1] S Hadi and I M Radjawane 2009 *Ocean Current*

[2] M F Azis. 2006 *Ocean Water Motion J.Oseana*

[3] S Rizal, Ichsan S., Muhammad, Taufiq I., and Mulyadi A W 2009 *Simulation of Baroclinic Flow Pattern in East Indonesia Waters with Three-Dimensional Numerical Model*

[4] W P Putri 2005 *Study of Tidal Dynamics in Bali Strait*

[5] T A Theoyana, Widodo S P, Anastasia R.T.D.K, and Purwanto 2015 *Characteristics of Tidal Flow in Badung Strait Bali*

Optimizing and Implementation Contactless Tag-Reader System for Smart Classroom and Laboratory Access

Rida Ariyanti[1], Elyas Palantei[2], Intan Sari Areni[3]
1,2,3 Electrical Engineering Department, Hasanuddin University, Makassar, Indonesia

E-mail: ridaariyantizainal@gmail.com, elyas_palantei@unhas.ac.id , intan@unhas.ac.id

Abstract. Advances in wireless technology support smart card technology became contactless technology. Wireless technology allowed users to access or controlling. However, existing contactless smart cards have limited coverage with low security levels. Therefore, this research perform optimization on user access further with better security assurance. This research implemented in classroom and laboratory access. Hardware that used consists of Smart cards and Raspberry Pi that serves as a host and card reader that could be interconnected with the microprocessor. On the smart card stored user data (registration number and name) and on the Raspberry server stored user data that could access the class or laboratory. User's data on smart cards were stored on chips that act like RAM (Random Access Memory). The chip was embedded in the Smart card in the same function as storage media. Raspberry that had function as server and microprocessor that provide access to registered users. This study produced classroom access system by using smart card with distance > 5cm with automatic access system for user registered on system. It was according to this research proposed about security and historical data. Reader in this research use wireless transmission system. Historical data stored on the server (Raspberry pi) which records the time in and out, and the duration of time while in the room. Additional technology in the classroom access system is equipped with an automatic door close system for double security in the classroom. The advantages of this system was using server and microcontroller in one package so that in terms of cost is cheaper than the smart card access technology available today.

1. Introduction

Smart cards were cards with chips as data storage that can be used for network access systems, storing values and other data[1]. Smart card technology were supported by the advancement of wireless communication technology made smart cards became contactless to facilitated user in access or controlling. Open data transmission over a wireless system proccess have a security challenge, namely user authentication and privacy [2]. Previous research on contactless smart cards focused on file access for printing data and evaluation in tags [4], [5], which then developed file access on cloud storage systems [6] but there was no guarantee of user safety. Research then develope in 2016 to ensure the security of data with remote systems and used dynamic (changing) ids and remote servers verify the legitimacy of users through open and unsafe communication channels [8]. The research has focused on authentication but has not access the system by using microcontroller. Therefore, the

Content from this work may be used under the terms of the Creative Commons Attribution 3.0 licence. Any further distribution of this work must maintain attribution to the author(s) and the title of the work, journal citation and DOI.
Published under licence by IOP Publishing Ltd

authors do research that implemented in everyday life in the academic field, namely access classroom and laboratory. Classroom and laboratory access systems that use smart cards with optimizations at a range of objectives to enable users for access with an high-class security algorithm, which is an automatic closed door. The host computer (in research using Raspberry) and card reader can be interconnected with the microprocessor [3].

The structure of this paper is related research on part 2. The system model presented in the next part 3 of the results is discussed in part 4 and finally the conclusions in part 5.

2. Releated works

The researches and implementation of smart cards has grown from year to year. Previous research had been conducted using contactless and IC smart cards as a medium for identity authentication, analyzing conditions and ways for cloud print systems, and researching key printer driver technology, data encryption, identity authentication to print centers by Wu Yun [4].

Another reseach was focused on smart cards with contactless interfaces. Using Mifare DESFire able to found some weakness of smart card that can cause vulnerability if protocol were not implemented properly. This method can be used by researchers to evaluate the security of protocol implementation on some types of smart cards [5]. Another research about Unified RB-DAC method Approach With Secure Authentication Using Smart Card Architecture for file access on cloud computing using smart cards. RBAC was a system where all users could access the specified data for a particular need, but can not specify the data allocation for each user. The RABC system incorporates the ACL (Access Control List) & ACM (Access Control Matrix) concept of the DAC with RBAC to improve the integrity of data in RBAC. Thus to achieve this, ACLs were add for all objects (data) and ACM for all objects in the system [6]. Related research Authentication of remote users with mechanisms by which servers verify the legitimacy of users through open and unsecured communication channels. The Password authentication using smart cards has become one of the commonly adopted methods for password protection during transmission by Zhengxian Gao, et al. [7].

Based on the previous research, this research was an optimization. Previous research had evolved using smart cards with contactless reader, but with limited distance and low security. The Improvement in this research was using the server and mikrokontroller in one package. In the field of academics, classroom or laboratory security is very important, so the system was develope by adding applications for the history user who uses the room, hours of entry, hours out and duration of users are in the room.

3. System Model

3.1. System Design

Raspberry Pi is a micro module computer that has a digital output port input such as microcontroller board. Excess Rasberry Pi compared to other microcontroller board that has a port / connection for display in the form of TV or PC Monitor, USB connection for Keyboard and Mouse and has 26 pin I / O digital [8].

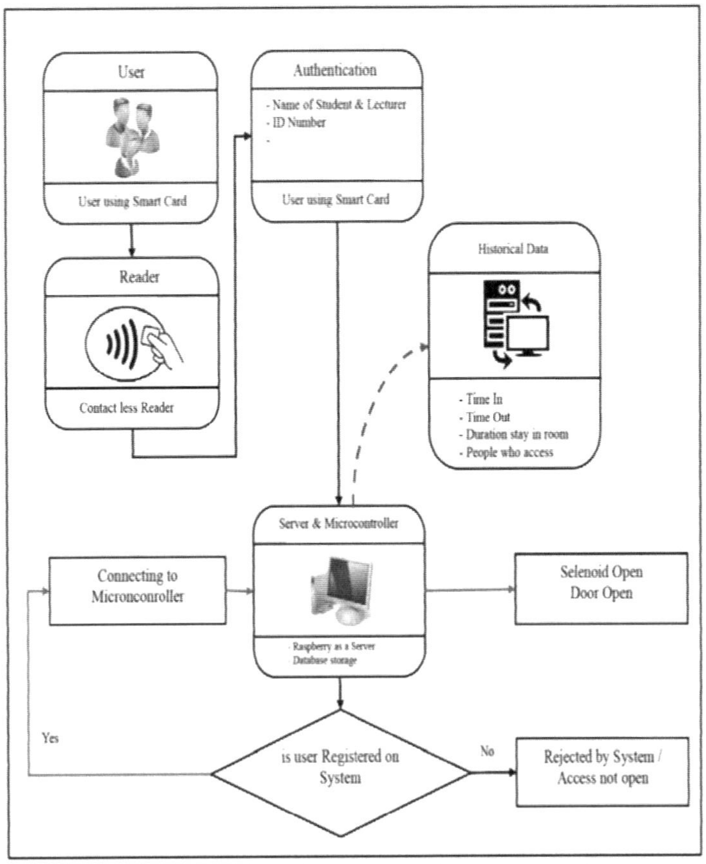

Figure 1. Algorithm Classroom and Laboratory Access

Table 1. GPIO Raspberry

The Function of GPIO in Raspberry			
Function	***Pin Number***	***Pin Number***	***Function***
3.3 V	1	2	5 V
12C0 SDA	3	4	DNC
12C0 SDL	5	6	GROUND
GPIO 4	7	8	UART TXD
DNC	9	10	UART RXD
GPIO 17	11	12	GPIO 18
GPIO 21	13	14	DNC
GPIO 22	15	16	GPIO 23
DNC	17	18	GPIO 24
SP10 MOSI	19	20	DNC
SP10 MISO	21	22	GPIO 25

The Function of GPIO in Raspberry			
Function	***Pin Number***	***Pin Number***	***Function***
SP10 SCLK	23	24	SPC10 CE 0 N
DNC	25	26	SPC10 CE 1 N

Room access system has several steps in development, namely:

- Input the information on a smart card (name, student/teacher identification number).
- Entry of a data base on the server which users can access which consists of the name and identification number.
- Access system programming with the detail time in, time out, duration. It proposed to accessing historical data can be displayed on screen. As a microcontroller, Raspberry was connected to the solenoid then the door can automatically open when the authentication process were completed.

GPIO pins can be configured as inputs. Connection of GPIO to a voltage higher than 3.3V was destroy GPIO blocks in SoC [9]. In addition to the input output on some pins GPIO has functions as serial communication such as I2C, SPI and serial communication UART [10].

3.2. System testing

The Figure 1. shows the Algorithm on classroom and laboratory access systems. Users who had been registered on the system and stored on the server access by using tag on the reader. Further authentication process whether the user can access or not. In accordance with the programmed procedure, if the user were not registered, the door would not open and there was a warning that the user were unidentified. Meanwhile, if the user were registered, the door that uses a selenoid and connected to the microcontroller / Raspberry will automatically open.

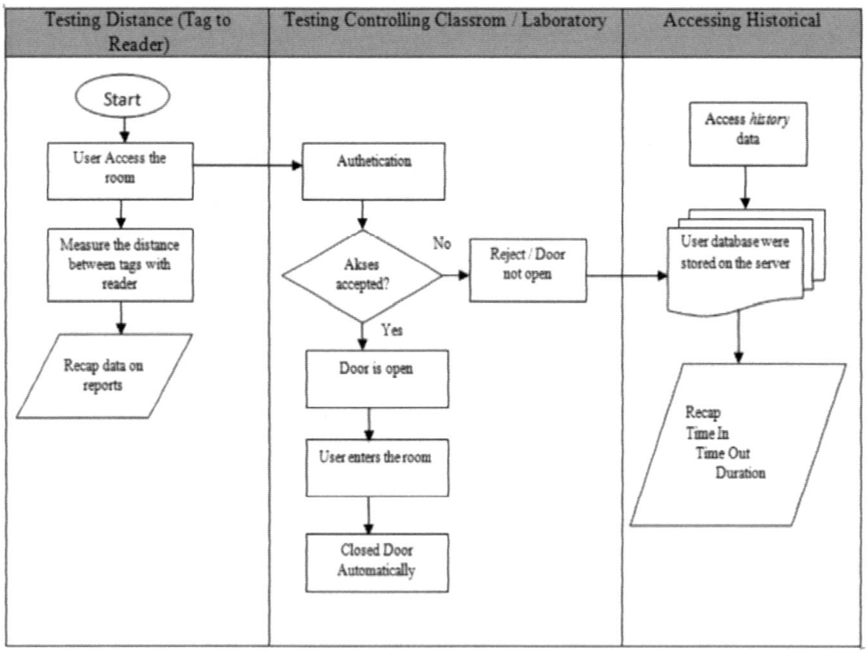

Figure 2. Testing System Diagram

Figure 2 shows a system test consist of 3 test scenarios were measuring the tag distance on the reader, the access system of the room where the moment after the user accessed the room, the door would be closed automatically. The last scenario was accessing historical data stored on the server. The data in the form of time in, time out and duration of users were in the room. This allowed the administrator or Person in charge to monitor the room. The security of the room will be ensured with additional features that automatically close the door when the user takes too long to enter the room.

4. Formatting the text

The The system testing process were verifying the distance between the reader and the tag.

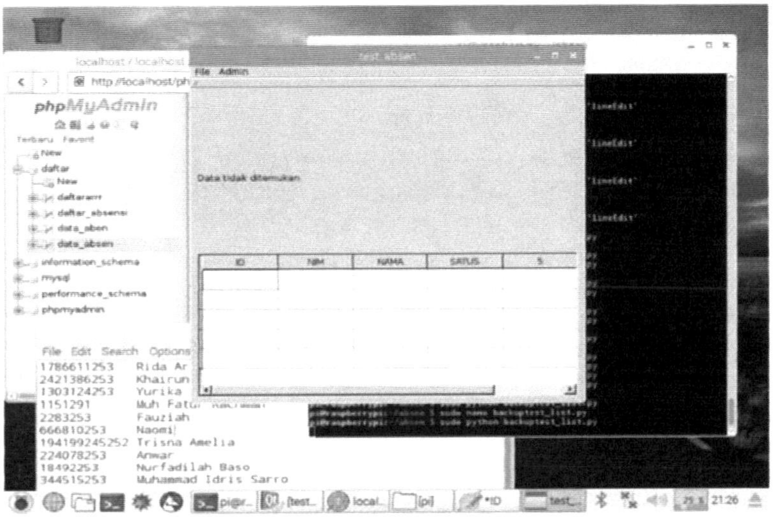

Figure 3. Accesses denied

Figure.3 shows system testing that users who did not have access to the system will not enter the room and the door will not open.

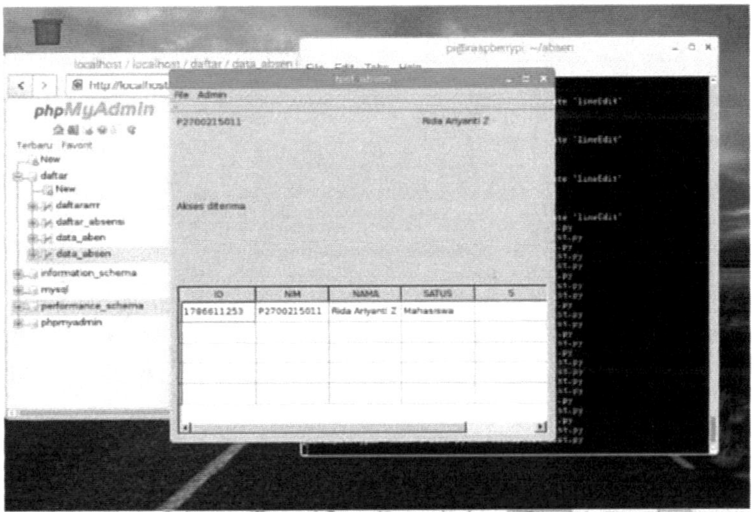

Figure 4. Accesses denied

Figure.4 shows system testing that users who have access to the system will enter the room and the door will open.
Based on the proposed of this research, the access room system using the tag is shown in the Table. 2.

Table 2. Test result

Distance (cm)	System Testing Results		Door open (Yes / No)
	Readable by System	Not Legible by System	
0.5	√	-	Yes
1	√	-	Yes
1.5	√	-	Yes
2	√	-	Yes
2.5	√	-	Yes
3	√	-	Yes
3.5	√	-	Yes
4	√	-	Yes
4.5	√	-	Yes
5	√	-	Yes
5.5	√	-	Yes
6	-	√	No.
6.5	-	√	No.
7	-	√	No.

In access distance tested the distance of tag to reader as shown in table 2 .Explains that the system could be accessed with a maximum distance of 5.5cm. The 100% security class system works by denied access for unregistered users.

Additional applications on this system were historical data about time in, time out and duration of users in the room can be recorded on the server. Thus, when needed the data can be accessed from the server. This is very important considering the security of the room such as classes and laboratories need a special security system. Thus, the optimization of the system related to the security of the room is added to the closed door automatically system algorithm when the user has authenticated.

5. Future research direction and Conclusion

The function of the reader in reading the tag, works well on the reader reading distance to the tag can reached 5.5 or > 5 cm. Software testing shows the system works based on proposed where the user registered on the system will be able to access the room and the door opens automatically. If the authentication process fails / the user is not registered, the door will not open and the access warning appears rejected. As the advantages of this system, data stored on the server in the form of historical time in time out, long in the room and logs of users who access can be seen on the display / monitor server. Future research will be done about detecting users within 100m of the reader.

Acknowledgments
This research support by Cyber Tech Community, Electrical Engineering of Hasanuddin University.

References

[1] M. Wassim Raad, Tarek Sheltami ,M. Sallout, "A Smart card Based Prepaid Electricity System", IEEE [Pervasive Computing and Applications, 2007. ICPCA 2007. 2nd International Conference].
[2] Alavalapati Goutham Reddy, Eun-Jun Yoon, Ashok Kumar Das, Kee Young Yoo, "Lightweight Authentication With Keyagreement Protocol For Mobile Network Environment Using Smart Cards", IEEE [The Institution of Engineering and Technology 2016].
[3] Agnes C. Noubissi, Julien Iguchi-Cartigny, Jean-Louis Lanet, "Hot Updates for Java Based Smart cards", IEEE, 2011 [ICDE Workshops].
[4] Wu Yun, "Research of Cloud Print Key Technology Base on Identity ". IEEE, 2012 [Third World Congress on Software Engineering].
[5] Matej Kacic, Petr Hanacek, Martin Henzl, Ivan Homoliak, "A concept of behavioral reputation system in wireless networks", IEEE, 2013 47th International Carnahan Conference on Security Technology (ICCST).
[6] B. Linkesh, S.Durga, E. Chowdary, Unified RB-DAC Approach With Secure Authentication Using Smart card Architecture, IEEE [2014 3rd International Conference on Eco-friendly Computing and Communication Systems].
[7] Shanu Gaharana, Darpan Anand, "Dynamic ID Based Remote User Authentication In Multi Server Environment Using Smart cards: A Review", IEEE, 2015, International Conference on Computational Intelligence and Communication Networks (CICN).
[8] Zhengxian Gao, Shou Hsuan Stephen Huang, Wei Ding, "Cryptanalysis of Three Dynamic ID-Based Remote User Authentication Schemes Using Smart cards", 2016, IEEE International Conference of Online Analysis and Computing Science (ICOACS).
[9] Qihao He; Bruce Segee; Vincent Weaver, "Raspberry Pi 2 B+ GPU Power, Performance, and Energy Implications", IEEE, 2016 International Conference on Computational Science and Computational.
[10] Vamsikrishna Patchava, Hari Babu Kandala, P Ravi Babu, "A Smart Home Automation technique with Raspberry Pi using IoT", IEEE, 2015, International Conference on Smart Sensors and Systems (IC-SSS).

The Effect of Grounding Resistance About Back Flashover on 150 KV Tranmission Network in Main Station of Sungguminasa - Tallasa (Makassar)

Naomi Lembang[1], SalamaManjang[2], IkhlasKitta[2]

Electrical Engineering Department[1], Post Graduate Program Of Hasanuddin University[2],Jl. Perintis Kemerdekaan KM. 10 Makassar, Sulawesi Selatan, 90245, Indonesia

E-mail: naomi_lembang@ymail.com[1], salamamanjang@gmail.com[2], ikhlaskitta@gmail.com[2]

Abstract. Lightningsurge on transmission network is a natural problem that affecting electricity voltage. This impact on the occurrence of equipment damages as result of *back flashover*. *Back flashover* causes an impulse wave due to the increase of voltage on the transmission and has a large grounding resistance value. There has been a disturbance in the transmission system SULSELRABAR causing black out on Tuesday, January 19th, 2016 at 14:13 pm. The disturbance occurred along with the rain and strong winds accompanied by lightning which caused the load loss of 746 MW. The research was conducted to determine the occurrence of direct lightning strikes on each tower, the effect of grounding resistance to *back flashover* on150 kV transmission network in main station of Sungguminasa - Tallasa, using modeling simulation of peak menara transmission network 150 kV in main station of Sungguminasa - Tallasa with software ATPDraw (Alternative Transient Program). The simulation results depicting one of the phase wire reach the peak current of 359.89 MA. In this condition occurs disruption that causing the impulse voltage and back flashover with disturbance current 22.65 kA. The large of grounding resistance has an average voltage exceeding the capacity limit from 1001.54 kV to 1042.31 kV. Based on the calculation, the impulse voltage obtained is 5.271 MV close to simulation result of ATPDraw(Alternative Transient Program) by 4.530 MV. The calculation of the current wave when there is lightning disturbance in the transmission line with a peak current of 60 kA and the wave front time of 1 set, obtained the highest value that is on the tower 39 the grounding value of 5.4 Ω, the grounding value obtained reflection coefficient of -0.937 and insulator voltage at 957.036 kV, the results of this research held the measurement of tower 39 using *Megger Det4tcr* and obtained the grounding value on the tower feet of 12.22 and 9.97 ohm. What makes this case interesting is that the installation is located in a wet-sand soil type, undertaken decrease grounding value of 39 of parallelization methods by adding 2 copper rods (*arde*) 5/8 inch diameter in 4 meters long and the addition of mixture of carbon and NaCl, the result of was able to reduce the magnitude of grounding resistance is 0.78 ohm, close to the calculation results of parallelization method is 0.0864 ohm.This method is more accurate because the disturbance of back flashover can be avoided by lowering the high grounding value due to lightning strikes and protecting other equipment, can be used in long-term, inexpensive and simple, very suitable applied of wet soil

1. Introduction

Lightning strike is a very serious threat in electrical systems where its network is widespread [1], disturbance for transmission that causing major loss for electricity service providers. On transmission network 150 kV, main station of Indonesia have problems with equipment damage caused by lightning strikes that immediately strike the ground wire and tower. At voltage exceeds or equal to critical limit voltage of isolator back flashover so that occurs *Back Flash Over* (BFO) and causing the form of impulse wave due to increase voltage of transmission network to the end of network of main station that can damages isolation. Therefore, that needs to be done is a study about the performance of transmission network 150 kV as result of lightning strikes using ATPDraw(Alternative Transient Program), The ATP(Alternative Transient Program) has been used to study the lightning performance of transmission lines [2]. to get the profile of voltage as it exceed or equal to critical limit of isolator back flashover or current result by using ATPDraw(Alternative Transient Program) tools. The result of point conditions BFO occurring in a transmission tower 150 kV Sungguminasa-Tallasa and using the theory of running wave to calculate lightning disturbances of tower that is BFO back flashover,
due to reflection of a travelling waves formed by the lightning strike, where the reflections from the tower base will arrive much sooner at the tower topthen reflections from adjacent towers [3].

Content from this work may be used under the terms of the Creative Commons Attribution 3.0 licence. Any further distribution of this work must maintain attribution to the author(s) and the title of the work, journal citation and DOI.
Published under licence by IOP Publishing Ltd

In order to reduce the likelihood of blackouts and the damage of transmission device as result of BFO(Back Flashover), then it conducted by lowering the grounding value with long-term durability. It conducted at point of BFO occurrence on the transmission tower of Sungguminasa-Tallasa.

To lowering the grounding value, then use grounding copper rods (*arde*) by grounded under the soil surface as deep as 4.20 meters in parallel close to the tower pole and the addition of NaCl + C (salt + charcoal) by measuring the grounding using parallelization method. Conductor connector of grounding system from electrical channel to salt can reduce grounding impedance simultaneously. The increase of voltage exceed the safe limit can be developed a continuous counter connected to the salt water can stabilize the potential profile [4].

Comparison of the addition of salt and bentonite to the grounding value by using rod type to lowering grounding value[5].So that the effect of grounding quantity is expected in the future to be considered to reduce the damage of transmission line equipment on main station 150 kV as result of the lightning surge.

2. Methods

The method of analysis in this study used 4 (four) problem-solving techniques with mathematical equations that will be simulated on ATPDraw(Alternative Transient Program), as well as mathematical equations to calculates the magnitude of grounding resistance to the voltage generated lightning disturbance on the isolator, and to calculate the grounding rod on tower feet, as follows:

2.1. Impulse Voltage

It is defined as a double exponential waveform which can be expressed by the equation:

$$(t) = Voe - at - e - bt \tag{1}$$

On the other hand, within some 30 m of the tower base (i.e. concentrated tower grounding system), the tower grounding impedance exhibits only current dependence, which can be modelled in accordance with guidelines provided in [6] and implemented in (Electromagnetic Transient Program) EMTP-ATP (Alternative Transient Program)software package by means of the MODELS language [7].

An analysis model of transmission line lightning strain 150 kV is simulated by using ATPDraw(Alternative Transient Program) software. The modelling of transmission line used in this simulation is using the historical data of lightning strike disturbance that happened at 150 kV transmission line especially in GI (Gardu Induk) Sungguminasa and GI (Gardu Induk) Tallasa on 19 January 2016.

Transmission line of the electrical system, especially in Tragi Bulukumba under interconnected GI Sungguminasa, GI Takalar, GI Jeneponto and GI Bulukumba were located on the line between the sea and the mountain that has many cumulonimbus clouds that produce lightning [8].

Voltage source is modelled by ATP(Alternative Transient Program) model (Alternating Current) AC3ph (3-phase source). Voltage source AC (Alternating Current) in ATPDraw(Alternative Transient Program) software is used peak amplitude of voltage system. The modelling of voltage source AC (Alternating Current) can be done by converting a voltage system of 150 $kV_{L-L(RMS)}$ to $V_{peak} = \sqrt{2} / \sqrt{3} \times V_{L-L(RMS)}$.

The transformer is modelled by a transformer component 3-phase ideal with a primary and secondary voltage ratio of 150 kV/20 kV = 7.5. As installed on field, the transformer group Yy load 20 kV connected to each transformer is modelled in ATPDraw (Alternative Transient Program) software using RLcY3 component. The following is the average of load percentage of each transformer installed and can be seen in Table 1.

Table 1. Capacity and transformer load installed on transmission line of Tragi Bulukumba

No.	Substation	Installed Capacity (MVA)	Load (%)
1	Sungguminasa	60	46
2	Tallasa	20	12
3	Tallasa	30	16
4	Jeneponto	20	8
5	Jeneponto	30	12
6	Bulukumba	20	17
7	Bulukumba	30	26

*Data source = PT PLN Persero. AP2B Sulselrabar

The load on this simulation is modelled by using RLcY3 component and the equations :

$$R = \sqrt{3}V^2/P \tag{2}$$

$$L = \sqrt{3}V^2/(2\pi f Q) \tag{3}$$

2.2. *Voltage on Disturbance (BFO)*

Strike on the ground wire or tower causing the increase of voltage that can cause BFO (Back Flashover) on the tower isolator, with the equation:

$$V_L = i \times R\varepsilon + L\frac{di}{dt} + V_M \tag{4}$$

2.3. *Lightning Disturbance on the Tower*

To calculate the lightning disturbance on the tower, i.e disturbance due to the back flashover, we use the theory of running wave, with equation [9]:

a. Calculates the possibility of the number of fire jumps

$$V_{50\%}\, 2\ \mu\text{det} = \left(K_1 + \frac{K_2}{t^{0,75}}\right) \times 10^3\ kV \tag{5}$$

1. Calculating ground wire surge impedance and coupling factor of ground wire corona radius :

$$R\ln\left(\frac{2h}{R}\right) = \frac{V}{E_o} \tag{6}$$

2. Surge impedance of a single ground wire :

$$Z_{11} = Z_{22} = 60\sqrt{\ln\frac{2h}{r}\ln\frac{2h}{R}} \tag{7}$$

b. Reflection coefficient on the base of the tower for R

$$d = \frac{R - Z_t}{R + Z_t} \tag{8}$$

Table 2. One data Tower and grounding resistance at Tragi Bulukumba GI Sugguminasa - GI Tallasa

No	Tower	Grounding(Ohm) = Rε
1	T. 39	5.4
2	T. 40	5.3
3	T. 46	4.6
4	T. 112	0.8
5	T. 114	0.6
6	T. 113	0.4

c. Voltage on isolator

This calculation is done for all the flash currents of 60 kA and the front of light wave that given 1μ.

$$V_i = e_o\Big\{(1-K)T + d\Big[\Big\{T - 2\Big(\frac{h_t}{c} - \frac{X_1}{c}\Big)\Big\} + (b - Ka)\Big(T - \frac{2h_t}{c}\Big)\Big] + d^2 b\Big[\Big\{T - 2\Big(\frac{2h_t}{c} - \frac{X_1}{c}\Big)\Big\} + (b - Ka)\Big(T - \frac{4h_t}{c}\Big)\Big] + d^3 b\Big[\Big\{T - 2\Big(\frac{3h_t}{c} - \frac{X_1}{c}\Big)\Big\} + (b - Ka)\Big(T - \frac{6h_t}{c}\Big)\Big]\Big\} \tag{9}$$

2.4. Grounding on Two Copper Rods on Tower in Transmission 150 kV GI Sungguminasa - GI Tallasa at Tower 39

Implementation of this grounding system is done on the tower SUTT 150 kV between Sungguminasa station and Tallasa main station at tower 39 located on Pallangga subdistrict which was held on 4 May 2017. In the implementation of parallelization system of this grounding installed 2 copper rods at a distance of 2 meters from each foot tower A and C with the addition of charcoal and salt. The giving of charcoal and salt is reduced gradually as shown in figure 7 in the parallelization measurement.

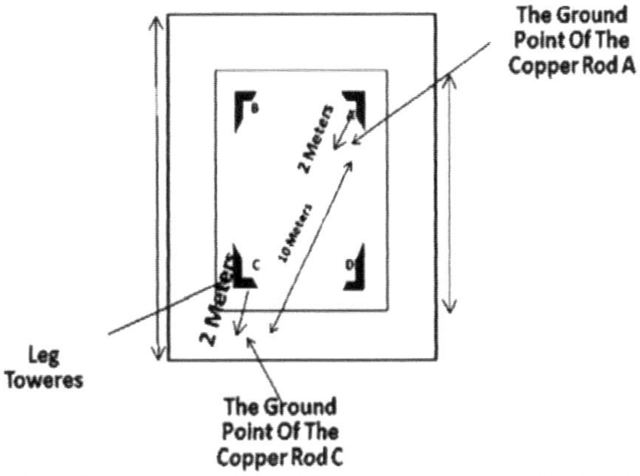

Figure 1. Grounding construction of tower transmission 150 kV (top view).

The resistance of tower foot 10 Ohm can be obtained by using one or more ground rods and/or counterpoise systems. **addition of a counterpoise combined with the effect of the sea water can help to reduce the voltage rise and then the grounding resistance at each tower [10].** The selection of a ground rod and/or a counterpoise system depends on the type of ground type resistance in which the transmission tower is located. Using a single ground rod, tower foot resistance is calculated using the following equation [11]:

$$R = (\rho / 2\pi L)\ \mathrm{In}\ (2L / d) - 1 \tag{10}$$

Measuring 2 copper conductor rods with a parallelization model price A is a multiple of ground rods, two rods is placed everywhere, given salt, the three given [11].

$$A = \sqrt{ar} \tag{11}$$

Types of soil.

Table 3. Resistance of soil type

NO	Types of Soil	The kind Resistance (ohm.m)
1	Marshland	10 up to 40
2	Clay and Fields	20 up to 100
3	Wet Sand	50 up to 200
4	Wet Gravel	200 up to 3000
5	Sand and gravel dry	<10.000
6	Rocky Ground	2000 up to3000
7	Sea water and fresh water	10 up to 100

*Source: PUIL 2000

3. Result and Discussion

In the analysis of the Back Flashover (BFO), special attention is given to the influences emanating from the insulator strings flashover characteristic and lightning statistics. The model could be applied to the transmission line as a whole or some of its portions, e.g. first several towers emanating from the substation or several towers crossing a mountain ridge [12]. Model analysis of transmission line 150 kV especially along the transmission line there are 81 towers located 26.43 km in GI Sungguminasa and GI Tallasa. The modelling of single diagram Tragi Bulukumba on Alternative Transient Program (ATPDraw). Table 4 is the calculation result of transformer capacity on the transmission line Sungguminasa –Tallasa.

Table 4. Transformer capacity installed on main station

No	S (MVA)	P (MW)	Q (MVAR)	R (Ohm)	L (mH)
1	46	37	27,72	18,745	79,597
2	12	10	7,44	69,841	296,564
3	16	13	9,54	54,467	231,282
4	8	7	5,04	103,098	437,785
5	12	9	7,02	74,019	314,307
6	17	14	10,44	49,772	211,344
7	26	20	15,3	33,962	144,211

*Data source = PT PLN Persero. AP2B Sulselrabar

3.1. The modelling of single diagram Tragi Bulukumba connecting GI Sungguminasa - GI Tallasa.

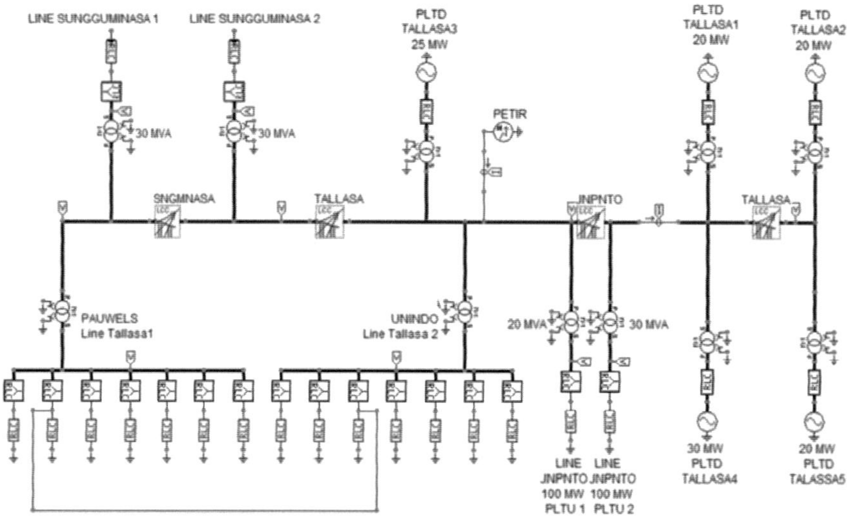

Figure 2. Simulation model of single diagram of Tragi Bulukumba connecting GI Sungguminasa - GI Tallasa after lightning strikes.

The shape of simulation result of front and end time wave of lightning is 0.15 s or 1/50 μs, the value of lightning peak current reaches 359.89 MA, resulting in a disruption to the electrical system and causes the occurrence of impulse voltage on Busbar and back flash over (BFO), The back flashover performance is estimated by means of several flashover models. They perform differently depending on the lightning waveform and investigated tower [13].

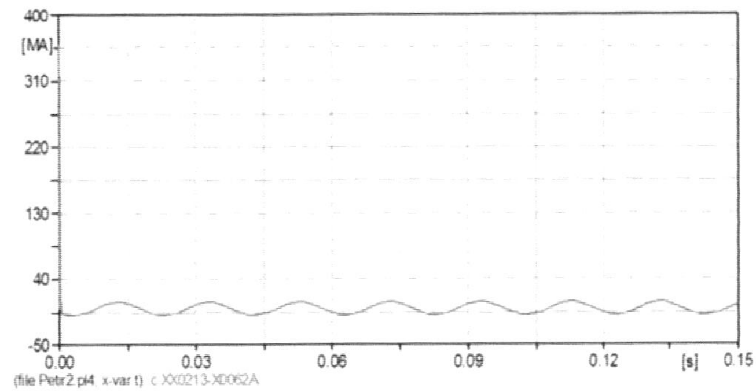

Figure 3.Model of lightning current type Heidler

3.2. Impact analysis the occurrence of back flashover (BFO) that affecting transmission line 150 kV GI Sungguminasa - GI Tallasa due to lightning strikes.

The simulation results for the current on transmission line 150 kV GI Sungguminasa - GI Tallasa is 22.65 kA and waveform in period T = 50 μs.

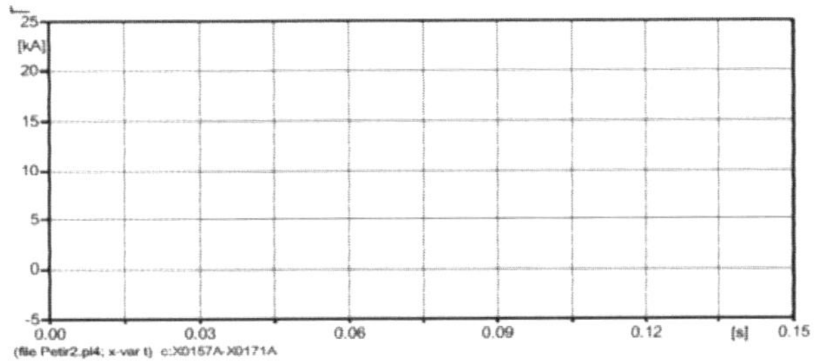

Figure 4.Current on electrical system of Tragi Bulukumba

Table 5 is a data tower and grounding resistance of Tragi Bulukumba under interconnected GI Sungguminasa, GI Takalar, GI Jeneponto and GI Bulukumba, there are several towers that have a large earth resistance value so that it can affect the occurrence of back flashover (BFO). Impact analysis of the occurrence of BFO (back flashover) is obtained the highest result of 1042.31 kV is in tower 39.

Table 5. Data tower and the magnitude of grounding resistance that causes back flashover (BFO).

No.	Tower	Ground (Ohm) = Rg	Measurement results (kV)	No.	Tower	Ground (Ohm) = Rg	Measurement results (kV)
1	T.38	5.2	1037.78	21	T.65	3.8	1006.07
2	T.39	5.4	1042.31	22	T.66	3.6	1001.54
3	T.40	5.3	1040.05	23	T.68	3.6	1001.54
4	T.41	5	1033.25	24	T.69	4.6	1024.19
5	T.42	5	1033.25	25	T.70	4.6	1024.19
6	T.43	5.2	1037.78	26	T.72	4	1010.60
7	T.44	4.2	1015.13	27	T.73	3.8	1006.07
8	T.45	4.4	1019.66	28	T.74	3.8	1006.07
9	T.46	4.6	1024.19	29	T.75	3.8	1006.07
10	T.47	4.6	1024.19	30	T.76	4	1010.60
11	T.48	4.4	1019.66	31	T.77	4.2	1015.13
12	T.49	4.4	1019.66	32	T.78	3.8	1006.07
13	T.50	3.6	1001.54	33	T.79	4.8	1028.72
14	T.51	3.6	1001.54	34	T.80	4.6	1024.19
15	T.52	3.8	1006.07	35	T.81	4.2	1015.13
16	T.53	3.6	1001.54	36	T.82	4	1010.60
17	T.60	3.6	1001.54	37	T.90	3.6	1001.54
18	T.61	3.6	1001.54	38	T.93	3.6	1001.54
19	T.62	3.8	1006.07	39	T.94	3.6	1001.54
20	T.64	3.8	1006.07	40	T.96	3.6	1001.54

3.3. The Analysis of Impulse Voltage Magnitude on Transmission Line 150 kV due to Back Flashover (BFO).

Simulation results of Figure 5 shows the impulse voltage values due to lightning strikes in GI Sungguminasa - GI Tallasa.

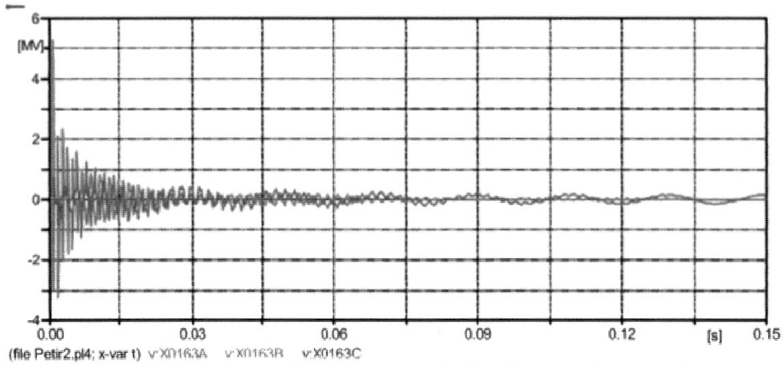

Figure 5. Impulse voltage due to lightning strikes on transmission lien GI Sungguminasa - GI Tallasa.

The simulation result on Tragi Bukulumba, the lightning current that is too high greatly affects the voltage on the electrical system. So the important role to resolve the lightning strikes is the grounding resistance of each tower, on average there is a large grounding resistance in some towers along GI Sungguminasa - GI Tallasa so that the voltage of back flash flashover (BFO) is strongly affects the disturbance of the electrical system. Hence, the conclusion that the very high lightning current in the electrical system of GI Sungguminasa - GI Tallasa makes R-phase becomes unstable which is increased by 5.271 MV and then the transition periode decrease to 190.59 kV and at the end period is decrease drastically to 3595 V. Lightning peak currents used is a simulation current of 22.65 kA and grounding 0.3 ohms so that the voltage in the phase wire is:

$$V = \frac{Z.I}{2} = \frac{0{,}4\,\Omega\,.\,22{,}65\text{ kA}}{2} = 4{,}530\text{ MV}$$

The impulse voltage obtained from the calculation of 5.271 MV approximate the simulation result of Alternative Transient Program (ATPDraw) is 4.530 MV. If seen from the calculation of the impulse voltage then it has exceeded BIL isolator is 750 kV. This indicates that it will cause a flashover, thus propagating in both opposite directions on the line toward the GI to both sides of transmission line. So that the relay will work read interruptions and tripped PMT/CB (Circuit Beaker) on both sides of the main station.

3.4. Impact Analysis of the Occurrence of Lightning Disturbance of Tower that Causing Back Flashover (BFO) with the Theory of Running Wave affecting transmission line 150 kV GI Sungguminasa - GI Tallasa due to a lightning strike.

In this method assumed that the lowest phase will experience the earliest fire jumps due to lightning strikes on the tower. In this calculation is done up to 1 μdet because at that time a negative reflection wave from adjacent tower (average gate length 300 meters) has reached the tower struck by lightning and this wave will reduce the voltage on the tower. Several models have been proposed for representing towers [14], [15], [16]. The simplest one is based on a single conductor distributed parameter line. Tower surge impedance values range from 100 to 300 ohms [17],

The line to be calculated is a line owned by PLN, i.e.Sungguminasa-Tallasa channel. This line is a dual air transmission line 150 kV with a configuration of wire and tower as shown in Figure 6. The data of wire and isolator are:

Radius of ground wire:	0.45 cm
Radius sub-conductors of phase wire:	1.45 cm
Total sub-conductors:	2
Distance between sub-conductors:	45.7 cm
High ground wire on the tower:	37.7 m
High phase wire C or A on the tower:	25.5 m
Large ground wire and phase wire:	7 cm
Distance length of isolator:	6.43 m

This calculation will see the effect of the magnitude of the grounding resistance to the voltage generated by the lightning disturbance of isolator. Therefore, we will use some grounding values obtained from the data of the State Electricity Enterprise, i.e. on voltage 5,4 Ω, 5,3 Ω, 4,6 Ω, 0,8 Ω, 0,6 Ω, and 0,4 Ω.
This calculation gradually and its stage is indicated by number given on the front.

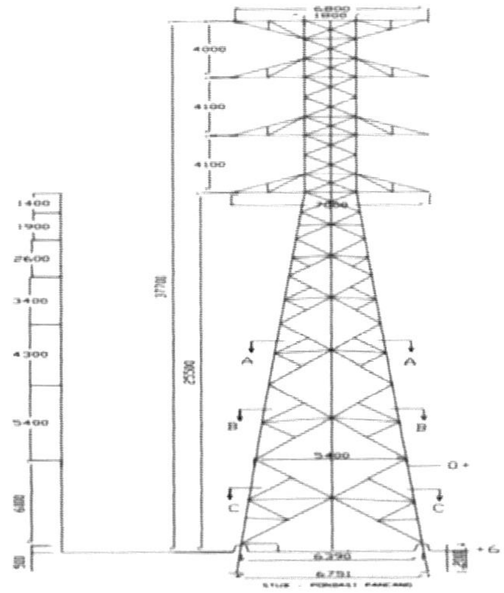

Figure 6.A configuration of wire and tower.

Table 6. The calculation result of lightning disturbance on the tower (calculation for C and A phases).

T (μdet)	I_o (kA)	*R* (Ω)	*d*	V_i (kV)
1	60	5,4	-0,937	957,036
		5,3	-0,938	954,967
		4,6	-0,946	937,819
		0,8	-0,99	842,694
		0,6	-0,993	838,454
		0,4	-0,995	834,238

The calculation result of highest voltage values on tower 39. The reflection coefficient is -0.937 and the voltage on the isolator is 957.036 kV, the greater of grounding value on the tower, it will increase the reflection coefficient value which means the possibility to occur back flashover will also be greater. This will cause the voltage value of isolator will also be greater. So, efforts made in the form of correcting the value of grounding resistance of foot tower by make parallelization modeling of ground rods with the addition of charcoal and salt. This is intended to reduce the grounding value of tower foot as well as to add lightning discharge point. The addition of ground point has a very maximum impact.

3.5. Grounding of Tower Feet and grounding Analysis of Two Copper Rods on the Tower Transmission 150 kV GI Sungguminasa - GI Tallasa on Tower 39.

Protection of transmission line against lightning disturbance using ground wire at tower foot to reduce tower foot resistance, to obtain tower foot resistance is less than 5 ohm. Then, conduct the grounding application of 2 copper rods by parallelization method with the addition of charcoal and salt, with a distance of grounding two meters from the tower feet A and C as shown 1. The installation of copper is done in square tower on the side 10 m. Two copper rods are installed 420 cm in deep from the tower foot A and C connected using BC (Bare Core) wire 50 mm in coupling at the tower foot, ground type of tower 39 is wet sand soil. Field corrosion test results in Poland show evidence of better corrosion protection by copper coated steel earth electrodes compared with zinc coatings made using different technologies [18].

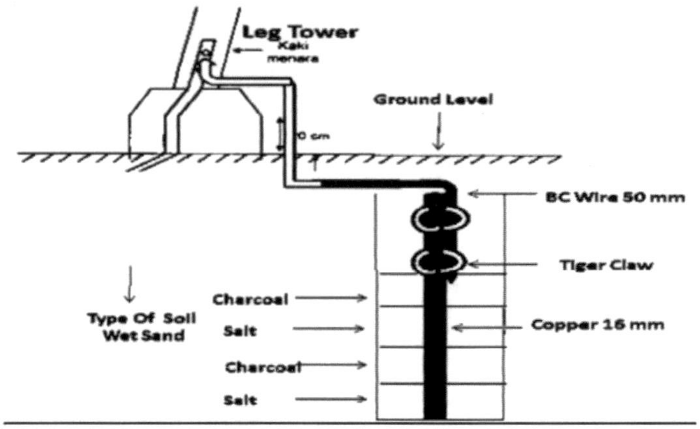

Figure 7.Ground rods of grounding from side.

Table 7.Measurement results of grounding on transmission Tower 39.

PLN Data	Leg of Tower (without Copper Charcoal and Salt)		Grounding of 1 Copper (Clasped Legs of Towers)		Parallelization Of Legs Coupled Tower OF Charcoal and Salt Addition
	Leg of Tower A	Leg of Tower B	Leg of Tower A	Leg of Tower B	Leg Of Tower A and B
5 ohm	12,22 ohm	9,97 ohm	1,24 ohm	1,74 ohm	0.78 ohm

Table 8. Calculation result of ground rods

1 Copper Rod	2 Copper Road
2,6071 Ω	0,0864 Ω

The calculation result of two copper rods with parallelization model installed perpendicular to the ground is 0.0864 Ohm. So the calculation has met the applicable requirements under 5 Ohm. This results showed that the configuration of rod electrode grounding with the addition of charcoal and salt is 0.78 can reduce the magnitude of grounding resistance. It means the more electrode is installed in the ground the smaller of value of grounding resistance, the longer of grounding copper to the ground touch the water the better the decrease of grounding value, and the addition of charcoal and salt affecting the decline in the grounding value because charcoal and salt contain chemicals in the soil particularly a number of soluble organic or inorganic substances need to be considered. To achieve an effective decrease in grounding resistance by grounded the copper deeper until it reaches where the salt and charcoal solutions are exposed to water or wet conditions. The objective of minimizing the grounding value of tower foot and increasing the point of lightning discharge, so that continuity of electricity service does not occur blackout during the rainy season due to lightning strike and can protect the transmission equipment 150 kV GI Sungguhminasa - GI Tallasa.

4. Conclusion

The results of this research can be concluded as follows; the simulation results of lightning strike one of the phase wires on transmission line GI Sungguminasa - GI Tallasa so that occurs a disturbance to the electrical system resulting in the impulse voltage on the busbar and back flashover (BFO). The calculation of 81 towers due to the magnitude of the grounding resistance on transmission line 150 kV GI Sungguminasa - GI Tallasa the average of voltage exceeds the predetermined capacity limit of the occurrence of back flashover (BFO) in isolator.

The calculation of running wave of reflection coefficient and the highest voltage is on tower 39 which means the possibility to occur back flashover will be greater already exceeds the limits of (Basic Impulse Level) BIL isolator 750 kV. There are several tower conditions that must be minimized the value of grounding, one of which is in tower 39.

This research is presented for a reduction of new grounding using copper or rod with a mixture of charcoal and salt. This method is more accurate because the disturbance of back flashover can be avoided by lowering the high grounding value due to lightning strikes and protecting other equipment, can be used in long-term, inexpensive and simple, very suitable applied of wet soil. For the content of charcoal need further research on the effect of each soil types. This method can last long, make the application will be more effective and efficient.

References

[1] Z. Reynaldo, P. Y. Eko, 10 October 2008, *Lightning Performance of Extra High Voltage 500 Kv Lines at East Java – Indonesia*, IPTEK, *The Journal for Technology and Science*, Vol. 19, No. 4, November 2008, 100.

[2] J.A Martinez, Member, IEE, and F Castro-Aranda, 2003, *Lightning Performance Analysis of Transmission Lines Using the EMTP*, Barcelona, Spain, IEEE, 300.

[3] P. Sarajcev, J. Vasilj, D. Jakus, 2015, *Method for estimating backlashover rates on HV transmissions lines based on EMTP-ATP and curve of limiting parameters*, University of Split, FESB, Dapartment of Power Engineering, R. Boskovica 32, HR-21000 Split, Croatia, Elsevier, Electrical Power and Energy systems 78 (2016) 127-137, 129.

[4] S. W. Georges, Member, IEEE, F. H. Slaoui, P. J. Lagace, Member, IEEE, X. D. Do, Senior Member, IEEE and J. Forth, Member, IEEE, 2003, *Evaluation of Grounding System Impedance ofa Near Sea Water HV Electric Power Installation,*1591.

[5] I.Janardana. *Perbedaan Penambahan Garam Dengan Penambahan Bentonit Terhadap nilai Tahanan Pentanahan Pada Sistem Pentanahan,* VOL, 4 No. 1 Januari – Juni 2005, 24.

[6] IEEE WG, 1996, *Modeling guidelines for fast front transients*. IEEE Trans. PowerDeliv. 11 (1) (1996) 493-506.

[7] L. Prikler, H.K. Hoidalen, 2009, *ATPDRAW version 5.6 for Windows9x/NT/2000/XP/Vista: User's6Manual,Norwegian University of Technology.Trondheim, Norway*.

[8] H.W. Dommel,1986, *Electromagnetic Transients Program*, Portland, Reference Manual,*Bonneville Power Administration.*

[9] T.S. Hutauruk, *1991, GelombangBerjalan Dan ProteksiSurja*, Jakarta, Erlangga, 132.

[10] S. Georges, F. Slaoui, P. J. LagacC X.D. Do, Senior Member IEEE J. Fortin, Member IEEE, 1999 IEEE , *Evaluation Of The Effect Of Salt Water And A Counterpoise On The Voltage Profile Of A HV Power Transmissions LineGrounding System In A Two Layer With High Resistivity Soil, Montreal, Canada,* , 1020.

[11] T.S. Hutauruk, 1991, *PengetanahanNetralSistemTenaga Dan PengetanahanPeralatan.* Jakarta, Erlangga, 141.

[12] SarajcevPetar, 12 May 2014,*Monte Carlo method for estimating backflashover rates on high voltage transmission lines.*University of Split, FESB, Dapartment of Power Engineering, R. Boskovica 32, HR-21000 Split, Croatia, Elsevier, *Electric Power Systems Research* 119 (2015) 247-257.

[13] Mackow, A., Kizilcay, M, *Mitigation methods to improve the lightning performance of hybrid transmission line.*Sigen, Germany,The work was supported by German TSO, Amprion.

[14] W.A. Chisholm and Y.L. Chow, September 1983, *I Lightning surge response of transmission towers*, i *IEEE Trans, on Power Apparatus and Systems,* vol. 102, nc. 9, pp. 3232-3242.

[15] M. Ishii et al, July 1991, *i Multistory transmission tower model for lightning surge analysis*, i *IEEE Trans, on Power Delivery,* vol. 6, no. 3, pp. 1327-1335.

[16] A.R. Hileman,1999, *Insulation Coordination for Power Systems*, Marcel Dekker.

[16]M. Robert, L. Marek, N. Radoslaw, 2010, *Influence of Climatic Conditions in China on Reliability of Power Earthing System*, IEEE, 1546.

[17] A. Imece, Chairman, January 1996, *1 Modeling guidelines for fast transients, IEEE Trans, on Power Delivery,* vol. 11, no. 1.

[18] M. Robert, L. Marek, N. Radoslaw, 2010, *Influence of Climatic Conditions in China on Reliability of Power Earthing System*, IEEE, 1546.

IOP Conf. Series: Journal of Physics: Conf. Series **1090** (2018) 012078 doi:10.1088/1742-6596/1090/1/012078

A Numerical Investigation of the Friction Contact of an Unfilled Styrene Butadiene Rubber by a Blade Sliding Indentation

B. Setiyana[1,2,a)], M. Khafidh[2], R. Ismail[1], J.Jamari[1] and D.J. Schipper[2]

[1]Laboratory for Engineering Design and Tribology,
Department of Mechanical Engineering, University of Diponegoro,
Jl. Prof. Soedharto, Tembalang, Semarang 59275, Indonesia

[2]Laboratory for Surface Technology and Tribology,
Faculty of Engineering Technology, University of Twente,
Drienerloolaan 5, Postbus 217, 7500 AE Enschede, Dutch (Netherlands)

[a)] Corresponding author: bsetiyana@yahoo.com

Abstract. Rubber is mostly modeled as a hyperelastic material and as a consequence, large deformation occur along friction contact against a rigid counterface. In general, Coefficient of friction (COF) of a contact surface consists of two components, i.e. adhesion and deformation (hysteresis). However, it is difficult to investigate the deformation component of COF analitically on the rubber sliding. By means of a rigid blade sliding indentation technique, this paper studies the friction contact phenomena on Unfilled Styrene Butadiene Rubber (SBR-0) numerically by using a Finite Element Analysis (FEA) in plane strain mode. By a given sliding speed, the FEA simulation is carried out with the various adhesion COF and sliding depth. The presented simulation output are stress, deformation and reaction forces. Results show that the deformation COF strongly depends on the sliding displacement. Finally, the overall COF highly increases and then decreases with respect to the sliding displacement and tends to indicate stick-slip phenomena.

1. Introduction

Elastomers or rubbers are used extensively in many industries because of their wide availability and low cost. They are also used because of their excellent damping and energy absorption characteristics, flexibility, resiliency, ability to seal against moisture and variable stiffness. Rubber is a very unique material. The unique properties of rubber are such that, it can undergo large deformations under load, sustaining strains of up to 500 percent in engineering applications and its load-extension behavior is nonlinear. Proper analysis of rubber components requires special material modeling and nonlinear finite element analysis tools that are quite different than those used for metallic parts. Due to the nonlinearity behaviour and large deformation under load, rubber is mostly modeled as hyperelastic material [1]. Therefore, it is difficult to investigate the contact phenomena from the rubber friction analitically.

Most of rubber or elastomer friction has been intensively studied experimentally in sliding or abrasion contact. There were some valuable results associated with a friction contact, i.e. a coefficient of friction (COF), stick-slip occurrence, abrasive wear etc. These phenomena play an important role in investigation of a contact force and rubber life time (in case on abrasive wear). By definition, the COF consists of an adhesion component and hysteresis (deformation) component [2,3]. The adhesion component relates to interfacial friction that depends on surface contact roughness, meanwhile, the hysteresis relates to internal friction of material that strongly depends on the degree of deformation and damping value.

Content from this work may be used under the terms of the Creative Commons Attribution 3.0 licence. Any further distribution of this work must maintain attribution to the author(s) and the title of the work, journal citation and DOI.
Published under licence by IOP Publishing Ltd

Rubber friction was mostly investigated experimentally as a contact between rubber and rigid rough surface [4,5]. Moreover, the friction or abrasion contact by using a single indenter was experimentally performed with a sphere, cone and blade indenter [6-10]. Along sliding contact, it obviously showed that the stick-slip phenomena occurred along friction contact and as a consequence, a periodic friction force and wear pattern were formed. The adhesion and hysteresis COF were experimentally estimated and showed that the overall COF was dominantly influenced by deformation component rather than the adhesion one [11]. A numerical investigation of the rubber friction by the blade indenter was performed to investigate the stress induced on the contact surface, meanwhile, the coefficient of friction (COF) occured was sligtly investigated [12].

By using a Finite Element Analysis (FEA), this study proposes dynamic or sliding contact in which a blade indenter tangentially slides over on the rubber surface. The material used is an Unfilled Styrene Butadiene Rubber (SBR-0) and is modeled as a hyperelastic rather than viscoelastic material because of a low damping value for unfilled rubber. The rigid blade indenter with a specified tip radius applies on the rubber surface with a constant sliding speed. The FEA output are von Mises stress on the rubber surface, deformation contour and reaction forces. The FEA simulation is carried out in various sliding depth and adhesion COF. Finally, the friction contact phenomena can be observed and the overall COF as well as the deformation COF can be developed.

2. Method

The finite element analysis of the present work was performed using a commercial finite element software package, ABAQUS 6.11 [13]. Regarding to hyperelastic model of the rubber, the material behaviour was obtained from experimental test that is required for FEA simulation and represented as Strain Energy Function (SEF) term. The SBR-0 (Unfilled Styrene Butadiene Rubber) with the Mooney-Rivlin SEF was used and the material was assumed as an incompressible material. The required data of SEF are adopted from Liang's experiment [12] that was performed by a uniaxial tensile test up to approximately 5.5 MPa stress and 300% strain in elastic condition.

Figure 1(a) shows a schematic illustration of a straight rigid blade sliding a rubber surface. A rigid blade indenter with 0.5 mm tip radius contacts on the rubber (elastomer) surface. Boundary conditions of the indentation system are depicted schematically in this figure as well. The rubber specimen of 10 mm high, 20 mm wide and 10 mm thick was modeled in the plane strain. The constant sliding speed of 5 mm/sec, maximum horizontal sliding of 2.0 mm with some various sliding depth and adhesion COF are applied. The chosen sliding depth are 0.5, 1.0 and 2.0 mm, and the given adhesion COF are 0.0, 0.5 and 1.5.

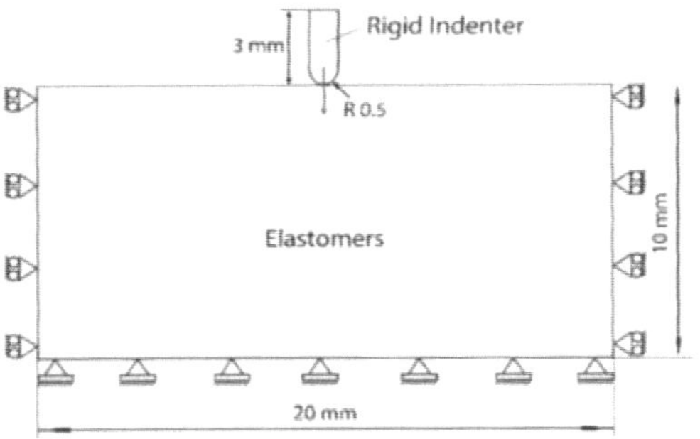

(a)

(b)

Figure 1. Initial model of the rubber sliding (a) Schematic model (b) FE model

IOP Conf. Series: Journal of Physics: Conf. Series **1090** (2018) 012078 doi:10.1088/1742-6596/1090/1/012078

The FEA simulation of the blade movement is started in rest position (initial state), then to be continued in move condition (sliding state) and finally in stop condition (final state). The FEA outputs are presented in the form of stress field and deformation contour in the initial state (before sliding), the sliding state and the final state (after sliding). From these results, the reaction forces in the horizontal as well as vertical direction can be developed. By the given various adhesion COF, the deformation and overall COF can be plotted in various sliding depth with respect to the sliding displacement.

3. Results and Discussions

Fig. 2 demonstrates the FEA output of the von Mises stress field and the deformation contour surface of SBR-0 by blade sliding indentation. The FEA output with the sliding depth 0.5 mm and adhesion COF 0.5 is depicted in these figures in the initial state, the sliding state and the final state respectively. These figures show that the highest contact stresses are located below at the indenter's blade tip, especially in the initial state because it is still in static condition. It can be seen that in the sliding state, the high stress regime tends to move toward to the sliding direction. In contrast, the high stress regime moves to the opposite of the sliding direction in the final state.

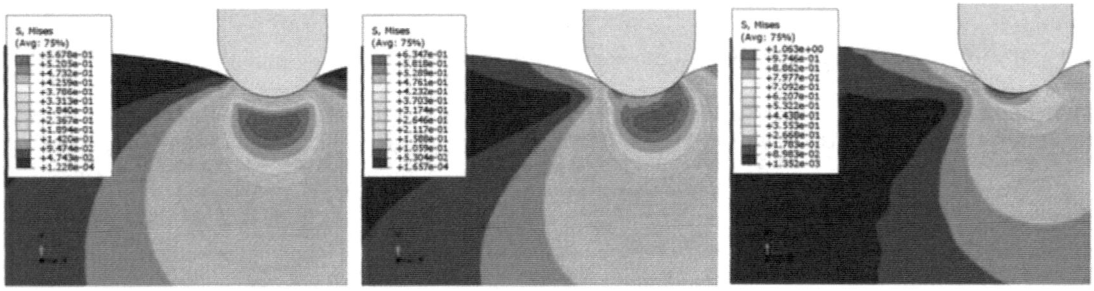

(a) (b) (c)

Figure 2. Stress field on the deformed model for sliding depth = 0.5 mm and adhesion COF = 0.5
(a) Initial state (b) Sliding state (c) Final state

The comparison of stress field on the deformed model in the sliding state is presented in Fig.3 for the various sliding depths, i.e. 0.5, 1.5 and 2.0 mm. It is clearly shown that the high stress area moves toward the sliding direction for all of the sliding state. Higher sliding depth yields to higher stress induced and located more away from the contact surface.

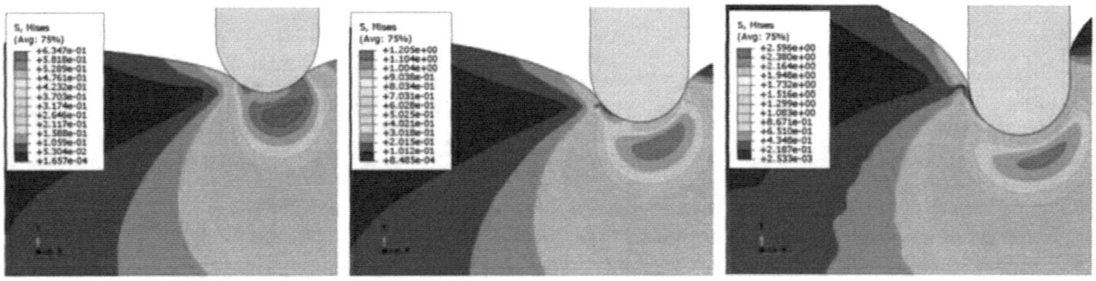

(a) (b) (c)

Figure 3. Stress field on the deformed model on sliding state for adhesion COF = 0.5
(a) Sliding depth 0.5 mm (b) Sliding depth 1.0 mm (c) Sliding depth 2.0 mm

The comparison of the stress field on the deformed model for the various adhesion COF, i.e. 0.0, 0.5 and 1.5 in the sliding state is presented in Fig.4. It is shown that the high stress area tends to move toward the sliding direction for higher adhesion COF. High adhesion COF yields to sligt increase of

IOP Conf. Series: Journal of Physics: Conf. Series **1090** (2018) 012078 doi:10.1088/1742-6596/1090/1/012078

the maximum stress.

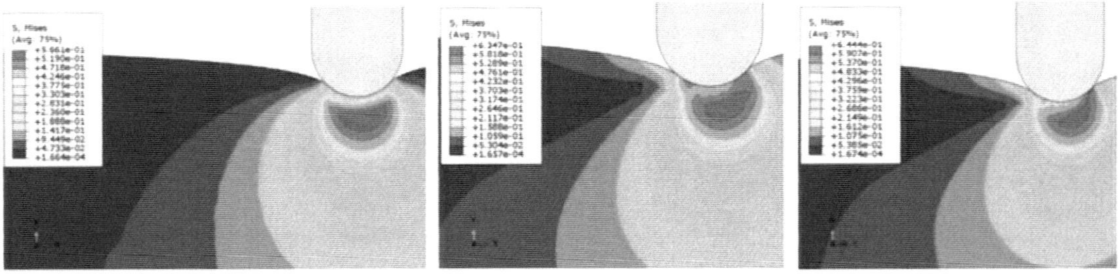

(a) (b) (c)

Figure 4. Stress distribution on the deformed model on sliding state for sliding depth = 0.5mm (a) Adhesion COF = 0 (b) Adhesion COF = 0.5 (c) Adhesion COF =1.5

The horizontal reaction forces as well as vertical reaction forces are depicted in Fig.5 and Fig.6 respectively. It can be clearly seen that horizontal reaction forces increase with respect to the sliding depth and adhesion COF. However, these reaction forces tend to decrease if the sliding displacement is extended over 1.5 mm, where it might be close to periodic forces.

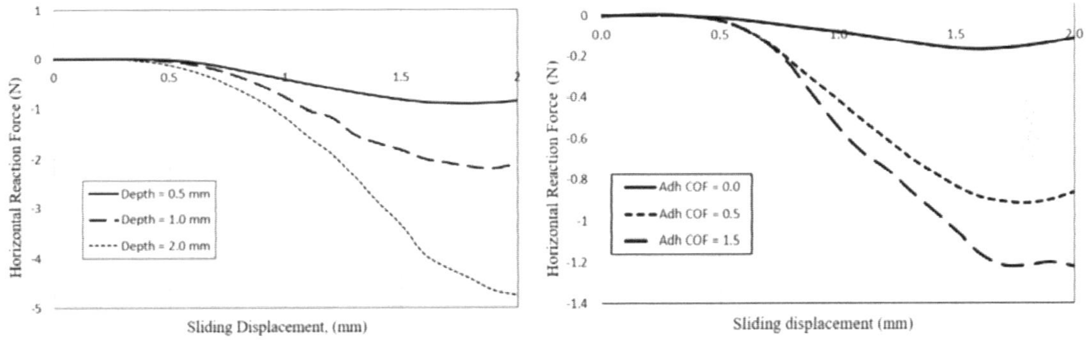

(a) (b)

Figure 5. Horizontal reaction force (a) by various sliding depth for adhesion COF = 0.5 and (b) by various adhesion COF for sliding depth 0.5 mm

The vertical reaction forces have relative constant value for low value of the adhesion COF and sliding depth, meanwhile, fluctuative values are obtained for the high value of the sliding depth as well as the adhesion COF. For high adhesion COF, it indicates any oscillation force that may also indicates any oscillation of the rubber surface along friction. In the contact between indenter tip and rubber surface, high adhesion causes high tangential force, therefore, stick occurrence is emerged in addition to slip occurrence. Thus, the stick-slip occurrence causes periodic or oscillation motion and this will lead to the emergence of oscillation forces. This phenomenon also occurs to the case of large sliding depth, that it causes high normal force, therefore, the tangential force is also high and the stick-slip occurrence is also emerged [14].

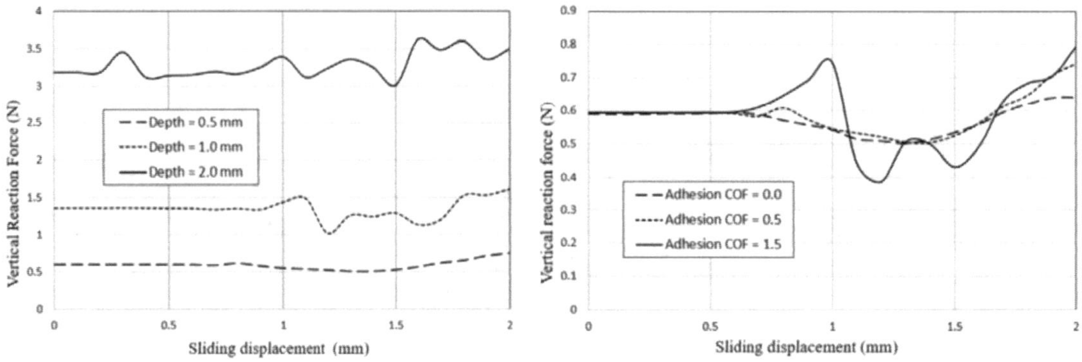

(a) (b)

Figure 6. Vertical reaction forces (a) by various sliding depth for adhesion COF = 0.5 and (b) by various adhesion COF for sliding depth 0.5 mm

Finally, the overall COF can be found by dividing the horizontal reaction forces with the vertical reaction forces. These results are depicted in Fig. 7 that is presented with respect to the sliding displacement.

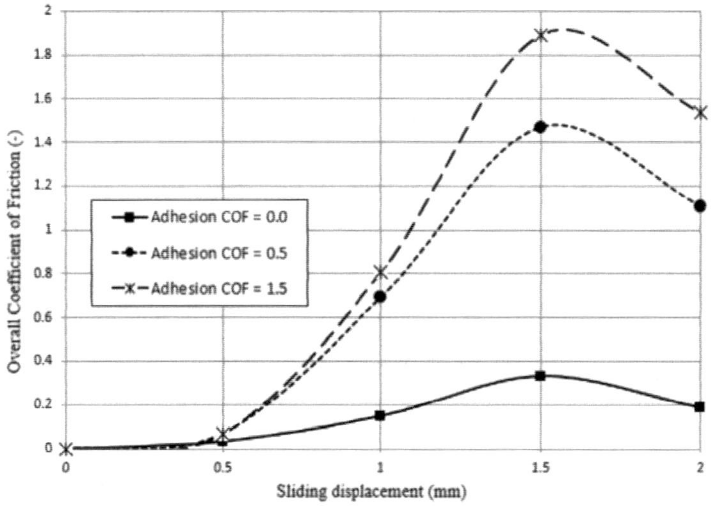

Figure 7. The overall coefficient of friction by given various adhesion COF for sliding depth 0.5 mm

Based on Fig.7, the overal COF highly increases and then decreases with respect to the sliding displacement. This overall COF tends to oscillatory change that might indicate the stick-slip occurrence along friction contact. These phenomena qualitatively are in accordance with the experimental results that the periodic value of COF occured along friction contact [6,15]. By the given constant adhesion COF, the overall COF is strongly influenced by the deformation COF component. In general, it indicates that the overall COF is dominantly influenced by the deformation component rather than the adhesion [11]. The deformation COF is developed from the friction force due to a tangential elastic force in the stick-slip contact. The deformation COF has fluctuative or periodic value that might follow the stick-slip theory. Theoretically, most of the stick-slip analysis is based on the static friction to represent a stick phase and kinetic friction to represent a slip phase [14].

4. Conclusion

This paper proposes a numerical investigation to identify the friction contact on the rigid blade sliding along unfilled rubber surface. Because of its hyperelastic behaviour, the elastic force has a dominant contribution on the friction force. And also, nonlinear behaviour of the rubber makes the sliding system difficult to be analitically investigated. Coefficient of friction (COF) is the main parameter of the friction contact that commonly consists of adhesion and deformation components. With the given constant adhesion COF, the deformation COF is strongly influenced by the sliding displacement and sliding depth. With the given adhesion COF, the overall COF highly increase and then decreases with respect to the sliding displacement. In general, the overall COF is dominantly influenced by the deformation component than the adhesion one. Qualitatively, the deformation COF tends to follow the stick-slip theory.

References

[1] MSC Software Whitepaper 2010 *Nonlinear Finite Element Analysis of Elastomer* (Santa Ana, USA: MSC Software Corporation)
[2] Johnson K L 1985 *Contact Mechanics* (Cambridge, UK: Cambridge University Press)
[3] Lafaye S 2008 *Wear* **264** 550-554
[4] Pinnington R J 2009 *Wear* **267** 1653–1664
[5] Vieira T, Ferreira R P, Kuchiishi A K, Bernucci L L B and Sinatora A 2015 *Wear* **328-329** 556–562
[6] Mane A, Loubet J L and Guerret C 2013 *Wear* **306** 149–160
[7] Coveney V and Menger C *Wear* **233-235** 702-711
[8] Zhang S W 2004 *Tribology of Elastomer, Tribology and Interface Engineering Series* (Amsterdam: Elsevier)
[9] Uchiyama Y and Ishino Y 1992 *Wear* **158** 141-155
[10] Liang H, Fukahori Y, Thomas A G and Busfield J J C 2009 *Wear* **266** 288-296.
[11] Wu Y P, Zhou Y, Li J L, Zhou H D, Chen J M and Zhao H C 2016 *Wear* **356-357** 1–8
[12] Liang H 2007 *Investigating the Mechanism of Elastomer Abrasion* (London, UK: PhD Thesis, University of London)
[13] ABAQUS 6.11 2011 *Standard User's Manual* (USA: Dassault Systems Simulia Corp.)
[14] Setiyana B, Ismail R, Jamari J and Schipper D J 2016 *Tribol. Online* **11,3** 512-518
[15] Rorrer R and Eiss N S 1995 *Tribol. Transaction* **38,2** 323-328

Design Project Management System Based on SOA Approach

Abba Suganda Girsang[1], Fahmi Jafar[2], Ahmad Nurul Fajar[3]

[1] **Computer Science Department, BINUS Graduate Program-Master of Computer Science, Bina Nusantara University, Jakarta, Indonesia 11480, Email : agirsang@binus.edu**
[2] **Computer Science Department, BINUS Graduate Program-Master of Computer Science, Bina Nusantara University, Jakarta, Indonesia 11480, Email:fahmijafar@gmail.com**
[3] **Information System Computer Science Department, BINUS Graduate Program-Master of Information Systems Management, Bina Nusantara University, Jakarta, Indonesia 11480, Email : afajar@binus.ac.edu**

ABSTRACT - Application architecture and information technology systems in the company should be able to support the company's policy that leads to their alignment with the business objectives. The alignment of the implementation of information technology systems to the needs and goals of the organization will be able to be answered by taking into account the integration in the development, so that the gap of these two things can be minimized. As a company engaged in IT consulting, project management system is one form of application systems that support core business objective of the company: to produce high-quality software. Nevertheless, the project management system should be able to work together with other supporting information systems, so that the flow of information within the company can run effectively. To achieve these objectives, required the analysis and design of an integrated system, which can be achieved through a service-oriented architecture based system. which includes the stages of requirements and analysis, design and development, and the last is the IT operations. The final result expected is the design of the service-oriented architecture based project management system which built through top-down approach that focuses on improving the effectiveness of business process.

1. Introduction

As a company engaged in IT consulting, project management system is one form of application systems that support core business objective of the software-vendor company: to produce high-quality software. Nevertheless, the project management system should be able to work together with other supporting information systems, so that the flow of information within the company can run effectively.

Data and information of the project stored in separate systems which could make it difficult for the parties involved in the project to collaborate in the project management process, as well as managerial parties in the company.

The information system which implemented in fragmented design could lead to difficulties in terms of synergies and integration of information related to the software project being built by the company to its customers.

To overcome these problems, required the analysis and design of a project management system which can be integrated to supporting systems while implementing business processes

Content from this work may be used under the terms of the Creative Commons Attribution 3.0 licence. Any further distribution of this work must maintain attribution to the author(s) and the title of the work, journal citation and DOI.
Published under licence by IOP Publishing Ltd

that have been run. System development approach that applies the principles of Service Oriented Architecture (SOA) is considered to be one of the solutions for the software-vendor company. With the SOA approach, business processes that have been implemented in the systems can be orchestrated from one system to another so that the flow of data and information can become more streamlined.

As in [1], Service Oriented Architecture is an approach in designing an application with the principle of reusing the components that already exists. SOA can act as an information technology architecture that can support a variety of applications to exchange data and participate in business processes.

With the basis of SOA in the development of the project management system, it is expected to obtain various benefits: the reusable service is established, application development become more efficient, as well as the increased collaboration between the developer of business unit application with the developer who developed the core business applications.

2. Related Works

A. Project Management

Project management is the application of knowledge, skills, tools, and techniques to project activities to meet project requirements [2]. As in [3], project management is the planning, organizing, directing, and controlling of company resources for a relatively short-term objective that has been established to complete specific goals and objectives.

Project management is defined by the PMBOK as the application of knowledge, skills, tools and techniques to project activities in order to meet stakeholder's needs and expectations from a project [4].

Project success has been defined to include the completion of: within the allocated time period, within the budgeted cost, at the proper performance or specification level, with acceptance by the customer/user, with minimum or mutually agreed upon scope changes, without disturbing the main work flow of the organization, and without changing the corporate culture [3].

B. Service-Oriented Architecture

SOA is an architectural style for building enterprise solutions based on services. More specifically, SOA is concerned with the independent construction of business-aligned services that can be combined into meaningful, higher-level business processes and solutions within the context of the enterprise [5].

As in [6], enterprise logic can be divided into two important halves: business logic and application logic. Each exists in a world of its own, and each represents a necessary part of contemporary organization structure. Business logic is a documented implementation of the business requirements that originate from an enterprises business areas. Business logic is generally structured into processes that express these requirements, along with any associated constraints, dependencies, and outside influences.

Application logic is an automated implementation of business logic organized into various technology solutions. Application logic expresses business process workflows through purchased or custom-developed systems within the confines of an organization's IT infrastructure, security constraints, technical capabilities, and vendor dependencies

IOP Conf. Series: Journal of Physics: Conf. Series **1090** (2018) 012079 doi:10.1088/1742-6596/1090/1/012079

3. Proposed Method

Framework on which to base this thesis starts from the observation of the background issues, especially related to project management system in a software-vendor company. Service lifecycle approach would be used during the process of this case study. Thus, it will achieve the objective of the case studies in the form of architectural design of service-oriented project management system.

A. Service Lifecycle Approach

The service lifecycle begins at inception (definition) and ends at its retirement (de-commissioning or repurposing). The service lifecycle enables service governance across its three stages: requirements and analysis, design and development, and IT operations . There are the three stages and the need for an enterprise service repository to enable service governance. **Requirements & analysis**: business initially identifies and prioritizes the business needs. **Design & development**: during the design phase, the business analysts work closely with the architect to hand off the business requirements. The architect is responsible for the high-level estimates, design, and handover to the development team. **IT operations**: IT operations is responsible for deploying, monitoring, and providing tier 1 support for all applications supported by IT. The approach described above illustrated in Fig. 1.

B. Project Management Business Processes

As one of the early stages of requirements and analysis will be carried out business processes mapping that will be included in the system in this case is the business processes related to project management of software development.

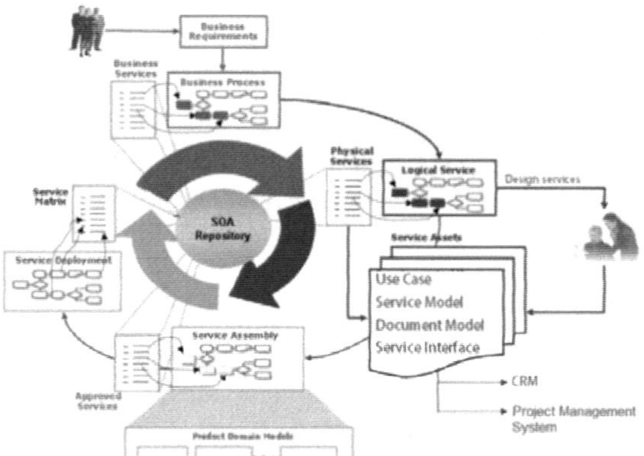

Fig. 1 Detailed Service Lifecycle Implementation [3]

1) Project Initiation

The project initiation phase is the 1st phase in the project management life cycle, as it involves starting up a new project. Project initiation is a procedure that must be done prior to a software development project is approved to work. The scope of this procedure include the determination of scope of work (software requirements) with customer, determination of the calculation of timeline and man-days, determination of a project cost up to the bid proposal to the customer and project initiation to project management department.

2) Project Management

Project management is a procedure that must be performed in managing a software development project according to standards of best practice so that the project is done it can provide satisfaction and more value to customers and meet the criteria for a successful project to improve the company's image and profits for the company.

3) Project Execution

Project execution is a procedure that must be done in managing software development projects undertaken in the company to the customer. The objective of the execution phase is to perform the work planned and approved during planning phase by developing the product or service that the project was commissioned to deliver. The process of software development project execution has special complexity compared to other projects, it would require a separate procedure. Characteristics of software that is intangible and detail the needs of new software generally can be identified clearly at the execution stage that would require a procedure of software development that are adaptive in order to accommodate the needs of customers and minimize project failures.

4. Analysis Result

A. Requirements and Analysis

To map the key business processes in this case study, will be described in diagrams that will help shape the business process model in high-level. Fig. 2 shows a high-level business architecture illustrated in a value-chain diagram. The value chain describes the full range of activities which are required to bring a product or service from conception, through the different phases of production (involving a combination of physical transformation and the input of various producer services), delivery to final consumers, and final disposal after use [7] [5].

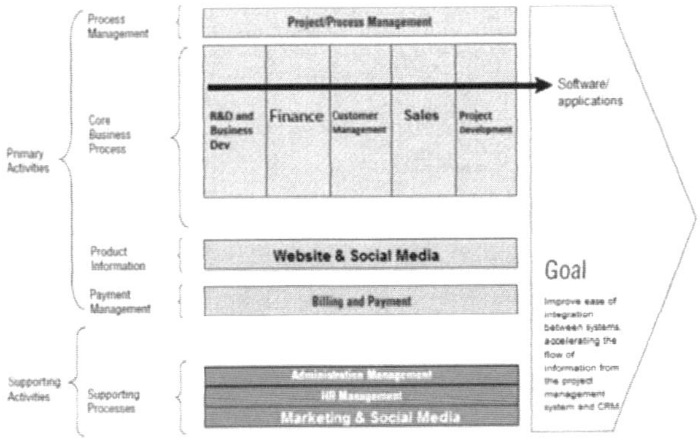

Fig. 2 Value-Chain Diagram

The main business processes and activities within the company are: Research and Development, Business Development, Finance, Customer Management, Sales and Project Development.

Two major business processes outlined in this study is related to Sales and Project Development, for both the business processes that directly relate to the application of project management and CRM systems, as well as the business processes that deliver value to the primary value chain in the company.

Business context diagram then can be depicted based on value-chain diagram above [5][8]. Business context diagram in Fig. 3 shows the four actors involved in the project management system, namely sales, project development, finance, and customer. Sales can initiate the project and obtain information regarding the current status of the project. Customers can get involved in the business process with stakeholders to update the data profile, and obtain data regarding the status of the project that is being done. Project development team can be involved through activities such as updating the status of the project, updating project expenditures (project expenses), as well as get the data item of unfinished work (pending task). Finance team can be involved through activities such as filling value of the project (project cost) as well as get the total cost of project expenditures.

Fig. 3 illustrates the scope of the interaction between the system (application) that can be bridged through the services, as well as its association with key business processes in the company. On the left and right of the conceptual architecture, there are systems (applications) that can interact with each other, CRM (customer relationship management) system, finance system, and project management system.

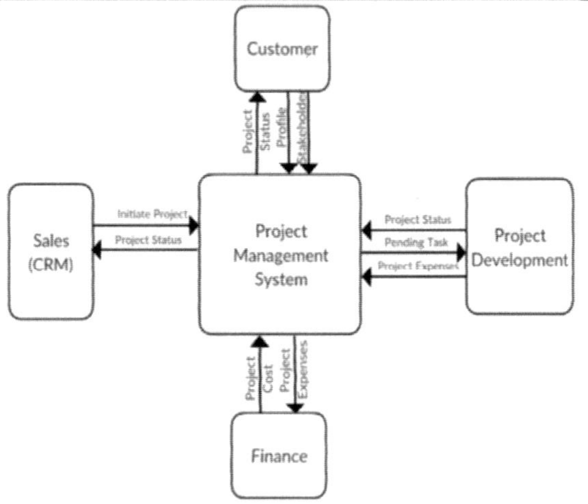

Fig. 3 Business Context Diagram

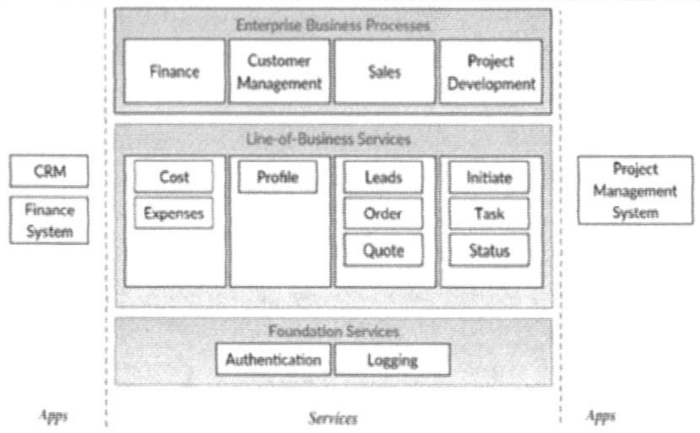

Fig. 4 Conceptual Architecture

At the bottom, there is a foundation service, service which supports the construction of the overall system, but does not implement business functionality.

B. Project Initiation

Business process model of project initiation deploys the activities in the beginning of a software development project in the company. At the end of the course of the process, after the Sales Department obtain confirmation of the project order from Customer, then the next activity is to initiate the project, which is done manually by communicating verbally or via a confirmation email to the Department of Project Management.

Sales Department will provide information and project documents to the project manager in the form of: customer data, stakeholders, project completion date, the value of the project (project valuation), budget (cost), and business requirement document.

Initiate project activity will be executed by utilizing the service, so that the project initiation process can be done automatically from the Sales Department (CRM application) to the Department of Project Management (Project Management System application). This can help avoid miscommunication between the department.

Based on the conceptual architecture diagram, services that can be used to initiate project activitiy are service projects which has initiate() operation, the operation to create a new project data in Project Management System.

C. Project Status

After starting the software development project, the process then continued with the implementation of business process monitoring and control over the course of the project development process. Process monitoring and control are run under the responsibility of the project manager to coordinate with the project team.

The project manager is assisted by the project team regularly provide status update regarding the progress of project, to the Sales Department, Customer, and directors (CEO).

Information update activity on the progress of project development in process monitoring and control can be executed through the service. Service that can be used in these activities is a service project that has updateProjectStatus() operation, which is an operation to update and get the status of project implementation in the Project Management System.

D. Project Expenses

The project manager will record expenses and control the use of cost so that it does not exceed the budget of the project cost which has been determined at the beginning of the project. Information update about the expenditure of the project is required to determine the amount of the costs incurred during the project development process, so that it can be monitored and controlled by project manager, finance department, as well as directors. Information update activity related to the project expenditure in monitoring and control can be executed through the service. Service that can be used in such activities is that finance service which has updateProjectExpenses() and viewProjectExpenses() operation. This is an operation to update and get the amount of the project cost in the Project Management System.

E. Design and Development

1) Service Model

From the stage of requirements and analysis, the services can be set up to handle the processes required. Services that formed include: projects, customer management, and finance management.

Fig. 5 Service Model

The function of the services in Fig. 5 are as follows. **Service "Project"**, it is used to handle activities that are in the process of the project, from the initiation of the project or create a project in the project management system. This service can also handle the process of updating the project progress status information, as well as view the status of the development of the project. **Service "Customer Management"**, is used to handle the processes associated with the customer, such as view customer profile and update customer profile. **Service "Finance"**, is used to handle the processes related to finance in a project, such as project expenses update, view project expenses, as well as view project cost.

2) Document Model

Based on processes and services that have been previously defined, can be determined the documents involved in the existing business processes. These documents can serve as input for the process as well as the output of the results of such processes, which passed through a service.

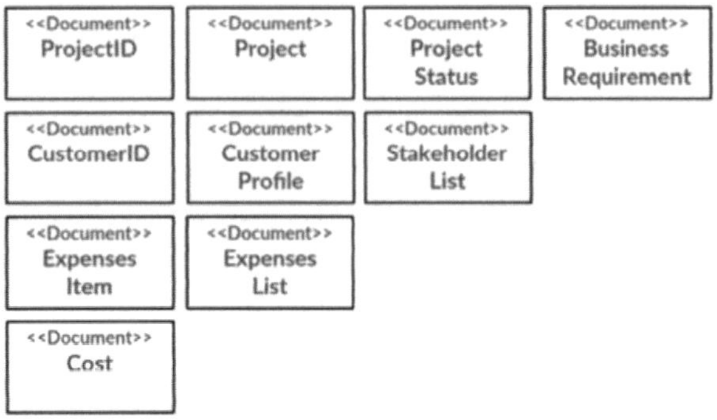

Fig. 6 Document Model

List of document in Fig. 6 is the required documents based on use case and sequence diagram. The document can be added in accordance with the addition of business processes. Document related to service "project" are : ProjectID (used as an input to obtain data Project), Project (the output of a service project), ProjectStatus and Business Requirement.

Document related to service "customer management" are : CustomerID (used as inputs to get CustomerProfile), CustomerProfile (the output of a customer management service to get Customer data), and CustomerList. Document related to service "finance", are : ExpensesItem (used as an input to register the expenses of a project), ExpensesList (an output of the operation to get the expenses data), and Cost (the output of the total budget of the project cost).

3) Service Definition

Based on the services and documents described above, the details of service's operation can be described along with the document involved, either as an input or output.

Service "Project" . Fig 7 shows that "project" service has three operations, that is: initiateProject(), updateProjectStatus(), and viewProjectStatus(). initiateProject() operation, requires business requirement document, CustomerID, and cost as input to create a project data in project management system. After the operation of the service successfully executed, it will produce a new project data in the project management system. **Service "Customer Management".** Fig. 8 illustrates that the customer management service has two operations, namely updateCustomerProfile(), and getCustomerProfile(). getCustomerProfile() operation requires CustomerID as input and will produce CustomerProfile as output. **Service "Finance".** Fig. 9 provides an illustration of the finance service which has three operations, namely updateProjectExpenses(), viewProjectExpenses() and viewProjectCost(). updateProjectExpenses() operation requires an expenses item document as input. viewProjectExpenses() operation requires ProjectID document to obtain a list of project expenses in a project. Similarly viewProjectCost() operation requires ProjectID document to get the total cost of a project.

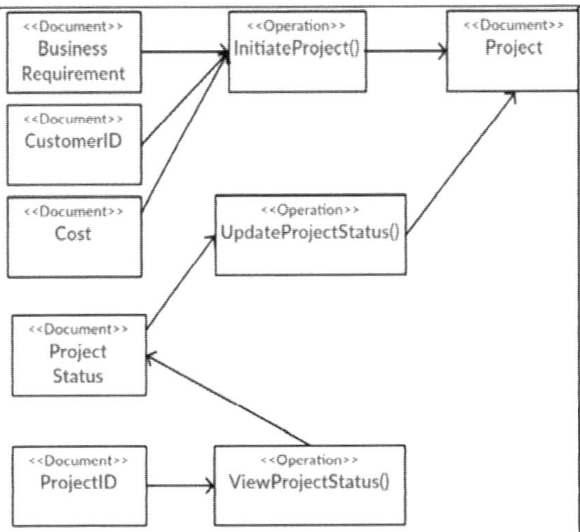

Fig. 7 Document and Operation - Service "Project"

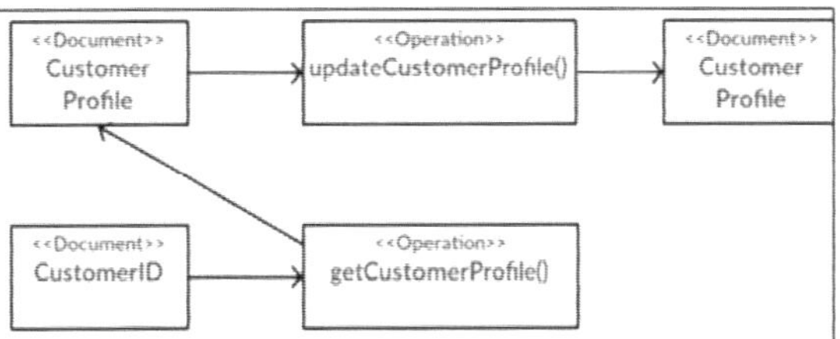

Fig. 8 Document and Operation - Service "Customer Management"

F. *IT Operations*

In the stage of IT Operations, the logical architecture design of services implementation and deployment will be using SOA Reference Architecture as shown in Fig. 10.

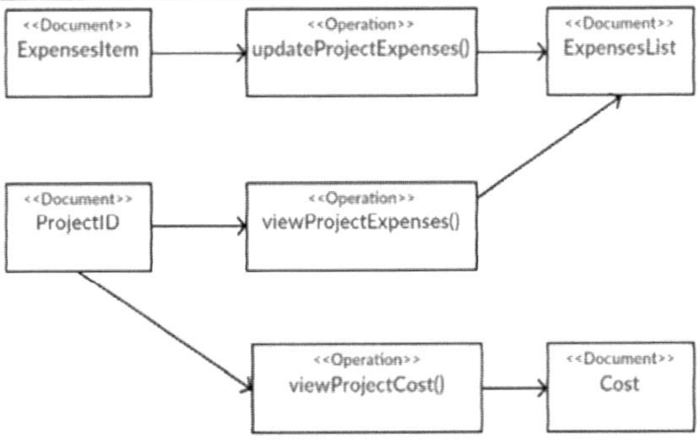

Fig. 9 Document and Operation - Service "Finance"

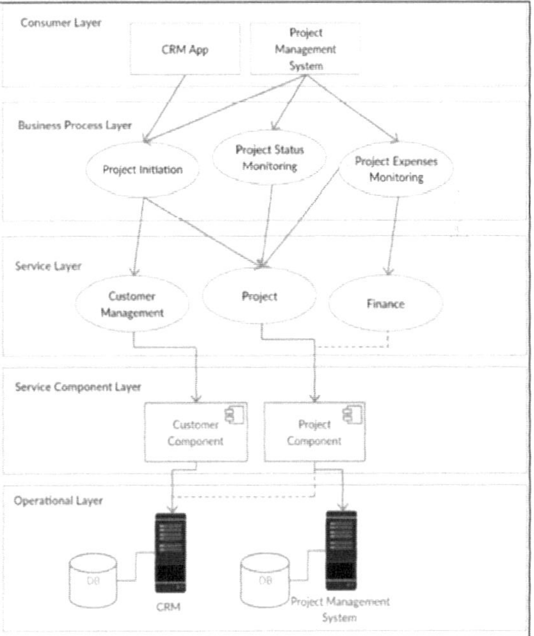

Fig.10 1 SOA Reference Architecture

There are five main layer in the architecture. In the operational layer, there are applications and modules that are used as source of data: CRM and Project Management System. The data from these two applications are used as a component in a service component layer, namely the customer and project component. In the service layer, there are services that have been designed before: project, finance, and customer management service. These services are the result of top-down analysis based on existing business processes, in which business processes are placed as an object in the business process layer.

In the consumer layer, there are users who interact directly with services, in this case is CRM application and project management system. Consumer of SOA can be developed and added to the other applications in the future, without having to build from scratch again to obtain data which related to the project because they can use the services that has been built.

G. Comparison of Manual-Based Process and Service-Based Process

To see how the service-oriented architecture can improve the effectiveness of the course of business processes, it can be compared between manual business processes and business processes which runs through a service-oriented architecture based system.

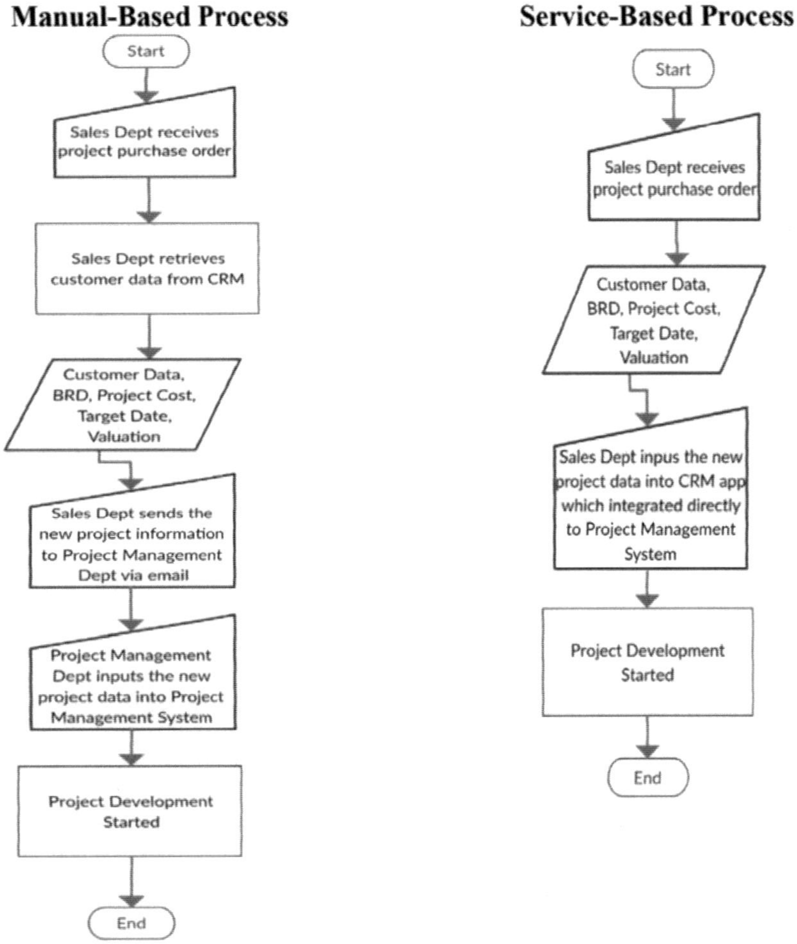

Fig. 11 . Project Initiation Process:Manual vs. Service

From the flow chart shown in Fig 11, manual-based process column, it appears that the project initiation process is done separately and not integrated between the CRM and Project Management System. The data transfer done from CRM to Project Management System which is bridged through the process of sending emails manually by the sales department.

But from the flow chart shown in Fig 11, service-based process column, it appears that the process of project initiation is done through an integrated CRM application interface directly with the Project Management System via an intermediary service. This can increase the effectiveness of project initiation process by reducing the stages which in manual process.

5. Conclusions

The architecture design of the SOA-based project management system successfully constructed based on project management business process in the company which has aligned with the business objectives. By using this architecture, the manual process can be cut which reduces the process time. Further, it can accelarate the flow information by integrating between the CRM application and project management systems. At last it can help to achieve the goal of project management business process in the company. Service lifecycle approach helps the process of designing a SOA-based project management system that can be developed into other business processes in the company. This project is built based on the current needs which relates directly to project management. Services has not been built all of department or mart in the company yet. In future, the development of all services bring the company grow continously and more agile.

References

[1] Hurwitz, J., Bloor, R., Baroudi, C., & Kaufman, M. Service Oriented Architecture for Dummies. United States: Wiley Publishing, 2007.

[2] Schwalbe, K., Information Technology Project Management 8e. United States: Course Technology, 2015.

[3] Kerzner, H., Project Management: A System Approach to Planning, Scheduling, and Controlling 11e. United States: John Wiley & Sons, Inc., 2013.

[4] Burke, R. Project Management: Planning and Control Techniques 4e. United Kingdom: John Wiley & Sons, Ltd, 2003.

[5] Rosen, M., Lublinsky, B., Smith, K. T., & Balcer, M. J. Applied SOA: Service-Oriented Architecture and Design Strategies. Indianapolis: Wiley Publishing, Inc., 2008.

[6] Rosen, M., “Use Context Diagrams to Bridge from Value Streams to Processes and Services,” A BPTrends Column., Oct. 2013.

[7] Kaplinsky, R., Morris, M., A Handbook for Value Chain Research, 2001.

[8] Reference Architecture for Service Oriented Architecture, OASIS, 2008

IOP Conf. Series: Journal of Physics: Conf. Series **1090** (2018) 012080 doi:10.1088/1742-6596/1090/1/012080

The *Hybrid* Model Algorithm on Sharia Insurance

[1]Rini Cahyandari, [2] Riva Lesta Ariany, [3]Sukono, [4]Yudha Satya Perkasa

[1]Department of Mathematics, Faculty of Science and Technology, UIN Sunan Gunung Djati Bandung

[2]Department of Mathematics Education, Faculty of Tarbiyah and Teacher Training, UIN Sunan Gunung Djati Bandung

[3]Department of Mathematics, Faculty of Mathematics and Natural Science, Universitas Padjadjaran

[4]Department of Physics, Faculty of Science and Technology, UIN Sunan Gunung Djati Bandung

[1]rini_cahyandari@uinsgd.ac.id

Abstract. Sharia insurance operational model was developed with the approach of trial and error by scholars and industry as a process to obtain the ideal format which can be used as a foundation for Islamic insurance operations. Models that are formed are: *mudharabah* model, *wakalah* model, and *hybrid* model. The development of the operational model of islamc insurance still continue because there are still many problems and Islamic issues that must be resolved. The *hybrid* model is the incorporation of the *wakalah* and *mudharabah* model. This model is mostly used in shariah insurance company in Indonesia because it is considered fair enough for policyholders and companies. In this paper, will do research on *hybrid* model by taking one example of Hajj travel insurance products, that is Mitra Mabrur Plus (AJB Bumiputera 1912) through the algorithm as follows: 1) translate the *hybrid* model scheme of Mitra Mabrur Plus into mathematical equations 2) drafting table of *hybrid* insurance Mitra Mabrur Plus model using mathematical equations on stage 1 and some initial value that set by the insurance company. This *hybrid* model table will show the details of the management of funds sourced from the contribution of the policyholders, including the quantity percentage of personal account, the percentage of *tabarru' account*, the percentage of *ujrah*, magnitudes of profit and benefit that be accrues to the policyholders. Through this table, are expected that management of the fund contribution is shariah insurance will be more transparent and fair for companies or policyholders.
Keywords: shariah insurance, hybrid model, personal account, tabarru' account, ujrah

1. Background

Insurance is an economic institution that aims to reduce risk, with combined within one management a number of objects that the numbers are fairly large, so that the overall losses can be foreseen within certain limits. **[2]**

Content from this work may be used under the terms of the Creative Commons Attribution 3.0 licence. Any further distribution of this work must maintain attribution to the author(s) and the title of the work, journal citation and DOI.
Published under licence by IOP Publishing Ltd

IOP Conf. Series: Journal of Physics: Conf. Series **1090** (2018) 012080 doi:10.1088/1742-6596/1090/1/012080

Other definitions about insurance on UU No. 12 of 1992 about Venture of Insurance according to Dahlan Siamat in his book about management of the financial institution are: "Insurance or the bearing is an agreement between two or more parties, with which the insurer committing yourself to the insured, by accepting the insurance premiums, to provide reimbursement to the insured because of the loss, damage or lost profits expected, or legal liability to third parties which may be suffered by the insured, arising from an event which is uncertain, or to provide a payment based upon died or life a person who paid". **[5]**

Overall, insurance in Indonesia can be distinguished into two types: conventional insurance and syariah insurance. Both have the same purpose, which is the management and mitigation of risks. The National Shariah Council MUI assign that definition of shariah insurance (*ta'min, takaful, tadhamun*) as an effort to protect each other and help each other among of peoples or party through investment funds in the form of assets or *tabarru'* which gives the pattern of returns to face certain risks through a contract that accordance with the shari'ah (Fatwa No. 21 MUI DSN/DSN-MUI/X/2001 about sharia insurance general guidelines).

Sharia insurance operational model that was already well known, that is: *wakalah* model, *mudharabah* model, *hybrid* model. According to the Fiqh terms, *mudharabbah* is a contract agreement between the two sides, one of which as a giver of the capital to the other so that developed, while its profits will be divided between them in accordance with the agreement that has been approved in advance **[6]**. Meanwhile, *wakalah* according to *fiqh* means submission, delegation, or the granting of a mandate. According to fatwa No. 52 DSN/DSN-MUI/III/2006 about *wakalah bil ujrah*, *wakalah* contract can be done between the insurance companies with policyholders. *Wakalah bil ujrah* contract for insurance, which is one of *wakalah* contract where the policyholder provides authority to the insurance company in exchange for awarding ujrah. Furthermore, the *hybrid* is the amalgamation of the *wakalah* model and *mudharabah* model, in which the magnitude of the fee *wakalah* revamped based on contributions and profit sharing on investment. This model is mostly used in shariah insurance company in Indonesia because it is considered fair enough for policyholders and insurance companies.

In this paper, will do research on *hybrid* model by taking one example of Hajj travel insurance products from AJB Bumiputera 1912, that is Mitra Mabrur Plus insurance. The stages of this research generally done through the algorithm as follows: 1) translate the *hybrid* model scheme of Mitra Mabrur Plus into mathematical equations 2) drafting the *hybrid* table of Mitra Mabrur Plus insurance.

2. Methodology

Stages of research conducted, including the following:

1. Visit and interview with actuary of AJB Bumiputera 1912 company about terms of insurance and Mitra Mabrur Plus insurance data.
2. Present the data from Mitra Mabrur Plus insurance in the form of table and scheme of *hybrid* model.
3. Translate the *hybrid* model scheme into mathematical equations.
4. Drafting the *hybrid* model table of Mitra Mabrur Plus insurance using mathematical equations in stage 3 and some initial value that set by the insurance company.

There are several terms that are often used in insurance, including:

1. Polis : The insurance agreement between the insurer and the policyholder.
2. Policyholders : A person or a group of people doing the alliance insurance contract (policy) with an insurance company.
3. Premium : An amount of money that is listed in the policy that approved by the policyholder, for paid to insurance companies according the agreement so that the policy remains active.
4. Personal Account : Collection of funds that belong to policyholders and payable if the agreement ends, the participant resigns, or dies.

5. *Tabarru' Account* : The collection of funds that intended by the policyholder as alms with the purpose to help each other and will be paid in the event of claims.
6. Contract : The alliance, the agreement, or conspiracy
7. Claims : Requests or demands for payment of benefits in accordance with the provisions that set in the polis.
8. Profit sharing : Profit sharing obtained proportionally based on the *nisbah* of profit Investment sharing that have been determined.
9. Surrender Value : The amount of money that have to be paid to policyholders, if the insurance agreement is terminated before the expiry of the agreement.
10. Rider : Conditions that attached to the insurance policy which gives additional benefits or restrictions.
11. Ujrah : Wages or rewards.
12. Hospital Bill : Rider (benefits) in the form of reimbursement of the cost ofhospital.
13. Critical Illness : Rider (benefits) in the form of sum assured in case of critical illness.
14. Death Coverage : An amount of money that are prepared by the insurance company as compensation if the policyholder dies.
15. Death Benefit : An amount of money that will be paid if the policyholder dies before and or after the period of the insurance agreement terminates.

Mitra Mabrur Plus insurance data are presented in tabular form as below:

Table 1 Mitra Mabrur Plus insurance data

MITRA MABRUR PLUS INSURANCE (AJB BUMIPUTERA 1912)						
		Year				
	Year of Polis	1	2	3	4	5
1	Contributions per year	1,200,000				
	Contributions per month (P)	100,000				
2	% Basis of *Tabarru'* (A)	7.47	7.47	7.47	7.47	7.47
3	% *Ujrah* (B)	19.35	11.84	9.34	9.34	9.34
4	% *Personal Account* (C)	73.18	80.69	83.19	83.19	83.19

Some initial value assignment, such:
Age of the Policyholder : 38 Years
Death Coverage : 6,000,000
Hospitall Bill : 600,000
40 Critical illness : 3,000,000

Mitra Mabrur Plus insurance data are presented in scheme form as below:

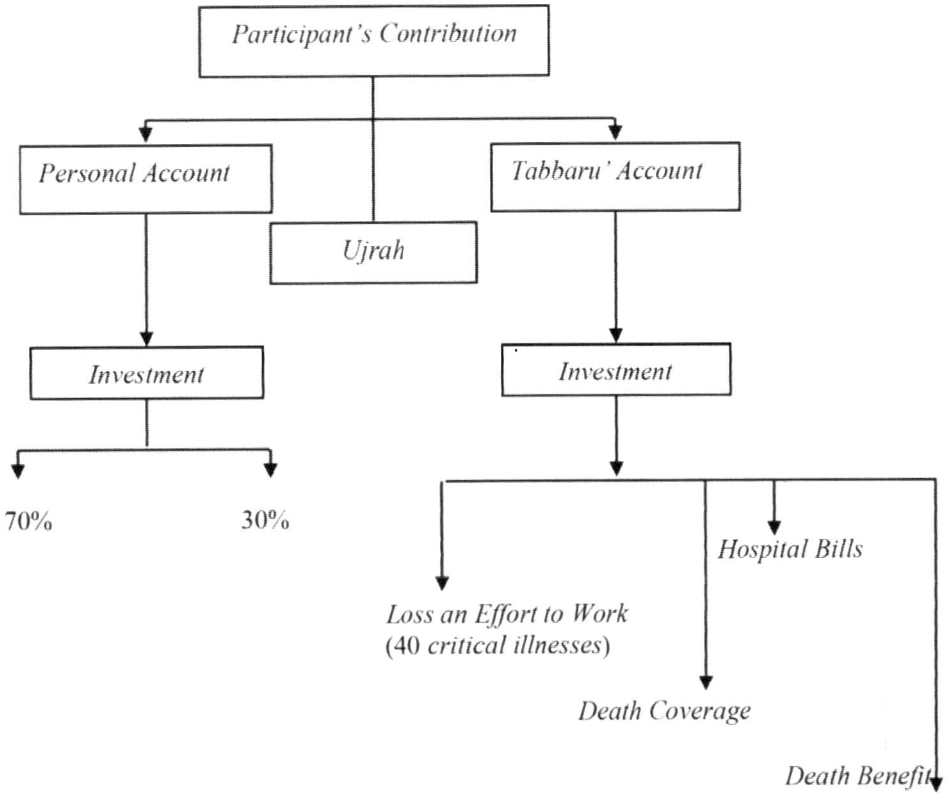

Figure 1 *Hybrid* model scheme of Mitra Mabrur Plus insurance

3. Result

Using the scheme in Figure 1 and the data in Table 1 as well as initial values which have been determined by the insurance company, then the obtained mathematical equations needed to compose the table of *hybrid* model from Mitra Mabrur Plus insurance as below:

Table 2 Mathematical equations for *hybrid* model scheme of Mitra Mabrur Plus Insurance

1	2	3	4	5	6	7	8	9	10
Month	***Age***	**Contribution**	***Tabarru' Account***	***Ujrah***	***Personal Account***	**Accumu lation**	***monthly profit***	***Yearly Profit***	***Surrender Value***
1	38	P	0.0747*P	0.1935 *P	0.7318*P	73,180	201		73,381
2-11	38	P	0.0747*P	0.1935 *P	0.7318*P	X_1	Y_1		X_1+Y_1
12	38	P	0.0747*P	0.1935 *P	0.7318*P	X_1	Y_1	0.05*(0.7318*P)	$(X_1+Y_1)+(0.05*(0.7318*P))$
13	39	P	0.0747*P	0.1184 *P	0.8069*P	(0.7318 *P) + (0.8069 *P)	Y_2		((0.7318*P) + (0.8069*P)) + Y_2
14-23	39	P	0.0747*P	0.1184 *P	0.8069*P	X_2	Y_2		X_2+Y_2
24	39	P	0.0747*P	0.1184 *P	0.8069*P	X_2	Y_2	0.05*(0.8069*P)	$(X_2+Y_2)+(0.05*(0.8069*P))$

25	40	P	0.0747*P	0.0934 *P	0.8319*P	(0.8069 *P) + (0.8319 *P)	Y_3		((0.8069*P) + (0.8319*P))+ Y_3
26-35	40	P	0.0747*P	0.0934 *P	0.8319*P	X_3	Y_3		X_3+Y_3
36	40	P	0.0747*P	0.0934 *P	0.8319*P	X_3	Y_3	0.05*(0.831 9*P)	(X_3+Y_3)+ (0.05*(0.8319*P))
37-47	41	P	0.0747*P	0.0934 *P	0.8319*P	X_4	Y_4		X_4+Y_4
48	41	P	0.0747*P	0.0934 *P	0.8319*P	X_4	Y_4	0.05*(0.831 9*P)	(X_4+Y_4)+ (0.05*(0.8319*P))
49-59	42	P	0.0747*P	0.0934 *P	0.8319*P	X_5	Y_5		X_5+Y_5
60	42	P	0.0747*P	0.0934 *P	0.8319*P	X_5	Y_5	0.05*(0.831 9*P)	(X_5+Y_5)+(0.05*(0.8319 *P))

advanced of Table 2,

1	2	11	12	13	14
Month	***age***	***40 critical Illness***	***death coverage***	***Hospital Bills***	***Death Benefit***
1	38	0.5*((P*12)*5)	(P*12)*5	0.1*((P*12)*5)	((P*12)*5) + 73,381
2-11	38	0.5*((P*12)*5)	(P*12)*5	0.1*((P*12)*5)	((P*12)*5) + (X_1+Y_1)
12	38	0.5*((P*12)*5)	(P*12)*5	0.1*((P*12)*5)	((P*12)*5) + ((X_1+Y_1)+(0.05*(0.7318*P)))
13	39	0.5*((P*12)*5)	(P*12)*5	0.1*((P*12)*5)	((P*12)*5) + (((0.7318*P)+(0.8069*P)) + Y_2)
14-23	39	0.5*((P*12)*5)	(P*12)*5	0.1*((P*12)*5)	((P*12)*5) + (X_2+Y_2)
24	39	0.5*((P*12)*5)	(P*12)*5	0.1*((P*12)*5)	((P*12)*5) + ((X_2+Y_2)+ (0.05*(0.8069*P)))
25	40	0.5*((P*12)*5)	(P*12)*5	0.1*((P*12)*5)	((P*12)*5) + (((0.8069*P)+(0.8319*P))+ Y_3)
26-35	40	0.5*((P*12)*5)	(P*12)*5	0.1*((P*12)*5)	((P*12)*5) + (X_3+Y_3)
36	40	0.5*((P*12)*5)	(P*12)*5	0.1*((P*12)*5)	((P*12)*5) + ((X_3+Y_3)+ (0.05*(0.8319*P)))
37-47	41	0.5*((P*12)*5)	(P*12)*5	0.1*((P*12)*5)	((P*12)*5) + (X_4+Y_4)
48	41	0.5*((P*12)*5)	(P*12)*5	0.1*((P*12)*5)	((P*12)*5) + ((X_4+Y_4)+ (0.05*(0.8319*P)))
49-59	42	0.5*((P*12)*5)	(P*12)*5	0.1*((P*12)*5)	((P*12)*5) + (X_5+Y_5)
60	42	0.5*((P*12)*5)	(P*12)*5	0.1*((P*12)*5)	((P*12)*5) + ((X_5+Y_5)+(0.05*(0.8319*P)))

In the table 2 above, showed that the contribution per month over the past five years are divided into three accounts, which is *tabarru' account* amounting to 7.47% per year, personal account 73.18% in the first year, 80.69% in the second year, 83.19% in the third year until the fifth year, and the leftover is *ujrah*. Next, assigned the initial value to surrender value amounted to 73.381 with gains in the first moth of 201. As for the calculation of the surrender value for the next months is obtained from the magnitudes of the accumulation and profit per month or per year. The benefit that be accrues to the policyholders, which is assured if critical illness by 50% of contributions per year and assured the hospital costs of 10% of the contributions per year.

Subtituting the P value of 100.000 to table 1, then resulting the *hybrid* table model as follows:

Tabel 3 *Hybrid* Model Table of Mitra Mabrur Plus Insurance

1	2	3	4	5	6	7	8	9	10	11	12	13	14
Month	*Age*	*Contribution*	*Tabarru' Account*	*Ujrah*	*Personal Account*	*Accumulation*	*Monthly Profit*	*Yearly Profit*	*Surrender Value*	*40 Critical Illness*	*Death Coverage*	*Hospital Bills*	*Death Benefit*
1	38	100,000	7,470	19,350	73,180	73,180	201		73,381	3,000,000	6,000,000	600,000	6,073,381
2	38	100,000	7,470	19,350	73,180	146,360	201		146,561	3,000,000	6,000,000	600,000	6,146,561
3	38	100,000	7,470	19,350	73,180	146,360	201		146,561	3,000,000	6,000,000	600,000	6,146,561
4	38	100,000	7,470	19,350	73,180	146,360	201		146,561	3,000,000	6,000,000	600,000	6,146,561
5	38	100,000	7,470	19,350	73,180	146,360	201		146,561	3,000,000	6,000,000	600,000	6,146,561
6	38	100,000	7,470	19,350	73,180	146,360	201		146,561	3,000,000	6,000,000	600,000	6,146,561
7	38	100,000	7,470	19,350	73,180	146,360	201		146,561	3,000,000	6,000,000	600,000	6,146,561
8	38	100,000	7,470	19,350	73,180	146,360	201		146,561	3,000,000	6,000,000	600,000	6,146,561
9	38	100,000	7,470	19,350	73,180	146,360	201		146,561	3,000,000	6,000,000	600,000	6,146,561
10	38	100,000	7,470	19,350	73,180	146,360	201		146,561	3,000,000	6,000,000	600,000	6,146,561
11	38	100,000	7,470	19,350	73,180	146,360	201		146,561	3,000,000	6,000,000	600,000	6,146,561
12	38	100,000	7,470	19,350	73,180	146,360	201	3,659	150,220	3,000,000	6,000,000	600,000	6,150,220
13	39	100,000	7,470	11,840	80,690	153,870	38,372		192,242	3,000,000	6,000,000	600,000	6,192,242
14	39	100,000	7,470	11,840	80,690	161,380	38,372		199,752	3,000,000	6,000,000	600,000	6,199,752
15	39	100,000	7,470	11,840	80,690	161,380	38,372		199,752	3,000,000	6,000,000	600,000	6,199,752
16	39	100,000	7,470	11,840	80,690	161,380	38,372		199,752	3,000,000	6,000,000	600,000	6,199,752
17	39	100,000	7,470	11,840	80,690	161,380	38,372		199,752	3,000,000	6,000,000	600,000	6,199,752
18	39	100,000	7,470	11,840	80,690	161,380	38,372		199,752	3,000,000	6,000,000	600,000	6,199,752
19	39	100,000	7,470	11,840	80,690	161,380	38,372		199,752	3,000,000	6,000,000	600,000	6,199,752
20	39	100,000	7,470	11,840	80,690	161,380	38,372		199,752	3,000,000	6,000,000	600,000	6,199,752
21	39	100,000	7,470	11,840	80,690	161,380	38,372		199,752	3,000,000	6,000,000	600,000	6,199,752
22	39	100,000	7,470	11,840	80,690	161,380	38,372		199,752	3,000,000	6,000,000	600,000	6,199,752
23	39	100,000	7,470	11,840	80,690	161,380	38,372		199,752	3,000,000	6,000,000	600,000	6,199,752
24	39	100,000	7,470	11,840	80,690	161,380	38,372	4,035	203,787	3,000,000	6,000,000	600,000	6,203,787
25	40	100,000	7,470	9,340	83,190	163,880	51,983		215,863	3,000,000	6,000,000	600,000	6,215,863
26	40	100,000	7,470	9,340	83,190	166,380	51,983		218,363	3,000,000	6,000,000	600,000	6,218,363
27	40	100,000	7,470	9,340	83,190	166,380	51,983		218,363	3,000,000	6,000,000	600,000	6,218,363
28	40	100,000	7,470	9,340	83,190	166,380	51,983		218,363	3,000,000	6,000,000	600,000	6,218,363
29	40	100,000	7,470	9,340	83,190	166,380	51,983		218,363	3,000,000	6,000,000	600,000	6,218,363
30	40	100,000	7,470	9,340	83,190	166,380	51,983		218,363	3,000,000	6,000,000	600,000	6,218,363
31	40	100,000	7,470	9,340	83,190	166,380	51,983		218,363	3,000,000	6,000,000	600,000	6,218,363
32	40	100,000	7,470	9,340	83,190	166,380	51,983		218,363	3,000,000	6,000,000	600,000	6,218,363

33	40	100,000	7,470	9,340	83,190	166,380	51,983		218,363	3,000,000	6,000,000	600,000	6,218,363
34	40	100,000	7,470	9,340	83,190	166,380	51,983		218,363	3,000,000	6,000,000	600,000	6,218,363
35	40	100,000	7,470	9,340	83,190	166,380	51,983		218,363	3,000,000	6,000,000	600,000	6,218,363
36	40	100,000	7,470	9,340	83,190	166,380	51,983	4,160	222,523	3,000,000	6,000,000	600,000	6,222,523
37	41	100,000	7,470	9,340	83,190	166,380	56,744		223,124	3,000,000	6,000,000	600,000	6,223,124
38	41	100,000	7,470	9,340	83,190	166,380	56,744		223,124	3,000,000	6,000,000	600,000	6,223,124
39	41	100,000	7,470	9,340	83,190	166,380	56,744		223,124	3,000,000	6,000,000	600,000	6,223,124
40	41	100,000	7,470	9,340	83,190	166,380	56,744		223,124	3,000,000	6,000,000	600,000	6,223,124
41	41	100,000	7,470	9,340	83,190	166,380	56,744		223,124	3,000,000	6,000,000	600,000	6,223,124
42	41	100,000	7,470	9,340	83,190	166,380	56,744		223,124	3,000,000	6,000,000	600,000	6,223,124
43	41	100,000	7,470	9,340	83,190	166,380	56,744		223,124	3,000,000	6,000,000	600,000	6,223,124
44	41	100,000	7,470	9,340	83,190	166,380	56,744		223,124	3,000,000	6,000,000	600,000	6,223,124
45	41	100,000	7,470	9,340	83,190	166,380	56,744		223,124	3,000,000	6,000,000	600,000	6,223,124
46	41	100,000	7,470	9,340	83,190	166,380	56,744		223,124	3,000,000	6,000,000	600,000	6,223,124
47	41	100,000	7,470	9,340	83,190	166,380	56,744		223,124	3,000,000	6,000,000	600,000	6,223,124
48	41	100,000	7,470	9,340	83,190	166,380	56,744	4,160	227,284	3,000,000	6,000,000	600,000	6,227,284
49	42	100,000	7,470	9,340	83,190	166,380	57,953		224,333	3,000,000	6,000,000	600,000	6,224,333
50	42	100,000	7,470	9,340	83,190	166,380	57,953		224,333	3,000,000	6,000,000	600,000	6,224,333
51	42	100,000	7,470	9,340	83,190	166,380	57,953		224,333	3,000,000	6,000,000	600,000	6,224,333
52	42	100,000	7,470	9,340	83,190	166,380	57,953		224,333	3,000,000	6,000,000	600,000	6,224,333
53	42	100,000	7,470	9,340	83,190	166,380	57,953		224,333	3,000,000	6,000,000	600,000	6,224,333
54	42	100,000	7,470	9,340	83,190	166,380	57,953		224,333	3,000,000	6,000,000	600,000	6,224,333
55	42	100,000	7,470	9,340	83,190	166,380	57,953		224,333	3,000,000	6,000,000	600,000	6,224,333
56	42	100,000	7,470	9,340	83,190	166,380	57,953		224,333	3,000,000	6,000,000	600,000	6,224,333
57	42	100,000	7,470	9,340	83,190	166,380	57,953		224,333	3,000,000	6,000,000	600,000	6,224,333
58	42	100,000	7,470	9,340	83,190	166,380	57,953		224,333	3,000,000	6,000,000	600,000	6,224,333
59	42	100,000	7,470	9,340	83,190	166,380	57,953		224,333	3,000,000	6,000,000	600,000	6,224,333
60	42	100,000	7,470	9,340	83,190	166,380	57,953	4,160	228,493	3,000,000	6,000,000	600,000	6,228,493

In the table 3, showed the details of the management of the fund contribution to policyholders during a five-year contract. Through this table could be seen nominal magnitudes each account, made per month or per year, and the benefits that become accrues of policyholders in case of claims during the year of polis runs or if the contract is stopped in the middle for some reason.

4. Summary and Advice

The *hybrid* model at the Mitra Mabrur Plus insurance considered good enough in terms of the mechanism. Shown at the percentage of the quantity of personal accounts that increased annually, and *ujrah* that was decreased. This conditions would be profitable for policyholders. Meanwhile, the percentage of quantity *tabarru' account* that stable during the contract years is profitable too for the insurance company because it has minimize loss that caused by fluctuations in investment result.

Hence, a *hybrid* model is considered fairer to policyholders as well as insurance companies. As research continued, it is advisable to perform this *hybrid* model optimization by changing the percentage of *tabarru' accounts* with the expectation of the raw quantity obtained as the ideal format that can maximize benefits to policyholders as well as profits for insurance companies.

Acknowledgment
Thanks to Mr.Windhu Wijatmoko, Drs, MM, AAIJ, FSAI as actuary of the AJB Bumiputera 1912 company that willing take the time to discuss about Mitra Mabrur Plus insurance, and Annisa Azizah, S.Si has helped directly in the writing of this paper. Thanks were also given to all who has helped directly or indirectly so that paper can be resolved properly.

References
[1] Cahyandari, R, et.al. *" A design of mathematical modelling for the mudharabah scheme in shariah insurance"*. IOP Conference Series: Materials Science and Engineering, Volume 166. 2017.
[2] Djojosoedarso, S. *"Prinsip-prinsip Manajemen Risiko dan Asuransi"*. Salemba Empat. Jakarta. 1999.
[3] Ghazali, P . L, et.al. *"Mathematical Modelling in Family Takaful"*. Journal of Applied Science, 11: 3381-3388. 2011a.
[4] Ghazali, P . L, et.al. *"Integration Model In Premium Life Table Of Family Takaful"*. Journal of Applied Science, 8(7): 3763-3776. 2012.
[5] Siamat, Dahlan. *"Manajemen Lembaga Keuangan"*. Lembaga Penerbit FE UI. Jakarta. 1999.
[6] Sula, M. S. *"Asuransi Syariah (Life and General) Konsep dan Sistem Operasional"*. Gema Insani. Jakarta. 2004.

Development of Ebola Vaccine Candidate by in Silico from Glikoprotein (GP) Gene of Ebola Zaire Virus

Yani Suryani[1], Opik Taupiqurrohman[1], Sri Aryanti[1], Moh. Nurul Subkhi[2], Epa Paujiah[3]

[1]Department of Biology, Faculty of Sciences and Technology, UIN Sunan Gunung Djati Bandung
[2]Department of Physics, Faculty of Sciences and Technology UIN Sunan Gunung Djati Bandung
[3]Department of Biology Education, Faculty of Tarbiyah dan Keguruan, UIN Sunan Gunung Djati Bandung
e-mail: yan_dikha@gmail.com

Abstract. Immunoinformatics is the one of bioinformatics divisions. The focus of this division is to design compounds of immune response activists or candidate vaccine in silico or computation. One type of immune response activator compound is a peptide. Peptides are small amounts of amino acid residues. The amount of amino acid residues that can activate the immune response ranges from 9-15 amino acids. Good candidate vaccine quality is shown affinity or strong bond between peptide and MHC (Major Histocompatibility Complex) is indicated by the value of energy of molecule-blocking process through low software. This research is conducted in Data and Process Laboratory of Biology Department, Faculty of Science and Technology, States of Islamic University Sunan Gunung Djati Bandung, in December 2016 until February 2017. The study amis to get candidate vaccine ebola peptide form 9 (nine) amino acid residues. The tool used in this research is the software which made based on the working principle of immune response. The software is SDS Workbench, IEDB-AR, Emboss, and CABSdock. The material used in this research is the sequence information of ebola virus glycoprotein. The results show that the peptides with FLYDRLAST (Fenilanalnin, Leusin, Tyrosine, Aspartic acid, Arginine, Leusin, Alanine, Serin Treonin) are potential candidate for the ebola peptide vaccine because of their high affinity values with MHC I, indicated by molecular-binding energy which is very low ie -1870.69 Kcal / mol.

Keywords: Glicoprotein, Immunoinformatics, *In Silico*, Zaire ebolavirus

1. Introduction

Ebola disease grows rapidly in Africa[1,2]. In human, this disease causes severe bleeding in the body even the death rate of patients reaches 90%[3,4]. Every year, cases and deaths from ebola disease continue to increase. In 2016, WHO recorded from 28,638 cases of ebola disease, the death rate reached 11,316 people[5].

The ebola disease is caused by five ebola species that have different mortality impacts on humans ie Zaire ebolavirus, Sudan ebolavirus, Tai Forest ebolavirus, Reston ebolavirus[6] and Bundibugyo ugolavirus[1,2,3]. The species of Zaire ebolaviruses are the most pathogenic

Content from this work may be used under the terms of the Creative Commons Attribution 3.0 licence. Any further distribution of this work must maintain attribution to the author(s) and the title of the work, journal citation and DOI.
Published under licence by IOP Publishing Ltd

species with mortality rates around 80%[3]. Prevention of this disease can be done with the vaccine[3].

The design of epitope/peptide based vaccine candidates using immunonformatics has been widely developed[4,7]. Epitopes are a minimal part of the antigen that can induce the immune system[8]. Epitope-based vaccines have many advantages such as more effective, more efficient, and the process is short for one to two years[9]. The immunoinformatics approach begins with the determination of the genome sequence and proceeds by predicting the antigen most likely to be used as a vaccine candidate from the genome sequence[10].

The ebola virus is composed of genomes consisting of nucleoprotein (NP), protein structural virion (VP) VP35, VP40, glycoprotein (GP), VP30, VP24 and RNA-dependent RNA polymerase (L)[3]. From the genome structure of the ebola virus, the glycoprotein gene (GP) is the most potential gene to be used as a source of ebola vaccine candidates. This is because GP in the ebola virus is the only surface protein of the virus and is very important in the process of attachment to the host cell and catalyze the occurrence of membrane fusion[2].

The immune response of the body is triggered by the entry of antigens or microorganisms into the body and is confronted by macrophage cells which will subsequently play the antigen presenting cell (APC). These cells will capture small amounts of antigens and are expressed to the surface of cells that can be recognized by T lymphocytes cells (Th or T helper). These Th cells will be activated and will activate other lymphocytes such as B lymphocytes or cytotoxic T lymphocytes. T cells only recognize immunogens which attaches to the MHC protein (Major Histocompatibility Complex) on the surface of another cell[19]. Proteins that interact with MHC is what can be used as a candidate vaccine because it can induce the immune system process.

One of the main conditions in designing epitope vaccines through the immunoinformatics approach is that there should be no part of the homologous candidate vaccine source sequence / similar to the human genome sequence, since it is feared that if the vaccine is made from a homologous candidate source with the human genome, the vaccine will cause the disease autoimmune[11]. The next important stage of vaccine candidate design is the determination of the antigen binding affinity of peptides / epitope on MHC molecules[7] Furthermore, peptide-MHC interaction analysis was done through molecular blocking process. This method is expected to obtain potential candidate vaccine peptide sequence.

2. Methods

The sequence of glycoprotein amino acid is obtained from the uniprot.org website. The sequence is then analyzed homologically with the human genome using SDSC (San Diego Supercomputer Center) Biology Workbranch - TFASTY. The quality of the protein bonds was then analyzed using MHC I using the program of IEDB-AR (Immune Epitope Database-Analysis Resources)[12,13]. The bonding and interaction analysis between MHC I-peptides and molecular-binding methods using the CABSdock program[14] and BIOVIA Discovery Studio Visualizer Version 4.5 was performed.

3. Results and Discussion

Homology or similarity can be seen based on statistical estimates called percent-likeness. The commonly used rule is that two sequences are said to be homologous if they have a> 30% equilibrium[16]. The result of homology analysis using SDSC Biology Workbranch program based on TFASTY showed that glycoprotein (GP) gene had a percent value of 25.15% with human genome (<30%). This shows that GP is can be used as a source of vaccine candidates.TFASTY is a part of fasta3 which contains a program of searching the DNA or protein sequence. TFASTY has a function to compare the amino or DNA sequences with DNA in the database.

The design of an epitope vaccine candidate is done by predicting the peptide sequence of the GP ebola gene that is strongly bound to MHC I. This prediction is made using the IEDB-AR program. The results of the analysis by IEDB-AR are IC50 values (Inhibitor Concentration 50) in nM. In the program there are provisions, ie peptides with IC50 values <50 nM are considered to have a high affinity, <500 nM (intermediate affinity) and <5000 nM (low affinity).

The IEDB-AR program uses three algorithm models to predict the IC50 value of Artificial Neural Network (ANN), Strabilized Matrix Method (SMM)[14] and Combinatory Library Sidney 2008 (comblib_sidney2008)[16]. ANN is a computer algorithm model with a simple processing system, where the system is able to transmit a signal if it receives a strong input signal[16]. In the SMM method, IC50 is calculated based on the MHC peptide position-specific score matrix[13]. It consists of a prediction process that resembles the work of neural networks with various coding schemes of specific bonding codes[13]. Comblib_Sidney2008 refers to a set of predictors (ie scoring matrices) derived from affinity measurements of peptide binding combination to the MHC alleles panel[16]. For the conclusion of the results, the IEDB-AR program displays them in a percentile rank, where small number percent rank values indicate a high affinity[17].. Here are 5 peptide sequences that have the smallest percentile rank of IEDB-AR analysis results in GP ebola genes (Table 1).

Table 1. The prediction results of peptide bond – MHC I

Gene of Ebola	**Allele**	**Residu Position**		**Length**	**Peptide**	**Persentil Level**	**Values of IC_{50} (nM)**		
		Early	**End**				**ANN**	**SMM**	***Comblib***
GP	HLA-A*02:01	40	48	9	FLYDRLAST	0,4	8,14	26,80	5,07x10-6
GP	HLA-A*02:01	12	20	9	FLLQLNETI	0,4	16,80	35,49	5,37x10-6
GP	HLA-A*02:01	25	33	9	ILFQRTFSI	0,5	10,17	28,39	1,1x10-5
GP	HLA-A*02:01	7	15	9	ALFCICKFV	0,5	22,50	26,01	3,23x10-5
GP	HLA-A*02:01	23	31	9	FLDPATTTS	0,8	334,08	359,01	7,11x10-6

Based on Table 1, peptides with the best affinity for MHC I are peptides in the order of FLYDRLAST (phenylalanine, leucine, tyrosine, aspartic acid, arginine, leucine, alanine, serine, threonine) since they have the smallest percentile rank value (0.4)[19] with IC50 value 8.14 (ANN) 26.80 (SMM) and 5,07x106 (comblid_sidney2008).

The molecular blocker is performed to visualize the affinity between FLYDRLAST peptide and MHC I. Through this method it is known that molecular interactions occur between specific FLYDRLAST peptides, specifically. Molecular tacking is done with the CABSdock program. Some of the things that are done on the molecular blocking process are the prediction of the active side bonding on the receptor structure, peptide modeling on the active side and the refinement of the protein-peptide bond[14]. Here are the results of FLYDRLAST peptide molecules with MHC I HLA * A 2: 1 allele (Figure 1).

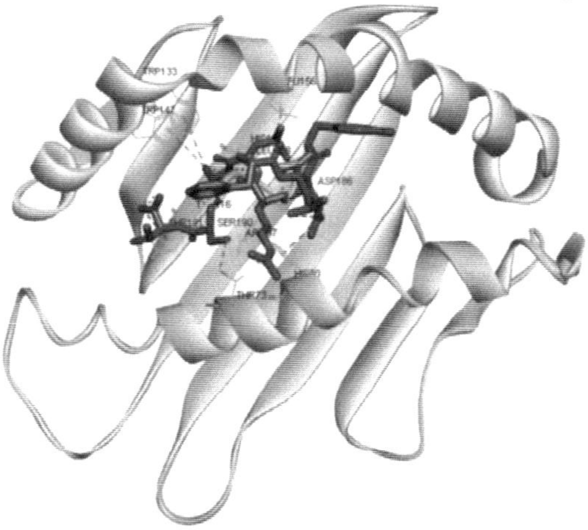

Figure 1. Molecule interaction between FLYDRLAST peptide and MHC I (PDB 2X4U Code)

Table 2. Non-Kovalen interaction between Amino Acide of Peptide FLYDRLAST and MHC I.

Amino Acid		**Distance (Å)**	**Type of Interaction**
Peptide	**MHC**		
ASP186	ARG97	3,97	Hydrogene bond, Electrostatics
SER190	THR73	3,13	Hydrogene bond
ASP186	ARG97	3,06	Hydrogene bond
ALA189	HIS114	2,85	Hydrogene bond
ASP186	HIS70	4,17	Electrostatics
THR191	TYR116	4,46	Electrostatics
ALA189	HIS114	3,28	Hydrophobic
LEU188	LEU156	3,68	Hydrophobic
ALA189	TRP133	4,78	Hydrophobic
ALA189	TRP147	4,27	Hydrophobic

The quality of molecular binding depends on the value of RMSD (Root Mean Square Deviation). Molecular binding qualities are categorized as good if RMSD values <3Å, medium quality when 3Å ≤ RMSD ≤ 5.5 Å and low quality when ≥5Å[14]. To find out the quality of the binding is done validation method. Validation is done by molecular binding on the three-dimensional structure of ligand receptor (peptide-MHC) contained in the database (GDP). The CABSdock program generates a RMSD value of 2.15 Å. This value indicates that the quality of molecular blockers is of good quality.

Figure 1 shows the visualization of the affinity that occurs between the FLYDRLAST peptide and MHC I. This affinity is clarified by the molecular interaction information between the peptide against MHC I in the range of 5 Å (Table 2). Figure 2 shows four hydrogen bonds, three electrostatic interactions and four hydrophobic interactions. Conventional hydrogen bonds are formed between SER190-THR73 and ASP186-ARG97. Hydrophobic interactions involving an alkyl group are formed between LEU188-LEU156, ALA189-TRP133 and ALA189-TRP147. Other interactions involving sigma-Pi or Pi-Pi orbitals are formed by ASP186-HIS70, THR191-TYR116 and ALA189-HIS114. The free energy produced by the bond between the FLYDRLAST peptide and the MHC I of the molecules is -1870.69 Kcal / mol. This value is very low, so the FLYDRLAST peptide has a strong affinity for MHC I [14]. A strong affinity between peptide and MHC I may induce cytotoxic T cells[11]. With induced cytotoxic T cells then the infected cells can be killed[18]. The activated immune system process indicates that the peptide is a potential epitope candidate for ebola vaccine.

4. Conclusion

The sequence FLYDRLAST (Phenylalanine, Leusin, Tyrosin, Aspartic Acid, Arginin, Leucine, Alanine, Serine, Threonine) is the most potential sequence of glycoprotein zaire ebolavirus for ebola vaccine

Acknowledgement
We thank to the IEDB (www.iedb.org) for providing open access to the ebola virus genome database and the CABSdock program (http://biocomp.chem.uw.edu.pl/CABSdock) that has facilitated the molecular interaction peptide-MHC and all immunoinformatics analysis programs used in this study.

References
[1] Jayanegara, A. P. (2016). Ebola Virus Disease – Masalah Diagnosis dan Tatalaksana. CDK-243, 43(8), 572– 575.
[2] Aditya, M. (2014). Ebola Hemorrhagic Fever : Clinical Management and Prevention. Juke, 4(8), 245–253.
[3] Dharmayanti, N., dan Sendow, I. (2015). Ebola : Penyakit Eksotik Zoonosis yang Perlu Diwaspadai. Wartazoa, 25(1), 29–38.
[4] Khan, M. A., Hossain, M. U., Rakib, S. M., dan Morshed, M. N. (2015). Epitope-based peptide vaccine design and target site depiction against Ebola viruses : an immunoinformatics study. Scandinavian Journal of Immunology, 82, 25–34. https://doi.org/10.1111/sji.12302.
[5] WHO. (2016). Ebola Situation Report January 20th 2016. Retrieved from http://www.who.int/csr/disease/ebola/situation-reports/archive/en/
[6] Agnandji, S. T. A., Huttner, A., Zinser, M. E., Njuguna, P., Bejon, P., Kremsner, P. G., … Siegrist, C. A. (2015). Phase 1 Trials of rVSV Ebola Vaccine in Africa and Europe. The New England Journal of Medicine, 374(17), 1647– 1660. https://doi.org/10.1056/NEJMoa1502924.
[7] Patronov, A., dan Doytchinova, I. (2013). T-cell epitope vaccine design by immunoinformatics T-cell epitope vaccine design by immunoinformatics. Open Biology, 3, 120– 139. https://doi.org/10.1098/rsob.120139
[8] Subroto, T., Hardianto, A., Kahari, A. A., dan Pradnjaparamita, T. (2013). Sintesis Tiga Peptida Bergugus Pelindung sebagai Prekursor Komponen Vaksin Influenza Universal. Jurnal Natur Indonesia, 15(2), 84–91.
[9] Guerra, R. E. S., Gomez, R. N., Alonso, D. O. G., dan Mendoza, S. R. (2014). An overview of bioinformatics tools for epitope prediction : Implications on vaccine development. Journal of Biomedical Informatics, 1–11. https://doi.org/10.1016/j.jbi.2014.11.003
[10] Rappuoli, R. (2003). Reverse vaccinology. Drug Discovery Today, 445–450. https://doi.org/10.1016/S1359-6446(03)02689-8

[11] Taupiqurrohman, O., Yusuf, M., Nuswantara, S., Subroto, T., Bioteknologi, P. S., Pascasarjana, S., … Padjadjaran, U. (2016). Potensi Gen Oncoprotein Human Papillomavirus Tipe 16 Sebagai Kandidat Vaksin Kanker Serviks Human Papillomavirus Type 16 Oncoprotein Genes as the Candidate of Cervical Cancer Vaccine. MKB, 48(35), 84– 91.

[12] Vita, R., Overton, J. A., Greenbaum, J. A., Ponomarenko, J., Clark, J. D., Cantrell, J. R., … Peters, B. (2015). The immune epitope database (IEDB) 3.0. Nucleic Acids Research, 43(D1), 405– 412. https://doi.org/10.1093/nar/gku938

[13] Zhang, Q., Wang, P., Kim, Y., Andersen, P. H., Beaver, J., Bourne, P., … Peters, B. (2008). Immune epitope database analysis resource Immune epitope database analysis resource. Nucleic Acids Research, 36, 513– 318. https://doi.org/10.1093/nar/gkn254

[14] Kurcinski, M., Jamroz, M., Blaszczyk, M., Kolinski, A., dan Kmiecik, S. (2015). CABS-dock web server for the flexible docking of peptides to proteins without prior knowledge of the binding site. Nucleic Acids Research, 43, 419– 424. https://doi.org/10.1093/nar/gkv456.

[15] Pearson, W. R. (2013). An Introduction to Sequence Similarity (" Homology ") Searching. Curr Protoc Bioinformatics, 1– 9. https://doi.org/10.1002/0471250953.bi0301s42.

[16] Fleri, W. (2016). T Cell Epitopes - MHC Class I Binding Prediction Tools Description IEDB Solutions Center. Retrieved from http://help.iedb.org/hc/en-us/articles/114094151691-T-Cell-Epitopes-MHC-Class-I-Binding-Prediction-Tools-Description

[17] Christy, J., dan Anand, A. (2014). Prediction of HLA-DRB1 * 15 Restricted CD 4+ Epitopes of Indian HIV-1 GAG Protein Using In Silico Approach. International Journal of Pharma and Bio Sciene, 5(4), 1212– 1222.

[18] Campbell, N. A., dan Reece, J. B. (2008). Biologi (8th ed.). Jakarta: Erlangga.

[19] Munasir, Z. (2001). Respons Imun Terhadap Infeksi Bakteri. Sari Pediatri, 2(4), 193– 197.

IOP Conf. Series: Journal of Physics: Conf. Series **1090** (2018) 012082 doi:10.1088/1742-6596/1090/1/012082

Appropriateness of Water Hyacint and Rice Husk as Filler Materials for Resin Polymer Insulation

Trisna Amelia Fitriah*[1], Salama Manjang[1] , Ikhlas Kitta[2]
Electrical Engineering Department[1], Post Graduate Program Of Hasanuddin University[2],Jl. Perintis Kemerdekaan KM. 10 Makassar, Sulawesi Selatan, 90245, Indonesia

Email: ameliafitriah26@yahoo.co.id[1], salamamanjang@gmail.com[2], ikhlaskitta@gmail.com[2]

Abstract .The use of natural fibers as a filler in the industry still continue to be developed, given its eco-friendly and contains material which is great. This research was conducted to find out the influence of boiled water and rice husk as filler material in epoxy resins. The making of the specimen using the method of hand lay-up with each variation of filler 10%-40% of the total weight of material test and the specimen without filler 0% for contrast. Polymer materials used i.e. epoxy resins. Testing is done namely pull-test, test volume resistivity and surface resistivity. Pull-test is performed using standard ASTM D-638 and resistivity test using standard ASTM D 257. The test results showed that the maximum volume resistivity values reached in the levels of water hyacinth 20% whereas on rice husk volume resistivity increase continues to occur until levels of 40% of the epoxy resins. Surface resistivity continues to increase in the levels of water hyacinth reaches 20% on epoxy resins as well as the addition of a filler rice husk. The tensile strength of the epoxy resins materials has decreased with the addition of water hyacinth and rice husk. The value of the maximum attainable tensile strength on the composition of the filler is 20%. With increasing levels of water hyacinth and rice husk in epoxy resins, the nature of the material brittle epoxy resins increase or an increase in the hardness of epoxy resin resulting in fragility.

1. Introduction

The development of industry with the advancement of technology has contributed greatly to the Indonesia's economy. On the other hand it also has an impact on the environment due to industrial waste as well as the increasingly intensive resource exploitation in industrial development. Material and energy efficiencies in utilization, processing and recycling result in competitive advantage and economic benefits. An environmental issue that must be considered one of them is pollution in water bodies. Eutrophication is one of the causes of contamination it is water pollution caused by the emergence of excessive nutrients into the aquatic ecosystem. Based on the above, the industrial development should be coupled with environmental management efforts in the form of waste handling that released or by searching for industrial raw materials environmentally friendly as an alternative material. material water hyacinth (Eichornia crassipes) and rice husk are alternative materials for high voltage isolation. Water hyacinthis a type of water plant that has a high growth rate, so it is considered as a weed that can disrupt aquatic ecosystems[1].

Content from this work may be used under the terms of the Creative Commons Attribution 3.0 licence. Any further distribution of this work must maintain attribution to the author(s) and the title of the work, journal citation and DOI.
Published under licence by IOP Publishing Ltd

Problems arising from the high population of water hyacinthis the decrease of the amount of water aquatic diversity, the occurrence of siltation, the decrease of water quality due to the decrease of oxygen due to decreasing the intensity of sunlight entering the water body, the increasing of disease vector, irrigation disturbance, transportation, and the decreasing of aesthetic value in water [1,2]. Lignocellulose is also contained in agricultural waste, one of which is rice husks. Rice husk is an agricultural residue whose availability is very abundant and can produce bio-ethanol because it contains lignocelluloses structure. Rice husks contain 24.3% hemicelluloses, 34.4% cellulose, and 19.2% lignin [3]. One example of a polymer that can be used as a base material of outer mounting insulator in the electrical field is resin epoxy. Resin epoxy material has several advantages including: epoxy fluid has a low viscosity that is easily mixed (thermoset stage) in the manufacture, but resin epoxy resin has a combination of: low viscosity properties, easy to form, low shrinkage, high adhesiveness, high mechanical properties, high electrical insulation, good chemical resistance (4). Cellulose and lignin is one of the criteria that demonstrate the power of fibre. Cellulose is the major constituent of the cell wall and including glucose polymer with 1.4-ß Glycoside bond in the chain length of the straight. The cellulose chains are connected by hydrogen bonds and van der walls [5].Lignin is a component that is very difficult didegradasi. Polymer composition comprising lignin aromatic units linkedby ether bonds andcarbo n-carbon. The main function of lignin in plants is strengthening the structure of plants [6]. Lignocellulose is also contained in agricultural waste, one of which is rice husks. When the rice husk is burned the ash content obtained is 17-26%, much higher than other fuels (wood 0.2-2%, coal 12.2%). Rice husks have an average high calorie value of 3410 kcal / kg and can be used as one of renewable energy sources [7]. rice husks are only used as primary or additional fuel in the brick making industry, seed planting medium, while the ashes are small for rubbing and others are thrown away. In fact, if the rice husk ash is heated above 500°C temperature during 105 minutes or more will give optimum amorphous silica content of at least 86% [8]. Isolator as one of the electrical equipment plays an important role in the process of electrical energy distribution. Popular outer mounting insulators developed today are polymer insulators, because polymeric materials have many advantages over porcelain and glass materials. Among these are high hydrophobic properties, have lower mass densities, are easy to manufacture because they do not require very high temperatures in the manufacturing process, and have a high degree of adhesion [9].

2. Experimental Procedures

2.1 Material

Resin epoxy used in this study is epoxy glue with a type of RTV (Room Temperature vulcanization) or vulcanized/harden at room temperature. It is obtained from PT. AVIA AVIAN, as for trademark is Avian 2 component consists of two parts, namely resin and hardener.

Water hyacinth and rice husks are used as filler for insulating material resin epoxy. Water hyacinthin question here is Latin Eichornia crassipes obtained from the waters of Makassar city while rice husk is obtained from Sidrap district. The composition of chemical compounds contained in water hyacinthand rice husk are used have been analyzed using XRF (X-Ray Fluoroscent) and the results can be seen in the following below.

Table 1. Chemical composition of rice husk ash

No	Coumpond Name	Pecentage (%)
1	SiO_2	97,99
2	P_2O_5	0,50
3	CaO	0,496
4	K_2O	0,491
5	MnO	0,268
6	Cl	0,179
7	Fe_2O_3	0,038
8	ZnO	0,0124
9	Rb2O	0,0076
1	SiO_2	97,99
2	P_2O_5	0,50

Table 2. Lignicellulose composition of *eceng gondok*

No	Coumpond Name	Percentage (%)
1	Lignin	3,97
2	Selulosa	23,54
3	Hemiselulosa	28,19

3. Material Preparation

Water hyacinthobtained from the waters of Makassar city before use is cleaned first by using running water to remove impurities, after cleansing leaves is dried for 7 days in open room that is not directly exposed to sunlight. After dry the leaves are refined and sieved to provide uniformity in the particles. Rice husks obtained from Sidrap district in the form of ash are first sieved to provide uniformity in the particles.

Resin epoxy is then mixed with filler each using a technique of hand lay-up mixing. The mixture is then inserted into the vacuum chamber to remove the air bubbles trapped in the mixture when mixing. The mixture is printed using a mold with a depth of 3 mm so that the material will be produced with a uniform thickness. After the material is hardened, it is conditioning on the material by keeping the material in the test chamber for 3 days with humidity reaching 85% and a temperature of 25oC. By doing this, no further diffusion of water vapor is expected to material from the humidity of the air around which the diffusion process is concerned may affect the data obtained. After the conditioning process is complete, the material is ready to be tested for characteristics. The material is prepared with different water hyacinthand rice husk levels to investigate the effect of addition of water hyacinthand risk husk on the characteristics of resin epoxy material. The composition of filler and resin epoxy tested can be seen in the following table :

Table 3. Sampling code and percentage of *water hyacinth*and rice husk ash levels

No.	**Sample Code of Water Hyacint**	**Percentage (%) (based on weight of *resin epoxy*)**	**No.**	**Sample Code of Rice Husk Ash**	**Percentage (%) (based on weight of *resin epoxy*)**
1	SEG 0	0	6	SSP 0	0
2	SEG 1	10	7	SSP 1	10
3	SEG 2	20	8	SSP 2	20
4	SEG 3	30	9	SSP 3	30
5	SEG 4	40	10	SSP 4	40

Eachcomposition is made up of 3 samples to provide more accurate data.

4. Characterization

The testing process is based on ASTM standard. The standard test used can be seen in the following table:

Table 4. Testing method

Name Of Testing	Method of testing
Tensile Strenght	ASTM 638
Surface resistivity	ASTM 257
Volume resistivity	ASTM 257

Testing is done as much as 1 to 10 times. The final value of each test type is the average of the tests performed so that the data obtained is more accurate. Testing of surface and volume resistivity and volume were performed using the Kyoritsuinsulation tester with electrode configuration and procedures based on the ASTM D257 standard. On surface and volume resistivity measurements, the voltage used by the test equipment is 5000, with ambient temperature of 25oC and humidity 81%. Testing of tensile strength was performed using tensile testing machine based on ASTM D 638 standard.

5. Results and Discussion

Volume resistivity

Volume resistivity of material has been tested and the results can be seen in the following figure:

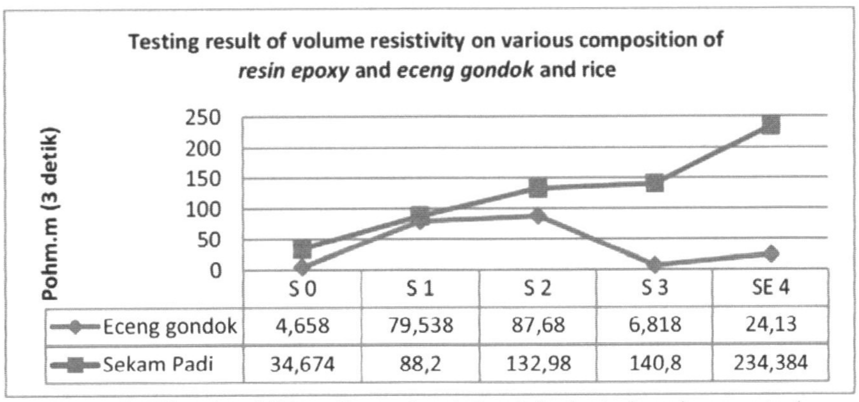

Figure 1. Effect the addition of water hyacinthand rice husks in resin epoxy to the volume resistivity of the material

In figure1. can be seen that the addition of water hyacinthin resin epoxy increases the volume resistivity of resin epoxy to the filler 20% composition. In addition, the addition of water hyacinthwill make the volume resistivity of resin epoxy to be decreased then not linear. In filler 20% composition the increased volume resistivity of Resin epoxy material is about 87.68% compared with resin epoxy without filler. The addition of rice husk in resin epoxy increased the volume resistivity of the resin epoxy to the filler 40% composition without decreasing. In the filler 40% composition the increased volume resistivity of resin epoxy material was about 234.384% compared with resin epoxy material without filler. There are several factors that can affect the volume resistivity of material especially if the material is tested at high-voltage DC impulse. Mineralogy of fillers, chemical bonds, molecular distributions and many other chemical factors will contribute to the resistivity of the material.

Surface resistivity

Surface resistivity of the material has been tested and the results can be seen in the following Figure 2

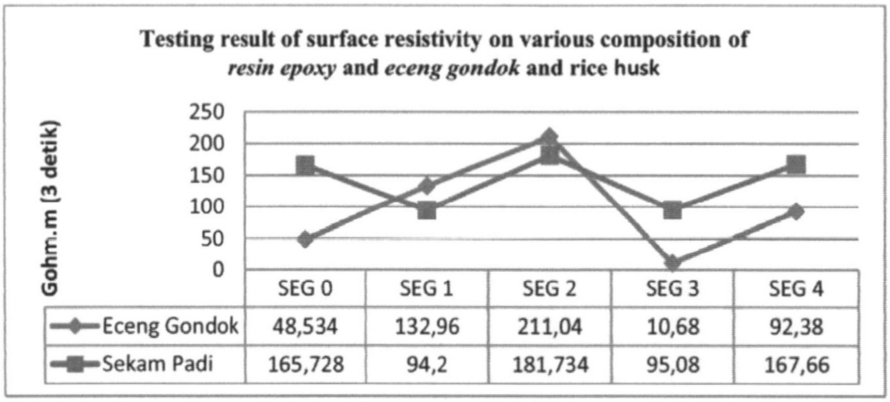

Figure 2. Effect the addition of water hyacinthand rice husks in resin epoxy to surface resistivity of the material

In figure 2 can be seen that with the addition of water hyacinthin the resin epoxy, the surface resistivity of resin epoxy is increased. Increased surface resistivity continues until the water hyacinthlevel reaches 20% in resin epoxy.

The addition of rice husk level to resin epoxy, surface resistivity of resin epoxy has decreased compared to resin epoxy without filler. Increased surface resistivity increases in the addition of filler 20%. The resistance value of a material will depend on the type of voltage/field applied to the material. In materials with high semi-conductive particle content, the resistance of resin epoxy will change depending on the size and type of voltage/field given. At a voltage magnitude/field given of material, the resistance value may change depending on the band gap width between the valence band and the conduction of the constituent compounds.

Tensile strength

The tensile strength of material tested can be seen in the following figure:

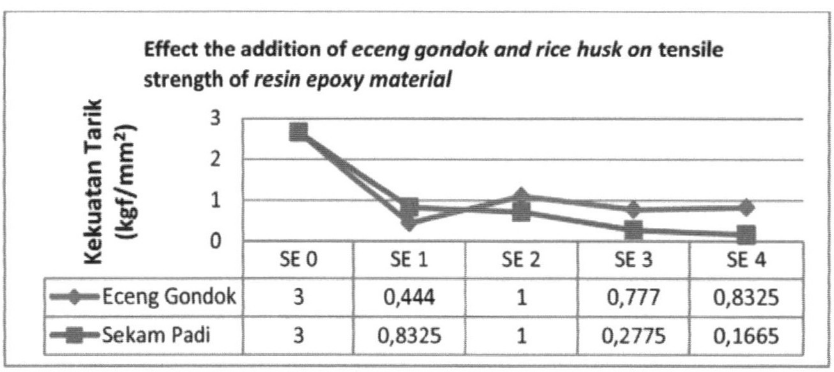

Figure 3. Effect the addition of water hyacinthand rice husks in resin epoxy to tensile strength of resin epoxy material

In figure 3 can be seen that there is a tendency to decrease the tensile strength of resin epoxy material, the addition of water hyacinth and rice husks. The maximum tensile strength value is achieved in the water hyacinth 20% composition of 1 kgf/mm2. In the composition, the tensile strength increase is 20% compared to the resin epoxy composition without filler. Also, figure 5 can be seen that initially there is a tendency to increase the tensile strength of the Resin epoxy material by increasing the water hyacinth content in the material until it reaches its peak in the water hyacinth 20% composition. After that there is a non-linear tensile change in composition above 30%, the tensile strength of the Resin epoxy material initially decreases to the water hyacinth 10% composition. After that there is an increase of tensile strength at 20% composition. It can be seen that with increasing levels of water hyacinth and rice husk in the resin epoxy, the elasticity of resin epoxy material is reduced or an increase in hardness of resin epoxy. The presence of some minerals in water hyacinth and rice husks in resin epoxy makes the stretching of interpersonal distance and the range of bond angles smaller between the monomers on Resin epoxy so that the tensile strength and elasticity between the monomers become weakened. As a result elongation of resin epoxy composite material and water hyacinthand rice husks decreases with increasing levels of filler in resin epoxy.

6. Conclusions

The conclusions that can be drawn from this research as follows:

1. In general, the utilization of water hyacinthand rice husk as filler in resin epoxy can improve some parameters of Resin epoxy material.
2. In general, the maximum increase of electrical and mechanical parameters is achieved at water hyacinth composition by 10 - 40% of the weight of resin epoxy
3. esting of surface resistivity and permittivity, the parameter values increase with increasing addition of water hyacinthand rice husks in the resin. There is characteristic irregularity in the result of tensile test. This may be due to an incomplete sample preparation caused by the nature of the resin epoxy material itself, which is difficult to be molded.

References

[1] Cheng, J, Xie, B., Zhou, J., Song, W., dan Cen, K. (2008). *Cogeneration of H2 and CHfrom Water Hyacinth by Two-Step Anaerobic Fermentation*. International Journal of Hydrogen Energy 35, 3029-3025. *4*

[2] Gunnarson, C. G. dan Peterson, C. M. (2007). *Water Hyacinth as aResource and Energy Production: a Literature Review.* Journal of Waste Management 27, 117-129.

[3] Soltani, N., Bahrami, A., Pech-Ganul, M. I., dan Gonzalez, L. A. (2015). *Review on the Physicochemical Treatments of Rice Husk for Production of Advanced Materials.* Chemical Engineering Journal 264, 899-935.

[4] Prasojo, Winarko Ari, Abdul Syakur, dan Yuningtyastuti, Analisis Partial Discharge pada Material Polimer Resin Epoksi dengan Menggunakan Elektroda Jarum Bidang, Penelitian, Universitas Diponegoro, Semarang, 2009.

[5] Perez, J., Dorado, J. M., Rubia, T., dan Martinez, J. (2002). *Biodegradation and Biological Treatments of Cellulose, Hemicellulose, and Lignin: An Overview*. Int. Microbiol 5, 53–63.

[6] Girisuta,B.(2007). *Levulinic Acid from Lignocellulosic Biomass*.Proefshrift University of Gronigen.

[7] Andy Chandra, S.T., M.M. 2012. Isolasi Dan Karakterisasi Silika Dari Sekam Padi. Lembaga Penelitian dan Pengabdian kepada Masyarakat Universitas Katolik.

[8] Hasyim Asy'ari (2008) "The Effect Of Industrial Pollutionto The Mechanical Performance Of Epoxy Resin Insulation Material Using Silicone Rubber and Rice Husk Ash As Filler". Jurnal Penelitian Sains & Teknologi, 9(1): 94-106.

[9] Manjang, Salama, and Masayuki Nagao. "Characteristics of high voltage polymer insulator under accelerated artificial tropical climate multi stress aging." *Electrical Insulating Materials (ISEIM), Proceedings of 2011 International Conference on*. IEEE, 2011.

[10] Kitta,Ikhlas.2015.Seminar Nasional Teknik Energi dan Ketenagalistrikan SNTEK 2015 Pengaruh *Fly Ash* Batubara Terhadap Perilaku *Hydrophobicyty* Isolator Polimer.

Institute of Physics
Dirac House, Temple Back
Bristol BS1 6BE UK

ISSN: 1742-6588
ISBN 978-1-5108-7300-1

CURRAN ASSOCIATES INC.
proceedings.com